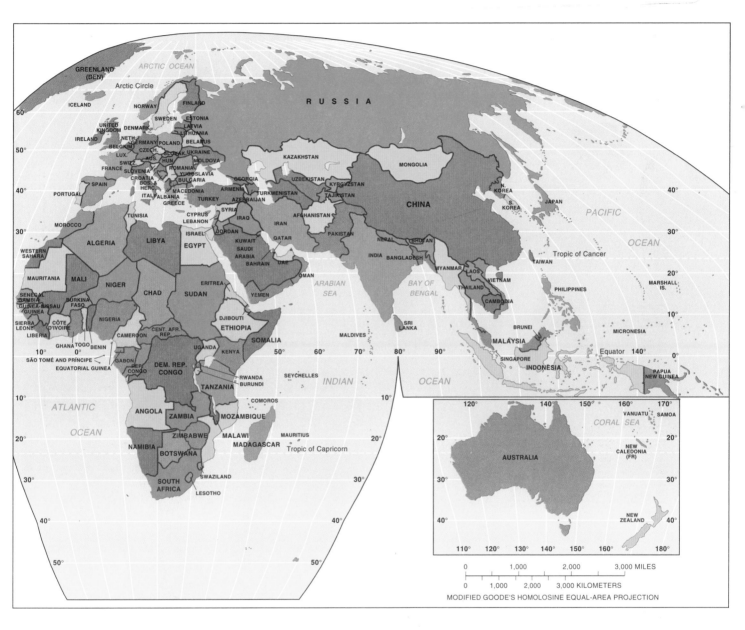

GREENLAND
(DEN)

ARCTIC OCEAN

Arctic Circle

ICELAND

NORWAY

60°

SWEDEN   ESTONIA
FINLAND

LATVIA

UNITED
KINGDOM   DENMARK   LITHUANIA

IRELAND

NETH.   GERMANY   POLAND   BELARUS

50°

BELGIUM

LUX.   CZECH.   SLOVAK   UKRAINE

SWITZ.   AUS.   HUN.   MOLDOVA

FRANCE   SLOVENIA   ROMANIA

CROATIA   YUGOSLAVIA

BOS. &   BULGARIA

40°   HERC.   MACEDONIA

SPAIN   ITALY   ALBANIA

PORTUGAL   GREECE   TURKEY

CYPRUS
TUNISIA   LEBANON
ISRAEL   SYRIA

30°   MOROCCO   JORDAN   IRAQ

ALGERIA   LIBYA   EGYPT   KUWAIT
SAUDI
ARABIA   BAHRAIN

WESTERN
SAHARA   QATAR

MAURITANIA   MALI   NIGER   OMAN

SENEGAL   CHAD   SUDAN   YEMEN
GAMBIA   BURKINA   ERITREA
GUINEA-BISSAU   FASO
GUINEA   NIGERIA   DJIBOUTI

SIERRA   CÔTE   CENT. AFR.   ETHIOPIA
LEONE   D'IVOIRE   REP.

LIBERIA   GHANA   TOGO   BENIN   CAMEROON   SOMALIA

SÃO TOMÉ AND PRÍNCIPE   UGANDA   KENYA

GABON   EQUATORIAL GUINEA   DEM. REP.
REP.   CONGO
CONGO   RWANDA

TANZANIA   BURUNDI

ANGOLA   ZAMBIA

ATLANTIC   MOZAMBIQUE

OCEAN   ZIMBABWE   MALAWI

NAMIBIA   BOTSWANA   MADAGASCAR

SOUTH   SWAZILAND
AFRICA   LESOTHO

R U S S I A

KAZAKHSTAN   MONGOLIA

GEORGIA   UZBEKISTAN   KYRGYZSTAN

ARMENIA   TAJIKISTAN

AZERBAIJAN   TURKMENISTAN   CHINA

AFGHANISTAN   N.
KOREA
IRAN   S.
KOREA   JAPAN

PAKISTAN

NEPAL   BHUTAN   TAIWAN

INDIA   BANGLADESH   MYANMAR   LAOS

VIETNAM

THAILAND   CAMBODIA

SRI
LANKA   BRUNEI
MALDIVES   MALAYSIA

SINGAPORE   INDONESIA

SEYCHELLES

COMOROS

MAURITIUS

PACIFIC

OCEAN

Tropic of Cancer

MARSHALL
IS.

PHILIPPINES

MICRONESIA

Equator   140°

PAPUA
NEW GUINEA

*ARABIAN
SEA*

*BAY OF
BENGAL*

INDIAN

OCEAN

Tropic of Capricorn

120°   140°   150°   160°   170°

VANUATU   SAMOA

CORAL SEA

20°   20°

NEW
CALEDONIA
(FR)

AUSTRALIA

30°   30°

40°   NEW   40°
ZEALAND

110°   120°   130°   140°   150°   160°   180°

0   1,000   2,000   3,000 MILES

0   1,000   2,000   3,000 KILOMETERS

MODIFIED GOODE'S HOMOLOSINE EQUAL-AREA PROJECTION

◼ The World Economy ◼

The WORLD ECONOMY

RESOURCES,

LOCATION, TRADE,

AND DEVELOPMENT

Third Edition

FREDERICK P. STUTZ, PH.D
San Diego State University

AND

ANTHONY R. de SOUZA, PH.D

PRENTICE HALL
Upper Saddle River, New Jersey 07458

Library of Congress Cataloging-in-Publication Data

The world economy : resources, location, trade, and development /
Frederick P. Stutz, Anthony R. de Souza. -- 3rd ed.
    p.   cm.
Includes bibliographical references and index.
ISBN 0-13-727769-5
    1. Economic geography   2. Economic history--1945-   I. Stutz,
Frederick P.   II.   Title.
HC59.D398   1998
330.9--dc21   97-35050
                                                          CIP

To Pam, My Wife, and to my children Christa, Tiffany, Derek, Janene, Michelle, and Weston—F. P. Stutz
For Nadia, Jason, and Sam—A. R. deSouza

Acquisitions Editor: *Daniel Kaveney*
Editor in Chief: *Paul F. Corey*
Editorial Director: *Tim Bozik*
Assistant Vice President of Production and Manufacturing: *David W. Riccardi*
Executive Managing Editor: *Kathleen Schiaparelli*
Assistant Managing Editor: *Lisa Kinne*
Marketing Manager: *Leslie Cavaliere*
Manufacturing Buyer: *Benjamin Smith*
Creative Director: *Paula Maylahn*
Art Director: *Joseph Sengotta*
Cover Art: *Sikkim, India—rice paddies in Ganges lowlands, George Holton/Photo Researchers, Inc.; Computer/Video Conference at NEC, Charles Gupton/Uniphoto Picture Agency; Sewing Upholstery, Billy E. Barnes/Stock Boston; Woman at Grocer's stall in market, Christopher Bissell/Tony Stone Images*
Cover Designer: *Joseph Sengotta*
Photo Researchers: *Beaura Ringrose/Karen Pugliano*
Editorial Assistant: *Margaret Ziegler*
Production Assistant: *Nancy Gross*
Art Editor: *Grace Hazeldine*
Editorial/Production Service: *Electronic Publishing Services Inc., NYC*

 ©1998 by Prentice-Hall, Inc.
Simon & Schuster/A Viacom Company
Upper Saddle River, New Jersey 07458

Previous edition copyright ©1994 by Macmillan College Publishing Company, Inc.
Earlier edition, entitled *A Geography of World Economy,* copyright ©1990 by Merrill Publishing Company.

Printed in the United States of America

10 9 8 7 6 5 4 3 2 1

ISBN 0-13-727769-5

Prentice-Hall International (UK) Limited, *London*
Prentice-Hall of Australia Pty. Limited, *Sydney*
Prentice-Hall Canada Inc., *Toronto*
Prentice-Hall Hispanoamericana, S.A., *Mexico*
Prentice-Hall of India Private Limited, *New Delhi*
Prentice-Hall of Japan, Inc., *Tokyo*
Simon & Schuster Asia Pte. Ltd., *Singapore*
Editora Prentice-Hall do Brasil Ltda., *Rio de Janeiro*

CONTENTS

PREFACE

The World Economy: Resources, Location, Trade, and Development(3rd edition) adopts an international perspective to examine how people earn a living and how the goods and services they produce are geographically organized. It also emphasizes conflicting arguments and theories essential for understanding a world economy in rapid transition. Designed around the themes of distribution and economic growth, this textbook explores the nature of the dynamic, international environment and key international issues that arouse the concern and interest of geographers. Among the issues discussed are population growth, economic development in underdeveloped countries, pollution and resource depletion, food production and famine worldwide, patterns of land use, economic justice, international business, social and economic development, and multinational and international commerce.

Most important, we have made a concerted effort to globalize the economic geography curriculum. The trend toward globalizing the curriculum at the college level is major. Some educators have asked if it is the latest education fad or if it is the initial phase of a long-lasting, important shift in higher education. If education is meant to equip and enable students to live their lives fully and in a meaningful manner, then it seems obvious to us that the answer has to be the latter. The World Economy is in a position to take advantage of the student demand to globalize the curriculum. Increasing the global community's economic and political interdependence, which is tightly linked to the technology of the extraordinary global communications revolution, necessitates a global education for the twenty-first century. As David Grossman, former director of the Stanford University program on international, cross-cultural education, has pointed out, "students are becoming citizens within the context of this global era in human history and this calls for competencies which traditionally have not been emphasized by schools." We attempt to emphasize these competencies in this book.

The Carnegie Report on Higher Education (1991) cogently sets forth the almost certain dangers that exist in the global future if such an emphasis is not adopted:

The World has become a more crowded, more interconnected, more volatile and more unstable place. If education cannot help students see beyond themselves and better understand the interdependent nature of our world, then each new generation will remain ignorant, and its capacity to live competently and responsibly will be dangerously diminished. (p. 42). It is easy to understand this point in light of the recent crises in Russia, the Middle East, and the Persian Gulf; in the economic unity being sought by the European Community (EU) and by Canada, the United States, and Mexico through the North American Free Trade Agreement (NAFTA); or by any number of political and economic crises that have occurred in the past few decades, which emphasize the crucial nature of a globally informed citizenry. That is why we argue that, among the myriad of other programmatic demands placed on universities across America, global education should have a high—perhaps the highest—priority.

The text progresses logically from one topic to another and is designed to be used in both geography and international business courses. We recommend that the chapters be read sequentially; however, because we wrote each chapter to stand on its own, the book can serve as a reference or as a refresher. Prepared as an introductory book on economic geography, international business, and international economies, the material can be read and understood without college-level prerequisites. Certain chapters can be omitted for short courses. For example, if the course is taught as a traditional economic geography course that emphasizes model development, Chapters 10, 11, and 12 may be omitted. If the book is used for courses in international business, world development, or world trade and economy, portions of Chapters 6, 7, and 8 may be omitted.

The World Economy offers specific pedagogical features, including new two- and four-color world maps that illustrate, unlike any competing book, the spatial nature of the world economy. This new book contains more than 400 updated maps and graphs, many of which are modified from Goode's Homolosine Equal Area Projections. We enlarged the maps, increasing the size of Europe for intelligibility. It was our opinion that these maps were the best pedagogical tools available to teach students the spatial pattern of the world economy. The maps incorporate recent data from the Statistical Abstracts of the United States, the World Bank, the Population Reference Bureau, the Encyclopedia Britannica, and other sources, several of which were not published until August 1993. In addition, this book includes chapter objectives, end-of-chapter summaries, key terms, suggested readings, many photographs, and box essays, which illustrate principles of the world economy. A special effort has been made to offer modern examples of classical models and to provide summary charts, tables, and supply-demand curves that explain the dynamics of the world economy. Most of all, this book encourages students to think through problems, by providing the information and concepts necessary to help them evaluate issues without subscribing or submitting to a particular set of values.

World Space Economy, by de Souza and Foust in 1979, and *A Geography of World Economy,* by de Souza in 1990, concentrated heavily on the economic geography of the United States and on national effects of international processes. *The World Economy* (2nd edition) took a much wider view, was 60% new, and enabled students to appreciate what is going on, not only in the United States but also elsewhere in the world economy. This continues the globalization trend and has a special new emphasis on international business patterns and dynamics. An insular view of the world is untenable in the 1990s, as discussed. The world is too much "with us" everyday.

Here are some of the new features of THE WORLD ECONOMY (3rd edition);

1. A new emphasis on global interdependence of countries, companies and cultures;
2. A new emphasis on flexible manufacturing, flexible labor, and the flexible economy, corporate downsizing and the new globalization of work
3. New emphasis on globalization of business, as well as globalization of culture, globalization of environmental problems and the globalization of telecommunications;
4. A new half chapter on telecommunications and information technology (IT) as a substitute for transportation—the Internet, data communicators, digital moble communications, low earth orbiting satellites, and telework;
5. A new emphasis on the transnational corporation and its new impacts on old theories of location, trade, and comparative advantage;
6. New emphasis on the declining role of traditional factors of production including natural resources and population;
7. A reduction of 14 chapters to 12 chapters and the addition of GIS, world wide web sights, statistical considerations, and exercises.

We are especially grateful to our colleagues who read the manuscript and offered useful comments. These colleagues are as follows: Patrick Alles, Indiana University-Purdue University at Indianapolis; Amy Glasmeier, The Pennsylvania State University; R. W. Jackson, The Ohio State University; Dr. J. Harold Leaman, Villanova University; Edward J. Malecki, The University of Florida; Beth Mitchneck, The University of Arizona; Gordon Mulligan, University of Arizona; Bruce W. Pigozzi, Michigan State University; Dr. Debra Straussfogel, The Pennsylvania State University; Paul Susman, Bucknell University; Goeffrey Martin, Southern Connecticut State University; and Patrick Alles, Indiana/Purdue University.

Senior reviewers for both the second and third editions of this book were: Bruce Pigozzi, Michigan State University and Edward Malecki, University of Florida.

We would like to especially thank James Rubenstein, author of *The Cultural Landscape: An Introduction to Human Geography,* for his help and inspiration. Sent Visser of Southwest Texas State University and Fred Stutz produced the Instructor's Manual. We would like to thank Geosystems, Columbia, Maryland, for cooperating with us in mutually designing the maps; Academy ArtWorks and Burmar Technical Corporation for rendering the noncartographic art; Anthony Calcara of Electronic Publishing Services Inc., NYC for managing all aspects of manuscript production; and the entire Prentice Hall staff, especially Dan Kaveney, Geoscience Editor at Prentice Hall, and Paul Corey, Editor in Chief, who made possible the successful completion of this book. Naturally, however, all mistakes and omissions are ours.

Finally, we wish to thank our families—Pamela, Christa, Tiffany, Derek, Janene, Michelle, and Weston Stutz; and Nadia, Sam and Jason de Souza—for their support and understanding during this busy time. We wish to especially thank Pamela Stutz for typing the drafts of this book in both 1993 and 1997 before it went to a commercial word processer.

F.P. Stutz
A.R. de Souza

Frederick Stutz
San Diego California
e-mail: stutz@mail.sdsu.edu
http://www.prenhall.com/stutz
http://rohan.sdsu.edu/faculty/fstutz

◼ The World Economy ◼

GLOBALIZATION OF THE WORLD ECONOMY: AN INTRODUCTION

OBJECTIVES

- To outline the globalization of the world economy
- To introduce the major problems of environmental constraints and disparities in economic development
- To indicate why economic geographers are interested in development problems and how geography can help to resolve these problems
- To understand the four principal political economies of the world
- To state the four questions important to understanding the world economy
- To acquaint you with the field of geography and, in particular, with the major paradigms and concepts of economic geography

New York Stock Exchange. New York City is still the financial capital of the world, having supplanted London; however, Tokyo is a major player. (A demographic and geographic explanation for the recent surge in the stock market is given in this text on pp. 94–102.) *(Photo: New York Convention and Visitors Bureau)*

The world economy is being transformed by a combination of technological and geopolitical forces. This combination is creating a globalization of culture, a globalization of the economy, and a globalization of environmental issues. Technological changes—improvements in transportation and communications—are reducing the friction of distance and barriers to worldwide exchange. The principal instruments of the globalization of culture are worldwide television, music, and consumption patterns. The principal instruments of globalization of the economy are multinational corporations, which through their activities are producing new efficiencies in production, distribution, and the use of the world's resources. The collapse of communism around the world and the reduction of state governmental intervention into corporate life have facilitated a globalization of the economy by way of the removal of institutional and trade barriers, which is leading to the success of market capitalism. This increased globalization of the economy helps explain the nature of competitive advantage of countries, states, and regions.

According to M. E. Porter (1990), regions first produce raw materials and export such factor-driven resources to aid in their development. Next, regions enter into an investment-driven stage of development whereby new technologies are capitalized in other locations. The third stage is innovation driven, whereby regions develop new technologies, new products, and new markets. In the final, wealth-driven stage, a country achieves high levels of affluence and consumption that slow down levels of innovation and investment.

Changes in world economy have been cultural, technological, political, and environmental. Growing transportation cost reductions have improved physical exchange due to a continuing low cost of energy. Advancement in computing power in telecommunications/digital systems provided by fiber-optic networks, satellite technology, and the ease, speed, quantity, and quality of information transactions has rearranged the global economic system (Chapter 4).

Because transportation and communication costs have fallen rapidly, many services and goods that were provided locally are becoming internationally mobile. World communication systems now allow for companies to subcontract their production planning and financial operations across continents, wherever price is cheapest and quality is the best. Information has become more mobile than ever before, and a new global economy exists in information transmission, corporate consultancy, cable television news, internet information services and software systems design, programming, and application. International finance has also become both global and computerized, and capital markets are now highly mobile, for all forms of marketable equities and securities, stocks, bonds, and currency transactions. The globalization of finance has been aided by financial deregulation—the removal of state controls over interest rates, tariffs, barriers into banking, and other financial services.

Around the world, countries have abandoned centrally planned models of development in favor of private free-market enterprise, lower barriers to trade, and a nonregulatory watchdog role for the state. At the consumer level, the cable internet (Web TV) system brings high speed internet access to the home. Cable television allows the flow of information in a whole new dimension. The new cable internet system is an industry leading a broad-band solution for consumers. It brings two-way multimedia communication to the home computer over the television cable. It speeds hundreds of times faster than a telephone line; it puts a consumer in touch with the whole world in real time (instantaneously). Services include distance learning, interactive services such as banking and securities trading, all parts of Netscape browser information retrieval, transactional shopping, entertainment on demand, and video.

GLOBALIZATION

Globalization refers to worldwide processes that make the world, its economic system, and its society more uniform, more integrated, and more interdependent. *Globalization* is the process of the economy becoming worldwide in scope. The globalization process is a useful way to explain why the movement of people, goods, and ideas within and among world realms are becoming more and more important to not only economic systems but also cultural, political, and environmental systems. Globalization is a process that involves a "shrinking" world because of the time it takes for a person, goods, or a piece of information to travel from one place to another. World citizens are exposed to a global culture, a global economy, and global environmental change on a scale that has never been seen before. In some regions, social and political problems, as well as economic problems, result from a tension between the processes promoting global culture, economy, and environment on the one hand, and a practice and preservation of local economic isolation, cultural tradition, and the localization of environmental problems on the other hand. We will now take a brief look at the three important dimensions of globalization that are occurring at an ever increasing rate in the world today: (1) globalization of culture, (2) globalization of economy, and (3) globalization of environmental change.

Globalization of Culture

Culture is the total learned way of life of a society. Culture can be defined as that body of beliefs, social forms, and material traits constituting a distinct social tradition of a large group of people. Customary beliefs involve religion, food, and attitudes toward population growth. Social forms involve language, which is the transmission of ideas through written symbols, signs, and dialects. Material traits involve food, clothing, and shelter. All of the world's peoples consume food, wear clothing, and construct shelter, but various cultural groups incorporate and produce these necessities in a variety of ways.

1. The globalization of culture is based on an increasing level of shared beliefs, social forms, and material traits worldwide.
2. Societies display fewer cultural differences than in the past.
3. Globalization of cultural preference is being created by improved telecommunication systems worldwide.
4. The penetration of global culture in different regions across the earth is taking place at different rates. All peoples do not share the same access to this globalization.

Globalization of Consumption

The survival of a culture's distinctive customary beliefs, social forms, and material traits is being threatened by the global diffusion of consumption preferences. For example, the world's young people enjoy wearing blue jeans, Nike shoes, consuming Coca-Cola and Pepsi, smoking Marlboro cigarettes, and eating McDonald's hamburgers. Consumption preferences in food, clothing, shelter, and leisure activities are displaying a globalization of culture today. According to James Rubenstein (1996), the globalization of culture is based primarily on diffusion of lifestyles and products from more developed countries, especially from the United States. Regardless of cultural tradition, according to Rubenstein, people around the world are inspired to drive an automobile, watch television, and own a house with a stove and indoor plumbing.

Students of globalization observe a new global landscape in shops, stores, restaurants, and service stations. Retail chains create recognizable logos and visual appearances that do not vary from one region to another. Customers recognize these logos and building designs in magazines in whatever landscape or part of the world they may find themselves in.

Telecommunications

Cultural groups in different regions share similar beliefs, social forms, and material traits, thanks to enhanced telecommunications. Because of cable television and international news services, we know a great deal about political and economic events happening worldwide

Japanese woman carries home the "original recipe." American foreign direct investment, culture, and products have saturated much of the world markets. American cultural and business influences are likely to increase in the future. As U.S. television programming becomes ubiquitous and the global information network becomes increasingly mature, American culture, the United States' most powerful export, will grow in popularity and power. This will have both good and bad effects, but it will broaden the linkage between America and other countries, especially for the younger generations. At the same time it will threaten and further alienate the United States from more conservative cultures in the third world, especially Islamic countries. Multinational corporations will continue to play an increasing role on the world scale. Many MNCs have grown to be larger in terms of annual sales than the economies of many countries. Additionally, MNCs are more informed and sophisticated than most governments and can act much faster because of less bureaucracy. *(Photo: Greg Davis, The Stock Market)*

within a few hours. Far away places are less remote and more accessible now than they were just 10 years ago. Through television programming and the Internet, we can reach into the countries and cities of the world's people that are far away or to remote databanks at universities and government agencies around the world.

Today, travel by jet aircraft is rapid compared to travel a century ago. We can learn of places far away via television and the Internet. We can receive pictures and messages worldwide at the touch of a button or the click of a mouse. We can communicate almost instantly with people in distant places around the world through desktop computers, cellular phones, and other telecommunication devices. We can instantly see people in distant places on the evening news broadcast on television (Figure 1.1).

This diffusion of global telecommunications has enhanced the globalization of culture through shared beliefs, social forms, and material traits. Africans, for example, have shifted away from traditional religions and animism and have adopted Christianity south of the Sahara and Islam in the countries sharing the Sahara Desert. The world's peoples still speak thousands of different languages, but English is becoming the world international language—*lingua franca*. More international communications and global business activities is conducted in English than any other language (see box: Globalization: An Emerging World Culture?).

Citizens in developed countries take for granted these telecommunication innovations, such as MTV—communications across oceans by telephones and fax and satellite communications via the cell phone. The world is being wired up into global networks of millions of personal machines interconnected to each

other by fiber-optic and satellite links. These networks allow essentially instantaneous communications to anyone else on the Net. That interchange can include mail, documents, books, pictures and photographs, voice and music, video and television images, and programs and film. Huge databases at universities, government agencies, and research institutes are in many cases accessible to anyone at no cost. The largest of these networks, the *Internet*, includes over 10,000 universities and colleges with databases of computer programs and other information.

What this change means is that we are quickly moving toward the time when anyone can get any information to almost anyone else at almost any time throughout the entire world. We are also increasingly moving information instead of people, and we are doing it almost instantaneously. This globalized network is transmitting data and will be the prime conduit for the capital commodity of the coming age: information. It is allowing the globalization of world culture, as new ideas sweep the globe and give extraordinary access of information to those who are connected.

This global information network is far more refined in the United States and Canada than any other region and therefore allows North America to project its culture to the rest of the world. The information network is the

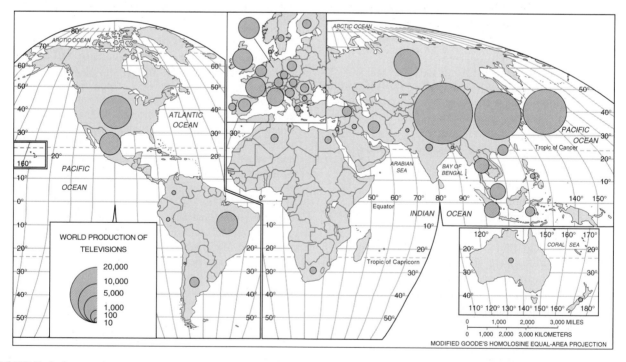

FIGURE 1.1

Manufacture of television receivers worldwide, 1990. In East Asian nations, there has been a great emergence in the production of consumer electronics within the last 10 years. In fact, Asia is the primary area of television manufacture, with more than two-thirds of the world's supply. The biggest single provider is China, which manufactures 22% of the world's total, followed by Japan and Korea, each having 14% of the world market share. *(Source: Data from United Nations, Industrial Statistics Yearbook, 1990, 1993b)*

GLOBALIZATION: AN

EMERGING WORLD CULTURE?

The world and its people are becoming interdependent at an ever increasing rate through trade and communications. Entertainment, especially television, modern technology, and multinational corporations have brought peoples into close contact. MTV is available in more than 100 countries now, bringing together performers and audiences from almost every culture. Although people live continents apart, they are exposed to common food, music, clothing, and most important, ideas and ideals. This culture of people who embrace these cultural norms is called *popular culture* and the movement to them is *cultural convergence*.

In India, Star TV, the new Asian satellite programming service, has been in service only since September 1993. Already, it has been cited as the chief inspiration for major changes in Indian culture, such as fashion and popular music, and has increased the demand for brand-name products. It has engendered vigorous new calls to liberate the Indian economy from a once dominant state control and to end the subjugation of women. *Baywatch* and *Santa Barbara* are the highest rated shows on Indian television. U.S. sitcoms, soap operas, and MTV are now broadcast from Pakistan throughout India, Myanmar, even to Beijing. Television is the most effective instrument ever invented for cultural diffusion. (It took 40 years for *Gone With the Wind* to be viewed by 21 million people. But in one night, 59 million people around the world watched the movie on TV.) The *Wall Street Journal* noted: "TV has become more significant than any other single factor in shaping the way most of us view our world."

By 1996, about 50% of the revenues of the seven largest U.S. movie studios came from the world economy outside the United States. The percentage is rising rapidly, as foreign countries present the greatest opportunities for increased sales outside the United States. By 1995, the five largest U.S. movie studios were owned by foreign multinational corporations (MNCs): MGM/UA, owned by an Italian MNC; 20th Century Fox, owned by an Australian MNC; Columbia Pictures, owned by a Japanese MNC; Universal Studios, also owned by a Japanese MNC. New movies made by the Walt Disney Corporation are also financed by Japanese investors. (This goes with global independence, an emerging world culture.)

Western clothing styles and fast food also represent an aspect of the new world culture. Blue jeans are sought after around the world, and U.S. companies exported more than 100 million pairs of them in 1993 alone. Jean-clad tourists in Russia are often asked if they are willing to sell the pants they are wearing. Most U.S. jeans sell at a much higher relative price in foreign countries than they do in the United States, because of their popularity and limited supply.

The food of the new world culture is convenience food. American fast food franchises have been set up around the world, including KFC, Burger King, Pizza Hut, and McDonald's. More than 50,000 Russians lined up at Pushkin Square to spend 6 rubles (valued at $9.00) for a Big Mac, Kartofel, and Koktel (hamburger, french fries, and milkshake) at the 1990 grand opening of McDonald's in Moscow. The huge restaurant employs 600 people and seats 1,000 customers. In Ho Chi Minh City, Vietnam, people jam the "California Burger."

In its quest to dominate the world palate, McDonald's occasionally accommodates local customs. Opening its first outlet in cow-revering India in 1997, the fast feeder offered New Delhi a beefless "Maharaja Mac" and vegetarian burgers. In 1997, the Swedish ski resort town of Lindvallen will launch a McFranchise right on the slopes—complete with ski-through window for on-the-go schussers.

Rock music, blue jeans, and fast food are all part of popular culture in the United States. As the intercontinental transmission of television becomes ubiquitous and the global information network becomes mature (allowing international subscribers access to huge amounts of U.S.-based knowledge, not only from television, but also from computer networks such as the Internet), American culture, the United States' most powerful export, will continue to grow in influence. People in less prosperous nations see these products as symbols of a wealthy nation that perhaps is choking on consumer goods. They worry that people may value wealth and material possessions more than they value the traditional family, the community, or even religion. The American cultural influence may cause a broadening link of commonality among younger generations of the developed world while at the same time threatening the spread of crime, drugs, divorce, and the cultural decay that troubles the west, alienating the United States from more conservative cultures.

Others suggest that American culture has helped to liberate the world because it speaks of democracy. After all, rock music, fast food, and blue jeans are available, not only to a privileged few, but to everyone. Others argue that American culture actually reflects a good deal of cultural diversity—that American culture is influenced by a wide variety of ethnic groups and national origins. After all, cultural exchange is not only one way. American musicians find inspiration in South African music, for example. Some British bands combine music styles from Latin America, Eastern Europe, and Africa. The Japanese group Chang Typhoon mixes traditional Japanese styles with reggae, funk, and salsa. French, Chinese, Mexican, and now Southeast Asian fast-food restaurants abound in the United States. Transnational corporations, communication, and entertainment have bridged geographical space and physical obstacles that once separated countries, cultures, and peoples. Mountain and ocean barriers were the archetypal symbols of the age behind. Communication satellites and world trade are the symbol of the age ahead!

---

perfect tool that will allow the sophisticated knowledge worker to mine the databases and knowledge bases of a huge number of sources. Power in the world today is shifting from being defined by money to being defined by information. This information age is exacerbating one of the major problems of the coming decades, the increasing disparity between have and have not nations and have and have not regions within countries (Figure 1.2).

The world contains handfuls of people here and there who are isolated and sheltered, who have never seen television, used a phone, or ridden in a motor vehicle. Access to communications and transportation of the information age is restricted by an uneven division of wealth worldwide. Even within a region, access may be restricted because of uneven distribution of wealth or because discrimination against a tribe, race, or women. Alvin Toffler, author of *Future Shock*, refers to the effect that information technology will have on the tempo of human activity and calls them "fast" and "slow" societies.

A determination to retain cultural traditions in the face of globalization in some cases leads to intolerance of people who display other beliefs, social forms, and material traits. Political, economic, and social disputes and unrest, wars, and conflict erupt in places such as Southern Europe, Central Africa and the Middle East, where different cultural groups have been unable to peacefully share the same geography. This telecommunication revolution that promotes cultural globalization has also permitted the preservation of cultural diversity in these cases.

Globalization of the Economy

The world economy is at work in creating a global cultural uniformity. Companies, societies, and individuals that were once unaffected by events and economic activity elsewhere now share a singular economic world with other companies, societies, and workers. The fate of an aerospace worker is tied to political change in Eastern Europe. The job of an auto worker in Detroit is related to the fall of the Mexican peso and the auto industry's investment in production plants in Mexico along the border. The globalization of the economy has meant that national and state borders and differences between financial markets has become much less important because of a number of trends: (1) the globalization of finance, (2) the increasing importance of transnational corporations, (3) global foreign direct investment from the *core regions* of the world—North America, Western Europe, and East Asia, (4) global specialization in the location of production, (5) globalization of the tertiary sector of the economy, (6) the globalization of the office function, (7) global tourism.

Globalization of Finance

In the past, companies had some difficulty moving small amounts of money from one country to another. International monetary exchanges frequently involved cumbersome procedures that could tie up the funds for weeks until all the paperwork had been approved.

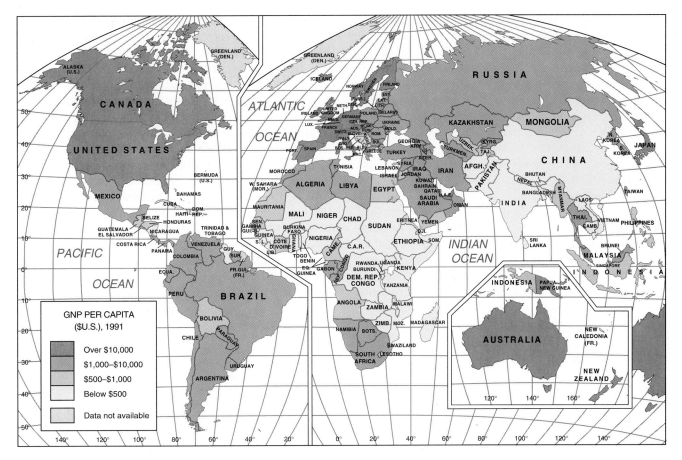

FIGURE 1.2

World variations and gross national product per person. Countries of the so-called First World, including North America, Western Europe, Japan, and Australia. These countries have gross national products per person at more than $10,000 per capita. Little data is presently available for the former Second World, including Eastern European socialist societies and the Soviet Union; however, now that the propaganda war is over, most Westerners realize that per capita incomes were falsely inflated. The remainder of the world consists of Latin America, Africa, and South and East Asia. Here, incomes are generally less than $2,500 per capita. (See color insert for more illustrative map.)

In the case of communist countries, no currency could be removed without government approval, and many governments prohibited large sums of money to move beyond their country borders.

Modern telecommunications and transportation allowed the technical aspects of moving money, materials, products, technology, and other economic assets—*factor flows*—around the world. Because of the Internet, companies can now organize economic activity around the world from a single point and can outsource (subcontract) white-collar banking services and credit services.

The telecommunications revolution has allowed a single global capital market. Computers can now monitor and trade in national currencies stocks, bonds, and annuities listed anywhere in the world instantaneously. Banks, financial houses, and corporations can operate worldwide partly because the decision centers that control the global economy—New York, London, and Tokyo—are all in different time zones. When Tokyo's

stock market closes at 3 p.m., it is 6:00 a.m. in London, only a few hours before the opening of the day's trading. The stock market will open in New York City at 9:00 a.m., when the London stock market is still open. When the New York Stock Exchange closes at 4:00 p.m., it is 6:00 a.m. the next morning in Tokyo, only hours before the opening of the Tokyo market for the following day. Consequently, banks and corporations can react immediately to changes in the value of commodities or gold on the world market, and the rate of exchange between the dollar, the pound, the yen, and other currencies. By 1995, over $900 billion was traded daily on the world's foreign exchange markets. This total is almost 100 times the average daily merchandise trade. In 1995, an additional $8 trillion worth of stock changed hands on the major stock exchanges.

Countries are expected to regulate their own stock exchanges, and the home countries of global banks are responsible for supervising the bank's financial operations.

Some companies attempted to avoid regulation and supervision by taking the assets offshore or to countries renowned for lenient banking regulations. An example of a major offshore banking center is the British Caribbean Colony of the Cayman Islands (population 25,000), holding assets of over $600 billion in 1995. Most assets are not held in banks, vaults, or leaded enclosures underground, but are held electronically in computers.

Another principal source of capital in banking is Tokyo, Japan, and the Tokyo stock market, which affects investments and jobs throughout North America and Europe. The Persian Gulf oil states are another important source of world capital, so the expensive Persian Gulf conflict raised interest rates on U.S. bank loans. Banking policy in Canada or Johannesburg, South Africa, affect charges of U.S. banks for precious metals, minerals, and primary products that flow to North America. Financial manipulations in Brazil, Argentina, and India affect the security of many U.S. elderly peoples' pension plans, which have been invested abroad (Figure 1.3).

Transnational Corporations

The globalization of the economy has been spearheaded by transnational corporations (TNCs), sometimes referred to as multinational corporations (MNCs), or multinational enterprises (MNEs). A transnational corporation may conduct research, operate industries, and sell products in many countries, not just where its headquarters are based. Most transnational corporations maintain their headquarters in one of the three regions of the core countries—North America (United States and Canada), Western Europe (especially Germany, France, the United Kingdom, and the Netherlands) and Japan. Most of the factories and firm locations of transnational corporations are within these three regions, as well as their headquarters. The United Nations in 1995 reported that transnational corporation employed 80 million in the core regions, with 20 million elsewhere (see Table 1.1).

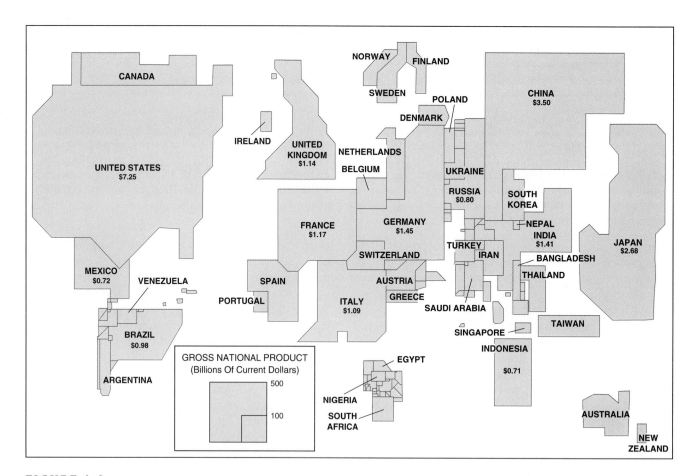

FIGURE 1.3

Relative sizes of the nations' economies, 1995. This map shows the size of each country's Gross National Product and not its actual land area. On this GNP cartogram, the whole African continent is smaller than Canada even though it has over 20 times the population of Canada. South America, China, and Russia also have small national economies relative to Japan, Europe, and the United States.

🔯

T a b l e 1.1
U.S. Multinational Corporations: Selected Examples, 1991

Company	Foreign Revenues (in billions of dollars)	Foreign Revenues as a Percentage of Total Revenues	Foreign Assets as a Percentage of Total Assets
Exxon	$78.1	75.9%	58.4%
IBM	40.4	62.3	54.3
General Motors	39.1	31.8	23.4
Mobil	38.8	68.1	56.6
Ford Motor	34.5	39.1	31.6
du Pont	17.1	44.8	39.8
Philip Morris	13.2	27.4	25.6
Xerox	8.6	44.3	27.3
Coca-Cola	7.4	64.0	46.8
Motorola	6.4	55.9	33.7
ITT	6.3	30.9	14.5
Goodyear Tire	4.7	42.9	40.2
PepsiCo	4.4	22.6	23.6
Woolworth	4.2	42.4	46.6
Monsanto	3.2	36.4	34.2
McDonald's	3.0	44.6	45.8

Source: "The 100 Largest U.S. Multinationals," Forbes (July 20, 1992), pp. 298–300.

Nature of Multinational Corporations

In 1970 the world's 15 richest nations were headquarters to 7,500 MNCs. However, by 1994 these same countries hosted 25,000 MNCs. By 1997, there were some 50,000 MNCs in the world controlling about 40 percent of all private sector assets, and accounting for a third of the goods produced for the world's market economies. The International Labor Office estimates that MNCs employ 100 million people directly, which is 4% of the employment in developed regions and 12% of employment in developing regions. Some countries such as Canada have extremely high proportions of total production by MNCs, especially in mining, manufacturing, and petroleum sectors of their economy. MNCs also play a disproportionately dominant role in other developed countries such as Belgium, France, the Netherlands, Italy, Great Britain, and Japan. Very large MNCs have sales of goods and services exceeding $100 billion annually (Figure 1.4). When the MNCs in various countries are ranked on the basis of GNP, or total sales, the top MNCs are larger than most countries' total economies. This is important to a small country's economy that can be affected by a decision of a global corporation (Figure 1.5). Today, multinational corporations control greater than half of total international trade simply via intracorporation transfers of components,

services, investments, profits, and managerial talent among their scattered plants and offices in various countries. Most of this intrafirm trade is not finished products and services, but components, subassemblies, parts, and semifinished products. (Multinational corporations and trade theory are discussed extensively in Chapters 10 and 11.)

Today, MNCs, not countries, are the primary agents of international trade, largely between and within their organizations. In so doing, MNCs change countries' reserves of natural resources in effect by moving human and physical capital and technology from one part of the world to the other, creating a new asset base, and allowing production and manufacturing to occur in outsourced locations where they may not have happened otherwise. A MNC will produce in a country where a set of characteristics taken together is more attractive: location, resource endowments, size and nature of market, political environment (Figure 1.6). Further, the MNC is able to use transfer pricing, the practice of price setting for goods and services provided by subsidiaries, so as to transfer taxable profits to the lowest tax country possible and minimize tax overhead—to shift its profits to countries that have the lowest tax rates (Figure 1.7).

The MNC is able to provide the developing world with new technology (called *technology transfer*). Today, MNCs develop profits from the foreign use of

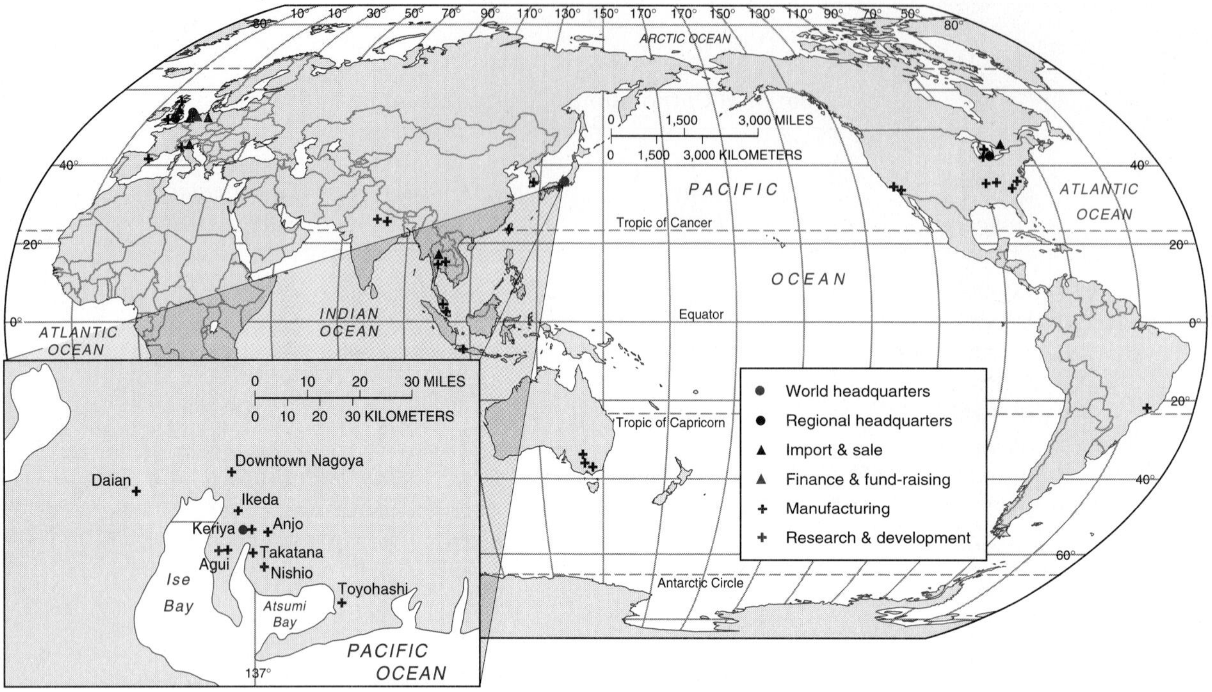

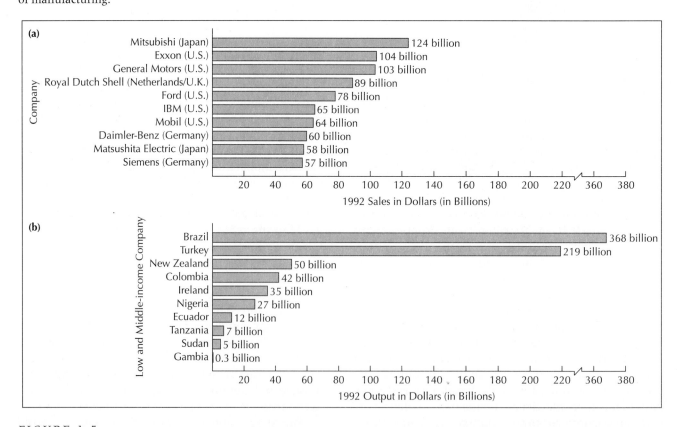

FIGURE 1.4

Nippon is a Japanese transnational corporation. Its world headquarters, research labs, and numerous factories are located in Nagoya, 100 miles west of Tokyo, Japan. Sales and manufacturing activities are scattered throughout North America, Europe, and the far East. Nippon makes air conditioners, heaters, and automobile and truck parts and demonstrates the globalization of manufacturing.

FIGURE 1.5

Many multinational corporations are larger than some national economies of world countries. The relative size of an MNC is important to small countries whose economies can be drastically affected by a decision a global corporation might make. *(Source: CIA World Fact Book, 1995; World Almanac, 1994)*

A Toy's Globalization Story: U.S. to Canada to China to Hong Kong to U.S.

① DESIGN/LICENSE
A U.S. company licenses a figure to a toy manufacturer in Ontario. The two companies design the product together.

② FINDING PRODUCERS
The Ontario company searches the world to find low-cost subcontractors to produce the toy. It settles on a factory in southern China where workers are paid about $2 a day to assemble parts from around the world.

③ INSPECTION
After the product is made, it is shipped to factories in Hong Kong or other parts of China where it is inspected by the buying agents–typically, the buyers for major retail chains like Wal-Mart, Kmart or Toys "R" Us.

④ SHIPPING
The Ontario company's Hong Kong subsidiary arranges for the product to be delivered to the "consolidators," the agents of the major retailers, who ship the product to the U.S. by boat.

⑤ ENTERING THE U.S.
The toy enters a major port like Long Beach, CA for customs clearance. From there it is forwarded to a retailer's distribution center.

⑥ INTO THE STORES
The toy arrives at local stores. The $8 retail price reflects the cost of shipping, marketing, materials and the U.S. company's licensing agreement. Wages paid to the workers in China make up only a tiny percent of the price.

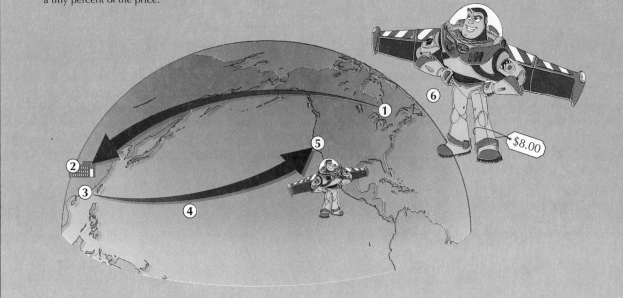

FIGURE 1.6
A toy's globalization story. Action figures are among the hottest sellers for children in toy stores across North America. How the product gets from a U.S. company to a retailer's shelf is a long, complicated process. This toy's story demonstrates the globalization of manufacturing and marketing.

technologies that they have developed in the home country but that are now used in the foreign-based plants under their own control by a foreign subsidiary. MNCs have a potential for making efficient allocations of the world's resources, but not in all instances does technology transfer trickle down to foreign-owned companies. The maquilladora industry in Mexico has been criticized for this.

Multinational corporations are able to compete on a world scale because they can operate with greater information efficiency and share that with their subsidiaries and branches throughout the world via the Internet and modern satellite and fiber-optic communication systems. This transnational communications ability is a tremendous advantage to the MNC in that it becomes aware of markets, products, labor, and business opportunities before other competing countries. Other advantages of the MNC are its large store of capital, technology, managerial skill, and overall scale economies.

Transnational corporations evolve in four stages. The first stage, demand abroad, was satisfied by export of a commodity from the corporation's home country to a new foreign market. In the second stage,

```
┌─────────────────────────────────────────┐
│              Germany                    │
│          (Tax Rate = 48%)               │
│                                         │
│ A software system is manufactured by    │
│ the parent firm at a cost of $2,000.    │
│ It is then sold to the firm's Irish     │
│ subsidiary for $2,000. German taxes     │
│ paid: $0.                               │
└─────────────────────────────────────────┘
                    │
                    ▼
┌─────────────────────────────────────────┐
│              Ireland                    │
│          (Tax Rate = 4%)                │
│                                         │
│ The Irish subsidiary turns around and   │
│ resells the same system to the U.S.     │
│ subsidiary for $2,500, earning a $500   │
│ profit. Irish taxes paid: $20.          │
└─────────────────────────────────────────┘
                    │
                    ▼
┌─────────────────────────────────────────┐
│           United States                 │
│          (Tax Rate = 34%)               │
│                                         │
│ The U.S. subsidiary sells the system at │
│ cost, for $2,500. No profit is earned.  │
│ U.S. taxes paid: $0. The Irish          │
│ subsidiary then lends money to the U.S. │
│ subsidiary for future expansion.        │
└─────────────────────────────────────────┘
```

FIGURE 1.7
Transfer pricing. Transfer pricing is a technique used by multinational corporations to shift profits to countries with low corporate tax rates and thus suffer a smaller total tax bite. The MNC in this case pays no taxes in Germany and the United States.

Foreign direct investment (FDI) indicates investment by foreigners in wholly owned factories that are operated by the foreign owner of the multinational corporation. U.S. transnational corporations are most likely to invest in Europe, Canada, and Latin America. Western European transnationals are most likely to invest in Eastern European, Russian, and African markets, as well as in North America. Japanese transnationals are most likely to invest in Asia and in North America.

Since the 1980s, governments in the three regions where transnational corporations are based, North America, Europe, and East Asia, have altered tax codes and regulations that formerly hindered transnational operations. For one, they have allowed transfer of pricing (see Figure 1.7). Other countries where transnational corporations wish to invest, especially in developing countries, have also changed laws from bureaucratic procedures to foster transnational operations within their borders. These changes have been successful in triggering economic growth and fostering FDI. In the 1980s, global FDI grew three times faster than world trade and four times faster than total world output, to a total of $2.5 trillion. This flow of FDI greatly increased and significantly rearranged the world's multinational manufacturing.

the transnational corporation established production facilities abroad to supply these new markets. Exports of the same item from the home country dropped. In stage three, the foreign production facilities supplied foreign markets other than the local market first serviced. In stage four, the foreign production facilities exported back to the home country products produced more efficiently. These four stages appeared first in the production of primary materials, including copper and oil, then in the evolution of international manufacturing, and are now in the international tertiary sector.

IBM corporate headquarters, Armonk, New York. Whether measured in terms of value of sales, value of assets, or number of employees, IBM is one of the largest MNCs in the world. The domestic activities of multinationals are only a part of their worldwide activities. They are the epitome of direct investors abroad. In the first half of the 1990s, however, IBM saw its market share for computer equipment drop due to aggressive policies of Japanese companies, and such software systems as Apple and Microsoft. *(Photo: Vergara Bob, IBM Corporation)*

In 1995, however, the United States and Canada, the European Union, and countries of East Asia, principally Japan, still accounted for 80 percent of the FDI, according to the United Nations. However, companies from an increasing number of developing countries have become active investors, including Brazil, Argentina, Malaysia, Indonesia, Mexico, Korea, Taiwan, and Singapore. In 1995, U.S. and Canadian firms accounted for 30% of the global FDI, and conversely, the United States was the world's largest recipient of FDI from other countries. FDI in the United States alone accounts for 12% of manufacturing jobs (Figure 1.8). FDI creates more new manufacturing jobs in the United States than American companies do. Foreign firms accounted for 23% of total exports from the United States in 1997 alone. Such FDI in America is carrying on research, introducing new technology and management techniques, and changing Americans' daily lives, both at work and in the shopping mall.

Globalization of Investment from the Core Countries

The globalization of the economy is increasingly centered in the core regions—North America, Western Europe, and Japan, as well as the Tigers of the Pacific Rim. From the world cities, or "command centers" in New York, London, and Tokyo, work orders are sent instantaneously to factory shops and research centers around the world. Manufacturing enterprises have located their production and assembly lines and lower-cost offices outside the high-cost core countries. For example, most U.S. sportswear companies, centered in New York and Los Angeles, have moved their production to Asian countries. Fila maintains headquarters in Italy but has moved 90% of its athletic sportswear to Asian countries as well. Mitsubishi corporate offices are, of course, in Japan, but increasingly its electronic components such as VCRs and even automobiles are produced in other Asian countries with lower labor costs. The countries in Latin America, Africa, and Asia

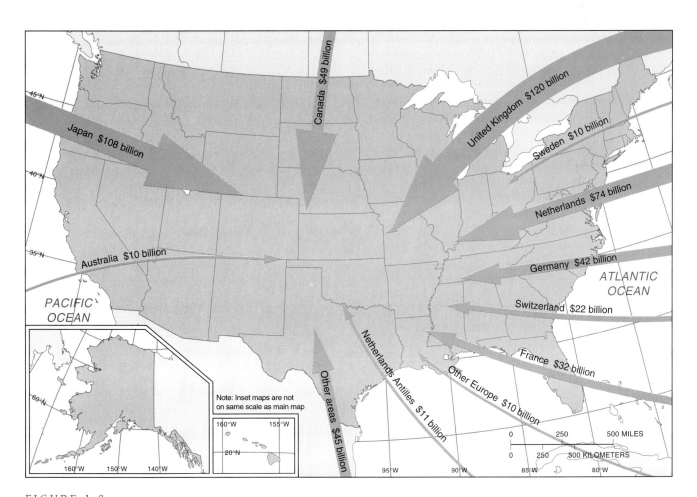

FIGURE 1.8
Foreign direct investment (FDI) in the United States comes primarily from Europe and Japan. Developed countries of the world are the primary investors in other developed countries. This diagram demonstrates the globalization of finance and the globalization of the world economy.

contain three-fourths of the world population and almost all of its population growth. Countries in these realms find themselves on the periphery of the world economy, or outer edge of global investment decisions made by transnational corporations.

Three trends demarcate foreign direct investments (FDI) in developing countries. First, the proportion of FDI that core countries are allocating to periphery countries is declining. Core countries are increasingly investing in one another. Second, FDI is becoming more geographically selective. Countries that are attracting the greatest FDI from the core countries are those that have chosen the export-led method of economic growth (discussed in Chapters 10 and 11). Export-led industrialization is characterized by countries welcoming foreign investment to build factories that will manufacture goods for international markets and employ local labor. Export-led policies rely on global capital markets to facilitate international investment and global marketing networks to distribute the products. The countries that have grown the fastest in recent decades have generally followed the export-led approach as opposed to the alternate approach, import substitution.

Import substitution policy protects domestic infant industries. This form of economic protectionism helped some countries industrialize in the past but involves economic risks. An import substitution country normally raises tariffs (taxes on imports) to protect a particular home industry. But all national consumers in that country subsidize the industry due to higher prices than the world market would suggest. Some protected industries will never be able to survive worldwide competition, so that in this way the population will go on subsidizing the industry indefinitely. Their economies are growing rapidly.

Four Asian nations attracted a larger proportion of FDI than others and grew at a 10% average per annum clip during the 1980s and early 1990s. They earned the nickname "The Four Tigers": Singapore, South Korea, Taiwan, and Hong Kong, which now belongs to China. They have reached core-level prosperity and have themselves become sources of capital and technology for the next level of Asian development.

A third characteristic of FDI in developing countries is that each core member has a majority share of the FDI in a group FDI in linked countries that have become its economic satellites. These link countries help explain world patterns of trade. For example, in recent decades, the United States has had a trade deficit in electronic consumer products with Hong Kong and South Korea. Table 1.2 shows that Hong Kong and South Korea fall within the Japanese group. Electronic factories in these countries are owned by Japanese corporations. Thus, the consumer electronics products that the United States and Canada import from Hong Kong

Table 1.2
Triad Foreign Investment Clusters

United States	European Union	Japan
Argentina	Brazil	Hong Kong
Bolivia	CIS states	South Korea
Chile	Croatia	Thailand
Colombia	Czechoslovakia	
Mexico	Hungary	
Panama	Poland	
Philippines	Slovenia	
Saudi Arabia	Yugoslavia	
Venezuela		

Note: Clusters identified by the UN Centre for the Study of Transnational Corporations, 1991. Czechoslovakia and Yugoslavia have each dissolved into several new states, but the new states remain within the designated spheres of economic influence.

and South Korea not only profit those countries, but profit Japanese corporations as well. In a similar manner, many of the imports from Latin America to Europe are products of U.S. corporations.

Developing countries compete with one another to attract FDI, and the greater degree to which such countries can improve their transportation, trade, and communications infrastructure, including roads, ports, highways, telephone systems, as well as education and political law and order, will attract new investment to their manufacturing base. The place of development within developing countries is increasingly being determined by their ability to attract foreign capital. If a developing country seeks foreign FDI, it must openly reveal national economic, political, and social conditions in order for the transnational corporations to do business there. Consistent enforcement and practice of the laws and statutes of the land are important. The U.N. center for the study of transnational corporations in 1995 stated that the substitution of private investment for foreign aid and foreign bank loans has a substantial liberalizing and democratizing influence on these countries in the world economy.

Global Locational Specialization of Work

Every location in the global economy can play a distinctive role, but more than ever this role is based on its particular accessibility and assets. Transnational corporations assess the economic and locational assets of each place in the world economy. Specialization in the location of global production shows a declining role of population and resources, the original factors of production in global development. Today, brain power has replaced muscle power, *transmaterialization* has changed the nature of resources, and *dematerialization* in the nature of products. Global transactions are replacing the older order of world trade. Input factors and components move intrafirm, final goods are fabricated close to the point of consumption, and national boundaries count much less than they did in the past global economy.

In the new global economy, transnational corporations maintain a competitive edge by correctly identifying geographic factors and the optimum location of each of the activities, including engineering systems, raw material extraction, production, storage, office function, marketing, and managing. Suitable places for each activity may be clustered in one country or may be disbursed in countries around the world. The resulting globalization of the economy has increased economic differences among places in a have and have-not world. Factories are closed in some locations and reopened in other countries (Table 1.3). Some countries become centers of technical research, whereas others become centers for low-skilled manual tasks. Changes in the geography of production have led to the spatial division of labor in which a region's labor specializes in particular functions. Transnationals decide where to locate in response to characteristics of local labor force, skill level, prevailing wage, attitudes toward unions, tariffs, and transportation rates. A transnational may close factories in regions with high wage rates and strong labor unions.

Table 1.3
Invented in the United States, Manufactured Elsewhere

U.S.-Invented Technology	U.S. Producers' Share of U.S. Market (%)*		
	1970	*1980*	*1992*
Phonographs	90	30	1
Color TVs	90	60	10
Audiotape recorders	40	10	0
Videotape recorders	10	1	1
Machine-tool centers	99	79	35
Telephones	99	88	25
Semiconductors	88	65	64
Computers	NA	94	74

*1–(value of imports / value of product shipments + value of imports)
Source: U.S. Department of Commerce, U.S. Industrial Outlook *(Washington, D.C.: Government Printing Office, 1993).*

Globalization of the Tertiary Sector

The globalization of service provision (the tertiary sector) and consumption is taking a growing share of international investment. Frequently, when people hear of foreign trade, they think of oil shipments and barges full of iron ore, bananas, or coffee. Cars from the Far East and clothing from India are no doubt shipped around the world. It is easy to imagine such tangible products being traded on international markets. The international tertiary sector, however, has helped transcend trade in primary products, natural resources, and manufactured goods. Trade in the tertiary sector—in the services—surpassed $1.5 trillion dollars in 1995, and its growth in the world economy is much more rapid than manufactured goods or natural resources. For example, U.S. service exports dwarf auto exports—over $200 billion versus $60 billion in 1996.

Business services that provided essential needs to transnational corporations in the United States and in other home lands provide these services continually as the transnational corporations venture out into the world arena (Figure 1.9). The international tertiary sector explains the globalization of legal counsel, business consulting, accounting, advertising, billing, and computer services. Many professionals, such as architects, physicians, software designers, and business consultants, market their skills throughout the world. Globalization of the tertiary sector includes welcoming tourists from foreign countries, helping business of the local economy, welcoming foreign students to local universities, and providing financial services, entertainment, and music recordings produced for global markets. Globalization of services displays the same trends as the globalization of manufacturing. Countries are losing their autonomy to increasingly global markets. The globalization of services operates in a world of a declining role of the nation state, but the continuing reinforcement of cultural differences at both national and regional scales.

The United States is by far the world's leading exporter of services and yearly shows huge trade surpluses. (When mixed with goods trading, however, the United States has a net deficit.) In education alone, the United States earned $10 billion in 1995. Other North American tertiary exports include real estate development, design and management, accounting, medical care, business consulting, computer software development, legal services, advertising, commodity brokerage and architectural design. Entertainment is by far the largest category of services for the United States. American movies, music, television programs, and home videos dominate the world's airwaves and account for a significant annual trade surplus.

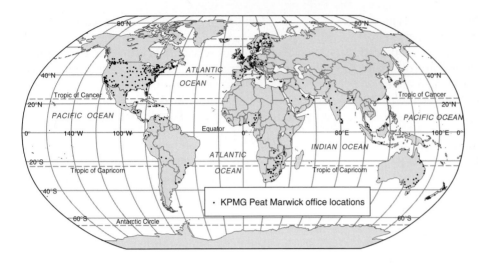

FIGURE 1.9
KPMG Peat Marwick, a joint Dutch and U.S. company, is the world's largest accounting firm. It has offices in all the major economic capitals in the world and demonstrates the globalization of finance in the world economy. *(Source: Courtesy of KPMG Peat Marwick)*

The mega service firms of North America are not really national and will respond in non-national ways. Large corporations are not international or transnational; they are non-national and will increasingly see economics and politics without a particular national flavor.

As the global transmission of television becomes ubiquitous and the intercontinental information network becomes mature (allowing international subscribers access to a huge amount of American-based knowledge) American culture, the United States' most powerful export, will grow in influence. This will have both positive and negative effects on the culture and the disposition of the world's peoples. It is likely to broaden the links of commonality among the younger generations of the developed world, especially those who are Internet and World Wide Web savvy. At the same time, it threatens to further alienate the United States from more conservative cultures.

By 1996, about 50% of the revenue of the seven largest U.S. movie studios came from outside North America. Additionally, foreign countries presented the greatest opportunities for increasing sales. Today, the four largest motion picture studios in the United States are foreign owned: MGM/UA, owned by an Italian MNC; 20th Century Fox, owned by an Australian MNC; Universal Studios, also owned by a transnational Japanese MNC; and Walt Disney Corporation is also financed by Japanese investors. Such an increasing dependence on foreign markets means that the treatment of foreign countries and their cultures is notably more sympathetic than when the ownership and intended audience were restricted to North America alone. Such shaping of culture is making North American movie goers more cosmopolitan.

The Global Office

The United States and other core countries of the world are exporting clerical work to developing countries. The developing countries have highly educated young people who are willing to work for lower wages than those in the core countries, but whose domestic economies cannot employ all of them. Transnational corporations require a large number of clerical assistants, computer operators and programmers, and telecommunication experts. The new telecommunications technology allows these corporations to access pools of skilled but inexpensive labor, wherever they may be found in the developing world.

U.S. companies have been moving clerical operations out of the large urban centers to save the cost of office space and commuting time for decades. Now, these same companies are relocating jobs around the world. For example, New Yorkers' bank credit card charges are processed in Mexico, and Boeing and Bechtel multinational corporations have located software development facilities that provide many jobs in Ireland. The time and cost of distance to faraway places for electronic transmission is almost negligible. The relocation of global office jobs to developing countries can boost national economies and will spread the globalization of cultural values and practices.

Perhaps no other country has as many skilled workers as India. Only in India does the production of university graduates outpace the economy. India generates more university graduates than the United States and Canada combined, some 40% of degrees in science or engineering. Multinationals still get what they came for—inexpensive labor. A circuit board designer in California earns between $60,000 and $100,000 a year, but makes only $25,000 per year in Taiwan and less than $10,000 a year in India or China. Today's Asian labor is as well trained as the West's, plus many employers find the work ethic and attitude to be superior. Major North American information economy transnationals are finding and making a home in India, which claims the second largest pool of English-speaking scientific talent after the United States and Canada. India boasts 100,000 software engineers and technicians. Hundreds of companies, many locally owned, supply software to western customers. An Indian engineer with five years experience earns $800 a month and a position in the upper middle class.

India's software development firms are connected to U.S. and European customers through a telecommunication network that was the brainchild of Indian software designers. The government of India provided all the requisite inputs for transnationals to hire local talent—training for manpower, high-speed data communications, red-tape–free systems, and a virtual red carpet for FDI. Yet infrastructure remains a problem. India has 9 million phone lines—one line per 100 people—and expects to increase that to 20 million phone lines by the year 2000. The price tag is $15 billion. There are 4 million people on the waiting list for phones in India, some of whom have been waiting for over five years.

Globalization of Tourism

By some measurements, tourism is the world's largest industry. The World Travel and Tourism Council estimated that 1995 travel and tourism activity generated $5 trillion in gross output, employed 250 million people, and produced 14% of the world's gross domestic product (GDP). Tourism is the leading earner of foreign exchange in many countries, and in 1995, tourism became the largest export industry of even the United States and Canada, earning a $15 billion surplus. Tourism is expected to grow by at least 6% per year worldwide, with higher rates reported for some developing countries, especially tropical regions or areas spectacularly endowed with wildlife for ecotourists (see Figure 1.10).

Individuals of the core countries account for not surprisingly the largest tourist expenditure. Europe is the most popular tourist destination, followed by the Americas and East Asia/The Pacific. When tourism is measured as a proportion of total national exports, the importance of this industry in the development of poor countries is clear. Turkey, for example, quadrupled the dollar value of its exports between 1980 and 1990, but the value of its tourist revenue increased three times as rapidly. By 1990, Turkey's revenue from tourism was 25% of the value of Turkey's export goods. Mexico is another country that bases its economic development largely on tourism. Mexico's foreign exchange earnings from tourism tripled between 1945 and 1955 and are

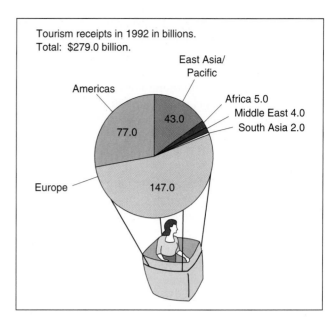

FIGURE 1.10

World tourism 1992 in billions of dollars. As a region, Europe is the largest international tourist destination because many small countries with affluent populations generate numerous world travelers. The United States is the country with the greatest number of arrivals and the greatest tourism expenditures.

among the world's highest today. Being relatively accessible to the world's largest tourist flow in North America has helped their tourist foreign exchange position. Thailand was one of the world's fastest growing economies from 1985 to 1995, yet despite Thailand's rapid industrialization, tourism to Thailand grew twice as fast. By 1995, tourism represented 10% of Thailand's gross domestic product and was by far the country's leading foreign exchange earner.

According to Edward Bergman (1996), the tourist potential of any location depends on the three A's: (1) the region's *accessibility* to the core countries; (2) the region's quality and number of *accommodations*; and (3) the region's cultural and physical *attractions*. Many world travelers, especially from northern, colder countries, find sunshine and a sandy beach to be the most attractive. The mountain scenery and recreation opportunities in the Rockies, the Alps, and even Mt. Fuji draw millions of tourists yearly. *Ecotourism* is growing in popularity; it is based on travel to distinctive, highly scenic regions of the world, with unusual natural environments or wildlife. The rich animal life in Tanzania and Kenya attracts millions yearly, as does the rich bird life on Lake Nicaragua, or the abundant sea life off the coast of Patagonia in Argentina. Antarctica is the fastest growing region for ecotourism today. Ecotourism provides capital needed to protect a country's natural environment as well as the biodiversity of rare animals and plants.

In sum, the countries best endowed for global tourism provide both natural and cultural attractions to their visitors, along with pleasant climates, good beaches, and attractive social and political milieus. Political stability is all important. For example, a favorite circuit for both Europeans and Japanese in the United States includes travel to the San Diego Zoo, the Grand Canyon, Disneyland, and San Francisco, and return.

Information Technology (IT) Driving Globalization

Communication improvements mean that a globalization of the world economy is moving forward rapidly to a point where people in any location can receive and send information to others in any other location almost at any time. World economy is increasingly moving towards moving information instead of people. The global network, which will export and import not products but information, will allow innovations to sweep the world at rapid rates. Power in the global economy will shift from being defined by money to being defined by access to information. This will exacerbate and increase the disparity between the have and have-not nations. "Fast" and "slow" societies in the future will be referring to the effect that information technology will have on the tempo of human affairs. This information-based economy means that relative success of individuals or groups is based on access to information, more than money or products, more than natural resources, labor pools, and other traditional metrics of power and wealth. A global information network of the information age is a pool that will allow a *knowledge worker* of the global economy to mine the databases and knowledge bases of all important information bases of the world. The world will be interdependent more than ever before, and this interdependent global network will facilitate the interchange of information among researchers and disciplines as never before. Information systems will be the facilitators of the knowledge-based world economy.

Real time information systems mean that information becomes available as it happens, or at least as soon as possible after software programs process it and make it available. Real time implies immediate accessibility so that everyone on the Net can seek critical information by accessing a computer screen. Real time is one of the essential differences between the world economy in the future and the world economy in the past. With real time information systems, more people will be making more decisions in a customized world economy, as the number of people who interface with customers becomes part of a self-managing business unit. Companies, individuals, and MNCs need feedback on their decisions as soon as it becomes available so that they can learn faster and make constant adaptions to meet customers' needs better than anyone else. The globalization of the world economy in the future will demand real time information systems so that world business decisions will be made with the minimum of bureaucracy. Rich companies need feedback immediately, not a month or a year from now when the accountants audit the books and determine everything in accordance with accounting principles.

The communications and IT revolution of the late 1990s is based on the networking of individual computers (see Chapter 4). The engine of change for the 21st century is based on the communications revolution of computers being linked to global networks of information databases and personal computers, linked to one another by fiber-optic and satellite communications. These communication networks, such as the Internet, allow almost instantaneous communication to anyone else on the network. Communication linkages can include photographs, voice and music, videos, television images and programs, films, documents, books, pictures, mail, and spreadsheets. With large commercial networks such as America Online or Compuserve, one can teleshop at 1,000 stores, make reservations at hotels in Europe, buy airplane tickets, monitor the weather and stock market, pay bills, and download large text files from newspapers, magazines, and even encyclopedias. Gigantic databases located at government agencies, international organizations such as the UN and the World Bank, Population Reference Bureau, and research institutions across the world are accessible to anyone at almost no cost.

Economic geographers study human-environmental problems, inventory and monitor resources, and provide expertise to better manage large corporations. Key questions include: What should be produced? How should it be produced? Where should it be produced? For whom should it be produced? *(Photo: ESRI Environmental Systems Research Institute, Inc.)*

Globalization of Environmental Problems

A constant list of new facts suggests changes in our environment. They are hints that something serious may happen around the globe. Consider these:

- Coastal areas have fewer beds of sea grass that house many marine species, while the algae epidemic is continuing to kill off more sea life, including coral reefs.
- Biologists are confused as to why sponges are quickly dying off near Greece, and why in France sea urchins are loosing their spikes.
- Weather extremes resulted in 500 deaths in Chicago in the summer of 1995, due to a heat wave. Two years earlier, the Midwest experienced its worst flood on record, while 1997 floods on the West Coast were unparalleled.
- A strange virus has been blamed for the recent death of 20,000 seals in the North Sea, while another virus caused a major epidemic among dolphins.
- We have seen a 90% decline in butterflies in the eastern United States, and the normally crowded routes southward to Mexico are nearly empty. The decline in these Monarch butterflies is due to their roosting trees being cut.
- Surprisingly, the North Atlantic has become rougher over the past 30 years. Recent studies indicate that waves are nearly 50% higher now than they were 30 years ago. Hurricanes are now more frequent.

Twenty years ago, a list of these occurrences would have sounded like something from an episode of *Twilight Zone* or *Star Trek*. Our job as informed citizens of the world must be to understand global environmental change as it affects the world economy, world regions, and realms (see Chapter 3).

Growth of the East Asian Economy

Many of the changes that went on in the world economy during the 1980s were fueled by the remarkable growth of East Asia. Although there were exceptions (i.e., the Philippines), these nations remained largely free from the debt crisis. In Japan and the *Four Tigers* (South Korea, Taiwan, Hong Kong, and Singapore), a new wave of products began to sweep across the economies of North America and Europe. Other East Asian nations, such as Malaysia, Indonesia, and Thailand, began to follow the example set by the leaders, developing extremely competitive, export-oriented industrial sectors.

The nations of East Asia have not been content to follow their natural comparative advantage in low-wage, labor-intensive production but have jumped right into value-added, sophisticated, high-tech production, which traditional economic theory says should have come much later in their development. They have been aided in part by developments in communication and information technology that allow MNCs to split up their production process and locate pieces of it around the globe (Figure 1.11). At the same time, however, they appear to have developed an alternative model of capitalism and international trade in which policymakers "govern" the market in ways that channel investment and consumption to strategic sectors, while still relying primarily on market forces and institutions to ensure that resources are allocated efficiently. These policies have posed a serious challenge to the economies of North America and Europe and will continue to generate friction and new developments well into the 1990s.

Why do U.S. businesses keep flocking to Asia, despite the risks? The potential for profit remains huge in markets that have leaped from developing-world to developed-world living standards in a single generation.

Despite the potential for temporary economic setbacks in debt-heavy countries such as Indonesia, Malaysia, or Thailand, growth in Asia should remain strong. Merrill Lynch estimates that East Asian economies will average 8% annual growth over the next 10 years, compared with about 3% for the United States and the rest of the industrial world.

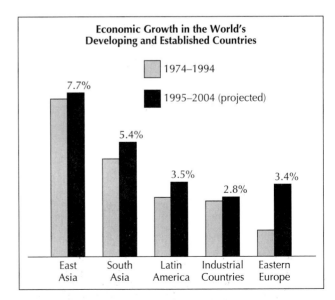

FIGURE 1.11
Economic growth in world's developing and established countries. *(Source: World Bank, 1997)*

Emissions from smokestacks, such as these in the Saar industrial region of Germany, increase acidic deposition and contribute to the buildup of carbon dioxide in the atmosphere. Until recently, carbon dioxide released from fuel combustion and deforestation were considered the chief contributors to the greenhouse effect. In the 1980s it was discovered that other trace gases—methane, ozone, chlorofluorocarbons, nitrous oxide—contribute to greenhouse warming on a scale comparable to that of carbon dioxide. *(Photo: United Nations)*

Collapse of Communism

A historic change was the collapse of the centrally planned economies of central and Eastern Europe. The inability of central planning to meet the expectations of consumers and address serious environmental problems led to tremendous frustration and alienation for the citizens of these nations. In addition, the ever widening disparities in the standards of living between citizens of nonmarket and developed market economies were visible proof of the failure of central planning. Given the political opening of "glasnost" in the former Soviet Union and the unwillingness of the Soviets to use military force to block change in its client states of central Europe, the administrative bureaucracy of central planning lost its legitimacy and, consequently, its ability to organize and motivate its economies.

The transition from central planning to market or quasi-market economies is filled with dangers and uncertainties. There are no handbooks or previous historical examples to draw lessons from, and as a result, many nations seem to be groping their way toward market economies in fits and starts. The transition process is expected to take a decade or more in most of these nations, and the variation among nations will be substantial in the degree to which their economies become centered around market institutions.

In general, reform must occur in four major areas. Although the four areas are inseparable, the first task is usually the stabilization of the macroeconomy through stopping both the movement toward hyperinflation and the continuous slump in production. Second, over the medium to longer term, markets must be created for goods and services as well as labor markets and financial markets. The third task is the privatization of some or all of the nation's shops, restaurants, hotels, factories, and other productive facilities. Finally, the states of central and Eastern Europe must redefine their role in the economy. This need extends from the creation of commercial codes for business, to the development of pension, health care, and education funding and systems, and the development of other social services and social and physical infrastructure. The last two chapters (11 and 12) in the text discuss these issues in much greater depth.

The impact of these changes on the rest of the world will be substantial. Just as the industrialization of East Asia has posed an economic challenge to traditional economic powers, there is likely to be a new challenge coming from central and Eastern Europe in the next decade or two. These nations have skilled labor forces, high rates of literacy, low wages, and bright futures if they overcome problems of political instability. In addition, in the short term, the world economy will feel these changes in the form of increased competition for the capital resources necessary for rebuilding. Developing countries have already begun to discover that many potential lenders and providers of development assistance have shifted their attention to the enormous capital needs of central and Eastern Europe.

Government Control Slows

A reduction of government intervention into economic life has produced globalization, especially since the fall of the Berlin Wall in 1989. Communist governments throughout Eastern Europe were ousted as political change brought on economic change. Most Eastern European countries abolished communist regimes and replaced them with non-communist coalition governments that were based on a platform of closer ties with the West and the establishment of a free-market economy. Farther east, former Soviet Republics asserted their right to sovereignty and declared independence shortly after the fall of the Soviet Union on December 25, 1991. By the beginning of 1992, Mikail Gorbachev and the formidable Soviet Union had fallen and had been replaced by 15 states attempting to develop market economies.

The Russian Parliament, Moscow. Boris Yeltsin is at the balcony of the parliament building, celebrating his rise to power. Demonstrators celebrate at the barricade after the news of the defeat of the coup that had attempted to return the USSR to hard-line communist rule. By 1994, the Russian people, for the first time, had elected a parliament and approved a new constitution. *(Photo: Impact Visual Photo & Graphics, Inc.)*

The underlying message for the world economy was clear. Communist and socialist economies, with non-democratic institutions, assure some degree of equity for their citizens but could not produce sustained growth for their economies. These systems are extremely destructive of the environment, compared to political or economic systems of a market economy with democratic institutions. The nuclear holocaust at Chernobyl devastated vast areas of the Ukraine and Belarus. The Aral Sea is vanishing, Lake Baikal is badly polluted, and toxic waste dumps are all over the former Soviet Union.

By 1992, Boris Yeltsin had come to power promising broad-ranging economic change, including removal of restrictions on trade and the privatization of enterprise. By 1996, more than 4,000 state factories had been sold to investors, and 80% of Russia's industrial workforce was employed in privatized factories. The agricultural privatization has progressed much more slowly, as individual farmers who receive plots of land were unable to purchase necessary equipment to be competitive with state-owned farms. The difficulty of conversion to a market economy has been substantial, and the threat of a return to authoritarianism and a state-run economy continues to loom on the horizon. The process of privatization has been grossly inequitable, and many of the former party bosses have violated their countrymen by claim of special privilege in the privatization process (see Chapter 12).

The globalization of China's authoritarian socialist economy has taken a slightly different approach. Here, the authoritarian state-run economy has allowed the encroachment of the market economy and the private sector to gradually reduce China's share of the total economy. The communist leadership role is to continue to provide order, stability, and basic human services. What the communist party considers truly threatening is not capitalism itself, but creeping irrelevancy. Ironically, that irrelevance is exacerbated by the greatest success. The economic reforms the communists themselves unleashed 20 years ago have produced economic success. The economic opportunities that unfettered the economy in December 1978 have made wealth a rival to the communist party for people's attention. Nowhere does the party's marginalization show up more clearly than with China's youth, especially in the booming coastal cities of the southeast. Top university graduates formerly considered party membership to be a career rite of passage required for advancement through the ranks.

No one can turn back the clock, but many wish to slow the rate of development long enough to examine its impact on people and on the environment. In the last three years, China's economy has expanded 40%. Production has helped China maintain that remarkable growth rate. However, sulfur dioxide emissions from coal furnaces have made Beijing one of the most polluted cities of the world and created upper bronchial and respiratory problems for many people.

Yet 50% of the state-owned enterprises are in the red. Returns from these enterprises dropped from 12% in 1985 to 2% in 1996. Though they contributed a decreasing share of industrial output, 48% today versus 80% in the late 1970s, they still employ three out of four urban workers, especially in the interior and northern cities. Some 320 million Chinese owe their livelihood to state firms. No wonder many argue in favor of maintaining state enterprises. But in many countries, inefficient state-owned enterprises are employment programs for the chronically unemployable (Figure 1.12).

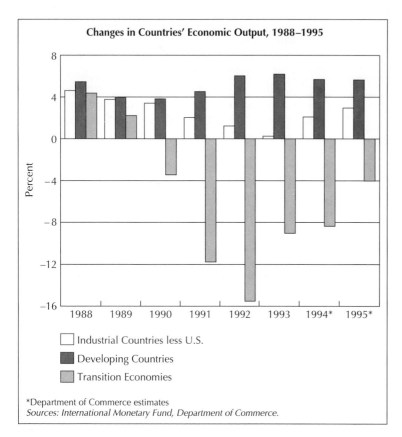

Changes in Countries' Economic Output, 1988–1995

☐ Industrial Countries less U.S.

■ Developing Countries

▨ Transition Economies

*Department of Commerce estimates
Sources: International Monetary Fund, Department of Commerce.*

FIGURE 1.12
Recessions in major industrial countries have been the primary reason why the world economic growth was slow in the 1990s. Recoveries are now underway in a number of these countries. Economic growth in the industrialized countries, excluding the United States, was 3% in 1995, compared to 2.1% in 1994. The European Union (EU) grew by 3% in 1996, up from 2% in 1994. Japan, while still in the early stages of recovery, grew by 2% in 1996, up from 1% in 1994. The biggest recession was in transition economies, including Russian and Eastern European countries, which still exhibit negative growth, in some cases, well into 1997.

Rise of East Asia

The 20th century began with a utopian view in Asia that communism could bring the best quality of life for everyone. The 20th century is ending with the understanding that a market-driven society brings the greatest prosperity to the greatest number of people. Economic shifts into market-driven economies in Asia between 1970 and 1998 reduced the number of desperately poor Asians from 400 million to 150 million, even while the populations of these countries—Indonesia, India, Malaysia, the Philippines, China, and Thailand—grew by two-thirds. Never before have so many people been raised from poverty in such a short period of time, thanks to free-market mechanisms. The free market, a Western idea, is a triumph of markets for both East and West (Figure 1.12).

The epitome of private enterprise in Asia are the "Four Tigers" (Singapore, South Korea, Taiwan, and Hong Kong). The Tigers demonstrated that poor countries could industrialize quickly and enter the world manufacturing scene by market economic policies and thus rise quickly on the ladder of economic development. Not too far behind are the achievements of Thailand, Malaysia, and Indonesia, which have also seen accelerated growth into the late 1990s. The high economic growth of these Asian countries is contrasted markedly with the economic stagnation of India, Pakistan, Bangladesh, and other Asian low-income countries that pursued policies of statism and isolationism, producing excessive bureaucratic barriers and tariffs and fostering inefficiency and slow growth. While India is changing its policies in this regard, in Africa, inefficient state-run manufacturing and agricultural enterprises continue to be the dominant mode of economic activity and thus penalize its citizens (Table 1.4).

The reduction in government-operated programs was not limited to the developing world. The 1980s and 1990s in North America and in Europe saw the planned reduction of "big government" with the benefits of laissez-faire market policies. Western conservatism was put into motion in the United States under Ronald Reagan, in England under Margaret Thatcher, and it proliferated to Germany and Scandinavia where the Christian Democratic parties strengthened their government position over Social Democrats and attempted to reduce the welfare state policies, social expenditures, and high tax rates of social governments. In France, however, the country's telecommunications, defense, utilities, airlines, railways, and banking remain government operated, in contrast to most of the remainder of European nations. In Latin America, economic programs took a tilt toward laissez-faire economic policies and allowed rapid progress of economic development for a few countries, notably Chile and Argentina.

◻

T a b l e 1.4

Economic Growth in the Giant Emerging Markets, 1976–1995

Item	Average Growth 1976–1985	Annual Percentage Change in Real GDP				
		1987	1989	1991	1993	1995[1]
World	3.4	4.0	3.4	0.9	2.3	3.6
Industrial Countries	2.8	3.2	3.3	0.8	1.3	3.0
Developing Countries	4.5	5.7	4.2	4.5	6.1	5.6
Big Emerging Markets[2]	6.0	7.8	4.3	5.0	8.5	6.3
Chinese Economic Area	7.9	11.3	4.5	7.8	13.1	8.7
China (12)	7.8	11.1	4.3	8.0	13.8	9.0
Hong Kong (11)	8.9	14.5	2.8	4.1	5.5	5.4
Taiwan (6)	8.6	12.3	7.6	7.2	6.1	6.4
India (29)	4.6	4.8	6.6	1.0	4.0	5.5
Indonesia (30)	5.7	4.9	7.5	6.9	6.5	7.0
South Korea (9)	8.0	11.5	6.4	9.1	5.5	7.5
Turkey (27)	4.0	9.3	0.3	0.9	7.5	3.0
South Africa (35)	2.1	–2.1	2.4	–1.0	1.2	3.0
Argentina (22)	–0.5	2.6	–6.2	8.9	6.0	5.5
Brazil (18)	4.0	3.6	3.3	1.2	5.3	4.0
Mexico (2)	4.3	1.8	3.3	3.6	0.4	–2.0
Poland (50)	1.8	2.0	0.2	–7.6	3.8	4.5
Other Developing Countries	0.3	0.0	3.9	3.1	–0.3	2.9
Eastern Europe and Former USSR[3]	3.9	2.9	2.4	–12.2	–10.3	–4.9

[1]Department of Commerce estimates
[2]Numbers in parentheses indicate the country's ranking for U.S. exports of manufactured goods in 1993
[3]Excludes Poland
Sources: International Monetary Fund, Department of Commerce.

The introduction of more than a billion workers from developing economies into the global labor pool will have a major impact. Developed countries such as Canada and the U.S. will have to decide whether to let them into the global economy or to keep them out. Letting them in is likely to cause some worker dislocations at home. But keeping them out is likely to cause massive unrest in parts of the world. Governments did not legislate globalization into being. It is the product of technological process. Governments can struggle against it, but history shows their effects will probably be futile. Nations that advance with globalization and open markets in the past have, for the most part, won out over those who struggle against this tide.

Framework of the World Economy

The world is full of problems—debt, unemployment, food shortages, environmental degradation—that are rooted in the structure and development of the world economic system. An understanding of the reasons for problems in the world economy begins by recognizing its domination by developed countries and the existence of an international economic order established as a framework for an international economic system. The term *world economy* refers to the capitalist world economy, a multistate economic system that was created in the late 15th and early 16th centuries. As this system expanded, it took on the configuration of a core of dominant countries with a periphery of dominated countries. The developed countries are in the industrial West, or the *First World*. The dominated countries, in the capitalist underdeveloped *Third World* in the South, are sometimes also called "developing" or, a bit more accurately, "less developed countries" (LDCs). Former

socialist countries of the East or countries of the formerly so-called *Second World* have become increasingly linked to the capitalist world economy.

Core/Periphery Model

Immanuel Wallerstein (1984) provides a useful classification of the world into a *core*, a *semi-periphery*, and a *periphery*. The world economy is an evolving market system in which a hierarchy of states develop based on economic development. The labels core, periphery, and semi-periphery are used to identify economic processes that operate at different levels of this world economy hierarchy. Core processes are economic forces that include high wages, high levels of urbanization, industrialism and post-industrialism, the quaternary sector of the economy, advanced technology, and a diversified product mix. The world periphery processes are low wages, low levels of urbanization, pre-industrial and industrial technology, and a simple production mix. Consumption is low. In between are states that are part of the semi-periphery where both sets of processes exist to a greater or lesser degree. The theory suggests that the semi-periphery countries are exploited by the core countries with regard to raw material and product flow, while at the same time exploiting periphery countries (see Figure 1.13).

This figure is based on GNP per country and GNP per capita. The countries of the core have high GNP per capita and large internal markets. These countries host a large number of internationally active multinational corporations and further define characteristics of their core status with high wages, high levels of urbanization, capital intensive production, and high levels of consumption. In contrast to the core countries are the countries with low GNP per capita. These countries have more political instability, lower wages, less labor-intensive production, and lower consumption rate. This categorization into core, periphery, and semi-periphery is theoretically more meaningful to describe the dynamics of the world economy than simply income or occupation.

The term *international economic system* refers to the system of geographically expanding and evolving capitalism and, in the world today, such underlying processes as the globalization of capital. By internationalization of capital is meant the export of capitalist production, banking, and services through direct investment by firms that create subsidiaries abroad. MNCs are the principal actors of this export.

The term *international economic order* refers to institutions such as those established after World War II to reflect the style of the core countries. Among these institutions are the World Bank, the International Monetary Fund (IMF), and the World Trade Organization (WTO). As the hegemonic power, the United States created institutions that were required to establish a liberal international economic order. Consequently, these institutions had a mandate to dismantle trade and currency restrictions of the interwar years and to facilitate capital mobility.

At any given time, the core of the world economy is dominated by one or more core states. In the postwar period preceding the 1970s, the United States was the

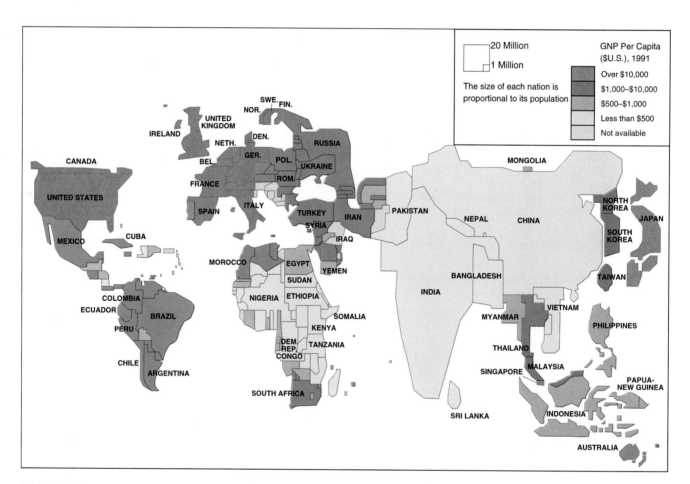

FIGURE 1.13

Categories of per capita Gross National Product on a cartogram of world population. Each country's area is proportional to its total population. GNP per capita are shaded. Africa, South Asia, and East Asia have the lowest GNP per capita, while Japan, Europe, and North America have the highest. The twin banes of North America in the 1990s—job insecurity and income inequality—are starting to lessen.

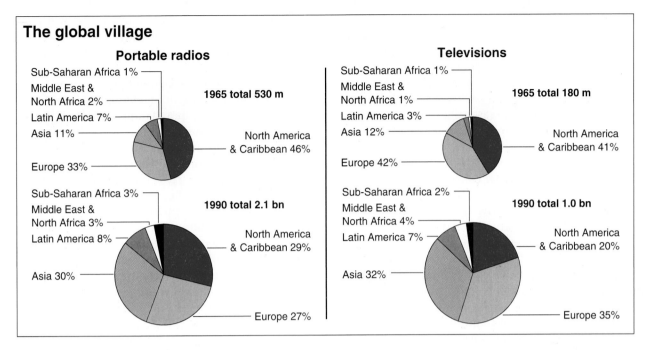

FIGURE 1.14

Radios have increased 400% between 1965 and 1990, while televisions have increased 500% during the same period. The rise of percentage ownership found in Asia is the most dramatic. A decade ago, only 10 million people used cellular phones, which were viewed as toys for the rich, but falling prices and the ability to call from anywhere has made wireless telephony the chat line of the global village. *(Data courtesy of The Economist Newspaper Group Inc., 1992. Reprinted with permission.)*

principal power. The relative decline of U.S. power that became evident after 1973 was triggered by intense competition from rival core states such as Japan and Germany. By the late 1970s, the world order created by the United States after World War II came to an end. Out of the old order came the tentative birth of a new one.

A major reason for the breakdown of the postwar world was a decline in the rate of profit of many firms in the industrial West. Faced with intense global competition, firms had to "automate, emigrate, or evaporate" (Magirier, 1983, p. 61). Some firms did go out of business, but others responded to the challenge to automate and especially to "go international." They became more international due in large measure to the speed of travel and the technology of information handling (Figure 1.14).

WORLD DEVELOPMENT PROBLEMS

World development implies progress toward desirable goals. It is a concept full of hope and enthusiasm, even though the consequences of the jolting and dislocating process can be horrendous for people when long-standing traditions and relationships break down. The purpose of development is to improve the quality of people's lives—that is, to provide secure jobs, adequate nutrition and health services, clean water and air,

cheap transportation, and education. Whether development takes place depends on the extent to which social and economic changes and a restructuring of geographic space help or hinder in meeting the basic needs of the majority of people (see Chapter 12).

Problems associated with the uncertainty and disorder of the development process occur at all scales, ranging from a Somalian villager's access to food and a modern clinic to the international scale of trade relations between rich and poor countries. Attempts to understand development problems at local, regional, and international levels must consider the principles of resource use as well as the principles surrounding the exchange and movement of goods, people, and ideas. This text, written from a geographer's perspective, discusses these principles within the context of the world's critical issues.

Table 1.5

Production Possibilities of PCs and Pizzas

Product	A	B	C	D	E
PCs ('000)	0	1	2	3	4
Pizzas (mil)	10	9	7	5	0

Two critical issues require immediate attention. One is the challenge to economic expansion posed by the environmental constraints of energy supplies, resources, and pollution (see Chapter 3). The other element is the enormous and explosive issue of disparities in the distribution of wealth between rich and poor countries, city and rural areas, wealthy and poor people, and men and women (see Chapter 12).

Environmental Constraints

The world environment—the complex and interconnected links among the natural systems of air, water, and living things—is caught in a tightening vice. On the one hand, the environment is being squeezed by the massive overconsumption and waste of consumer culture and its ethos of "trying to keep up with the Joneses." On the other hand, the environment is being squeezed by the poor people in developing countries who destroy their resource base in order to stay alive. The constraints of diminishing energy supplies, resource limitations, and environmental degradation are three obstacles that threaten the possibility of future economic growth and life itself.

Energy is the key to the long-run sustainability of human life. The oil shock brought to the attention of many Americans the possibility of a world drained of energy. In 1973, OPEC raised oil prices, and these prices continued to rise until 1981 (Figure 1.15). This action

dealt a blow, but not a fatal blow, to the world economy. Without substitutes, proven oil reserves at current rates of extraction are projected to be exhausted in the next 40 years.

There is already a significant poor-world energy problem. Oil is an unaffordable luxury for more than 50% of the world's population who cook and heat with fuelwood, charcoal, animal wastes, and crop residues. Even during the years of falling energy prices in the 1980s, developing countries obtained more than 40% of their energy from noncommercial sources. In countries such as India, Haiti, Indonesia, Malaysia, Tanzania, and Brazil, fuelwood collection is a major cause of deforestation—one of the most severe environmental problems in the underdeveloped world.

The fragility of the environment poses the most formidable obstacle to the economic process. Are there limits to growth? Is the world overpopulated? Some of our present activities, in the absence of controls, may lead to a world that will be uninhabitable for future generations. Topsoil is being lost because of overcultivation, improper irrigation, plowed grassland, and deforestation. Water tables are falling. In the United States, parts of the Ogallala water basin under the Great Plains are at least half depleted. Forests are being torn down by lumber companies and by people trying to keep warm or cook their food. Water is being poisoned by domestic sewage, toxic chemicals, and industrial wastes. The waste products of industrial regions are

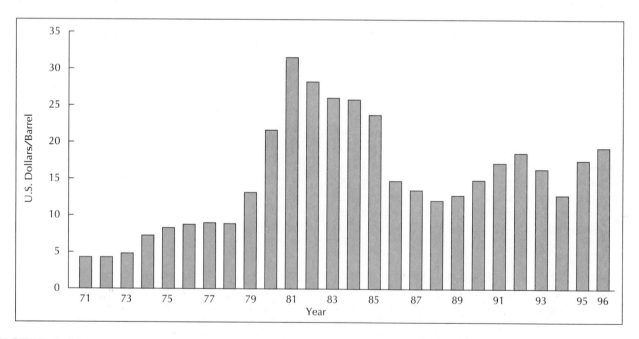

FIGURE 1.15

Average U.S. price per barrel of oil, 1971–1996. The price per barrel of oil peaked in 1981 with the Arab Oil Embargo. Since then, the price of a barrel of crude has dropped from $34 per barrel to a low of $12 per barrel in 1988. In the summer of 1993, a barrel of crude oil on the world market was approximately $20, effectively ruling out synthetic fuels, solar energy, and geothermal energy as sources of power because of the associated higher development and production costs. By 1994, a barrel had dropped to $15. There has been a steady rise in a barrel of crude oil on the world market, which by 1997 reached $24.

also beginning to threaten the world's climate. Accumulated pollutants in the atmosphere—carbon dioxide, methane, nitrous oxide, sulfur dioxide, and chlorofluorocarbons—are said to be enhancing a natural *greenhouse effect* that may cause world temperatures to rise. Recent warming of the Pacific Ocean just 0.25 degrees caused violent weather disruptions in South America, with billions of dollars worth of damage. El Nino, as this ocean warming effect is called, has caused torrential rainfall in Mexico and southern California in 1993, resulting in flooding, mudslides, and massive loss of life. Chlorofluorocarbons, which are used as aerosol propellants and coolants and in a variety of manufacturing processes, are blamed for damaging the earth's ozone layer. Ozone protects life from ultraviolet radiation given off from the sun. Yet another hazard to the environment is the fallout from nuclear bomb tests that took place in the 1950s and 1960s and from nuclear power reactor accidents such as those at Three Mile Island, Pennsylvania, and Chernobyl, Ukraine.

Environmental management must include policies regulating energy and other natural resources and policies curtailing activities that threaten the health and well-being of people. The shift to planning, which is already underway, is essential to ameliorate the problems caused by the economic system. It is also essential if the extraordinarily resilient and persisting system is to survive. These questions are further explored in Chapter 3.

Disparities in Wealth and Well-Being

Poverty afflicts the few in predominantly rich countries—hunger and malnutrition among families in Appalachia, bankrupt farmers on the Minnesota prairie, unemployed factory workers in Detroit, and single mothers on welfare in New York. This kind of poverty, that of poor individuals and families, is an issue of considerable importance. However, there is another kind of poverty that affects all but the few in poor countries; it is *mass poverty.* This is the most important world development problem of our time. You cannot doubt this assertion when you see the halt and maimed sidewalk people of Bombay, insistent begging children in the streets of Mexico City, or women and children carrying firewood on their backs in the countryside north of Nairobi. Mass poverty is intolerable and a critical issue that we must grapple with and try to overcome.

Who are the world's poor? They are the 15 million children in Africa, Asia, and Latin America who die of hunger every year. They are the 1.2 billion people, or 24% of the world's population, who do not have access to safe drinking water. They are the 1.4 billion without sanitary waste disposal facilities. They are the 3 billion people—more than 50% of the world's population—

who live in 43 countries in which the per capita income was less than $350 in 1990 (Figure 1.16). These people are caught in a *vicious cycle of poverty.*

The poor of the world are overwhelmingly the people of developing countries that have failed to keep up with the economic levels of the West since the beginning of the modern colonial period in the 16th century. During the boom that followed World War II, the gross national product (GNP) (i.e., the total domestic and foreign output claimed by residents) of the developed countries rose from $1,250 billion to $3,070 billion. That increase was three and one-half times the GNP, $520 billion, of the underdeveloped countries in 1972. Although per capita real income rose from $175 in 1952 to $300 in 1972, the per capita real income in developed countries rose from $2,000 to $4,000. Developed countries enjoyed 66% of the world's increase, whereas half of the world's population in underdeveloped countries (excluding China) made do with one-eighth of the world's income. By 1982, the national income of the United States (235 million people) was about equal to the total income of the Third World (more than 3,000 million people). In that year, 43% of Third World countries had national incomes amounting to less than one-thousandth that of the United States. The national incomes of 89% of developing countries were not equal to one-hundredth that of the United States.

The developing world is far from a homogeneous entity; that is, there are enormous differences among developing countries. *Physiological density* is one way in which the developing world varies. For some countries, a small amount of arable land and a large number of people can help create poverty. Mexico had a per capita GNP of $2,290 in 1990; Bangladesh had $180 in 1990. There are also huge differences in wealth within countries and among people. In India in 1990, 14% of the population of the Punjab was below the official poverty line. By contrast, more than half the population (60%) in the state of Bihar in northeast India fell below the poverty line. Household income is also uneven. In Bangladesh, the richest 20% of the population gets six times as much as the poorest 20%; in Brazil, the richest get more than 30 times as much as the poorest (Table 1.6). Land ownership is the ultimate indicator of economic inequality in the Third World. For example, in Bangladesh, 22% of the population owns 75% of the land; in Brazil, 1% owns 40%.

With the debt crisis of the 1980s, the United States finally discovered it had a real stake in the prosperity of the developing world. The inability of some countries to make payments on their debt placed the financial structures of the United States and some European nations in jeopardy. Many U.S. banks, including some of the largest, would technically be insolvent if their loans

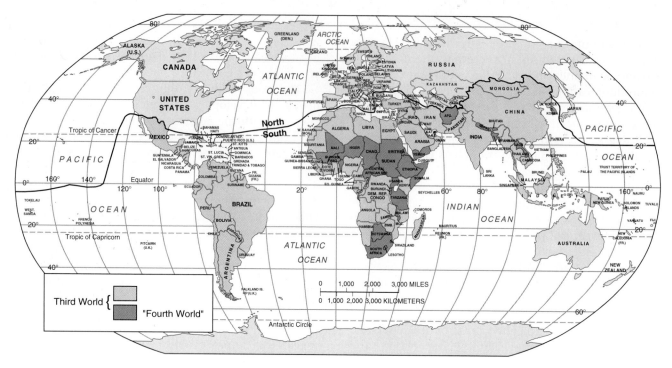

FIGURE 1.16

The Third and "Fourth" Worlds. The Third World consists of Latin America, Africa, and South and East Asia. These are the areas of the world that are the least developed. The Fourth World consists of the least developed of the Third World nations with the lowest per capita incomes, usually below $350. The countries included are grouped in the Sahel area of Africa, South Africa, and in scattered locations throughout South and East Asia. The strong north/south division of developed versus undeveloped nations is striking. The First World, or the most developed countries, includes Anglo-America and Europe. Former communist and socialist countries of Eastern Europe and the Soviet Union appear as more developed than the Third World and were formerly called Second World countries. While this designation is no longer appropriate as many of these former Soviet countries look toward the European community for economic ties and membership, their per capita GNPs are closer to Third World nations. We now understand that per capita incomes for the former Soviet Union and Eastern European nations were for a long time overstated for propaganda purposes.

❖

Table 1.6

Selected Socioeconomic Indicators of Development, 1997

Country	*(1)* Per Capita GNP, 1997 (in Dollars)	*(2)* Life Expectancy at Birth, (Years) M	F	*(3)* Infant Mortality per 1000 Live Births	*(4)* Adult Literacy Rate (%)	*(5)* Daily Per Capita Calorie Supply	*(6)* Per Capita Energy Consumption*
Japan	39,640	77	83	4	99	2848	3484
United States	26,980	73	79	7	99	3666	7794
Canada	19,380	75	81	6			
Brazil	3640	70	76	48	78	2709	897
Mauritania	480	50	53	101	17	2528	114
Haiti	250	48	52	48	5	1911	51
India	340	59	59	75	43	2104	226
Bangladesh	240	58	58	77	33	1925	51
Ethiopia	100	46	48	120	5	1658	20
Mozambique	80	44	48	118	28	1632	84

*Kilograms of oil equivalent.
Source: Statistical Abstract of the United States, 1997; PRB, 1997.

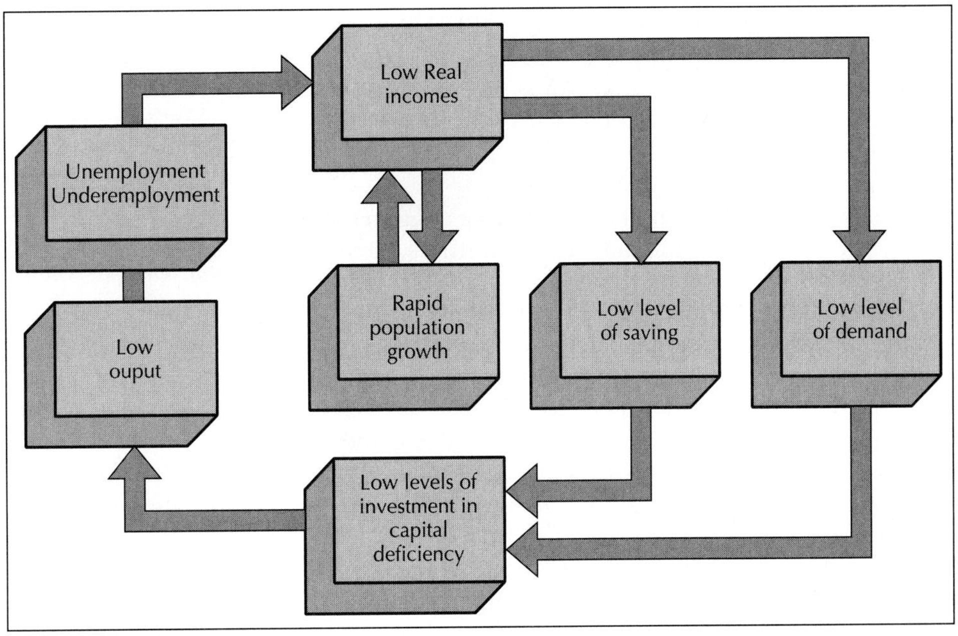

FIGURE 1.17
(a) The cycle of poverty for Third World countries reinforces itself. Most Third World nations have low per capita income, which leads to a low level of saving and a low level of demand for consumer goods. This makes it very difficult for these nations to invest and save. Low levels of investment in physical and human capital result in low productivity for the country as a whole, which leads to underemployment and low per capita income. In addition, many of these countries are faced with rapid population growth, which contributes to low per capita incomes by increasing demand without increasing supply or output. (b) Well-being and health are circular and mutually reinforcing. Better health and nutrition support better economic productivity, better economic productivity supports obtaining a better education, and a better education leads to a delay in marriage and a desire to have fewer children.

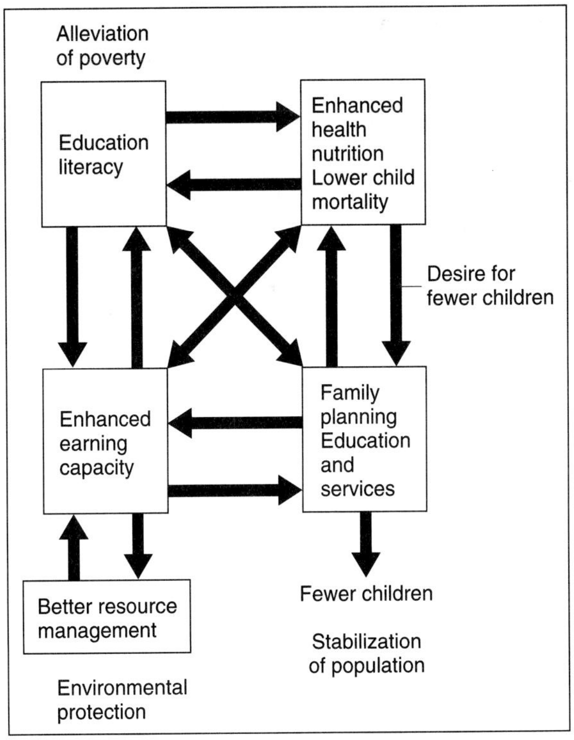

(a)

(b)

to developing nations were declared in default. This led to enormous pressure to resolve the immediate problems of the debt crisis, many of which were directly related to the poor economic performance of the economies of the debtor nations.

Unfortunately, for many of the debtors, the solution sometimes proved to be more painful than the problem itself. Under the guidance of the IMF and other international agencies, stringent limits were placed on the economic policies of debtors, with the result that a majority of citizens in these nations often found themselves worse off. The goals of *IMF conditionally*, as it came to be called, were to restore growth, reduce central government involvement in the economy, and expand the exports of goods and services, while reducing imports so that the debtor would have sufficient earnings of foreign revenue to make payment on the interest and principal of their debt.

There is little evidence that these policies helped to restore economic growth, but they did result in export surpluses that made debt servicing easier. As a consequence, the mid-1980s saw a remarkable reversal in the flow of financial resources—from the flow of the 1970s in which rich nations provided net financial transfers to poor nations to assist in the development effort, to a flow from poor to rich. (See tables in World Bank's *World Development Report*, 1992, p. 50.) In 1978, net financial transfers from rich to poor of long-term lending was $33.2 billion. By 1988, the flow from poor to rich was $35.2 billion; the money went largely for interest payment and principal debt repayment. Needless to say, these transfers have become a serious obstacle to further economic development in poor countries where capital and financial resources are scarce and every dollar lost had repercussions through the economy (Figures 1.17a and 1.17b).

WORLD ECONOMIC CONDITIONS

World economic growth is up. World GDP growth is expected to be about 3.5% in 1995, up from 3% in 1994 and 2.3% in 1993. This is the fastest rate of economic growth since 1988 (Figure 1.18).

Economic recovery in industrial countries is underway. Recessions in major industrial countries have been the primary reason that world economic growth was slow in the 1990s (see Figure 1.12). Recoveries are now underway in a number of these countries, and economic growth in the industrial countries, excluding the United States, is likely to be close to 3% in 1995 compared to 2.1% in 1994. The European Union could grow close to 3% in 1995, up from 2% in 1994. Japan, while still in the early stages of recovery, grew 2.2% in 1995, up from 1% in 1994.

Developing countries will continue strong economic growth. Developing countries comprise about 40% of the market for U.S. exports. Overall, their GDP should average 5.6% growth in 1995, the same as in 1994.

Asia will grow the fastest. Among all the world's regions, Asia has been experiencing the strongest economic growth, over 8% annually for the last several years. In 1995, this growth will slow somewhat to around 7%. The principal factor in the slowdown is China, which after growing at double-digit rates since 1992, will slow to 9%.

Growth for some important countries in Latin America should be in the 4% to 5% range. Argentina has had four years of growth averaging 7% annually and the outlook for 1995 is for a respectable growth rate of 5.5%. Brazil has had growth of 5% in each of the last two years and, despite some uncertainties, should grow between 3% and 5% in 1995.

The short-term outlook for Mexico is for a modest contraction on the heels of the peso devaluation. As of early March 1995, the government of Mexico was estimating that GDP would decline about 2% in 1995. President Zedillo's economic recovery plan aims to provide the necessary short-term measures to cope with the financial crisis and to enable the country to continue its economic expansion over the longer term.

THE FOUR WORLDS

The division of the world into First World, Second World, Third World, and even Fourth World is laden with theory. For Westerners, the First World includes the countries of Western Europe, North America, Australia, New Zealand, and Japan. Some would even include the Republic of South Africa and Israel. The Second World was represented by the Soviet Union and Eastern Europe; whether this designation will remain seems doubtful. The Third World consists of the remaining countries in Latin America, Africa, the Middle East, and Asia. The Fourth World is really a subset of the Third World. These are the scarcest countries on earth with less than $150 per capita per year. These are the so-called "basket cases." Through the eyes of the "free world," the Second, Third, and Fourth Worlds are associated with negative and undesirable traits (i.e., lack of freedom, poverty) in contrast to the positive and desirable characteristics of First World countries (see Figure 1.16). The strong geographic regionalism of countries, north to south, is striking.

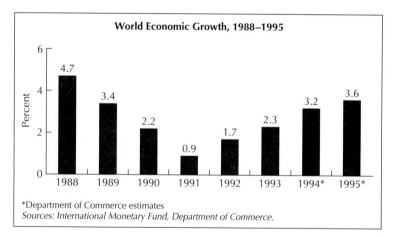

*Department of Commerce estimates
Sources: International Monetary Fund, Department of Commerce.

FIGURE 1.18
World economic growth is increasing. World GDP growth was 3.5% in 1995, up from 3% in 1994, and 2.3% in 1993. This is the fastest rate of economic growth since 1988. The global economy is in a strong recovery. Expanding foreign markets are the chief factor for improved export growth worldwide. U.S. trade deficits will remain large, but world economic conditions favor North American exports. The United States is in the eighth year of an expanding economy, partly because the Cold War's end has reduced the federal deficit (from $290 billion in 1992 to $60 billion in 1997) and thus interest rates by shifting money away from defense spending. At the same time, convert countries to capitalism have craved U.S. products as never before, and U.S. exports reached almost $700 billion by the end of 1997. American companies and workers experienced job restructuring in the early 1990s (see Chapter 8). Now these companies are more highly competitive and far ahead of their European and Japanese competition in using computers and telecommunications systems. The result has been increasing efficiency, corporate profits, and stock prices. Joblessness is 11% in Germany, 10% in Canada, 6.5% in the United Kingdom, 4.9% in the United States, and 3.3% in Japan in 1997.

Fewer Manufacturing Workers: The percent change in manufacturing workers, 1990 to 1995, showed Sweden (−3.6%), U.K. (−3.2%), Italy (−2.8%), France (−2.5%), Netherlands (2.1%), Belgium (−1.8%), Canada (−1.0%), U.S. (−0.6%), Japan (−0.5%), and Norway (−0.4%). The photo shows T.V. assembly line, Bloomington, Indiana. With television sales down 8% since 1997 and retail prices down 31% since 1991 for color televisions, this RCA plant has announced plans to relocate to Mexico, citing the inaffordability of the $10 an hour wage ($19 with benefits factored in). The wage rate in Mexico for T.V. assembly is one- tenth the U.S rate. *(Photo: McGuire Studio/Thomson Consumer Electronics Inc.)*

Because there is no one indisputable reality, it is desirable to be aware of different economies, systems, and world views. For this reason, we have attempted to select and organize the material for this text from a broad base of inquiry. The focus is on a comparative approach in which different perspectives are explained and contrasted. Looking at the world through different ideological lenses better enables us to meet the challenge of world development problems. The way in which a society answers the four questions of economic geography depends on its ideology.

THE FOUR QUESTIONS OF THE WORLD ECONOMY

The basic problem of economic geography deals with *scarcity* in space and the methods to overcome it so that a fair distribution of resources can occur (see Figure 1.16). The problem of scarcity does not strike people in the United States or Canada in quite the same way as it does people in the developing world. We are used to seeing grocery stores and shopping malls loaded with food and consumer goods, but it is quite the opposite situation in much of the rest of the world. Scarcity in space has been a problem for humankind from the earliest days, and much of the remainder of this text aims at describing the processes whereby scarcity has originated and how the world can overcome it.

The *four basic questions of economic geography* and the world economy arise for each country and each economic system in an effort to overcome their society's scarce resources. The answers to the questions will vary by the type of political economy—capitalist, command economy, traditional economy, or some hybrid form.

Question 1: With human and capital resources available, exactly *what* should be produced and at what level or scale of production?

Question 2: Given what should be produced and production levels, exactly *how*, or with what combination of techniques, labor and capital inputs, and other resources, should the output be produced?

Question 3: Once it is agreed upon what will be produced and how it will be produced, *where* should the output be produced? A related question is: Once spatially dispersed economic activity exists, why is it here?

Question 4: After producing the output at geographic locations from the resources available, exactly which groups should receive what share of the goods and services? That is, *who* will receive the goods?

The field of economics centers on question No. 1; international business centers on question No. 2; political economy centers on question No. 4. By far, the core of economic geographic studies through the decades has centered on question No. 3. A shift in the field is starting, which emphasizes the need to answer Nos. 2 and 4. Also note that these four questions have been stated in the "normative," or the third level of inquiry. That is, they ask what is the best answer to these questions. As is the case with all sciences, most of our study is at the first and second level of inquiry; that is, we try to first describe, then explain, economic phenomena in this text. Only then can we hope to answer the normative questions. (These questions are discussed explicitly in Chapters 8 and 9.)

The Factors of Production

We have all heard of the *factors of production*. However, what does this phrase really mean, given the four previously listed questions, which are basic to all the economic geographies of the world's countries? The factors of production include land, labor, capital, and entrepreneurial skills (management). *Land* can be divided up into not only the units of physical geographic space that exist across the earth, but also into the amount of *raw materials* extracted from the earth. *Labor* is the human, physical input from individuals required in the production process. *Capital* is the machines, tools and infrastructure, storage, transportation, and distribution facilities used by labor to produce the goods from the spatially dispersed raw materials. (Here, capital or *real capital* is an economic resource. *Money capital* is not a resource and is not included in capital.) *Entrepreneurial skill*, or management, is the coordinating effort used to combine land, resources, labor, and capital to produce goods and services.

The interdependencies of economic geography are that the factors of production, land, resources, labor, capital, and entrepreneurial skills can be substituted for one another in the final production process to attain variations of the four questions. For example, the production of goods and services is still a possibility in nations that have almost no land or no resources by substituting capital and labor with the addition of increased transportation costs. These are the factors that account for the prodigious east Asian and Japanese economies of the 1990s.

Generally speaking, the lack of two or more factors of production can relegate a country to great levels of depravation. The amount of interchange between the factors of production determine how the output will be produced. *Compensation* is made in the form of *rent*, which is paid for the use of someone else's land; *wages*, which are paid for the use of labor; *interest*, which is paid for the use of capital; and *profits*, which are the reward paid for entrepreneurial ability.

The factors of production are employed differently within each political economy. No two countries employ the same mix of the factors of production to produce the same products. Different areal sizes, amounts of technology and capital, entrepreneurial skills, and quantities of labor, as well as differing political, cultural, and historic factors, mean that each country must answer *what is produced? how is it produced? where is it produced?* and *to whom will it be given?* in different ways. Each country will dictate the amount of individual freedom, the amount of social support, and the process of distribution in answering these basic questions, based on their political, cultural, and economic philosophies. These political economic systems are discussed next.

POLITICAL ECONOMIES

Briefly, we discuss the political economic systems of the world: *pure capitalism*; the *command economy* of Marxism; *mixed political economy*, including fascism and socialism; and the *traditional economy* of developing nations.

Capitalistic Economies

Pure, or laissez-faire, capitalism is the economic geographic system that is characterized by the private ownership of the economic means of production and minimum governmental intervention. In such a system, sometimes called *market economy*, a series of competitive markets and prices are used to coordinate economic geographic activity and final production levels. Freedom of enterprise and choice is the distinguishing factor in this

Egyptian farmer tilling the soil at the oasis of El Faiyum. This field is being prepared for growing cotton, to meet a worldwide demand for cotton clothing. In the future, the poorer countries of the world will have to rely on agriculture to raise their standards of living and to supply the capital they need to create industries. Agricultural production, therefore, must be increased. Many developing countries have placed their limited capital into the manufacturing sector, thus providing less than enough food for their hungry people. This capital is needed to purchase equipment, fertilizers, and new and better seeds, and to develop vast irrigation schemes and land reforms. Some developing countries, such as Egypt, have grown a disproportionate amount of nonfood crops for the export revenue it generates. *(Photo: United Nations)*

type of economy. In pure capitalism, each individual, or firm, seeks to maximize his or her own self-interest.

Now, we must answer our four basic questions: What will be produced? How will it be produced? Where will it be produced? For whom will it be produced? Only profitable products will be produced, based on market demand and price. Price will be maintained by the utility and value of the good, based on consumers maximizing self-interests. How and where the goods are produced is based on labor and technology efficiency and the lowering of production costs through wise selection of locations. The most efficient producers, based on capital inputs, transportation, and the division of labor, as reflected in low prices and quality of goods through competition, are the survivors. Their production processes and locations will dictate how and where goods will be produced in a market economy.

Pure capitalism features two groups of decision makers—private households and businesses. The mechanisms that operate to bring households and businesses together are the *resource market* and the *product market*, or the *circular flowing market system* (Figure 1.19). The *demand curve* shows the schedule of total goods that all consumers are willing to buy at each price (Figure 1.20).

The *law of supply* conversely states that the level of prices and the quantity of goods supplied are directly related. As prices increase, sellers offer more goods to be purchased because they can make a greater profit. The *supply curve* then is the schedule that relates the

total amount of a good or service that all sellers, together, are willing to sell at each possible price. To the seller, it is the price multiplied by the total good or service sold that equals earnings. From earnings, production costs must be subtracted to yield net income or profit. The amount of money remaining after costs are subtracted from total income is the principal factor that gives suppliers a willingness to produce more or less of a commodity or service.

Unlike the command economic system, if the gap between costs and earnings is small or negative, producers may change production to something that yields a higher return in the marketplace over the long run. However, if the difference between earnings and costs is large or producers become more efficient, they will most likely put more resources into the production of their goods or services in order to obtain greater total sales. The point at which supply and demand curves meet is called the *equilibrium price*—the price and quantity where supply is equal to demand at a particular price. The equilibrium price does not leave suppliers with unsold goods or consumers with unfilled demand.

The Command Economy

Almost the exact opposite on the political economy compass from pure capitalism is that of the *command economy*, sometimes known as Marxism or communism.

FIGURE 1.19
Circular flow in the capitalist economy. The circular flow in the capitalist economy involves a resource market where households supply resources to businesses, and where businesses provide money income to households. It also consists of the product market where businesses manufacture and produce goods and services for households, while households provide money revenue from their wages and income to consume such goods and services. In the resource market, shown in the upper half of the diagram, households are on the supply side and businesses are on the demand side. The bottom half of the diagram shows the product market; households are on the demand side and businesses are on the supply side. (*Source: Adapted and modified from McConnell and Brue, 1993, p. 59*)

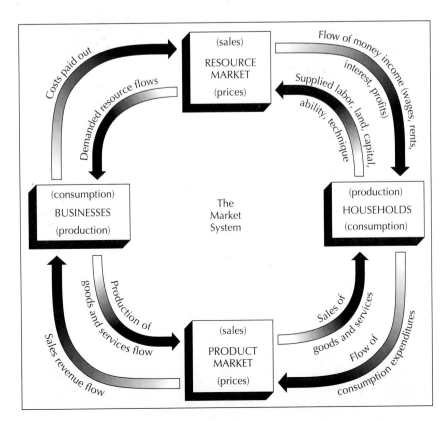

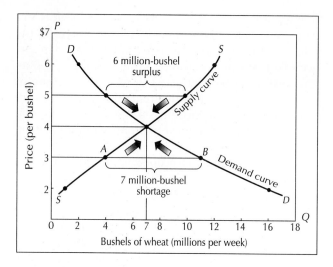

FIGURE 1.20

The market equilibrium price and quantity for wheat in the United States, 1997. On January 15, 1997, the price per bushel of wheat in the United States was $4. The equilibrium quantity was 7 million bushels of wheat per week. At $3 per bushel, farmers will only supply 4 million bushels per week (point A). However, because of the low price a whopping 11 million bushels per week is demanded (point B). Therefore, there is a shortage of 7 million bushels per week. The shortage means that some consumers will pay more than $3 per bushel, and they will bid higher in the marketplace. Gradually, their higher bids will bring more wheat farmers into the market so that both price and quantity are bid up to $4 per bushel and 7 million bushels. A different scenario in which farmers offer wheat at $5 per bushel brings a demand of only 4 million bushels of wheat per week. But farmers are willing to provide 10 million bushels per week at the $5 price. The 6 million bushel surplus at $5 per bushel will not occur because the price is unrealistically high. Prices will fall, thus increasing the demand by consumers and reducing the supply by farmers back to the equilibrium price.

It is a form of totalitarianism, in which the means of production and all property and resources are owned and controlled by a central government that serves as the economic planning agency. All prices, quotas, levels of production, plant location and size, and any other major decision regarding the use of the means of production and resources, as well as the distribution of output and price levels, are set by the central bureaus of the command economy.

Production units, plants, and firms are owned by the government, and production quotas and output targets are established externally, not on the basis of market factors, demand, and supply but according to central directives. Consumer goods frequently are produced secondarily to capital goods in such economies, or an emphasis is placed on heavy industry, military might, and collectivized farming on huge state-run farms.

How are the four critical questions of economic geography answered in the command economy? The answers to the "how," "what," "where," and "for whom" questions posed alone are determined purely on the basis of the state leaders' choice, not on market factors. In the last chapter, we present, in economic terms, the reasons that the Soviet economy failed and the challenges Russia and other former Soviet republics face in the future.

Although it is difficult to find pure capitalism or a pure command economy anywhere in the world today, Hong Kong and the United States' economic geography comes close to capitalism, followed by Canada, many of the countries of Western Europe, and Japan, as well as some Latin American and Asian countries. The command economy of the former Soviet Union and that of the present-day Peoples Republic of China were certainly the largest experiments of this nature in history. Other countries have tried to imitate the policies of the former Soviet Union and China, including Vietnam, North Korea, Cuba, Nicaragua, and a number of African countries.

With the fall of the Soviet Union in 1991 and the breakup of its republics, the command economy, as a political economic form of geography, has certainly received a damaging blow. Most political economists now agree that the command economy is a failure.

Mixed Economic Systems

Most economic and political economies of the world are positioned somewhere between the extremes of pure capitalism and the pure command economy. The U.S. economy, for example, leans toward pure capitalism but has important modifications that help to stabilize the economy. The government is an active agent in redistributing wealth, providing basic services, and producing certain goods that would not be otherwise available because of their money-losing nature. The U.S. Post Office and the Army are examples.

The American form of government-supported capitalism has spawned large economic enterprises in huge MNCs. While China and the former Soviet Union approximated the command economy, important market mechanisms were retained to help determine prices and levels of production. Private ownership of land and resources has been allowed by Deng Xiaoping in China for 20 years, and vestiges of private ownership had been creeping into Soviet society at approximately the time of its collapse in 1991. Russia, the former Soviet republics, China, and most of Eastern Europe are now moving their economies toward a more market-oriented capitalistic system, and the difficulties of such a shift are addressed in Chapter 12.

Bill Clinton at the G-7 Economic Summit, Denver, June 1997. Antagonists say that the pure capitalist system exaggerates income distortion that creates political instability. The question for capitalism is: Is the common good best served by the uninhibited pursuit of self-interest? Financial compensation for CEOs at 350 of America's largest companies jumped in 1997 to 1.5 million, up 25% over the three previous years. Investment bankers can earn 200 times more than school teachers. In Los Angeles, 5,000 city employees earn less than $7.50 per hour, which is the 1997 poverty line. Social stability may require some balance between those at the top end and those at the bottom end of the payroll. Those at the bottom should not feel that the deck is stacked irreversibly against them. *(Photo: Mark Reinstein, Uniphoto Picutre Agency)*

Private ownership and the market system, which are typical of U.S. economic geography, do not always fit together. This is the case with central planning by military dictatorships in Latin America. Private ownership of property and the means of reduction are allowed, but at the same time, the economy is highly controlled. Nazi Germany was also typical of this approach.

In former Yugoslavia, now involved in late stages of a civil war, market socialism was the rule of the day. Public ownership of land and resources was regulated by market mechanism, which relied on free markets and supply and demand to regulate economic geography.

To some extent, the Canadian, British, French, and Scandinavian socialist economies also resemble hybrid, or mixed, systems. Sweden is a good example, where although 95% of land and resources are under private ownership, the government has set economic standards and prices and strongly affected the redistribution of income by taxing individuals who make more than $50,000 at the 80% level. A health network such as Sweden's, from "cradle to the grave," is highly desired by other nations. The United States has been unwilling to pay for such a system.

In addition, the Japanese economic system is an example of a hybrid system, which involves private ownership. It also receives a great deal of government support and regulation, which produces a growth economy in which government favors those businesses that are the most profitable and successful.

The Traditional Economy

Most of the world's population falls into another political economy known as the *traditional economy*. Here, price, exchange, distribution, and income are regulated by many years of convention, culture, and custom. In India, Southeast Asia, Middle America, and most of Africa and South America, economic and social caste and heredity define economic roles of individuals and upward mobility, or lack thereof. There is a strong fatalism among members of society; that is, it is widespread

Jean Chretian, Prime Minister of Canada. Four years after the United States and Canada plunged into the North American Free Trade Agreement with Mexico, the spoils of a continental-scale economy have benefited Canada the most. The U.S. auto industry is moving parts and vehicles across the Canadian and Mexican borders in greater volume, helping it drive down costs and compete against foreign rivals. Canadians are most enthusiastic about quickly expanding NAFTA to Latin America. The Canadian economy is booming, largely because of trade, some 80% of which is now with the United States, compared with 60% one decade ago. *(Photo: Thierry Orban, Sygma)*

thinking that economic conditions are locked and will not improve, and that one's individual position in society cannot change. Religious and cultural values frequently dictate the societal norms as well as economic aspirations, and the status quo is self-perpetuated.

Most of the traditional economies are poor by Western standards, by the standards of the command economies, or by the mixed economic systems of Asia. In traditional economies, the "what," "how," "where," and "for whom" questions are answered with the observance of culture, habit, and custom, tempered by a strong dose of pragmatism.

In future chapters, we discuss economic development of developing countries, traditional economies, and the difficulty they face in making a choice as to which economic model to use. These countries need a model that will result in the greatest welfare to their populations, in simultaneous growth, and in compatibility of their societal and cultural norms.

WHAT SHOULD BE PRODUCED?

The first question that a society faces is the "what" question: What should be produced? We address this question briefly here, as well as in future chapters (e.g., Chapters 7 and 10). Each economy faces scarcity and limited amounts of resources; each society is faced with trade-offs. If more of one good is produced, less of another good will be available, given a fixed set of resources. A factory can produce cruise missiles or airplanes, but not both at the same time. One axiom of economic geography is that, for an economy operating at nearly full production, more of product A can be produced only at the expense of, or by producing less of, product B. *Opportunity cost* is the amount of product B we must give up to produce a unit of product A. Within the limits of their resources, needs, and time frame, societies must trade off resources and decide "what" shall be produced, over and over again.

Each of us is faced with similar trade-offs and opportunity costs. The question of whether to save for graduate school or buy a new car may be a difficult choice. We may choose to live near our downtown office on expensive land but have a short commute, saving time. Or we may choose to live in a distant suburb on a cheaper piece of land (per unit area), consume more of it than the downtown condo would require, but have a farther (and more costly from an opportunity-cost perspective) commute.

Production Possibilities Curve

In order to demonstrate the opportunity costs societies face, consider an economy that produces only two commodities—PCs and pizzas. Figure 1.21 gives

information on the opportunity costs of producing PCs and pizzas (see Table 1.5 on page 27).

The trade-off between PCs, shown in thousands, and pizzas, shown in millions, can be seen. At point A, 11 million pizzas can be produced, but all the resources have been put to this effort and none are left for PCs. At point B, the number of pizzas has been reduced to 9 million, which allows 1,000 PCs to be produced with available levels of technology and resources. Finally, at point E, 4,000 PCs can be produced, but no resources are left for pizzas. While no society is a two-commodity economy, the example shows various opportunity costs, or trade-offs, of goods and resources within the society that can be compromised to produce more of one commodity than another. To gain more of a good, in this case pizzas, we must pay more because the resources used to make pizzas and PCs are not perfectly interchangeable. To produce more PCs,

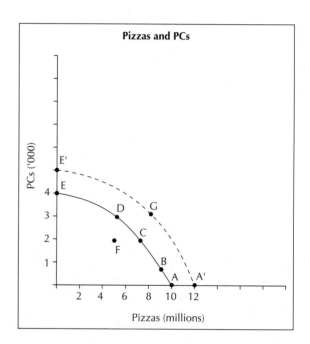

FIGURE 1.21
The production possibilities curve of PCs and pizzas. The curve ABCDE describes a schedule of trade-offs for a two-commodity society. The society can choose to produce only PCs at the cost of producing no pizzas (point E), or it can choose to produce only pizzas and no PCs at point A. Points B, C, and D along the curve suggest trade-offs between the commodities. As a society chooses to produce a large amount of one or the other commodity, the opportunity cost goes up and a smaller amount of the remaining commodity can be produced. Point G outside the curve suggests a level of production that has not yet been attained by the available resources and the technology of the society, but towards which it is moving. Point F under the curve suggests a possible condition of unemployment or underutilization of resources for the society.

we must change equipment and technology and retrain individuals.

Figure 1.21 shows us that for any pair of goods and services, resources are not perfectly interchangeable. Consequently, to increase the output of one commodity, there must be a reduction in the output of one or more other commodities. Societies, therefore, must choose what products and services will be produced because there will be a reduction in the number of other products and services.

Answering the "How" Question

The dotted line in the production possibilities curve (see Figure 1.21) represents economic growth. The only way we can move from the solid line to the dotted line is to increase the amount of resources, acquire more labor, or increase the level of employed technology, meaning more efficient machinery or technical skills of the workforce, to develop new ways of using the resources to a maximum.

Moving from curve E-A to E'-A' answers the question of "how" production takes place. The goal of all societies is to move from present production levels, through a growth process, to higher production levels in the future. The problem is selecting a strategy that produces the most benefits for the society as a whole. This involves techniques to measure increased benefits from higher production and answering the question "for whom," so that a society can measure the benefit of distributing goods to social groups. These factors, used to determine "how" production takes place and "who" receives the goods, depend on the type of political economy—capitalism, command economy, mixed systems, or a traditional system, as previously described.

Future Versus Present Economic Growth

One of the basic choices that economic managers must make is a production mix that favors either capital goods, which are devoted to future production, versus consumer goods, which satisfy a present want and need in society. (Capital goods are goods that are used to produce other goods, whereas consumer goods are goods that are used by individuals for their own needs and wants, in the short run.) After World War II, the Soviet Union and the United States made different choices along the production possibilities curve. The Soviet Union decided to put an emphasis on capital goods for long-term economic growth and emphasized truck factories, large electric plants, mechanized agricultural farms, and a heavy military arsenal. In the United States, the emphasis was on producing consumer goods for a victorious country that had put up with the shortages and inconveniences of war. Automobiles, television sets, single-family homes, and a variety of clothing and shoes flooded the markets in America. In the Soviet Union, resources were shifted away from consumer goods toward capital goods, whereas in the United States, a larger population of resources were shifted away from capital goods and toward consumer goods.

During the next 45-year period, the people of the Soviet Union suffered inconvenience and shortage as the command economy rebuilt its production base. By 1989, the people of the Soviet Union, informed of their lack of consumer goods as compared to the West by Radio Free Europe and ubiquitous television transmission, finally grew tired of constantly being exploited and of suffering from low standards of living. The population of the United States, however, enjoyed a high

FIGURE 1.22
Two countries with opposite production trade-offs. A country's present choice of substitution between "goods for the future" and "goods for the present" in its production possibilities curve will help determine the economy's future growth and the production possibilities curve's new location. Country X has chosen a greater proportion of consumer goods, or goods for the present, at the expense of capital goods, or goods for the future, shown on the left. Country Y has chosen a greater proportion of capital goods, or goods for the future, as opposed to a modest proportion of consumer goods,

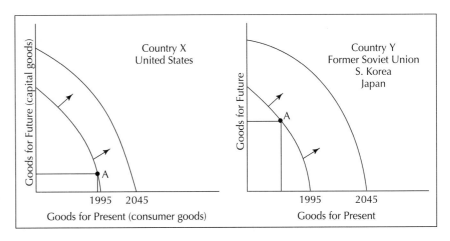

or goods for the present, as shown on the right side. Country X's production choice results in a growth curve, which is moderate and upward to the right. Country Y's selection of an aggressive future-oriented economy results in a greater rightward shift to the production possibilities curve, suggesting greater overall economic growth.

consumption level and a standard of living that was unparalleled by any society in the world. At the same time, the United States did not avoid producing capital goods and a strong economic base. Both the Soviet Union's and the United States' decisions were appropriate for the resource levels available and for the desires of the political economy of each.

The pleasure of buying a new car now and skipping graduate school may be sacrificing future economies. Developing countries today have had to make the same choices, often with disappointing results. Many times, the population has pressured the government to fulfill demands for consumer goods at the expense of capital goods needed to develop a society in the 20th century that will be competitive in the world market. All too often, the result has been that the productivity of the country has been in areas that did not allow it much future growth, leaving the country underdeveloped, with dim prospects for the future. The country opted for export, nonfood cash crops, such as cotton or sugar, and used the revenue to buy guns.

Now, observe Figure 1.22. On the left-hand side, country X has chosen a disproportionate commodity mix, emphasizing consumer goods over capital goods. Note point A for country X. In this hypothetical example, country X's growth during the next 50-year period is shown, by the outward curves and arrows, to be moderate at best. Compare the production possibilities curve of country X to country Y on the right side. Country Y chose to answer the "how" question of production by emphasizing capital goods over consumer goods. Note point A on country Y's production possibilities curve. Fifty years later, the overall growth of both capital goods and consumer goods are substantially more than that of country X. Country Y's emphasis on production for present consumption has sacrificed its production for its future generations, because of its slower growth rate. Country Y has sacrificed, in the short-run, consumer goods and private consumption, so that its long-run growth is faster. Another way of stating the differences is that country X has sacrificed the opportunity costs of more growth in the future to satisfy a present demand for consumer goods.

Rwanda and the Sahel of Africa

In Western society, from Western Europe to the United States and Canada and westward to Japan, rightward shifts of the production possibilities curve have always been the rule. The economy has, in the long-run, always grown in western society, either at a low rate or at a high rate. In the poorest parts of the world, the production possibilities curve may actually be shifting to the left, however. Recent famine in the Sahel of Africa, especially in the east, near the horn of Africa, including countries of Ethiopia, Chad, Sudan, and Somalia, indicate

Baku, the oil capital of the former Soviet Union. At the end of 1991, the communist system and the USSR collapsed. The Commonwealth of Independent States (CIS) was formed, and Mikhail Gorbachev was removed from power. Further, state-assisted industry and a subsidized economy has completely imbalanced the Russian continent, requiring them to receive financial aid from their former capitalist enemies. The advantages of Baku's geographical position as a meeting place of the trade routes from Russia, Persia, Central Asia, and the Caucasus make the oil fields on the Caspian Sea an important economic factor. The city of Baku is the capital of the Republic of Azerbaijan, with a population of 2 million people. American oil companies are now flooding into Baku because of recent new oil discoveries. *(Photo: Sebastiao Salgado, Contact Press)*

that neither present consumer goods nor capital goods may be growing from year to year. Because of drought, ecological destruction, encroaching deserts (desertification), overgrazing, and as we have learned through Rwanda and Operation Restore Hope in Somalia, rampant warlords commandeering food supplies and defending feudal territories, production has dwindled in almost all categories of goods. For example, the per capita GNPs in each of these countries mentioned has decreased by an average of 10% from 1980 to 1993.

The importance of *production possibility analysis* is obvious for developing countries, as well as for highly developed countries. First, any product that is produced has an opportunity cost and will lower the output of other products in the society that could have been produced. Politicians have called this a "zero sum game."

The sum of all commodities produced can neither expand nor contract in the short run. One must "rob from Peter to pay Paul," so to speak. Second, if a country uses resources inefficiently, a lower level of societal gain will accrue, and levels of unemployment, underemployment, and underutilization will develop.

THE GEOGRAPHIC PERSPECTIVE

World development problems are inherently geographic. Many development problems have geographic solutions or, more accurately, partial solutions. They certainly raise the important geographic questions of location and accessibility, relationships between settlements and land use, changing transport and communication linkages, efficient flows of commodities, and the spread of ideas and innovations. They also raise questions about center-periphery relations at a hierarchy of scales ranging from a farmer's access to a market to connections between developed and developing worlds.

The Field of Geography

Geography uses a distinctive language—the *language of maps*. A map reveals human excitement, wonder, and concern for spatial relations. Geographers have systematically recorded what is where on the face of the earth. Modern geographers also use the map as a research tool to ask and answer questions about spatial relations.

The spatial dimension is central to geography. It is also central to the internationally interdependent character of our lives. Events in one place have a direct and immediate impact on events in other places. An increase in the price of OPEC oil swells the coffers of countries such as Kuwait and Saudi Arabia, but impoverishes other countries such as Tanzania and Bangladesh that feel the pinch much more than the United States or Japan. To keep their trucks moving, poor countries must pay for oil in hard currencies that leave less money for fertilizers, schools, hospitals, and new development projects.

Trends in Economic Geography

Economic geography is concerned with the spatial organization and distribution of economic activity, the use of the world's resources, and the distribution and expansion of the world economy. In its infancy, economic geography was called commercial geography, which developed during the era of European exploration and discovery from the 15th century through the 19th century. There was much excitement and adventure then. Commercial geographers were on sea voyages, and their reports brought masses of factual information about other lands to merchants and government officials.

The term "economic geography" was coined in the United States in 1888. Twelve years later, Ellen Stemple authored a book with that title, and by the end of World War I, economic geography was a respected part of the discipline. In fact, in 1925, a new journal, *Economic Geography*, began publication.

As a distinct field of study, economic geography was affected by three major themes of geography: (1) human-environmental relations, (2) areal differentiation, and (3) spatial organization. Although all three thematic approaches have always been present, a human-environmental emphasis flourished largely by itself until the 1930s; areal differentiation was most influential from the late 1930s to the late 1950s; and spatial organization has since emerged as the dominant approach. To better understand the fundamental concepts of geography that we examine later in the text, it is useful to first take a brief look at these three approaches.

Geographers sought to explain variations in economic development in terms of *environmental determinism*, or *environmentalism* for short. For example, Ellsworth Huntington in 1924 argued that certain areas of the globe stimulated human efficiency more than others. The industrial countries, in Huntington's view, had the most "stimulating climates," whereas most developing countries had "difficult climates."

Few, if any, geographers who study human-environmental relations now claim that the physical environment is the sole determinant of people's economic behavior. Geographers now place emphasis on human adaptation and adjustment to potentialities in the environment. They attempt to discover how particular groups of people, especially in a local area, organize their thoughts about the environment and how to come to grips with it (Porter, 1965).

Economic geographers were disappointed in their encounter with environmentalism and its underlying preoccupation with race. The main reaction to the period of excessive environmental determinism was to adopt the view that all geographic phenomena were unique and that theory building was of little value. Hence, for a period between the late 1930s and the late 1950s, the primary focus of concern was *areal differentiation*—differences rather than similarities—among places.

The areal differentiation concept, which led to some of the great regional writing on which much of the present academic status of geography was built, led geographers to overlook the need for comparative studies. Areal differentiation dominated geography at the expense of *areal integration*. In the 1950s and 1960s, geographers such as Peter Haggett from Bristol, UK, and others, such as Brian Berry, Bill Garrison, Richard Morrill, Waldo Tobler, John Nystuen, and Art Getis, from the University of Washington, scorned the unique approach. They argued that economic geography needed to become a theoretical subject, which it did.

This theme in geography came to the fore in the decade of the 1960s. It has done a great deal to help geographers think in new ways about geographic distributions and spatial relations. *Spatial organization* is concerned with how space is organized by individuals and societies to suit their own designs. It provides a framework for analyzing and interpreting location decisions (e.g., market versus raw material location, accessibility versus transportation costs) and spatial structures (e.g., land-use patterns, industrial location, settlement). The popularity of the organization-of-space theme was influenced by governments that were subsidizing geographic research, especially for planning and policy-oriented studies. It was also influenced by the *quantitative revolution*, with its emphasis on quantification and experimentation with a wide range of statistical techniques. A more important emphasis, however, was on the formulation of hypotheses, data collection, and the search for theory. This empirical approach aimed at building theory was called *logical positivism*. Economic geographers found some of the theories they were looking for in the social and biological sciences, including location theory.

Location theory attempts to explain and predict geographic decisions that result from aggregates of individual decision making. The main aim of location theorists is to integrate the spatial dimension into classical economic theory. The origins of location theory stem from the work of Johann Heinrich von Thünen on agricultural location in the 1820s and subsequent contributions to industrial and settlement theory by Alfred Weber and Walter Christaller. They developed *normative* (i.e., theory-building) models relating to business and industry in a world of pure competition that assumed entrepreneurs are completely rational and attempt to maximize profits with perfect knowledge of the cost characteristics of all locations. This image of an entrepreneur became known as "economic person"; that is, an omniscient, rational individual who is driven by a single goal—to maximize profits.

In the 1970s and 1980s there were at least three departures from location theory. First, *behavioral geographers* criticized location theory for its emphasis on abstract patterns of land use and maximization of profits. They questioned the relevance of location theory for understanding location decisions and spatial structures in the real world.

Second, *phenomenological geographers* rejected the perception of the "economic person" and the notion that we live in an "objective world." Life takes on meaning only through individual experiences and needs; for example, resources have no existence apart from human wants. The phenomenological approach in geography is based on a relationship between the observer and the observed. The scholar looks at a problem from the subject's viewpoint, or "lived world." Phenomenology took a step back from the progress of location theory of the 1960s by rejecting most theory. Because experience in the field was individualistic, there was no need for models or theory.

Third, *structuralist geographers* charged that traditional theories of spatial organization obscure more than they reveal. In their view, location theories are narrowly conceived and blind to historical facts—designed primarily to serve the goals of those who wield power in the economic system. These geographers believe that a structuralist view can provide a more precise set of ideas about the world economy by recognizing and drawing attention to the power relations of societies. They see the relations of places in the context of the world's political economy. They recognize a contemporary reality—the disadvantageous situation of most people with respect to the control and use of resources—and expose the structure that preserves and intensifies that situation. Structuralist geographers acknowledge and analyze prevailing value systems, a topic largely ignored by most economic geographers (Figure 1.23).

A number of radicals have been influenced by phenomenology, as previously discussed, and reject notions of objectivity or the existence of theory in economic geography, because to them, all experience is personal. These *deconstructionists*, descendants of postmodern and new-age mentalities, have become *humanistic*. They seek to view the world through the eyes of the exploited population that they seek to study. Some of the research on the homeless population, the inner-city unemployed, or undocumented temporary workers are examples.

A major thrust has been made by counterculture radicals who now call themselves *radical humanists*, arguing that labor, the environment, and the means of production have for centuries been dominated by the white, male power establishment. This power structure has been, and continues to be, they argue, exploitive, greedy, and debilitating to women, members of minority groups, foreign cultures (especially those of the Third World), and also to the environment. Radical humanists propose that this new understanding of "politically correct" research thinking and writing, with its undergirding plot of exploitation and domination, replace neoclassical location theory and analysis as the principle paradigm of future economic geography.

The majority of research in economic geography today remains location theory and analysis. It aims to understand "what" products and services are produced and "how" they are produced (that is, with what combination of resources), as well as "where" they are produced and "why there?" The radical humanist's principle aim is to expose the fallacy of currently accepted answers to these three questions and, most im-

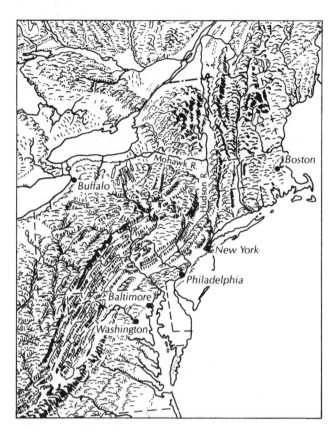

FIGURE 1.23
The Hudson-Mohawk corridor and the relative location or situation of New York. New York has a good harbor, was centrally located among the other 12 colonies, and was connected to the interior of America by water routes. These three locational and accessibility factors meant that it became the primate city of America rather than Philadelphia, Boston, or Washington.

portant, to answer the "for whom" question in a politically correct manner.

Economic geography has been characterized by major changes in thinking. These shifts reflect a need to deal with new realities of the world. However, to attempt to reduce the mosaic of views of the economic geographer's task to one or two general views would be misleading. Nonetheless, many geographers would agree that the theme of spatial organization is particularly valuable in helping us to understand world development problems. This theme receives more attention than others in this text.

Some Fundamental Concepts of Space and Location

Geographers examine space at multiple levels. There is a hierarchy of spatial perspectives, from personal space to international space. Personal space is the familiar "close at hand" world of the individual. International space is the entire world—those vast areas controlled more by governments and institutions than by individuals.

Agglomeration and *accessibility* are extensions of the concept of distance. For example, a shopping center reduces the distance consumers must travel to purchase goods by clustering (agglomeration) many stores at one point. A city itself is a clustering strategy, reducing aggregate distance among residential, business, and recreational functions, making them more accessible to the population.

Relative location is measured by the cost, in both money and time, required to overcome it. These costs are referred to as the *friction of distance*. The handicap of distance has declined historically because of transport improvements, but it may increase in the future with a rise in the cost of oil. In general, the retarding effect of distance is less in developed countries with their modern and well-managed transport systems than it is in underdeveloped countries. At the same time, geographic space is much more "sticky" in traffic-clogged New York or Washington, D.C., than it is in rural Wyoming.

Concrete qualities of space are referenced to specific points or areas on the earth's surface. Any location requires a fixed reference point. Geographers deal with two kinds of location: absolute and relative. *Absolute location* (site) is position in relation to a conventional grid system, such as latitude and longitude or street addresses. *Relative location* (situation) is position with respect to other

Vancouver is at the terminus of Canada's transcontinental transport routes, including TransCanadian Highway and both Transcontinental Railway Lines. It is also at the mouth of two river canyons, the Fraser and the Thompson. It has grown, due to its site and situation, to become Canada's second-ranking city in western Canada, as well as an attractive place to live because of the temperate climate created by its proximity to the marine influences of the Pacific Ocean. It acts as a super port for the exportation of coal and ores to Japan at Robert's Bank, 22 miles to the south; this advantage has added to its urban influence. *(Photo: Porterfield/Chickering, Photo Researchers, Inc.)*

locations. It is a measure of connectivity and accessibility, and it usually changes over time. The concept of relative location is of greater interest to economic geographers than absolute location.

To illustrate the importance of relative location, consider the position of New York City. The absolute location of New York (40°45' north latitude and 74° west longitude) tells us nothing about the city if we are interested in understanding why it became one of the world's great cities and ports. In 1820, four main ports on the northeastern seaboard competed for trade between the United States and Europe: Boston (pop. 61,000), Baltimore (pop. 63,000), Philadelphia (pop. 64,000), and New York (pop. 131,000). But New York's slight edge over its rivals became unassailable after 1825. Why? Geographers find the answer in the relative location of New York. The Appalachian Highlands represented a cost and time barrier (high friction and distance) between the resources of the American interior and the return flow of manufactured goods from Europe. New York is at the mouth of the Hudson River, which is almost at sea level all the way to Albany, where it is joined by the Mohawk River that cuts through the Appalachians (Figure 1.23). In 1825, the Erie Canal was completed linking the Hudson-Mohawk corridor with Lake Erie. An advantageous location relative to a primary traffic artery provided a major impetus for New York's explosive growth during the 19th century. By 1840, New York, with a population of 349,000, was nearly three and one-half times the size of its closest rival, Baltimore.

The concept of relative location is vital to our understanding of the integration of the world economy. *Spatial integration*—the linking of points of production (absolute location)—was mandatory for the development of the economic system. It involved, through the construction of transport networks, the transformation of absolute space into relative space. Growth of the world economic system, proceeding at different rates in various regions, determines the relativity of geographic space.

A *spatial process* is a movement or location strategy. Geographers are interested in movements, such as the flow of raw materials to processing plants, the distribution of finished products from manufacturing plants, and the trade in commodities. Also of interest are location strategies, such as the decision of Chagga farmers on the slopes of Mount Kilimanjaro to grow coffee, the choice of the most accessible place for a new school serving children in villages of a rural area in India, and the decision of Japanese companies to locate assembly plants near the U.S. market in cities along the U.S.–Mexican border.

Spatial structure is the internal organization of a distribution—the location of the elements of distribution with respect to each other. Spatial structures limit, channel, or control spatial processes. Because they are the result of huge amounts of cumulative investment over years and centuries, large alterations to the spatial structures of towns, regions, or countries are difficult to make.

Spatial structure and spatial process are circularly causal. Structure is a determinant of process, and process is a determinant of structure. For example, the existing distribution of regional shopping centers in a city will influence the success of any new regional shopping center in the area.

Flows of goods, people, and information are collectively known as *spatial interaction*. The amount of spatial interaction tends to decline with distance (Figure 1.25). This rule, called the *distance-decay effect*, holds for all sorts of things and all sorts of geographic scales. Information people have about places declines with distance; at larger geographic scales, air-passenger traffic is subject to the same effect.

The amount of interaction between places is also a function of their size. Big places generate more information, people, and goods than small places. New York

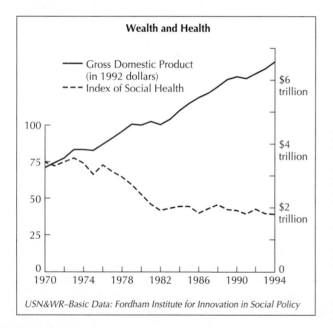

Wealth and Health

— Gross Domestic Product (in 1992 dollars)
- - - Index of Social Health

USN&WR–Basic Data: Fordham Institute for Innovation in Social Policy

FIGURE 1.24

Wealth and health. Between 1970 and 1994, the Gross Domestic Product in the United States doubled, but certain measures of social well-being dropped. Five of 16 social problems improved over the 25 years, including infant mortality, drug abuse, high school dropouts, elderly poverty, and food stamp coverage. Ten problems became more acute: child abuse, teen suicides, elderly health costs, child poverty, health insurance coverage, housing costs, the rich/poor gap, real wages, homicides, and alcohol-related road deaths. Gains from rapid technological change, the rise of the service sector, and globalized markets for North American products have penetrated deeply but unevenly into the national economy.

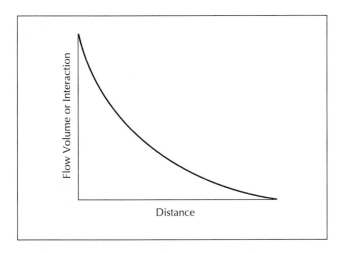

FIGURE 1.25

The distance-decay effect. The greater the distance traveled, the fewer the number of trips that will occur. This rule of transportation is true not only for air travel, but also for pipeline movements, telephone traffic, and automobile traffic. Is it true for your own pedestrian travel as well?

and Chicago have more interaction than New Orleans and Cincinnati, even though both pairs of cities are about the same distance apart (see Chapter 4).

Geographic Research in Aid of Development

A basic function of geographic research is to influence planning or organized action and the development of policy. An equally important function is to influence scholars, teachers, and students in the field and in the other social and environmental sciences.

The sort of knowledge and advice geographers provide their important constituencies fits under three headings: (1) human-environment, (2) spatial organization, and (3) the inventorying and monitoring of research. In the human-environment area, geographers provide information about the best match between the environment and the product realized from the environment. They provide information about how the biophysical environment, ecosystem management, and systems of livelihood are linked. In the area of spatial organization, geographers provide information about the kinds of spatial structures that favor processes of development and the procedures that can bring these structures into being. Inventorying research involves the collection and analysis of information, especially for planning the use of human and natural resources. Monitoring research provides information about change—whether certain changes are harmful to people, to the resource base, or both, and whether the goals of development are being achieved.

Economic Geographers

Economic geographers—university researchers; local, state, and federal government (e.g., the Aid for International Development Agency, the U.S. State Department); or private industry (e.g., Exxon, IBM, or McDonald's)—work to solve local, national, and international problems in a variety of settings. Some geographers have produced research in aid of international development in the developing world. The work of Iowa professor Gerard Rushton in developing a series of health care centers in India, the work of New York professor Jeffery Osleeb in tracking the spread of disease in Africa, the work of Tennessee professor Bruce Ralston, in improving transportation efficiency in Africa, and the work of California professor Janet Franklin in checking desertification in Africa are but a few of the hundreds of examples that exist of academic-based research. Many geography departments have multimillion-dollar annual research budgets in aid of development supplied by both private foundations as well as local, state, and federal governments.

For each score of economic geographers working at the international level, and usually funded by the U.S. government or international governments, there are hundreds working at the national and state level with funding from both the federal and state governments, as well as from the private sector. Researchers at the city and regional level surpass the national level by a factor of 10. Here, local governments and businesses provide the support and needed research agendas.

Geographers do not always solve development problems directly, but their skills and ways of thinking can help to resolve critical world issues. Our major responsibility as teachers and students is to try to understand the world and to explore the causes of that reality. The world, developed and underdeveloped, poses challenges of immense importance and complexity. In this exciting age of change, we can help to meet these challenges.

SUMMARY AND PLAN

The changing state of the world arouses the interest of the geographer. World development problems raise geographic questions, and the actual development involves varying degrees of change across geographic space. In this chapter, we present the geographer's perspective. We provide a definition of the field and introduce the main concepts geographers use to interpret and explain world development problems at a variety of scales, ranging from small areas and regions to big chunks of the world.

This text describes how geographers study and analyze the world economy and how they attempt to resolve world development problems. The varied perspectives that geographers use to approach principle economic, location, and allocation problems are addressed. Alternative world views, as derived from four basic political economies, are also a focus of discussion.

The following chapters of this text, which progress in logical sequence, are organized around the themes of distribution and economic growth. Chapters 2 and 3 deal with population and resources, the prime variables in economic geography. To understand the effects people and resources exert on the world economy, we need to learn about the principles of location. Chapters 4 to 9 supply the foundation stones required

to understand these principles: information technology and transportation, agricultural land use, urban land use, cities as service centers, and industrial location from the standpoint of firms and world regions.

The two chapters on industrial location provide a link between the development of businesses and the development of places—a link vital to the issues discussed in the remaining chapters. Chapters 10 and 11 deal with the expanding world of international business—its operations, environments, and patterns. The final chapter examines the geography of development and illustrates how economic growth and resource use can combine to create a world of uneven and unequal development.

KEY TERMS

absolute location
absolute space
accessibility
areal differentiation
areal integration
capital
circular flowing market system
command economy
comparative advantage
concrete space
demand
development
distance-decay effect
entrepreneurial skill
environmental determinism
factors of production
four questions of economic geography
four worlds
Four Tigers
friction of distance
GDP

GNP
greenhouse effect
human-environmental relations
interest
international economic system
international economic order
internationalization of capital
isotropic surface
labor
land
law of demand
law of supply
location theory
logical positivism
mixed economic systems
money capital
opportunity cost
phenomenological approach
physiologic density
point, line, area
product market

production possibilities analysis
profits
purchasing power parity
pure capitalism
radical humanist
raw materials
real capital
relative location
rent
resource market
spatial organization
spatial structure
spatial process
spatial interaction
spatial integration
structuralist geographers
Third World debt crisis
traditional market
transnationals (or MNCs)
vicious cycle of poverty
world economy

SUGGESTED READINGS

Anderson, J., Brook, C., and Cochrane, A. 1996. *A Global World*. New York: Oxford University Press.

Berry, B.J.L., Conkling, E., and Ray, M. 1997. *The Global Economy and Transition*. New York: Prentice Hall.

Barnes, T., and Gregory, D. 1997. *Reading Human Geography*. New York: John Wiley.

Bingham, R. D., and Hill, E. W., eds. 1997. *Global Perspectives on Economic Development*. New Brunswick, N.J.: Center for Urban Policy Research.

Frank, A. G., 1980. *Crisis in the World Economy*. London: Heinemann.

Gaile, G. L., and Willmott, C. J., eds. 1989. *Geography in America*. New York: Merrill/Macmillan.

Gould, P. 1985. *The Geographer at Work*. London: Routledge and Kegan Paul.

International Institute for Environment and Development and World Resources Institute. 1988. *World Resources 1988–89*. New York: Basic Books.

Janelle, D. G., ed. 1992. *Geographical Snapshots of North America*. New York: Guilford.

Johnston, R. J., Taylor, P. G., and Watts, M. J., eds. 1996. *Geographies of Global Change*. New York: Blackwell.

Martin, G. J., and James, P. E. 1991. *All Possible Worlds: A History of Geographical Sales*. New York: John Wiley & Sons.

Massey, D., and Jess, P. 1996. *A Place in the World: Places, Cultures, and Globalization*. New York: Oxford University Press.

Porter, M. E. 1990. *The Competitive Advantage of Nations*. New York: The Free Press.

Rubenstein, J. M. 1996. *An Introduction to Human Geography*, 4th ed. New York: Macmillan.

Sayer, A., and Walker, R. 1993. *The New Social Economy: Reworking the Division of Labor*. Cambridge, Mass.: Blackwell.

World Resources 1996-1997, 1996. New York: Oxford University Press.

WORLD WIDE WEB SITES

USGS: NATIONAL MAPPING PROGRAM
http://www.usgs.gov
This site provides accurate and up-to-date cardiographic data and information for the United States. These data products and information provide a framework of spatial information needed by Federal, State, and Local Government agencies, as well as the private sector to deal with such problems as conserving our natural resources, identifying and mitigating hazards, defining and studying ecosystems, and supporting economic development.

**USGS: THE GEOGRAPHIC NAMES
INFORMATION SYSTEM**
http://www.usgs.gov
The (GNIS), developed by the USGS in cooperation with the U.S. board on Geographic Names, contains information about almost 2 million physical ands cultural geographic features in the United States. The GNIS is the official repository of domestic geographic names in the United States.

WORLD TRAVEL GUIDE
http://www.wtgonline.com/country/
Maps, climate graphs, and country information.

YAHOO
http://www.yahoo.com
regional, country, or US states; very extensive information about all geographic aspects of places; Society & Culturelinks to cultural information about countries; links to many other sites.

YAHOO ALTA VISTA WEB PAGES
http://av.yahoo.com/bin/
Many links, such as to the CIA world fact book, with diverse information.

**ESRI—ENVIRONMENTAL SYSTEMS RESEARCH
INSTITUTE**
homepage: http://www.esri.com
jumpstation: http://www.esri.com/services/
jumpstation/jumpstation.html
Links to government sites, commercial sites and educational sites with geographic information.

CIA PUBLICATIONS AND HANDBOOKS
http://www.odci.gov/cia/publications/pubs.html
World factbook with diverse geographical data about countries.
USGS http://www.usgs.gov/
Topographic maps, other maps, physical geographical data.

US BUREAU OF THE CENSUS
http://www.census.gov
For a good introduction to desktop mapping and the latest news about TIGER and other data sources, visit:
http://www.wessex.com

TWO CARTOGRAPHIC SITES

**NATIONAL GEOGRAPHIC MAP MACHINE
[SHOCKWAVE]**
http://www.nationalgeographic.com/ngs/maps/
cartograhic.html

COLOR LANDFORM ATLAS OF THE UNITED STATES
http://fermi.jhuapl.edu/states/states.html
For those interested in map information, National Geographic's Map Machine and the Landform Atlas of the United States, pro-vided by Ray Sterner of the Johns Hopkins University Applied Physics Laboratory, are effective places to start. Map Machine's Atlas allows users to click on a world map or on continent or country menus to retrieve country maps, with concise information and flags. There are also selected area maps available, created from weather satellite data, as well as political and physical maps and a Macromedia Shockwave enhanced world map that allows users to view a world map interactively.

**INTERNATIONAL GOVERNMENT'S META-PAGES—
NORTHWESTERN UNIVERSITY**
http://www.library.nwu.edu/govpub/resource/internat/
Mike McCaffrey-Noviss of the Northwestern University Library Government Publications department provides no-nonsense, utilitarian pointer pages to over 60 country government pages and nearly 100 International Government Organizations (IGOs). Sites included are the official government or organization sites; pages are updated biweekly, and aim for breadth of coverage. Depth of coverage of both government and IGO sites varies by country. International regional and topical pages are forthcoming.

**FINDING INFORMATION ON THE INTERNET:
A TUTORIAL**
http://www.lib.berkeley.edu/TeachingLib/Guides/
Internet/FindInfo.html
The Teaching Library of the University of California, Berkeley makes the content of its Internet Workshops freely available to the public via this site. The Tutorial includes an introduction to the Internet, a glossary of terms, things to know before searching the World Wide Web, how to create search strategies, and how to refine your topic and identify the search tools to fit your needs. There are also well designed pages on how to construct and refine searches for Infoseek, Hobbot, and AltaVista.

**FEMINISM AND GLOBALIZATION: THE IMPACT OF
THE GLOBAL ECONOMY ON WOMEN AND
FEMINIST THEORY—*GLOBAL LEGAL STUDIES
JOURNAL***
http://www.law.indiana.edu/glsj/vol4/nol/toc/html

**GLOBAL LEGAL STUDIES JOURNAL"INDIANA
UNIVERSITY SCHOOL OF LAW**
http://www.law.indiana.edu/glsj/glsj.html
Indiana University's *Global Legal Studies Journal*, a bi-annual "peer-reviewed interdisciplinary journal focusing on the intersections of global and domestic legal regimes, markets, politics, technologies, and cultures," is highlighted by a symposium issue each fall. The latest symposium issue concerns feminism and globalization of market forces.

THE XEROX PARC MAP VIEWER
http://mapweb.parc.xerox.com/map/
Since June 1993, the Map Viewer, provided by the Xerox Palo Alto Research Center, has served interactive web-based maps of the world. Maps are available in three projections (equirectangular, sinusoidal, ellipse) and views may be zoomed in or out. The world map has viewable features, including borders and rivers; note, however, that political boundaries are based on 1985–90 data. The United States map is much more detailed and includes features such as borders, rivers, roads, and railroads.

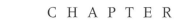

C H A P T E R

2

POPULATION AND THE WORLD ECONOMY

O B J E C T I V E S

- To describe and account for the world distribution of human populations
- To examine the economic causes and consequences of population change
- To describe the major demographic and economic characteristics of a population
- To describe and explain economic migrations, past and present
- To consider how changing demographics affects the business firm
- To apply demographic techniques to plan regional urban economic change

Crowds of people gather at the Syracuse University Carrier Dome sports complex. *(Photo: Michael J. Okoniewski, The Image Works)*

Human beings are the most important element in the world economy. People are not only the key productive factor, but their welfare is also the primary objective of economic endeavor. People are the producers as well as the consumers of goods and services. As world population continues to grow by leaps and bounds, we face the critical question of whether there is an imbalance between producers and consumers. Does population growth prevent the sustainability of development? Does it lead to poverty, unemployment, and political instability?

To help answer these questions, this chapter examines the determinants and consequences of population change for developed and developing countries. It analyzes population distributions, characteristics, and trends. It also reviews competing theories on the causes and consequences of population growth. It concludes with an application of demographic methods to the process of regional economic change.

POPULATION DISTRIBUTION

In 1997, there were more than 5.9 billion people in the world; these people were very unevenly distributed. Such contrasts in population distribution are studied on different scales of observation. *Population distribution* refers to the arrangement, spread, and density of people. On a macroscale, we examine broad geographic areas, such as continents, countries, and regions. On a microscale, we look at small areas, such as population variations within cities. In this discussion, the emphasis is on population distributions across large geographic areas.

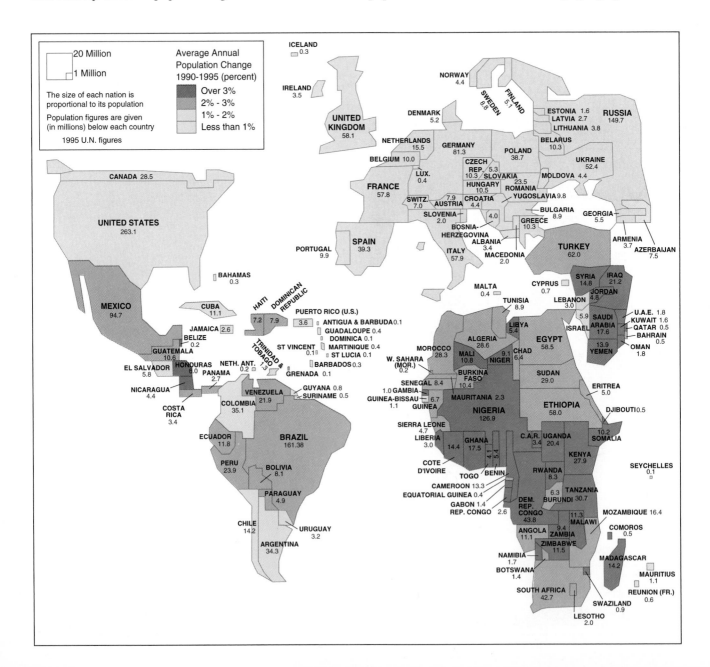

Population Size

A comparison of the world's population by continents shows that Asia's population is the largest, as it has been for several centuries. In 1997, Asia contained 3.6 billion people or 61% of the world's population. Europe (581 million) and Russia (149 million), along with the other republics of the Soviet Union (135 million), were home to about 16%, Africa (734 million) to 12%, Latin America (480 million) to 8%, North America (295 million) to 5%, and Oceania (28 million) to less than 1%. The populations of the developing world—Africa, Asia (excluding Japan and Russia), and Latin America—accounted for three out of every four humans alive in 1997. Of the 4.6 billion in the developing world, 63% lived in Asia, 16% lived in Africa, and 11% lived in Latin America.

Given such large variations among continents, it is not surprising that national population figures show even more variability (Figure 2.1). Ten out of the world's nearly 200 countries account for two-thirds of the world's people (Table 2.1). Five countries—China, India, the United States, Indonesia, and

FIGURE 2.1

Cartogram of 1995 world population. This map shows the area of each country in proportion to its population. Geographic space has been transformed into population space. Asia dominates the map, especially China, with its 1.2 billion people, and India, with its 900 million people. Europe is much larger than on a normal map, where countries are sized in proportion to their geographic space. Both South America and Africa show up much smaller than their normal size, because their populations are relatively small.

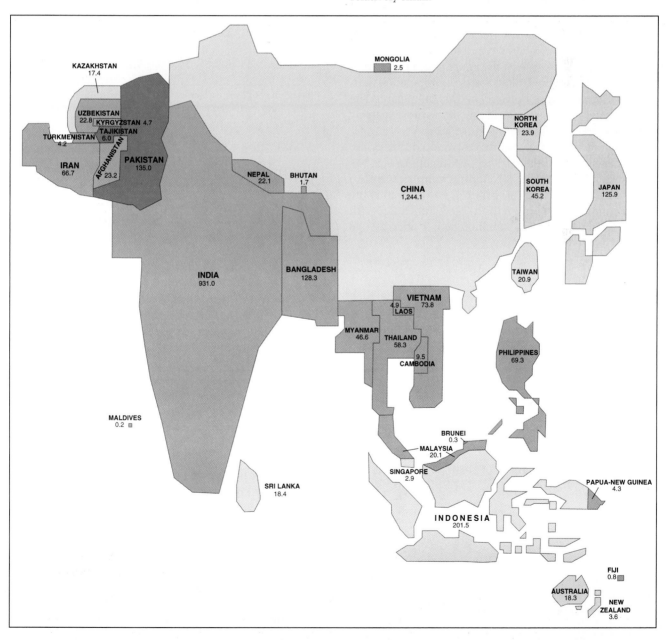

Brazil—contain half of the world's population. Approximately 21% of all people live in China, 16% in India, 5% in the United States, and 3% each in Indonesia and Brazil. Six of the top 10 countries in population size—China, India, Indonesia, Japan, Bangladesh, and Pakistan—are in Asia. Only 3 of the 10 most populous nations are considered to be developed (the United States, Russia, and Japan).

Population Density

Because countries vary so greatly in size, national population totals tell us nothing about crowding. Consequently, population is often related to land area. This ratio is called population density—the average number of people per unit area, usually per square mile or kilometer. Several countries with the largest populations have relatively low population densities (Table 2.2). For example, the United States is the fourth most populous country, but in 1997 it had a population density of only 72 people per square mile. If the entire world population were placed inside the United States, its population density would be roughly equivalent to that of England. The United States and Canada form one of the more sparsely populated areas of the world (Figure 2.2).

Many of the world's most crowded countries are small city-states or islands, such as Singapore (12,347 people per square mile). Excluding countries with a very small area (such as Singapore), Bangladesh is the world's most crowded nation (see Table 2.2). In Bangladesh, more than 100 million people are crowded

◫

Table 2.1

1997 The Ten Most Populous Countries and Canada

Country	Growth rate	Population in 1997 in Millions	In 2025*
China	1.1	1,238	1,570
India	1.9	970	1,385
United States	0.6	268	335
Indonesia		204	276
Brazil	1.7	160	213
Russia	−0.5	147	131
Japan	0.2	126	121
Pakistan	2.9	138	233
Bangladesh	2.0	122	180
Nigeria	3.1	107	232
Canada		30	37

Source: Population Reference Bureau, 1997, World Population Data Sheet.
*estimated

◫

Table 2.2

The Ten Most Densely Populated Countries, U.S., Canada*

Country	Population (Millions)	Area (Sq Miles)	Population Density
Bangladesh	111.4	55,600	2004
Taiwan	20.8	12,460	1672
South Korea	44.3	38,020	1165
Netherlands	15.2	14,410	1055
Japan	124.4	143,750	865
Belgium	10.1	11,750	855
Rwanda	7.7	10,170	759
Sri Lanka	17.6	25,330	696
India	882.6	1,269,340	695
Israel	5.2	8020	653
U.S.	265.2	3,792,575	71
Canada	30.2	3,851,787	8

*Based on countries with at least 5000 square miles of territory.
Source: Population Reference Bureau, 1996, World Population Data Sheet.

into an area the size of Iowa. Three of the top 10 densely populated countries—the Netherlands, Japan, and Belgium—are developed, whereas another three—South Korea, Taiwan, and Israel—are newly industrializing countries (NICs). The remainder are clearly less developed nations, reminding us that the relationship between density and development is a complex one.

Contrary to popular opinion, not all crowded countries are poor. But what explains the fact that many people in the Netherlands or Singapore live well on so little land? What part of the explanation lies in their industrious people and their ability to adapt to change? What part of the explanation lies in their history of trade or their relative locations? Singapore is on one of the great ocean crossroads of the world. But being on a crossroads has worked no similar miracle for Panama. In 1997, Singapore had a per capita income ($23,370) that was nearly seven times that of Panama ($1680).

Population density is a potentially valuable abstraction, but it conceals much variation. Egypt had a reasonably low figure of 144 people per square mile in 1997, but 96% of the population lives on irrigated, cultivated land along the Nile Valley where densities are extremely high. Similarly, in the United States there are densely populated and sparsely populated areas. Large areas to the west of the Mississippi are essentially devoid of people, whereas the Northeast is densely settled. The island of Manhattan, for example, has a density that is virtually the same as Hong Kong.

Most people are concentrated in but few parts of the world (see Figure 2.2). Four major areas of dense settlement are (1) East Asia, (2) South Asia, (3) Europe, and (4) the eastern United States and Canada. In addition, there are minor clusters in Southeast Asia, Africa, Latin America, and along the U.S. Pacific coast.

FACTORS INFLUENCING POPULATION DISTRIBUTION

Is there a reason for the massing of people in some areas? One possible explanation is physical environment. People tend to concentrate along edges of continents, at low elevations, and in humid midlatitude and subtropical climates. Lands deficient in moisture, such as the Sahara Desert, are sparsely settled. Few people live in very cold regions, such as northern Canada, arctic Russia, and northern Scandinavia. Equatorial heat and moisture, as in the Congo and Amazon basins, appear to deter settlement. In addition, many mountainous areas—whether because of climate, thin stony soils, or steep slopes—are inhospitable habitats. There are, however, many anomalies relative to population distribution and physical environment correlations. For example, more people inhabit highland than lowland environments in many Latin American and East African locales due to oppressive heat at lower elevations. For example, Mexico City is at an elevation of 7,000 feet.

Caution must be exercised in ascribing population distribution to natural elements alone. Furthermore, to hold that natural elements control population distribution is deterministic. Certainly climatic extremes, such as insufficient rainfall, present difficulties for human habitation and cultivation. However, given the forces of technology, the deficiencies of nature increasingly can be overcome. Air-conditioning, heating, water storage, and irrigation are examples of the extensive measures that technology offers to residents of otherwise harsh environments.

If physical environments alone cannot explain population distribution, what other factors are involved?

Human distributions are molded by the organization and development of economic systems. They are influenced by cultural traits, which also affect demographic components of fertility, mortality, and migration. Such social disasters as war may alter population distribution on any scale. Social and political decisions, such as tax policies or zoning and planning ordinances, are eventually reflected on the population map. None of these factors, however, can be considered without reference to historical circumstance. Present population distribution is explicable only in terms of the past. For example, the 19th-century Industrial Revolution made British coal sites major centers of population concentration and economic growth. Populous areas in Britain associated with coal include the Birmingham-Coventry district, Bradford-Leeds, Stoke-on-Trent, Manchester, South Wales, and Glasgow in Scotland. The influence of coal upon population distribution in Britain is still strong, yet its significance is waning as other sources of energy free industry and people from the coalfields.

Changing Population Patterns

Between 1995 and 2025, many countries in Eastern Europe and the former Soviet Union—including Bulgaria, Belarus, Hungary, Romania, Russia, and the Ukraine—are expected to lose population, continuing the present pattern. Russia is projected by the United

Mexico City is a spectacular example of a primate city. But most of its residents dwell a world apart from the glitter of the high-rise corridor. By 1994, Mexico had a population of 92 million, and Mexico City arguably comprised one-quarter of that amount. It is difficult to calculate the true size of Mexico City because of the myriad of squatter settlements on the outskirts of town without water, sewers, or streets. For the first time, Mexico City is showing signs of leveling off and even declining in population due to the urban pollution and squalor it has generated. *(Photo: Tatiana Parcero, Mexican Government Tourism Office)*

FIGURE 2.2
Population dot map of the
world. This map shows
population clusters within each
country. Population density is
shown on this map as a
concentration of dots, with one
dot representing 10,000 people.
Population density is defined as
the number of phenomena
occurring within a real unit.
Population density worldwide is
normally expressed as the
number of people per square
mile. Population density in East
Asia, notably China and Japan,
as well as South Asia, including
Bangladesh, India, and Pakistan,
is extremely high. Population
density in Northern Asia, Africa,
and South and North America is
quite low, comparatively
speaking. Three major and two
minor areas of world population
concentration occur. These are:
(1) East Asia; (2) South Asia;
(3) Europe; (4) Northeastern
United States and Southeastern
Canada; (5) Southeast Asia,
especially the country of
Indonesia and the island of
Java.

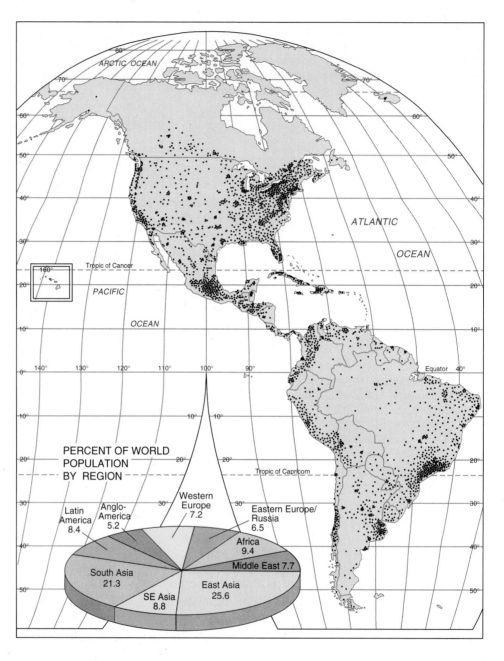

Nations to decline by 7% from the current 147 million to 137 million during this period. Additionally, the populations of certain highly developed countries, including Germany, Denmark, Italy, Spain, Sweden, Norway, and Japan, are projected to decline over the three decades. The United States is the principal exception to this trend for developed countries. The population is projected to grow from 265 million to 330 million (see Figure 2.8).

Developing countries will experience population growth, however. For example, Africa is projected to double between 1995 and 2025, from 728 million to 1.5 billion. The population of Asia is expected to grow from its current 3.5 billion to 5 billion, a 40% increase.

The decline in fertility in developing countries does not mean zero population growth. The demographic momentum in China is projected to expand the 1.2 billion population in 1995 to over 1.5 billion by 2025, even though China's fertility rate is below replacement level. South Korea, Singapore, and Thailand are at or below replacement levels presently. On the other hand, in Latin America, the fertility rate is 2.8%.

In India, there has been a dramatic decline in the fertility rate, which stands presently at 3.8%. It is here in South Asia, including Bangladesh and

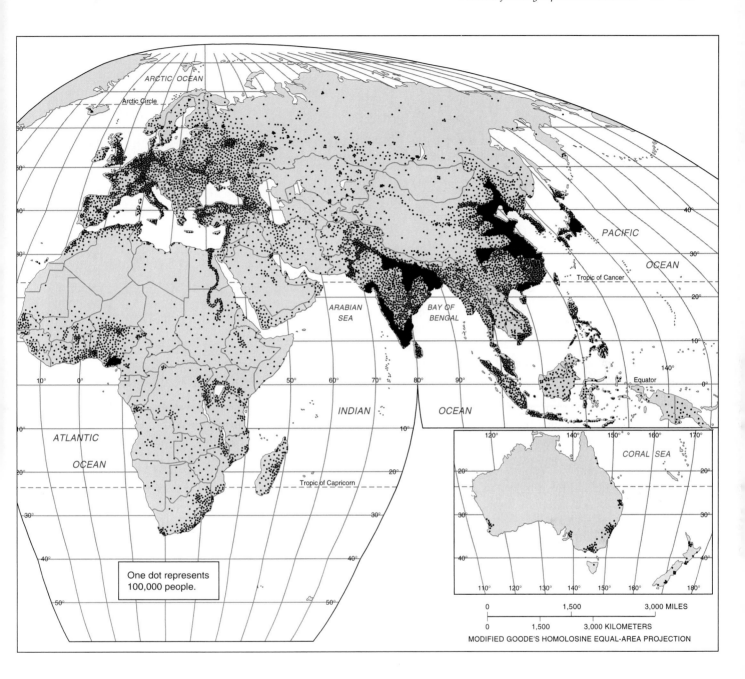

One dot represents 100,000 people.

MODIFIED GOODE'S HOMOLOSINE EQUAL-AREA PROJECTION

Pakistan, and in the Middle East and Africa—the least developed countries—where the demographic outlook is the most disturbing. These regions' fertility as a group was 6.6% in 1960, but only 6.5% by the early 1990s.

Urbanization

The city is the most impressive and forceful expression of humankind's struggle with nature. It is a built environment—a giant resource system—that has little regard for the physical environment. As a center of commodity production and final consumption, it exerts a major influence on population distribution. Population concentrations reach their most extreme form in cities.

Modern urban growth is the result of agricultural, industrial, and transport revolutions of the late 18th and 19th centuries in Europe. The agricultural revolution allowed farmers to produce a surplus of food needed for growing nonagricultural populations. The Industrial Revolution spurred the mass movement of surplus population away from the countryside and into the emerging factories in cities. The transport revolution permitted the cheap and fast distribution of the goods required by an expanding urban population. These three developments also increased the size of the trade areas that acted as markets for the goods and services of the growing towns.

The outcome of these revolutions was an increase in the size of urban areas and in the level of urbanization. *Urban growth* is the increase in the size of city populations. Cities grow through natural increase—the excess of births over deaths—and through in-migration. *Urbanization* refers to the process through which the proportion of population living in cities increases. An urban area may be defined on the basis of numbers of residents. In the United States, any place with at least 2,500 residents is classified as urban. There are no universal standards, however. Countries have developed their own criteria for differentiating urban from rural places.

Urban growth and urbanization occur simultaneously in most countries, but their rates vary. Indian cities are growing by about 3.6% annually, but the percentage of total population in urban places is increasing slowly, from 20% in 1970 to 26% in 1996. Brazil, with a similar urban growth rate, is experiencing a much faster rate of urbanization. The percentage of total population living in urban areas increased from 46% in 1970 to 76% in 1997.

World urbanization has increased dramatically since 1800. In 1800, some 50 million people—about 5% of the total population—lived in urban areas. By 1997, more than 2.5 billion people, about 43% of the total population, lived in cities.

Levels of urbanization vary widely among regions and countries of the world (Table 2.3 and Figure 2.3). In 1997, developed countries were 75% urbanized, whereas developing countries were only 35% urbanized. Latin America has become a predominately urban continent since the end of World War II, but Asia and Africa are still overwhelmingly rural.

The degree of urbanization in a country usually corresponds to its level of economic development. In general, developed countries have the highest urbanization and income levels (see Figures 2.3 and 2.4), but the relationship between town-dwelling population and the level of income per person does not imply that urbanization equals development. (Compare Figures 2.3 and 2.4 with Figure 1.3.) Urbanization in the developing world is not being accompanied by a rapid increase in prosperity. High urban growth rates are diluting capital resources and reducing living standards. They are generating congestion, slums, employment problems, and deteriorating services.

To understand the role of rapid population growth in urban problems, it is necessary to distinguish between the increasing proportion of urban to total population (urbanization) and the absolute growth of

◈

T a b l e 2.3

Proportion of Population in Urban Areas

Region	1950	1970	1985	1996	2000
World	29	37	41	45	50
Developed countries	54	67	72	75	76
Underdeveloped countries	17	25	31	35	39
Africa	16	23	30	31	33
Latin America	41	57	69	71	73
East Asia	17	27	29	36	38
South Asia	16	21	28	30	32
North America	64	74	74	75	75
Former Soviet Union	39	57	66	68	70
Europe	56	67	72	74	75
Oceania	61	71	71	71	72

Source: Population Reference Bureau, 1996.

urban population (urban growth). Urbanization increased during the economic expansion of developed countries from 17% in 1875 to 26% in 1900. It also increased at about the same rate in the Third World between 1950 and 1975; the rise was from 17% to 27%. This comparison suggests that economic development rather than population growth is the main determinant of urban growth. But Third World urban population increased at an annual rate of 4.1% between 1950 and 1975, far faster than the rate of 2.8% a year in developed countries between 1875 and 1900. Urban growth rates reflect national growth rates. In the developing world, urban growth is paralleled by high rates in rural areas. Natural increase accounts for 60% of the urban population growth. Rural-to-urban migration accounts for the remaining growth.

The most striking feature of urbanization in the developing world is the growth of huge primate cities—cities that dominate the urban landscape of a nation. In less developed nations, a *primate city*, a major center of growth, is typically the political capital as well as the center of industry, finance, and commerce. In 1950, only three cities in underdeveloped countries ranked in the top 10 in the world (Figure 2.5). By 1990, the number had grown to 8 of the top 10. Twenty-one of the largest 25 cities in the world are in developing countries (Table 2.4). Compared with developed countries, the projected rates of big city growth are very rapid indeed (Table 2.5). A large part of the population of many developing countries is concentrated in a primate city. At least one in five people

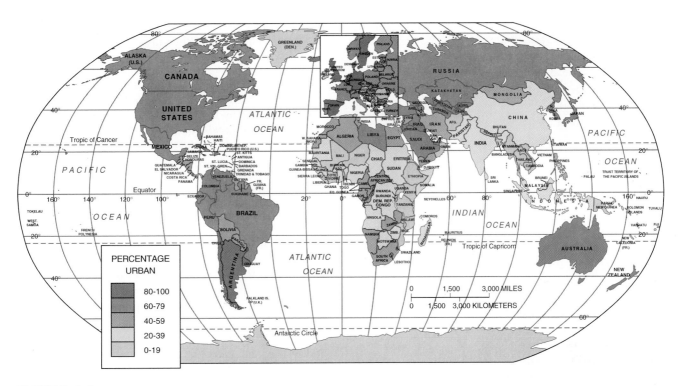

FIGURE 2.3
World urbanization. The level of urbanization varies worldwide and is dependent on a number of factors, including economic development. If the level of economic development is low, a large portion of the population is engaged in hunting/gathering and agriculture. These activities are not conducted in cities. As a society becomes more developed, a greater proportion of the population is engaged in manufacturing and the service industries, which are normally concentrated in cities. The most urbanized countries include Northwestern Europe, North America, Australia, and New Zealand. Least urbanized countries are located in East Asia, South Asia, and Africa.

in Argentina, Iraq, Peru, Chile, Egypt, South Korea, Mexico, and Venezuela live in a primate city. At the present time, cities such as Lima, Bangkok, Baghdad, and Buenos Aires account for more than 40% of the total urban population of their respective countries (Figure 2.6).

FIGURE 2.4
Relationship between per capita GNP and urbanization. The percentage of population living in urban places generally increases with per capita GNP. Ethiopia and other low-income countries of the world have a low percentage of population living in urban places. Japan, Germany, Canada, and the United States are examples of countries with relatively high GNP per capita and a high proportion of population living in urban places— close to 80%. Some countries such as the United Kingdom and Hong King have high percentages of their populations living in urban places, but moderate per capita GNPs. This is due to a space availability problem. Why do countries in Latin America have a higher percentage of population living in urban places than their per capita GNP would suggest? *(Source: Data from Population Reference Bureau, 1987)*

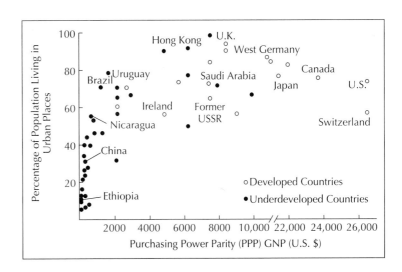

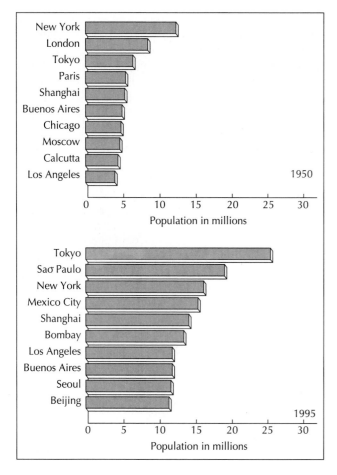

FIGURE 2.5
The 10 most populous world cities in 1950 and 1995. In 1950, only one city, New York City, exceeded 10 million people, but by 1995 at least 10 cities exceeded 10 million. Tokyo had become the world's largest city with more than 25 million people. In 1950, 7 of the top 10 world cities were found outside the third World, but by 1995, 7 of the top 10 cities were Third World cities. *(Source: Population Reference Bureau, 1997)*

POPULATION GROWTH OVER TIME AND SPACE

To this point, we have set the world stage in demographic terms, but we now need to put population growth into the context of changes over time and variability across space. The chief force affecting world population distribution was once migration. This was true until European transoceanic movements were disrupted during the economic depression of the 1930s. Today, the main cause is natural increase—the excess of births over deaths. Human population is increasing, and threatens to go on increasing. Each year an additional 95 million people inhabit the earth. The major impetus to world population growth comes from developing countries, in which more than three-fourths of humankind dwell. With more than 4 billion people already and another billion expected by the year 2000,

Table 2.4
The World's Twenty-Five Largest Cities, 1996

	Population (millions)	Average Annual Growth Rate 1990–96 (percent)		Population (millions)	Average Annual Growth Rate 1990–96 (percent)
Tokyo, Japan	26.8	1.41	Tianjin, China	10.7	2.88
Sao Paulo, Brazil	16.4	2.01	Osaka, Japan	10.6	0.23
New York, United States of America	16.3	0.34	Lagos, Nigeria	10.3	5.68
			Rio de Janeiro, Brazil	9.9	0.77
Mexico City, Mexico	15.6	0.73	Delhi, India	9.9	3.80
Bombay, India	15.1	4.22	Karachi, Pakistan	9.9	4.27
Shanghai, China	15.1	2.29	Cairo, Egypt	9.7	2.24
Los Angeles, United States of America	12.4	1.60	Paris, France	9.5	0.29
Beijing, China	12.4	2.57	Metro Manila, Philippines	9.3	3.05
Calcutta, India	11.7	1.67	Moscow, Russian Federation	9.2	0.40
Seoul, Republic of Korea	11.6	1.95	Dhaka, Bangladesh	7.8	5.74
Jakarta, Indonesia	11.5	4.35	Istanbul, Turkey	7.8	3.67
Buenos Aires, Argentina	11.0	0.68	Lima, Peru	7.5	2.81

Source: United Nations (U.N.) Population Division, World Urbanization Prospects, 1997 Revision (U.N., New York, 1997), Table A.12, pp. 132–139, and Table A.14, pp. 143–150.

Table 2.5

Projected Populations, Percentage of Total Urban Populations, and Number of Cities Larger than 4 Million People

Region	2000		
	Population (Millions)	Percentage of Urban	Number of Cities
World	681	21.7	79
Developed countries	167	16.5	20
Underdeveloped countries	514	24.2	59
Africa	74	20.4	12
East Asia	154	23.0	14
South Asia	199	25.8	23
Latin America	123	28.6	12

Source: U.N. Food and Agricultural Organization, 1996. The State of Food and Agriculture, 1994.

how will the developing world manage? How will the vast population increase affect efforts to improve living standards? Will the developing world become a permanent underclass in the world economy? Or will the reaction to an imbalance between population and resources be waves of immigration and other spillovers to the developed countries?

The *demographic transition* describes the pattern of population change experienced historically by the now developed countries of Europe, North America, and Asia, and it is the most pervasive theory with regard to what we can expect in the future from other countries. In its basic format, the theory suggests that all societies were characterized by high birth and high death rates until roughly the time of the Industrial Revolution. Beginning in Europe and then spreading to North America, the death rate began to decline as an accompaniment to the rising standards of living that went along with the Industrial Revolution. The drop in the death rate was not at first matched by a decline in the birth rate, and the result, of course, was an increase in the rate of population growth (Figure 2.7). This is the phenomenon that gave rise to the term "population explosion"—it is the *transition* from high birth and death rates to low birth and death rates. This latter phase results from the eventual response of the birth rate to downward pressure. Industrialization, which first lowered the death rate, ultimately produces the motivations for families to limit family size, and over time, the birth rate drops enough to once again reach parity with the death rate, and then population growth slows or stops.

The current rapid growth rate of the world population is a recent phenomenon. It took from the emergence of humankind until 1850 for the world population to reach the billion mark. The second billion was added in 80 years (1850–1930), the third billion was added in 30 years (1931–1960), the fourth in 16 years (1960–1976), and the fifth in only 11 years (1977–1987). Although the overall rate of population growth is slowing down, a sixth billion will be added before the year 2000 (Figure 2.8).

The trains are crowded in Bangladesh. Here, in one of the world's most fertile lands, the combination of high population density and the concentration of land into fewer and fewer hands contributes to continuing poverty and hunger. By 1994, India had a population of 900 million and a birth rate of 31 per 1,000 population per year. Due to its 2.1% natural increase, its doubling time is 34 years. However, India's infant mortality rate is still rather high, at 91 deaths per 1,000 births, while that of North America is only 8. In 1993, India's per capita GNP was $330 (in U.S. dollars). By 1994, Bangladesh had a population of 120 million people and a natural increase of 2.4%, giving it a doubling time of only 29 years. In 1993, its per capita GNP was $220 (in U.S. dollars). *(Photo: World Bank)*

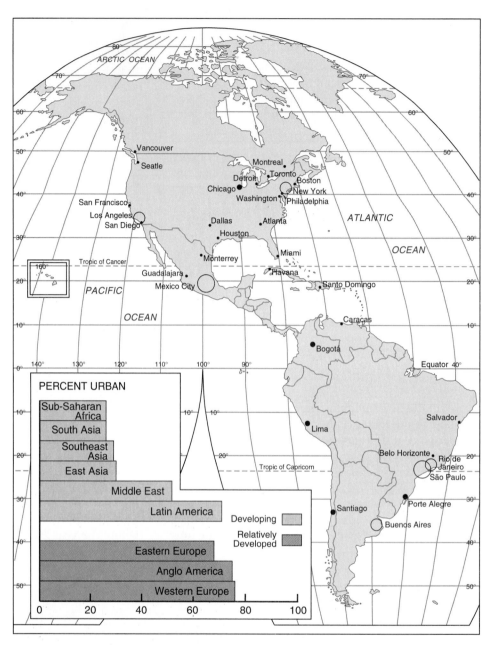

FIGURE 2.6
Major cities of the world. Cities with populations of more than 2 million are shown on this map. While urbanization of a country usually increases with per capita income, more than 10 million cities are located in the developing world, especially in East and South Asia. Europe has more than its proportional share of large cities, while Africa has a noticeable lack of large cities. The growth of large cities in developing countries suggests the overall rapid growth of populations and the migration from rural areas to the city as more jobs are offered in the manufacturing sector and fewer labors are required in the agricultural sector.

The immediate cause for the surge in the growth of the world population is the difference between the crude birth rate and the crude death rate. The crude *birth rate* is the number of babies born per 1,000 people per year, and the crude *death rate* is the number of deaths per 1,000 per year. For example, the U.S. birth rate in 1996 was 15 and the death rate was 9. During that year, the growth rate was 15 minus 9, or 6 per 1,000, which is a percentage *rate of natural increase* of 0.6.

Like all living things (and some that are inanimate, such as your savings account), human populations have the capacity to grow exponentially (1, 2, 4, 8) rather than arithmetically (1, 2, 3, 4), which is why population can increase so rapidly. Thus, the historical pattern of population growth shown in Figure 2.8 does

indeed look like an explosion. We can express this intuitively by talking about the *doubling time*—the number of years that it takes a population to double in size, given a particular rate of growth. For example, at an annual increase of 0.7% per year, the doubling time for the U.S. population is 100 years. As growth rates increase, doubling times decrease sharply. At 1.5%, the rate of world increase in 1997, the doubling time is 46 years. The rate of growth in Mexico is 2.2% and the doubling time is 32 years. In general, the doubling time for a population can be determined by using the *rule of seventy*, which means that you divide 70 by the average annual rate of rate. Thus, for Mexico, 70 divided by the growth rate of 2.2 equals a doubling time of 32 years.

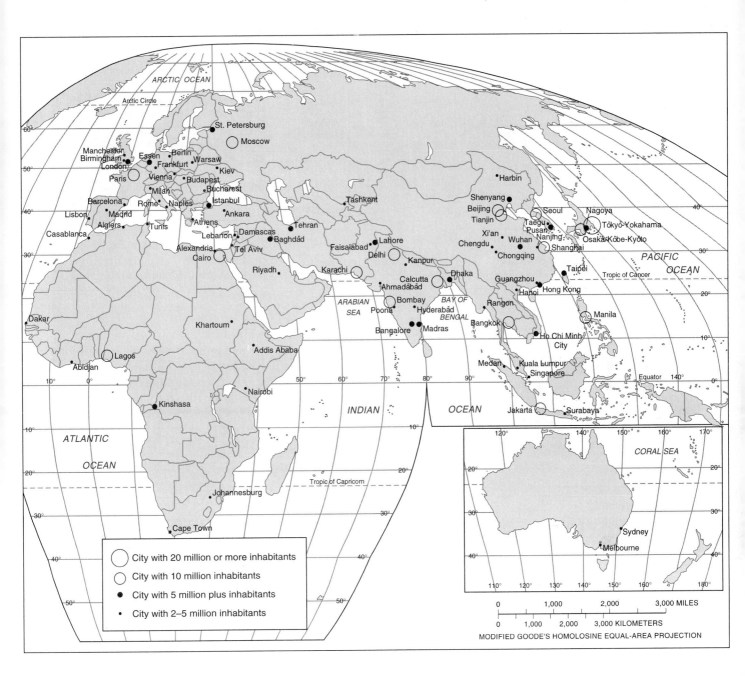

Most of the world's population growth is occurring in the developing world (Figure 2.9). Of all the continents, Africa has the fastest rate of growth. In 1996, the population of Africa was growing by 3% per year. For Kenya, with a *fertility rate* of nearly seven births per woman, the rate was 2.7%. At that rate of increase, Kenya's 1996 population of 28 million will double in just 25 years.

Rapidly declining death rates and continued high birth rates are the cause of this explosion. Death rates have been falling to fewer than 10 deaths per 1,000 people each year in Asia and Latin America, and to about 13 per 1,000 in Africa. Birth rates are changing less spectacularly (Figure 2.10). They are highest in Africa (41 births per 1,000 people annually), Latin

America (26 per 1,000), and Asia (24 per 1,000). These latter figures compare with birth rates of 11 per 1,000 in Europe and 15 per 1,000 in North America.

After accelerating for two centuries, the overall rate of world population growth is slowing down. In 1992, population was growing at 1.7% a year, down from a peak of 2% in the late 1960s. The rate of growth declined to about 1.5% in 1997 and 1% in 2020. However, the absolute size of population will continue to increase because the size of the base population to which the growth rate applies is so large (Figure 2.11).

The United Nations projects world population at 6.1 billion in 2000 and 8.5 billion in 2025. Almost all of this increase will occur in the developing coun-

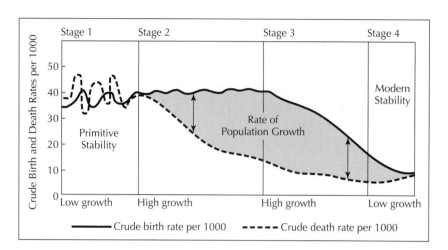

FIGURE 2.7
The demographic transition. Four stages of demographic change are experienced by countries as they develop from primitive to modern. Stage 1: Both high crude birth rates and crude death rates occur in the early stages of civilization. The resulting crude natural increase is quite low. Stage 2: Declining crude death rate, but a continually high crude birth rate, results in an increasing population growth rate. Stage 3: Crude death rates remain low, while crude birth rates start to decline rapidly, resulting in a rapid but declining growth rate. Stage 4: Both crude death rates and birth rates are low, resulting in a low natural increase.

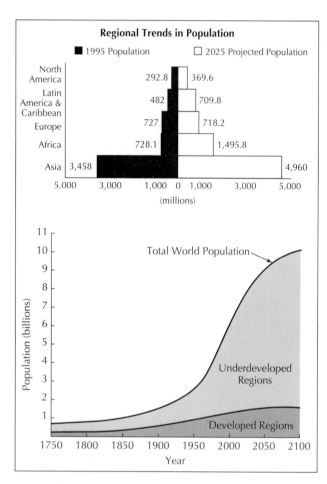

FIGURE 2.8
A major upturn in the logistic growth curve occurred around 1950 with important improvements in agriculture and an increasing food supply. In addition, medical science breakthroughs, such as polio and chickenpox vaccines, and improvements in nutrition substantially reduced death rates. Population growth rates are expected to slow by 2050 as the above improvements in science and nutrition permeate the Third World.

tries. The largest absolute increase is projected for Asia, reflecting its huge population base. Future population growth will further accentuate the uneven distribution of the world's population. In 1997, 80% of the world's population lived in the developing world, but by the year 2025, the proportion will increase to 84% (Figure 2.8).

POPULATION PROCESSES

We discussed the fact that the rate of natural increase for a country is measured as the difference between the birth rate and the death rate. Births and deaths represent two of the three population processes; the third is migration. Every population displays some combination of these three processes as the determinant of its pattern of growth. In general, we can express the relationship among them using the *demographic equation*:

Population at Time 2 = Population at Time 1
+ Births − Deaths + In-migration − Outmigration

Thus, in order to understand the current demographic situation in any country or region of the world, we must understand all three of these basic demographic processes. The population of any geographic area changes through the interaction of three demographic variables: births, deaths, and migrations. The difference between births and deaths produces natural increase (or decrease) of a population. Net migration is the difference between immigrants and emigrants. Natural increase usually accounts for the greatest population growth, especially in the short run. However, in the long run, migration contributes far more than the number of people moving into an area because the children of immigrants add to the population base.

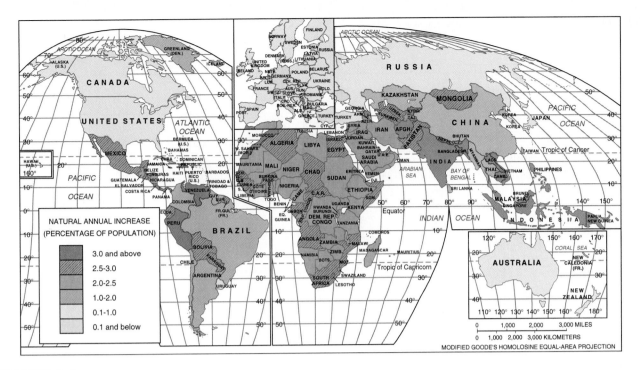

FIGURE 2.9

Natural annual increase in world population, 1997. The natural annual increase is the rate of growth of a population for a particular year. The fastest growing areas of the world include Africa, Central America, and Southwest Asia. Here growth rates exceed 2.5%, with a number of countries in Central America and Africa actually exceeding 3% per year. Three percent growth per year does not seem like a high growth rate; however, it indicates total population doubling time for a country of only 23 years. With a 2% growth rate, a country would double in 35 years. For a 1% growth rate, a country may double in 70 years. Natural increase in Europe as a whole is only 0.2%, and several countries have declining rates of growth. (See Color Insert 1 for more illustrative map.)

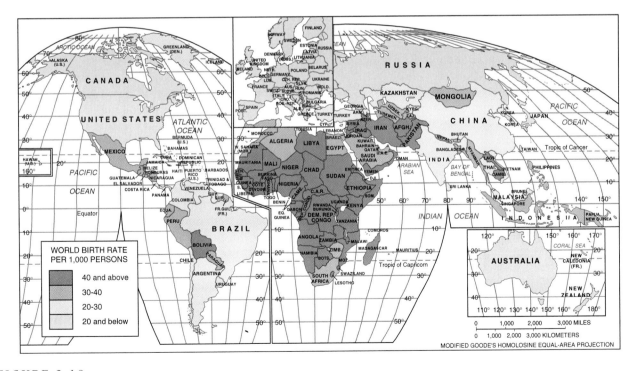

FIGURE 2.10

World birth rate, 1997. The world birth rate is the number of children born each year per 1,000 people who live in a country. The world birth rate closely mirrors the natural annual increase. *Crude* birth or death rates mean that we are counting a country's population as a whole and not studying separate rates of birth for women of different age groups, or by income or racial character. Again, Central America, Africa, and Southwest Asia lead the world in the rate of crude birth. (See Color Insert 1 for more illustrative map.)

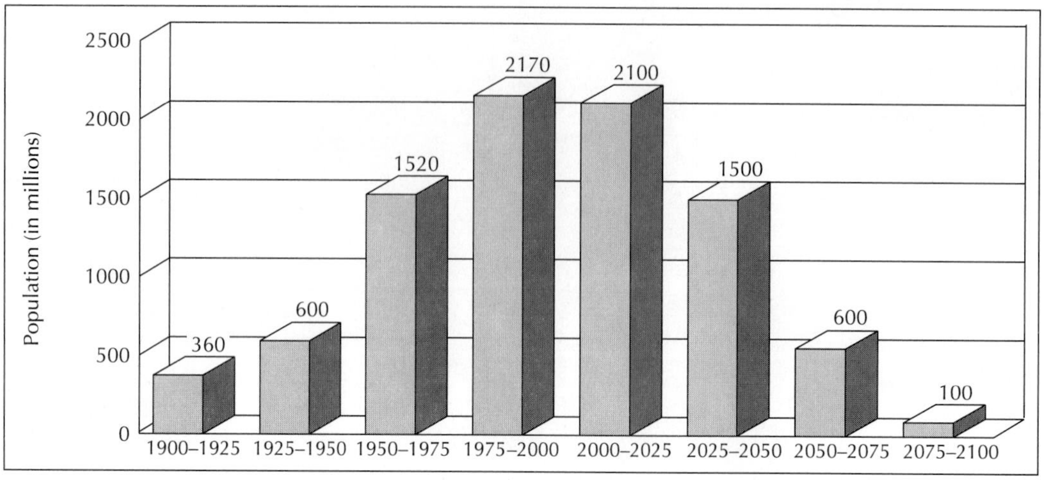

FIGURE 2.11
Net additions to the world's population between 1900 and 2100. The world's population has increased exponentially in the 20th century. The period from 1975 to 2000 will show the largest quarter decade of increase, with more than 2 billion people being added to the world's population, followed by the period from 2000 to 2025. From there, the world's population should slow down in rate of increase. *(Source: Weeks, 1994, p. 35)*

Fertility and Mortality

Earlier in the chapter, we mentioned the demographic transition as a theory that describes the fertility and mortality changes that have occurred in the developed societies and that may provide some guidance for future demographic trends in the less developed world. Thus far, only North America, Europe, and East Asia have moved completely through the demographic transition. Elsewhere in the world, largely in the Southern Hemisphere, death rates have dropped dramatically, especially since the end of World War II, but birth rates remain well above death rates, resulting in the high rate of population growth we have been describing.

It is important to remember that death rates do not drop evenly as an economy develops over time. *Infant mortality rates* (deaths to babies under age 1) tend to drop earliest and quickest. It was not uncommon in premodern societies to find an infant mortality rate in excess of 200 infant deaths per 1,000 live births—20% or more of all babies died before reaching their first birthday. Nowhere in the world today are rates that severe, but the highest rates (averaging 100 infant deaths per 1,000 live births) are found in sub-Saharan Africa (Figure 2.12).

Because the drop in the death rate disproportionately affects the very young, it acts exactly like an increase in the birth rate—more babies survive to grow to adulthood. Life spans likewise increase (Figure 2.13). One of the reasons the very young are more affected is that, as the death rate drops, it does so initially because communicable diseases are brought under control, and the very young are particularly susceptible to such diseases. The control of communicable disease

has the serendipitous economic side effect of reducing the overall illness level in society, thus promoting increased labor productivity. Workers miss fewer days of work, are healthier when they do work, and are able to work productively for more years than when death rates are high. Eventually, as death rates drop, the timing of death shifts increasingly to the older ages, to the years beyond retirement when the economic impact on the labor force is minimal.

Although death rates have declined throughout the world, it is still true that mortality is lowest in the Northern Hemisphere (especially in northwest Europe and in Japan) and highest in the Southern Hemisphere (especially in sub-Saharan Africa). A virtually identical pattern prevails in the world with respect to fertility (Tables 2.6 and 2.7).

Fertility levels fell first in Western Europe, followed quickly by North America, and more recently by Japan, and then the remainder of Europe. In all of those areas of the world, reproductive levels are near, or even in some countries below, the level of generational replacement. Elsewhere in the world, however, birth rates remain at much higher levels, although in China and Southeast Asia the birth rates are dropping very quickly. There has been a modest decline in South Asia, the Middle East, and much of Latin America, but few signs of a decline have yet been seen in most of sub-Saharan Africa. The seeming intransigence of birth rates in Africa has baffled many analysts, but Australian demographers (Caldwell and Caldwell, 1990; Caldwell, Orubuloye, and Caldwell, 1992) have argued that the explanation lies in the very different way in which African society tends to be organized.

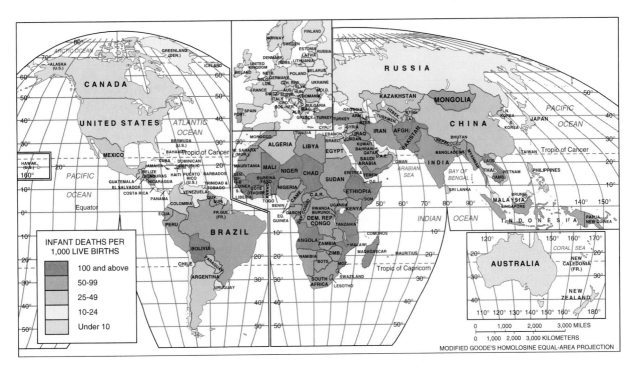

FIGURE 2.12

Infant deaths per 1,000 live births, 1997. Infant mortality rates are the number of deaths of infants under 1 year of age per 1,000 live births for a country. World infant mortality rates average approximately 75 per 1,000 for 1997. The Unites States, for example, averaged 7 deaths per 1,000 live births, but several European nations had lower infant mortality rates. As with natural income increase and world birth rates, infant deaths per 1,000 live births are related to development. Infant deaths per 1,000 live births are highest in Africa and Southwest Asia, more than 100 in many cases. (See Color Insert 1 for more illustrative map.)

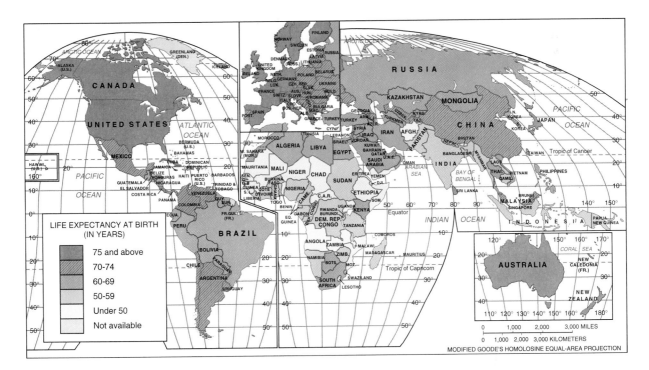

FIGURE 2.13

Life expectancy at birth. Babies born in 1997 are expected to live an average of 64 years. Life expectancy, however, ranges from the high 70s in northwestern European nations and in North America and Australia to the low 40s in some African nations. (Sierra Leone's life expectancy is only 33 years, while Iceland's and Japan's are 77 years.)

Table 2.6
Decline in Total Fertility Rate, 1981–1997

	1981	1985	1990	1997
Africa	6.4	6.3	6.2	5.6
Latin America and Caribbean	4.4	4.2	3.5	3.0
Asia (excluding China)	5.5	4.6	4.1	3.0
China	2.3	2.1	2.3	1.8
Developed Countries	2.0	2.0	2.0	1.6

The core of African society is its emphasis on ancestry and descent. In religious terms, this is usually reflected by a belief in the intervention of ancestral spirits in the affairs of the living. In social terms, the emphasis is reflected in the strength of ties based on the family of descent—the lineage. A person's true spiritual home, and in some societies the physical home for a lifetime, is the lineage, rather than the conjugal family (Caldwell and Caldwell, 1990, p. 119.)

The consequence of the African family structure is that children continue to be economic assets to the family (the lineage) even though the resulting high rate of population growth is very difficult for each nation to cope with.

Overall, then, almost all the world's nations are experiencing more births than deaths each year, with the biggest gap being found in the less developed nations and the narrowest difference existing in the more developed nations. In addition to these patterns of natural increase, many areas of the world are also impacted by migration.

Increasing life expectancy in a region or country is an important indicator of social progress. According to the World Health Organization, between 1980 and 1995, overall life expectancy increased from 61 to 65 years. In developed regions, life expectancy is 71 years for males and 79 years for females; in less developed

Table 2.7
Crude Birth and Death Rates for Selected Countries, U.S., Canada

Country	Crude Birth Rate	Crude Death Rate	Annual Rate of Increase (%)	Doubling Time (Years)	% Married Women Using Contraception
World	25	9	1.6	43	56
Developing Nations					
Average (excluding China)	32	10	2.2	31	43
Egypt	30	8	2.3	31	—
Kenya	44	10	3.3	21	33
Madagascar	46	13	3.3	21	17
India	29	10	1.9	36	41
Iraq	45	8	3.7	19	18
Viet Nam	30	7	2.3	30	65
Haiti	42	19	2.3	30	18
Brazil	25	8	1.7	40	77
Mexico	28	6	2.2	31	65
Developed Nations					
Average	12	10	0.3	264	66
United States	15	9	0.5	127	71
Canada	13	7	0.6	116	75
Japan	10	7	0.6	267	64
Denmark	13	12	0.1	533	78
Germany	10	11	−.1	—*	75
Italy	10	10	0.0	—*	—
Spain	10	9	0.1	630	59

*Population is declining.
Source: Data from 1996 World Population Data Sheet, Population Reference Bureau, Inc.

In Mali, women often have to travel several miles from their villages to gather firewood. Their heavy workloads are one of the factors that contribute to high fertility; the children are needed to help with house chores and farm work. As of January 1994, Mali had a population of 10 million and a natural increase of 3%, giving it a doubling time of 23 years. As is the case in most African countries south of the Sahara, the infant mortality rate of 111 per 1,000 live births is relatively high. Their per capita GNP in 1993 was $200 (in U.S. dollars). *(Photo: Ian Steele, United Nations)*

regions, 62 years for males and 65 years for females; in least developed regions, the values are 51 and 54 respectively. The gap in life expectancy among regions is expected to widen from 35 years in 1995 to 37 years by the year 2000 (Table 2.8).

The most dramatic reversal of life expectancy has occurred in Russia. Life expectancy for males in Russia fell from 65 years in 1987 to 57 years in 1995. Life expectancy for females in Russia declined from 75 years in 1987 to 71 years in 1995. The gap in life expectancy between men and women widened from about 10 years in 1989 to nearly 15 years in 1995. Statistics show that the rise in mortality is mainly due to diseases of the heart and circulatory system, plus increased crime and social unrest.

Migration

Migration is a purposeful movement involving a change of permanent residence. It is a complex phenomenon that raises a lot of questions. Why do people move? What factors influence the intensity of a migratory flow? What are the effects of migration? What are the main patterns of migration?

CAUSES OF MIGRATION

Most people move for economic reasons. They move to take better-paying jobs or search for jobs in new areas. They also move to escape poverty or low living standards. Some people move because of cultural pressures or adverse political conditions. Others move to fulfill personal dreams. Whatever the motive, migrants seek generally to better themselves. The causes of migration are sometimes divided into "push-and-pull" factors. Push factors might be widespread unemployment, population pressure, shortage of land, famine, or war. Hunger in Sweden in the 1860s is a good example of a push factor. In the half century between 1861 and 1910, more than 1 million Swedes moved to the United States. In the late 1970s and early 1980s, the various communist purges in Vietnam, Kampuchea, and Laos "pushed" out approximately 1 million refugees, who resettled in the United States, Canada, Australia, China, Hong Kong, and several countries in Western Europe. Pull factors may be free agricultural land, the "bright lights" of a developing world primate city, or a booming economy. The rich oil-exporting countries in the Middle East act as a pull factor for millions of immigrants seeking employment. In Kuwait, nearly 80% of the workforce was composed of foreigners at the time of Iraqi invasion in 1991.

Migrations can be voluntary or involuntary. Most movements are voluntary, such as the westward migration of pioneer farmers in the United States and Canada. Involuntary movements may be forced or impelled. In forced migration, people have no choice; their transfer is compulsory. Examples include the African slave trade and the deportation or "transportation" of British convicts to the United States in the 18th century. In impelled migration, people choose to move under duress. In the 19th century, many Jewish victims

◈

T a b l e 2.8
Growth by 2025

	Population in 1995	Growth by 2025		Population in 1995	Growth by 2025
California	31,589,000	56%	Kansas	2,565,000	21%
New Mexico	1,685,000	55%	New Jersey	7,945,000	20%
Hawaii	1,187,000	53%	Delaware	717,000	20%
Arizona	4,218,000	52%	Minnesota	4,610,000	20%
Nevada	1,530,000	51%	South Dakota	729,000	19%
Idaho	1,163,000	50%	Louisiana	4,342,000	18%
Utah	1,951,000	48%	Washington, D.C.	554,000	18%
Alaska	604,000	47%	Nebraska	1,637,000	18%
Florida	14,166,000	46%	Missouri	5,324,000	17%
Texas	18,724,000	45%	Mississippi	2,697,000	17%
Wyoming	480,000	45%	Vermont	585,000	16%
Washington	5,431,000	44%	Rhode Island	990,000	15%
Oregon	3,141,000	39%	Maine	1,241,000	15%
Colorado	3,747,000	39%	Wisconsin	5,123,000	15%
Georgia	7,201,000	37%	Connecticut	3,275,000	14%
North Carolina	7,195,000	30%	North Dakota	641,000	14%
Montana	870,000	29%	Massachusetts	6,074,000	14%
Virginia	6,618,000	28%	Illinois	11,830,000	14%
UNITED STATES	**262,755,000**	**28%**	Indiana	5,803,000	13%
Tennessee	5,256,000	27%	Kentucky	3,860,000	12%
South Carolina	3,673,000	27%	New York	18,136,000	9%
New Hampshire	1,148,000	25%	Iowa	2,842,000	7%
Maryland	5,042,000	24%	Michigan	9,549,000	6%
Oklahoma	3,278,000	24%	Ohio	11,151,000	5%
Arkansas	2,484,000	23%	Pennsylvania	12,072,000	5%
Alabama	4,253,000	23%	West Virginia	1,828,000	1%

of the Russian pogroms elected to move to the United States and the United Kingdom without the immediate lash of fear.

THE ECONOMICS OF MIGRATION

Many reasons exist for migration, including religious discrimination, cultural differences, and political factors. This section examines migration of a voluntary nature, due to economic purposes. This category of migration comprises the largest single category of human migrations throughout history. Consider two regions as shown in Figure 2.14. One region, Region A on the right, is a highly industrialized country (i.e., the United States or Germany). The region on the left, Region B, is a less developed region (i.e., Mexico, Morocco, or Greece). A labor market exists in each, and the developed region is shown on the right side as a downward sloping demand curve. The quantity of labor is measured on the horizontal axis and the price of labor is measured on the vertical axis. As the price increases on the vertical axis, the demand for labor decreases, and as the price decreases, the demand for labor in the developed country increases. The supply curve S1 in the developed country increases upward to the right, suggesting that a greater supply of labor is available at higher prices. As the price for labor increases, individuals who would not care to work at lower wages now come into the marketplace. They substitute work at higher prices for staying home and taking care of children, going to school, or being in retirement or semi-retirement.

In order to facilitate the analysis for the less developed country, such information is shown on the left side of Figure 2.14. Instead of measuring the quantity of labor from zero to the right for the less developed country, we now measure it from zero to the left. Price remains on the vertical axis. The supply curve in the less developed country slopes upward to the left and the demand curve slopes downward to the left. In this manner, we get a

FIGURE 2.14
Migration and wage differentials. The quantity of labor is measured on the horizontal axis, and the price of labor is measured on the vertical axis. As the price increases on the vertical axis, the demand for labor decreases in both the developed country and the less developed country. The supply curves of both developed and undeveloped countries slope upward, suggesting a greater supply of labor is available at higher prices. Because the equilibrium price of labor is higher in the developed country than in the undeveloped country, labor migrations occur from less developed to developed, or from B to A, to take advantage of higher wages. The greater the wage differential, both the greater the flow and the longer the distance of flow. The wage rate differential between San Diego and Tijuana, for example, is 8 to 1 for comparable occupations. *(Source: Adapted from Stutz, 1992c)*

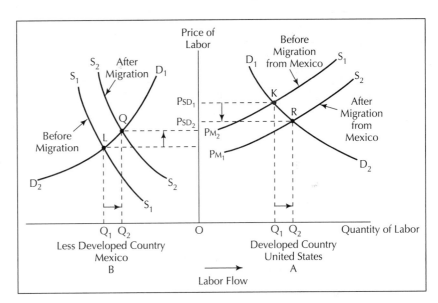

back-to-back set of supply and demand curves for a less developed country and developed country.

Observe the equilibrium position of the supply and demand for labor in the developed country before migration occurs. This occurs at point K. Also note the equilibrium position for the supply and demand of labor in the less developed country. This equilibrium position occurs at L. In classical *labor migration theory*, there is an assumption that information about job availability and labor wages differentials is available and known by workers. Labor in the less developed country finds out about jobs available in the developed country at higher wage rates. Now because the equilibrium price in Region A is quite a bit higher than that in Region B, labor migrates from Region B to Region A to take advantage of higher wages. The greater the differential, the greater the flow and the longer distance the flow can operate. In Region A, extra labor is now coming into the region, which is used to working for lower wages. Because the extra labor is supplied over and above the indigenous supply and the labor is used to working at lower wages, the supply curve moves downward toward the right. The new equilibrium is R. In addition, because the labor pool has left Region B, the supply of labor is reduced. The supply curve moves upward and to the right in Region B, thus raising the equilibrium price to Q. The new equilibrium price in Region A, at R, is at a lower level than it was prior to labor migration. Thus, migration will continue as long as there is a difference between the wages of Region A and Region B, which exceed a cost associated with migration. In the case of flows from Mexico to the United

States, most categories of employment are paid two to five times the rate in Mexico. Consequently, the flows both on a daily basis (Figure 2.15) and on a longer-term basis (Figure 2.16) continue to occur at high levels.

In classical migration theory, transportation costs and other costs associated with moving an individual or family are included, such as selling a property and purchasing one in the new region. The costs of labor in the different regions will not be exactly equalized. But classical trade and migration theory tell us that the long-run price of labor in the two regions should come into close harmony with one another. Relocation and similar migration costs should be split up over the period of work remaining in the life of the mover.

However, when we observe real world labor movements and price differentials, we find that wage rates do not seem to be converging between regions. Major discrepancies occur in the wages paid in various regions of the United States and Europe, as well as in South America and India. If classical theory held, the problems of the world would not exist with regard to variations based on labor rates. There would be simply less difference per capita in incomes. One major reason that major labor differential rates exist is because of the imperfections in the migration system knowledge. Many workers in the less developed countries do not know that jobs they may be qualified for in more highly developed countries even exist. As an individual contemplates changing locations and even countries, he is beset by a series of social factors, including lack of friends and knowledge and the feeling of uneasiness in his new setting. Consequently, the largest number of

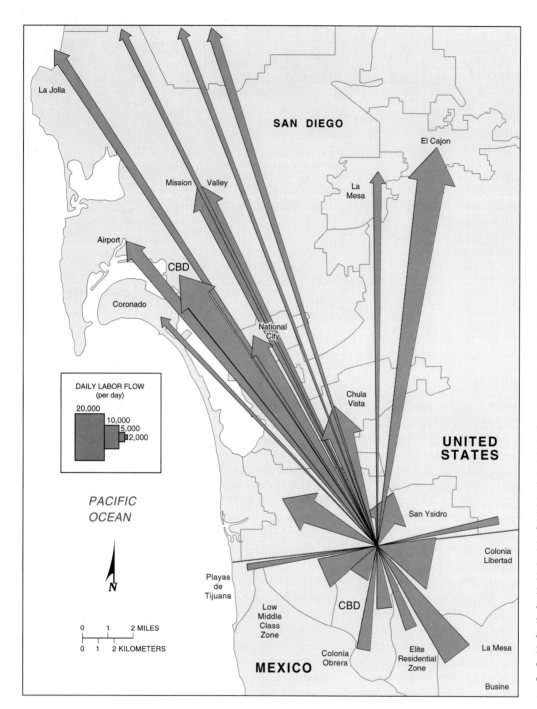

FIGURE 2.15
Mexican Commuters—Labor flow from Mexico. Two hundred thousand daily migrations occur from Tijuana, Mexico, to San Diego, California. The labor flow (25 to 50 percent of total flow) is caused by wage differentials. Pay for the same type of job in San Diego is eight times higher than in Tijuana. Laborers in Tijuana must obtain a green card or work permit to work in the United States. Most laborers leave their home early to get in line at the border crossing gate. Eighty percent of the laborers who live in Tijuana and work in San Diego return home to Mexico daily. Sweat shop manufacturing, auto repair, landscaping and agriculture work, and service job categories associated with restaurants and hotels comprise the chief labor categories. *(Source: Stutz, 1992d, p. 92)*

international labor migrants are young males who do not have families to relocate.

Cultural differences also exist. The cultural shock of a developed nation, especially when one does not have the resources to live adequately or does not speak the native language, presents social problems. Institutional barriers also exist, such as the status of immigration or the length of time allowed in the host country. Blacks, Puerto Ricans, Mexican Americans, and women have encountered such resistance in the past in the United States in their search for improved working conditions and wages. Consequently, we cannot expect that economic forces will lead to a total smoothing of wage inequities throughout a country or throughout the world. At best, only a small portion of the population in a low-wage region is sensitive to and has the ability to gain access to and higher pay from the developed nation. Therefore, there will continue to be a discrepancy in per capita earnings between less developed countries and developed countries and between depressed regions and economically healthy regions within countries.

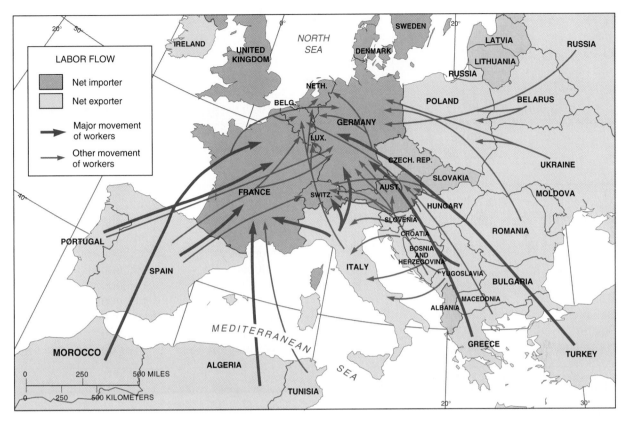

FIGURE 2.16

Guest worker labor flows in Europe. Guest worker labor flows occur principally from North African and Southern European nations, where wages are low, to Northwestern European nations, where population is aging and labor rates are much higher. Here, labor flow is for much longer durations than a single day, as in the case of Tijuana to San Diego. Labor flow is influenced by several factors, including distance, similarity of culture and language, and cooperative agreements between exporting and importing countries regarding rules of migration and labor occupation.

Finally, the more urbanized the region is at the destination, compared to that of the origin, the more attractive it is for potential migrants. The availability of work and wage rates account for major labor flows throughout the world, from countries lacking in jobs and high rates to countries with jobs available at relatively higher wage rates. Major labor flows occur (1) from Mexico and the Caribbean to the United States and Canada; (2) from South American countries to Argentina, Venezuela, and Peru; (3) from North Africa and southern European nations to northwestern Europe; (4) from Africa and Asia to Saudi Arabia; and (5) from Indonesia to Malaysia, Singapore, and Australia. Migrants vary by age. Young adults are most likely to be migrants because of their desire for an improved life and energy in overcoming travel and hardships. Young adults also have fewer ties to their home towns (Table 2.9).

BARRIERS TO MIGRATION

The intensity of a migration flow is reduced by would-be migrant characteristics, political restrictions, and distance. In the late 19th century, British sociologist E. G. Ravenstein (1885, 1889) studied migration in England and concluded that most people move short distances and that the frequency of moves declines with distance. All movement costs, except long-distance moves are more expensive than short-distance ones. Subsequent studies have modified some of Ravenstein's generalizations on the migration process. The concept of intervening opportunity, for example, holds that people's perception of a faraway place's comparative advantage is changed when there are closer opportunities. In steamship days, many British emigrants chose South Africa rather than crossing another ocean to Australia.

Almost all countries regulate the flow of immigration. The United States limits immigration to approximately 600,000 people annually. In addition, an estimated 5,000,000 people enter the United States illegally and live in a half-world of constant threat. Billions of dollars are spent annually to police the borders of the United States, and much of this money is used to try to keep Mexicans out.

▩

T a b l e 2.9

1997 Undocumented Nation

(All figures are estimates)

Top 10 Countries of Origin of Illegal Immigrants			*States with the Largest Illegal Immigrant Resident Population*		
State	*Illegal immigrants in U.S.*	*% of U.S. total*	*State*	*Illegal immigrant population*	*% of nationwide total*
1. Mexico	2,700,000	54.1%	1. California	2,000,000	40.0%
2. El Salvador	335,000	6.7%	2. Texas	700,000	14.1%
3. Guatemala	165,000	3.3%	3. New York	540,000	10.8%
4. Canada	120,000	2.4%	4. Florida	350,000	7.0%
5. Haiti	105,000	2.1%	5. Illinois	290,000	5.8%
6. Philippines	95,000	1.9%	6. New Jersey	135,000	2.7%
7. Honduras	90,000	1.7%	7. Arizona	115,000	2.3%
8. Bahamas	70,000	1.4%	8. Massachusetts	85,000	1.7%
9. Poland	70,000	1.4%	9. Virginia	55,000	1.1%
10. Nicaragua	70,000	1.4%	10. Washington	52,000	1.0%

Source: Immigration and Naturalization Service, 1997.

CHARACTERISTICS OF MIGRANTS

Some countries have higher rates of migration than do others—both into and within the country. In general, the countries that have long histories of migration, such as the United States, Canada, and Australia, have higher migration rates in the modern world than do other countries, such as China, where migration is far less common. When people do move, they are far more likely to be young adults than they are any other group. In some parts of the world, women are more prone to migration than men, and in other parts of the world, men are more likely to be migrants, although the increasing ease of transportation and communication may be breaking down these gender differences. Regardless of gender, it is the young who most frequently move.

CONSEQUENCES OF MIGRATION

Migration has demographic, social, and economic effects, due especially to the fact that migrants tend to be young adults and often the more ambitious and better educated members of a society. Obviously, the movement of people from Region A to Region B causes the population of A to decrease and B to increase. Because of migratory selection, the effects are more complicated. If the migrants are young adults, their departure increases the average age, raises the death rate, and lowers the birth rate in Region A. For the region of immigration, B, the opposite is true. Thus, A's loss and B's gain is accentuated in the short run. If migrants to

Region B are retirees, their effect is to increase the average age, raise the death rate, and lower the birth rate in the region of reception. Arizona, for example, has attracted a large number of retirees resulting in a higher-than-average death rate.

Conflict is a fairly frequent social consequence of migration. It often follows the mass movement of people from poor countries to rich. There were tensions in Boston after the Irish arrived, and the same tensions have come with recent migrants—Cubans to Miami and Puerto Ricans to New York. Social unrest and instability also follow the movement of refugees from poor countries to other poor countries. Generally, poor migrants have more difficulty adjusting to a new environment than the relatively well educated and socially aware. But migrants, on frequent occasion, do suffer from guilt. Many migrants to the United States feel they should go back to their home country to share in its tasks and problems.

The economic effects of migration are varied. With few exceptions, migrants contribute enormously to the economic well-being of places to which they come. For example, guest workers were indispensable to the economy of West Germany prior to reunification. Without them, assembly lines would have closed down, and patients in hospitals and nursing homes would have been unwashed and unfed. Without Mexican migrants, fruits and vegetables in Texas and California would go unharvested and service in restaurants and hotels would be nearly nonexistent.

In 1997, the U.S. border patrol arrested more than 3 million persons passing illegally from Mexico into the United States. Each day along the 2,000 mile (3,750 km) frontier, Mexicans attempted to pass through the 3-meter-high grilled fence separating the two countries to enter their neighboring country to the north. Most people are captured, returned to Mexico, and freed the same day; however, the majority of these Mexicans will try again and eventually be successful. This clandestine immigration makes important problems for the United States. This photo shows a portion of the fence separating Tijuana, Mexico, on the left, from San Diego, on the right. Children born to undocumented women in the

United States immediately become legal citizens. Many women make the perilous journey and stay with friends until the delivery date. The migration's most important direct result is the money migrants send home to Mexico—$2.5 billion to $4.0 billion yearly. That figure is equal to half the yearly foreign direct investment in Mexico. *(Photo: J. P. Laffont, Sygma)*

In the short run, the massive influx of people to a region can cause problems. The U.S. Sunbelt states have benefited from new business and industry but are hard-pressed to provide the physical infrastructure and services required by economic growth. In Mexico, migrants to Mexico City reduce the standard of living in their competition for scarce food, clothing, and shelter. Despite massive relief aid, growing numbers of refugees in the developing world impoverish the economies of host countries.

Emigration can relieve problems of poverty. If the Irish and Italians had remained at home in the 19th century, they would have added to the population pressure. External migration relaxed the problem of poverty and may have averted revolution in Jamaica and Puerto Rico in the 1950s and 1960s. However, emigration can also be costly. Some of the most skilled and educated members of the population of Third World countries migrate to developed countries. Each year, the income transferred through the "brain drain" to the United States amounts to billions of dollars, although billions of dollars are also sent back home in the form of remittances to family members who stayed behind.

PATTERNS OF MIGRATION

To examine patterns of migration, it is helpful to consider migration as either external (international) or internal (within a country). It is also convenient to subdivide external migration into intercontinental and intracontinental, and internal migration into interre-

gional, rural-urban, and intermetropolitan. International migrations, so important in the past, are now far exceeded by internal population movements, especially to and from cities.

The great transoceanic exodus of Europeans and the Atlantic slave trade are spectacular examples of intercontinental migration. In the five centuries before the economic depression of the 1930s, these population movements contributed strongly to a redistribution of the world's population. It has been estimated that between 9 and 10 million slaves, mostly from Africa, were hauled by Europeans into the sparsely inhabited Americas. The importance of the "triangular trade" of Europe, Africa, and the Americas can hardly be exaggerated, especially for British economic development. Africans were purchased with British manufactured goods. They were transported to plantations where they undertook production of sugar, cotton, indigo, molasses, and other tropical products. The processing of these products created new British industries. Plantation owners and slaves provided a new market for British manufacturers whose profits helped further to finance Britain's Industrial Revolution.

The Atlantic slave trade was dwarfed by the voluntary intercontinental migration of Europeans. Mass emigration began slowly in the 1820s and peaked on the eve of World War I, when the annual flow reached 1.5 million. At first, migrants came from densely populated northwestern Europe. Later they came from poor and oppressed parts of southern and eastern Europe.

Between 1840 and 1930, at least 50 million Europeans emigrated. Their main destination was North America, but the wave of migration spilled over into Australia and New Zealand, Latin America, Asia, and southern Africa. These new lands were important for Europe's economic development. They offered outlets for population pressure and provided new sources of foodstuffs and raw materials, markets for manufactured goods, and openings for capital investment (Figure 2.17).

Since World War II, the pattern of intercontinental migration has changed. Instead of heavy migratory flow from Europe to the New World, the tide of migrants is overwhelmingly from developing to developed countries. Migration into industrial Europe and continued migration to North America is spurred partly by widening technological and economic inequality and by rapid rates of population increase in the developing world.

The era of heavy intercontinental migration is over. Mass external migrations still occur, but at the intracontinental scale. In Europe, forced and impelled movements of people in the aftermath of World War II have been succeeded by a system of migrant labor. The most prosperous industrial countries of Europe attract workers from the agrarian periphery. France and Germany are the main receiving countries of European labor migration. France attracts workers especially from Spain, Italy, and North Africa (see Figure 2.16). West Germany draws workers from Italy, Greece, and Turkey. Migrant workers from southern Europe usually have low skills and perform jobs unacceptable to indigenous workers. Similarly, thousands of Mexicans, many of whom are illegal aliens, find their way to the United States each year to work (see Figure 2.15).

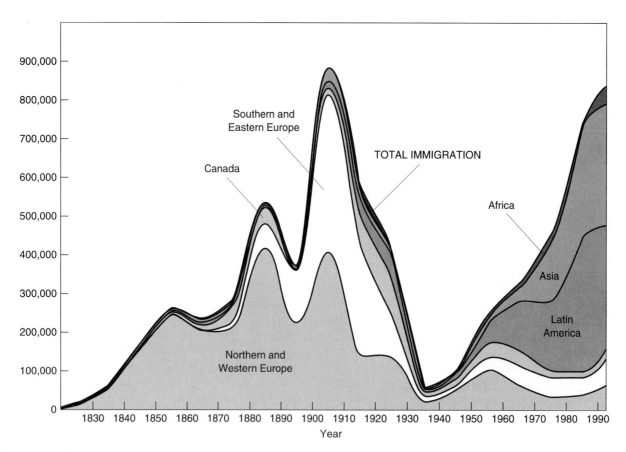

FIGURE 2.17

Immigration to the United States by region of the world. European countries provided more than 90% of all immigrants to the United States during the 1800s. Up until 1960, Europeans continued to provide more than 80% of the total migration to the United States. But since the 1960s, Latin America and Asia have supplanted Europe as the most important source of immigrants to the United States. During the Great Depression and WWII years, immigration was at an all-time low. (The curves have been smoothed by using a five-year moving average.) In 1997, 950,000 people came to the United States legally, and California was the top destination. Add the taxes that an immigrant and his or her descendants are likely to generate over their lifetimes; then subtract the cost of the government services they are likely to consume. The result is that each new immigrant yields a net gain to the government of $80,000. States and cities lose $25,000, while the impact on the federal treasury is $105,000.

The system of extraterritorial migrant labor also exists in the developing world, most notably in Africa, where laborers move great distances to work in mines and on plantations. In West Africa, the direction of labor migration is from the interior to coastal cities and export agricultural areas. In East Africa, agricultural estates attract extraterritorial labor (Figure 2.18). In southern Africa, migrants focus on the mining-urban-industrial zone that extends from southern Zaire in the north, through Zambia's Copper Belt and Zimbabwe's Great Dyke, to South Africa's Witwatersrand in the south.

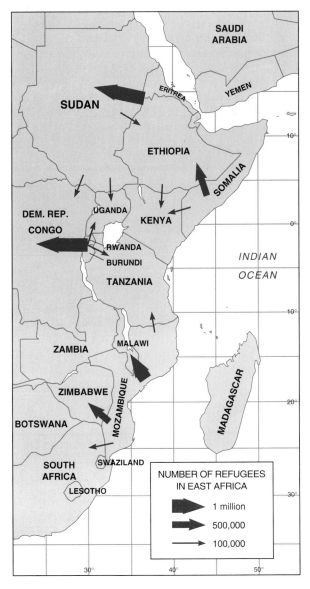

FIGURE 2.18
Arrows represent involuntary migration in East Africa by graduated line segment. Eritreans, Ethiopians, Somalis, and Sudanese have been forced to move because of raging civil wars. To escape civil war, Hutus have been forced to migrate from Rwanda, and residents of Mozambique have fled to Malawi and other neighboring states.

In the modern era, the refugee problem has swung from Europe to the developing world, and then back to the Balkan states of eastern Europe. Refugee generating and receiving countries are concentrated in Africa (6 million people), Southeast Asia (4 million), and Latin America (2 million). The causes of refugee movement include wars or ideology (e.g., Vietnam, World War II, Afghanistan); racial and ethnic persecution (e.g., South Africa, Bosnia-Herzegovina); economic insufficiency increased by political intervention (e.g., Ethiopia, Chad); and natural and human-caused disasters (e.g., Belize hurricane of 1961, Bhopal chemical accident of 1984).

Colonizing migration and population drift are two types of interregional migration. Examples of colonizing migration include the 19th-century spontaneous trek westward in the United States and the planned eastward movement in the USSR beginning in 1925. General drifts of population occur in almost every country, and they accentuate the unevenness of population distribution. Since World War I, there has been a drift of black Americans from the rural South to the cities of the nation's industrial heartland. Since the 1950s in the continental United States, there has been net out-migration from the center of the country to both coasts and a shift of population from the Frost- and Rust-Belt states to the Sunbelt (Figures 2.19 and 2.20).

The most important type of internal migration is rural-urban migration, which is usually for economic motives. The relocation of farm workers to industrial urban centers was prevalent in developed countries during the 19th century. Since World War II, migration to large urban centers has been a striking phenomenon in nearly all developing countries. Burgeoning capital cities, in particular, have functioned as magnets attracting migrants in search of "the good life" and employment.

In highly urbanized countries, intermetropolitan migration is increasingly important. Although many migrants to cities come from rural areas and small towns, they form a decreasing proportion. Job mobility is a major determinant of intercity migration. So, too, is ease of transportation, especially air transportation. For intermetropolitan migrants from New York, the two most popular destinations are Miami and Los Angeles.

POPULATION STRUCTURE

Except for total size, the most important demographic feature of a population is its age-sex structure. The age-sex structure determines the needs of a population; therefore, it has significant policy implications. A fast-growing population implies a high proportion of young people under the working age. A small proportion of

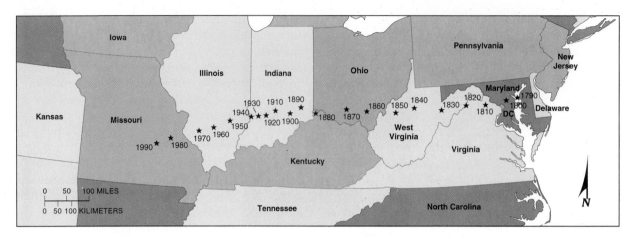

FIGURE 2.19
The population center of the United States. Since the first U.S. census in 1790, the center of population of the United States has moved steadily west and, at a somewhat slower pace, south. The giant leap was made in 1960 when both Hawaii and Alaska were added to the United States as the 49th and 50th states. Until this point, the population center of the United States paralleled Interstate 70. For the first time in history, the population center of the United States moved west of the Mississippi River in 1980. In 1990, the center of the population was at Rolla, Missouri, 100 miles southwest of St. Louis. The new pattern seems to have shifted in a more southerly direction, with a large population attraction to the southwestern United States, especially California. The new population center seems to be drifting southwest, paralleling Interstate 44.

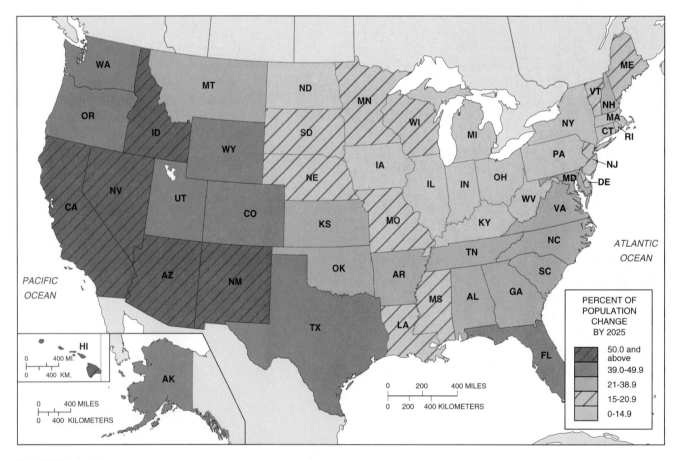

FIGURE 2.20
Percent of population change in the United States 1990 to 2025. The most rapidly growing states between 1980 and 1990 were located in the West and South. States of the Northeast, South, Midwest, and Great Plains grew at a much lower rate, or even lost population in some cases. Texas, California, and Florida accounted for almost 50% of the population growth during this period, and each registered at approximately a 25% growth rate. A few less populated states, such as Arizona and Nevada, actually had population growth rates at much higher levels (data for above map from Table 2.8).

workers results in a smaller output per capita, all else being equal. A youthful population also puts a burden on the education system. When this cohort enters the working ages, a rapid increase in jobs is needed to accommodate them. By contrast, countries with a large proportion of older people must develop retirement systems and medical facilities to serve them. Therefore, as a population ages, its needs change from schools to jobs to medical care.

The age structure of a country is often examined through the use of *population pyramids*. They are built up in 5-year age groups, the base representing the youngest group, the apex the oldest. Population pyramids are compared by expressing male and female age groups as percentages of total population. The shape of a pyramid reflects long-term trends in fertility and mortality and short-term effects of "baby booms," migrations, wars, and epidemics. It also reflects the potential for future population growth or decline.

Three representative types of pyramid may be distinguished (Figure 2.21). One is the squat, triangular profile. It has a broad base, concave sides, and a narrow tip. It is characteristic of developing countries having high (even if declining) birth rates, such as exemplified by Mexico in 1990. In fact, the number of Mexicans working or looking for work will have more than doubled between 1980 and 2000, from 20 million to 42 million (Figure 2.22). What does Mexico's population future portend? Will the Mexican economy be able to support such an increase, or will the potential for out-migration be enhanced?

In contrast, the pyramid for the United States in 1990 describes a slowly growing population. Its shape is the result of a history of declining fertility and mortality rates, augmented by substantial immigration.

With lower fertility, fewer people have entered the base of the pyramid; with lower mortality, a greater percentage of the "births" have survived until old age. As a result, the U.S. population has been aging, meaning that the proportion of older persons has been growing. The pyramid's flattened chest reflects the "baby dearth" of the depression years when total births dropped to about 2.5 million from an average of close to 3 million a year. At the time, the fertility rate dropped close to 2.1, which is the level that leads to a stable population if maintained indefinitely. The bulge at the waist of the pyramid is a consequence of the *baby boom* that followed World War II. In the mid 1950s, the fertility rate increased to 3.8 and the number of births each year exceeded 4 million. After 1964 there was another "baby bust." By 1976, the fertility rate had fallen to 1.7, a level below replacement. Members of the baby-boom generation, however, were having children in the 1980s. Thus, even though the birth rate is lower than ever before, the U.S. population continues to grow from natural increase as well as from immigration.

A few developed countries have very low rates of population growth—in some cases *zero population growth* (ZPG) or *negative population growth* (NPG). They have low birth rates, low death rates, and in some cases, net out-migration. France is one of these countries. Because of very low fertility for a long time, the country is experiencing very slow population growth, and although there is a steady stream of foreigners (especially Algerians) being let into the country, France works hard to limit immigration. Population decline is an economic concern to many European countries. Who will fill the future labor force? Is the solution the immigration of guest workers from developing countries?

Chinese children in Shanghai welcoming visitors from the United States. By January 1994, China's population reached 1.2 billion, or more than one-fifth of the world population. Considering its Third World status and its huge size, it has a remarkably low 1.2% average natural increase. One child per family is the norm, and it is enforced by the government with a series of economic sanctions. While the per capita income is only $370 (in U.S. dollars) per person, it represents one of the largest markets in the world. The United States and other developed countries are presently trying to open the Chinese market and, therefore, have given China a most favored nation trading status in spite of its poor human rights record. *(Photo: Paolo Koch, Photo Researchers, Inc.)*

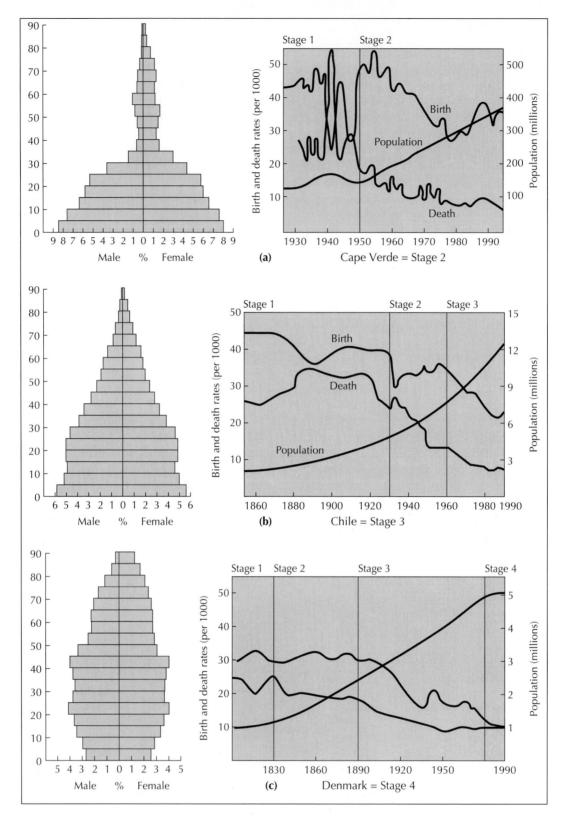

FIGURE 2.21

(a) The population pyramid (left) and the demographic transition (right) for Cape Verde. The vertical axis measures age, and the horizontal axis measures population by gender. (b) Chile entered stage 2 of the demographic transition in 1930 when death rates abruptly declined. It entered stage 3 in approximately 1960 when birth rates began to decline. It has yet to enter stage 4, and the population continues to rise dramatically. Because of its status as a developing country, it has a relatively broad-based pyramid. Population members younger than 15 and older than 65 are dependent populations. (See Color Insert 1 for more illustrative map.) (c) The population pyramid and demographic transition for Denmark are shown. The shape of the so-called pyramid is quite different than that of Chile. Because of lower birth rates, the population pyramid is more in the shape of a spark plug. Denmark, which entered this stage in 1830 and stage 3 in 1890, entered stage 2 of the population pyramid 100 years earlier than Chile. In the 1970s, Denmark entered stage 4, and its current population growth is zero.

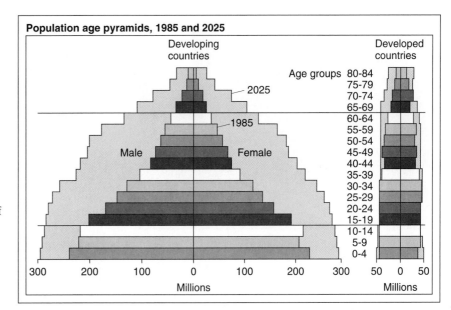

FIGURE 2.22
Population pyramids for developed and developing countries projected to the year 2025. Even with the supposition of a continuing gradual decline in the fertility rate, the populations of developing countries will grow immensely. *(Source:* Population and the Environment: The Challenges Ahead, *United Nations Population Fund, 1991)*

DEMOGRAPHIC CHARACTERISTICS

Demographic characteristics are those qualities of humans that we often label as *human capital*, including educational attainment, labor force participation, occupation, and income. However, these characteristics, which are often called *achieved characteristics*, are often confounded by the existence of *ascribed characteristics*, with which we are born and which may affect our ability to achieve our desired level of living. These ascribed characteristics include race and ethnicity (often mixed up with religion) and gender. These interrelationships are shown in Figure 2.23.

Educational Attainment

Probably no characteristic is more important to demographic trends than education level. Education has a profound set of effects on all types of human behavior, including fertility, mortality, and migration. Education also influences almost every aspect of a person's life chances. Virtually without exception, the more educated a person, the fewer children the person is likely to have, the longer the person is likely to live, and the more likely the person is likely to migrate. Educational attainment has increased significantly over time in the United States.

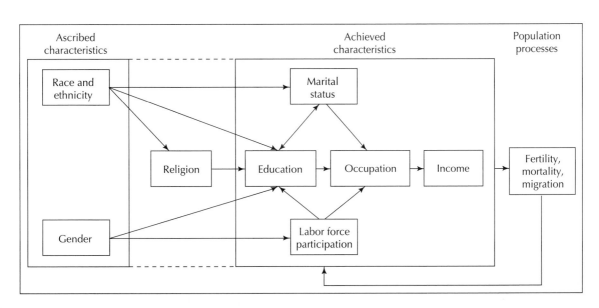

FIGURE 2.23
Population characteristics affect population processes and are also affected by them. *(Source: Weeks, 1992, p. 262)*

Third World female children are less likely to receive a good education than female children in more developed countries. It is much more difficult for these children to achieve a prominent economic or political position in their society, compared to opportunities available to women in the developed economies. This scene shows an elementary school for girls in Manila, Philippines. Operated by the Immaculate Conception Sisters, the school was built by the parents of the children and the nuns of the nearby convent. *(Photo: United Nations)*

In the developed countries, nearly all adults are literate (although there is some controversy about this in the United States), and indeed, the majority of adults are high school graduates. But in less developed nations, large numbers of adults are not literate. World Bank data indicate that 50% of the Algerian adult population is illiterate, as is 62% of the adult population in Haiti, 70% in Pakistan, and nearly all adults (88%) in Somalia. "Less developed" and "uneducated" go together, hand-in-hand.

On the more positive side, there is a clear relationship in the United States between education and income. To be sure, knowledge of the relationship is what spurs many students to stay in college. Table 2.10 shows the relationship as of 1991 in the United States. You can see that males with a postgraduate degree who were working full-time, year-round were earning an average of $49,093 in 1991, nearly $30,000 a year more than the men who had not graduated from high school. Income for women is lower, but you can see that the pattern is the same.

Labor Force Participation

The population of a country is conventionally divided into two parts: the economically active and the economically inactive. Economically active people are the productively employed and temporarily unemployed. They compose what is known as the income-earning labor force. Men are still dominant in the paid labor

force but the proportion of men to women is changing slowly. In 1950, women accounted for 31% of the world's income-earning labor force; in 1985, the figure was 35%. But the absolute number of women engaged in wage and employment increased considerably. Between 1975 and 1985 more than 100 million women joined the labor force, and in 1985 an estimated 676 million women were gainfully employed.

Official labor-force statistics are deceiving, however. They exclude adults and children in the informal sector, whose work may involve begging, shoe shining, selling handicrafts, prostitution, drug dealing, or petty theft. They also ignore the invisible, unpaid work of men, women, and children. Women workers are especially invisible to official enumerators, in particular, in the case of women who do agricultural work. In 1970, Egypt's census revealed that only 3.6% of women did agricultural work, but local surveys discovered that up to 40% of women were actually involved in planting, tilling, and harvesting. In Africa, women generally do the lion's share of such work. Women's agricultural work is underestimated in the developed world as well.

Women are also engaged in other forms of unpaid work—cooking food, feeding infants, washing and mending clothes, and collecting water and firewood. A woman in a Pakistani village spends approximately 60 hours per week on domestic work. In the developed world, women who are "just housewives" work an average of 56 hours per week.

The wage-earning labor force engages in thousands of different kinds of activities that may be classified into five sectors, as follows:

Primary activities, including agriculture, mining, quarrying, forestry, hunting, and fishing
Secondary activities, including manufacturing and construction, public works, utilities

◈

T a b l e   2 . 1 0

Annual Median Income for Year-Round, Full-Time Workers Age 25 and Older: United States, 1991

Educational Attainment	Males	Females
Less than High School	19,654	13,816
High School Graduate	26,515	18,323
Some College	31,566	22,227
College Graduate	39,115	28,042
Postgraduate	49,093	33,771

Source: U.S. Bureau of the Census, 1992. Educational Attainment in the United States: March 1991. Current Population Reports, Series P-20, No. 462, Table 8.

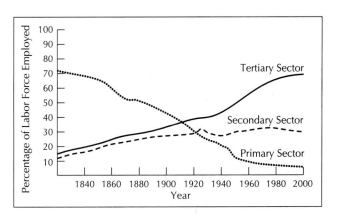

FIGURE 2.24
The changing composition of the U.S. labor force. Between 1820 and 1990, the primary economic sector of the economy fell from more than 70% of the labor force to less than 2%, while the secondary sector increased from 11% to 32% in 1980. Since 1980, the secondary sector has fallen to 28%. The tertiary sector has risen from 15% in 1820 to more than 70% today. Interestingly, in 1916, each of the three sectors of the economy accounted for approximately 33% of the labor force.

Tertiary activities, including most services, retail commerce, wholesaling, personal services, restaurant, hotel, repair, and maintenance

Quaternary activities, transportation and communications, information services, producer services, finance, insurance, real estate, administrative services

Quinary activities, medical care, research, education, arts, recreation (Figure 2.24).

Economic development alters labor-force composition. As the U.S. economy grew, the proportion of the labor force in secondary and tertiary activities increased at the expense of primary activities, particularly agri-

culture (Figure 2.25). Compared with the United States, the share of the labor force in agriculture in the contemporary developing world is high.

The ratio of industrial to agricultural workers provides a measure of a nation's economic advancement and power. As expected, developing countries have a low ratio of industrial to agricultural workers. Although manufacturing industries are moving to the developing world, most people have been barely touched by the iron embrace of the industrial age that is more than 200 years old. The majority of people continue to live their lives by rhythm of the seasons, not of machines.

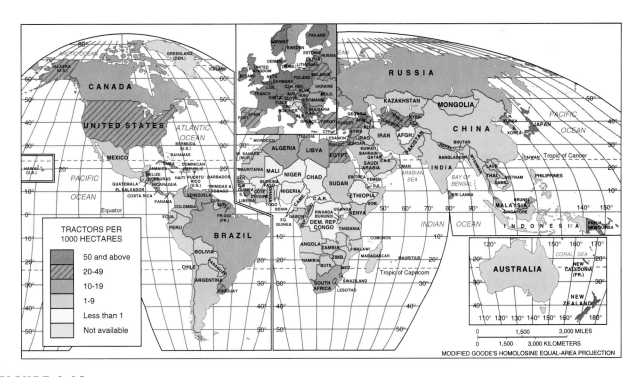

FIGURE 2.25
Farm workers in developed countries have more tractors than their counterparts in less developed countries. Once such farmer can produce enough food to feed 50 to 100 others, allowing these people to work in the secondary and tertiary sectors of the economy. A fourth sector is sometimes identified, the quaternary sector, which comprises those occupations dealing with the information economy. Which developing countries are well supplied with tractors?

UNEMPLOYMENT

Unemployment means able and willing workers with no jobs. In the United States, the jobless rate was 3.4% in 1971, 9% during the recession of 1974–1975, and 5.7% in 1997. Levels of unemployment are much higher for women, blacks, and the young than for adult males. Unemployment of young people (24 years of age and under) was 16% in 1997. Unemployment figures, however, underestimate the problem. People who have failed to find work and stop trying are excluded from the statistics. Moreover, the figures fail to account for adults, especially women, who want formal employment but who never obtain it because of domestic responsibilities or the lack of acceptable occupations.

In developing countries, official unemployment is a phenomenon of an urban sector composed of a minority of the population. Although unemployment rates are officially around 10%, an estimated 25% of those living in developing world cities survive outside the mainstream economy, and a further 750 million are unemployed or underemployed in the rural sector. People who work less than a specified number of hours are not underemployed. Underemployment is the gap between the amount of labor required for a decent living and the actual payment given for work; thus, underemployment is an index of poverty (Figure 2.26b).

Income

Per capita income is the most familiar index to measure a population's relative development, its economic well-being, and its capacity to consume. The distribution of total GNP among selected countries is striking. The wealth produced by the United States and Canada alone is nearly five times greater than the combined GNPs of Africa and Latin America, which have about three and one-half times the population of North America. In 1997, North America had a per capita income of $25,580. The corresponding figures for Africa and Latin America were $630 and $3,270, respectively.

However well income per capita may seem to reflect international disparities, this measure should be viewed with caution. First, it fails to indicate ways in which incomes are distributed through the strata of societies. In most underdeveloped countries, wealth is concentrated in a small elite. Even within developed countries, major inequities in the internal distribution of income exist. A second problem stems from the fact that countries measure income per capita in a variety of ways. Finally, the per capita income indicator fails to take into account economic activities outside national monetized accounting systems in developing countries. Street vending, for example, which plays an important role in the lives of many people in the developing world, is not officially accounted for. For these reasons,

per capita income has only limited utility, and often serves only to publicize the problem of poverty.

To delve more deeply into the problems of poverty, social scientists have constructed multidimensional measures of human well-being. An example is the Population Crisis Committee's (1987) *Human Suffering Index* constructed from 10 variables: income per capita, average annual inflation rate, average annual growth of labor force, average annual growth of urban population, infant mortality rate, daily per capita calorie supply, percentage of population with access to clean drinking water, energy consumption per capita, adult literacy rate, and personal freedom/governance. Each of 130 countries was assigned a score of zero (high) to 10 (low) for each variable, and these scores were added together to form an index. Countries ranged from Mozambique (most human suffering) to Switzerland (least human suffering) (Figure 2.26). The *Human Suffering Index* is a useful descriptive measure of the differences in living conditions among countries.

ECONOMIC GROWTH AND ECONOMIC DEVELOPMENT

The debate over population growth and economic development historically has been three-cornered. In the first corner, arguing that population growth stimulates development, you will typically find nationalists—people seeking freedom for their country from economic and political exploitation by more powerful nations. A frequent corollary of nationalism is the idea that more people will bring more productivity and greater power. Another form of nationalism is that which appeared as the official United States position at the 1984 International Population Conference in Mexico City—in any free-market system, population growth stimulates demand and, thus, helps the economy.

In the second corner, you will find Marxists (and others, as well) arguing that social and economic injustices result simultaneously from the lack (or slowness) of economic development and the (erroneous) belief that there is a population problem. The Marxist position maintains that no cause-and-effect relationship exists between population growth and economic development; that is, poverty, hunger, and other social welfare problems associated with lack of economic development are a result of unjust social and economic institutions, not population growth.

Finally, in the third corner, you find those who have historically antagonized the Marxists, namely the *neoMalthusians*. They are, of course, latter-day advocates of the thesis that population growth, unless checked, will wipe out economic gain. The difference between Malthus and the neoMalthusians is that Malthus was opposed to birth control, whereas neoMalthusians are

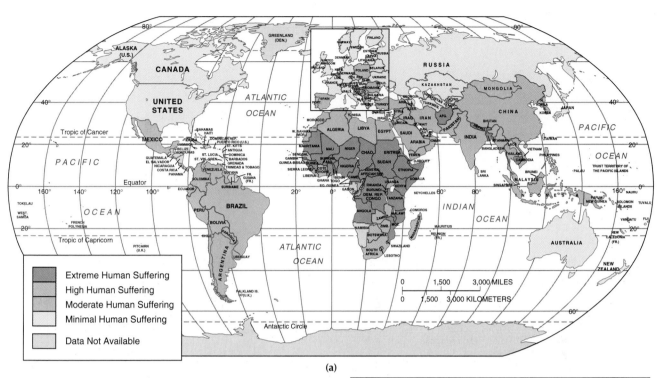

(a)

FIGURE 2.26

(a) The human suffering index. The human suffering index measures income per capita, average annual inflation rate, average annual growth of labor force, average annual growth of urban population, infant mortality rate, daily per capita calorie supply, percentage of population with access to clean drinking water, energy consumption per capita, adult literacy rate, and personal freedoms. On this basis, human suffering appears most severe in the developing world, especially in Africa and portions of South Asia. (b) U.S. unemployment rate by education level. *(Source: Based on Population Crisis Committee data, 1987)*

strong advocates of birth control as a preventive check to population growth (Figure 2.27).

Impact of Population Size on Economic Development

As a population grows larger, the ability to garner resources for development may grow progressively smaller. This is true for individual nations just as it is true for the entire world. Although we can conjure up images of standing room only at the point at which all economic activity most certainly would have to stop, in reality, the limit is far less than that. But how much less? This is a question that is still puzzling, yet has been the object of a good deal of scrutiny as researchers have tried to define an optimum population size for the earth or for a particular country. In trying to determine an optimum size, we ask how large a population can be before the level of living begins to decline.

It is widely recognized that there are economies of scale associated with size; that is, too few people may

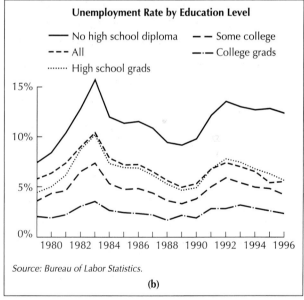

(b)

retard economic development as surely as too many people might. The world is much better off economically with 5 billion people than it was with 1 billion. General Motors can produce a car far more cheaply than you could build one, precisely because they sell so many cars that they can afford the expensive assembly plants that reduce production costs per car. Although larger is sometimes more economical, a population may grow too large to be efficient or so large that, at a given level of living, it will exhaust resources. When it reaches that point, it is said to have exceeded its carrying capacity, or the size of population that

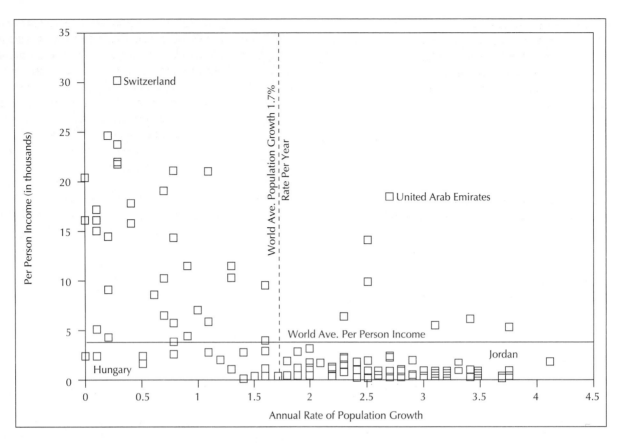

FIGURE 2.27
Per person income and annual rate of population growth. In 1990, as in previous years, there was a high rate of population growth associated with low levels of income. *(Source: Weeks, 1994, p. 384)*

could theoretically be maintained indefinitely at a given level of living (Figure 2.28).

The carrying capacity will vary according to which level of living you might choose for the world's population. The lower that level, the greater the number of people that can be indefinitely sustained. However, if the desired level of living is too high, you may well exceed the carrying capacity and start draining resources at a rate that will lead to their exhaustion. Once you have done that, you lower long-run carrying capacity. For example, if you and everyone else in the world were content to live at the level of the typical South Asian peasant, then the number of humans that the world could carry would be considerably larger than if everyone were trying to live like the General Motors board of directors. Indeed, it is highly doubtful that the world has enough resources for 5 billion people to ever approach the successful business executive's level of living.

Impact of Age Structure on Economic Development

A rapidly growing population has a young age structure. This means that a relatively high proportion of the population is found in the young ages. Two important economic consequences of this youthfulness are that the age structure affects the level of dependency, and it puts severe strains on the economy to generate savings for the investment needed for industry and for the jobs sought by an ever increasing number of new entrants into the labor force.

A high rate of population growth leads to a situation in which the ratio of workers (people of working age) to dependents (people either too young or too old to work) is much lower than if a population is growing slowly. This means that in a rapidly growing society, each worker will have to produce more goods (that is, work harder) just to maintain the same level of living for each person as in a more slow-growing society. This may seem like an obvious point. The father of six children will have to earn more money than the father of three just to keep his family living at the same level as the smaller family. But it goes deeper than that. A nation depends at least partially on savings from within its population to generate investment capital with which to expand the economy, regardless of the kind of political system that exists. With a very young age structure, money gets siphoned into taking care of more people (buying more

FIGURE 2.28
Population growth, poverty, and environmental degradation are not separate issues. They are very much interrelated. Numerous connections exist between unchecked population growth and social and environmental problems. These factors tend to lock the poor into the vicious cycle of illiteracy and squalid conditions that define absolute poverty.

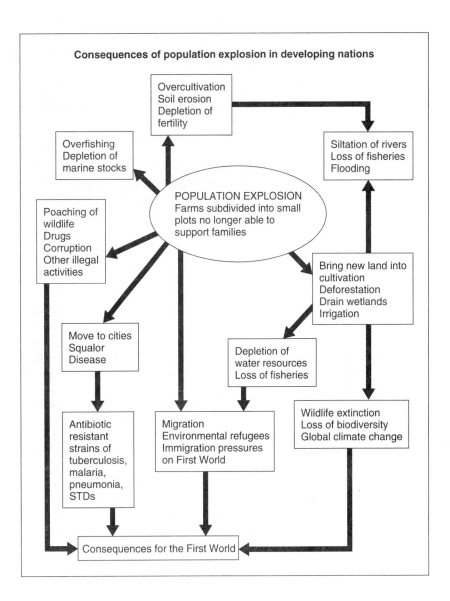

Consequences of population explosion in developing nations

food, and so on) rather than into savings per se. A very old age structure may also be conducive to low levels of saving, because in the retirement ages, people may be taking money out rather than putting it in.

In a growing population, the number of prospective entrants into the labor force is also growing every year, as each group of young people matures to an economically active age. If economic development is to occur, the number of new jobs must at least keep pace with the number of people looking for them. The expansion of jobs is, of course, related to economic growth, which in turn relies on investment and may be harder to generate with a young age structure.

In countries like Pakistan and Mexico, for example, the workforce will grow at about 3% per year (from 1985 to 2000). In contrast, growth rates in the United States, Canada, and Spain will be closer to 1% per year, Japan's workforce will grow just 0.5%, and Germany's workforce (including the Eastern sector) will actually decline (Johnston, 1991).

When the rate of labor-force growth is slow, the new job entrants can simply step into the places vacated by people dying or retiring. As the rate of population growth increases and the age pyramid flattens out at the bottom, the ratio of new seekers to those leaving the labor force goes up rapidly. Bradshaw and Frisbie (1983) calculated that in Mexico, from 1990 to 2000, the economy will have to create 330 new jobs for every 100 people leaving through death or retirement just to maintain current levels of employment. In the United States, only 159 new jobs per 100 people leaving the labor force are projected to be needed during the same time period. Can the Mexican economy work that much harder than the United States economy?

Maybe. If not, the pressure on Mexicans to migrate to the United States will, of course, continue.

Demographic Economic Forecasting: Companies and Cities

Our analysis throughout the chapter has been focused either on countries (where development occurs) or on individuals (who are the producers and consumers of the goods and services involved in economic development). We conclude the chapter with an overview of applying demographic insights to: (1) the *firm*—the organization in which people produce the goods and services required for a society to improve its overall level of well-being, and (2) the *city*—the organization that plans and governs a society so it can function smoothly and efficiently.

Demographic Insights for the Firm

Businesses can be likened in many ways to the human beings who staff them—they are born, they grow and mature, they die, and, in between birth and death, some of them migrate. Each of these demographic processes is of crucial importance in economic geography because the number, size, and distribution of firms in an area helps to determine the economic well-being of the region. The study of these processes has been variously labeled as "organizational ecology" (see, for example, Hannan and Freeman, 1989) or more recently *firmography*, meaning the demography of the firm.

The organizational ecology approach has focused especially on the birth and survival of firms, relating both processes to the density of similar organizations. New, innovative types of businesses, created (by definition) in an area in which very few such firms exist, have a fairly high mortality rate. However, as the number of similar firms increases, the survivability of all increases because they come to be viewed as more legitimate members of the community and the demand for their goods or services goes up. In a sense, then, there is safety in numbers—an advantage to agglomeration that is well known to regional scientists. However, density may increase beyond a critical point that is conceptually similar to the carrying capacity. This will vary from one type of business to another, but when that point is reached, the competition for resources (such as customers) will discourage the birth of new companies and will speed up the demise of some existing companies (Hannan and Carroll, 1992). Just as some humans survive the plague while others die, some businesses survive the competitive process, especially those that are well embedded in the community network (Baum and Oliver, 1992) (see Chapter 8: Business Reengineering).

The regional pattern of births, deaths, and migration is of special interest to geographers because it brings into play the issue of land use patterns and transportation linkages. A recently constructed database in The Netherlands has allowed Dutch geographers to track the birth, death, and movement of firms in that country with interesting results. Data reveal that the migration of firms is fairly limited and exhibits distinct distance decay. Firms that move are likely to go only a short distance, especially out of the central city into the suburbs, but rarely do they make interregional moves. Furthermore, firms that are growing (the "healthier" firms) tend to be located closest to the major urban or suburban centers as well as near the major transportation arteries. (Chapters 8 and 9 discuss the location and behavior of firms.) These are not surprising findings, but they help to support the planning

processes of urban policymakers. We now turn to the demographics of the firm of today.

Firm target marketing

Target marketing (TM) is a micro-geographic GIS system that uses consumer as well as census data to accurately classify every household in the United States into one of 50 unique *market segments*. Each market segment consists of households that are demographically similar, are at similar points in the life cycle and share common interests, purchasing patterns, financial behavior, and needs for products and services.

To give flexibility in consumer segmentation strategies, TM can assign each of the 50 market segments to one of nine groups. Each group contains segments with similar characteristics or habits, giving the ability to simultaneously target many segments that will respond alike to product, service, or market efforts.

TM can be used in many ways:

Customer Profiling. Identify the purchasing behaviors and lifestyles of customers and prospects to develop the most profitable marketing strategy possible.

Product and Service Demand. Determine the products and services consumers use and need, then design an effective cross-selling strategy.

Market Analysis. Pinpoint where consumers with the highest potential for sales and profits are located, and target as few as five to 15 households.

Media Selection. Select the advertising media that will entice customers and prospects to respond to promotions.

Direct Marketing. Maximize the response and return on direct marketing investment, and obtain a higher response rate while mailing fewer pieces.

TM is built by combining georeferenced, consumer demand data files with the most current census data. Over 100 unique characteristics for more than 160 million consumers are used, including demographic, socioeconomic, housing, purchasing activity, consumption patterns, and consumer financial information from credit files.

The data is then statistically clustered at the ZIP + 4 level of geography. Over 23 million ZIP + 4s, each consisting of only five to 15 households, are used in the TM clustering process. This gives the ability to target the lowest possible level of geography of any segmentation system. From this clustering process, 95 unique and homogeneous sub-segments are identified.

TM can then classify every household in the United States into one of 50 market segments based on the unique characteristics of its ZIP + 4. This extensive segmentation process makes TM an accurate segmentation and targeting system for any market, of any size and shape, anywhere in the United States and Canada. TM gives the ability to precisely quantify, locate, and target the most profitable customers and prospects.

TM shows what percentage of total customer households are in each market segment. It then compares this composition to the total geographic market area, allowing one to calculate customer penetration per segment. Indices help determine the propensity of each segment to be a customer. Target marketing software can answer questions in the following areas:

Retail

How to develop a market-oriented customer database and use it to cross-sell current customers and acquire new ones (Figure 2.29).

How to determine which markets and specific store sites offer the greatest potential for success (Figure 2.30).

How to define a realistic trade area for each store.

How the addition of a new site will impact existing store sales.

How to increase customer traffic and store sales by tailoring "store types" and merchandise mix to the distinctive needs of each store's customer base.

How to measure the effectiveness of marketing programs.

Restaurants

How to develop a market-oriented customer database to attract new customers and increase frequency and per ticket sales among current ones.

How to identify the markets and restaurant sites that offer potential for successfully introducing a new restaurant concept (Figure 2.31).

How to use drivetime or daytime population as key components in defining the trade area.

How to minimize the adverse impact of cannibalization when adding new restaurant sites within a market.

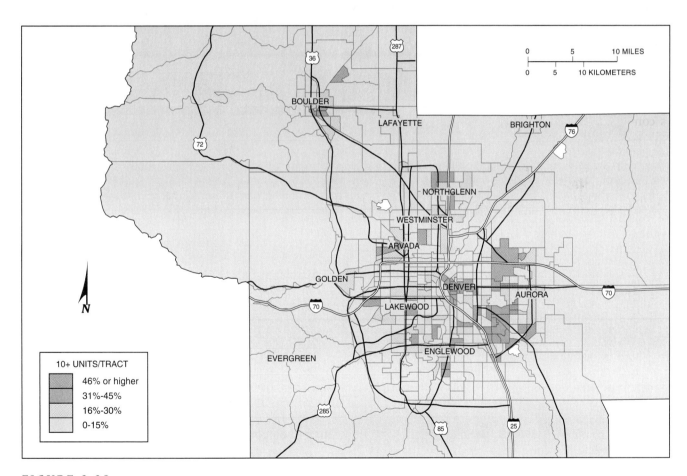

FIGURE 2.29

Duds and Suds in Denver and Boulder, Colorado. Duds and Suds, a laundromat chain serving beer and snacks, gravitates to apartment districts. This company was interested in finding census tracts with high proportions of apartments and, therefore, households without built-in laundry facilities. The census tracts, with high proportions of apartment units, are identified and are clustered in the new residential suburbs of Aurora in East Denver, in central Denver, and in the Colorado University area of Boulder.

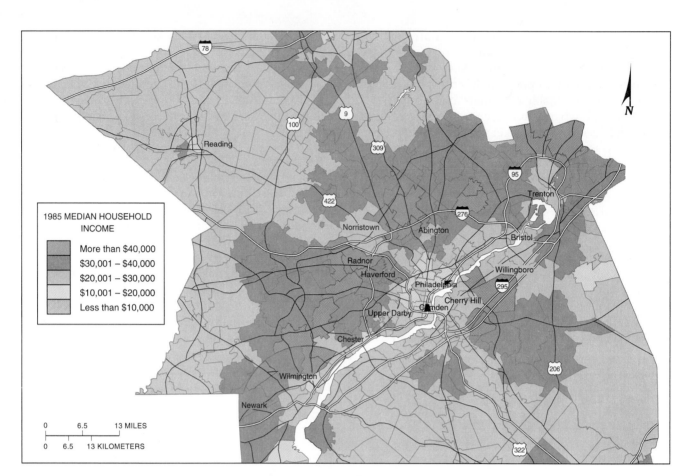

FIGURE 2.30
Location for Upscale Gyms in Philadelphia, Pennsylvania. Market-area analysts and planners for firms nationwide frequently want to know the median household income of census tracts in a metropolitan area, such as Philadelphia, shown here. High-rent, medium-rent, and low-rent neighborhoods can be targeted for various products. In this case, an upscale and expensive fitness club chain, called Fitness Advantage, selected the census tracts where median household income was more than $50,000, shown in dark blue. These locations were optimum locations for their service clubs.

How to segment and target current and potential customers in terms of specific restaurant patronage behavior.

Real Estate
How to realistically define the trade area for a site.
How to quantify the sales potential for a shopping center trade area and demonstrate the value for a diverse array of prospective tenant businesses.
How to define the impact and opportunities posed by competition located within the trade area.
How to strengthen long-term tenant relationships through joint target marketing efforts.

Banking
How to accurately identify, locate, quantify, and target customers to maximize existing financial relationships and acquire new ones.

How to analyze markets and evaluate sites in order to make the best decisions regarding branch expansion, consolidation, mergers, and acquisitions (Figure. 2.32).
How to increase commercial revenue by segmenting and targeting profitable small and mid-sized businesses within a market.
How to more quickly and economically satisfy regulatory compliance requirements of CRA and HMDA.
How to employ Branch Classification Analysis to optimize local marketing programs, match product mix to market demand, and establish a sustainable competitive advantage.

Insurance
How to segment and target the best current and potential policyholders in terms of premiums, claims,

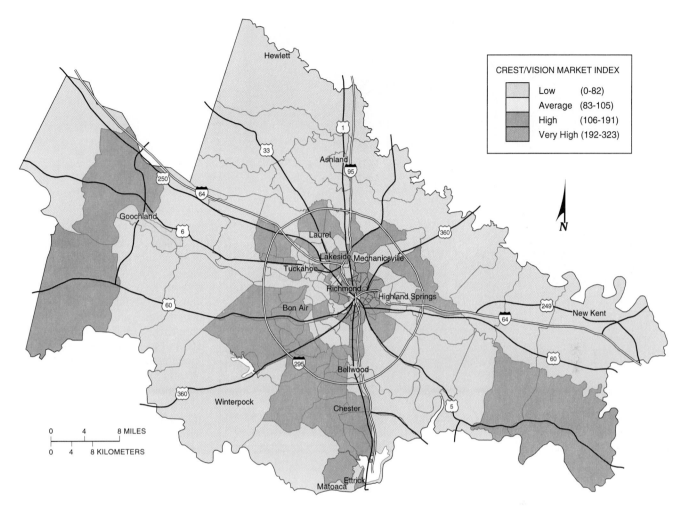

FIGURE 2.31
Fried chicken fast-food locations in Richmond, Virginia. This map shows the potential sales locations in the greater Richmond, Virginia, area for a fried chicken restaurant chain, The Golden Skillet. An eating-habit scale was used, along with a market-segmentation index. Areas shaded in dark blue are the most likely to patronize a Golden Skillet fast-food chicken restaurant. Downtown Richmond and nearby neighborhoods rank the highest, whereas suburbs farther away rank somewhat high. The eating-habit scale and the market-segmentation index are based on 70 U.S. census sociodemographic and economic variables that are keyed to each census tract.

policy relationships, lapse experience, or other revenue and profit criteria.

How to more effectively and efficiently cross-sell new products and services to policyholders.

How to increase the number of qualified sales leads generated from compiled list mailings while reducing cost per conversion.

How to estimate a market share and determine where resources should be allocated to tap unrealized policyholder potential and revenue growth.

How to identify, segment, locate, and target high-potential business prospects for property/casualty, group life/health, pension, annuity, and risk management sales.

Investments

How to segment and target the best current and potential investors in terms of net asset value, original investment amount versus potential, investment objective, distribution channel, or other revenue and profit criteria (Figure 2.33).

How to increase the lift of direct marketing efforts by targeting segments most likely to respond to a specific offer.

How to effectively and efficiently cross-sell additional funds and services to existing investor base.

How to target a proprietary message and the medium to maximize impact and response among prime potential investors.

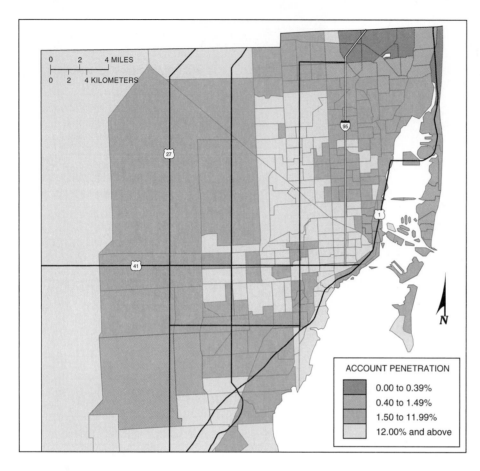

FIGURE 2.32
Sales penetration for Atlantic Federal Savings in the Miami, Florida, area. Where should Atlantic Federal Savings open new branches? Savings-account penetration is shown on this map with darker areas in northern Miami Beach and the city of Miami showing the highest levels of penetration. If a tract has 1,000 households and 30 accounts, the penetration is 3%. A small dot shows Atlantic Federal Savings branches. Areas of south Miami look promising for new branches.

How to use investor profile to increase response and conversion for direct mail programs.

Telecommunications

How to identify, profile, and target current customers for successfully cross-selling and upselling additional services.

How to more accurately project demand for a service area in order to allocate capital resources, define sales territories, and establish objectives.

How to segment and profile business customers for more successful cross-selling, upselling, and new customer acquisition (Figure 2.34).

How to obtain a clearer understanding of the characteristics and needs of retail and business customers for new product development and strategic planning.

How to increase the number of qualified leads through targeted consumer and business direct marketing programs.

Cable TV

How to increase penetration among homes passed-up by segmenting and screening current subscriber base.

How to reduce cancellation by targeting retention-based programs to subscribers who are most likely to disconnect.

How to increase revenue per subscriber by identifying and targeting those subscribers most likely to sign up for the pay-per-view, premium channels, and cable radio.

How to maximize local advertising revenues by linking subscriber profiles to local advertiser target audience potential.

How to more accurately project market demand and growth for planning and allocating strategic resources for new services such as interactive TV and Web TV.

Utilities

How to segment and profile residential customer base to better understand their needs in terms of appliance ownership, lighting, home improvements, home office equipment, and participation in demand side management or conservation programs.

How to use this knowledge to increase response rates and decrease cost-per-response of direct marketing programs.

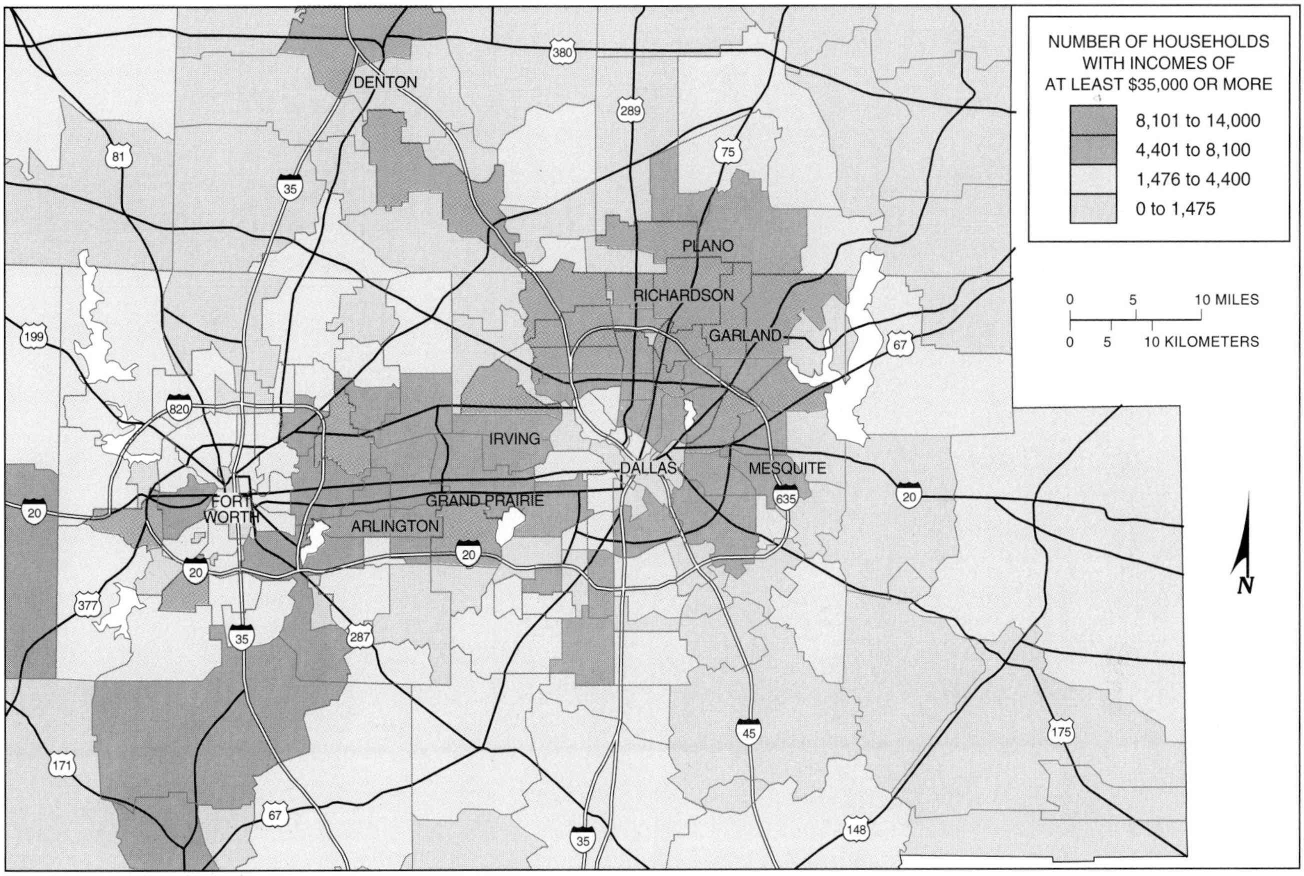

Firm Target Marketing 91

FIGURE 2.33
Direct mail advertising in Dallas, Texas. A direct mail advertising firm wanted to identify the high-rent districts of the Dallas–Fort Worth–Denton area. They planned to advertise and mail to every household a time-share and vacation sales promotion, centered in Vail, Colorado. Census tract groups shown here have average income of $25,000 or more. Dark readings show, for example, census tract groups where at least 8,000 to 14,000 households earn more than $25,000.

How to more accurately forecast future service area growth and demands for strategic allocation of resources.

How to use a Commercial and Industrial Targeting System to identify, quantify, locate, and target the best customers and prospects.

How to use this system to determine the market potential of new programs and be cost-effective.

How to best position the company with potential new commercial and industrial businesses.

Consumer Goods

How to identify, segment, and profile customers in terms of demographics, socioeconomics, lifestyle, and media usage (Figure 2.35).

How to develop targeted programs to acquire new customers, increase sales volume, and boost market share.

How to develop more targeted and cost-effective co-op advertising and direct mail promotions.

How to strengthen relationships with retailers and help them increase sales of the product.

Nonprofit Organizations

How to segment, profile, and target donors to increase average dollar donations and the percentage response generated by direct mail programs.

How to develop programs that are tailored to the lifestyle and behavior of donors.

How to increase the response and cost-effectiveness of direct mail programs that utilize a wide array of compiled lists.

How to be more successful in recruiting volunteers for a cause.

The Firm of the 21st Century

The *demographics of the firm* are being affected by two major factors—technology and generation. The

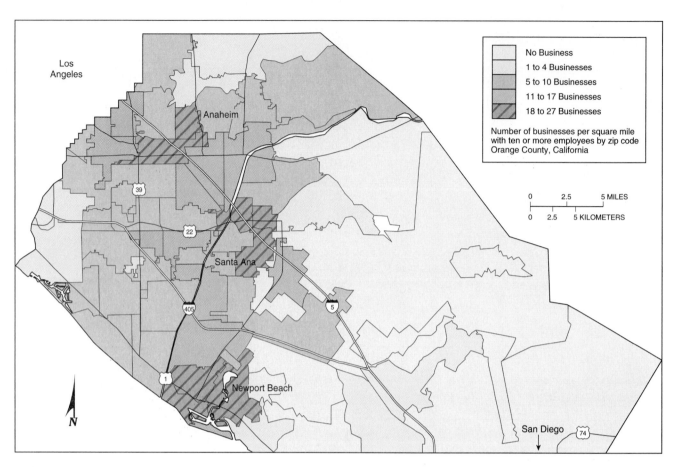

FIGURE 2.34

Xerox potential sales areas in Orange County. Potential sales areas for Xerox machines and equipment are shown here. The problem was to identify census tracts that had numerous businesses and, therefore, clients for Xerox Corp. The blue areas in the center of the map are clustered on Newport Beach. The top two blue shaded areas are in the Irvine Ranch Industrial Park. The more businesses per square mile with 10 or more employees, the larger the potential sales area for Xerox Corp.

technological change means the restructuring or reengineering of the workplace environment. In the past, industrial revolutions have always meant automating old jobs. Today, we are in the midst of the greatest industrial revolution in the workplace since the Industrial Revolution itself in the 18th century. The introduction of the new technology of the microcomputer is ushering in the information age. The information revolution will affect the workforce in every firm across North America, Europe, and Japan in the near future. The majority of the workforce today in this developed world are primarily white collar, administrative, and clerical workers performing repetitive functions that computers will replace.

The next major change in the demographics of the firm is generational in that North American *baby boomers*, those born between the years 1945 and 1965 (approximately 90 million in total), are moving into their peak spending and productivity years, near the year 2000. Baby boomers are assuming power structure positions in business and politics, bringing with them

the information technologies of the future. The result is the greatest job restructuring in history, ushering in a new era of prosperity and spending. The New York Stock Exchange, which stands at over 6,800 today, is projected by many to hit 10,000 by the year 2010 because the coming decades will mean greater spending, greater growth, and greater change in how the developed world works, lives, and where it lives. The downsizing of white-collar labor functions in the United States is a suggestion that this revolution is underway. Many of the best paid jobs of the past era have been eliminated in middle management categories in the 1990s recession. Many of the highest paid bureaucratic office functions went as well, and in the next decade, clerical office functions are likely to be largely eliminated (see Chapter 4: Transportation and Communication, and Chapter 8: Business Reengineering).

Lifetime employment with a single company is becoming a thing of the past. Terminating salaried workers who earn good pay and benefits and hiring them back as *temps* (temporary workers) is the pat-

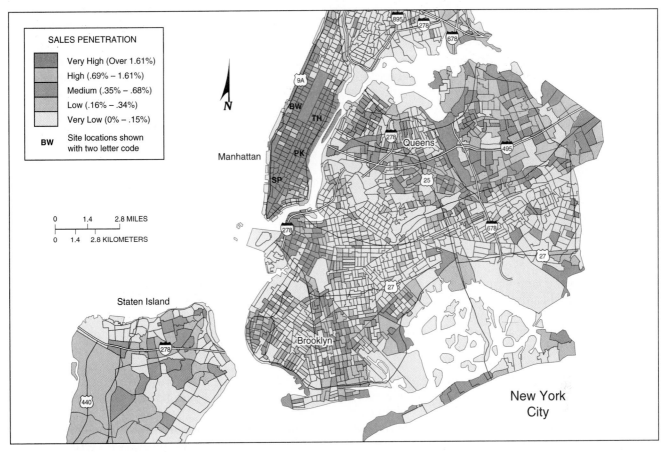

SALES PENETRATION

Very High (Over 1.61%)
High (.69% – 1.61%)
Medium (.35% – .68%)
Low (.16% – .34%)
Very Low (0% – .15%)

BW   Site locations shown
      with two letter code

0   1.4   2.8 MILES
0   1.4   2.8 KILOMETERS

Manhattan

Staten Island

Queens

Brooklyn

New York
City

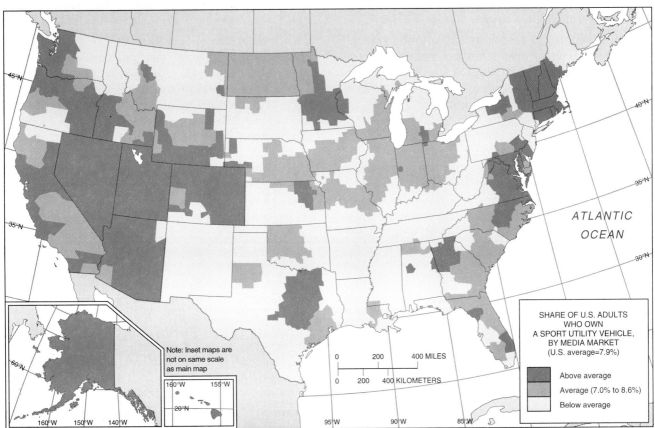

Note: Inset maps are
not on same scale
as main map

ATLANTIC
OCEAN

SHARE OF U.S. ADULTS
WHO OWN
A SPORT UTILITY VEHICLE,
BY MEDIA MARKET
(U.S. average=7.9%)

Above average
Average (7.0% to 8.6%)
Below average

0   200   400 MILES
0   200   400 KILOMETERS

FIGURE 2.35 (top) True Value Hardware Store Penetration in New York City. (bottom) SUV Nation. Sport utility vehicles were once bought for traveling off-road. Now they have become one of the most popular vehicles in North America. One driver in 12 in 1997 owned a SUV, double the 1992 rate. Most drivers are upscale, 40-something, driving their kids to the field. New England is SUV country, as is the western United States.

tern of the firm today. Fleeting are the days of corporate responsibility for a worker's life, career, and
retirement. The new employment pattern emerging
today is also a shift from an industrial, standardized
assemblyline (Fordist) economy to an informational,
customized, post-Fordist *flexible economy*. We are
changing from a nation of gigantic firms that are slow
to respond to market demands, to a nation of small,
quickly reacting firms, operating in niche markets
and catering to an increasing customized need of a
wealthier society. A shift is occurring from a society
patterned around stable families, farms, and factories
to a more mobile, multi-career, fast-changing urban
and ex-urban society in which the basic unit is the
individual, not the large firm.

The hierarchial family has been diminishing in importance for some time. The family structure is changing along with the demographics of the firm. Like the
family, the firm is becoming a network of individuals
rather than a hierarchy. Families and firms must adapt
and keep up with the rapidly changing world of multiple careers, rapid personal growth, flexible work
hours, more work based in the home, relying on
telecommunications, rather than face-to-face contact,
and the need for nonstop learning. These are the
changes that are revolutionizing every phase of the demographics of the firm today.

A restructuring of our society today from a *standardized economy* to an *information economy* is as important as
the shift from nomadic herding and foraging to an agricultural society (circa 10,000 b.c.) or a shift from an agricultural society to an industrial society (circa 1800). It is
not surprising that people are angry and confused with
this restructuring. The protectionists of our society are
defending the old ways, fighting to preserve them harder
than ever. The siren of protectionism understandably
holds a certain lure for workers whose worry is that their
jobs are threatened by foreign competition. This is especially true among unskilled employees whose job security is undermined by the low cost of labor outside the
North American and other core countries. These pervasive economic fears that the United States is losing its
workers abated somewhat by 1998, not the least because
unemployment, at less than 5%, reached a 10-year low.

Some asserted that NAFTA would take millions of
jobs from the United States to Mexico. The truth has
turned out to be somewhat the opposite. Increased
cross-border commerce spurred by NAFTA has added
jobs on both sides of the border. For example,
California—and San Diego in particular—has benefited enormously from rising trade with Mexico. In 1994,
1995, and 1996, the first three full years under NAFTA,
California exports to Mexico soared by 20% per year to
$40 billion in sales. Sales to Mexico created more than
150,000 jobs in California, including an estimated
42,000 in San Diego County.

A worker revolution in the core countries is inevitable when the changes are fundamental in the
world economy, like those facing the work revolution
of firms today. The world will never fully return to the
social values and structures of the past, no matter how
many politicians run on a platform of protectionism.

THE GREAT (BABY) BOOM AHEAD

Economists have never displayed much skill in predicting recessions. However, historians have found one
thing common to most periods just before a downturn—the emergence of a widespread belief that this
time we have found the magic elixir to produce an extended period of prosperity.

The National Association of Business Economists
recently surveyed its members and found few who see
anything but sustainable growth and low inflation
ahead. Although all business cycles eventually end, almost all of the 44 economists who responded to the
survey expect economic expansion to continue for the
foreseeable future.

In March 1997, the current expansion was six years
old. It is the third-longest recovery on record, and if a
recession does not strike in the next three years, it will
become the longest peacetime expansion, overtaking
the 92-month recovery in the 1980s. Just after the turn
of the century, if growth continued apace, the expansion would breach the 106-month record set in the
1960s when the Vietnam War gave the economy a
boost. If that is still a bit far off to contemplate seriously,
there is still little to suggest that the economy will deviate anytime soon from its path of moderate growth
and low inflation. The Federal Reserve seems to have
hit on just the right monetary policy formula, with unemployment low but wage and price increases showing
no acceleration.

By continuing to invest in new equipment and
squeezing more efficiencies from their workforces,
companies have managed to maintain profit growth.
Still, to the extent that bumps along the economic road
have been smoothed over, stability has come at a substantial cost. Improvements in productivity in recent
years have come in part from wide-scale layoffs, creating a new level of anxiety in the workforce. Among
business and political leaders, there are those who
argue that stability has been bought at a price of foregone growth and opportunity.

Indeed, some of the same forces that are dampen-
ing the economy's ups and downs, most notably the
globalization of competition, are speeding the *life cycles* of
products and technology in other industries (see
Chapter 9). This post-Fordist revolution will hallmark
personalized service in the coming decades. Reliance on
information technologies and telecommuting will allow

a major population shift to small towns and cities within 100 miles of major metropolitan areas. These *exurbs* will reduce the costs of living and doing business and increase the quality of life.

The so-called baby boom is the surge of a new generation that peaks about every 40 years. The baby boom generation is the largest in world history, 90 million strong (see Figures 2.36a and 2.36b).

The age at which this generation enters the workforce is called the "innovation swell"; the age at which this generation, 48 years later, peaks in its spending is called the "spending swell"; and the age at which the generation moves into its power position, 65 years later, is the "power swell."

As the baby boomers entered the workforce, they brought new social and technological ideas and entrepreneurial spirit. This pressure forced older, mature industries and companies to restructure themselves and to make large investments at all levels to train the new workers. Initially, this new boomer generation was not very productive and had low earnings and savings rates. In the 1960s and 1970s, the baby boomers entered the workforce, 90 million strong, including more working women than ever before. This large swell entering the workforce required a huge investment in capital stock and infrastructure—office space, desks, training programs, computer terminals, parking garages, not to mention cafeterias and clothing stores.

The baby boomers redefined the workplace, causing social and technological change, but their conformist, civic-minded bosses were not accustomed to such change. A few of the upstart new entrepreneurs included Bill Gates (Microsoft), Michael Dell (Dell

FIGURE 2.36

The cycles in birth rates in the United States 1940–1995. The largest generation by far is the baby boom generation of the 1940s, 1950s, and early 1960s, about 90 million strong. It comprises more than two-thirds of our workforce. By 1995, it was three-fourths of the workforce, where income and spending power is created. The relative timing and size of baby boom generations of Europe, the United States, and Japan. The United States had a far larger baby boom than either Europe or Japan. Canada's demographics are parallel to the United States, adding to its overall baby boom effect. The peak in the North American baby boom is 37 years of age in 1996. The spending dimension of a generation peaks in spending around age 49, so North America has a good 12 years of boom ahead for the economy. There are three booms here. The middle one, the biggest, is the famed baby boom generation, marked by annual birth rates that reach nearly 4.5 million babies in the late 1950s. The previous boom was more of a boomlet, with birth rates rising to a little over 3 million in the early 1920s. The third boom is made up of children of the baby boomers and won't be as big a boom, because baby boomers are having fewer than two children per couple. This enormous bulge in the birth rate that began after WWII sets the stage for the wave of the future just ahead. *(Source: From* The Great Boom Ahead: Your Comprehensive Guide to Personal and Business Profit in the New Era of Prosperity *by Harry S. Dent, Jr. Copyright © 1993 Harry S. Dent, Jr. and James V. Smith, Jr. Reprinted with permission of Hyperion)*

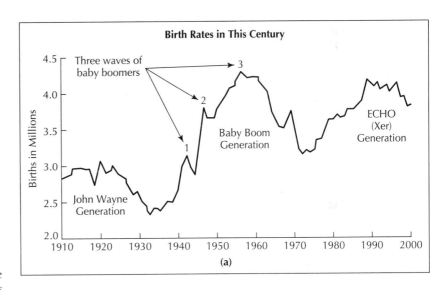

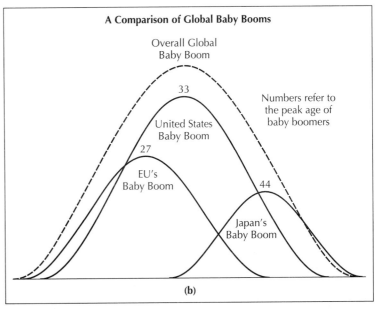

Computers), Steven Jobs (Apple), and Eric Anderson (Netscape). The result of the influx of baby boomers was new products, new services, and new technologies in niche markets, improving service and reducing service delivery times.

The innovation swell of the baby boom generation established the industries and markets for the future. The new technologies, products, and markets moved through a life cycle called an "S" curve, or a *logistic growth curve* (see Figures 2.37a to 2.37d). A new product takes many years and a difficult trial period before it is accepted. The first third of a logistic growth curve is adaptions from 0.1% to 10% of the population. The market may be suspicious of the products, or the product may be overpriced and not user friendly. When market saturation reaches 10%, adoptions mushroom into the mainstream and the growth from 10% to 90% of the market takes only as long as the initial growth from 0.1% to 10% took.

In 1900, less than 0.1% of urban families could own a car, and by 1914, only 10% owned automobiles. In the next 14 years, however, automobile ownership exploded, and by 1928, 90% of urban families owned an automobile. The telecommunication industry is about to enter its powerful 10% to 90% period in the coming years, much like the automobiles did in the Roaring 20s.

Large retooling and training investments that were needed to absorb the baby boomers into the job market required consumers to pay a high "inflation tax" to finance them. It took a large investment to launch new technologies and industries. Technological innovations are expensive to start up, and there is a correlation between inflation and such revolutions. It was no accident that America saw the highest inflation in U.S. history in the late 1970s as the largest generation in history, the baby boomers, went through their early job years. During this time period, the most powerful innovations and technologies in history started to emerge—the digital revolution from personal computers to fax machines to cellular phones.

Now that the baby boomer swell has entered the labor force, present labor force growth is low

FIGURE 2.37
(a) The common S curve for consumer adoption patterns for any innovation. As the proportion of adopters increases along the S curve, the long-run average cost curve of a product or service declines. (b) The long-run average cost curve declines as the S curve starts its rapid upward growth. (c) In the adoption process, the initial product cost curve declines, only to be challenged by a new firm producing a product or a new product innovation. Initially, the new product's cost curve is higher than the original cost curve, but eventually at point E, it drops below the initial product cost curve and forces the initial product out of the market. (d) Product cost curves for mainframe and microcomputers. Microcomputers and new technology have all but supplanted mainframe computers, the old technology, for most applications in decentralized environments. *(Source: see Figure 2.36)*

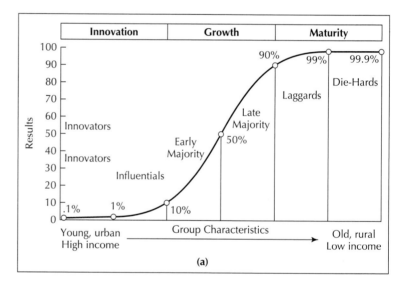

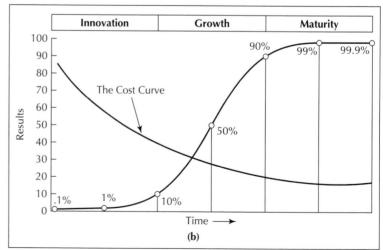

(Figure 2.38). Inflation is also consequently low, and the consumer price index (CPI), compared with labor force growth, shows that inflation will continue to stay low throughout the coming decades. Throughout the coming decades, labor force growth will be very slow. What this means for business demographics is that companies will not have to train masses of new workers and charge them the "inflation tax." However, the present workforce must be redeployed so that everyone becomes more productive to keep up with the high growth rates of the industry and market demand. This will be possible because of the microelectronics revolution, which allows workers to leverage information technologies and become more efficient in their work patterns (see Chapter 8: Business Reengineering).

After the restructuring of the North American economy from manufacturing industries to service industries, especially information sectors, productivity is expected by some to return. A new generation of workers entering the job market drives inflation up by bringing high cap-

italization costs associated with the innovation, resulting in high investments and technological retooling and training. Disinflationary periods occurred in the 1980s and 1990s. Thus, turmoil was then followed by a period of high productivity when the effect of those technological investments was felt, driving inflation back down.

The baby boom is over, but the baby boom consumers are just moving into their most important spending years. According to the Bureau of Labor Statistics, family spending follows a predictable life cycle that results in maximum spending between age 46 and 50 (Figure 2.39). The great number of baby boomers moving as consumers in and out of their peak spending years causes booms and busts in the economy. The baby boom spending wave began in the 1980s. North American baby boomers will move into their peak years of total spending for durable goods around the year 2010, according to economist Harry Dent (1995), creating the biggest business boom in the history of the world (Figure 2.40).

The baby boom spending swell is changing the demand for products and services and changing how we work and live. The innovation of the baby boom generation is requiring a new *customized/flexible economy* requiring creativity in products and services in an unprecedented surge in productivity. This business trend is propelled by the advancement of baby boomers into their power years when they have the business decision-making capacity to change organizations and accommodate new technologies to service their different skills and lifestyles. The combination of the microcomputer revolution and the telecommunications revolution, with the individualistic taste of the baby boom generation, means that there will be growth in all segments of the economy, especially those companies that can offer customization and flexibility to individual's demands and needs. This means high quality and high value added products and services delivered rapidly, custom produced flexibly with fast response and delivery.

Flexible Economy

The *customized/flexible economy* is the specialized economy of today, the rough opposite of the standardized economy of yesterday. In the *standardized economy*, the idea was to reduce an assemblyline process into its simplest parts, and have an individual or teams of people accomplish an isolated part. These people would never see

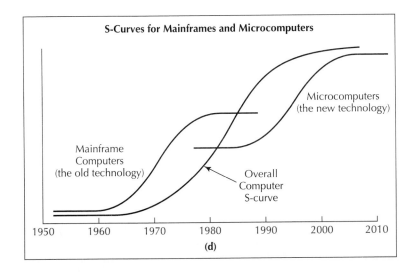

The Dual-Cost Curve Leading to Lower Prices

Cost Per Unit

Initial Product Cost Curve

New Product (or new product innovation) Cost Curve

Higher cost, initially

Lower cost eventually

A
B
C
D
E

Time →

| Opinion leaders buy at Point A | Influentials buy at Point B | Early majority buy at Point C | Late majority buy at Point D | Laggards buy at Point E | Die-hards don't buy at any price! |

(c)

S-Curves for Mainframes and Microcomputers

Microcomputers (the new technology)

Mainframe Computers (the old technology)

Overall Computer S-curve

1950 1960 1970 1980 1990 2000 2010

(d)

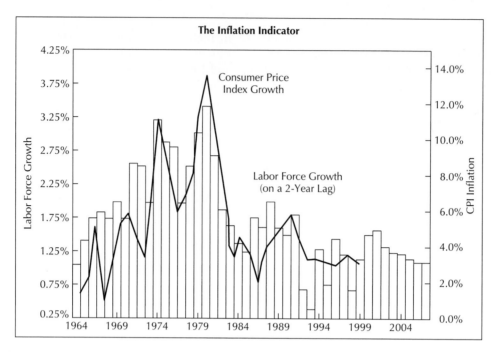

FIGURE 2.38

The rate of baby boomers entering the labor force is a tool that predicts inflation, which will stay low throughout the coming boom. This figure shows the consumer price index as the mirror image of labor force growth. There is a high correlation between the two. The lower productivity and higher investments required to incorporate such a generation (the baby boomers) peak about 2 years after entry into the workforce. Many factors explain inflation, but the rate of baby boomers coming into the labor force is the best predictor. The economy borrows from itself, or its own consumers, by raising the prices of goods and services to accommodate the lower productivity as a large number of laborers enter the workforce. *(Source: Figure 2.36)*

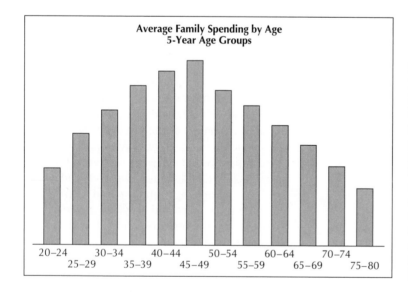

FIGURE 2.39

Peak spending years. According to the U.S. Bureau of Labor Statistics, when people are between the ages of 45 and 50, they are at their peak spending years. This histogram shows how people spend money through various 5-year periods of their life. People in their teens don't have much money to spend, and then as they move through their 20s, 30s, and 40s, they earn and spend more and more. But once they hit their early 50s, they begin to save money in anticipation of retirement. They don't spend as much as they used to. According to the Department of Labor, spending actually hits its high point at age 49. *(Source: Figure 2.36)*

the whole process. That was the assemblyline (Fordist) approach, and it required a growth of a hierarchy of managers to coordinate and control all of the over-specialized functions. The new approach in the flexible economy is somewhat the opposite. Individuals work in small, self-managing teams focused on a particular need or problem solution of a market or customer, and they get to know that market or customer's

needs better than large organizations of the past (Figure 2.41).

These small teams are multifunctional and cannot afford to become overspecialized in a lengthy, time-consuming, and costly bureaucratic process. These teams must be flexible enough to do whatever it takes to solve a marketing problem. The small teams are thus more efficient and less time consuming than the as-

FIGURE 2.40

When U.S. births are lagged forward 47 to 49 years and are compared with the best measurement of the performance of the U.S. economy (the Standard & Poor's 500 Index), there is a remarkably high correlation. Pulling two facts together: (1) there are many people about to become 49 years old, and (2) people who are 49 years old spend more money than anyone else, resulting in more people spending more money than ever before in national history. This chart is the same as the birth rate chart shown in Figure 2.36a, but simply moved forward 49 years to show when the baby boomers will peak in their spending years. The theory is that if there are more people than ever

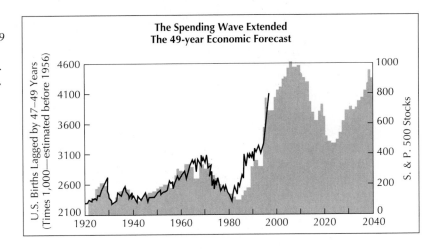

spending more money than ever, good things will happen to the economy, because private sector spending traditionally drives the U.S. economy. As the previous generation's 49-year-olds grew in numbers, beginning in the mid-1940s, the economy expanded, and the stock market moved ahead steadily for roughly 30 years. Beginning in 1970, as the number of 49-year-olds began to decline, the stock market fell along with their numbers. In the late 1980s, as the number of big spenders again began to increase, up went stocks, creating the biggest bull market in U.S. history. That bull market continues into 1997. But look at what is coming! The biggest move has not yet begun. The single most powerful economic force to move through any society in the history of the world has not yet hit its peak. *(Source: See Figure 2.36)*

semblyline approach because customization, fast response, and niche markets are essential.

The Dell Computer Corporation has designed its newest factory without room for inventory storage. The Chrysler Corporation can increase vehicle production without building new factories. The General Electric Company expects to save millions of dollars by buying spare parts for its plants over the Internet.

Indeed, even high-technology companies have not managed to immunize themselves from traditional inventory bloat. Another computer maker, the Digital Equipment Corporation, stumbled badly in 1996 when it was caught with thousands of unsold personal com-

puters that were built before a plunge in memory chip prices. When PC prices followed chip prices lower, Digital sustained heavy losses, forcing a revamping that cost 7,000 jobs.

Nevertheless, technology proponents say the world is in the beginning of a process that will accelerate with the rapid adoption of the Internet as the primary means of commercial communication (see just-in-time delivery, Chapter 8).

General Electric, the diversified industrial and financial giant, is converting its entire supply chain from paper to the Internet, a shift executives say will save hundreds of millions of dollars a year. General

FIGURE 2.41

The standardized economy will give way to the flexible, customized economy in the future. An entire new economy will be built on top of the old. The flexible economy will be driven by investments in flexible software and will use information as a fuel. The result will be customized products and services with short turn-around times, quick delivery, and low levels of stocks and warehousing. Products will become more affordable than our standardized products and services of today. The microcomputer industry and a broad range of new niche products and services will increasingly dominate the North American economy in the coming decades. Microtechnologies increasingly will allow the world's developed countries to custom design, market, and produce products and services at lower cost.

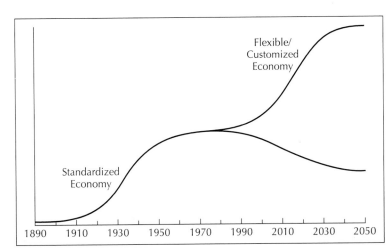

American Generation X-ers are beset with an uneven impact of the flexible economy on different layers of the workforce. At the top, *tier I*, where 25% of North Americans work, is the technology- and service-oriented economy. Here, the salaries are strong, and there are not enough well-trained workers to fill some specialized jobs such as software programming. The technology sector is growing at approximately 40% per annum, and the Information Technology Association of America estimates that high tech companies need 200,000 new workers per year but cannot immediately find them. For this tier, salaries are going up faster than inflation. For the middle tier, *tier II*, companies are outsourcing production to nonunion subsidiaries. This tier of the workforce, whether unionized or not, and whether blue collar or white collar, is holding even and eking out small gains by holding more than one job or by having more members of the household go to work at temporary jobs without benefits. At the lowest tier, *tier III*, of the workforce, companies continue to close expensive, aging factories and relocate production units to less expensive areas of North America, but also Mexico and Asia. These pressures have hit the textile industry of the Carolinas, just as they have taken a toll on the steel, auto, and machine tool industries of the upper Midwest. The least skilled workers in tier three are not able to receive wage increases to keep up with inflation because of an abundant supply of immigrants from Latin America and the Far East and from former welfare recipients seeking work. The trend in the overall economy is to drive costs down and profits up. Real economic growth for the United States as a whole is moving along at a 5% annual rate and unemployment is falling below the critical 5% level for the first time in more than a quarter decade. The inflation that haunted the economy throughout the 1970s is in check. Deficits have shrunk from $290 billion five years ago to an estimated $60 billion for 1998. *(Photo: Ron Chapple, FPG International)*

Electric's lighting division has already begun using the Internet to solicit bids for spare parts for its factories overnight instead of over two weeks. The solicitations go out over the global network to a selected group of suppliers, who reply with their bids the next day.

The epitome of this customization is the microelectronics industry. Small teams require less overhead and have more flexibility to meet customers' demands. Functional specialization or Fordism is a thing of the past for most corporations of the future. There will be experts and information providers that support multi-functional front-end teams, and these people will have to specialize even more in their functions and expertise. More and more firms will require multi-functional people with an emphasis on achieving customized results. The firm of the future will not have specialization of production, but specialization of knowing products, knowing the competition, knowing the business, and most of all knowing the customers.

Today, small firms are growing at the expense of large companies. It is easy to see the crumbling of the giant bureaucratic organizations, including Sears, GM, and IBM. These companies led in their respective fields, but corporate bureaucracy so slowed implementation that other companies beat them to the market with the fruits of their research. The firms of the future will be small, working with multi-functional teams and individuals, but networked together to be flexible and to be able to shift in the market quickly. Speed is everything for the firm of the future in the flexible economy. In the traditional economy, or the standardized economy, size was everything, but in the flexible economy, speed counts far more. In this information-driven age of customized products and flexible manufacturer and services, entrepreneurial spirit, and adaptability are more important than vertical integration, hierarchical bureaucratic structures, and top down command and control systems.

An example of small, flexible systems is the Internet, which allows computer owners to link up with other large computers worldwide to make work more efficient. The Internet is the most efficient information system for individuals and teams communicating around the world, and the number of Internet users is growing exponentially. The Internet allows users with sets and subsets of data to network together to share information, to acquire data, and entertain themselves. The Internet has a life of its own and functions almost organically.

Migration to Exurbs

Another major spatial change is occurring due to the demographics in the firm. A geographical shift of the population is occurring from the urban area as the workforce begins moving to the exurbs, or outer suburbs, and back to smaller towns and communities. The move to the exurbs is being propelled by (1) retirees, (2) those seeking relief from the high-cost suburbs, and (3) those desiring a greater community or native lifestyle. These shifts will be made possible by the *electronic cottage*—the increased power of computers and telecommuting in ways that have not been used in the past. This power will be employed to decentralize today's firms through the communication revolution—the moving of information rather than people. As baby boomers move into their peak spending years, bolstered by two-income family earnings, they will spend more than ever before on high quality durable goods and convenience items, and less on the standardized, mass produced items and services of the past. Just as in flexible manufacturing, baby boomer consumer demand will center on the high quality, high choice niche markets. Labor patterns of the future will focus on personal freedom and an ability to control one's time schedule and work location, health, and enjoyment, rather than only securing high income and retirement benefits, which in themselves are disappearing fast in the new information economy. Such new workers in the firms of the year 2000 will not so much be "doing their own thing" as "controlling their own time."

The X generation (children of the baby boomers) will provide business growth opportunities for housing, office buildings, restaurants, all matter of personal and business services, home conveniences, entertainment, and more. Small town exurbs will become boom towns, and as many as 80 million North Americans could shift outside metropolitan areas by the year 2010 (Figure 2.42).

Such boom towns will provide a slashed cost of living, especially for food and real estate needs, but ample opportunity to start up a *carbon copy business,* which is a successful business idea transplanted from a major city to the exurbs that share common consumer demographics and lifestyles with the city. Starbucks, Video Discount, Charles Swabb Investment, Fitness Advantage, Woody's Woodfired Pizza are examples. The *carbon copy business* will be an important exurb model for small firms, while improving the quality of life for individuals. Such firms in the exurbs suggests four important principles of the coming consumer demographics of the firm, according to economist Harry

Canadian Generation X-ers have a pragmatic idealism and a desire to succeed in this globalized economy that is as intense as the fire that sent Peace Corps workers abroad during the baby boom generation of the 1960s. A globalized, interconnected economy is likely to be the X-ers best insurance for a peaceful world for their children. The X-ers in North America are part of a globalized generation that shares values across borders, moving ideas and products with the freedom of E-mail. For the X-ers worldwide, age and ambition are the issues. Nearly 50% of the world's population is under 30 years of age. Given the X-ers great ambitions for themselves and their countries, the prospects for the two essential ingredients for peace—capitalism and democracy—seem to have a good chance to succeed worldwide. The odds of sustainable peace and prosperity, according to the lessons of history, are slim. The X-ers will be the first generation in a century to choose their own identity, accept others as they are, and do what interests them. That is because around the world, economic and political liberalization is catching on. Less war equals more money. At the end of 1997, none of the planet's 180 nations was shooting at each other.

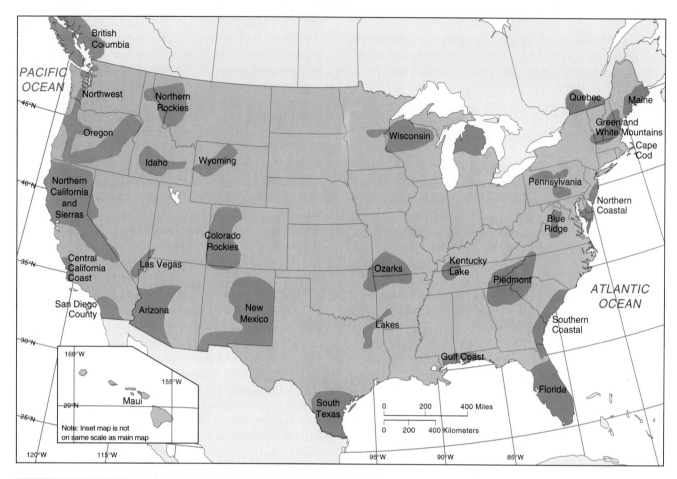

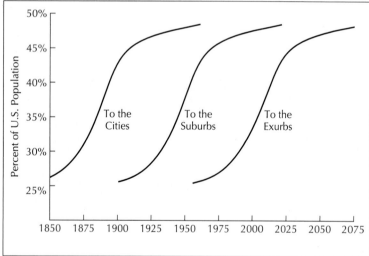

FIGURE 2.42
(top) The leading retirement exurb communities showing rapid growth in the late 20th century. Most are rural communities filled by exurbanites from medium and large cities. (bottom) Small town exurbs will become boom towns, and as many as 80 million North Americans could shift outside metro areas by 2010. The move to the exurbs is being propelled by retirees, those seeking relief from high cost cities, and those desiring a rural lifestyle. *(Source for [bottom] see Figure 2.36)*

Dent (1995), associated with the information revolution: (1) customization of products, (2) carbon copy businesses on a small scale, (3) distance networking many small businesses through computers, and (4) leveraging information technologies to exploit the growth of small towns (see Chapter 7: Smart Cities).

Demographics for City Planning

Without proper planning, modern cities would wander aimlessly and be unable to ensure a high quality of life for their residents. This last section describes demographics used in the planning process by urban policymakers. The example is taken from San Diego, California, but represents similar efforts in 100 other cities throughout the world.

The population of the San Diego region increased by more than half a million people between 1980 and 1990, making it one of the fastest growing metropolitan areas in the United States. In 1984, the San Diego region reached 2 million residents. During 1987, the region experienced a growth rate of 3.6%, bringing the total population to 2,328,328 residents by the end of the year and to 2.5 million by 1990. San Diego County is currently the fifteenth largest metropolitan area in the nation in terms of population and the fifth largest legal city. It is the fourth largest county in the nation (behind Los Angeles, California; Cook, Illinois; and Harris, Texas). Overall, the county also experienced the third largest numeric increase in population between 1980 and 1988, behind Los Angeles, California, and Maricopa County in Arizona. It is evident that such rapid development creates many problems in the provision of physical and social infrastructure at acceptable cost, in preserving the quality of the environment, in safeguarding the safety of the populace, and in many other areas.

The San Diego Association of Governments (SANDAG) is a quasi-government agency consisting of the County of San Diego in the southwestern corner of California, as well as 18 cities located therein. SANDAG forecasts population, housing, and economic activities for the entire San Diego region and for the smaller geographic areas within it. Locally, this product is known as the Regional Growth Forecast (RGF). The RGF has a wide variety of uses and applications and is based on a large computer geographic information system (GIS). It helps determine the need for transportation systems and the size and location of public facilities such as fire stations, schools, hospitals, and sewage and water treatment plants. The RGF is also used to assess water and energy demands for county agencies and geographic areas and can help predict the future quality of the region's air based on developmental aspects of land uses and population growth. Local governments that do not have a large planning and GIS capacity make use of the RGF and other products of local technical assistance from SANDAG as they evaluate housing needs for their constituencies and update their general and community plans. Uses of the RGF are summarized in Figure 2.43.

There are two phases to the RGF process, and four major models are used to obtain the projected population and land-use values. The first phase uses the Demographic and Economic Forecasting Model (DEFM), which produces a forecast for the San Diego region as a whole. The second phase employs three allocation models to disaggregate the RGF tabulations to each of the geographic sub-areas in the county.

Table 2.11 is an example report: it gives subregional forecast model outputs. This information can be retrieved and printed out for standard geographic areas

FIGURE 2.43
The San Diego Association of Governments produces a Regional Growth Forecast based on population, housing, employment, and income data from the U.S. Census and from locally generated surveys. Economic policy functions and historical data series, as well as land-use policy assumptions, are also used to produce the Regional Growth Forecast. The Regional Growth Forecast has a wide variety of uses and applications and is based on a large geographic information system (GIS). It helps determine the need for transportation systems and the size and location of public facilities, such as fire stations, schools, hospitals, and sewage and water treatment plants. *(Source: Parrott and Stutz, 1992, p. 248)*

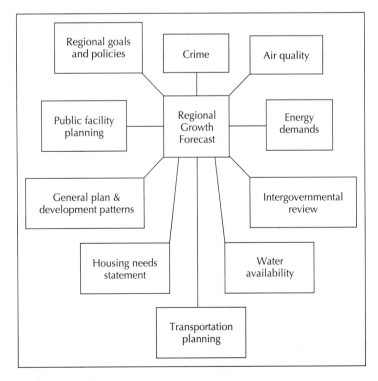

▨

T a b l e 2.11
SANDAG Sub-Regional Forecast Model Outputs

Population
Total population
  Household
  Group quarters
    Civilian
    Military

Employment
Total employment
  Civilian
    Basic
    Agriculture (SIC 1–9)
    Mining (SIC 10–14)
    Manufacturing (SIC 20–39)
    Transportation (SIC 40, 42)
    Wholesale (SIC 50, 51)
    State and federal government (92)
    Hotel (SIC 70)
    Basic military*
  Local serving
    Retail trade (SIC 52–59)
    Retail services (SIC 72, 74–88)
    Business services (SIC 73, 89)
    Construction (SIC 15–17)
    Finance, insurance and real estate
    Local government (SIC 93, 94)
    Local serving transportation
  Uniformed military†

Occupied housing units
  Total occupied units
  Single family
  Multiple family
  Mobile homes
Persons per household

Land use
Total acres
  Developed
    Single family
    Multiple family
    Mobile homes
    Basic
    Local serving
    1986 Freeway
  Vacant developable
    Low density single family
    Single family
    Multiple family
    Mixed use
    Local serving
    Industrial
  Unusable
  Redevelopment/infill acres
    Single family to multiple family
    Single family to mixed use
    Single family to local serving
    Multiple family to mixed use
    Multiple family to local serving
    Single family intensification
    Multiple family intensification

*All military persons at their place of work, excluding persons living on-base in barracks or on-board ships. Civilian persons working on military bases are included in the State and federal government category.
†Basic military + military group quarters.
SIC: Standard Industrial Classification
Source: Parrott and Stutz, 1992, p. 251.

such as traffic analysis zones (TAZ), census tracts, or entire cities, or for any user-defined geographic area. An example of output from the projection programs is Figure 2.44, which shows total change in population by grid cell between the base year 1986 and the year 2010.

Summary

Because people are the most important element in the world economy, it is essential to learn about population distribution, qualities, and dynamics. After considering the variable distribution of populations and their de-

mographic, cultural, and economic characteristics, we examined the processes of population change. The components of population change are migration and natural increase. The principal force affecting world population distribution used to be migration; now it is natural increase.

Although the population growth rate is falling, the world's population is projected to increase for decades to come, due to the large momentum built into the vast and youthful population of the Third World. As a result, there is considerable interest in the question of whether and how population growth affects economic growth. We compared and contrasted a number of viewpoints,

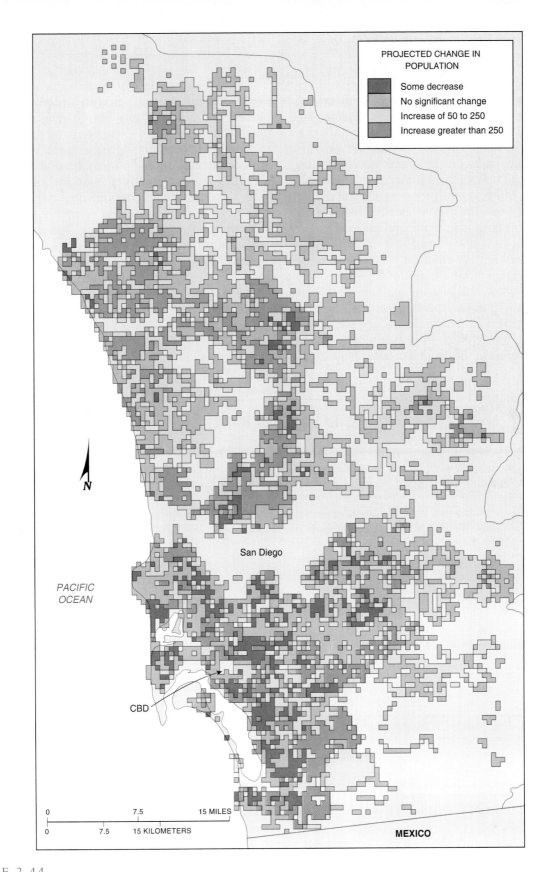

FIGURE 2.44

One of the outputs of the Urban Development Model is the projected change in San Diego County's population by one-quarter square mile grid scale. This map shows the change in population between 1986, the base year, and the year 2010. Downtown San Diego is located 25 kilometers north of the Mexican border and next to San Diego Bay. The greatest increase in population is projected to occur in the southern regions of San Diego, near the Mexican border, and in areas well to the north of downtown. These locations represent the greatest amount of available developable space. Unshaded areas to the east of San Diego represent mountainous terrain, while the large open area 45 kilometers north of the Mexican border represents a naval air station and environmentally sensitive wildlife habitats. *(Source: Parrott and Stutz, 1992, p. 252)*

including conservative and progressive. Some were optimistic; others, pessimistic. Most population experts believe that overpopulation deters growth, depletes natural resources, and destroys the environment. Hence, efforts must be made to slow population growth. Organized fertility-reduction programs, however, must be combined with development strategies that give poor people more control over the processes of social change that affect their lives. The organizational ecology of the firm examines its birth, migration, and survival.

North America is likely headed for a strong economic boom between 1998 and 2007–2008. The economy is driven by predictable cycles of family spending in the durable goods and housing markets. Each new generation follows a similar spending pattern. The baby boom, as the largest generation in history, will create the greatest economic boom ever as it moves into its peak spending years in the coming decade, 2000–2010.

This boom will be like the Roaring 20s, with vigorous economic growth, increased productivity, wages gains, low inflation and interest rates, rising saving, and falling debt ratios. Computers and the Internet will race into mainstream affordability and application just as cars, electricity, and phones did in the Roaring 20s, creating a revival of productivity and jobs. Consumers will see their discretionary income rise, and as mortgage rates fall, housing will become more affordable. We will also see a 60-year cycle shift from large metropolitan areas back to smaller towns and exurban areas. Lifestyle shifts and mid-life crises that baby boomers are going to be experiencing in the coming decade show eight types of towns and cities that will boom: vacation, college, classic, revitalized business towns, exurbs, emerging new cities, and large growth cities. Finally, population projections have valuable roles to play in the planning and survival of cities.

KEY TERMS

achieved characteristics
ascribed characteristics
baby boom
baby bust
birth rate
capital
carrying capacity
death rate
demographic equation
demographic transition
doubling time
fertility rate
geographic information system (GIS)
human capital
human suffering index

infant mortality rate
labor force
labor migration theory
law of diminishing returns
limits to growth
migration
natural increase
negative population growth
neoMalthusians
optimum population size
organizational ecology
physiological density
population bomb
population composition
population distribution

population hurdle
population pyramid
primary activities
primate city
"push-and-pull" factors
rate of natural increase
regional growth forecast (RGF)
rule of seventy
secondary activities
tertiary activities
unemployment
urban growth
urbanization
zero population growth

SUGGESTED READINGS

Hannan, M., and Carroll, G. 1992. *Dynamics of Organizational Populations*. New York: Oxford University Press.

Hannan, M., and Freeman, J. 1989. *Organizational Ecology*. Cambridge, MA: Harvard University Press.

Knox, P. L. 1994. *Urbanization: An Introduction to Urban Geography*. Englewood Cliffs, NJ: Prentice Hall.

Parrott, R., and Stutz, F. P. 1992. Urban GIS applications. In Maguire, Goodchild, and Rhind (eds.), *GIS: Principles and Applications*. London, England: Longman.

Rubenstein, J. M. 1996. *An Introduction to Human Geography*, 4th ed. New York: Macmillan.

Stutz, F. P. 1992. Urban and regional planning. In T. Hartshorn (ed.), *Interpreting the City: An Urban Geography*. New York: John Wiley & Sons.

Stutz, F. P. 1992. Maquiladoras branch plants: Transportation—labor cost substitution along the U.S./Mexican border. In *Snapshots of North America*. Washington, DC: 27th Congress of the International Geographical Union Official Book.

Stutz, F. P. 1992. Labor shed of Tijuana in relation to the U.S. mexican border. In T. Hartshorn (ed.), *Interpreting the City: An Urban Geography*. New York: John Wiley & Sons.

Stutz, F. P., Parrott, R., and Kavanaugh, P. 1992. Charting urban space-time population shifts using trip generation models. *Urban Geography*, 13(5):468–475.

Weeks, J. 1996. *Population: An Introduction to Concepts and Issues*, updated 5th ed. Belmont, CA: Wadsworth.

World Resources Institute, 1996. *World Resources 1996-97*. New York: Oxford University Press.

W O R L D   W I D E   W E B   S I T E S

DEMOGRAPHY & POPULATION STUDIES
http://lcweb.loc.gov/homepage/lchp.html
Library of Congress; links together libraries & online catalogs.
http://econwpa.wustl.edu/EconFAQ/EconFAQ.html
Resources for Economics on the Internet.
http://lambik2.rri.wvu.edu/spacestat
SpaceStat—Software for Spatial Data Analysis.
http://www.geom.umn.edu/docs/snell/chance/welcome.html
The CHANCE Database.
http://www.worldbank.org
World Bank.
http://www.lib.virginia.edu/gic
GIS and other map generation; NCGIA link, among others.
http://mpas.esri.com
Mapping on the internet.

UPDATED US POPULATION ESTIMATES [.ZIP]
http://www.census.gov/population/www/estimates/popest.html
The US Census Bureau, in association with the Federal-State Cooperative Program for Population Estimates (FSCPE), has recently released updated population estimates on the national, state, and county level. Updated national estimates are available on a monthly basis, and annual estimates are available for 1990- February 1997. On the state level, estimates have recently been updated through 1996 for demographic components. On the county level, 1990 and 1996 estimates are now available. All files are available as ASCII text, with some of the larger ones compressed in .zip format. Documentation and layout of files is available.

TWO SEARCHABLE INTERNATIONAL DATABASES

U.S. CENSUS INTERNATIONAL DATABASE
http://www.census.gov/ftp/pub/ipc/idbnew.html

UNITED NATIONS INFONATION [FRAMES]
http://www.un.org/Pubs/CyberSchoolBus/infonation/
The US Census Bureau and United Nations provide excellent, interactive searchable databases of international statistical information. The Census Bureau's IDB is highlighted by IDB Online Access, which allows the user to select one of 26 demographic, ethnic, and economic variables and any or "all [of] the countries in the world," and receive a table for that variable for the latest year, selected individual years, or a range of years. Selected variables are available by urban/rural residence. Data can be displayed as text, spreadsheet, or in user-configurable format. IDB is a product of the Census Bureau's International Programs Center. InfoNation data, gleaned from the United Nations Statistics Division, allows the user, through a frames-based interface, to select up to seven of 185 countries and four of 37 variables. Retrieval is for the latest year available only, providing a snapshot view of country information. InfoNation provides geographic and social indicators as well as demographic and economic. Together, IDB and InfoNation are a formidable resource. Note that because of the use of different sources, data may vary for the same variables across the two databases.

USA COUNTIES 1996—OREGON STATE UNIVERSITY GOVERNMENT INFORMATION SHARING PROJECT
http://govinfo.kerr.orst.edu/usaco-stateis.html

OSU GOVERNMENT INFORMATION SHARING PROJECT
http://govinfo.kerr.orst.edu/
Oregon State's well-known Government Information Sharing Project has recently added the US Census Bureau's USA Counties 1996 database to its arsenal of interactive demographic, economic, and educational databases. USA Counties 1996 allows the user to choose from nearly 3,500 variables in 26 major subject categories from age to health to wholesale trade. Single county profiles can be retrieved for multiple variables; conversely, the user can retrieve multiple county, single variable profiles.

SOURCEBOOK OF CRIMINAL JUSTICE STATISTICS 1995
http://www.cs.wisc.edu/scout/report/archive/scout-961129.html#3

SOURCEBOOK OF CRIMINAL JUSTICE STATISTICS 1995
http://www.albany.edu/sourcebook/

UNIFORM CRIME REPORTS 1990-1993
Abstract: http://www.cs.wisc.edu/scout/report/archive/scout-961129.html#3
Site: http://www.lib.virginia.edu/socsci/crime/index.html
http://govinfo.kerr.orst.edu/usaco-stateis.html

THE ARGUS MAP VIEWER
http://www.argusmap.com
The ARGUS Map Viewer is being hailed as the first geographic viewer for the Internet by allowing you to interactively browse and explore the Web and to link to information and site using maps with high quality vector graphic.

U.S. GAZETTEER
http.//tiger.census.gov/cgi-bin/gazatteer
Identifies states, cities, and counties in the U.S. Can also search for 5-digit zip codes.

THE CENSUS BUREAU
http://www.census.gov
The Census Bureau Web site was designed to enable "intuitive" use and is intended to be visually appealing, concise, and quick loading. It was designed so users can effectively locate and utilize the resources the site has to offer, such as the "Population Clock" and its small search engine.

THE INTER-UNIVERSITY CONSORTIUM FOR POLITICAL AND SOCIAL RESEARCH
http://www.icpsr.umich.edu
ICPSR provides a wealth of data on National, State, & Local Elections; Census Information; Political Behavior Attitudes; Poll & Survey Data.

LIBRARY OF CONGRESS
http://www.loc/.gov
Astounding variety of resources and exhibits.

THE YELLOW PAGES: GEOGRAPHY
http://theyellowpages.com/geography.html
Lists links of geography sites and search tools.

MAPQUEST
http://www.mapquest.com
Maps of cities and regions in the U.S. and the world.

RESOURCES AND ENVIRONMENT

OBJECTIVES

- To describe the nature, distribution, and limits of the world's resources
- To examine the nature and extent of world food problems and to make you aware of the difficulties of solving them
- To describe the distribution of strategic minerals and the time spans for their depletion
- To consider the causes and consequences of the "energy crisis" and to examine present and alternative energy options
- To examine the nature and causes of environmental degradation
- To compare and contrast "growth-oriented" and "balance-oriented" lifestyles

Mining coal—the world's most abundant fossil fuel.
(Photo: International Labour Office)

Our prosperity depends on the availability of natural resources and the quality of the environment. Yet as the 20th century draws to a close, there is growing concern that human economic activities in developed countries, and increasingly in developing countries, are depleting resources and irreparably degrading the physical environment. How did we get into this situation? What can be done to effectively manage resources and protect the environment?

There has been an endless debate about how the world got into trouble as a result of industrial and agricultural activities, particularly among resource optimists and pessimists. *Resource optimists* believe that economic growth in a market economy can continue indefinitely; they see "no substantial limits . . . either in raw materials or in energy that the price structure, product substitution, anticipated gains in technology and pollution control cannot be expected to solve" (Notestein, 1970, p. 20). In contrast, *pessimists* assert that there are limits to growth imposed by the finiteness of the earth—by the fact that air, water, minerals, space, and usable energy sources can be exhausted or overloaded. They believe these limits are near and, as evidence, point to existing food, mineral, and energy shortages and to areas now beset by deforestation and erosion. To pessimists, a world with a projected population of 10 billion in the year 2100 is unthinkable. Population and economic growth ought to stop.

Some scholars think that the long-run debate on resources and the environment, which waxes and wanes according to general economic conditions, is counterproductive, evading practical issues that demand our immediate attention. What we need to do, according to geographer Thomas Vale (1985), is to keep our purposes in mind and try to understand how to achieve our ends. If our purpose is to create a habitable and sustainable world for generations to follow, how can we redirect present and future output to serve that end? One solution is to transform our present *growth-oriented lifestyle*, which is based on a goal of ever-increasing growth, to a *balance-oriented lifestyle* designed for harmony and endurance. A balance-oriented lifestyle would include an equitable and modest use of resources, a production system compatible with the environment, and appropriate technology. The aim of a balance-oriented world economy is maximum human well-being with a minimum of material consumption. Growth occurs, but only growth that truly benefits people. However, what societies, rich or poor, are willing to dismantle their existing systems of production to accept a lifestyle that seeks satisfaction more in quality and equality than in quantity and inequality? Are people programmed for maximum consumption by a value system constantly reinforced through advertising willing to change their ways of thinking and behaving?

This chapter, which discusses growth-oriented versus balance-oriented philosophies of resource use, deals with the complex components of the population-resources issue. Have population and economic growth rates been outstripping food, minerals, and energy? What is likely to happen to the rate of demand for resources in the future? Could a stable population of 10 billion be sustained indefinitely at a reasonable standard of living utilizing currently known technology? These are the salient, critical questions with which this chapter is concerned.

RESOURCES AND POPULATION

Popular perception in the industrial West appreciates the need to reduce population growth but overlooks the need to limit economic growth that exploits resources. We suffer from a view of limitless resources, which threatens our affluent way of life. We are liquidating the resources on which it was built. Developing countries are aggravating the situation. Their growing populations put increasing pressure on resources and the environment, and their governments aspire to affluence through Western-style urban industrialization that depends on the intensive use of resources.

Poor countries do not have the means for running the high-energy systems manifest in the industrial West. Even by conservative estimates, a middle-class "basket of goods" requires six times as much in resources as a basket of essential or basic goods. The expansion of GNP through the production of middle-class baskets means that only a minority of people in poor countries would enjoy the fruits of economic growth. Resource constraints prevent the large-scale production of middle-class goods for the growing populations of the developing countries.

However, measures of material well-being (e.g., per capita incomes, calories consumed, life expectancy) show that people in many countries are better off today than their parents were. But there are problems with this optimistic assessment. These improvements are based on averages; they say nothing about the distribution of material well-being. Another difficulty is that the world may be achieving improvements in material well-being at the expense of future generations. This would be the case if economic growth were using up the world's resource base or environmental carrying capacity faster than new discoveries and technology could expand them.

Carrying Capacity and Overpopulation

The population-resources problem is much debated, particularly during periods of economic shortages and rising prices. Pessimists believe that the world will enter a stationary state known as *carrying capacity*, which is the population that can be supported by available resources. They point to recurring food crises and famines in Africa as a result of overpopulation. On the other hand, optimists believe in the saving grace of modern technology. Technological advances in the last 200 years have raised the world's carrying capacity, and future technical innovations as well as *transmaterialization*—the substitution of new raw materials for old—hold the promise of raising carrying capacity still further.

Economist Ernest Schumacher, whose book *Small Is Beautiful* (1973) was written during a period of inflation and rising oil prices, doubted the feasibility of a completely technical solution to the population-resources problem. Technologies may be unable to cope with the matter of supplying growing populations with raw materials for existence in light of the *second law of thermodynamics*. This law holds that the amount of energy in the universe is fixed, but the amount of work that can be derived from that energy is irreversibly diminished. For example, once gas is burned in the engine of a car, its value as a source of useful energy is gone forever.

The answer to the population-resources problem also depends on the standard of living deemed acceptable. To give people an essential basket of goods instead of a middle-class basket of goods would roll back resource limitations. The establishment of a basic goods economy depends on our capacity to develop alternatives to the high-energy, material-intensive production technologies characteristic of the industrial West. Already, there are outlines of a theory of productive resources suited to the needs of a basic goods economy. Some of the main ideas are: (1) the adoption of a sun-based organic agriculture; (2) the use of renewable sources of energy; (3) the use of appropriate technology, labor-intensive methods of production, and local raw materials; and (4) the decentralization of production in order to increase local self-reliance and minimize the transport of materials. These productive forces would minimize the disruption of ecosystems and engage the unemployed in useful, productive work. Trends indicate that economies that produce essential goods for human consumption face neither unemployment nor overpopulation. Moreover, secure supplies of basic goods provide a strong motivation for reducing population size.

Optimum Population

The best possible world would be one with an optimum population that permits progressive improvements in human well-being; however, what is optimum for one country may not be for another. Furthermore, governments of rich industrial countries set the terms for what is optimal or suboptimal in relation to resources. Yet it

Water for Chad. Water is an important ingredient to sustain human life. Fifty percent of the world's people do not have adequate, clean water. Villagers in Chad are delighted as the water pours out of a new water system they have worked together to construct. The system is part of an anti-desertification project funded by the United Nations Development Program and the U.S. government. Acute water shortage in many parts of the world requires solutions that will be costly, technically difficult, and politically sensitive. Water scarcity contributes to the impoverishment of many countries in east and west Africa, threatening their ability to increase food production fast enough to keep pace with modern population growth. *(Photo: Ruth Massey/UNDP, United Nations)*

is only through the operation of the world economy that wealthy countries can appropriate the resources necessary to support the large numbers of people who enjoy middle-class lifestyles.

TYPES OF RESOURCES AND THEIR LIMITS

All economic development comes about through human resources (e.g., human labor, skills, and intelligence). In order to produce the goods and services people demand in today's global economy, we need to obtain natural resources. What are natural resources and what are their limits?

Resources and Reserves

Natural resources have meaning only in terms of technical and cultural appraisals of nature and are defined in relation to a particular level of development. *Resources*, designated by the entire box in Figure 3.1, include all substances of the biological and physical environment that may someday be used under specified technological and socioeconomic conditions. Because these conditions are always subject to change, we can expect our determination of what is useful to also change. For example, uranium, once a waste product of Canada's radium mining of the 1930s, is now a valuable ore. Taconite ores became worthwhile in northeast Minnesota only after production of high-grade, non-magnetic iron ore declines in the 1960s.

At the other end of the extreme are reserves, designated by the upper left-hand box in Figure 3.1. *Reserves* are quantities of resources that are known and available for economic exploitation with current technologies at current prices. When current reserves begin to be depleted, the search for additional reserves is intensified. Estimates of reserves are also affected by changes in prices and technology. *Projected reserves* represent estimates of the quantities likely to be added to reserves because of discoveries and changes in prices and technologies projected to occur within a specified period, for example, 50 years.

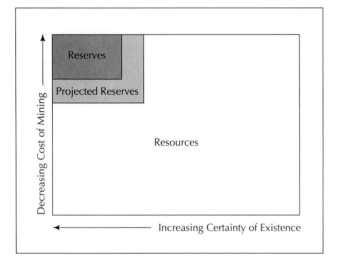

FIGURE 3.1

Classification of resources. Resources include all materials of the environment that may someday be used under future technological and socioeconomic conditions. Reserves are resources that are known and available with current technologies and at current prices. Projected reserves are reserves based on expected future price trends and technologies available.

Renewable and Nonrenewable Resources

Resources may be classified in various ways (Figure 3.2), and a major distinction is between nonrenewable and renewable resources. *Nonrenewable resources* consist of finite masses of material, such as fossil fuels and metals, which cannot be used without depletion. They are, for all practical purposes, fixed in amount. This is because they form slowly over time. Consequently, their rate of use is important. Large populations with high per capita consumption of goods deplete these resources fastest.

Many nonrenewable resources are completely altered or destroyed by use; petroleum is an example. Other resources, such as iron, are available for recycling. Recycling expands the limits on the sustainable use of a nonrenewable resource. At present, these limits are low in relation to current mineral extraction.

Renewable resources are those resources capable of yielding output indefinitely without impairing their productivity. They include *flow resources* such as water and sunlight and *stock resources* such as soil, vegetation, fish, and animals. Renewal is not automatic, however; resources can be depleted and permanently reduced by misuse. Productive fishing grounds can be destroyed by exploitation. Fertile topsoil, destroyed by erosion, can be difficult to restore and impossible to replace. The future of agricultural land is guaranteed only when production does not exceed its maximum sustainable yield. The term *maximum sustainable yield* means maximum production consistent with maintaining future productivity of a renewable resource.

In our global environment, the misuse of a resource in one place affects the well-being of people in other places. The misuse of resources is often described in terms of the *tragedy of the commons* (Hardin, 1968). This metaphor refers to the way public resources are ruined by the isolated actions of individuals. We appear to be unwilling to use a minimum share of a resource. People who fish, when there is no rule of capture law, are likely to try to catch as many fish as they can, reasoning that if they don't, others will. Similarly, dumping waste and pollutants on public waters and land or into the air is the cheapest way to dispose of worthless products. There is an apparent unwillingness to dispose of these materials by more expensive means unless mandated by law.

Sometimes resources are unavailable, not because they are depleted but because of politics. Resources are under the control of sovereign nation-states. Many wars between countries in the 20th century have been resource wars. For example, the conflict in the Persian Gulf, from the Iraqi invasion of Kuwait in August 1990 to their retreat in March 1991, was an attempt by Iraq to control a larger share of the world's known oil resources that was defeated by the United States and its allies wishing to maintain their oil supplies. Another

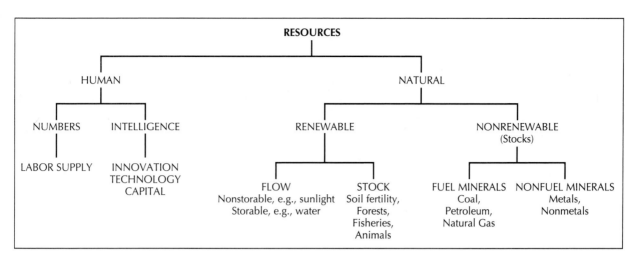

FIGURE 3.2

Typology of resources. Renewable resources are those resources capable of replenishing themselves or being replenished by individuals. Stocks of forests, animal populations, and fish are examples. Flow resources, such as sunlight and water, are also of the renewable variety. Nonrenewable resources including fuel minerals and non-fuel minerals, once used up, are for all practical purposes gone forever. Thus, increasing price and different levels of technology mean that the search for new deposits continues.

example of the politics of resources also comes from the Middle East. Here, fierce, atavistic national rivalries make water a conflict-laden determinant of external policies. While some parts of the Middle East are blessed with adequate water, most of the region is insufficiently supplied. Some observers predict that political tension over the use of international rivers, lakes, and aquifers in the Middle East may escalate to war in the next few years.

FOOD RESOURCES

Thanks to scientific advances in farming, world food production has been increasing faster than population. Despite these advances, which averted large-scale starvation in the 1970s and 1980s, millions of people still go hungry daily. With demand for food expected to grow at 3 to 4% per year over the next 20 to 30 years, the task of meeting that need will be more difficult than ever before. A record explosion in the world's population coupled with the problem of poverty threaten the natural resources on which agriculture depends. To make matters worse, environmental degradation perpetuates poverty, as degraded ecosystems diminish agricultural returns to poor people (Figure 3.3).

Nutritional Quality of Life

The gulf between the well-fed and the hungry is vast. Average daily calorie consumption is 3,300 in developed countries and 2,100 in developing countries. The largest number of calories available is in Ireland and the lowest number available is in Ghana (Table 3.1). But these are average figures. There are people in Ireland on the breadline and people in Ghana with plenty to eat. Averages mask extremes of *undernutrition*—a lack of calories—and overconsumption.

Even with a high calorie satisfaction, people may suffer from *chronic malnutrition*—a lack of enough protein, vitamins, and essential nutrients. The most important measure in assessing nutritional standards is the daily per capital availability of calories, protein, fat, calcium, and other nutrients. In the world today, the sharpest nutritional differences are not from country to country or from one region to another within countries. They are between rich and poor people. The wretched of the earth carry the major burden of hunger.

Causes of the Food Problem

Hunger among the poor of the world is often attributed to deforestation, soil erosion, water-table depletion, the frequency of droughts, and the impact of storms such as hurricanes. Although the environment does have a bearing on the food problem, it has limited significance compared to the role of societal conditions.

POPULATION GROWTH

Population growth is one cause of the food problem. However, presently, at the global level, there is no food shortage. In fact, world food production has been growing steadily from 1961 to 1994. Even by the year 2010, according to the United Nations Food and Agricultural Organization (FAO), production increases, assuming continuing high investments in agricultural

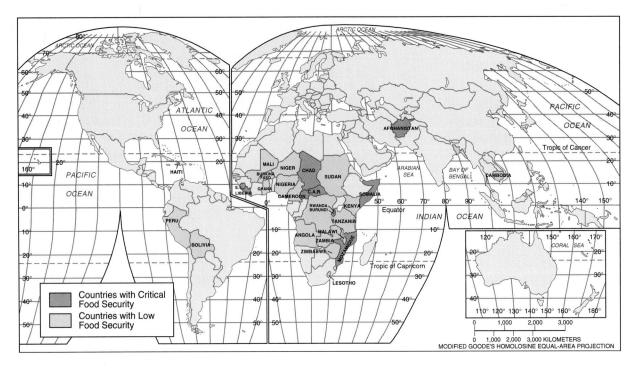

FIGURE 3.3

Developing countries with low or critical food security indexes, 1990–1992. Africa remains the continent most seriously challenged by food shortages. Fifteen countries in the region are facing critical food emergencies. Of the 27 countries with household food security problems, 22 are in sub-Saharan Africa.

◈

Table 3.1

Calorie Intake and Calorie Requirement Satisfaction

	Highest Calorie Intake			Lowest Calorie Intake		
Country	Calorie Intake per Person per Day	Percentage of Requirements	Country	Calorie Intake per Person per Day	Percentage of Requirements	Food Aid*
Ireland	4054	162	Ghana	1573	68	51
Denmark	4023	150	Chad	1620	68	23
East Germany	3787	145	Mali	1731	74	44
Belgium	3743	142	Kampuchea	1792	81	0
Bulgaria	3711	148	Uganda	1807	78	33
Yugoslavia	3642	143	Mozambique	1844	79	486
United States	3616	137	Burkina Faso	1879	79	51
Czechoslovakia	3613	146	Haiti	1903	84	136
UA Emirates	3591	N.A.	Bangladesh	1922	83	1304
Libya	3581	152	Guinea	1987	86	32

*In thousands of metric tons.
Source: World Bank, 1991.

research, will be sufficient to meet effective demand and rising world population. However, commentators such as Lester Brown are more pessimistic about future world food production. They argue that food produc-

tion will be constrained by the shrinking backlog of unused agricultural technology, the limits to the biological productivities of fisheries and rangelands, the fragility of tropical and subtropical environments, the increasing scarcity of freshwater, the declining effectiveness of additional fertilizer applications, and by social disintegration in many developing countries. The best prospects for production increases are in the temperate zone nations such as the United States, Australia, and Europe.

The success of global agriculture from 1961 to 1994 has not been shared equally. In Africa, per capita food production has not been able to keep up with population growth. By contrast, Asia and to a lesser extent Latin America have experienced tremendous success in per capita terms.

The food and hunger problem is most severe in Africa. Fifteen countries are experiencing exceptional food emergencies. Of the 27 countries with food-security problems, 22 are in Africa (Figure 3.4). Indeed, famine, the most extreme expression of poverty, is now mainly restricted to Africa. The fact that famine has been held at bay for decades in Latin America and Asia suggests that famine can be eliminated. But how? Certainly, bringing an end to Africa's wars would go a long way toward eradicating famine. But peace is not in itself a sufficient condition for removing acute hunger. Appropriate policies and investments are needed to stimulate rural economic growth that underpins food security and to provide safety-net protection for the absolute poor.

The pace of urbanization in the developing countries has also contributed to the food problem. In recent decades, millions and millions of people who previously lived in rural areas and produced some food have relocated in the urban areas, where they must buy food. As a result of urbanization, there is a higher demand for food in the face of lower supply.

MALDISTRIBUTION

The problem of food distribution has three components. First, there is the problem of transporting food from one place to another. Although transport systems in developing countries lack the speed and efficiency of those in developed countries, they are not serious impediments under normal circumstances. The problem arises either when massive quantities of food aid must be moved quickly or when the distribution of food is disrupted by political and military conflict.

Second, serious disruptions in food supply in developing countries are traceable to problems of marketing and storage. Food is sometimes hoarded by

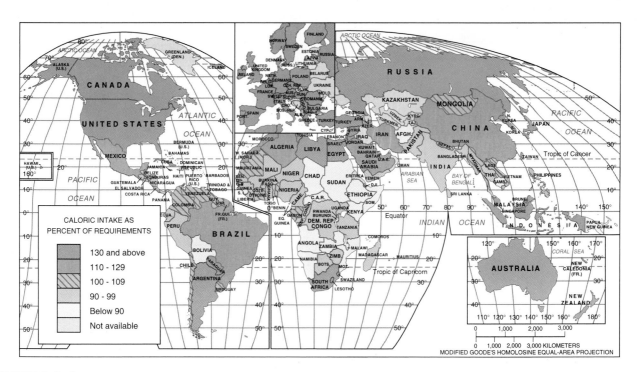

FIGURE 3.4

Caloric intake as a percentage of adult daily requirements. Highly developed regions of the world receive, on the average, 130% of the daily caloric requirements (2,400 calories per day) set by the United Nations Food and Agricultural Organization (FAO). Some countries in South America, South Asia, and many countries in Africa receive less than 90% of the daily caloric requirements needed to sustain body and life. Averages must be adjusted according to age, gender, and body size of the person and by regions of the world. Although it appears from this map that the great majority of world populations are in relatively good shape with regard to calories per capita (food supply), remember that averages, which are used in this case, tend to overshadow destitute groups in each country that receive less than their fair share. Again, the situation is most severe in the Sahel or center belt of Africa. (See Color Insert I for more illustrative map.)

merchants until prices rise and then sold for a larger profit. Also, much food in the tropics is lost due to poor storage facilities. Improvements in storage facilities would provide people, especially in rural areas, with security against disasters, including a breakdown in communications.

A third aspect of the distribution problem is in the inequitable allocation of food. Only North America, Australia, and Western Europe have large grain surpluses. But food grain is not always given when it is most needed. Food aid shipments and grain prices are related. Thus, U.S. food aid was low around 1973, a time of major famine in the Sahel region of Africa, because cereal prices were at a peak.

CIVIL UNREST
Devastating examples of depriving food to secessionist areas include the Ibadan government in Nigeria starving the Biafrans and the Addis Ababa government in

Third World farmers, such as these in Indonesia, depend on high rice yields. Rice is the staple food for more than one-half the world's population. While rice and other grains supply energy and some protein, people must supplement grains with fruits, nuts, vegetables, dairy products, fish, and meat in order to remain healthy. *(Photo: World Bank)*

Ethiopia starving the Eritreans into submission, with 6 million people dying in the process. Such conflicts divert resources from civilian use and complicate the stability of the governments, creating famine and prohibiting the flow of developmental aid (See box, The Anatomy of Starvation).

ENVIRONMENTAL DECLINE
As population pressure increases on a given land area, the need for food pushes agricultural use to the limits, and marginal lands, which are subject now to *desertification* and *deforestation*, are brought into production. Removal of trees allows a desert to advance, because the wind break is now absent. The cutting of trees also lowers the capacity of the land to absorb moisture, which diminishes agricultural productivity and increases the chances of drought. Desertification and deforestation are symptoms as well as causes of the food problem in developing countries. Natural resources are mined by the poor to meet the food needs of today; the lower productivity resulting from such practices is a concern to be put off until tomorrow.

GOVERNMENT POLICY AND DEBT
In many developing countries, government policies have emphasized investment in military equipment at the expense of increasing agricultural production. In addition, some governments in Africa have provided food at artificially low prices in order to make food affordable. However, this practice robs farmers of the incentive to farm. Farmers cannot make a living from such low commodity prices.

The average debt of many developing countries runs into the billions. In 1996, aggregate debt of African countries stood at $200 billion. Simply put, African countries have no surplus capital to invest in their infrastructure or food production systems. Instead, they have to install austerity, reducing levels of government services in support of economic growth, particularly agricultural growth.

POVERTY
The inequitable allocation of food is related to poverty, the major cause of the hunger problem. Food goes to markets that can afford it, not to where it is needed most. Where food is produced is immaterial as long as costs are minimized and a profitable sale can be made. Thus, in the midst of hunger, food is exported for profit. If Americans or Canadians are prepared to pay more for meat than many Mexicans, then it is not surprising that the market fails to include the poor.

STRUCTURE OF AGRICULTURE
Closely associated with poverty as a cause of hunger in developing countries is the structure of agriculture, including land ownership. Land is frequently concen-

trated in a few hands. In Bangladesh, less than 10% of rural households own more than 50% of the country's cultivable land; 60% of Bangladesh's rural families own less than 2%. In fact, many of them own no land at all. They are landless laborers who depend on wages for their livelihoods. But without land, there is often no food.

In recent decades, agriculture in developing countries has expanded. This expansion is in the export sector, not in the domestic food-producing sector, and it is often the result of deliberate policy. Governments and private elites have opted for modernization through the promotion of export-oriented agriculture. The result is the growth of an agricultural economy based on profitable export products and the neglect of those aspects of farming that have to do with small farmers producing food for local populations.

Increasing Food Production

There is broad agreement that yield increases will be the major source of future food production growth. Through the year 2010, the FAO optimistically estimates that increase yields will amount to about 55% of production growth in developing countries, with the expansion of arable land accounting for an additional 21% and increased cropping intensity for another 13%. The result of these methods of increasing food supply would be to put additional pressures on land and water resources and contribute significantly to anthropogenic sources of greenhouse gases.

EXPANDING THE CULTIVATED AREA
The world's potential available land for cultivation is estimated to be about twice the present cultivated area.

🔷

T a b l e 3.2
Projected Reserves of Selected Strategic Minerals

Resource	Static Index*	Exponential Index (Years)	U.S. Consumption as a Percentage World Total†	Percentage of U.S. Consumption Imported	Sources of Major Resources
Aluminum	100	31	41	97	Guinea, Australia, Brazil, Jamaica
Chromium	420	95	19	73	South Africa, Zimbabwe, Finland
Cobalt	110	60	32	95	Zaire, Zambia, Canada
Copper	36	21	33	27	Chile, United States, Zambia, Canada, former USSR
Gold	11	9	26	31	South Africa, former Soviet Union, United States
Iron	240	93	28	22	former Soviet Union, Brazil, Australia, India
Lead	26	21	25	16	United States, Australia, Canada
Manganese	97	46	14	100	former Soviet Union, South Africa, Australia
Mercury	13	13	24	57	Spain, former Soviet Union, Algeria
Molybdenum	79	34	40	0	United States, Chile, Canada
Nickel	150	53	38	68	New Caledonia, Canada, Cuba
Platinum	130	47	31	92	South Africa, former Soviet Union, Zimbabwe
Silver	16	13	26	64	United States, Canada, Mexico
Tin	17	15	24	72	Malaysia, Indonesia, Thailand, China
Tungsten	40	28	22	68	China, Canada, United States, South Korea
Zinc	23	18	26	69	Canada, United States, Australia

* Static Index refers to the number of years reserves will last to 80% depletion with consumption growing at current rates.
† Exponential Index refers to the number of years reserves will last to 80% depletion with consumption growing at 2.5% per annum.
Source: U.S. Bureau of Mines, 1997, pp. 5–6.

❖

THE ANATOMY OF

STARVATION

"Restore Hope," the U.S. mission to Somalia, was a challenge to the cessation of starvation that still plagues the African nation. What follows is a look at the anatomy of starvation and what can be done to counterbalance it in order for the body to recuperate.

SELF-CANNIBALIZATION

1. When the body does not receive adequate food and nutrition, the first thing it loses is strength.
2. The body begins to feed off itself, and much water is lost, causing the body to shrink.
3. The body's loss of water causes distended, swollen stomachs, seen in starving persons.
4. A baby can withstand only a loss of 25% of its average weight and still survive. Beyond that, death comes within 30 to 50 days as the body's organs shrink and waste away.
5. The endemic diseases of Africa—malaria, river blindness, sleeping sickness—compound the problem.
6. Self-cannibalization leads to disorientation as people lose hope. Parents may even steal food from their children as they panic and become disoriented.

BABIES AND YOUNG CHILDREN

1. Babies born to starving mothers have low survival rates and are usually underweight. The baby's survival is frequently based on the availability of breast milk, which starving mothers cannot produce. Babies that do survive may have birth defects, including brain damage.
2. With proper care after delivery, babies born to starving mothers can become healthy and live without permanent brain damage.

Some of these African mothers can do little to save their children's lives. Children are suffering the complications of undernutrition and malnutrition. World per capita food production figures from 1970 to 1997 are disappointing. Asia has managed to increase production fast enough to stay well ahead of population growth, but Latin America and the near East, including Southwest Asia, the Middle East, and North Africa, have barely managed to stay even with the population. The worst report is from Africa, where population growth is outstripping food production year by year. Per capita food production in Africa has been falling steadily for the last 25 years. Worse declines have come in Angola, Botswana, Gabon, Mozambique, Rwanda, Somalia, Ethiopia, and Sudan. Many factors contribute to these fluctuations, including drought, changing world prices, and civil and ethnic unrest. Hundreds of thousands of tribal members have been killed and even more made homeless recently in the countries of Rwanda and South Africa. *(Photo: Andrew Holbrooke, The Stock Market)*

3. For older children, months of starvation can be overcome, but it is common for children deprived of food in their early years to have brain damage and chronic disease.

RESTORING HOPE AND HEALTH

1. In the refugee camps, the worst cases are children and adults whose intestines have withered away. They cannot digest food; therefore, they must be fed intravenously.
2. An IV supplying carbohydrates is used immediately. The first organ to be resuscitated is the brain.
3. The next step is to provide patients with a tiny amount of liquid to be taken orally. The liquid is a solution of water, sugar, and salt.
4. After several days, patients are fed a nutritious soup, comprised of ground maize, sunflower oil, sugar, vegetable oil, and bean flour.
5. Weeks later, milk and solid foods are administered.

Vast reserves are theoretically available in Africa, South America, and Australia, and smaller reserves in North America and in the former Soviet Union. However, many experts believe that the potential for expanding cropland is disappearing in most regions because of environmental costs and the cost of developing infrastructure in remote areas.

About half of the world's potentially arable land lies within the tropics, especially in sub-Saharan Africa and Latin America. At least 45% of this land is under forest in protected areas, and about 72% of the potential cropland suffers from soil and terrain constraints as well as excessive dryness. In Asia, two-thirds of the potentially arable land is already under cultivation; the main exceptions are Indonesia and Myanmar. South Asia's agricultural land is almost totally developed.

The expansion of tropical agriculture into forest and desert environments contributes to deforestation and desertification. Since World War II, roughly half of the world's rain forests in Africa, Asia, and Latin America have disappeared. Conversion of this land to agriculture has entailed high costs, including the loss of livelihoods for the people displaced, the loss of biodiversity, increased carbon dioxide emissions, and decreased carbon storage capacity. *Desertification*—the growth of human-made deserts—threatens about one-third of the world's land surface and the livelihood of nearly a billion people. Many of the world's major rangelands are at risk (Figure 3.5). The main factor responsible for desertification is overgrazing, but deforestation (particularly the cutting of fuelwood), overcultivation of marginal soils, and salinization caused by poorly managed irrigation systems are also important influences. Deforestation and desertification are destroying the land resources on which the development of the developing countries depends.

RAISING THE PRODUCTIVITY OF EXISTING CROPLAND

The quickest way to increase food supply is to raise the productivity of land under cultivation. Remarkable increases in agricultural yields have been achieved in developed countries through the widespread adoption of new technologies. Corn yields in the United States are a good example. Yields expanded rapidly with the introduction of hybrid varieties, herbicides, and fertilizers. Much of the increase in yields came through successive improvements in hybrids.

The approach for increasing yields in developed countries has been adopted in developing countries. This technology-package approach is known as the *Green Revolution*, in which new high-yielding varieties of wheat, rice, and corn are developed through plant genetics.

The Green Revolution is a major scientific achievement, but it is not a panacea. It depends on new seeds. It depends on chemical fertilizers, pesticides, and herbicides, which have contaminated underground water supplies as well as streams and lakes. It depends on large-scale, one-crop farming, which is ecologically unstable because of its susceptibility to pestilence. It depends on controlled water supplies, which have increased the incidence of malaria, schistosomiasis, and other diseases. It is restricted primarily to wheat, rice, and corn hybrids that are low-grade protein foods. It is confined mainly to a group of 18 heavily populated countries, extending across the subtropics from South Korea in the east to Mexico in the west (Figure 3.6). It is also benefiting countries that include 18% of the world's land surface and that are home to 56% of the world's population.

Politically, the Green Revolution promises more than it can deliver. Its sociopolitical application has been largely unsatisfactory. Even in areas where the Green Revolution has been technologically successful, it has not always benefitted large numbers of hungry

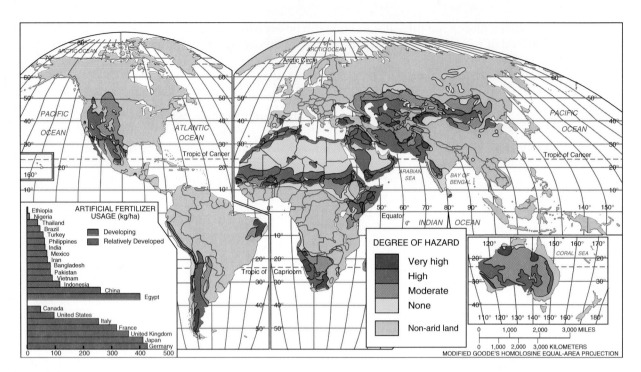

FIGURE 3.5

World desertification. The main problem is overuse by farmers and herdsmen. Approximately 9% of the earth surface has lost its topsoil due to overuse of lands by humans, creating desertification. An additional 25% of the earth's surface is now threatened. Topsoil is being lost at a rate of approximately 30 billion metric tons a year. Approximately 20 million acres of agricultural land are wasted every year to desertification by agricultural overuse. Plants have a fibrous root system, which holds the soil in place. When plants are uprooted by overplowing or by animals, the plants that stabilized shifting soil are removed. When the rains come, water erosion can wash away the remaining topsoil. Sand particles, which are heavier than the humus, remain behind and start to form dunes.

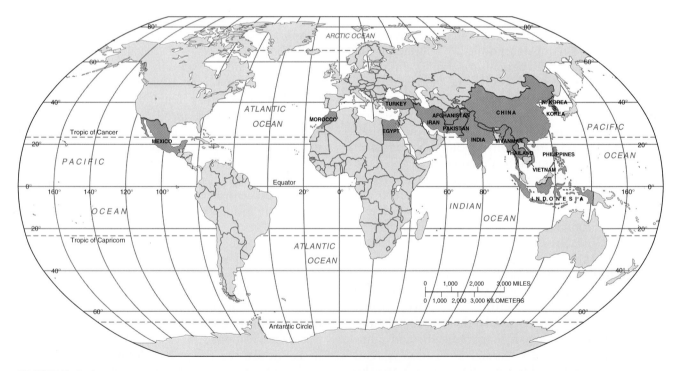

FIGURE 3.6

The chief benefitting countries of the Green Revolution. Countries indicated in blue represent these countries. The Green Revolution was the result of plant scientists genetically developing high yielding varieties of staple food crops such as rice in East Asia, wheat in the Middle East, and corn in Middle America. By crossing "super strains" that produced high yields with more genetically diverse plants, both high yield and pest resistance were introduced.

people without the means to buy the newly produced food. It has benefitted mainly Western-educated farmers, who were already wealthy enough to adopt a complex integrated package of technical inputs and management practices. Farmers make bigger profits from the Green Revolution when they purchase additional land and mechanize their operations. Some effects of labor-displacing machinery and the purchase of additional land by rich farmers include agricultural unemployment, increased landlessness, rural-to-urban migration, and increased malnutrition for the unemployed who are unable to purchase the food produced by the Green Revolution.

The Green Revolution is not winning its battle against hunger because it focuses on food production. The world food problem is not so much one of food production, but of food demand in the economic sense. Unfortunately, the Green Revolution does nothing to increase the ability of the poor to buy food.

The Green Revolution has helped to create a world of more and larger commercial farms alongside fewer and smaller peasant plots. However, given a different structure of land holdings and the use of appropriately intermediate technology, the Green Revolution could help developing countries on the road toward agricultural self-sufficiency and the elimination of hunger. Intermediate technology is a term that means low-cost, small-scale technologies "intermediate" between primitive stick-farming methods and complex agroindustrial technical packages.

Creating New Food Sources

Expanding cultivated areas and raising the productivity of existing cropland are two methods of increasing food supply. A third method is the identification of new food sources. There are three main ways to create new food sources: (1) cultivate the oceans, or mariculture; (2) develop high-protein cereal crops; and (3) increase the acceptability and palatability of inefficiently used present foods.

CULTIVATING THE OCEANS
Fishing and the cultivation of fish and shellfish from the oceans is not a new idea. At first glance, the world seems well supplied with fisheries because oceans cover three-fourths of the earth. However, fish provide a small proportion—about 1%—of the world's food supply.

Between 1950 and 1987, consumption of fish from the oceans increased at a more rapid rate than the growth of population and even exceeded beef as a source of animal protein in some countries. However, since 1987, fish caught by commercial fishing fleets leveled off and declined as a result of overfishing (Figure 3.7). Overfishing has been particularly acute in the North Atlantic and Pacific Oceans.

Countries such as Iceland and Peru, whose economies rely heavily on fishing, are sensitive to the overfishing problem. Between 1970 and 1975, Peru's catch of its principal fish, the anchovy, declined by over 75% because of overfishing. In order to prevent continued overfishing, the Peruvian government nationalized its fish-meal production industry. The Peruvian experience demonstrates that the ocean is not a limitless fish resource, as did the quest for blue whales a century earlier.

Commercial fishing fleets employ sophisticated techniques but catch what nature has provided, much like hunters and gatherers. An alternative approach is to follow the example of animal husbandry by devising methods for commercial fish farming. *Mariculture*, or fish farming, is now expanding rapidly and accounts for 5% of the world's fish caught yearly. The cultivation of high-quality food fish such as trout and salmon is big business in Norway, Japan, and other fishing countries.

HIGH-PROTEIN CEREALS
Another source of future food production rests in higher protein cereal crops. Scientists are experimenting with various high-yield, high-protein cereal crops in the hope that development of hybrid seeds will be able to help the protein deficiency of people in developing countries who do not have available meats from which to gain their protein needs, as do people in developed countries.

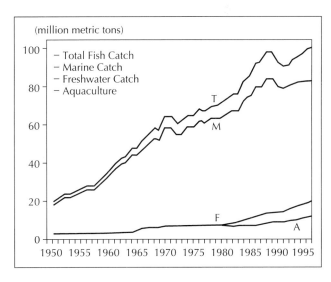

FIGURE 3.7
The global fish catch has been increasing over the last 20 years. Fish, crustaceans, and mollusks reached 100 million metric tons in 1989 and then decreased to 85 million tons in 1990, according to figures from the Food and Agricultural Organization (FAO) of the United Nations. Reduced quotas for some North Atlantic fish species, due to conservation techniques, were partially responsible for the 1990 decline. (*Source: World Resources Institute, 1993, p. 179*)

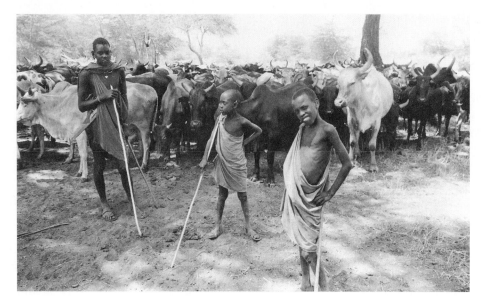

The Kenyan rangelands on which these herders' cattle graze are in jeopardy. With growing grazing pressures, more than 60% of the world's rangelands and at least 80% of African, Asian, and Middle Eastern rangelands are now moderately to severely desertified. About 65 million hectares of once productive land in Africa have become desert during the last 50 years. *(Photo: World Bank)*

Fortification of present rice, wheat, barley, and other cereals with minerals, vitamins, and protein-carrying amino acids is an approach that also deserves attention. This approach is based on the fortified food production in developed countries and stands a greater chance of cultural acceptance because individual food habits do not necessarily need to be altered. But developing countries rely on unprocessed, unfortified foods for 95% of their food intake. Large-scale fortification and processing would require major technological innovation and scale economies to produce enough food to have an impact on impoverished societies.

MORE EFFICIENT USE OF FOODS

In many developing countries, foods to satisfy consumer preferences as well as religious taboos and cultural values are becoming limited. The selection of foods based on social customs should be supplemented with information concerning more efficient use of foods presently available. An effort should be made to increase the palatability of existing foods that are plentiful.

Fish meal is a good example. Presently, one-third of the world's fish intake is turned into fodder for animals and fertilizer. Fish meal is rich in Omega 3 fatty acids and amino acids necessary for biological development. However, in many places, the fish meal is not used because of its taste and texture.

Another underused food resource is the soybean, a legume rich in both protein and amino acids. Most of the world's soybeans wind up being processed into animal feed or fertilizer and into production of nondigestible industrial materials. In addition, world demand for tofu and other recognizable soybean derivatives is not large. By contrast, hamburgers, hot dogs, soft drinks, and cooking oils made partially from soybeans are more acceptable. New, widely acceptable food products need to be created from the soybean in developing countries, not only because of its nutritional value but also because it will supplement dwindling supplies of more attractive foods.

A Solution to the World Food Supply Situation

As we have emphasized, there is a widely shared belief that people are hungry because of insufficient food production. But the fact is that food production is increasing faster than population, and still there are more hungry people than ever before (Figure 3.8). Why should this be so? It could be that the production focus is correct, but soaring numbers of people simply overrun these production gains. Or it could be that the diagnosis is incorrect—scarcity is not the cause of hunger, and production increase, no matter how great, can never solve the problem.

The simple facts of world grain production make it clear that the overpopulation/scarcity diagnosis is incorrect. Present world grain production can more than adequately feed every person on earth.

Ironically, the focus on increased production has compounded the problem of hunger by transforming agricultural progress into a narrow technical pursuit instead of the sweeping social task of releasing vast, untapped human resources. We need to look to the policies of governments in developing countries to understand why people are hungry even when there is enough food to feed everyone. These policies influence the access to knowledge and the availability of credit to small farmers, the profitability of growing enough to sell a surplus, and the efficiency of marketing and distributing food on a broad sale.

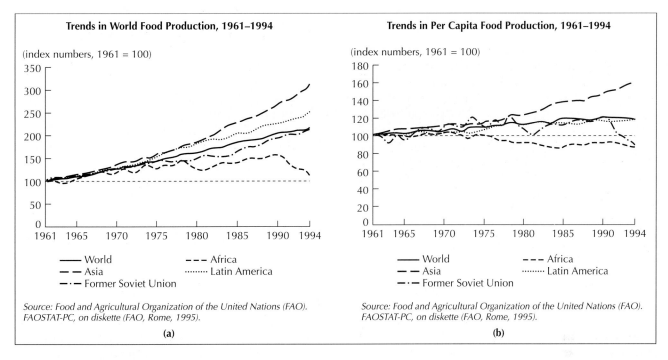

Trends in World Food Production, 1961–1994

(index numbers, 1961 = 100)

Source: Food and Agricultural Organization of the United Nations (FAO).
FAOSTAT-PC, on diskette (FAO, Rome, 1995).

(a)

Trends in Per Capita Food Production, 1961–1994

(index numbers, 1961 = 100)

Source: Food and Agricultural Organization of the United Nations (FAO).
FAOSTAT-PC, on diskette (FAO, Rome, 1995).

(b)

FIGURE 3.8
(a) Trends in world food production, 1961–1994. The precipitous drop in food production in the former Soviet Union.
(b) Trends in per capita food production, 1961–1994. Both the former Soviet Union and Africa show declining per capita food production.

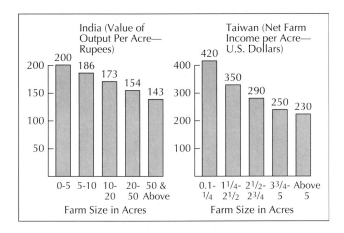

FIGURE 3.9
Small farm efficiency in India and Taiwan. Small, carefully formed plots are more productive per unit area than large estates throughout the world because they use fewer costly inputs. Each unit of land is more intensively worked, and there is a higher density of laborers per acre.

The fact is that small, carefully farmed plots are more productive per unit area than large estates because they use fewer costly inputs (Figure 3.9). Yet, despite considerable evidence from around the world, government production programs in many developing countries ignore small farmers. They rationalize that working with bigger production units is a faster road to increased productivity. Often, many small farms is the answer. In the closing years of the 20th century, many agricultural researchers, having gained respect for traditional farming systems, agree with this conclusion.

NONRENEWABLE MINERAL RESOURCES

Although we can increase world food output, we cannot increase the supply of minerals. A mineral deposit, once used, is gone forever. The term *mineral* refers to a naturally occurring inorganic substance in the earth's crust. Thus, silicon is a mineral whereas petroleum is not, since the latter is of organic origin. Although minerals abound in nature, many of them are insufficiently concentrated to be economically recoverable. Moreover, the richest deposits are unevenly distributed and are being depleted.

Except for iron, nonmetallic elements are consumed at much greater rates than elements used for their metallic properties. Industrial societies do not worry about the supply of most nonmetallic minerals, which are plentiful and often widespread. There is no foreseeable world shortage of nitrogen, phosphorus, potash, or sulfur for chemical fertilizer[3] or of sand, gravel, clay, or dimension stone for building purposes. Those commodities the industrial and industrializing countries do worry about are the metals—the raw materials of economic power.

Location and Projected Reserves of Key Minerals

Only five countries—Australia, Canada, South Africa, the United States, and the former Soviet Union—are significant producers of at least six strategic minerals vital to defense and modern technology (Figure 3.10). A larger number of mainly developing countries are major producers of between one and six minerals required by modern industry. Of the major mineral-producing countries, only a few—notably the United States and the former Soviet Union—are also major processors and consumers. The other major processing and consuming centers—Japan and western European countries—are deficient in strategic minerals. Compared to the former Soviet Union, the United States lacks several important metallic ores, including chromium and manganese (Table 3.2).

How good is the world supply of strategic minerals? Our knowledge of world mineral reserves is summarized by the U.S. Bureau of Mines. The table indicates the number of years 16 strategic minerals will last under two assumptions: (1) the number of years reserves will last with consumption growing at current rates (Column 2), and (2) the number of years reserves will last with consumption continuing to grow exponentially (Column 3). Data in Column 3 are more realistic than those in Column 2; they indicate that most of the key minerals will be exhausted within 100 years and some will be depleted within a few years. Column 4 shows U.S. consumption as a percentage of world total. Except for molybdenum, an alloying element, domestic production is insufficient to cover production (Column 5).

The United States is running short of domestic sources of strategic minerals. Its dependency on imports has grown steadily since 1950; prior to that year, the country was dependent on imports for only four designated strategic minerals. If measured in terms of percentage imported, U.S. dependency increased from an average of over 50% in 1960 to over 80% in 1995.

Minerals projected as future needs by the United States are unevenly distributed around the world. Many of them, such as manganese, nickel, bauxite, copper, and tin (Figure 3.11) are concentrated in the former Soviet Union and Canada and in developing countries. Whether these critical substances will be available for U.S. consumption may depend less on economic scarcity and more on international tensions and foreign policy objectives.

Is the U.S. technological society gravely threatened by the degree to which domestic supplies must be supplemented by importing them from other countries? The need for a country to import a particular material does not mean that the material is unavailable at home. We have merely scratched the solid crust of the earth for the materials we need. Given an economical source of ultraterrestrial energy (solar radiation) and even more ingenious methods of extraction, who can say how much our

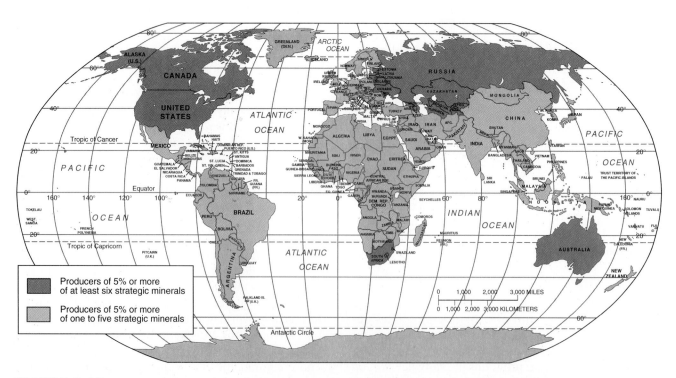

FIGURE 3.10
Major producers of strategic minerals. *(Source: Based on data from U.S. Bureau of Mines, 1986)*

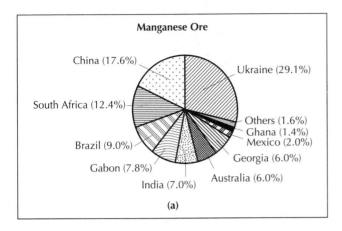

Manganese Ore

China (17.6%)
Ukraine (29.1%)
South Africa (12.4%)
Others (1.6%)
Ghana (1.4%)
Mexico (2.0%)
Brazil (9.0%)
Georgia (6.0%)
Gabon (7.8%)
Australia (6.0%)
India (7.0%)

(a)

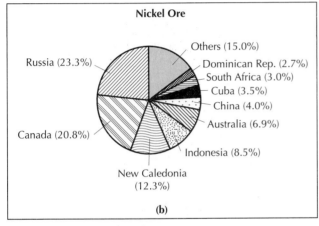

Nickel Ore

Russia (23.3%)
Others (15.0%)
Dominican Rep. (2.7%)
South Africa (3.0%)
Cuba (3.5%)
China (4.0%)
Australia (6.9%)
Canada (20.8%)
Indonesia (8.5%)
New Caledonia (12.3%)

(b)

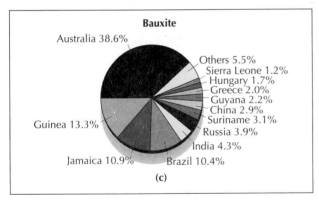

Bauxite

Australia 38.6%
Others 5.5%
Sierra Leone 1.2%
Hungary 1.7%
Greece 2.0%
Guyana 2.2%
China 2.9%
Suriname 3.1%
Russia 3.9%
India 4.3%
Guinea 13.3%
Brazil 10.4%
Jamaica 10.9%

(c)

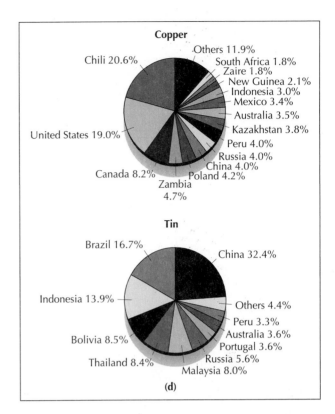

Copper

Chili 20.6%
Others 11.9%
South Africa 1.8%
Zaire 1.8%
New Guinea 2.1%
Indonesia 3.0%
Mexico 3.4%
United States 19.0%
Australia 3.5%
Kazakhstan 3.8%
Peru 4.0%
Russia 4.0%
China 4.0%
Canada 8.2%
Poland 4.2%
Zambia 4.7%

Tin

Brazil 16.7%
China 32.4%
Indonesia 13.9%
Others 4.4%
Peru 3.3%
Australia 3.6%
Bolivia 8.5%
Portugal 3.6%
Russia 5.6%
Thailand 8.4%
Malaysia 8.0%

(d)

FIGURE 3.11

(a) Manganese ore. (b) Nickel ore. Both of these ores are important to the steel industry as alloying metals. The former Soviet Union countries of the Ukraine and Georgia produce 34% of the world's supply of manganese ore, while China produces another 18%. South Africa, Brazil, Gabon, and India each produce a sizable share. Russia produces the largest proportion of nickel ore, followed by Canada, New Caledonia, Indonesia, and Australia. (Source: U.S. Bureau of Minerals Yearbook, 1993.) (c) World production of bauxite (aluminum ore), 1992. Australia produces almost 40% of the world's supply of bauxite. Guinea in West Africa produces another 13%, while Jamaica provides the largest source of bauxite in the Western Hemisphere. Most of the bauxite from Jamaica is exported to North America. (d) World production by country of copper and tin, 1992. The Americas are well endowed with copper, providing almost 50% of the world's supply. Copper is North America's most plentiful mineral proportional to its use. China produces almost a quarter of the world's supply of tin, while another quarter comes from South America and 30% is produced by three Southeast Asian countries.

mineral needs could eventually be obtained domestically? The environmental cost of "moving mountains" to win these commodities would be another matter, however.

Solutions to the Mineral Supply Problem

Affluent countries are unlikely to be easily defeated by mineral supply problems. Human beings, the ultimate resource, have developed solutions to the problem in the past. Will they in the future? Although past experience is never a reliable guide to the future, there is

no need to be unduly pessimistic about the exhaustion of minerals as long as we use our brainpower to develop alternatives.

If abundant supplies of cheap electricity ever became available, it might become possible to extract and process minerals from unorthodox sources such as the ocean. The oceans, which cover nearly three-quarters of the earth, contain large quantities of dissolved minerals. Salt, magnesium, sulphur, calcium, and potassium are the most abundant of these minerals and amount to over 99% of the dissolved minerals. More valuable minerals that are also present include copper, zinc, tin, and silver. Some minerals such as bromine and magnesium are being obtained electrolytically from the oceans at the present time.

Much more feasible than mining the oceans is devoting increased attention to improving mining technology, especially to reducing waste in the extraction and processing of minerals. Equally feasible is to utilize technologies that allow minerals to be used more efficiently in manufacturing. Also, if social attitudes were to change, encouraging lower per capita levels of resource use, more durable products could be manufactured, saving not only large amounts of energy but large quantities of minerals too.

Reusing minerals is still another option to our mineral problems. Every year in the United States and other affluent countries, huge quantities of household and industrial waste are disposed of at sanitary landfills and open dumps. These materials are sometimes called "urban ores" because they can be recovered and used again. For years, developed countries have been recycling scarce and valuable metals such as iron, lead, copper, antimony, silver, gold, and platinum, but large amounts of scrap metals are still being wasted. Although we could recover a much greater proportion of scrap, this is unlikely when prices are low or when virgin materials are cheaper than recycled ones.

None of these options will solve our mineral problems completely. Probably the best solution to our mineral problems will come about by substitution or *transmaterialization*. Minerals used in the manufacturing process are periodically replaced by new ones. This idea is illustrated in Figure 3.12, which shows material demand changes for selected minerals over a span of 50 years. Without doubt, one of the most significant developments in our quest to solve the problem of dwindling metallic elements is the trend toward replacing them with advanced engineering nonmetallic materials such as composites—products made of such materials as graphite or glass embedded in plastic—or ceramics—products made of common materials such as clay and silicon.

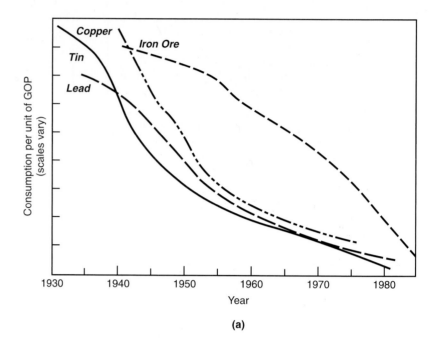

(a)

FIGURE 3.12

(a) The consumption of lead, tin, copper, and iron ore per unit of GDP for the United States, 1930–1985. *Transmaterialization* is the process whereby natural materials from the environment are systematically replaced by higher quality or technologically more advanced materials linked to new industries—glass fibers, composites, ceramics, epoxies, and smart metals. (b) Transmaterialization of asbestos, chromium, and manganese per unit of GDP. (c) Transmaterialization to increase consumption of aluminum, platinum, rare earths, and titanium per unit of GDP. *(Source: Adapted from L. M. Waddell and W. C. Labys, Transmaterialization: Technology and Materials' Demand Cycles, Morgantown: West Virginia University, 1988)*

Environmental Impact of Mineral Extraction

Mineral extraction has a varied impact on the environment, depending on mining procedures, local hydrological conditions, and the size of the operation. Environmental impact also depends on the stage of development of the mineral—exploration activities usually have less of an impact than mining and processing mineral resources.

Minimizing the environmental impact of mineral extraction is in everyone's best interest, but the task is difficult because demand for minerals continues to grow and ever-poorer grades of ore are mined. For example, in 1900 the average grade of copper ore mined was 4% copper. By 1973, ores containing as little as 0.53% copper were mined. Each year more and more rock has to be excavated, crushed, and processed to extract copper. In fact, the immense copper mining pits in Montana, Utah, and Arizona are no longer in use because foreign sources, mostly in the developing countries, are less expensive.

Open-pit mines and quarries amount to a small fraction of the total area of the United States. In gen-

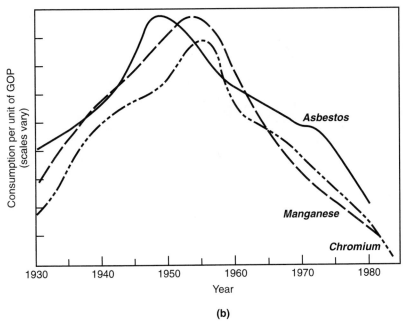

(b)

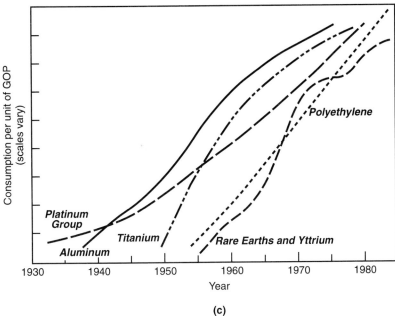

(c)

eral, their impact on the environment is local, but as long as the future remains technological and materialistic, the demand for minerals is going to increase. Lower and lower quality minerals will have to be used and, even with good engineering, environmental degradation will extend far beyond excavation and surface plant areas.

Energy

The development of energy sources was crucial for economic development (Figure 3.13). Today, commercial energy, which accounts for over 80% of all en-

ergy use, is the lifeblood of modern economies. Indeed, it is the biggest single item in international trade. Oil alone accounts for about one-quarter of the volume of world trade.

Until the energy shocks of the 1970s, commercial energy demands were widely thought to be related to population growth and rising affluence. Suddenly, higher prices brought energy demands in the industrial countries to a virtual standstill. The increase in oil prices even surprised energy analysts. It signaled a change in the control of the oil market from the international oil companies headquartered in the United States to the capitals of the producing countries.

There have been many attempts to account for the transition of power within the international oil market. Was it the international oil companies's scramble for higher profits? Was it a result of the U.S. government's effort to preserve American leadership in the international oil market? Or was it greed on the part of the oil sheiks? Whatever the answer, Americans came face to face with the "energy crisis," especially during the winter of 1976–1977. A combination of below normal temperatures and a shortage of fuel was devastating. Thousands of factories were cut back to "plant protection" levels and had to shut down, and more than 3 million workers were laid off. Americans appreciated that this was not "just another crisis." They learned firsthand that when energy fails, everything fails in an urban-industrial economy.

During the 1980s, energy demand forecasts were consistently excessive, resulting in excess capacity in energy industries. Oil prices decreased from $30 in 1981 to $15 in 1987, shaking predictions made in 1980 that costs would soar to more than $40 per barrel. OPEC, once considered an invincible cartel, lost oil sales between 1980 and 1987. Its share of world oil output dropped from 57% in 1975 to 30% in 1985 as non-OPEC countries expanded production. Many developing countries, strapped by heavy energy debts, were relieved to see prices falling. Oil-exporting developing countries, such as Mexico, Venezuela, and Nigeria, which came to depend on oil revenues for an important source of income, were hurt the worst.

Slaves   Domestic animals   Windmills/water wheels   Steamship

FIGURE 3.13
Throughout human history the advance of technological civilization has been tied to the development of energy resources.

Although prices rose to $25 per barrel during the Persian Gulf War of 1991 due to reduced production and a nervous world market, prices fell to $15 per barrel again by 1994 due to world recession and reduced demand. To be sure, there is no world oil crisis now, but what of the next 20 to 40 years? Will there be ways to keep oil expensive around the year 2015 (Figure 3.14)?

Energy Production and Consumption

Most commercial energy produced is from nonrenewable resources, and most renewable energy sources, particularly wood and charcoal, are used directly by producers, mainly poor people in the developing countries. Although there is increasing interest in renewable energy development, commercial energy is the core of energy use at the present time. Only a handful of countries produce several times more commercial energy than they consume. If we take petroleum consumption and production as an example, the main

energy surplus countries include Saudi Arabia, Mexico, Iran, Venezuela, Indonesia, Algeria, Kuwait, Iraq, Libya, Qatar, Nigeria, and the United Arab Emirates. Saudi Arabia is by far the largest producer of petroleum and has the largest proven reserves. Nearly one-half of all African countries are energy paupers. Several of the world's leading industrial powers—most notably Japan, many western European countries, and the United States—consume more energy than they produce (Figure 3.15).

The United States leads the world in total energy use, but leaders in per capita terms also include Bahrain, Qatar, and the United Arab Emirates, Canada, Norway, Sweden, Japan, Australia, and New Zealand (Figure 3.16). With around 5% of the world's population, the United States consumes roughly one-quarter of the world's energy. By contrast, developing countries consume about 30% of the world's energy but they contain about 80% of the population. Thus there exists a striking relationship between per capita energy consumption and level of development (Figure 3.16).

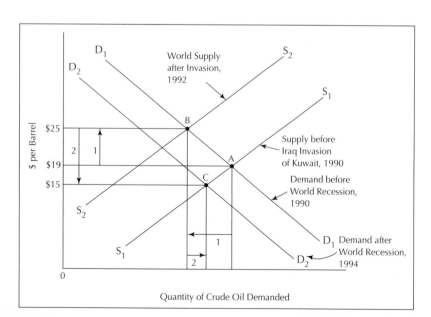

FIGURE 3.14
The Kuwait invasion by Iraq in 1990 had the immediate effect of knocking out world supply of oil by Kuwait and Iraq. With these supplies absent in the world flow of oil, as shown by S2 above, the supply of crude oil demanded was reduced and prices rose (shift 1, from equilibrium point A to point B). Prices also rose in expectation of higher oil prices, even before a reduced supply hit the refineries. The oil price per barrel, which hit $35 during the Arab Oil Embargo of 1973 and 1978 and $25 a barrel during the Kuwait invasion, has dropped back down to $15 as of January 1994 due to world recession and low levels of demand (shift 2, from equilibrium point B to point C).

The map of world petroleum trade also reinforces the image of the developed world as the all-consuming energy sink. Most developing countries receive a meager energy ration, well below levels consistent even with moderate levels of economic development. Commercial energy consumption in developed countries has been at consistently high levels, whereas in developing countries it has been at low but increasing levels (Figure 3.18).

Oil Dependency

Because Americans were seriously affected by the 1973 Arab oil embargo, and because imported oil as a proportion of total demand increased from 11% in the late 1960s to 50% in the mid-1970s, political leaders called for a national policy of oil self-sufficiency to end U.S. dependency on uncertain suppliers of petroleum. For example, the Reagan administration in the mid-1980s proposed the development of energy sources on most federal land and off-shore sites, filled a 500-million-gallon strategic oil reserve, and encouraged the development of nuclear power. But at the same time, air and water pollution regulations were relaxed, tax credits for home energy conservation expenditures were ended, a bill to compel firms to build energy-efficient appliances was vetoed, and fuel-economy standards for new cars were delayed, then passed. Following the Persian Gulf crisis in 1990, the

FIGURE 3.15
The production and consumption of crude oil by major world regions in millions of metric tons, 1995. The developed market economies of Europe, North America, and Asia, especially Japan, consume a far greater proportion of energy resources than they produce. Conversely, the Middle East, especially the Persian Gulf region, produces the most crude oil of any world regions but consumes only one-sixth of its production. Latin America, Africa, and the former Soviet Union consume less crude oil than they produce. *(Source: British Petroleum, 1995)*

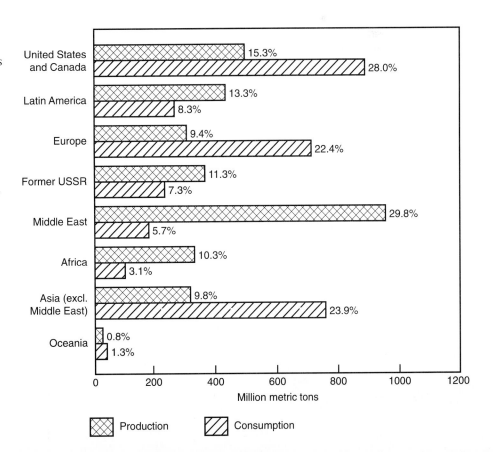

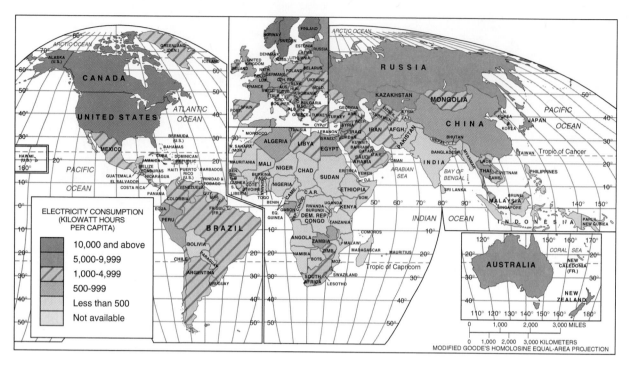

FIGURE 3.16

World per capita electricity consumption. The United States, Canada, and the Scandinavian countries consume more electricity per capita than any other countries. When the electricity usage of the United States, Canada, Europe, and Russia is combined, 75% of electricity usage in the world is accounted for, but only 20% of the people. By comparing this map to the map of crude petroleum proven reserves, the deficit areas of the world such as Europe and Japan, which have high energy needs but low fossil fuel resources, can be seen. In addition, there are areas in the world with high fossil fuel resources but low energy needs, such as the Middle East countries, Mexico, Venezuela, Indonesia, Argentina, Algeria, Nigeria, and China. (See Color Insert I for more illustrative map.)

U.S. Congress imposed stricter fuel standards on cars manufactured in future years.

These conflicting policies did not help to end U.S. dependency on foreign oil. In fact, they downgraded federal efforts to encourage American households and companies to conserve fossil fuels. High-priced oil from OPEC temporarily reduced U.S. consumption of oil (by 39% between 1973 and 1986), but in the mid 1980s, dependence on cheap imported oil began to rise again in response to lower prices.

Nonetheless, U.S. industry did become more energy efficient in the 1980s. Between 1983 and 1990, industries reduced their share of total U.S. energy consumption from 40% to 36%. In terms of conservation efforts, however, the United States lags behind

A Shell/Esso production platform in Britain's North Sea gas field. British oil exploration was stimulated by a dramatic increase in the price of oil in the mid-1970s and early 1980s, as well as by a recovery of oil prices in the late 1980s. Britain's North Sea oil and gas investment may keep the country self-sufficient until the end of the 20th century. *(Photo: Shell Oil Company)*

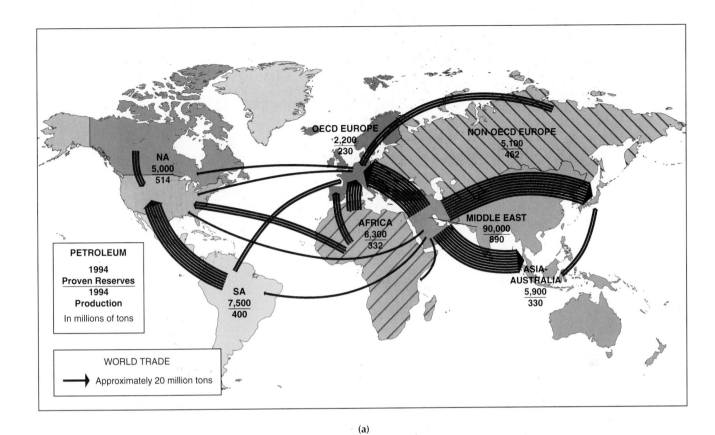

(a)

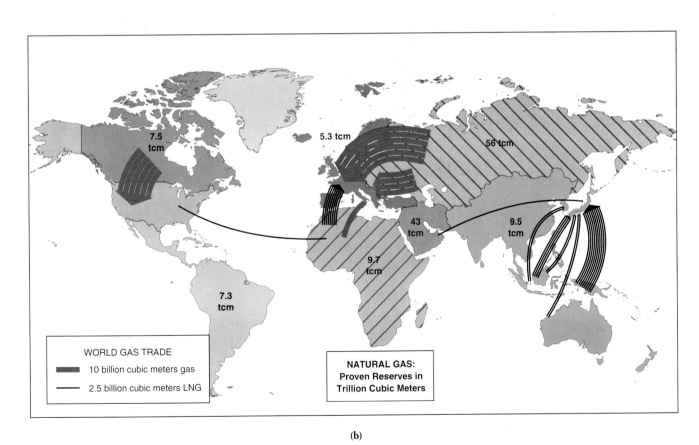

(b)

FIGURE 3.17

(a) World oil production and reserves at year's end, 1994. (b) World natural gas production and reserves at year's end, 1994.

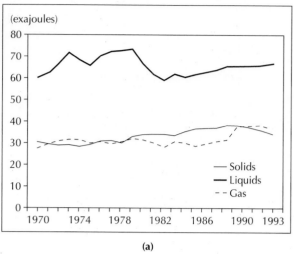

(a)

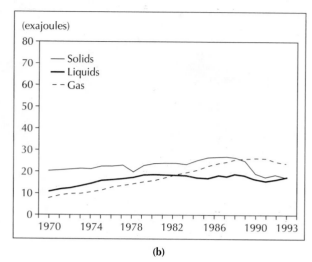

(b)

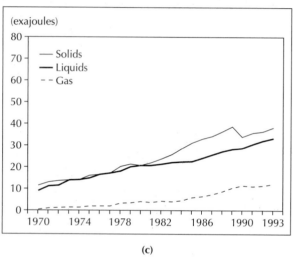

(c)

FIGURE 3.18

Differences in commercial energy consumption, 1970–1993. (a) OECD (developed) Countries; (b) Eastern Europe and the Former Soviet Union; (c) All developing countries. *(Source:* 1993 Energy Statistics Yearbook, *United Nations Statistical Division, New York, 1995)*

Japan and western Europe, where energy is more expensive. It also lags behind the former Soviet Union, whose per capita consumption is greater than Europe's despite a lower standard of living.

Oil dependence varies widely around the world, from virtually 100% in some African countries—for example, Sierra Leone—to 35% in Canada and 17% in China. The United States imports about 40% of the oil it consumes, but only a small proportion comes from the Middle East. Japan, Italy, and France are much more dependent on Persian Gulf oil.

Production of Fossil Fuels

The Arab oil embargo stimulated fossil fuel production in the United States and throughout the world. In the United States, coal production increased in the mid and late 1970s in response to the increase in the price and decrease in the supply of OPEC oil. In the 1980s, the apparent decline of the OPEC cartel and the entry of important new producers and others, such as Saudi

Arabia, to put oil on the world market forced a reduction in U.S. oil production (Figure 3.19). Had it not been for the war between Iran and Iraq and, later, the impact of the Persian Gulf conflict, the drop in oil prices might have had more drastic consequences on U.S. production and on the economies of the oil-patch states of Texas, Oklahoma, and Louisiana.

The embargo made the United States and other developed countries aware of their dependency on imported oil and on the world distribution of fossil fuel reserves. The United States is richly endowed with coal but has only modest reserves of oil and natural gas. Over 65% of the world's oil resources are located in the Middle East. This figure is much higher than it was 10 years ago because of a reevaluation of the region's oil reserves. Other large reserves are found in Latin America, primarily Mexico and Venezuela (13.5%), and in the producing countries of the former Soviet Union and Africa (Figure 3.20). Natural gas, often a substitute for oil, is also unevenly distributed, with nearly 40% in the former Soviet Union and 32% in the Middle East (Figure 3.21).

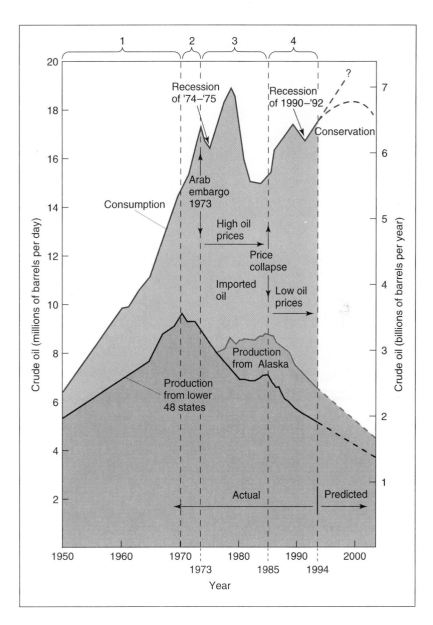

FIGURE 3.19
Oil production and consumption in the United States, 1950–2010. Four stages are observable: (1) Up to 1970, discovery of new reserves allowed production to match increasing consumption. (2) From 1970 to 1973, lack of new oil discoveries caused production to decline while consumption continued to increase, causing great oil imports and triggering the oil crisis of 1973. (3) In the late 1970s and early 1980s, high prices created increased production and lower consumption; also, the Alaskan oil fields came into production. (4) From 1980 onward, oil prices declined sharply due to the collapse of OPEC. *(Data from:* Statistical Abstracts, *U.S. Department of Commerce, 1996)*

Future Energy Policy

To ensure that an energy crisis does not recur, the United States must carefully consider energy sources and their end uses, paying special attention to reducing the amount of industrial and home energy inefficiency (Figure 3.22). It should also embrace a firm energy policy that conserves remaining oil and gas while renewable forms of energy are being developed, and that improves environmental quality. In addition, recurrent crises can also be prevented if the United States takes full account of world energy considerations. Energy is a worldwide problem that cannot be solved by confrontation among groups of countries.

Who was to blame for the energy crisis of the 1970s? The big oil companies were at least partially responsible in their maneuvers to restrict production and imports in the manner of monopolies, oligopolies, and global cartels. The large, often excessive profits enjoyed by the oil companies were obtained at high cost to Americans, who saw no appreciable improvement in national energy self-sufficiency.

Who suffered from the energy crisis? Certainly most people in developed countries (especially those on fixed incomes), many small firms, and even independent oil companies. Non–oil-producing developing countries such as Jamaica, Guyana, Mali, and Zambia were hurt much worse. They could not afford to spend foreign exchange on expensive oil imports.

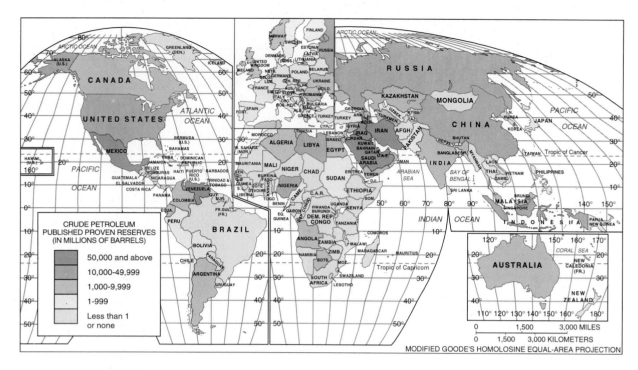

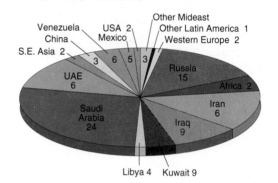

CRUDE PETROLEUM RESERVES
(Percentage of world reserves totalling
approximately 1 trillion barrels)

FIGURE 3.20
Proven oil reserves, 1995. Two-thirds of the world's proven reserves are located in the Middle East. Latin America is the second most important world region for proven reserves.

Adequacy of Fossil Fuels

In the next few decades, energy consumption is expected to rise significantly, especially because of the growing industrialization of developing countries. Most of the future energy production to meet increasing demand will come from fossil energy resources—oil, natural gas, and coal. The question people are asking these days is: How long can fossil fuel reserves last, given our increasing energy requirements? Estimates of energy reserves have increased substantially in the last 20 years, and therefore there is little short-term concern over supplies; consequently, energy prices are relatively low. If energy consumption were to remain more or less at current levels, which is unlikely, proved reserves would supply world petroleum needs for 40 years, natural gas needs for 60 years, and coal needs for at least 300 years. Although the size of total fossil fuel re-

sources is unknown, they are finite, and production will eventually peak and then decline.

Oil: Black Gold

Most of the world's petroleum reserves are heavily concentrated in a few countries, mostly in politically unstable regions. Although reserves increased by 43% between 1984 and 1994, most of this increase is attributed to new discoveries in the Middle East. Regionally, however, reserves have been declining in important consuming countries. For example, reserves in the former Soviet Union declined by 10% between 1984 and 1994. They also declined by 14% in the United States during the same period. Despite new discoveries, Europe's reserves are likely to be depleted early in the new century. Moreover, exports of oil from Africa and Latin America will probably cease by 2025. The Middle East will then

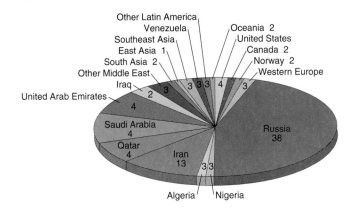

NATURAL GAS RESERVES
(Percentage of world reserves totalling
approximately 5 quadrillion cubic meters)

FIGURE 3.21
Major natural gas fields are located in North America and in Eurasia. The major concentrations of natural gas do not necessarily coincide with those of oil.

be the only major exporter of oil, but political events there could cause interruptions in oil supplies, creating problems for the import-dependent regions of North America, Western Europe, and Japan/Far East.

Natural Gas

The political volatility of the world's oil supply has increased the attractiveness of natural gas, the fossil fuel experiencing the fastest growth in consumption. Natural gas production is increasing rapidly, and so too are estimates of proven gas reserves. Estimates of global gas reserves have increased during the last decade, primarily due to major finds in Russia, particularly in northwestern Siberia, and large discoveries in China,

South Africa, and western Australia. Reserves have also been increasing in Western Europe, Latin America, and in North America. Gas production will eventually peak, probably in the first two or three decades of the 21st century. As a result, gas supplies will probably last a bit longer than oil supplies.

The distribution of natural gas differs from that of oil. It is found in countries that have yet to find oil. It is also more abundant than oil in the former Soviet Union, Western Europe, and North America, and less abundant than oil in the Middle East, Latin America, and Africa. A comparison of Figures 3.15 and 3.23 shows that natural gas also differs in its pattern of production and consumption. Because of the high cost of transporting natural gas by sea, the pattern of production is similar to that of consumption. The high

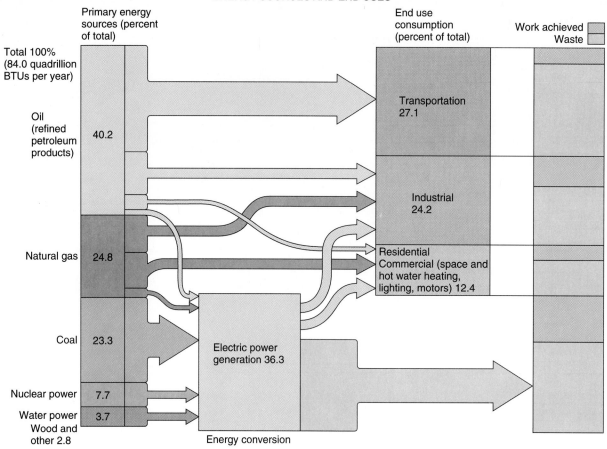

FIGURE 3.22
Energy sources and end uses.

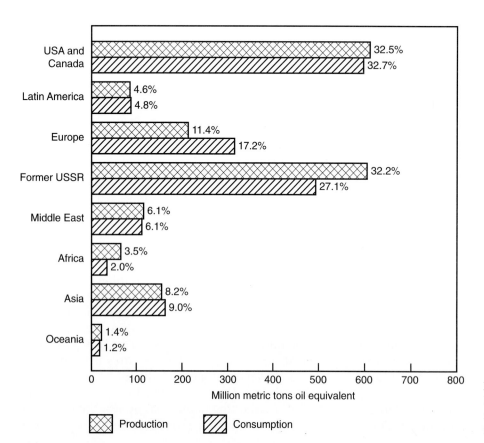

FIGURE 3.23
Production and consumption of
natural gas by major world
regions, 1995.

cost of moving gas by ocean-going vessels makes it difficult to share this resource between have and have-not countries.

Coal

Coal is the most abundant fossil fuel, and most of it is consumed in the country in which it is produced (Figures 3.24 and 3.25). Use of this resource, however, has been hampered by inefficient management by the international coal industry, the inconvenience of stor-

ing and shipping the fuel, and the environmental consequences of large-scale coal burning.

The principal fossil fuel in North America is coal (Figure 3.26). With the exception of the former Soviet Union, the United States has the largest proved coal reserves. Coal constitutes 67% of the country's fossil fuel resources, but only a small fraction of its energy consumption. It could provide some relief to the dependence on oil and natural gas (Figure 3.27).

The use of coal, however, presents problems that the use of oil and natural gas do not, making it less

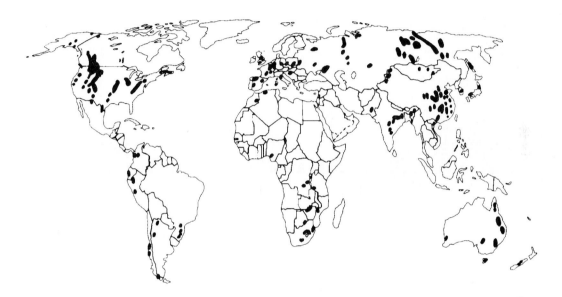

FIGURE 3.24
Major coal deposits. The United States, China, and Russia lead the world in major coal deposits. Australia, Canada, and Europe also have major deposits.

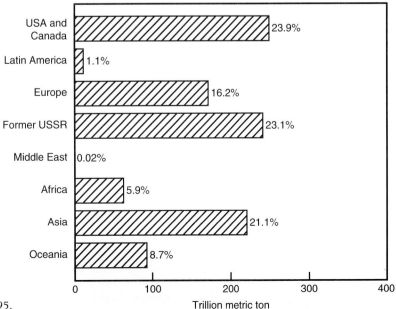

FIGURE 3.25
Proven coal reserves by major world regions, 1995.

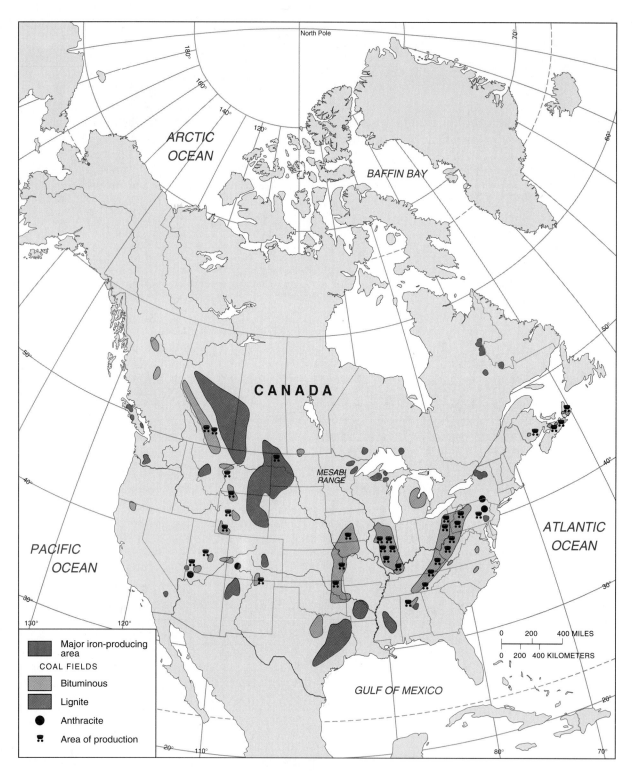

FIGURE 3.26
Major iron producing areas and coal fields of North America. *(Source: Fisher, 1992, p. 121)*

desirable as an important fossil fuel. These problems are as follows:

1. Coal burning releases more pollution than other fossil fuels, especially sulfur. Low grades of coal, called bituminous coal, have large supplies of

sulfur, which, when released into the air from the burning of coal, combines with moisture to form acid rain (Figure 3.28).

2. Coal is not as easily mined as oil or natural gas. Underground mining is costly and dangerous, and open-pit mining leaves scars difficult to re-

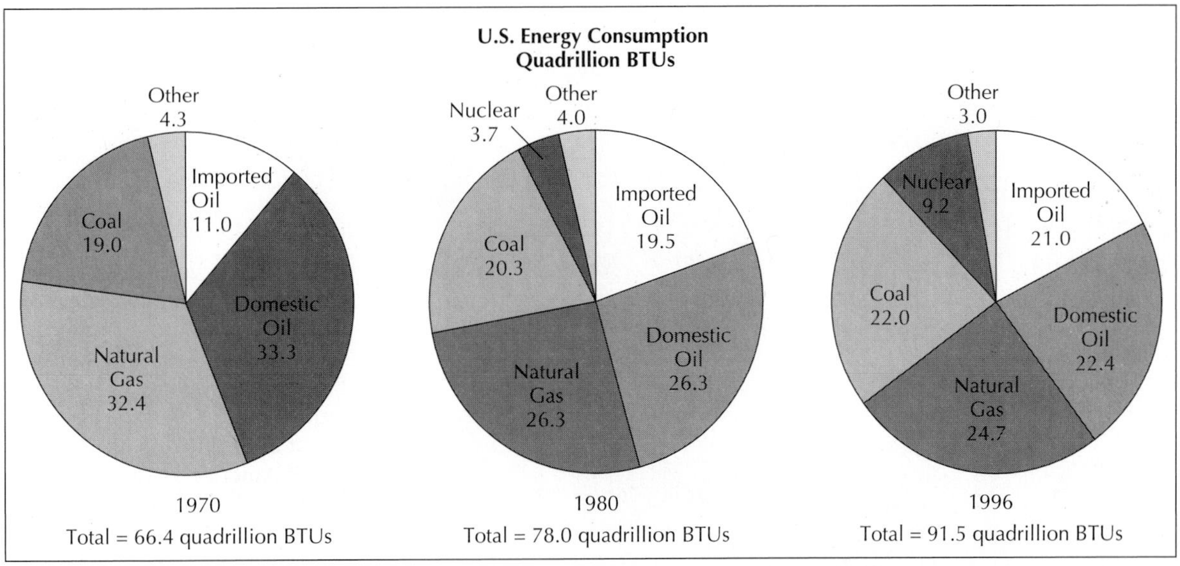

FIGURE 3.27
U.S. energy consumption in BTUs. The three principal sources of fossil fuels are coal, natural gas, and petroleum. From 1970 to 1991, the proportion of energy use depending on coal increased from 19% to 23%, while that of natural gas declined from 32% to 25%. Domestic oil consumption declined from 33% to 22%, but it has been supported by imported oil, which has increased from 11% to almost 18% of total energy consumption in the United States. Another dramatic change, besides that of reliance on imported oil, is the increase of nuclear power, which was less than 1% in 1990 to more than 9% in 1991. The number of total BTUs jumped substantially between 1970 and 1980 by almost 14% in a 10-year period; however, the increase between 1980 and 1991 was only approximately 4.5%.

habilitate to environmental premining standards.

3. Coal is bulky and expensive to transport, and coal slurry pipelines are less efficient than oil or natural gas pipelines.

4. Coal is not a good fuel for mobile energy units such as trains and automobiles. Although coal can be adapted through gasification techniques to the automobile, it is an expensive conversion and it is not, overall, well adapted to motor vehicles. Coal could be a substitute for gasoline via electricity. Electric cars will be more prevalent in the future.

ENERGY OPTIONS

The age of cheap fossil fuels will come to an end. As societies prepare for that eventuality, they must conserve energy and find alternatives to fossil fuels, especially alternatives that do not rape the environment. How viable are the options?

Conservation

According to U.S. utility companies and the government, the way to reduce the gap between domestic production and consumption in the short run is for consumers to restrict consumption. Energy conserva-

tion is a good thing because it stretches finite fuel resources and reduces environmental stress. Conservation can substitute for expensive, less environmentally desirable supply options and help to buy time for the development of other more acceptable sources of energy.

Many people believe that energy conservation means a slow-growth economy; however, data indicate that energy growth and economic growth are not inextricably linked. In the United States, from the early 1870s to the late 1940s, GNP per capita increased sixfold, whereas energy use per capita only slightly more than doubled. Energy efficiency, the ratio of useful energy output to total energy input, increased steadily throughout this century, partly as a result of industries installing better equipment. Even greater improvement can be expected in the new century. According to the U.S. Department of Energy, American firms could manufacture the same amount of product with 50% less energy than at present.

Alternative Energy Options

NUCLEAR ENERGY
One option is nuclear energy, the form which is currently in use commercially—*nuclear fission*. But nuclear fission causes many frightening issues, which became clear to all of us after the nuclear accidents at the Three

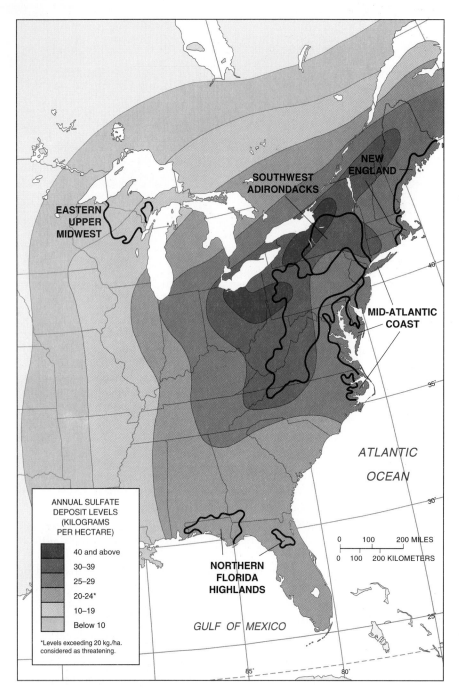

FIGURE 3.28
Annual acid rain deposit levels in North America. When sulfur is released into the atmosphere from the burning of coal and oil, it combines with moisture to create acid rain. The worst inflicted areas of acid rain in North America occur downwind from the principal polluting regions of the industrialized Midwest. Ohio, western Pennsylvania, and northern New York State are the areas most heavily inflicted with acid rain deposits. Acid rain and snow deposits are also well documented in Europe, which is in a belt of prevailing wind coming from the west, as is the United States. The industrial regions of Central Europe, especially the Ruhr Valley in northern Germany, have created strongly acidic precipitation in much of Scandinavia and Eastern Europe. Sulfur, released into the atmosphere from the burning of coal, combines with water vapor to produce sulfuric acid. Such acid creates substantial air pollution and etches away at limestone buildings, monuments, and markers on the earth. Acid precipitation can also kill plant and animal life, especially aquatic life. Literally thousands of lakes in Sweden and Norway no longer support the fish they once did. Local governments dump large amounts of powdered limestone into these water bodies to neutralize the acid.

Mile Island plant in Pennsylvania in 1979 and at the Chernobyl station in the Ukraine in 1986.

These issues range from environmental concerns caused by radiation to problems of radioactive waste disposal. Early radioactive wastes were dumped in the ocean in drums that soon began leaking. Likewise, many sites throughout the United States have contaminated groundwater supplies and leak radioactive wastes. Most present and future sites suffer from *NIMBY* ("not in my backyard") and *LULU* ("locally undesirable land uses") effects. One hotly discussed strategy is to store much of the nation's nuclear waste at

Yucca Mountain, Nevada, miles and miles away from major towns and cities. Another problem associated with the use of nuclear energy is the danger of terrorists stealing small amounts of nuclear fuel to construct weapons, which, if detonated, would wreak world havoc. The last problem associated with nuclear energy is its high-scale economy. Each plant costs billions of dollars to build and needs elaborate engineering and backup systems, as well as precautionary measures.

Nuclear power is less acceptable in North America than in some western European countries and Japan (Figure 3.29). In Japan, two-thirds of the electrical

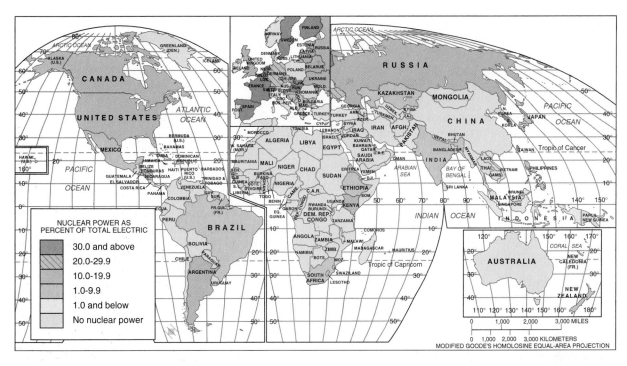

FIGURE 3.29

Nuclear power as a percentage of total energy use. The most important areas of nuclear energy production in the world include Western Europe and Japan. These are areas that have a relatively small amount of fossil fuels to satisfy local demand for energy. In Europe, for example, France, Germany, and the Scandinavian countries of Sweden and Finland produce more than 50% of their electrical energy from nuclear sources. Nuclear power is much less prevalent in the developing nations of the world because of extremely high-scale economies, or start-up costs, and the need for expensive uranium fuels. (See Color Insert I for more illustrative map.)

The large hyperbolic cooling tower and reactor containment dome of the Trojan nuclear power plant in Rainier, Oregon, add to the tranquility of this night scene. Safety issues surrounding the use of nuclear energy are fraught with turmoil. Most OECD countries expanded their nuclear energy production during the last 20 years, with France and Japan in the lead. Expansion of nuclear capacity had slowed by 1994 because of cost concerns and the chilling effects of the accidents at Three Mile Island in Pennsylvania and at Chernobyl in the Ukraine. New energy sources, such as geothermal, solar, biomass, and wind energy have increased and now provide up to 5% of total primary energy requirements in Australia, Austria, Canada, Denmark, Sweden, and Switzerland. *(Photo: U.S. Department of Energy)*

energy produced comes from nuclear power plants. Belgium, France, Hungary, and Sweden produce more than one-half of their energy from nuclear power plants, while Finland, Germany, Spain, Switzerland, South Korea, and Taiwan are also major producers and users. In the United States and Canada, countries that are less dependent on nuclear energy, the eastern portions of these countries are more receiving of nuclear power plants than western portions. For example, New England draws most of its electrical power from nuclear power plants. Interestingly, some countries have decided to throw in the towel on nuclear power generation because of high risk and high cost. For example, Sweden is phasing out its nuclear plants. The phase-out began in 1995 and is expected to be completed by the year 2010.

Nuclear fusion has the potential to be a solution to the environmental concerns of nuclear fission. If research on this new technology is successful, nuclear fusion would provide limitless amounts of very cheap electricity and pose no radiation dangers. The raw material for nuclear fusion is the common element hydrogen. At the present time, the U.S. government does not permit the use of nuclear fusion for the generation of electricity.

GEOTHERMAL POWER

The development of geothermal power holds promise for the future in several countries that have hot springs, geysers, and other underground supplies of hot water that can easily be tapped. The occurrence of this renewable resource is highly localized, however. New Zealand obtains about 10% of its electricity from this source, and smaller quantities are utilized by other countries such as Italy, Japan, Iceland, and the United States.

If the interior of the earth's molten magma is sufficiently close to the surface (i.e., 10,000 feet), underground water may be sufficiently warm to produce steam that can be tapped by drilling geothermal wells. Geothermal energy is most producible in giant cracks or rifts in the earth's tectonic plate structure that occur in earthquake or active volcanic areas around the Pacific Rim. Wyoming and California are noted examples.

HYDROPOWER

Another source of electric power, and one that is virtually inexhaustible, is hydropower—energy from rivers. Developed countries have exploited about 50% of their usable opportunities, the former Soviet Union and Eastern Europe about 20%, and developing countries about 7%. In developed countries, further exploitation of hydropower is limited mainly by environmental and social concerns. In developing countries, a lack of money and markets for the power appears to be the main obstacle.

One of the main problems of constructing dams for hydropower is the disruption to the natural order of a watercourse. Behind the dam, water floods a large area, creating a reservoir; below the dam, the river may be reduced to a trickle. Both behind and below the dam, the countryside is transformed, plant and animal habitats are destroyed, farms ruined, and people displaced. Moreover, the creation of reservoirs increases the rate of evaporation and the salinity of the remaining water. In tropical areas, reservoirs harbor parasitic diseases, such as schistosomiasis. Since the construction of the Aswan High Dam, schistosomiasis has become endemic in Lower Egypt, infecting up to one-half of the population. An additional problem is the silting of reservoirs, reducing their potential to produce electricity. The silt, trapped in reservoirs, cannot proceed downstream and enrich agricultural land. The decrease in agricultural productivity from irrigated fields downstream from the Aswan High Dam has been substantial.

The long-term hydrological, ecological, and human costs of dams easily transform into political problems on international rivers. An example is Turkey's Southeast Anatolia Project, which envisages the construction of 22 dams and 19 hydroelectric power plants. Because the project is being developed on two transboundary rivers—the Tigris and Euphrates—problems and disputes have arisen with two downstream users—Syria and Iraq—whose interests the project affects.

SOLAR ENERGY

Like river power, solar energy is inexhaustible. In the 1970s and early 1980s, solar energy caught the public eye through the publicity of the relatively few solar homes and buildings constructed in the United States. Large-scale utilization of solar energy, however, still poses technical difficulties, particularly that of low concentration of the energy. It is estimated that the energy of the sun's direct rays at sea level is slightly more than 1 horsepower per square meter. So far, technology has been able to convert only slightly more than 30% of solar energy into electricity; however, depending on the success of ongoing research programs, it could provide substantial power needs in the 21st century.

Solar energy's positive aspects are that it does not have the same risks as nuclear energy, nor is it difficult, like coal, to transport. It is almost ubiquitous, but varies by latitude and by season. In the United States, solar energy and incoming solar radiation are highest in the southwest and lowest in the northeast. In July, the highest readings for incoming solar energy are in Nevada, California, and Arizona, and in December they are in Nevada, Utah, Arizona, and New Mexico.

Passive solar energy is trapped rather than generated. It is captured by large glass plates on a building or house. The greenhouse effect receives short-wave radiation from the sun. Once the rays penetrate the glass, they are converted to long-wave radiation and are trapped within the glass panel, thus heating the interior of the structure or water storage tank.

The other way of harnessing solar energy is through *active solar energy*. One type of this form of solar energy is generated with photovoltaic cells made from silicon. A bank of cells can be wired together and mounted on the roof with mechanical devices that maximize the direct sun's rays by moving at an angle proportional to the light received. Another type of active solar energy system is a wood or aluminum box filled with copper pipes and covered with a glass plate, which collects solar radiation and converts it into hot water for homes and swimming pools.

Cost is the main problem with solar energy. High costs are associated with the capturing of energy in cloudy areas and high latitudes. But unlike fossil fuels, solar energy is difficult to store for long periods without large banks of cells or batteries. Currently, solar energy production is more expensive than other sources of fuel.

To promote the development of innovative energy supplies when the Arab oil embargo hit in the 1970s, the U.S. government offered tax incentives, including tax deductions for solar units mounted on housetops. Although this tax deduction offset the high costs of constructing solar energy systems, maintenance and reliability soon become a problem. Families that move lose their investment, because most systems installed are rarely recoverable in the sale price of homes.

Scientists continue to study ways to supply entire cities with solar power. In the future, this may well be possible. A major test for solar energy is *California's Solar 1*, a power plant in the Mojave Desert, 200 miles north of Los Angeles. It has 1,818 classroom-sized mirrored panels that follow the direct rays of the sun. The mirrors reflect sunlight into a boiler tank. The water is heated, creating steam to run turbines.

WIND POWER

The power of the wind provides an energy hope for a few areas of the world where there are constant surface winds of 15 mph or more. The amount of wind power available at the 50-meter height (or 164 feet), which is thought to be a favorable height for large wind machines, is measured in megawatt hours per square meter per year. The greatest majority of *wind farms* in the United States are in California. However, wind machines are an expensive investment, and the initial cost plus the unsightliness of the wind machine has ended most wind farm projects. Wind farm potential in

California has never matched expectations, and wind farming is presently a stagnated industry.

BIOMASS

Still another form of renewable energy is biomass—wood and organic wastes. In 1980, biomass accounted for about 14% of global energy use. For Nepal, Ethiopia, and Tanzania, more than 90% of total energy comes from biomass. The use of wood for cooking—the largest use of biomass fuel—presents enormous environmental and social problems because it is being consumed faster than it is being replenished. Fuelwood scarcities—the poor world's energy crisis—now affects 1.5 billion people and could affect 3 billion in the new century unless corrective actions are taken.

With good management practices, biomass is a resource that can be produced renewably. It can be converted to alcohol and efficient, clean-burning fuel for cooking and transportation. Its production and conversion are labor intensive, an attractive feature for developing countries that face underemployment and unemployment problems. But the low efficiency of photosynthesis requires huge land areas for energy crops if significant quantities of biomass fuels are to be produced.

Many countries have expanded their use of biomass for fuel since the 1973 oil embargo. The United States is using more wood-fired boilers for industrial and domestic purposes and is producing gasohol from corn. In 1985, 20 developing countries formed the Biomass Users Network (BUN), which proposes, among other things, to convert unprofitable export crops such as sugar cane into biomass fuel for local consumption.

ENVIRONMENTAL DEGRADATION

On some days in Los Angeles, pollution levels reach what is called locally a *level three alert*. People are advised to stay indoors, cars are ordered off the highways, and strenuous exercise is discouraged. In Times Beach, Missouri, which is some 50 miles south of St. Louis, dioxin levels from a contaminated plant became so high that the EPA required the town to be closed and the residents to be relocated. Around Rocky Flats, nuclear wastes of plutonium have degraded the soil so that radioactivity levels are five times higher than normal. In New England, acid rain has become so bad that it has killed vegetation and fish in rivers and lakes.

Environmental awareness came to the forefront following the publication in 1962 of Rachel Carson's *Silent Spring*. It awakened people to rising levels of pollution at all scales, from the local to the global. In 1969, the U.S. government enacted a piece of landmark legislation, the National Environmental Policy Act (NEPA), requiring environmental impact reports of major government improvements. This event was followed by Earth Day in April 1970. Dennis Hayes organized a movement at the Harvard Law School as America's wake-up call to the consequences of rapid industrialization. In 1971, California was the first state to enact legislation to guard and preserve the environment with the California Environmental Quality Act (CEQA). Massachusetts was next, followed by other states. These pieces of state environmental legislation did for the states and private projects what NEPA did for government projects across the country. In 1990, Earth Day was celebrated again and organized by Dennis Hayes. This time it was a global event, involving 250 million people in 150 countries.

Two years later came the Earth Summit in Rio de Janeiro. This meeting in 1992 was supposed to highlight linkages between environmental and economic growth issues. Instead, it highlighted "dueling hemispheres." Developed countries were more interested in environmental issues, the developing countries in economic concerns. In general, for the developing countries to be willing to talk about the environment, developed countries must first be willing to talk about economics. Only when the developed countries give the developing countries what they want, a fair deal es-

pecially in the sphere of trade and capital transfer, will there be any real hope for the global management of global environmental issues.

Environmental problems, caused mainly by economic activity, may be divided into four overlapping categories: (1) pollution, (2) wildlife and habitat preservation, (3) nonrenewable natural resources, and (4) environmental equity.

Pollution

Pollution is a discharge of waste gases and chemicals into the air and water. Such discharge can reach levels that are sufficient to create health hazards to plants, animals, and humans as well as to reduce and degrade the environment. The natural environment has the capacity to regenerate and cleanse itself on a normal basis; however, when great amounts of gases and solids are released into it from industrial economic activity, recycling and purification needs are sometimes overwhelmed. From that point on, the quality of the environment is reduced as pollutants create health hazards for humans, defoliate forests, inundate land surfaces, reduce fisheries, and burden wildlife habitats.

Exxon-Valdez cleanup. Fifteen thousand birds were the first victims of the ecological catastrophe caused by the running aground of the Exxon-Valdez in March 1990, creating one of the largest oil spills in history. Five thousand sea otters soon followed suit in the Prince William Strait near Alaska. This oil spill actually increased the U.S. GNP because billions of dollars are being spent on cleanup, but the resource loss did not show up as negative income in the accounting system. A new world system of accounting must be practiced that includes natural resource loss, such as oil and timber. This system would incorporate the value of biological resources and biodiversity itself, where possible, as well as the costs of genetic resources, degraded watersheds, eroded soils, and depleted energy supplies. *(Photo: Bill Nation, Sygma)*

AIR POLLUTION

Air pollutants, the main sources of which are illustrated in Figure 3.30, are normally carried high into the atmosphere, but occasionally, and in some places more than others, a temperature inversion prevents this from occurring. Usually, temperature decreases at the rate of about 3.5 degrees F. per 1,000 feet, called the *normal lapse rate*. Because air is cooler at higher elevations, warm air on the surface of the earth rises and dissipates through normal wind patterns. However, when a temperature inversion occurs, the normal lapse rate is disrupted up to an elevation of 3,000 to 4,000 feet above the ground. Inversions are literally caps, preventing the escape of pollutants. Under these conditions, the earth and its inhabitants are under an even greater risk. These conditions promote the formation of smog that blocks out sunshine, causes respiratory problems, stings the eyes, and creates a haze over cities such as Los Angeles, Mexico City, Tokyo, and Milan.

Air pollution gives rise to different concerns at different scales. Air pollution at the local scale is a major concern in cities because of the release of carbon monoxide, hydrocarbons, and particulates. Air pollution at the regional scale is exemplified by the problem of acid precipitation in eastern North America and Eastern Europe (Figure 3.30). At the global scale, air pollution may damage the atmosphere's *ozone layer* and contribute to the threat of *global warming*.

The earth's protective ozone layer is thought to be threatened by pollutants called *CFCs (chlorofluorocarbons)*. When CFCs such as freon leak from appliances such as air conditioners and refrigerators, they are carried into the stratosphere where they contribute to ozone depletion. As a result of the 1987 Montreal Protocol, developed countries must stop using CFCs by the year 2000 and developing countries by the year 2010. Scientists hope that this international agreement will effectively reduce ozone depletion.

Concern about global warming is much the same as our worry about ozone. The burning of fossil fuels in ever greater quantities increases the amount of carbon dioxide in the air, which makes the atmosphere more opaque, reducing thermal emission to space. Anthropogenic heat-trapping gases, such as carbon dioxide, warm the atmosphere, enhancing a natural *greenhouse effect* (Figure 3.31). Since the 1890s, the average temperature of the earth's surface has increased by 2 degrees F. This increase in temperature may or may not

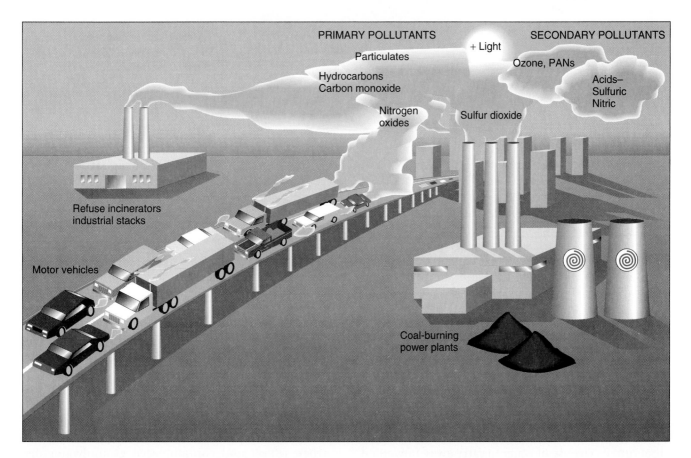

FIGURE 3.30
The primary sources of major air pollutants. Industrial processes are the major sources of particulates; transportation and fuel combustion cause the lion's share of the other pollutants.

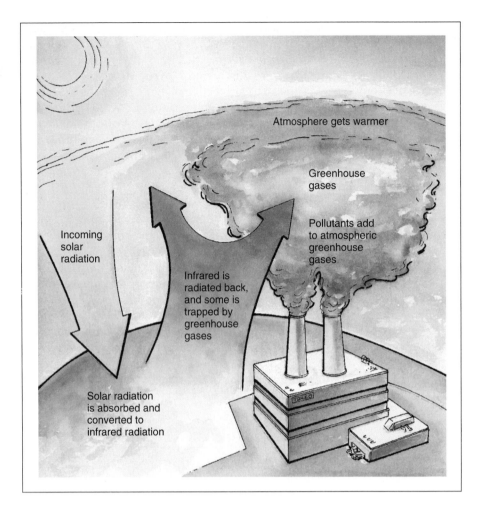

Atmosphere gets warmer

Greenhouse gases

Pollutants add to atmospheric greenhouse gases

Incoming solar radiation

Infrared is radiated back, and some is trapped by greenhouse gases

Solar radiation is absorbed and converted to infrared radiation

FIGURE 3.31
The greenhouse effect. Incoming solar radiation is absorbed by the earth and converted to infrared radiation. Greenhouse gases act as an insulator and trap the infrared heat, raising the temperature of the lower atmosphere. The overall effect is global warming.

be human-induced, however. There are divergent views on the issue. Nonetheless, even if the observed global warming is consistent with natural variability of the climate system, many scientists agree that it would be socially irresponsible to delay actions to slow down the rate of anthropogenic greenhouse gas buildup. For example, continued warming would increase sea levels, disrupt ecosystems, and change land-use patterns. While agriculture in some temperate areas may benefit from global warming, tropical and subtropical areas may suffer.

WATER POLLUTION

Fresh water is perhaps the most important resource. Although there is more than enough water to meet the world's needs now and in the future, the problem is that when we use water, we invariably contaminate it. Major wastewater sources that arise from human activities include municipal, mining, and industrial discharges, as well as urban, agricultural, and silvicultural runoff. The use of water to carry away unwanted waste material is an issue that has come into prominence, because water is being used more heavily than

ever before. As humans multiply, problems of water utilization and management increase. These problems are most acute in developing countries, where some 1 billion people already find it difficult or impossible to obtain acceptable drinking water. But water is also an issue in developed countries. In the United States, for example, the major water management problem through most of the 20th century focused on acquiring additional water supplies to meet the needs of expanding populations and associated economic activities. Recently, water management has focused on the physical limits of water resources, especially in the West and Southwest, and on water quality. Passage of the Clean Water Act in 1972 has resulted in improvement in water quality of streams that receive discharges from specific locations or *point sources* such as municipal waste-treatment plants and industrial facilities. Efforts to improve water quality in the 1990s have also emphasized the reduction of pollution from diffuse or *non-point sources* such as agricultural and urban runoff and contaminated groundwater discharges. These sources of pollution are often difficult to identify and costly to treat.

Wildlife and Habitat Preservation

Wildlife and habitat preservation for plants and animals called *renewable natural resources* are in danger throughout the world. These natural environments are critical reserves for endangered species of plants and animals. Wildlife, forestlands, and wetlands, including lakes, rivers and streams, and coastal marshes, are subject to acid rain, toxic waste, pesticide discharge, and urban pollution. They are also endangered by encroachment of land development and transportation facilities worldwide (Figure 3.32). The demand for tropical hardwoods, such as Philippine mahogany, has already removed 50% of the tropical hardwoods necessary for the maintenance of the earth's ozone layer. In the United States alone, expanding economic activity has consumed forests and wetlands, depleted topsoil, and polluted at a rapid rate. Even certain species of plants and animals have been reduced, including the grizzly bear, American bison, whitetail deer, prairie dog, gray wolf, brown pelican, Florida panther, American alligator and crocodile, and a variety of waterfowl and tropical birds, such as California's Least Bel Viro and Gnat Catcher.

The problem of wildlife and habitat preservation is exacerbated by the need for economic gain. For example, the populations of the Asian tiger and African elephant, each desired for a certain portion of its pelt, tusk, or skin, have been depleted. A variety of other questions beset wildlife managers and environmental farmers. Should farmers be permitted to drain swamps in Louisiana to farm the land, thus removing the habitat for American alligators? Should forest fires started by lightning be allowed to burn themselves out, as has been the practice on western U.S. forests and rangelands? The enormous Yellowstone National Park fire in 1990 brought this practice into question, because it removed not only tourists but wildlife habitats of 90% of the park's animals. The trade-off of residential lands versus wetlands, wildlife migration versus forest management, highway safety versus habitat preservation, and conservation versus real economic development and growth of the U.S. economy are difficult issues that bring to mind the opportunity costs and production possibilities curve presented in Chapter 2. It is difficult to select the best alternative.

Nonrenewable Natural Resource Management

Mineral and geologic resources and extraction, as discussed in this chapter, represent an enormous set of environmental management questions. As we saw, the deposits of petroleum, iron ore, copper, bauxite, zinc, and other metallic deposits are nonrenewable. Present economic activity depletes world resources, and the trade-off again besets us. Preservation of nonrenewable resources is essential for long-run economic growth. Piles of tailings, deep scouring in the earth's surface from open-pit mining, pollution of groundwater supplies and streams, enormous garbage heaps, the filling of canyons with solid waste, and oil spills, which are becoming more numerous, are but a few of the nonrenewable natural management concerns that must be addressed in the 1990s.

Some nonrenewable resource experts estimate that at least 80% of the conservation effort remains to be tapped. Cars are currently on the market that get double the average of 25 miles per gallon, and Peugeot, Toyota, Volkswagen, and Volvo all have prototypes that get from 70 to 100 mpg. To guarantee sales levels, manufacturers are waiting for government mandates, consumer demand, or fuel shortages before putting these models into full-scale production. GM began marketing its electric car in California in 1998. Of course, carpooling with just one person in addition to the driver is an easy way to double the mpg, because auto occupancy in North America is about 1.2 people per vehicle trip. Metals recycling is another way to conserve nonrenewable natural resources. Energy conservation is already saving $100 billion per year in oil imports in North America and half that much in metals recycling. Advances in conservation have been extremely cost effective as well as environmentally benign.

FIGURE 3.32
An inverse relationship exists between human population size and the survival of species worldwide. Uncertainty about the extent of decreasing biodiversity is reflected in the width of the species curve. *(Source: Mesoule, "Conservation: Tactics for a Constant Crisis,"* Science, 253, *1991, p. 744)*

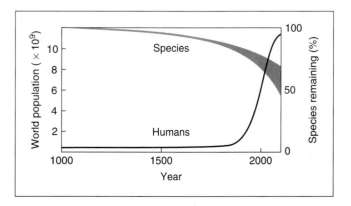

Environmental Equity and Sustainable Development

Economic development policies and projects all too often seem to carry as many costs as benefits. Build a dam to bring hydroelectric power or irrigation water, and fertile river bottomlands are drowned, farming people are displaced, waterborne diseases may fester in the still waters of the reservoir, and the course of the river downstream to its delta or estuary is altered forever. Dig a well to improve water supplies in dry rangelands, and overgrazing and desertification spread outward all around the points of permanent water. Mine an ore for wealth and jobs, and leave despoiled lands and air and water pollution. Build new industries, shopping malls, and housing tracts, and lose productive farmland or public open space. Introduce a new miracle crop to increase food production, and traditional crop varieties and farming methods closely synchronized to local environments disappear. Is it possible to reckon the costs and benefits of "progress"? Can we develop a humane and decent society, one that encourages both equity and initiative, a society capable of satisfying its needs without jeopardizing the prospects of future generations? How do we go about creating a sustainable society?

The term "sustainable development" was introduced in 1987 in *Our Common Future*, a report by the World Commission on Environment and Development—the Brundtland Commission, named after the Norwegian prime minister who headed the commission. There, sustainable development was defined as "development that meets the needs of the present generation without compromising the ability of future generations to met their own needs." Since then, "sustainable development" has become a buzzword around the world as commissions, industry, local communities, and grassroots activists present their own agendas and priorities for the future. Most agree with the Brundtland Commission's definition of sustainable development and accept its focus on the importance of long-range planning, but as a policy tool it is vague, providing no specifics about which needs and desires must be met and fulfilled and how. How do we, rich and poor alike, move from the perception of environmental problems to policies that will create a sustainable society? (See Table 3.3.)

In the debate on sustainable development, two different emphases have clearly emerged. In the industrialized countries of North America, Europe, and Japan, the emphasis has been on long-term rather than short-

Ships are loaded with timber for export to Japan at the Weyerhauser Docks, Longview, Washington. The timber industry represents a major economic activity in the northwestern United States and western Canada. Key issues center around how sustainable forest practices can contribute to both ecological restoration and soil quality. Loss of ecosystem habitats and endangered species, such as the Northern spotted owl, continue to raise questions about the logging industry. In dry areas, stripping of land vegetation for fuel wood leads to wind and water erosion. Worldwide overexploitation accounts for 10% of the degradated soils of the world. Africa has the highest percentage of degradated soils from this cause. The loss of tropical forest is currently a significant environmental and developmental issue as well. Loss of tropical forest diminishes biodiversity, contributes to climatic change by releasing stored carbon into the atmosphere, and often results in serious soil degradation, sometimes rendering the land unfit for future agriculture. *(Photo: Martha Tabor, Impact Visuals Photo & Graphics, Inc.)*

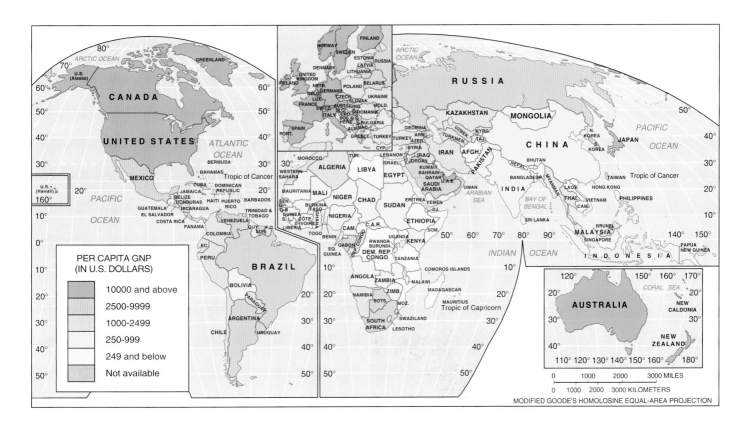

ABOVE: The annual GNP per capita exceeds $15,000 in most relatively developed countries, compared to less than $1000 in most developing countries. Several petroleum-rich countries in southwestern Asia have relatively high GNPs per capita, although by other measures they may rank among the developing countries. The difference in annual GNP per capita between relatively developed and developing countries has grown since the early 1980s.

BELOW: Physiological density refers to the number of people per unit area of agricultural land. The higher the physiological density, the greater the pressure people exert on the land and food resources of a country. Geographers use physiological density rather than population density as a true measure of the population pressure on the land because it shows the potential availability of needed food resources.

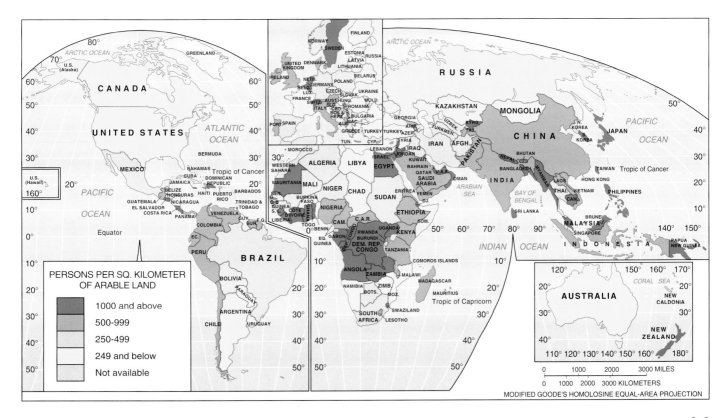

1.1

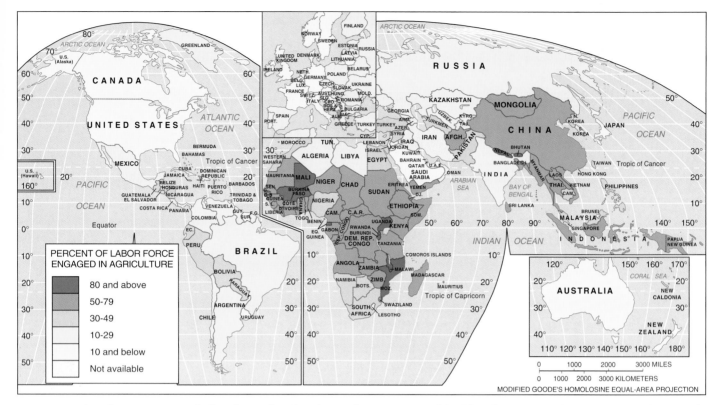

ABOVE: Note that the more economically advanced nations have a lower portion of the labor force engaged in agriculture. The United States, Canada, Europe, and Australia are examples. The most important priority for people throughout the world is to secure enough food to survive and prosper. Sixty percent of all workers in the world are engaged in agricultural pursuits; less than 4% of the workers in Europe and Anglo-America are engaged in agriculture. Most of these people earn their living working in offices, stores, and factories.

BELOW: The natural annual increase is the rate of growth of a population for a particular year. The fastest growing areas of the world include Africa, Central America, and Southwest Asia. In these countries, growth rates exceed 2.5%, with a number of countries in Central America and Africa actually exceeding 3% per year. Three percent growth per year does not seem like a high growth rate; however, it indicates a total population doubling time for a country to be only 23 years. The natural increase in Europe is only 0.2 percent.

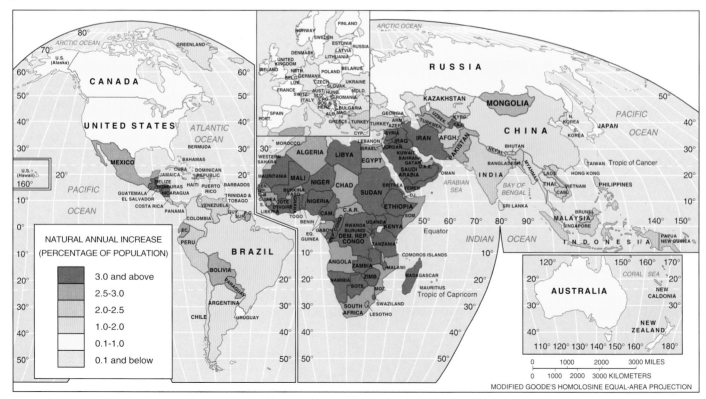

1.2

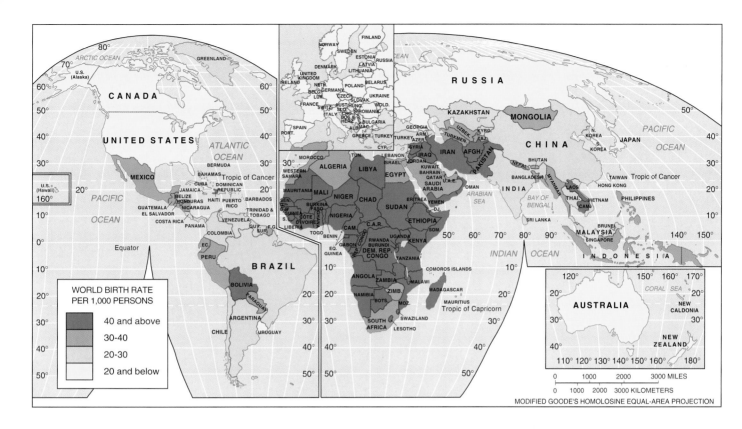

ABOVE: The world birth rate is the number of children born each year per 1000 people that live in a country. The world birth rate closely mirrors the natural annual increase. Crude birth rate measures a country's population as a whole and does not study separate rates of birth for women of different age groups, or by income or racial character. Again, Central America, Africa, and Southwest Asia lead the world in crude birth rate.

BELOW: Infant mortality rates are the number of deaths of infants under the age of 1 year per 1000 live births for a country. World infant mortality rates averaged approximately 75 per 1000 for 1993. The United States, for example, averaged 8 deaths per 1000 live births, but several nations had lower infant mortality rates. As with natural annual increase and world birth rates, infant deaths per 1000 live births are related to development.

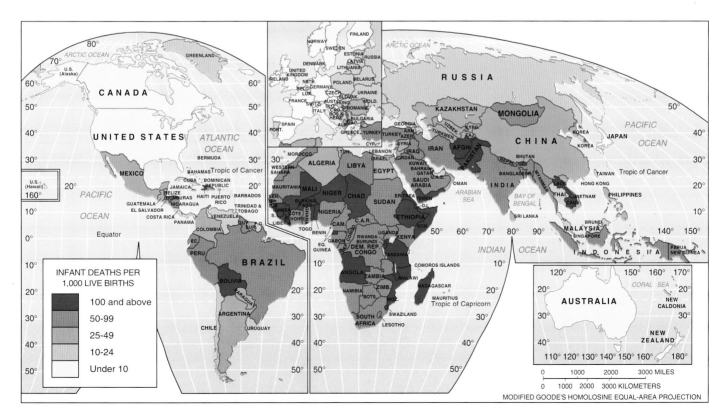

1.3

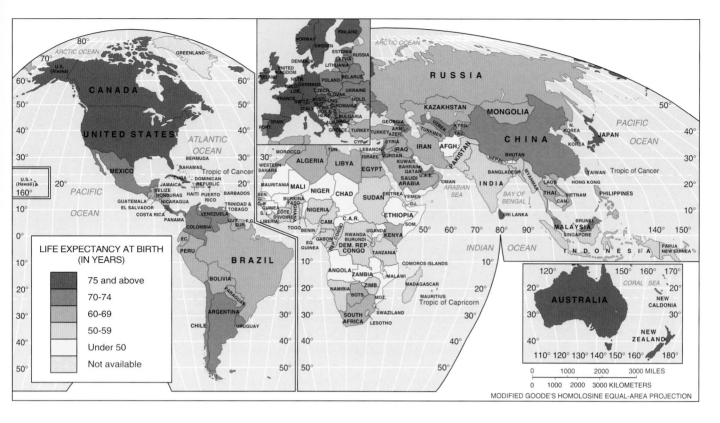

LIFE EXPECTANCY AT BIRTH
(IN YEARS)

- 75 and above
- 70-74
- 60-69
- 50-59
- Under 50
- Not available

MODIFIED GOODE'S HOMOLOSINE EQUAL-AREA PROJECTION

ABOVE: Babies born in 1997 are expected to live an average of 64 years. Life expectancy, however, ranges from the high seventies in Northwestern European nations, North America, and Australia to the low thirties in some African nations.

BELOW: Between 1980 and 1990, the most rapidly growing states were located in the West and the South. States of the Northeast, South, Midwest, and Great Plains grew at a much lower rate, or even lost population in some cases. Texas, California, and Florida accounted for almost 50% of the population growth during this period, and each registered an approximately 25% growth rate.

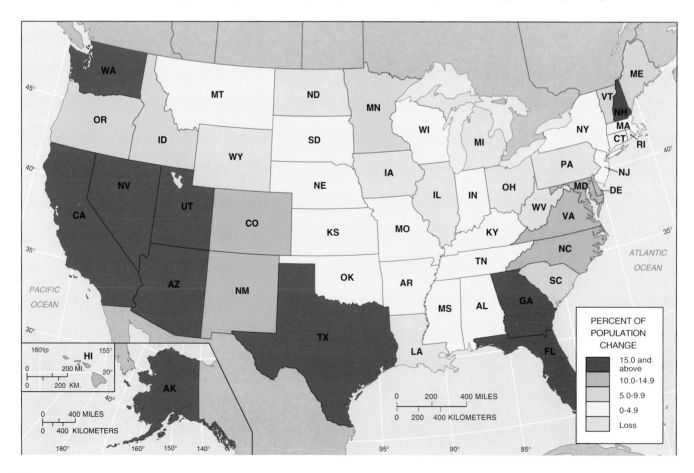

PERCENT OF POPULATION CHANGE

- 15.0 and above
- 10.0-14.9
- 5.0-9.9
- 0-4.9
- Loss

1.4

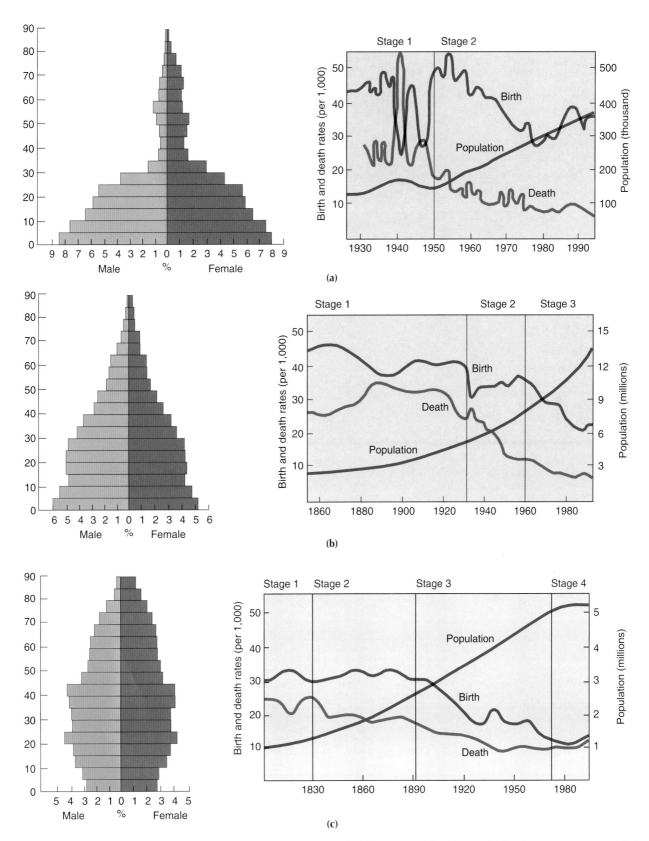

(a) The population pyramid (left) and the demographic transition (right) for Cape Verde. The vertical axis measures age, and the horizontal axis measures population by gender. (b) Chile entered stage 2 of the demographic transition in 1930 when death rates abruptly declined. It entered stage 3 in approximately 1960 when birth rates began to decline. It has yet to enter stage 4, and the population continues to rise dramatically. Because of its status as a developing country, it has a relatively broad-based pyramid. Population members younger than 15 and older than 65 are dependent populations. (See Color Insert 1 for a more illustrative map.) (c) The population pyramid and demographic transition for Denmark are shown. The shape of the so-called pyramid is quite different than that of Chile. Because of lower birth rates, the population pyramid is more in the shape of a spark plug. Denmark, which entered this stage in 1830 and stage 3 in 1890, entered stage 2 of the population pyramid 100 years earlier than Chile. In the 1970s, Denmark entered stage 4, and its current population growth is zero.

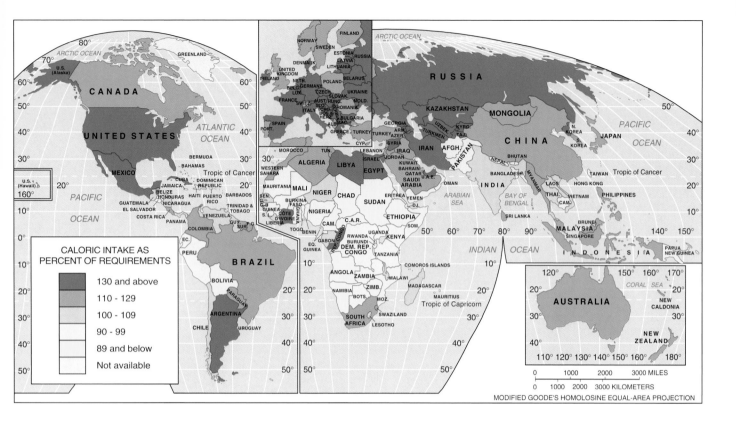

ABOVE: Highly developed regions of the world receive, on the average, 130% of the daily caloric requirements (2400 calories per day) as set by the U.S. Food and Agricultural Organization (FAO). Some countries in South America and south Asia, and many countries in Africa, receive less than 90% of the daily caloric requirements needed to sustain body and life. Averages must be adjusted according to age, gender, and body size of the person and regions of the world. Averages also hide destitute subpopulations.

BELOW: The greatest number of proven petroleum reserves exists in countries of the Middle East surrounding the Persian Gulf, in particular the Arabian Peninsula. The area includes more than 50% of all proven world reserves. Mexico and Venezuela also show a disproportionately large share of proven reserves.

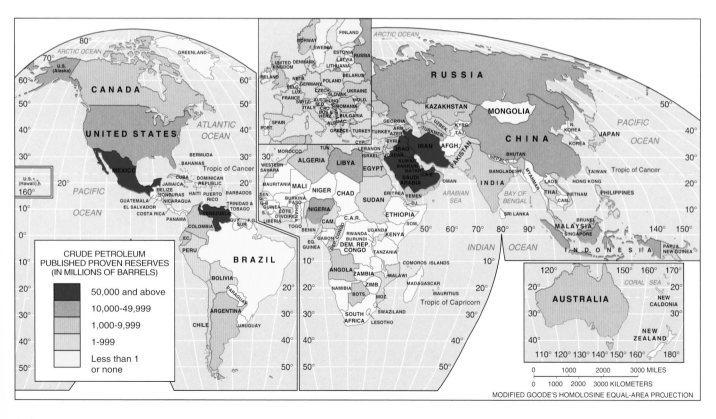

1.6

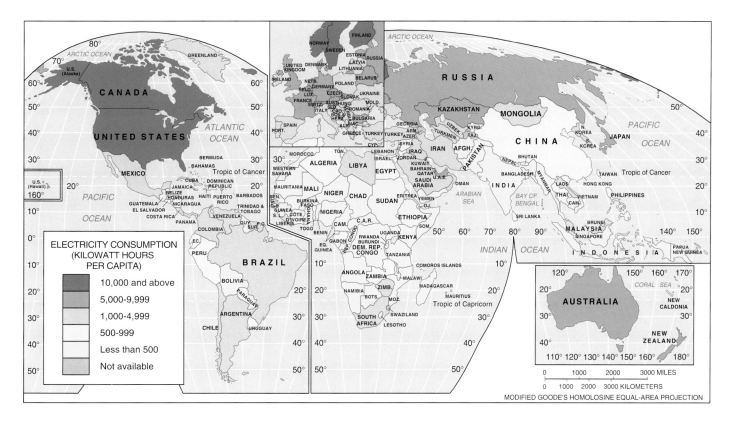

ABOVE: The United States, Canada, and the Scandinavian countries consume more electricity per capita than any other countries. When the electricity usage of the United States, Canada, Europe, and Russia are combined, 75% of electricity usage in the world is accounted for, but these same countries account for only 20% of the world's population. By comparing this map to the map of crude petroleum reserves, deficit areas of the world, such as Europe and Japan, can be seen.

BELOW: The United States, followed by the former Soviet Union and China, produce more coal than any other countries. Australia, India, and South Africa also produce substantial amounts of coal. Since coal is bulky and more costly to transport than liquid oil, less coal is shipped internationally to supply world energy demands. Alternative energy sources, such as wind power, geothermal, and solar-based approaches offer hope, however, until radical breakthroughs in energy technology occur, coal will remain America's greatest energy source.

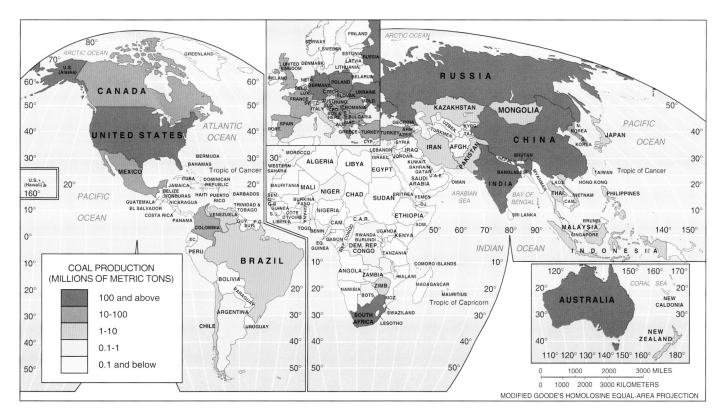

1.7

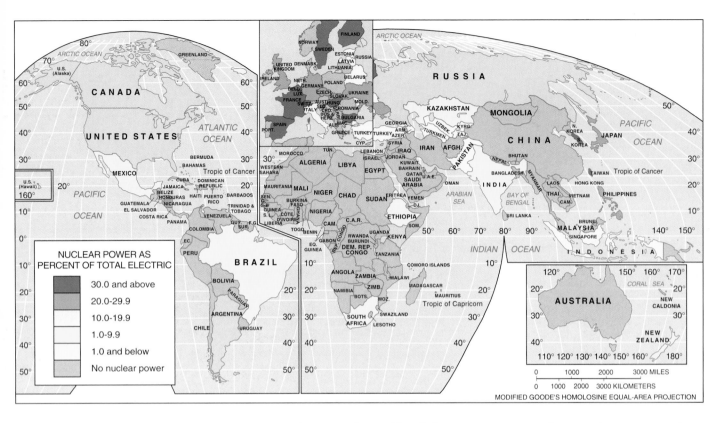

NUCLEAR POWER AS PERCENT OF TOTAL ELECTRIC

30.0 and above
20.0-29.9
10.0-19.9
1.0-9.9
1.0 and below
No nuclear power

MODIFIED GOODE'S HOMOLOSINE EQUAL-AREA PROJECTION

ABOVE: The most important areas of nuclear energy production in the world include Western Europe and Japan. These are areas that have a relatively small amount of fossil fuels to satisfy local demand for energy. In Europe, for example, France, Germany, and the Scandinavian countries of Sweden and Finland produce more than 50% of their electrical energy from nuclear sources. Nuclear power is much less prevalent in the developing nations of the world because of extremely high-scale economies, or start-up costs, and the need for expensive uranium fuels. Public sentiment seems to be against future use of nuclear power. Besides the possibility of melt-down, the problems of nuclear waste are immense.

BELOW: Areas of dominant influences (ADIs) are based on radio, television, and newspaper advertising. ADI for cities approximates the breaking point between cities based on the law of retail gravitation model as applied to adjacent pairs of cities.

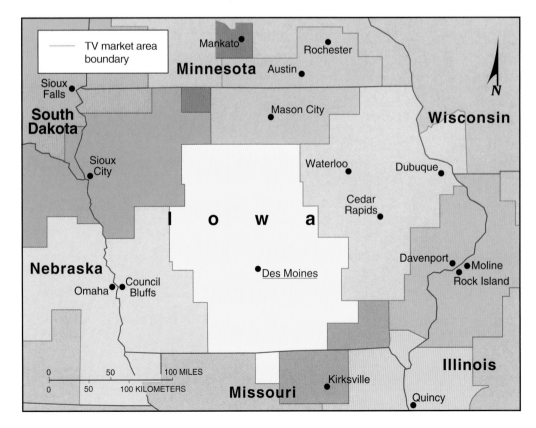

1.8

🔲

T a b l e 3.3

Incidence of Household Environmental Problems in Accra, Jakarta, and São Paulo

Environmental Indicator	Incidence of Problem (percentage of all households surveyed)		
	Accra, Ghana	*Jakarta, Indonesia*	*São Paulo, Brazil*
Water			
No water source at residence	46	13	5
No drinking water source at residence	46	33	5
Sanitation			
Toilets shared with more than 10 households	48	14–20	<3
Solid waste			
No home garbage collection	89	37	5
Waste stored indoors in open container	40	27	14
Indoor air			
Wood or charcoal is main cooking fuel	76	2	0
Mosquito coils used	45	28	8
Pests			
Flies observed in kitchen	82	38	17
Rats/mice often seen in home	61	82	25

Note: Sample sizes were as follows: Accra, *N* = 1,000; Jakarta, *N* = 1,055; and São Paulo, *N* = 1,000. Missing values, never more than 3% of the sample, are omitted. Missing values refers to cases where the question under consideration was not answered. It includes both "no response" and "not applicable." In some cases, questions were slightly different for the three cities, which accounts for the use of intervals. *From: McGranahan & Songsore, "Health, Wealth, and the Urban Household," Environment, July–Aug 1994, p. 9.*

term growth, and on efficiency. The emphasis has been on economics: If today we rely on an incomplete accounting system, one that does not measure the destruction of natural capital associated with gains in economic output, we deplete our productive assets, satisfying our needs today at the expense of our children. There is something fundamentally wrong in treating the earth as if it were a business in liquidation. Therefore, we should promote a systematic shift in economic development patterns to allow the market system to internalize environmental costs. By getting prices right, we can move toward a sustainable future.

This Western emphasis on the economic aspects of sustainable development has been criticized in Africa, Asia, and Latin America. Critics from the less-developed world accuse environmentalists from the industrialized world of dodging the issues of development without growth and the redistribution of wealth. While critics from the developing world may believe in the power of markets to distribute goods and services efficiently, they argue that social, political, and psychological influences are too pervasive for economics to provide answers to all our problems. They are more focused on human dignity than efficiency. Such views, from 52 individuals of 34 countries, were presented by participants in the 2050 Project. Many criticized the West's materialism and excessive consumption of resources. Sixto Roxas of the Philippines stated "we must not only de-materialize consumption and production, we must, as in the far distant past, resacralize everything: nature, community, consumption, production, governance, science and technology." Instead of seeking wealth and prosperity for all, Margaret Maringa of Kenya asserted that in her desired future, "no person has more or less than he or she needs for basic survival." Many of the participants in the 2050 Project put concern with basic human needs first, ahead of environmental concerns. Let us work toward a sustainable future, they said, but let us do so by ensuring food, shelter, clean water, health care, security of person and property, education, and participation in governance for all. An extension of this sentiment was a desire to protect basic values as well—to respect nature rather than dominate it, and to use the wisdom of indigenous groups, elders, and tribal leaders to reexamine current, mostly Western, structures of government, sources of knowledge, and the relationships that people have with the environment.

Is the West ready to listen to those whose lives reflect a different set of values and priorities? Surely there are many paths to a sustainable future, each determined by individualized priorities of what is desired and therefore worthy of sustaining. Surely too, in following those paths, we all must recognize that the future is constrained by physical resources that are often finite or whose availability is difficult to determine. Finally, we must realize that no region can achieve sustainability in isolation. A desirable and sustainable future will be the result of many policy changes, some small and at the local level, others international and far reaching. Desires for the future both unite and divide us. Deep ideological divides must be overcome, but if we accept that the futures of rich and poor are inextricably linked, perhaps we will achieve the humility necessary for compromise. There are millions of positive images of the future. Our responsibility is not to choose among them—that belongs to future generations. Our responsibility is to leave them social and natural resources that will allow them to make these images a reality.

Environmental Production Possibilities

Figure 3.33 shows the production possibilities curve again. This time, dollars worth of economic goods, both capital and consumer, is shown on the vertical axis, and dollars worth of pollution abatement is shown on the horizontal axis. Each society is confronted with getting one at the expense of the other. Currently, developing societies are not in on the pro-

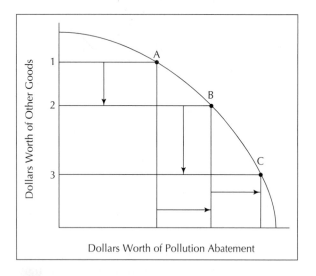

FIGURE 3.33
Production possibilities curve of pollution abatement cost. If a society adopts pollution abatement, other goods and services will have to be cut back. An initial level of a pollution abatement at point 1 can be purchased for a small reduction in the dollars worth of other goods and services shown by the intersection on the production possibilities curve at point A. Increasing levels of pollution abatement will require greater proportions of sacrifice in the dollars worth of other goods and services in a society at points B and C.

duction possibilities curve, meaning that they are receiving the total worth of other goods and spending 0 dollars worth of pollution abatement. United States and European nations have moved to position *B* on the production possibilities curve, meaning that they have sacrificed certain dollars worth of other goods, both capital and consumer, to achieve a level on the pollution abatement axis. Pollution abatement has been expensive. Economic goods must be sacrificed to protect the environmental commons that everyone can enjoy. What will be the level of trade-off? In the early stages, if a nation has no pollution abatement and wants to achieve some, it may be able to trade $1 worth of goods for $2 worth of pollution abatement. The production possibilities curve is fairly elastic at the upper level. At middle levels, $1 of other goods can be traded equally for $1 of pollution abatement. At lower levels of the curve, which are below *B*, there is a so-called law of *increasing cost*. Simply put, a dollar's worth of sacrifice of other goods will only buy 50 cents worth of pollution abatement. The point at which society selects the trade-off is based on economic, social, cultural, and political values and perceptions, and it is not an easily resolved issue.

From a Growth-Oriented to a Balance-Oriented Lifestyle

It appears unlikely that energy availability will place a limit on economic growth on the earth; however, drastic changes in the use of energy resources seem certain. The ultimate limits to the use of energy will be determined by the ability of the ecosystem to dissipate the heat and waste produced as more and more energy flows through the system.

In countless ways, energy improves the quality of our lives, but it also pollutes. As the rate of energy consumption increases, so too does water and air contamination. Sources of water pollution are numerous: industrial wastes, sewage, and detergents; fertilizers, herbicides, and pesticides from agriculture; and coastal oil spills from tankers. Air pollution reduces visibility; damages buildings, clothes, and crops; and endangers human health. It is especially serious in urban-industrial areas, but it occurs wherever waste gases and solid particulates are released into the atmosphere.

Pollution is the price paid by an economic system emphasizing ever-increasing growth as a primary goal. Despite attempts to do something about pollution problems, the growth-oriented lifestyle characteristic of Western urban-industrial society continues to widen the gap between people and nature. "Growthmania" is a road to nowhere. It is easy to see why. If the U.S. economy grew at a 5% annual growth rate, by about the year 2110, it would reach a level 50,000% higher than the present level. Problems of acquiring, processing, and disposing of materials defy imagination (Figures 3.34 and 3.35).

Many argue that we must transform our present linear or growth-oriented economic system into a balance-oriented system (Table 3.4). A balance-oriented economy explicitly recognizes natural systems. It recognizes that resources are exhaustible, that they must be recycled, and that input rates must be reduced to levels that do not permanently damage the environment. A balance-oriented economy does not mean an end to growth, but a new social system in which only desirable low-energy, high-labor growth is encouraged. It requires a deemphasis on the materialistic values we have come to hold in such high esteem. If current resource and environmental constraints lead us to place

FIGURE 3.34
Stages in the life cycle of environmental problems. Sewage and water treatment are in the final stages of environmental control, while indoor air pollution and urban sprawl are in the early stages.

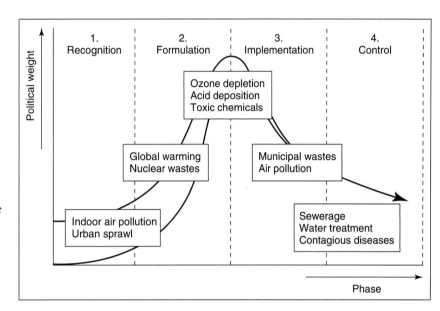

FIGURE 3.35
Acid deposition. Emissions of sulfur dioxide and nitrogen oxides react with the hydroxyl radicals and water vapor in the atmosphere to form their respective acids, which come back down either as dry acid deposition or, mixed with water, acid precipitation. Various effects of acid deposition are noted.

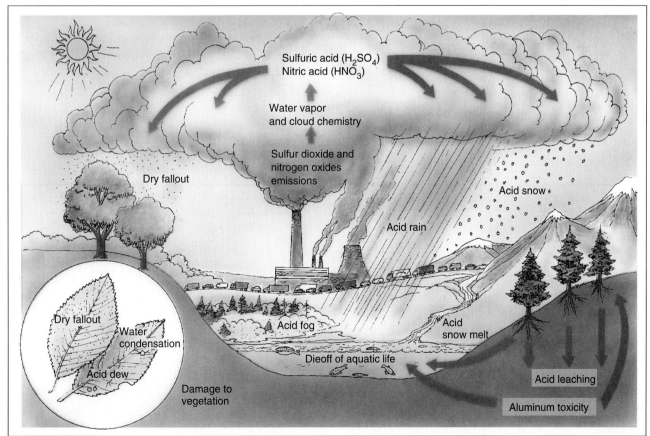

■

T a b l e 3.4

Benefits That May Be Gained by Reduction
and Prevention of Pollution

1. Improved human health
    Reduction and prevention of pollution-related
      illnesses
    Reduction of worker stress caused by pollution
    Increased worker productivity

2. Improved agriculture and forest production
    Reduction of pollution-related damage
    More vigorous growth by removal of stress due
      to pollution
    Higher farm profits, benefitting all agriculture-
      related industries

3. Enhanced commercial and/or sport fishing
    Increased value of fish and shellfish harvests
    Increased sale of boats, motors, tackle, and bait
    Enhancement of businesses serving fishermen

4. Enhancement of recreational opportunities
    Direct uses such as swimming and boating
    Indirect uses such as observing wildlife
    Enhancement of businesses serving vacationers

5. Extended lifetime of materials and less cleaning
      necessary
    Reduction of corrosive effects of pollution,
      extending the lifetime of metals, textiles,
      rubber, paint, and other coatings
    Cleaning costs reduced

6. Enhancement of real estate values

a higher premium on saving and conserving than on
spending and discarding, then they may be viewed as
blessings in disguise.

S UMMARY

We introduce this chapter by restating the resources-
population problem. It is possible to solve resource
problems by: (1) changing societal goals, (2) changing
consumption patterns, (3) changing technology, and
(4) altering population numbers. In the Western world,
much of the emphasis is on technological advancement
and population control.

Following a review of renewable and nonrenewable resources, we explore the question of food resources. The food "crisis" is essentially a consequence
of societal goals. Food production is increasing faster
than population growth, yet more people are hungry.
Socioeconomic conditions offer a more cogent expla-

nation of why this is so than either population growth
or environmental factors. In the course of transforming agriculture into a profit base for the wealthy, the
Third World poor are being forced out of the production process.

Unlike food, which is replenished by the seasons,
nonrenewable minerals and fossil fuels, once used, are
gone forever. We discuss some of the alternatives to
fossil fuels and point to energy conservation as a potent alternative with potential that remains to be fully
exploited. In conclusion, the comparison between
growth-oriented and balance-oriented lifestyles un-
derscores the importance of quality concerns as they
relate to economic growth.

C h a p t e r 3

APPENDIX A

W HY GIS?

Geographic Information Systems (or GISs) are becoming
widely established in the commercial, government, and
education sectors. The term "GIS" frequently is used to
describe a number of applications and systems. For the
beginner, the term may cause a great deal of confusion
because it appears to have a wide variety of definitions
and cover very different subject areas.

This section gives an introduction to the development of GISs, how GIS can be defined, and a short history of its evolution.

Introduction

Many organizations are now spending large amounts of
money on GISs and on geographic databases. Predictions
made by reputable firms suggest billions of dollars will be
expended on these items over the next decade. Why
should this be true now when only a few years ago such
spending was a rarity?

There are two obvious answers to this question.
The first is that the costs of computer hardware needed
to do a particular job are decreasing rapidly, and so
reaching a wider and wider audience. But more important still is that geography (and also the data that
describe it) is part of our everyday world; almost every
decision we make is constrained, influenced, or dictated by some fact of geography.

Fire trucks are sent to fires by the fastest routes
available. Central government grants are often handed
out to local governments on the basis of the population in each area, and we study diseases partly by mea-

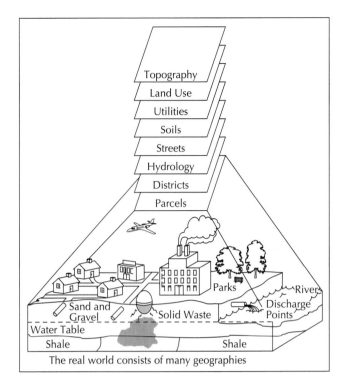

Topography
Land Use
Utilities
Soils
Streets
Hydrology
Districts
Parcels

Parks
River
Sand and Gravel
Solid Waste
Discharge Points
Water Table
Shale
Shale

The real world consists of many geographies

Geographic information systems store geocoded information about the earth's surface in data layers. Each layer, a coverage, represents a different aspect of human or environmental information. The layers can be viewed individually, or in combination, and each piece of information has an xy-coordinate, so it can be retrieved individually or in combination with other data.

suring where they are prevalent and how rapidly they are spreading. In principle, then, there is not only a need, but also an opportunity for GISs; from these have come the rapid growth in their popularity.

Such generalized explanations do not help you, our reader, to know why and how a GIS can help you. To bring you the benefits, we need to show you how to achieve results with a GIS.

First, however, it is vitally important to have some understanding of what a GIS actually is and what it can be used for—the latter is limited only by your imagination. We provide this understanding of GIS in several different ways.

What Is a GIS?

The use of GISs has grown dramatically in the 1980s to become commonplace in many businesses, universities, and governments. They are now used for an amazingly wide range of applications. As a result, there are many different definitions of what a GIS is and what it can or should do. We will describe a GIS as follows:

> A system of hardware, software, and procedures designed to support the capture, management, manipulation, analysis, modeling, and display of spatially referenced data for solving complex planning and management problems.

Although this definition is accurate, comprehensive, and widely accepted, we suspect that it does not help the newcomer to GIS. Instead, we can use a simpler definition for GIS as follows:

> A computer system that can hold and use data describing places on the earth's surface.

Spatial Operations

Many widely used computer programs—such as spreadsheets (e.g., Lotus 1-2-3), statistics packages (e.g., SPSS), or drafting packages (e.g., AutoCAD)—can handle some simple data of this kind (i.e., geographical or spatial data).

Why, then, are they not usually thought of as a GIS? The generally accepted answer is that a GIS is only a GIS if it permits spatial operations on the data. As a simple example, consider the file of data in the table below.

The table shows the (very) approximate number of people working on all aspects of GIS in each of these centers of activity in the field in 1989.

Name	Latitude	Longitude	GIS Population	Longitude	Latitude
London	80N	0	51	0	80N
Zurich	25N	8E	47	8E	25N
Utrecht	40N	5E	52	5E	40N
Redlands	50N	117W	34	117W	250N
Santa Barbara	50N	119W	34	119W	50N
Orono	30N	69W	45	69W	30N
Buffalo	30N	78W	42	78W	30N

Text and graphic images provided courtesy of Environmental Systems Research Institutes, Inc.

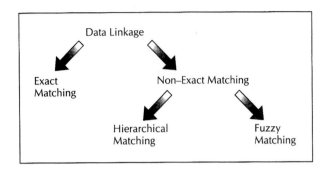

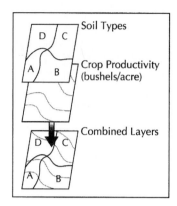

If we ask, "What is the average number of people working on GIS in each location?," this is an aspatial query; answering it does not use the stored value of latitude and longitude, describing where places are in relation to each other, and it is easily computed by many programs.

If, however, we ask, "How many people are working in GIS in the major centers in Western Europe?," "Which centers are within 1,000 miles of each other?," or "What would be the shortest route if I had to visit all of these centers?," these are spatial queries and can only be answered using the latitude and longitude data and other information, such as the radius of the earth. Such questions are readily answered by GISs.

Data Linkage

It is also the case that GIS can usually link different data sets together. Suppose we want to know the death rate due to cancer among those people under 10 years old in each county. Suppose also that (as usual) we have the numbers of people of this age in each county in one file and the numbers of death in the group for each county in another file. We need to combine, or link, the two data files. After this is done, division of one figure by the other for the same county gives the desired answer.

If this seems trivial—and scarcely in need of a GIS— it is not always so. Consider the different ways in which data sets may need to be linked together.

Exact matching is when you have information in one computer file about many geographical features (e.g., counties) and additional information in another file about the same set of features. The operation to bring them together is easy and is achieved through use of a key that is common to both files—the county name. Thus, the record in each file with the same county name is extracted and the two are joined together and stored in another file.

Sometimes, however, some information is available for more detailed geographical areas than is other information. Typically, for instance, you get frequently collected (e.g., finance or unemployment) data only for large areas and infrequently collected data (e.g., census) for very small areas. If the smaller areas "nest" (i.e., fit exactly within) the larger ones, then the solution is to use hierarchical matching: Add the data for the small areas together until the grouped areas match the bigger ones and then do an exact match.

On many occasions, however, the small areas do not match the larger ones. This is especially true when you are dealing with environmental data. Crop boundaries are usually defined by the edges of fields, and these rarely match the boundary between types of soil. If we wish to answer questions such as, "What soils are the most productive as far as wheat is concerned?," we need to overlay the two data sets and compute what crop productivity exists on each and every type of soil. In principle, this is like laying a map on tracing paper over another and noting the combinations of soil and crop productivity.

The important point is that a GIS can do all these operations because it uses geography or space as the common key between the data sets—information is linked together if it relates to the same space as does another set of information.

Why is data linkage so important? Consider the situation in which you have two data sets for a given area, such as income for every county in the country and the average cost of housing for the same area. Each data set may be analyzed or mapped individually. Alternatively, they may be combined: one such combination exists. If, however, we have 20 data sets covering the county, rather than two, we have more than one million possible combinations. Not all of these may be meaningful (e.g., soil type and unemployment), but we will be able to tackle many more tasks than if the data sets were kept separate. By bringing them together, we add value to our database. To make this a reality, we need a GIS.

Generic Questions That a GIS Can Answer

Thus far, we have described a GIS in two ways—through formal definitions and through its ability to carry out spatial operations and to link data sets together using space as the common key. We can, however, also describe what a GIS is by listing the type of questions that it can (or should be able to) answer. If we stand back far enough from a particular application, we can see there are five generic questions that a sophisticated GIS such as ARC/INFO can tackle.

The first of these generic questions seeks to find out what exists at a particular location. We can describe the location in different ways, for example, by place name, post or zip code, or geographic coordinates, such as latitude and longitude.

The second question is the converse and involves hunting through geographic space to find where certain conditions are satisfied, for example, a fishing lake without a public telephone, but no more than 50 miles from home.

The third question may involve both of the first two, but seeks the differences between the results for the two moments in time.

Questions four and five are more sophisticated. In asking question four, we may wish to know whether there is a cluster of deaths due to cancer among residents around a nuclear power station. Just as important, we will wish to know how many anomalies there are that do not fit the pattern and where they are located.

Finally, question five seeks to determine what happens, for example, if we add a new road to the network, or if a leak of a toxic substance occurs into groundwater; by its nature, answering such a query requires both geographic and other information (and possibly even scientific laws).

A good GIS (especially ARC/INFO) can be used to answer all five types of query; however, some systems take much longer than others to produce a result, and

Questions	Type of Task
What is at...?	Inventory and/or monitoring
Where is...true/not true?	Inventory...
What has changed since...?	Inventory...
What spatial pattern exists?	Spatial analysis
What if...?	Modeling

some are difficult to use. Indeed, many GISs at present have very limited capability to carry out spatial analysis or modeling.

Some Applications of GIS

The first applications of GIS varied between different parts of the world, depending on the local needs. Hence, in mainland Europe, the main effort went into the building of land registration systems and environmental databases. In Britain, however, the greatest expenditure in the 1980s was into systems for the utility companies and in the creation of a comprehensive topographic database for the whole country (mostly derived from maps at 1/1,250 and 1/2,500 scale).

In Canada, an important early application was in forestry for planning the volume of timber to be cut and access paths to the timber and in reporting all this to the provincial governments. In China and Japan, there has been heavy emphasis on monitoring and modeling possible environmental changes—unsurprising because of the catastrophic effects of flooding, earthquakes, and other natural hazards in these countries.

In the United States, all of these applications have also been important, but one other that deserves special mention is the use of GIS technology in the *TIGER* (Topologically Integrated Geographical Referencing) project by the U.S. Bureau of Census. This project produced a computerized description of the geography of the United States to facilitate taking and reporting the 1990 census and cost about $170 million. TIGER probably represents the largest-but-one collection of geographic data yet made—and ARC/INFO can read TIGER files.

The resulting geographic data files are on sale and can be used for a variety of purposes. When combined with demographic or other data from the 1990 U.S. Census, they can be used to target mail to suitable customers, underpin car guidance systems, and much else.

The most important point to note is that all these applications have been carried out using similar software and techniques. Thus, GISs are general-purpose tools.

The largest collection of geographic data yet assembled is the amazing volumes of satellite imagery collected from space. Unlike much other (vector) geographic data, these come in raster (or grid) form—small square areas of ground are each represented by one or more numbers that describe the properties of the ground area. Until recently, such data were invariably analyzed using special-purpose software, often on special hardware.

What a GIS Is Not

A GIS is not simply a computer system for making maps, even if these are drawn at a variety of scales and on a variety of different projections and in different colors. Maps are important to a GIS because much information is stored in map form that needs to be converted into computer form; maps are also important as an effective means of demonstrating results.

The basis of a map—the coordinate system on which it is based—is also the framework on which all the nongeographical data are hung. However, not all geographic information comes from maps (e.g.,meteorological data) and not all results are produced in this form (e.g., statistical summaries).

Even more important, a true GIS never holds a map in any conventional sense. Thus, we would never hold the area shown as a road on a small-scale map—we want to hold the data rather than the pictorial representation of data. In this case, we would hold the centerline of the road and a note of the true width or type of route.

From this information, we would compute the plotted width as appropriate for any given map. Equally, we would not usually hold lots of bird's-eye views of Mount St. Helens; we would hold a grid of heights of the ground and compute the particular view we wished, drawn in a way to suit a particular purpose and to please a particular individual.

In short, a GIS does not hold maps or pictures—it holds a geographic database from which we can (if we wish) produce the images, although we can also act on the data in other ways. To remove any confusion, ARC/INFO does not use the term "map" to refer to data sets but calls these "coverages."

It will be very clear from this that the database concept is central to GIS; in this, the GIS differs from simple drafting or computer mapping systems. Their role is only to produce good graphic output; thus, to produce a computer map, it is not strictly necessary that all lines meeting in a junction have the same coordinates. If the errors are small, the user will not be able to see them. Even small errors of this kind would, however, wreak havoc if the area of a polygon had to be computed.

Likewise, it is not essential that all the real-world objects, entities, or features making up a digital map are coded as such; provided that all of the lines are coded as to be drawn with a particular pen, it does not matter that the four lines making up the sides of a house are unrelated inside the computer. Once they are drawn, they will look to the human eye to form the real-world feature.

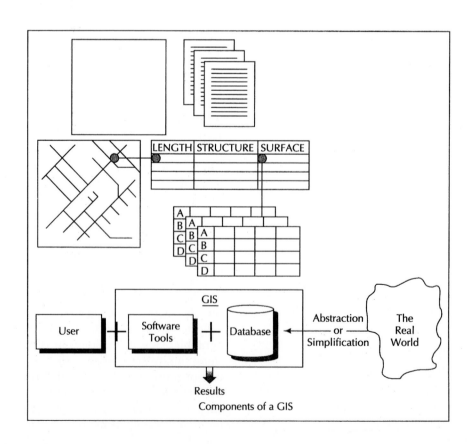

Components of a GIS

However, if we want to go beyond just making pictures, such as if we wish to study water flow through a river system, we need to know three pieces of information about every feature stored in the computer: what it is, where it is, and how it is related to other features (e.g., which roads link together to form a network). Database systems provide the means of storing a wide range of such information and updating it without the need to rewrite programs as new data are entered.

All contemporary GISs incorporate a database system. In ARC/INFO, ARC handles the "where the features are" information while INFO is the database that handles the descriptions of what the features are and the relationships between features.

The Components of GIS

From everything said earlier, we see that a GIS is made up as shown in the preceding figure.

Some people would argue, however, that the user becomes part of the GIS whenever complicated analyses have to be carried out, such as spatial analyses and modeling. These usually require skills in selecting and using tools from the GIS toolbox and intimate knowledge of the data being used.

At present and for years to come, off-the-shelf and general-purpose GISs will rely on users to know what they are doing—pressing a button is not enough. *Getting Started With ARC/INFO*, the primer on which this article is based, cannot make you an expert, but it is designed to teach basics and good practice; you need to read more detailed textbooks and to explore the GIS to become one.

Source: David Rhind, "Why GIS?" ARC News, October 1992, 11(3):1–4.

C h a p t e r   3

APPENDIX B

GIS SUITABILITY MAPPING AND MODELING

John Lyle and Frederick Stutz (1983) produced a land-use plan for San Diego County using suitability mapping. Land-use suitability mapping is a GIS technique that can help find the best location for a variety of land-use developmental actions, given a set of goals and other criteria. The mapping technique is based on environmental and human processes, and it analyzes the interactions among three sets of factors: location, land-use development actions, and environmental effects. The technique can yield three types of maps: (1) a map showing what land use will cause the least change in environmental processes; (2) a map showing qualitative predictions of environmental impacts of proposed land-use developments, given certain land-use developmental actions to be carried out and specific environmental actions to be controlled; and (3) a map showing the most and least suitable locations for those land-use actions.

At the heart of environmentally sensitive, systematic land planning and in the pivotal position between analyses and the definition of alternatives, mapped models, or suitability maps exist. A *suitability map* assesses the ability of each increment of land under study to support a given use. As the technique is now developing, the assessment of suitability is commonly based on predictions of the results likely to come about if a certain development is placed on a particular piece of land. Thus, in suitability mapping, environmental impact analysis and land planning can be effectively merged into a spatial decision support system (SDSS).

The suitability mapping process, then, begins with a collection of information forming coherent descriptions to these three component sets of factors, and a means of defining the connections among them. These three sets of information—describing locations, developmental actions, and environmental effects—form the three legs of the tripod on which the suitability mapping process is built.

It is important to recognize that we do actually need all three. Since suitability mapping has become more widespread, efforts to simplify and hasten the process have often resulted in failure to take into account either developmental actions or environmental effects. Too often, locational maps are simply overlaid without definition of either the uses or environmental factors being considered. This can lead to meaningless conclusions. It happens partly because the collection on locational information is the most visible, usually more extensive and complex than the other two. This collection is essentially a map file showing geographic distribution of the locational variables that interact with developmental actions and environmental effects.

Generally, at the most basic level, these include the horizontal layers of the biosphere, starting with bedrock composition and other geological factors and proceeding upward through soil types, hydrology plant and animal communities, and microclimates. To this we often add human contributions to the environment, such as existing land uses, transportation routes, accessibility, and political and social boundaries.

Any or all of these land factors can be almost indefinitely expanded. Soil types can be grouped by various characteristics, such as bearing capacity, expansion potential, or porosity. Or they can be divided into capability classes according to the U.S. Soil Conservation Service system. Hydrology might include groundwater basins and their capacity and recharge areas. Often, streams and tributaries are mapped to second, third, or fourth orders. Sometimes rates of flow are included. Man-made factors can likewise vary in their level and type of detail. Some developmental actions for which environmental effects can be estimated appear in the table below.

The suitability modeling process is briefly described as follows:

Given:

- Environmental effects to be minimized (stated in terms of transformations to be controlled) in priority order
- Land-use developmental actions to be carried out (e.g., residential)

Find:　　　Most or least suitable locations

Steps

1A. List environmental processes or transformations related to given developmental actions.

1B. List developmental actions related to given environmental transformations.

1C. List effects (outputs) related to environmental transformations in 1A.

2A. Match list of developmental actions in 1B with those given and record those that match.

2B. Match list of effects (outputs) in 1C with those given and record those that match.

3A. List locational variable, relative importance, and attribute sensitivity range related to developmental actions in 2A.

3B. Determine where interactions occur between locational variables and transformations in 2A and 3B.

3C. Assign a range to each variable according to its interactions; ranges must total 100%.

3D. Assign a score to each attribute based on its sensitivity range.

4. Print weighted map based on these scores.

Source Tools

Developmental action key charts linking process to development

Transformation key charts

Transformation key charts

Computer

Computer

Transformation key charts

Computer

Scoring matrix

Transformation key charts

Database, plotter

These totals give the relative importance of each land variable in providing the related environmental effect in the transformation key chart. Once this is done, the scores are simply inserted in the modeling program to be totaled for each grid cell, and the results printed by the computer. The map that resulted from the matrix just described is shown in the photo. In this case, the purpose was to accurately identify specific lands with agricultural potential in a single watershed within the San Diego Coastal Plain, where urban development is progressing rapidly and agricultural uses are in direct competition with residential and recreational uses.

Models for urban suitability and fire hazards at the regional scale were derived by similar means. In these examples, each cell represents the full 1,000-foot square. The urban suitability model is broad and inclusive, encompassing the consideration of a range of variables, while the fire hazard model is more limited in scope. The urban suitability map can provide a basis for urban growth policies and infrastructure locations and other long-range planning decisions. The fire hazard map can also play a role in these or be used for such practical purposes as locating fire stations or areas where shingle roofs should be banned or fire-retardant plants planted.

Developmental Actions for Which Environmental Effects Were Estimated

---

Cutting	Energy generation
Filling	Energy consumption
Excavation	Fertilizer application
Dredging	Groundwater extraction
Soil retention	Off-road vehicle use
Retaining walls	Active recreation
Dams	Passive recreation
Channels	Automobile operation
Walkways	Solid waste disposal
Demolition	Shoreline protection structures
Sewage disposal	Vegetation removal
Outdoor lighting	Vegetation introduction
Roof drainage	Animal species removal
Site drainage	Animal species introduction
Fences	Settling and debris basins
Paving	Spreading grounds
Pest control	Power transmission lines
Weed control	Power transmission structures
Irrigation	Building foundations
Aquaculture	Building superstructures
Soil cultivation	

Source: Lyle and Stutz, 1983.

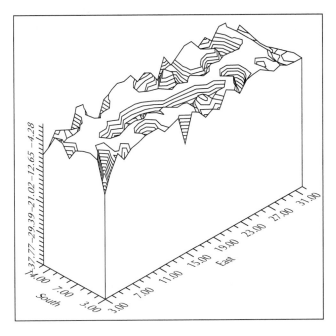

REGIONAL SCALE
URBAN SUITABILITY MODEL

≡ MOST SUITABLE

▓ LEAST SUITABLE

San Diego GIS Land Use Masterplan. The urban suitability model. The lightest shaded areas represent the most suitable areas for urban development; the darkest shaded areas are least suitable. *(Source: Lyle and Stutz, 1983)*

Frederick Stutz and Stuart Aitken located a California beltway with GIS. The highest areas on this block diagram of San Diego county show the most suitable location for a freeway. The sinks represent areas of physical or social (community cohesion) hazard. Caltrans has purchased this "plateau" land and is constructing the freeway. *(Source: Aitken and Stutz, 1993)*

Suitability mapping is a type of GIS spatial decision support system that gives ratings to all pieces of land in a study area based on human and physical land characteristics. It helps answer question three of the world economy: Where is the best location for economic activity X? *(Photo: Evans and Sutherland)*

KEY TERMS

acid rain
balance-oriented lifestyle
biomass
California Environmental Quality Act (CEQA)
carrying capacity
conservation
deforestation
depletion curves
desertification
energy
fossil fuels
geothermal energy
Green Revolution
growth-oriented lifestyle

intermediate technology
limits to growth
marine fisheries
maximum sustainable yield
mine tailings
minerals
National Environmental Policy Act (NEPA)
NIMBY and LULU effects
nonrenewable resource
normal lapse rate
Organization of Petroleum Exporting Countries (OPEC)
overpopulation
price ceiling

recycle
renewable resource
reserve
reserve deficiency minerals
resource
second law of thermodynamics
solar energy
stationary state
strategic minerals
tragedy of the commons
transmaterialization
triage
undernutrition
wind farm

SUGGESTED READINGS

Cutter, S. L., Renwick, H. L., and Renwick, W. H. 1991. *Exploitation, Conservation, Preservation: A Geographic Perspective on Natural Resource Use*, 2nd ed. New York: Wiley.

Durning, A. B., and Brough, H. B. 1991. *Taking Stock: Animal Farming and the Environment*. Worldwatch Paper 103. Washington, D.C.: Worldwatch Institute.

Ehrlich, A. H., and Ehrlich, P. R. 1987. *Earth*. New York: Franklin Watts.

Elsom, D. 1992. *Atmospheric Pollution: A Global Problem*, 2nd ed. Cambridge, Mass.: Blackwell.

Feshbach, M., and Friendly, A., Jr. 1992. *Ecocide in the USSR*. New York: Basic Books.

Goldsmith, E. P. B., Hildyard, N., and McCully, P. 1991. *Imperiled Planet: Restoring Our Endangered Ecosystems*. Cambridge, Mass.: MIT Press.

Goudie, A. 1992. *The Nature of the Environment*, 3rd ed. Cambridge, Mass.: Blackwell.

Goudie, A. 1993. *The Human Impact on the Natural Environment*, 4th ed. Cambridge, Mass.: MIT Press.

Harrison, P. 1992. *The Third Revolution*. New York: St. Martin's Press.

International Institute for Environment, Development and World Resources. 1988. *World Resources 1988–89*. New York: Basic Books.

Kemp, D. D. 1990. *Global Environmental Issues: A Climatological Approach*. New York: Routledge.

Knight, C. G., and Wilcox, P. 1975. *Triumph or Triage? The World Food Problem in Geographic Perspective*. Washington, D.C.: Association of American Geographers.

Lappe, F. M., and Collins, J. 1980. *Food First*. London: Abacus.

Mannion, A. M. 1991. *Global Environmental Change: A Natural and Cultural Environmental History*. New York: John Wiley & Sons.

Nebel, B. J. and R. T. Wright. 1996, *Environmental Science*, 5th ed. Upper Saddle River, N.J.: Prentice Hall.

Stutz, F. P. 1995. Environmental Impacts of Urban Transportation. In S. Hanson (ed.), *The Geography of Urban Transportation*. New York: Guilford.

WORLD WIDE WEB SITES

CONSERVATION DATABASES—WCMC
http://www.wcmc.org.uk/cis/index.html
The World Conservation Monitoring Centre, whose purpose is the "location and management of information on the conservation and sustainable use of the world's living resources," provides five searchable databases. Users can search by country for threatened animals and plants (plants are available for Europe only), protected areas of the world, forest statistics and maps, marine statistics and maps, and national biodiversity profiles (twelve countries only at present). Information is drawn from several sources, and database documentation varies from resource to resource.

STATE OF THE WORLD'S FORESTS 1997—FAO
[.PDF, 200P.]
http://www.fao.org/waicent/faoinfo/forestry/
SOFOTOC.htm
The United Nations Food and Agriculture Organization provides this book which "presents information on the current status of the world's forests, major developments over the reporting period (1995–1997), and recent trends and future directions in the forestry sector." SOFO provides information on global forest cover, including estimates for 1995, change from 1990, and revised estimates for forest cover change from 1980 to 1990.

ARID LANDS NEWSLETTER—UNIVERSITY OF ARIZONA
http://ag.arizona.edu/OALS/ALN/ALNHome.html
The Office of Arid Lands Studies at the University of Arizona's College of Agriculture supports the Arid Lands Newsletter web site. This site contains a browsable archive of current and past issues of the semiannual newsletter. Six newsletters are listed, covering the following topics: desertification, constraining geographical borders, computing and the web, biodiversity conservation, desert architecture, and deserts in literature.

NCGE REMOTE SENSING TASK FORCE
http://www.oneonta.edu/~baumaps/ncge/rstf.html

SEA FLOOR MAPS AND DATA
http://www.ngdc.noaa.gov/mgg/announcements/announce/predict.html
Declassified military data originally developed by NOAA from the U.S. Navy's GEOSAT and European Space Agency satellites are now available on the Internet.

ENVIRONMENTAL SYSTEMS RESEARCH INSTITUTE (ESRI)
http://www.esri.com
This home page contains a variety of valuable information on GIS technology. Among its offerings are free versions of Arc View software for Windows, maps, images and commercial data ses, online books and conference proceedings and secure electronic ordering. Use any popular browser to access the page.

GIS "HOT LIST"
http://gis.queensu.ca/pub/gis/doc/gissites.html

HOW FAR IS IT?
http://gs213.sp.cs.cmu.edu/prog/dist
Provides distance information, as the crow flies.

"MAPPING THE WORLD BY HEART"
http://www.world std.com/~mapping
David Smith's home page has links to worldwide geographical and educational resources, such as time zone maps and other related global information. They advise that these maps work best with a graphics browser such as Netscape.

NATIONAL SCIENCE FOUNDATION
http://www.nsf.gov
The National Science Foundation promotes and advances scientific progress in the United States by competitively awarding grants for research and education in the sciences, mathematics, and engineering.

SCIENCE & THE ENVIRONMENT
http://www.voyagepub.com/publish
Science & Environment is the award winning magazine that takes its readers globe-trotting to discover the state of the environment around the world. Each issue contains 80 stories filled with colorful photographs, maps, and graphics. Topics include Biodiversity & Wildlife; Alternative Energy & Fuels; Marine Ecology; Waste Management & Recycling; Clean Air; Clean Water; Health; Population and Agriculture.
This site was created by Voyage Publishing.

SOCIAL SCIENCE DATA ARCHIVES
http://ssda.anu.edu.au

THE NATIONAL COUNCIL FOR THE SOCIAL STUDIES
http://www.ncss.org/on-line

THE PENNSYLVANIA STATE UNIVERSITY DEPT. OF GEOGRAPHY WEB PAGE
http://www.gis.psu.edu/generalhtml/psgresources.html
Educational materials, graphics, and maps of Pennsylvania.

DESTINATIONS
http://www.lonelyplanet.com
This travel guide includes data with interactive maps and color photos.

ENVIRONMENTAL PROTECTION AGENCY
http://www.epa.gov
This site provides everything you ever wanted to know about environment and material resources.

GEOPEDIA
http://www.geopedia.com
Geopedia Online contains key information on every country of the world. Each country profile provides facts and data on geography, climate, people, religion, language, history, and economy making it ideal for students of all ages. A new environmental and global locator server developed by the European Commission is available on the World Wide Web at:
http://enrm.ceo.org
Further information on the information locator service can be found at:
http://www.g7.fed.us/gils.html
The Economic Research Service of USDA has launched a new and improved web site for research and data. Please stop by for a visit at:
http://www.econ.ag.gov

CHAPTER

4

TRANSPORTATION AND COMMUNICATIONS IN WORLD ECONOMY

OBJECTIVES

- To develop an understanding of modern transportation and communication systems
- To consider the impact of transport costs on locational patterns
- To demonstrate the relationship between transport and economic development
- To examine communications innovations and online computer networks
- To examine several new developments in transportation policy
- To consider recent innovations in transport development of metropolitan areas

Traffic in Calcutta, India. *(Photo: World Bank)*

For most of human existence, economic development was tied to natural conditions. People occupied narrowly circumscribed areas that were mostly isolated from other groups of people. Gradually, improvements in the efficiency and flexibility of growing transportation systems changed patterns of human life. Control and exchange became possible over wider and wider areas and facilitated the development of more elaborate social structures.

The course of human history was changed when European capitalism and its overseas progeny laid the foundations for technological culture. From the 16th century onward, there were great revolutions in science and trade, great voyages of discovery, and a consequent increase in productive, commodity, and financial capital. Capitalism required a world market for its goods; hence, it broke the isolation of the natural economy and of feudal society. The engine that drove this economic expansion was accumulation for accumulation's sake. In an effort to increase the rate of accumulation, all forms of capital had to be moved as quickly and cheaply as possible between places of production and consumption. To annihilate space by time, some of the resulting profits of commerce were devoted to developing the means of transportation and communication. "Annihilation of space by time" does not simply imply that better transportation and communication systems diminish the importance of geographic space; instead, the concept poses the question of how and by what means space can be used, organized, created, and dominated to facilitate the circulation of capital (Harvey, 1985, p. 37).

The transformation in transportation technology, together with capitalist development, served to integrate isolated producers. The integration of production points into a national or international economy does not change their absolute location (*site*), but it does alter their relative location (*situation*). Transport improvements increase the importance of relative space. The progressive integration of absolute space into relative space means that economic development becomes less dependent on relations with nature and more dependent on relations across space.

Most people no longer live in spatially restricted societies. Whether they live on farms or in cities, they can travel from place to place, communicate with each other over long distances, and depend on goods and information that come from beyond their immediate environment. Geographers refer to movements of goods, people, and ideas—by means ranging from walking to digital telephone networks—as spatial interaction.

Improvements in transportation promote spatial interaction; consequently, they spur specialization of location. By stimulating specialization, better transportation leads to increased land and labor productivity as well as to more efficient use of capital. As

societies abandon self-sufficiency for dependency on trade, wealth and income rise rapidly.

Trade occurs when time and money required to move goods over geographic space are within limits to permit local specialization. The amount of trade is related to the location of specialized production, the cost and time it takes to overcome the friction of distance, and the demand for goods. Production costs set the savings or additional wealth derived from local specialization and scale economies and influence the distance separating related activities. Specialization and trade may increase as long as production-cost savings exceed transport costs. For some activities, diseconomies occur at low levels of specialization; hence, production takes place at many locations. For other activities, concentration of production at a few locations is generally more profitable.

Transportation determines the utility or worth of goods. In today's world, almost nothing is consumed where it is produced; therefore, without transport services, most goods would be worthless. Part of their value derives from transport to market. Transport costs, then, are not a constraint on productivity; rather, trans-

Container cargo handling at the Maersk Line Terminal, Port Newark, New Jersey. Containerization has greatly improved the operation, management, and logistics of conventional ocean-going freight. The impact of the container evolution has gone far beyond shipping and international trade alone. Newly designed cellular vessels have much faster ship turnaround times in ports, as well as improved cargo-handling productivity at ports. An expanded interface between water and land transportation has occurred. Container trains have also enhanced the economy and scale of rail transportation. *(Photo: Port Authority of New York and New Jersey)*

port increases the productivity of an economy because it promotes specialization of location.

Transportation is a key for understanding geographic patterns. How does the geographic allocation of transport routes affect development? How do transport networks shape and structure space? How do they modify location? What is the impact of transport costs and transit time on the location of facilities? How is information technology (IT) changing the way we live, work, and conduct business? This chapter provides answers to these questions in discussions on transport costs, routes, and networks; transport development; transportation and communications innovation; and metropolitan concerns in transportation policy.

TRANSPORT COSTS IN THE WORLD ECONOMY

One of the major forces structuring the spatial organization of production is "the tyranny of distance": the fact that all movement costs. Societies have made a tremendous investment in both human and natural resources to overcome the friction of distance. Although transport innovations have reduced circulation costs, locational costs still exert a powerful influence on patterns of production. For that reason, we need to consider the following questions: What is the true form of transport costs? What determines specific transport rates? What effects do international regimes for shipping and aviation have on transport costs? What is the impact of transport costs on location? For what industries is transit time more crucial than cost?

General Properties of Transport Costs

Alfred Weber's industrial location theory emphasizes the cost of moving materials and finished products from place to place. Initially, it makes two normative assumptions about transport costs in order to concentrate on the idealized effects of distance. These assumptions are that: (1) transport costs are a linear function of distance, and (2) transport costs are exclusively a function of distance (zero distance equals zero cost). In reality, transport costs are much more complex.

Actual transportation costs can be categorized as either *terminal costs* or *line-haul costs* (Figure 4.1). Terminal costs must be paid regardless of the distance involved. They include the cost of preparation for movement, loading and unloading, capital investment, line maintenance, and other kinds of costs that are not a function of distance. Line-haul costs, in contrast, are strictly a function of distance. For example, fuel costs are proportional to the distance a load must be moved.

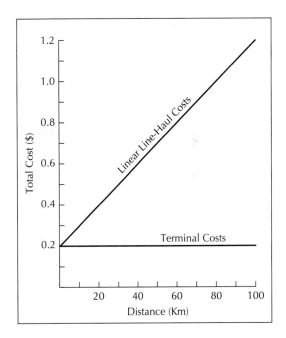

FIGURE 4.1
Terminal and line-haul costs. Terminal costs are also fixed costs. Line-haul costs incurred "over the road" are also variable costs.

Terminal costs are fixed in the short run, but are altered by technological changes. International shipping provides an example. In 1960, most merchant ships were general cargo vessels, and high loading and unloading costs as well as long turnaround times were involved in their operation. Since that time, more and more ships have been built to handle specialized cargoes. The first of the new ships were oil tankers, followed by vessels designed to carry ores, grains, and containers. The reduction in handling costs in ports and the rapid turnaround of the ships more than offset the cost of building specialized handling facilities (e.g., gantry cranes costing more than $2 million).

Recent developments in cargo handling have changed the location and appearance of ports. In the Netherlands, for example, old, enclosed dock systems of ports such as Rotterdam have been joined by new, deep-water terminals such as Europoort. In the 1960s, oil terminals capable of handling tankers weighing more than 250,000 tons were constructed at Europoort. Subsequently, additional deep-water facilities have been developed to handle trade in grain, coal, and ore. The most recent development has been the provision of container-handling facilities and roll-on/roll-off terminals.

The world's first containerized service tied trucks and ships together in 1956. McLean, a U.S. trucking firm, organized this operation at Sea-Land and used converted tankers on trips from the port of Newark, New Jersey, to Houston, Texas. Florida was added to this route in 1957 and Puerto Rico in 1958. By the mid-1960s, Sea-Land initiated service by new cellularized containerships from New York to Europoort, Bremen, and Grangemouth in Great Britain. Three years later, trans-Pacific service was established from Oakland, California, to Hong Kong, Taiwan, and Singapore. By the early 1970s, numerous carriers entered into the containership business.

At first, the greatest appeal of the containership was its speed and economy in port. Moreover, it facilitated the multimodal transport of goods. For example, commodities from Japan and other Pacific Rim countries could be transported economically to Europe via North America. Later, container operations sought to speed up the ocean voyage as well—top usable speeds increased from 15 knots in the 1950s to 33 knots in the 1970s. In addition, the oil crisis of the 1970s led to efforts to make more economical use of fuel to stabilize line-haul costs.

With the emergence of a new international division of labor, ports continue to modernize their methods of handling cargo as they compete with each other for shares of global commodity traffic. In developed countries served by many ports, competition has decreased the relevance of the traditional concept of the port hinterland (i.e., the area served by the port). On the West Coast of the United States, for example, ports in California, Oregon, and Washington compete fiercely for the mounting trade with the Pacific Rim. In underdeveloped countries and in marginal zones within developed countries, limited port systems still serve particular hinterlands.

CARRIER COMPETITION

Competitive differences in transport media account for variations in terminal and line-haul costs (Figure 4.2). Trucks have low terminal costs partly because they do

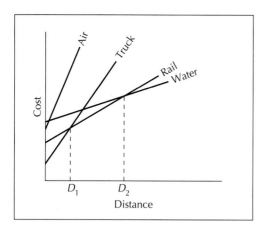

FIGURE 4.2

Variations in terminal and line-haul costs for air, truck, rail, water, and pipeline. Why do you suppose pipeline and water have the highest terminal costs? Why do air and truck have the highest line-haul costs?

not have to provide and maintain their own highways, and partly because of their flexibility. If provisions for parking are adequate, they can load and unload almost anywhere. However, trucks are not as efficient in moving freight on a ton-kilometer basis as are railroad and water carriers. Of the three competing forms of transport, trucks involve the least cost only out to distance D_1. Railroad carriers have higher terminal costs than truck carriers, but lower than water carriers, and a competitive advantage through the distance D_1-D_2. Water carriers, such as barges, have the highest terminal costs, but they achieve the lowest line-haul costs, giving them an advantage over longer distances.

CURVILINEAR LINE-HAUL COSTS

Thus far, line-haul costs have been portrayed as a linear function of distance. Actual line-haul costs, however, are curvilinear (Figure 4.3). As the graph illustrates, line-haul costs increase with distance, but at a decreasing rate. The distance from O to D_2 is twice the distance from D_1 to D_2, but does not involve twice the cost (C_1 to C_2). As the distance increases, the average cost per kilometer constantly decreases. This characteristic of actual transportation costs is often called "economies of the long-haul," which occurs for at least three reasons. First, terminal costs are the same regardless of the length of the trip. As line-haul costs increase, terminal costs become proportionally less of the total. Second, line-haul rates are lower for longer hauls. Short hauls by rail, for example, are moved by "local trains"; longer hauls are moved by "through" freight trains, which stop less frequently and operate more efficiently. Third, tapered line-haul costs prevent rates from restricting long-distance hauls. Rates would soon become high enough to prevent traffic if they increased in direct proportion to distance.

STEPPED FREIGHT RATES

Theoretically, every station along a line from a given origin should pay a different rate based on its actual distance from the origin, but computing large numbers of rates is both time-consuming and expensive to administer. Consequently, zonal-rate systems are a common feature of transport companies. For example, railroads group stations into areas and charge a single rate for all stations within the same zone. In general, group rates are set in relation to control points, often the largest centers in each zone, thus reinforcing the urban dominance of these centers.

Railroads have stepped freight rates that retain the tapering principle and favor long-haul movements (Figure 4.4). The total transport bill per unit of delivered material from station R to Y should be only slightly more than to X. However, because of the nature of the rate zones, Y pays a much higher price than X. Station Z pays the same price as Y because it is in the same rate zone. Station X has a competitive advantage over Y, but

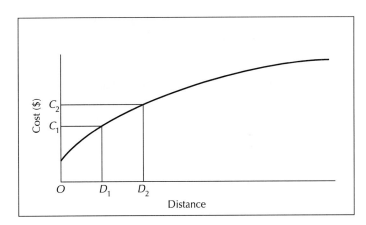

FIGURE 4.3
Curvilinear line-haul costs make the longer trips less costly per mile.

Y does not have a cost advantage over Z despite the greater distance of Z from R. Historically, cities such as Chicago, St. Louis, and some Missouri and Ohio River crossings have occupied and benefited from strategic positions in railroad rate groupings.

Commodity Variations in Transport Rates

The transportation cost curve varies according to differences in transport modes—but what determines specific rates? One set of factors pertains to the nature of the commodity. Factors that enter into any determination of commodity rates include: (1) loading and packaging costs, (2) susceptibility to loss or damage, (3) shipment size, (4) regularity of movement, (5) special equipment and services, and (6) elasticity of demand.

LOADING AND PACKAGING COSTS
Of particular importance in determining the reasonableness of rates is the weight density of a commodity. Light, bulky commodities usually incur higher freight charges per carload or shipload than heavy, compact articles. This explains why rates generally favor "knocked-down" or "set-up" commodities. For example, parts for an automobile are shipped at a much lower rate than for a finished car.

A low weight-density factor is not the only reason why some commodities load cheaply. Ability to load commodities compactly must also be considered. Articles of odd shape, such as furniture, may not load efficiently. Sometimes containers cannot be filled without damage to commodities. Melons, for example, cannot be loaded more than a few layers deep without crushing those on the bottom. Furthermore, some articles cost more to load. Rubber latex can be piped to ship rapidly and cheaply, whereas television sets must be handled with care to avoid damage. Coal requires little advance preparation for shipment, but furniture requires special crating and packing that add to terminal costs (Figure 4.5).

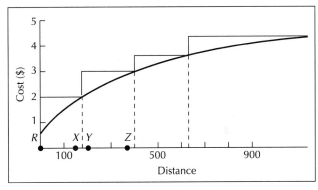

FIGURE 4.4
Stepped freight rates render ease of billing, but give advantages to cities located near the end of a flight zone.

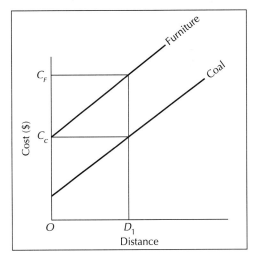

FIGURE 4.5
Variations in loading and packaging costs.

DAMAGE AND RISK VARIATIONS
Commodity rates reflect susceptibility to loss or damage. Except in the case of ocean carriers, most transport companies are liable for loss and damage during transit and must assume a greater risk for some commodities than for others. Sand, gravel, brick, and iron ore are not easily damaged, but fresh fruit and vegetables,

television sets, and china run a high risk of damage because of perishability or fragility.

SHIPMENT SIZE

Some commodities are shipped in bulk, whereas others must be carried in small quantities. Railroads charge higher rates for less-than-carload lots than for carload shipments. Rates are lower for commodities transported in volume over a period because carriers can better organize operation and handling methods and reduce costs. A high volume of a single commodity lowers line-haul costs per ton. Striking examples are fully loaded trains carrying only coal or iron ore.

REGULARITY OF MOVEMENT

If traffic moves regularly, carriers can operate at lower costs and should charge lower rates. Schedules are worked out more easily, and vehicle and labor needs can be planned. Irregularity of movement increases rates. This is often true of seasonal movements of fruits, vegetables, and wheat. Many railroad cars must be supplied over a short period. These either have to stand idle much of the time or be diverted from other routes, disrupting regular service.

SPECIAL EQUIPMENT AND SERVICES

The type of equipment required for a commodity affects freight rates. Commodities that require refrigerator cars are more expensive to transport than articles that can be carried by ordinary boxcars. Some commodities require special services. For example, refrigerator cars may have to be precooled before being loaded with shipments of fresh fruits and vegetables, and carriers may charge extra for the costs incurred.

ELASTICITY OF DEMAND

Previous factors related to the relative cost of transporting different commodities, but the *elasticity of demand* for transportation must also be considered. Defined in general terms, the elasticity of demand is the degree of responsiveness of a good or service to changes in its price.

Carriers generally charge what the market will bear. Often goods with a very high value per unit of weight, such as television sets, are able to bear a higher transportation rate than goods with a very low value per unit of weight, such as coal (Figure 4.6). Thus, the left-hand graph illustrates transportation price inelasticity for television sets. An increase in the rate from P_1 to P_2 produces only a slight change in the quantity of shipments. Coal, however (in the right-hand graph), exhibits a great change in the quantity of shipments, with only slight change in the transportation rate (P_1 to P_2).

The value of an article does not completely determine a commodity's ability to bear a higher freight rate. Transportation services are not purchased simply for the sake of consuming ton-kilometers. Transportation

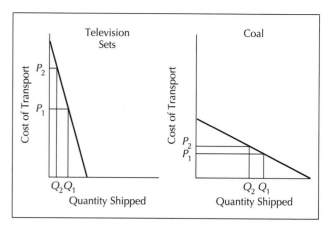

FIGURE 4.6
Demand elasticity for transportation. Higher valued television sets are more valuable and less elastic than quantities of coal. A large price increase to ship television sets results in a small reduction in the quantity shipped. A small price increase to ship coal results in a large reduction in the quantity shipped.

is a means of distributing localized commodities. The most localized commodities can usually bear higher rates within the framework of loading, shipment size, and damage and risk characteristics.

Freight Rate Variations and Traffic Characteristics

The characteristics of carriers and routes form a second set of factors determining specific transportation rates. Important factors include: (1) *carrier competition,* (2) *route demand,* and (3) *backhauling.*

CARRIER COMPETITION

An absence of competition between transport modes means a carrier can set rates between points to cover costs, and in the absence of government control, a carrier may set unjustifiably high rates. Intermodal competition or government regulation reduces the likelihood of such practices. The effect of competition between carriers is to reduce rate differences between competitors. For example, the opening of the St. Lawrence Seaway in 1959 resulted in lower rail freight rates on commodities affected by low water-transport rates.

ROUTE DEMAND

An important factor influencing the cost of haul is traffic density. High demand for transportation over a particular route can lower transportation rates. High demand lowers both line-haul and terminal costs per unit. The air shuttle between New York and Washington, D.C., is an example. Demand for this trip is so high that rates per passenger can be much lower than rates for trips over routes of similar distance where demand is low. High volume lowers the terminal costs per passenger. Fully loaded aircraft means lower line-haul costs per passenger.

BACKHAULING

Many carriers face heavy demand only in a specific direction. Consider the large volume of produce shipped from Florida to New York. Trucks must often return empty for the next load. The cost of the total trip, however, is used to determine the transportation rate. Because carriers must make return trips anyway, they are willing to charge very low rates on the backhaul. Any revenue on backhaul is preferred to returning empty. Rates are higher where there is little or no possibility of backhauling; most such runs occur in the transportation of raw materials from resource points to production points. An example is the railroad that carries iron ore pellets from Labrador to the port of Sept Iles, Quebec. This railroad may be likened to a huge conveyor belt that operates in one direction only. By contrast, the distribution of finished products generally involves traffic between many cities, creating a reciprocal flow and lower rates.

Regimes for International Transportation

In the international arena, transport rates and costs are affected by the nature of the regime governing the transport mode. To illustrate, let us consider the contrasting regimes of civil aviation and shipping. The international regime for aviation is dominated by the authoritative allocation of resources by states. By contrast, the international regime for shipping has been shaped by market-oriented principles. These different regimes were established by the industrialized countries. The regime for civil aviation developed in the early 20th century and reflects a concern for national security. The regime for shipping evolved over more than 500 years and has been more concerned with facilitating commerce than with security.

CIVIL AVIATION

The fundamental principle governing aviation is that states have sovereign control over their own air space. From this principle, rules and procedures have developed that permit countries to regulate their routes, fares, and schedules. As a result, many countries, developed and developing, have secured a market share that is more or less proportional to their share of world airline traffic. Developing countries have been able to compete with companies based in the industrialized world on an equal footing. Air India, Avianca, and Korean Air Lines can challenge Delta, Air France, and British Airways.

SHIPPING

The international regime for shipping has left developing countries in a weak position with regard to establishing and nurturing their own merchant fleets. In a world of markets, few underdeveloped countries have much influence when it comes to setting commodity rate structures. Lack of control over international shipping is an important area of concern in the Third World's quest for development.

The Rhine River is the main avenue of freight traffic for Europe. It enters the North Sea at Rotterdam, the world's busiest port. Although the Rhine River is heavily polluted from industrial wastes and chemical spills, it is still a scenic river and, along with castles such as this one, accounts for a large amount of tourism. The castle, Burg Katz, at Goarshausen recently sold to a Japanese businessman for $2.2 million, who will turn the historic landmark into a luxury hotel. *(Photo: Archive Photos)*

Although the regime for shipping is characterized by the market, the market is inherently unfair—it favors developed countries over developing countries. Hence, LDCs are faced with rate structures that work against them, inadequate service, a perpetuation of center-periphery trade routes, and a lack of access to decision-making bodies. Those LDCs generating cargoes such as petroleum, iron ore, phosphates, bauxite/alumina, and grains cannot penetrate the bulk-shipping market, which is dominated by the vertically integrated MNCs based in developed countries. Cartels of shipowners, known as liner conferences, set the rates and schedules for liners (freighters that ply regularly scheduled routes).

Developing countries have attempted to change the international rules of shipping. They want to generate fleets of a size proportional to the goods generated by their ports. Their accomplishments have been limited, however. The UNCTAD Code of Conduct for Liner Conferences, which was adopted in 1974, was rejected by the United States. The Liner Code gives developing country carriers a presumptive right to a share of the market; however, proposals to eliminate flags of convenience have not been accepted. Flags of convenience assume little or no real economic link between the country of registration and the ship that flies its flag. They inhibit the development of national fleets, but for shipowners they offer a number of advantages, including low taxation and lower operating costs. Liberia and Panama are the most important open-registry, or flags-of-convenience, countries. Flags of convenience are used mainly by oil tankers and bulk-ore carriers controlled by MNCs.

Transport and Location

TRANSPORT COSTS AND LOCATION THEORY
With a knowledge of actual transport costs, it is possible to examine their implications on modifications of Weber's industrial location theory (Figure 4.7). Consider Weber's solution for one pure raw material (i.e., a raw material that loses no weight in processing) localized at M and sold as a finished product at MKT (Figure 4.7a). Terminal costs are zero and line-haul costs are linear. What happens to the Weberian solution when terminal costs are added? The solution is given in Figure 4.7b. We must always pay at least one set of terminal costs because either the raw material or the finished product must be moved. At either the mine or the market, one set of terminal costs is paid, but at any intermediate location, two sets of terminal costs must be paid. This raises total transportation costs by an amount equal to one set of terminal costs. Thus, mine or market locations have a clear advantage over intermediate points in terms of terminal costs.

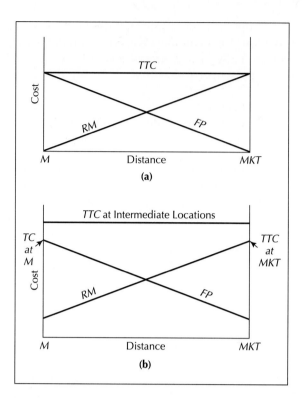

FIGURE 4.7
Weber's model: the effects of terminal costs. M is site of raw material; MKT is the market; RM is raw material procurements costs; FP is finished product distribution costs; TC and TTC are total transportation costs.

Curvilinear line-haul costs also favor mine or market locations (Figure 4.8). For simplicity, the diagram eliminates terminal costs. It shows that curvilinear line-haul costs favor the long haul. Shipping the raw materials from M to D_1 costs C_1; shipping the finished product from D_1 to MKT costs C_2. Shipping the raw material all the way

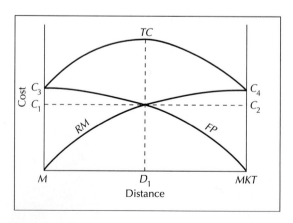

FIGURE 4.8
Weber's model: the effects of curvilinear line-haul costs favor the plant at the source of raw material (RM) or at the market (MKT), but not in between.

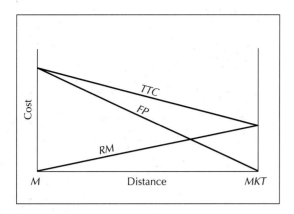

FIGURE 4.9
Weber's model: variations in commodity rates. Here, raw materials are cheaper to ship than finished products. Thus, total transport costs (TTC) are lower at the market (MKT).

to MKT, however, only involves a cost of C_4. Total transport costs are minimized at either mine or market.

Commodity rates influence the location of economic activities. As a very general rule, transport rates are lower for raw materials than for finished products favoring market locations (Figure 4.9). The graph retains linear transport costs for simplicity and shows that the lower raw material transport rate minimizes total transport costs (TTC) at the market.

Although tapering freight rates usually disfavor processing at intermediate locations because of additional terminal costs, these must be paid anyway at necessary transshipment or *break-of-bulk* points where a change in carrier must occur. This fact helps to explain why processing often takes place in port cities. Oil and sugar refineries, for example, often lie at tidewater. Iron and steel plants, the biggest and most visually impressive of all industrial establishments, are also attracted to coastal locations. The Ijmuiden works of Hoogovens in the Netherlands and the Mizushima works of Kawasaki steel on the north shore of the Japanese Inland Sea have deep-water access to ore and coal from international sources and can dispatch finished products to distant markets (Figure 4.10).

Thus far, we have examined the effects of transport rates on location in terms of *freight-on-board* (FOB) pricing. Consumers pay the plant price plus the cost of transportation; those close to the plant pay less than more distant consumers (Figure 4.11). However, many producers adopt a pricing policy known as *cost-insurance-freight* (CIF)

FIGURE 4.10
Market areas: (a) with linear line-haul costs; (b) with curvilinear line-haul costs.

pricing. In this pricing strategy, each consumer pays production costs plus a flat markup to cover transportation charges (Figure 4.12). Each consumer is charged a CIF price at C_1. Consumers from A to B are charged more than the actual cost of transportation. Consumers from X to A and B to Y are charged less than the actual cost. Close-to-plant consumers pay the distribution costs of more distant consumers.

What effect does CIF pricing have on the market area of producers? The FOB prices of producers A and B are shown in Figure 4.13. The market-area boundary is at X_1 with FOB pricing. If B adopts a CIF pricing strategy, the market-area boundary shifts to X_2. Producer A can, of course, counter by also adopting CIF pricing. Thus, each consumer would pay the same price to each producer, and price competition would disappear. Producers would then be forced to compete through advertising or other means. For finished products, such as clothing, CIF pricing tends to be the rule rather than the exception in the United States.

TRANSIT TIME AND LOCATION
Weber's industrial location theory and its modifications emphasize the cost of moving materials to the plant

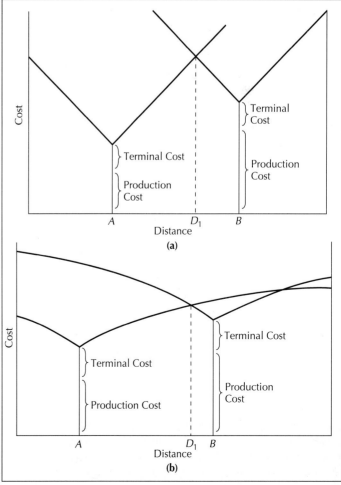

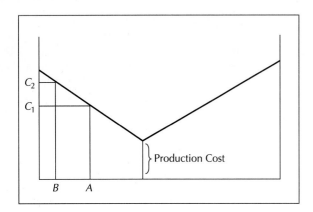

FIGURE 4.11
Freight-on-board pricing.

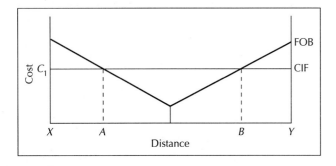

FIGURE 4.12
Freight-on-board and cost-insurance-freight pricing.

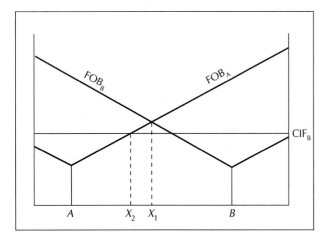

FIGURE 4.13
The effect of cost-insurance-freight on market-area boundaries.

The Mizushima works of Kawasaki Steel on the north shore of the Japanese Inland Sea has deepwater access for coking coal and iron ore from international sources. Major coking coal supply sources are Australia and Canada; major iron ore supply sources are the Philippines and Australia. *(Photo: Kawasaki Steel Corporation)*

and finished products to consumers. Transport costs are of crucial importance for industries that are raw material seekers and market seekers, but they are of little importance for industries dealing in materials and final products that are of very high value in relation to their weight. This is especially so for high-tech firms.

High-tech firms rely on input materials from a variety of domestic and foreign sources; thus, the advantages of locating a plant near any one supplier are often neutralized by the distance separating them from other suppliers. Their markets also tend to be scattered. Transport is a factor of some locational significance for these firms, but transit time is more crucial than cost. High-technology firms require access to high-level rapid-transport facilities to move components and final products, as well as specialized and skilled personnel. For this reason, they are often attracted to sites near major airports with good national and international passenger and air-cargo facilities. Concentrations of high-tech firms and research and development facilities are located in Silicon Valley near San Francisco, along the M4 motorway from Slough to Swindon to the west of London's Heathrow Airport, and in Tsukuba Science City, situated some 35 miles northeast of the center of Tokyo and 25 miles northeast of the Narita International Airport.

Transport Improvements and Location

Transport innovations have reduced circulation costs and fostered the new international division of labor. They have encouraged the decentralization of manufacturing processes in industrialized countries, both

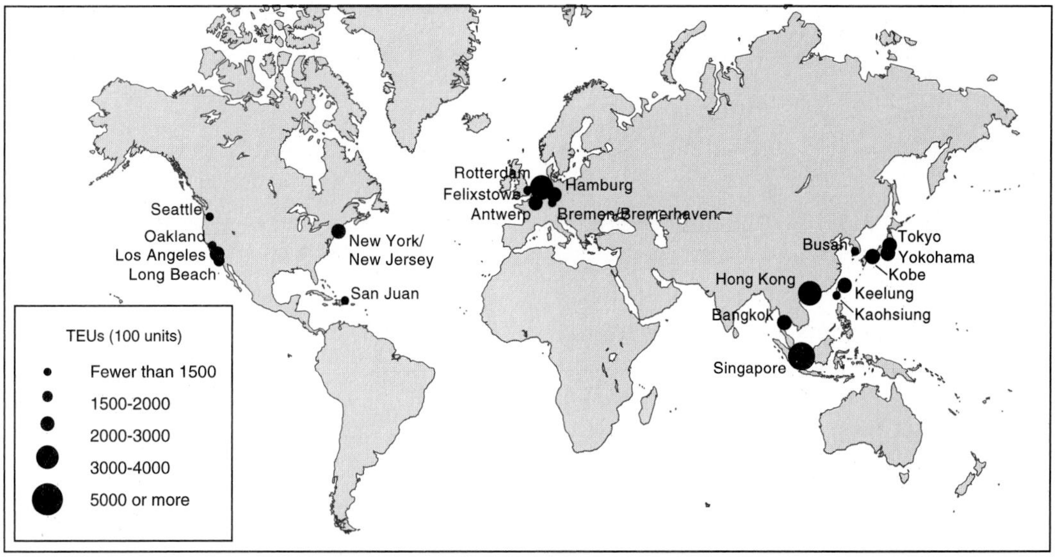

FIGURE 4.14

Most of the world's container ports are located in East and Southeast Asia and in Europe. New York, Los Angeles, and Vancouver represent major container ports in North America. Such ports are "hub ports" and act as major centers where container traffic splits into feeder flows to and from centers within the hub's respective hinterland. *(Source: Fleming and Hayuth, 1994, p. 10)*

from major cities toward suburbs and smaller towns, and from central regions to those more peripheral. They have also encouraged the decentralization of manufacturing processes to those LDCs with a free-market ideology and an abundance of weakly unionized, low-wage labor.

The "container revolution" (Figure 4.14) and bulk-air cargo carriers have enabled MNCs based in the United States, Japan, and Western Europe to locate low-value–added manufacturing and high-pollution manufacturing processes "offshore" in more than 80 Third World free-trade zones. Almost one-half of these zones are in Asia, including Hong Kong, Malaysia, and South Korea. Free-trade zones are areas where goods may be imported free of duties for packaging, assembling, or manufacturing and then exported. These global workshops are geared to export markets, often with few links to the national economy or the needs of local consumers. They tend to be located near ports (e.g., La Romana, Dominican Republic), international airports (e.g., San Bartola, El Salvador), and in areas virtually integrated into global centers of business (e.g., Mexico's northern border or maquila zone) in Frederick Stutz's 1992 study of Tijuana–San Diego.

Routes and Networks

Movements of goods, people, and information are highly channeled, but routeways do not exist by themselves. Rather, they are organized into networks. These individual networks—shipping lanes, railroads, high-

ways, pipelines—service transport demand and bind regions together.

Networks as Graphs

A network is a highly complex system, and each different type has its own special characteristics. Networks differ in terms of density, shape, type of commodity or information carried, and type of flow (either continuous or intermittent). These widely varying characteristics make networks difficult to describe, evaluate, and compare. In order to uncover the basic spatial structure of networks, geographers reduce them to the level of graphs.

A network idealized as a *graph* consists of two elements of geographic structure: (1) a set of vertices (V) or nodes that may represent towns, railroad stations, or airports; and (2) a set of edges (E), lines, or links that may represent highways, railroads, or air routes. The reduction of a network to a system of vertices and edges illustrates topological position only (Figure 4.15). The location of vertices is considered in terms of their relative position on the graph regardless of their absolute location. Distance between vertices is determined in terms of intervals, not route length.

Network Connectivity

We can evaluate the *connectivity* of a network simply by considering the system of edges and vertices. Connectivity means the ease of moving from one place to

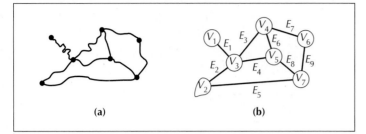

(a) **(b)**

FIGURE 4.15
Reduction of a network (a) to a graph (b). *(Source: Haggett and Chorley, 1969, p. 5)*

another within a network. Some networks are more successful in achieving ease of movement than others. The degree of success is known as the efficiency of the network, a property that we must be able to measure in order to compare networks.

The *beta index* is one of the simplest measures of network connectivity. It expresses the ratio between the number of edges in a system and the number of vertices in that system:

$$\beta = \frac{E}{V}$$

When the number of edges to vertices is large, the beta value is large, indicating a well-connected network. Conversely, more vertices than edges signifies a poorly connected network. A sequence of seven four-point graphs illustrates how we can measure differences in connectivity (Figure 4.16). In the simplest case, there are four unconnected vertices; therefore, the beta value is zero. The index reaches unity when all vertices are connected by the same number of edges. It exceeds unity when there are more edges than vertices. Although the maximum value in Figure 4.16 is 1.75, larger graphs would yield higher beta values. High beta values imply that an economy is advanced and can afford bypass links around intervening places.

The beta index may be used to compare the structure of networks. For example, a nomogram portraying beta values for the railroad networks of several countries indicates that the index is high in developed countries, such as France, and low in underdeveloped countries, such as Ghana (Figure 4.17). In addition to comparing several different networks at the same time, the beta index may also be used to compare a single network as it changes over time.

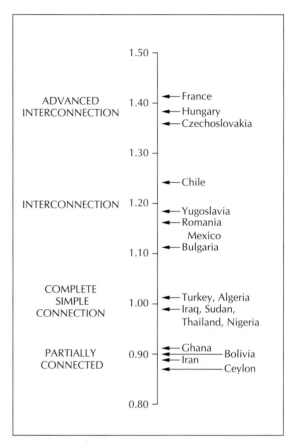

FIGURE 4.17
Beta values for the railroad networks of 18 countries. *(Source: Kansky, 1963, p. 99)*

Measures of network connectivity, however, have low discriminating power. For example, with a ratio like the beta index, the same value may be obtained for two networks having patterns that are not at all alike.

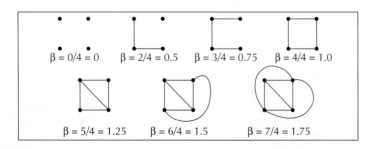

$\beta = 0/4 = 0$ $\beta = 2/4 = 0.5$ $\beta = 3/4 = 0.75$ $\beta = 4/4 = 1.0$

$\beta = 5/4 = 1.25$ $\beta = 6/4 = 1.5$ $\beta = 7/4 = 1.75$

FIGURE 4.16
Beta values for seven four-vertex graphs. How is economic development related to connectivity and beta values for a country?

NETWORK ACCESSIBILITY

The search for indices of higher discriminatory power led to the development of graph-theoretical measures of *accessibility*. These include the accessibility index and the *Shimble or dispersion index*. The accessibility index measures the shortest paths from each vertex to every other vertex; the dispersion index computes the accessibility of a network as a whole.

The accessibility of a vertex is

$$A_i = \sum_{j=1}^{n} d_{ij}$$

where d_{ij} is the shortest path from vertex i to vertex j. For a hypothetical four-point network, the accessibility index for $V_1=3$, $V_2=4$, $V_3=4$, and $V_4=5$ (Figure 4.18). These values correspond to our intuitive notions about vertex accessibility. The most accessible place is V_1, the least accessible is V_4, and V_2 and V_3 have intermediate accessibility. If the index were applied to the U.S. railroad system, we would find that Chicago has a low accessibility index compared to New York or Los Angeles, both of which are peripheral. When we deal with graphs much larger than a four-point network, the accessibility value for a vertex cannot be obtained by visual inspection. Computers are employed for large graphs, but for medium-size graphs we can use a shortest-path matrix. A matrix is an array of numbers ordered in rows and columns. The shortest-path matrix for the four-point network in Figure 4.18 is illustrated in Table 4.1.

The Shimble or dispersion index is defined as

$$D = \sum_{i=1}^{n} \sum_{j=1}^{n} d_{ij}$$

For the four-point network in Figure 4.18, the value of the dispersion index is 16 (Table 4.2). This value defines the graph's compactness in terms of all the paths within it and can be used to compare one network with another. For example, when the dispersion value is

◼

Table 4.1
A Shortest-Path Matrix

To:	V_1	V_2	V_3	V_4	Row Sum
From:					
V_1	0	1	1	1	3
V_2	1	0	1	2	4
V_3	1	1	0	2	4
V_4	1	2	2	0	5

◼

Table 4.2
Connectivity Matrix and Dispersion Value

To:	V_1	V_2	V_3	V_4	Row Sum
From:					
V1	0	1	1	1	3
V2	1	0	1	2	4
V3	1	1	0	2	4
V4	1	2	2	0	5
				Total	16

known, it can be used as a standard against which to measure the impact of new links on total accessibility.

Dispersion values are only an initial step in evaluating a transport network. Routes must be considered in terms of numerous criteria: characteristics of modes of transportation (e.g., carrier capacity, cost, frequency of service, speed); vehicular capacity; technical quality (e.g., surface, curvature, gradient); and stress (e.g., overuse of certain links). The cost of building, improving, or maintaining links must also be taken into account, as well as the route's effectiveness in meeting given objectives. For example, the cost of improving a road to meet the capacity for peak demand must be balanced against the cost of congestion, time loss, and deterioration of the route from overuse.

Density and Shape of Networks

Graph theory is particularly useful in measuring accessibility, but it neglects important aspects of network structure. For example, graph theory fails to consider network density or shape.

NETWORK DENSITY

By *network density* is meant the total number of route miles per unit area. This measure may be considered in several dimensions of space.

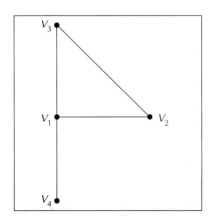

FIGURE 4.18
A four-vertex network, abstracted as a graph.

Examination of topographic maps reveals strong differences in road density at the local level. Villages have a denser pattern than the surrounding countryside, and downtown areas of cities have denser street patterns than suburbs. In the central area of Detroit, about 50% of the land is devoted to roads, but 3 kilometers from the central business district this figure drops to 34%. Distance-decay gradients, however, vary directly with city size. As city size increases, the need for interaction increases, and the proportion of land devoted to transport needs increases in linear fashion.

A study of the road pattern in Minneapolis–St. Paul by John Borchert (1961) provides much information on distance-density gradients for urban transportation networks. Because there was a high correlation between number of road junctions and road length, Borchert counted the number of road junctions on the map. Counting the number of junctions per unit area was much less tedious than measuring the number of miles of street per unit area. Borchert found a strong association between population density, as measured by the number of single-family dwellings, and network density, as measured by the number of intersections. In replicating Borchert's study in Dar es Salaam, Tanzania, in 1969, de Souza obtained similar findings.

At the regional level, variations in network density are closely related to patterns of uneven development. In 1970, de Souza calculated the road density in Tanzania by measuring the length of the main and secondary roads in grid squares of approximately 740 square kilometers. The appropriate length was then assigned to the center of each square, and isolines were drawn to join places of equal density. Road density is positively associated with areas of high population density and of intense commercial activity. For example, high road densities occur around the primate city of Dar es Salaam and provincial towns such as Tanga and Iringa. They also occur in zones of export agriculture such as the cotton-growing area to the south of Lake Victoria and the coffee-growing area on the well-watered slopes of Mount Kilimanjaro.

At the world level, the distribution of network density is highly skewed (Table 4.3). Norton Ginsberg (1961) calculated that a few countries have dense networks and many countries have sparse networks. Nearly two-thirds of the countries have distributions below the world mean. The distribution of countries with high and low densities is related to levels of economic development.

The greatest concentration of surface transportation facilities appears in Western Europe, the United States and southern Canada, Japan, and in western parts of the former Soviet Union. In these regions, road and rail densities are so high that virtually no place is inaccessible. Somewhat less dense networks are found in parts of Uruguay, Argentina, eastern Brazil, eastern Australia, India, and Pakistan, and in parts of the Mediterranean Middle East. Most LDCs are poorly served by roads and railways; for example, the vast tropical heartlands of South America and Africa and the interior of China are not easily accessed.

An Amtrak conventional passenger train at Harper's Ferry, West Virginia. New magnetic levitated trains will shuttle passengers between American cities at over 300 mph. Using far less energy and time than automobile and air travel, one will go by train from Los Angeles to San Francisco in an hour and a half, or between Washington, D.C., and Boston in less than an hour. Maglevs are twice as fuel efficient as cars and four times as efficient as airplanes, while producing no pollution themselves. Trains are considerably more comfortable than cramped aircraft and, as the French and Japanese trains have shown, could grow to be a significant portion of the public transportation market. *(Photo: Jerome Wexler, Photo Researchers, Inc.)*

◙

T a b l e   4.3

Distribution of Routine Density

Route Media	Roads	Railroads
Number of countries compared	126	134
World mean density, km/100 km²	10.3	0.95
Maximum density, km/100 km²	302.0	17.90
Minimum density, km/100 km²	0.0	0.00
Percentage of countries below world mean	64	67

Note: Although these data are for the late 1950s, they are still an effective representation of world-scale network patterns.
Source: Ginsburg, 1961, pp. 60, 70.

NETWORK SHAPE

There is a striking contrast between developed and developing countries with regard to the shape and orientation of transport networks. In LDCs, these features are a reflection of their colonial history. Resulting networks often have a strong directional focus; they resemble drainage systems that converge on coastal ports. For example, railroad development in Brazil linked the port of Rio de Janeiro with Sao Paulo and the export-producing areas inland. In Argentina, railroad development centered on Buenos Aires, and in Uruguay, on Montevideo. Port cities served as transshipment points for the export of primary products and the distribution of imported finished goods. Therefore, the networks of developing countries are typically fan-shaped. They distort and sharpen geographic and social inequalities because of an inadequate number of interlinkages.

In developed countries, the shape of transport networks is a fuller lattice, which allows a more even distribution of places by offering a degree of internal interchange. For example, Britain was crisscrossed with a dense network of main and branch railroads as early as 1900. However, major routes converged on London. This tendency strengthened after World War II when the government nationalized and modernized the rail network.

Location of Routes and Networks

Spatial interaction is movement on the earth's surface and depends on the existence of a demand-supply relationship. If a demand-supply relationship exists between two unconnected places and is profitable, then it is probable that a transport route will be constructed.

Choosing the actual location for a new route is a political task, but the information used to aid decision making is based on economic principles. Of critical importance in deciding where to construct a new route is the balance between fixed and variable transport costs. Fixed costs are construction costs. Variable costs are operating costs that depend on the length of the routeway and the volume of traffic flowing along it.

A demand-supply relationship explains many, but not all, transport patterns. It does not, for example, account for the geographic pattern of Roman roads in Britain. Roman roads were built across the island to meet the government's need for fast communication between centers of civil administration in the south and east and defense lines and fortresses in the north and west. Postroads were constructed with little regard for construction costs and local economic needs.

Minimum-Distance Networks

As an example of how costs influence the geographic pattern of routes in a network, consider two extreme minimum-distance solutions to a problem in which a demand-supply relationship exists between five towns (Figure 4.19). In Figure 4.19a, the fixed costs of road construction are low, but operating costs are somewhat

TGV Express Train in France. In 1983, the Trans Grande Vitesse was introduced by the French Railway with service between Paris and Lyon at speeds of up to 200 mph on an entirely new track. By 1994, it had captured millions of new passengers from the highways and from domestic airlines. Proposed high-speed trains from Naples to Milan, from Lisbon to Marseille, from Bordeaux to Glasgow via the Channel Tunnel, and from Geneva to Amsterdam are to be opened by the year 2005. Despite these advances, interurban rail will continue to occupy a subordinate role in the United States. But in China, Russia, and India, where private car ownership is low, the railway still carries the bulk of interurban traffic. *(Photo: Rapho-Beaune, Photo Researchers, Inc.)*

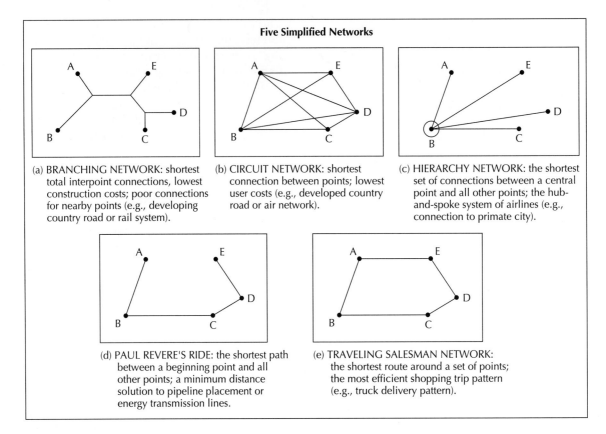

FIGURE 4.19

Five graphs representing five simplified networks; each depicts a class of transport problems in its simplest form. *(Source: Adapted from Haggett, 1967, p. 31, and Fellman, Getis, and Getis, 1992, p. 68)*

higher. The resulting network minimizes construction costs and is called a *least builder's cost network*. This network has a lower degree of connectivity and is less convenient to users. In Figure 4.19b, operating costs are low, but fixed costs are high. The resulting network is maximum benefit to users because each of the five towns is directly linked to every other town. This is the *least user's cost network*. In Figure 4.19c, the *hierarchical network* shows the shortest set of connections between a central city and all other points. In Figure 4.19d, the network displays the shortest path between a beginning point and all other points in the system. It is called the *Paul Revere Network*. In Figure 4.19e, the shortest network interconnecting a set of points, where the origin and destination point are identical, is given. This is called the *traveling salesman network*.

The railroad pattern in the United States can be understood partly in terms of least-cost-to-use and least-cost-to-build motives. The least-cost-to-use network is characteristic of the eastern half of the country where cities are clustered and transport demands are high. In the west, where railroads preceded settlement and cities are scattered, the least-cost-to-build solution dominates. Two other examples of least-cost-to-build

networks include Amtrak's National Rail Passenger System (Figure 4.20 top) and the French Master Plan for the future TGV Network (Figure 4.20 bottom).

Deviations From Straight-Line Paths

Most transportation routes deviate from a direct straight-line connection. There are two main types of distortion from the straight-line connection; they are positive and negative deviations.

Positive deviations occur when routes are made longer in order to increase traffic. They are constructed to pick up as many settlements as possible. At one time, this type of deviation was common in developed countries. Adherence to on-line settlement, however, declined in importance with economic development. An extreme example is the U.S. Interstate Highway System. When it was built, many small towns, villages, and hamlets were bypassed in favor of larger settlements that generated the bulk of the traffic. In developing countries, linking as many settlements as possible played hardly any part in the construction of routes during the colonial period. The length of the routes

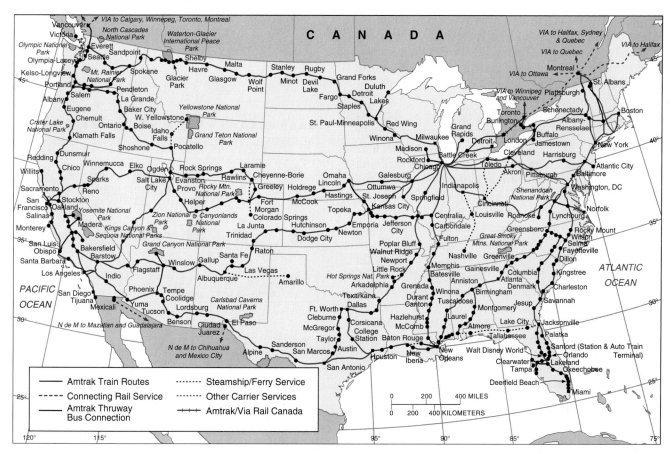

FIGURE 4.20
(top) Amtrak's national rail passenger system. (bottom) Europe's planned new rail network. As part of a worldwide mission to encourage more travelers onto rail from other modes, railway companies are developing improved passenger and freight information services. Real time information at stations and on trains is now a reality in some quarters, with more services to come. High speed passenger transport applications are driving technology innovation in the rail industry. Expansion of the European HSR program is back on track, with eyes now on extensions into eastern Europe. Similarly, Japan is pushing ahead with its HSR extensions. With Korea and even the U.S. now taking high speed seriously, HSR procurements are looking up. Vehicle positioning technology widely used in the automotive industry is now gaining a foothold on the railways. Amtrak's recent use of Qualcomm-supplied GPS coverage is just one example. Spain has gone down a similar track for some of Catalonia's freight and passenger lines. With moves to shift more freight onto railways, positioning technologies are becoming increasingly important to support tracking and tracing functions.

was geared to a pattern of economic growth based on export production. Areas where export production was insignificant or where there was little demand for imports were bypassed by road and rail. As a result, these places stagnated and lost ground.

Negative deviations arise from the need to reduce the distance traveled through high-cost areas. A transport route is often distorted from a reasonably straight line because different areas have different building-cost characteristics. For example, differential building costs may be the result of terrain difficulties, such as the presence of a mountain barrier. But economic development necessitates tearing down physical barriers to exchange. In the 19th century, a great deal of trade between the east and west coasts of the United States was diverted via the Cape route—a diversion that added some 9,000 miles to the overland distance. When the time factor came to be more highly valued, the degree of distortion was reduced with the construction of the Panama Canal. Today, east-west surface transportation across the United States is hardly distorted at all by overland barriers.

Development of Transport Networks

Historically, the development of transport networks has reflected and induced settlement, industrialization, and urbanization. The impact of transport networks on regional economic development is demonstrated in the 1963 stage model of network change in underdeveloped countries created by Edward Taaffe, Richard Morrill, and Peter Gould. Studies in Nigeria and East Africa provided the basis for their model. They defined the extension of transport in underdeveloped countries explicitly in terms of penetration from the coast. "[Transport links reflect] (1) the desire to connect an administrative center on the sea coast with an interior area of political and military control; (2) the desire to reach areas of mineral exploitation; [and] (3) the desire to reach areas of potential agricultural export production" (Taaffe, Morrill, and Gould, 1963, p. 506).

African Transport Development

The Taaffe, Morrill, and Gould model illustrates how the interplay between the evolution of a transport network and urban growth is self-reinforcing (Figure 4.21). The ideal-typical sequence begins the first stage when early colonial conquest creates a system of settlements and berthing points along the seacoast. Gradually, a second stage evolves with the construction of penetration routes that link the best located ports to the inland mining, agricultural, and population centers. Export-based development stimulates growth in the interior, and a num-

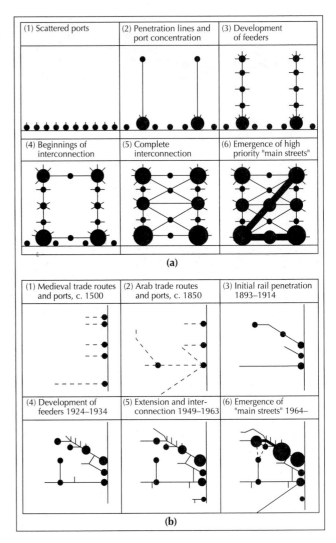

FIGURE 4.21
An idealized sequence of transportation development: (a) the Taaffe, Morrill, and Gould model, 1963; and (b) an adaption to East Africa by Hoyle, 1973.

ber of intermediate centers spring up along the principal access routes. This process results in the third stage of transport evolution—the growth of feeder routes and links from the inland centers. By the fourth stage, lateral route development enhances the competitive position of the major ports and inland centers. A few nodes along the original lines of inland penetration (i.e., N_1 and N_2) become focal points for feeder networks of their own, and they begin to capture the hinterlands of smaller centers on each side. The fifth stage evolves when a transport network interconnects all the major centers. In the sixth and last stage, the development of high-priority linkages reinforces the advantages of urban centers that have come to dominate the economy.

According to Taaffe, Morrill, and Gould, high priority linkages in underdeveloped countries are likely to emerge not along export-trunk routes, but along routes connecting two centers concerned with internal exchange activities. Nigerian geographer Akin Mabogunje (1980) questioned the logic of this assumption in that the sixth stage arises from earlier phases in which export-trunk routes receive the greatest attention. He argued that this contradiction can only be explained by the fact that the model is grounded in the history of transport development in developed countries. Thus, stage six of the model emphasizes the difficulty of transforming a colonial network into one suitable for more self-centered development (McCall, 1977).

The idealized model of network change describes one typical sequence of development. It shows that a transport network has the short-run purpose of facilitating movement, but that its fundamental effect is to influence the subsequent development and structure of the space economy through the operation of geographic inertia and cumulative causation. The term geographic inertia refers to the tendency of a place to maintain its size and importance even after the conditions originally influencing its development have changed, ceased to be relevant, or disappeared. The term cumulative causation refers to the process by which economic activity tends to concentrate in an area with an initial advantage. The stage model, therefore, illustrates how a space economy roots itself ever more firmly as initial locational decisions that shaped the system are subsequently reinforced by other decisions. The result is a concentrated and polarized pattern of development.

The evolution of transport systems is linked to the process of development. For thousands of years, most people walked and carried the goods they consumed. This time-consuming mode of transportation greatly limited the movement of commodities until after 1500, when major improvements in transportation began to take place.

At the time of Columbus, a mercantile revolution occurred in which nation-states replaced the chained economies of feudal society. The ideology of merchant capital, mercantilism, points to foreign trade as the source of a country's enrichment. To mercantilists, there existed a finite amount of trade in the world; consequently, the goal was to obtain the largest share. Mercantilism, therefore, was rooted in commercial expansion, for which improved transport was vital. Commercial expansion first arose in Portugal and Spain; then, early in the 16th century, it became a characteristic of Western Europe. From many rival harbors, especially Antwerp, Amsterdam, and London, a commercial network spread out to embrace parts of sub-Saharan Africa and its adjacent Atlantic islands, the Americas, and parts of Asia.

Opposition to mercantilism increased in the late 18th century, gradually giving rise to a climate of economic liberalism. This new climate was especially evident in Britain, where economic life was being transformed by improvements in farming techniques and by the Industrial Revolution. From 1800 to the present, capitalism promoted the development of cheap and rapid forms of transport in an evolutionary process spurred by competition among the various transport modes.

Southeast Asian Transport Development

P. J. Rimmer (1977) identified an alternate transport development strategy from his experiences in Southeast Asia. He termed it a hybrid transport system because the colonial powers' transport and development system is superimposed through the colonization process on the economic and cultural system of the less developed region. The result is that the colonial power exercises almost complete control over the exchange of goods internationally as well as the indigenous and imported dimensions of the transportation and economic system that results. Rimmer generalized four stages in the development of transport and trade between developed and less developed countries:

1. A *precontact stage* in which no transportation or trade exists between the developed country or the Third World country. Within the Third World country, a rudimentary transportation system exists.
2. Beginnings of *colonialism stage*. Initial contact is made between regions, but the developed country's role in the social, political, and economic life of the undeveloped country is minimal. However, permanent trading posts and garrisons are established.
3. *High colonialism.* In this stage, the European power develops a set of railways, roads, and inland transportation systems with its trading post now established as the new capital of the developing region. Diversification of economic activity begins an importance of manufactured goods, and exportation of natural resources moves into higher levels of intensity.
4. The *neocolonial stage.* Although no radical adjustments are made from earlier stages, a greater modernization and diversification occurs in the Third World country in terms of transport and economic development.

A fifth stage, which could be called *independence and codominance*, should now be added to the Rimmer model to coincide with Australia, New Zealand, and American developmental history. Here, a Third World nation or colony has developed into a regional economic and transport power on its own and has received full independence. With diversification and increased levels of international trade and transportation infrastructure development, links to the original developed country are strengthened and each region has developed a strong sense of identity as well as its own intensive levels of comparative advantage, as well as highly developed internal transportation networks.

FLOWS IN NETWORKS

Networks are constructed to carry flows of goods, people, and information. Geographers call this flow *spatial interaction*. Spatial interaction is the movement of goods, people, and ideas between areas, countries, cities, and even places within cities. These flows represent the exchange of supplies and demands at different locations. The flows are on networks. Geographic space is built with different availabilities of resources, commodities, information, opportunities, and populations. Because of this fact, movement occurs as a technique to satisfy human wants and needs with natural resources available and, finally, with the knowledge and skills that can convert the natural resources into usable items.

The California geographer, Waldo Tobler, coined the term *distance decay*, which he called "The First Law of Geography: Everything is related to everything else, but nearer things are more related to each other than are distant things." Distance decay is a geographic concept that describes the attenuation or reduction in the flow, or movement, between places with increasing distance between them (Figure 4.22).

Most food shipments, natural resource flows, and commodity movements occur within regions and within countries, rather than between them. The underlying principle of distance decay is the *friction of distance*. There are time and cost factors associated with extra increments of distance for all types of flow or movement. For individuals, the out-of-pocket costs of operating a vehicle or truck are combined with the cost of a person's time. The California Department of Transportation calculates the cost of traffic jams by basing a person's time at $6 per hour. With longer distance, it is more expensive to ship commodities because of labor rates of drivers and operators, as well as over-the-road costs of vehicle operation. Telephone calls and parcel deliveries, likewise, can be more expensive for more distant locations.

Distant decay and the friction of distance are not exerted in a linear pattern. From observation of Figure 4.22, we can determine that interaction between places is related in an inverse way to the square of the distance separating the two places.

The Gravity Model

Newton's law of gravitation states that any two objects in space attract one another according to a force that is proportional to the product of their masses and inversely proportional to the square of the distance separating them. Thus, Newton's law of gravitation can be expressed as the force of attraction F, which is equal to Mi, the mass of the first body, times

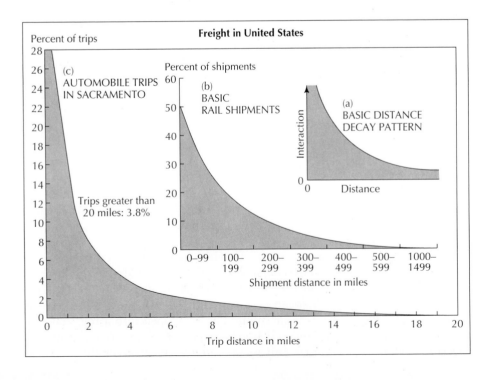

FIGURE 4.22
The distance-decay principle:
(a) the basic pattern; (b) rail freight trips in the United States, 1990;
(c) auto trips in Sacramento, 1991.
(Source: Modified from Fellman, Getis, and Getis, 1992, p. 64)

Mj, the mass of the second body, divided by the distance separating i and j, which is given as:

$$F = \frac{Mi \times Mj}{dij^2}$$

where K is the gravitational constant. Force equals mass of i times mass of j divided by distance between i and j.

This so-called *gravity model* has been applied to flows between two points in the transportation literature for more than 50 years. Geographers, planners, and engineers have realized that the gravitational model is useful in a social context to explain flows between two places, i and j. The flow, or interaction, Iij, is equal to the mass or population, Pi, of the first place, times the mass or population, Pj, of the second place, divided by the distance between the two, raised to a power given by:

$I_{ij} = P_i \times P_j$ divided by d_{ij} raised to the power 2

$$I_{ij} = \frac{P_i \times P_j}{d_{ij}^2}$$

(*Iij* equals population of i, times population of j, divided by distance between i and j squared.)

The gravity model has had perhaps the greatest use of any model in the field of transportation studies because it can be used to predict flows between two points for commodities, people traveling, information, telephone traffic, and radio and television messages. Frequently, distance is substituted by travel time or travel cost over a network, rather than by the straight-line path separating two places. Places have included, at one scale, continents and countries, and at the other scale, states or regions within countries, and even zones within cities. The gravity model, if properly weighed through regression analysis, has an extremely broad applicability.

The gravity model can be summarized by two principles: (1) larger places have a greater drawing power for flows of commodities, individuals, and information than flows to smaller ones; and (2) places that are more distant have a weaker attraction for one another than closer places. The gravity model is limited to flows between places, taking two places at a time. Using the principles of the gravity model, William J. Reilly (1931) developed the law of retail gravitation, which described the breaking point between two spheres of influence, of the border marking the outer edge of the trade areas of cities or regions.

The Law of Retail Gravitation

In the *law of retail gravitation*, Reilly (1931) used the gravity model concept to identify the breaking point between two cities (Figure 4.23). The assumption is that people will travel to the most rewarding destination, based on nearness of the place and its size. In this case, as in the case of the gravity model, the size of the destination is a measure of the rewards, or value available. The model is given by the formula, *BP* equals *dij* divided by 1 plus the square root of P_2 divided by P_1, where

$$BP_1 = \frac{d_{ij}}{1 + \sqrt{\dfrac{POP_2}{POP_1}}}$$

BP = Distance from the center of city 2 to the breaking point or boundary between city 1 and 2
d_{12} = the distance between city 1 and city 2
P_1 = the population of city 1
P_2 = the population of city 2

Here, the law of retail gravitation has been applied to find the breaking point between San Diego and Los Angeles. The two cities are approximately 120 miles apart, downtown to downtown. The population of the San Diego urban region is a little bit short of 3 million. The population of Los Angeles is 12 million, more or less depending on which subcities are included in the Los Angeles metropolitan region. The calculations show that the breaking point between the spheres of influence of

FIGURE 4.23
The law of retail gravitation applied to the breaking point between San Diego and Los Angeles. The breaking point (BP) between spheres is 40 miles north of San Diego's downtown and 80 miles south of Los Angeles' downtown.

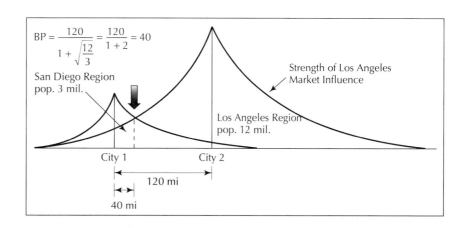

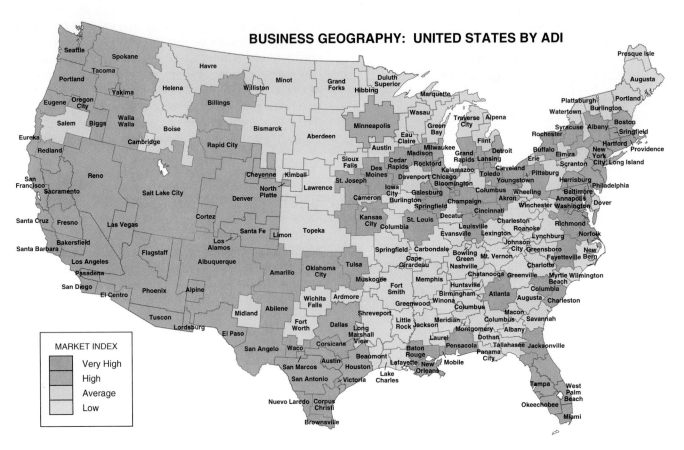

FIGURE 4.24
Areas of dominant influence (ADI). ADI is based on radio, television, and newspaper advertisements. For U.S. cities, ADI approximates the law of retail gravitation model if it were applied to adjacent pairs of cities. *(Source: Stutz, 1994, p. 232)*

San Diego and Los Angeles occur at 40 miles north of San Diego. This point coincides fairly closely with the boundary between San Diego County and Orange County and with a change of advertising media.

Most people south of the line receive the *San Diego Union* newspaper and listen to San Diego television channels for regular broadcasting and the evening news. They also follow the San Diego Padres baseball team, San Diego Chargers football team, and the San Diego State University Aztecs athletic teams.

The majority of the people north of the breaking point, in Orange County, receive the *Los Angeles Times* newspaper, tune-in predominantly to Los Angeles television stations, and follow the myriad of Los Angeles professional athletic teams, including the Angels, the Dodgers, the Lakers, the Clippers, and the Kings. Here, UCLA and USC have a strong following.

Figure 4.24 shows *areas of dominant influence* (ADI) for the United States as a whole in 1994, based on radio, television, and newspaper advertising. This remarkable map, created from data provided by nationwide advertising companies, approximates the results of the law of retail gravitation if it were applied to every city pair adjacent to one another (Stutz, 1994).

IMPROVED TRANSPORT FACILITIES

Prior to the development of railroads, overland transportation of heavy goods was slow and costly. Movement of heavy raw materials by water was much cheaper than by land. For this reason, most of the world's commerce was carried by water transportation, and the important cities were maritime or riverine cities.

To bring stretches of water into locations that needed them, canals were constructed in Europe beginning in the 16th century, with the height of technology represented by the pound lock developed in the Low Countries and northern Italy. Until the 19th century, canals were the most advanced form of transportation and were built wherever capital was available. Road building was the cheap alternative where canals were physically or financially impractical.

The most active period of canal building coincided with the early Industrial Revolution. The vast increase in manufacturing and trade fostered by the canals paved the way for the Industrial Revolution (Mantoux, 1961). The canals were financed by central governments on the Continent and by business interests in England, where a complex network was built during the last 40 years of the 18th century and the first quarter of the 19th century. Somewhat later, artificial waterways were constructed in North America. They supplemented the rivers and Great Lakes, the principal arteries for moving the staples of timber, grain, preserved meat, tobacco, cotton, coal, and ores (Figure 4.25).

At sea, efforts before the Industrial Revolution concentrated on expanding the known seas and on improving ships (e.g., better hulls and sails) to allow for practical transport over increasing distances. By 1800, or a few decades later in the case of the technology of sail, the traditional technology of transport reached its ultimate refinement. Subsequently, the rapid expansion of commerce and industry overtaxed existing facilities. The canals were crowded and ran short of water in dry periods, and the roads were clogged when traffic in wet periods destroyed the surface on which it

moved. The result was an effort to utilize mechanical energy as the motivating power.

The invention of the steam engine by James Watt in 1769 paved the way for technical advances in transportation. Its application to water in 1807 and to land in 1829, through the development of the locomotive, heralded the era of cheap transportation. The steamship reduced the cost of transportation by water, but the locomotive had a revolutionary effect on land transportation. In England, the railway served existing markets and provided urban populations with an excellent system of freight and passenger transportation. In the United States, the railroad was an instrument of national development; it preceded virtually all settlement west of the Mississippi, helped to establish centers such as Kansas City and Atlanta, and integrated regional markets. In developing countries, railroads linked export centers more firmly to the economies of Europe and North America.

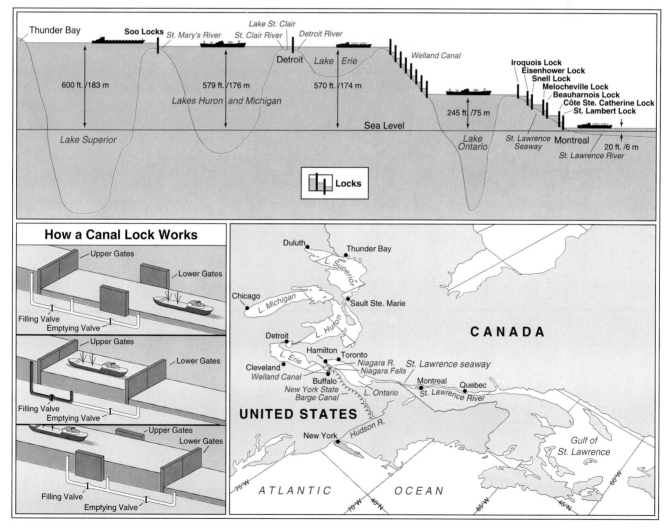

FIGURE 4.25
The St. Lawrence Seaway.

Eventually, mechanical power was used for localized urban transportation. Until the late 19th century, cities were mainly pedestrian centers requiring business establishments to agglomerate in close proximity to one another. This usually meant about a 30-minute walk from the center of town to any given urban point; hence, cities were extremely compact. The transformation of the compact city into the modern metropolis depended on the invention of the electric traction motor by Frank Sprague. The first electrified trolley system opened in Richmond, Virginia, in 1888. The innovation, which increased the average speed of intraurban transport from about 5 to more than 15 miles per hour, diffused rapidly to other North American and European cities, as well as to cities in Australia, Latin America, and Asia.

The 19th century was a time when the road was reduced to a feeder for the railroad. Road improvements awaited the arrival of the automobile. In the United States, heavy reliance on the automobile is a cross between a love affair with the passenger car and a lack of alternatives. In cities such as Denver and Los Angeles, roughly 90% of the working population travels to and from work by car; in the less auto-dependent cities like New York, cars still account for two-thirds of all work-related trips. By comparison, in Europe, where communities are less extensively suburbanized and average commuting distances are half those of North America, only about 40% of urban residents use their cars. In Tokyo, a mere 15% of the population drives to work.

In the developing world, technical developments in transportation have created a crisis—the result of a mismatch between transportation infrastructure, services, and technologies and the need for mobility of the majority of the population (Replogle, 1988). Governments that favor private car ownership by a small but affluent elite distort development priorities. Importing fuels, car components, or already assembled cars stretches import budgets thin. Similarly, building and maintaining an elaborate highway system devours enormous resources. The 1960s and 1970s saw a road-building boom in many LDCs, to the detriment of railroads and other forms of transport. With insufficient resources for maintenance, many roads in LDCs are in disrepair. In cities, bus systems and other means of public transportation are also in a poor state, meeting only a small proportion of transportation needs. Often, the poor cannot afford public transportation at all. Walking still accounts for two-thirds of all trips in large African cities like Kinshasa and for almost one-half the trips in Bangalore, India. Pedestrians and traditional modes of transportation are increasingly being marginalized in the developing countries (Figure 4.26).

In developing world cities, the reverse seems to be true. Most individuals cannot afford the operation or purchase of an automobile, and intercity areas are badly congested with inefficient public transportation, which is grossly inadequate in the inner areas. The inner areas are badly congested with a mixture of motorized traffic, pedestrian traffic, and animal-drawn carts and wagons.

Transport changes in the last 175 years have not been confined to railroads and roads. At sea, ships equipped first with steam turbines and then diesel engines facilitated the rapid expansion of international trade. In addition, the opening of the Suez Canal in 1869 and the Panama Canal in 1914 dramatically reduced the distance of many routes (Figure 4.27). The trend in ocean shipping today is not so much increased size and speed of vessels as it is increased specialization. Bulk carriers of oil, grain, and ores are replacing break-bulk vessels. Containerships, which tie trucks and ships together, have be-

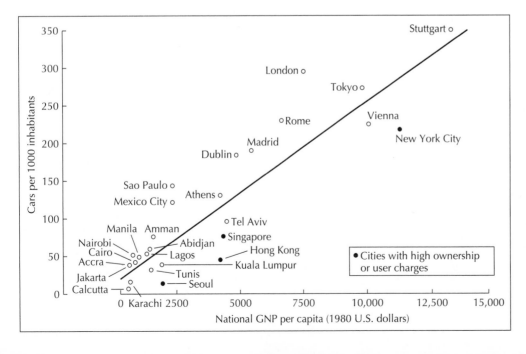

FIGURE 4.26
Private car ownership in selected major cities.
(Source: Hoyle and Knowles, 1993, p. 75)

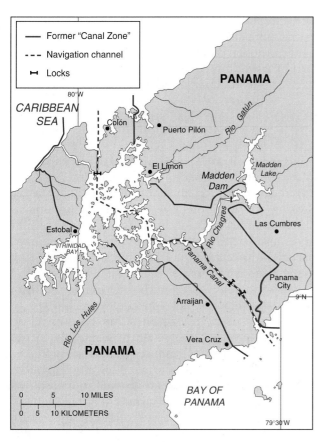

FIGURE 4.27
The Panama Canal, with 18,000 transits in 1997, is approaching its capacity. Yet the canal has a growing strategic and commercial importance.

41% between 1882 and 1900. Between the 1870s and 1950s, improvements in the efficiency of ships reduced the real cost of ocean transport by about 60%.

Cheaper, more efficient modes of transport widened the range over which goods could be shipped economically. For example, the international trade in iron ore was negligible in the 19th century, but by 1950 exports of iron ore accounted for just under 20% of world production, and by 1967 for 36% (Manners, 1971, pp. 348–349). Much of the trade is accounted for by long-haul traffic, from Venezuela to the United States or from Australia to Japan. Similarly, steam coal, exported from Alaska, Colombia, South Africa, Australia, and China is shipped, on average, 4,000 miles to its destination.

Cheap transportation also contributed to the growth of cities. It enabled cities to obtain food products from distant places and facilitated urban concentration by stimulating large-scale production and geographic division of labor. Furthermore, transportation improvements changed patterns of urban accessibility. North American cities have grown from walking- and horse-car-scale cities (pre-1800–1890), to electric streetcar cities (1890–1920), and finally, to dispersed automobile cities in the recreational automobile era (1920–1945), the freeway era (1945–1970), edge city era (1970-1990), and the exurb era (1990-present) (Figure 4.28).

come the basic transoceanic carrier. Planes have ousted passenger liners and trains as the standard travel mode for long-distance passengers. The shipment of cargo by air, however, is still in its infancy. Only perishable, high-value, or urgently needed shipments are sent by air freight.

Cost-Space and Time-Space Convergence

Transport improvements have resulted in what geographers call *cost-space* and *time-space convergence*—that is, the progressive reduction in cost of travel and travel time between places.

Transport improvements have brought significant cost reductions to shippers. For example, the opening of the Erie Canal in 1825 reduced the cost of transport between Buffalo and Albany from $100 to $10 and, ultimately, to $3 per ton. Railroad freight rates in the United States dropped

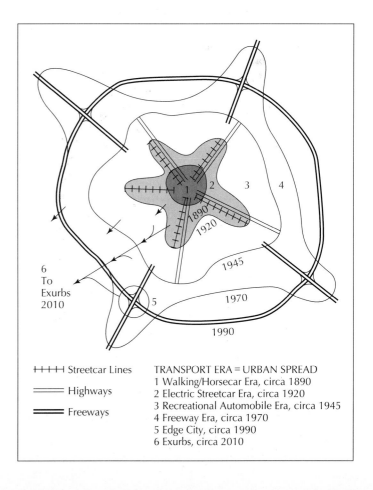

FIGURE 4.28
Stages of metropolitan growth and transport development in a North American city.

├─┼─┼─┤ Streetcar Lines
═══════ Highways
━━━━━━━ Freeways

TRANSPORT ERA = URBAN SPREAD
1 Walking/Horsecar Era, circa 1890
2 Electric Streetcar Era, circa 1920
3 Recreational Automobile Era, circa 1945
4 Freeway Era, circa 1970
5 Edge City, circa 1990
6 Exurbs, circa 2010

Developments in transportation have also cut travel time extensively—to where relative distances between places melt away. The travel time between Edinburgh and London, a distance of 640 kilometers, decreased from 20,000 minutes by stagecoach in 1658 to under 60 minutes by airplane today (Figure 4.29). Time-space convergence was marked during the period of rapid transport development; for example, in the 1840s, travel time between Edinburgh and London was longer than 2,000 minutes by stagecoach, but by the 1850s, with the arrival of the steam locomotive, the travel time had been reduced by two-thirds, to 800 minutes. By 1988, the rail journey between Edinburgh and London took 275 minutes. When the line was electrified in 1995, travel time was reduced to under 180 minutes.

Air transportation provides spectacular examples of time-space convergence. In the late 1930s, it took a DC-3 between 15 and 17 hours to fly the United States from coast to coast. Modern jets now cross the continent in about 5 hours. In 1934, QUANTAS/Empire Airways planes took 12 days to fly between London and Brisbane. Today the Boeing 747 SP is capable of flying any commercially practicable route nonstop. The result is that any place on earth is within less than 24 hours of any other place, using the most direct route. Reduction in travel times between London and Paris have likewise been dramatic (Figure 4.30).

Transportation Infrastructure

Transportation and communication infrastructures allow countries to specialize in production and trade. This regional division of labor is comparable to the task division of labor among its workers. The transportation and communications infrastructure of a country influences its internal geography. In some countries, transportation and communications are slow and difficult. Some regions are totally inaccessible (Figure 4.31). Most of Southeast Asia, South America, South Africa, and Asia have poorly developed infrastructures.

Fast and efficient transportation systems release capital for productive investment and allow the development of natural resources, regional specialization of production, and internal trade among regions. Even though India appears to be well connected, it is a country whose economic growth is harnessed by an inadequate transportation and communications infrastructure. Passenger and commodity traffic on India's roads has increased 30-fold since independence in 1948. Roads are overcrowded and 80% of villages lack all-weather roads. The government plans to connect every village of 500 people or more with all-weather roads by the year 2000. Such accessibility would allow regional specialization of production and cash crop farming, especially of high value crops such as fruits and vegetables that spoil quickly.

Conversely, the well-developed infrastructures of North America and Europe bespeak their level of economic development. However, most countries of the world have simple transport networks penetrating the interior from ports along the ocean. These railroads and highways are called tap routes. Tap routes are the legacy of colonialism or the product of neocolonialism. Such routes facilitate getting into and out of a country, but they do not allow for internal circulation, or circu-

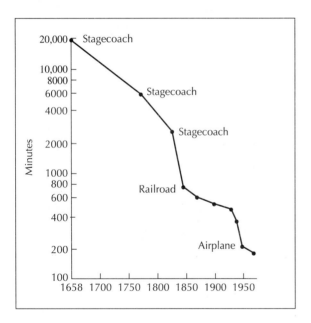

FIGURE 4.29
Time-space convergence between London and Edinburgh.
(Source: Adapted from Janelle, 1968, p. 6)

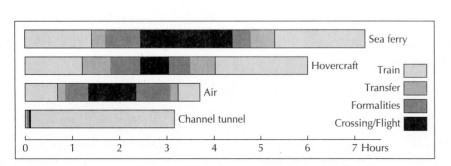

FIGURE 4.30
Journey times between London and Paris for ferry, air, hovercraft, and channel tunnel trails. *(Source: Hoyle and Knowles, 1993)*

The Concorde travels at supersonic speeds between London and New York, but this status symbol does not fly cheaply. For example, it uses about four times as much fuel per passenger as wide-bodied jets. In this era of increasing fuel costs, there appear to be limits to the economical annihilation of space by time. *(Photo: British Airways)*

lation between countries in the same region. South America, Africa, and Asia are examples of continents where the better roads are tap routes.

The transmission of information is as important as the dissemination of goods and materials. An abundance of information availability reveals economic development and political liberty. Home mail delivery is a good start and is available only in the world's core countries. The provision of telephones is another measure of communications infrastructure. The United States and Canada have almost 50 percent of the world's telephones. Yet together, they comprise only 4 percent of the world's population. Africa has less than 1 percent of the world's telephones and comprises 15 percent of the world's population.

The information superhighway, based on the Internet, is providing—through cable, television, and satellite—information, entertainment, and even video conferencing, as well as a great range of additional services, to homes equipped with cable TV. Japan, the United States, and Europe consume 95% of such service presently (see Communications Improvements later in this chapter).

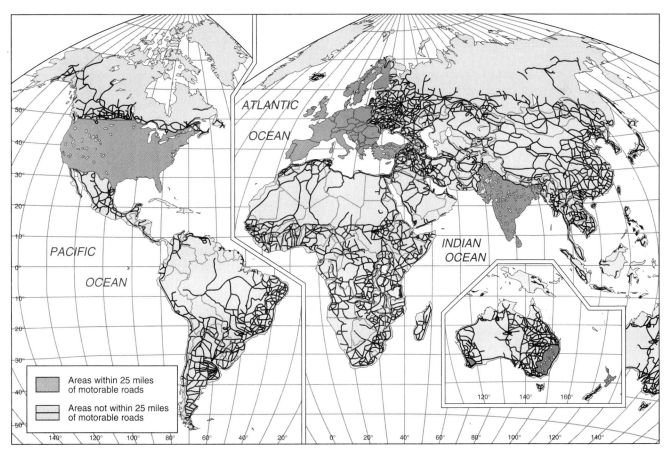

FIGURE 4.31
World road and highway density.

Transportation policy

Well-established transportation policy and regulation continued until the 1970s. The purpose of transportation national policy and regulation of air lines and rail carriers was to ensure quality control, protect companies and customers by closely guarding entry and exit from the market, and establish quality and safety control standards throughout the industry. During this period, providers were not only providing basic transportation service but also meeting a social obligation. For example, Britain's Road Traffic Act of 1930 introduced a system of licenses and rates that effectively regulated the sporadic and unsafe market for bus services in that country. In addition, the Railway Act in Britain provided benefits of inexpensive travel to all sectors of the population.

Deregulation and Privatization

By the mid-1970s the international *theory of contestable markets* required that there be free entry of new transportation operators into the market to ensure efficiency and welfare maximization. The move toward *privatization* had begun. Regulation, on the contrary, was criticized as increasing inefficiency, limiting competition, and raising prices to consumers. The Swedish railways, for example, were deregulated by 1968, and British trucking also was deregulated in that year. U.S. domestic airlines began deregulation in the mid-1970s. In Great Britain, in 1980, the Transport Act removed all controls on bus service and express service between cities. The 1985 Transportation Act deregulated local bus service inside and outside greater London. By the mid-1980s, the British government had sold nationalized transportation companies, including the National Freight Corporation, the British Transport Docks Board, the British Transport Hotels, Seaspeed Ferries, British Airport Authority, the National Bus Company, the Scottish Bus Group, and many other municipally owned companies.

Suddenly, in 1990 and 1991, a collapse of the communist economic system in central Europe and the former Soviet Union occurred. This collapse began a new phase in transportation deregulation and privatization, which has continued unabated throughout the world. The state-owned Soviet Unified Transport System was soon dismantled, and privatization set into Eastern Europe and the former Soviet Union. In New Zealand, road freight deregulation has been completed, while in Australia interstate bus services have been deregulated.

Other privatization moves in developed countries have included Air New Zealand, Japanese National Railways, and Air Canada. In each case, an attempt has been made to increase efficiency and reduce public debt by balancing budgets.

Privatization and *deregulation* have been hampered in developing countries because of a lack of foreign exchange to purchase necessary spare parts and replacement equipment. Sri Lanka, for example, has deregulated all bus routes, while communist China has deregulated long distance coach service, and fares are now allowed to vary for the first time. Nigeria has followed suit, by privatizing Nigeria Airways and its National Shipping Line, while Singapore privatized Singapore Airlines in 1985 and has offered to privatize its mass transit corporation.

Deregulation of the U.S. Airlines

The Civil Aeronautics Board (CAB) of the United States regulated the U.S. airline industry from 1938 until recently. During most of this period, the CAB's goal was to preserve the 16 trunk line airlines that existed in 1938 and to provide good service at fair prices with a high level of quality control. More recently, the 16 companies were reduced to 11 companies by mergers.

Because of the rapid economic growth of the United States, air passenger traffic increased 1,000% between 1950 and 1970. The move had been from military planes to propeller planes and finally to modern jet aircraft. Airfares remained almost constant because of the lower cost of operating more efficient planes. However, the oil embargo of 1973–1974 and 1978 increased pressure for *deregulation* for U.S. domestic air services. In 1978, the United States Airline Deregulation Act limited the CAB's route licensing powers (eventually phasing them out) and its fare controls.

Domestic U.S. airlines are now open to any carrier that might venture into the market. The most important result of airline deregulation has been more competitive fares and survival of the most efficient companies. The development of a hub-and-spoke network has been a cost-saving measure. Most direct flights have been reduced, and now air service requires at least one stopover in an airline hub city, unless the city pair are very large American cities. Service from smaller cities is directed into larger city airports or hubs and then linked to final destinations by direct flights.

Privatization and deregulation have kept fares down. In 1976, only 15% of passengers on domestic air routes used discount coach fares; by 1987, 90% of the passengers used discounted tickets. However, as average fares have fallen on long-haul routes, fares on short routes have risen. Load factors have increased substantially with a hub-and-spoke system and the number of flights has declined, leading to lower overall costs.

After deregulation, a musical chairs occurred within the airline industry as eight former local service airlines

(Hughes Airwest, North Central, Frontier, Ozark, Piedmont, Southern, Texas International, and US Air) grew rapidly and began to compete at regional levels. In the early 1980s, Alaska Air, Air Cal, Air Florida, PSA, and Southwest joined the competition. However, things had economically toughened by 1984 and the number of regional or national competitors had been reduced to only 10 airlines by 1988 and 7 by 1992 by financial failures, mergers, or takeovers. By 1991, Brannif, Continental, and Eastern were bankrupt and had stopped their operations, while National and Western had been taken over by Pan Am and Delta, respectively.

Hub-and-Spoke Networks

In order to remain competitive, the airlines that did survive the shake-out of the 1980s restructured their networks so that they could reduce direct flights between most city pairs and operations could be more efficient and cost-effective with the hub-and-spoke network model. Hubs serve central locations that collect and redistribute passengers between sets of original cities. For example, it requires N (N-1) connections to provide direct linkages between N places and the network. The number of connections required to link these N places can be reduced to N-1 if a hub is established and direct flights are provided between the hub and the other nodes or places in the system. Therefore a hub-and-spoke network provides a certain level of economy. Extremely large passenger volumes are funneled through hubs, and this allows the airlines to fly larger and more efficient aircraft and to offer more frequent flights between major hubs, increasing load factors (Figure 4.32).

However, *hub-and-spoke networks* can provide disadvantages, especially to the travellers who find their trips more lengthy, frequently with a change of planes, and fewer direct flights available. Also, great congestion is created at the main hub cities, and this affects efficiency both in the air and on the ground. It is important for airlines to make careful decisions as to the location and exact number of hubs so that their operation is competitive with other airlines. There have been a large number of optimum hub location studies in the literature. These mathematical optimization approaches have attempted to capture the real world realities of air passenger networks and the design problems that face most airlines (O'Kelly and Lao, 1991).

Not all cities have fared equally well. Of the 183 U.S. airports, 137 increased their number of flights between 1977 and 1988 by more than 50%. In contrast, the remaining 46 airports showed a precipitous decline on the order of 10% to 15% because of the order of deregulation. Resulting from this mad scramble to reduce fares and elevate efficiency for mega-

hubs, Atlanta, Chicago, Dallas, and Denver have surfaced, each serving as a major connection for two or more airlines in Salt Lake City and Minneapolis–St. Paul. Memphis and Detroit have also entered the market as major hubs. For example, American Airlines utilizes Dallas–Fort Worth and Chicago as national hubs, while Continental uses Houston, New York, Denver, and Cleveland. Delta Airlines' major hubs are Atlanta, Cincinnati, and Dallas; whereas for Northwest, Detroit, Memphis, and Minneapolis–St. Paul serve as hubs. Chicago is ranked well above all other nodes in the United Airlines network, and Pittsburgh and Charlotte are identified as major hubs for US Air (Shaw, 1993).

Transportation of Nuclear Wastes

Nuclear wastes produced during the fission process include reactor metals, such as fuel rods and assemblies, coolant fluids, and gases found in the reactor. Fuel rods are the most highly radioactive waste found on earth today, and their dangerous level of radioactivity requires that they be transported to a special site for storage. The storage site must be located so that the radioactive material will not contaminate the ground water or the biosphere in any way. Geologic stability is a must. Most experts support deep underground geologic disposal in salt domes or other rock formations. Presently, the United States has 100 sites where radioactive waste has been temporarily stored. These include: Hanford, Washington; Livermore, California; Beatty and Las Vegas, Nevada; Idaho Falls, Idaho; Los Alamos and Albuquerque, New Mexico; Amarillo, Texas; Weldon Springs, Missouri; Sheffield, Illinois; Paducah, Kentucky; Oakridge, Tennessee; Aiken and Barnwell, South Carolina; and Niagara Falls and West Valley, New York. Recently, however, the federal government has selected Yucca Mountain, Nevada, as a permanent facility for the storage of high-level nuclear wastes from commercially operated power plants in America (Figure 4.33).

Special containers on rail and truck will be used, and the routes will avoid high population areas. One of the factors responsible for the selection of Yucca Mountain is the so-called *NIMBY effect* (not in my back yard). The problem with nuclear waste deposition is that, because no official or engineer can assure that each site will be completely safe, local residents want no possibility of an accident. Another more regional response to the removal of nuclear spent fuel is the so-called NIMTOO effect. NIMTOO stands for "not in my term of office." Politicians are quite sensitive to their constituencies, pleas, and concerns for safety, and usually vote to remove nuclear wastes and power plants from their districts.

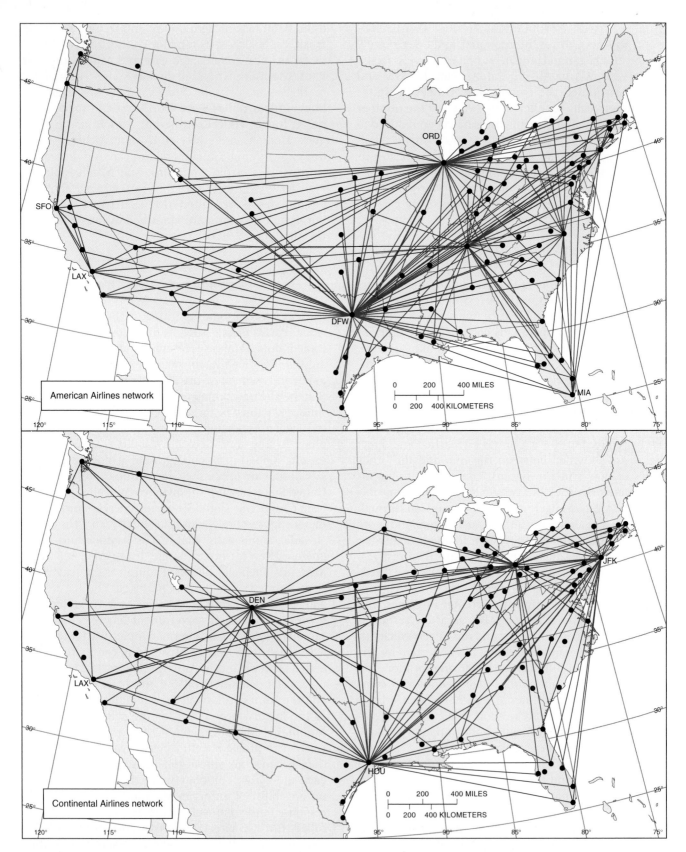

FIGURE 4.32
Hub-and-spoke networks for two major U.S. air carriers, 1993. (These figures were first published in the *Journal of Transport Geography*, Vol. 1, No. 1, March 1993, pp. 51–54, and are reproduced here with the permission of Butterworth-Heinemann, Oxford, UK.)

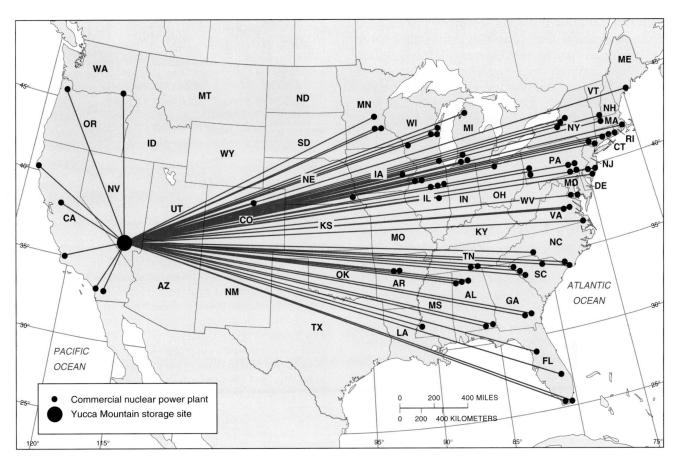

FIGURE 4.33
High-level nuclear waste from commercial power plants will be transported to Yucca Mountain, Nevada.

PERSONAL MOBILITY IN THE UNITED STATES, 1990

This section documents only one aspect of urban transportation problems and solutions, the personal mobility changes in the United States during the last 30 years. Solutions to the tremendous increase in personal mobility in the developed world include new road construction, rail-based rapid transit systems, traffic management support programs, and transportation coordination.

Personal mobility in the United States is at its highest point in history, with individuals making more and longer trips and owning more vehicles. Three factors account for this greater level of mobility in general. The first is the overall increased performance of the national economy. People have more money to spend on transportation, and this has led to greater automobile ownership and greater travel distances. The second is related to the increasing growth of cities and their spread over the surrounding countryside through a low-density urban expansion. On the average, dis-

tances between home and job have increased. Only recently has there been an increase in efforts to combine mixed-use activities in the same location and, thereby, reduce travel. A third major reason for increased mobility is the changing role of women in the workforce. Many more women own their own vehicles, have entered the workforce full-time, and have increased their travel demands during the last 30 years.

Because of increased mobility, individuals have benefited in the social and economic sense, but society as a whole may have felt the negative consequences. New concerns about rising levels of air pollution, congestion on the freeways, and the movement of goods are being posed. The new mobility has created a set of problems that have more difficult answers. Two techniques to address the issue of greater congestion and slower average speeds are being forwarded. One is to increase volumes on present roadways through *intelligent vehicle highway systems* (IVHS), and the other is to reduce travel demand by planning a land-use mix in localities so that trip origins and destinations are less separated. This approach is called *transit oriented development* (TOD). However, before we discuss these two measures of reducing congestion and increasing the greater volumes of flow, we first examine trends occurring in personal mobility in the United States.

The *Nationwide Personal Transportation Survey* (NPTS) has been conducted approximately every 7 years for the past 30 years. The results displayed here give important information regarding mobility of the U.S. population. Between 1969 and 1990, the number of households, the number of drivers, the number of workers, the number of vehicles, the number of vehicle trips, the household vehicle miles traveled (VMT), the number of person trips, and the person miles of travel have all increased at a much faster rate than the population (Table 4.4). During this period, the population increased by 21%; households increased at a rate of 49% in the same period; licensed drivers, 58%; workers, 56%; household vehicles, 128%; household vehicle trips, 82%; household vehicle miles traveled (VMT), 82%; person trips, 72%; and person miles to travel, 65%.

In addition, the number and percent of households that did not own a vehicle dropped from 20% in 1969 to only 9.2% in 1990, a 33% reduction (Table 4.5). However, during the same period, the number of households with three or more vehicles increased by 535%. During this period, the population increased by 21% and the number of households increased by 49%, but the number of household vehicles increased by 128% (Table 4.4).

The *journey-to-work trip*, both in terms of total miles of travel and in terms of number of trips, continued to account for the largest proportion of travel by U.S. households between 1969 and 1990. Both

Rush-hour traffic fills the northbound lanes of the San Diego freeway in California. Rapid growth of automobile ownership in Western cities in the second half of the 20th century has been met by unparalleled congestion. Greater levels of wealth in Western countries, more drivers, more auto ownership, and more women entering the labor force have contributed to high levels of car ownership. In California, auto ownership approaches one car for every person, whereas in cities worldwide the figure rarely exceeds 10 per 100, and much lower figures than that for rural areas. *(Photo: Spencer Grant, Photo Researchers, Inc.)*

annual vehicle miles traveled and annual number of vehicle trips per household increased by 22% (see Figure 4.34). Vehicle trip lengths, which had been decreasing slightly between 1969 to 1983, showed slight increases in 1990. The home-to-work trip av-

◼

Table 4.4

Summary Statistics on Demographic Characteristics and Total Travel, 1969, 1977, 1983, and 1990 NPTS

	1969	1977	1983	1990	*Percent Change* 69–90	69–90
Licensed Drivers (000)	102,986	127,552	147,015	163,025	2.2	58
Male	57,981	66,199	75,639	80,289	1.6	38
Female	45,005	61,353	71,376	82,707	2.9	84
Workers (000)	75,758	93,019	103,244	118,343	2.1	56
Male	48,487	55,625	58,849	63,996	1.3	32
Female	27,271	37,394	44,395	54,334	3.3	99
Household Vehicles (000)	72,500	120,098	143,714	165,221	4.0	128
Household Vehicle Trips (000,000)	87,284	108,826	126,874	158,927	2.9	82
Household VMT (000,000)	775,940	907,603	1,022,139	1,409,600	2.9	82
Person Trips# (000,000)	145,146	211,778	224,385	249,562	2.6	72
Person Miles of Travel# (000,000)	1,404,137	1,879,215	1,946,662	2,315,300	2.4	65

Source: National Personal Transportation Survey, U.S. Department of Transportation, 1992.

🔯

Table 4.5
Number of Households by Vehicles Available, 1969 and 1990 NPTS

Number of Vehicles Available	(Thousands)		Percent Change	
	1969*	1990	69–90†	69–90‡
No Vehicle	12,876 (20.6%)	8,573 (9.2%)	–1.9	–33
One Vehicle	30,252 (48.4%)	30,654 (32.8%)	0.1	1
Two Vehicles	16,501 (26.4%)	35,872 (38.4%)	3.8	117
Three or more vehicles	2875 (4.6%)	18,248 (19.5%)	9.2	535
All Households	**62,504**	**93,347**	**1.9**	**49**
All Household Vehicles	**72,500**	**165,221**	**4.0**	**128**
Vehicles Per Household	**1.16**	**1.77**	**2.0**	**53**

*The 1969 survey does not include pickups or other light trucks as household vehicles.
† Compounded annual rate of percentage change.
‡ Percentage change rate.
Source: Household and vehicle data.

eraged about 11 miles, while social and recreation trips averaged almost 12 miles per trip. Other family or personal business trip lengths averaged 7.5 miles, and shopping trips averaged 5.1 miles in 1990. For all purposes, however, the average vehicle occupancy declined substantially from 1977 to 1990, an average of 16% for all trip purposes. The average vehicle oc-

cupancy for a home-to-work was 1.1; for all purposes, 1.6; for shopping, 1.7; for other family business, 1.8; and for social and recreation, 2.1. These factors of decline in vehicle occupancy are explained partially by the increased number of vehicles per household and the decrease in average household size in the United States during this period.

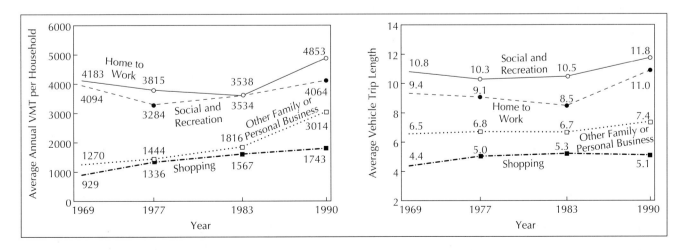

FIGURE 4.34
U.S. household vehicular travel for selected trip purposes, 1969, 1977, 1983, and 1990 NPTS: (a) average annual vehicle miles traveled (VMT), and (b) average vehicle trip length.

Commuter traffic in Bejing. For half of the world's population, the bicycle is a principal means of transportation. In China, less than one person in 1,000 owns an automobile. Public buses are cheap and moderately efficient, but human power is still very important for intraurban movement by foot and bicycle. Pedestrian movement dominates urban areas in terms of the number of trips. If the price of gasoline or the price of parking doubled, would you be willing to travel by bicycle to get to work or school? *(Photo: Paolo Koch, Photo Researchers, Inc.)*

The distribution of journey-to-work trips by mode showed that public transit usage had dropped from 8.4% in 1969 to 5.5% nationwide in 1990. Conversely, the automobile accounted for 83% of all travel in 1969 and 91.4% of all travel by 1990. However, the average journey-to-work trip length increased 7% between 1983 and 1990, from 9.9 miles to 10.4 miles (see Figure 4.35). The average travel time for commuting trips declined for the same period, however, from 22 minutes in 1969 to 19.7 minutes in 1990, a reduction of 10% during the 21-year period. Consequently, the journey-to-work is somewhat longer in recent years but average speeds are greater because of the improvement of highways; interstates, freeways, and beltways stretching into suburbia; and the development of employment centers near belt-ways in suburban locations. Lastly, an increase in automobile travel and a decrease in public transit has shortened average commute times as well.

Because of the rapid escalation in average vehicle price, Americans retain their vehicles for a longer period. For example, in 1969, 42% of household automobiles were 2 years old or less; by 1990, only 16% were 2 years old or less. During the same period, the number of automobiles that were 10 or more years in age increased from 6% in 1969 to 30% by 1990. The usage of older vehicles, which have been shown to burn energy less efficiently and cleanly than newer vehicles, contributes to the energy and air pollution problems that the United States now faces.

Mobility and Gender

Personal mobility changes in the United States by gender between 1969 and 1990 have been even more dramatic than increases in trip length, vehicle miles traveled, and automobile ownership. Travel by women between 1983 and 1990 has increased 50% from 6,382 annual miles to 9,528 miles per female driver (see Figure 4.36 and Table 4.6). The average annual miles per licensed male driver also increased dramatically between 1969 and 1990 by 46%. The most dramatic age period for both male and female was the 16

FIGURE 4.35
Average commute trip distance by mode, 1969, 1977, 1983, and 1990 NPTS.

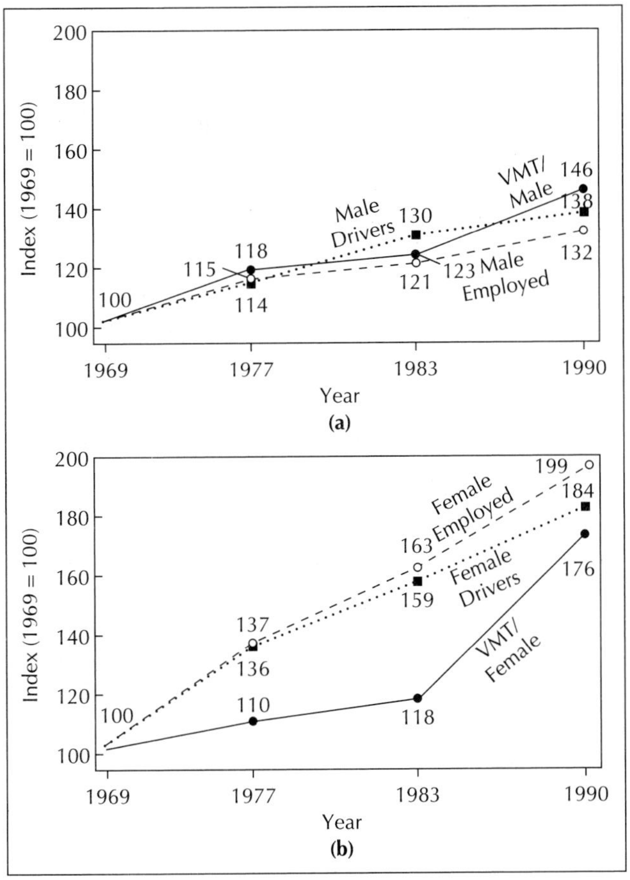

FIGURE 4.36
Changes in annual miles of travel (a) per male drivers, (b) per female drivers, for 1969, 1977, 1983, and 1990 NPTS. Recently, travel by females has increased much faster than travel by males because of increased car ownership by females and more females entering the labor force.

to 19 age period, where a 75% to 106% increase occurred, followed by females in the 20 to 34 age period, where a 100% increase occurred in average annual miles driven per licensed driver. The enormous increase for females stems from increased automobile ownership due to a more affluent society and a greater number of women who entered the labor force requiring the use of an automobile. The increase in age groups 16 to 19 and 20 to 24 also reflects an affluent society with younger drivers receiving greater use of the family car and, in many cases, their own vehicle ownership. Finally, in 1969, the distribution of annual miles driven by males was 73% and by females 27%. However, by 1990, although travel by males still accounted for the greatest proportion, 65%, the amount accounted for by female drivers increased to 35% of annual miles driven in the United States.

Due to the overall increases discussed and the number of trips females made for family trips and personal business as well as work-related business, they took, as

Table 4.6
Average Annual Miles per Licensed Driver by Driver Age and Sex, 1969, 1977, 1983, and 1990 NPTS

Age	1969	1977	1983	1990	Percent Change 69–90*	69–90†
Male						
16–19	5461	7045	5908	9543	2.7	75
20–34	13,133	15,222	15,844	18,310	1.6	39
35–54	12,841	16,097	17,808	18,871	1.9	47
55–64	10,696	12,455	13,431	15,224	1.7	42
65+	5919	6795	7198	9162	2.1	55
Average	**11,352**	**13,397**	**13,962**	**16,536**	**1.8**	**46**
Female						
16–19	3586	4036	3874	7387	3.5	106
20–34	5512	6571	7121	11,174	3.4	103
35–54	6003	6534	7347	10,539	2.7	76
55–64	5375	5097	5432	7211	1.4	34
65+	3664	3572	3308	4750	1.2	30
Average	**5411**	**5940**	**6382**	**9528**	**2.7**	**76**

*Compounded annual rate of percentage change.
†Percentage change rate.
Source: Driver's estimate of annual miles driven, including driving in all vehicles (personal and commercial).

a group, 70% more trips in 1990 than in 1983. The number of trips females took for family and personal business between 1983 and 1990 grew more than any other category, 37%. At the same time, 16% more work trips were taken by women because of increased employment outside the household.

Lastly, travel by individuals 65 years of age and older compared to all age groups between the years 1983 and 1990 is shown in Table 4.7. Similar to trip-making for the entire population, individuals 65 years and older took approximately 6% more trips over the 7-year period. While this increase is comparable to all age groups, the average trip length of the population 65 years and older increased 19.4% between 1983 and 1990, but the average trip length of all ages increased only 9.2%. When the average trip length is multiplied by the average number of person trips, the resulting average annual number of person miles of travel for individuals 65 years and older has increased by almost 26% during a 7-year period, a significant increase.

Transit Oriented Development

Two major transit developments are being suggested for the year 2000 to accommodate travel demands in the city. The first of these is called TOD (transit oriented development). The second is called IVHS (intelligent vehicle highway systems). We first look at TOD and then examine IVHS. The idea behind TOD is to develop design guidelines to help the community to reduce dependency on automobiles and the distance of travel demands. Mixed use of residential employment, commercial and recreational, will bring trip-ends closer to the individual. This is the goal of TOD. When neighborhoods are designed for the comfort and enjoyment of people and their needs instead of just for people and their cars, it is reasoned that traffic congestion as well

as the quality of life can be improved. At least two factors help reduce travel distances in Europe. One is higher density communities with shorter transportation and travel demands, and the other is very high costs of automobile ownership and gasoline expense. The European communities have employed concepts of TOD for generations, but the United States, with its low density urban sprawl and inexpensive land and gasoline, is just now coming to grips with the concept.

The TOD describes a scenario of land use with a compact pattern of mixed use designed to encourage pedestrian trips and support public transit. The emphasis is to provide public places and commercial traveler needs within walking distance of home. TOD presumes an understanding of links between land use and transportation and air quality. It is only one of the delicate long-term solutions to congestion in U.S. cities. TOD has come about through the passage, in 1992, of the Federal Intermodal Surface Transportation Efficiency Act (called Ice Tea), which targets coordinated community planning as a prerequisite to the obtaining of federal transportation funds.

As we have just seen with the National Personal Transportation Survey of 1990, longer commuting distances from homes in the suburbs have contributed to a dramatic increase in vehicle miles driven and in the production of air pollutants. In addition, it is almost impossible to provide good transit service to these distant suburbs. Expectations for transit oriented development are based on traffic surveys and forecasts, home price comparisons, regional air pollution and congestion modeling, ridership and consumer surveys, and other types of attitudinal techniques. All of these measurements indicate that a certain proportion of America's urban population would tolerate a pedestrian-oriented development that could reduce congestion, increase densities, and create more transit-oriented neighborhoods.

Because suburban communities in the outer city or edge city have lower densities, even trips to schools, stores, parks, and to visit friends have become more

❖

T a b l e 4.7

Travel by Individuals 65 Years and Older Compared to All Age Groups, 1983 and 1990 NPTS

Age	Average Annual Person Trips			Average Annual Person Miles of Travel			Average Trip Length		
	1983	1990	Percent Change*	1983	1990	Percent Change*	1983	1990	Percent Change*
65 and older	672.3	713.5	6.1	4447.5	5596.4	25.8	6.7	8.0	19.4
All ages	977.9	1042.4	6.6	8483.9	9670.6	14.0	8.7	9.5	9.2

*Percentage change rate.
Source: NPTS, 1990.

lengthy, and walking trips have become inconvenient and unsafe. This parallels the growing lack of the sense of community that has pervaded the orientation of the suburbs. The elimination of front porches, the preponderance of two-car garages dominating the facade of single-family homes, and the location of neighborhood services well beyond the reach of pedestrians have practically eliminated the chances of one meeting with a neighbor.

Naturally, the new emphasis on TOD would require amended zoning in the local region and changed general plans to ensure a greater mix of land uses, limited auto-oriented uses, the establishment of pedestrian scale walks and entrances. Further, there would be zoning for facilities that parallel public thoroughfares, for sidewalks and plantings, and for a minimum rather than a maximum width of streets, and a through traffic speed. The development of alleys and rear parking and the discouragement of the use of cul-de-sacs are some additional features of TOD.

Intelligent Vehicle Highway Systems

During the next 30 years, traffic volume in the United States is expected to double. Yet each year, some 135 million drivers spend about 2 billion hours stuck in highway traffic. An estimated $46 billion is lost by American drivers trapped in traffic delays, by detours, and by wrong turns. However, IVHS could greatly ameliorate this situation. The Clinton Administration is encouraging the development of high-tech highways that will use IVHS technology.

"Smart" cars are equipped with microcomputers, video screens, and other technologies that combine to take the frustration out of driving. Through the use of in-vehicle computers and navigation systems, drivers are guided step-by-step to their destinations. Fast, accurate information allows drivers to avoid accidents and congestion, while simultaneously offering information on restaurants, hotels, attractions, and emergency services—all available with the touch of a screen.

"Smart" highways are created by installing vehicle sensing systems. These sensing systems monitor traffic volume, speed, and vehicle weights. This information helps traffic engineers and transportation planners regulate signals to control traffic flow and plan new roads. The radar-based collision warning systems will automatically signal a car to brake to avoid collisions. If successful, cars could someday safely travel faster and closer together, thus allowing more vehicles to use the road at the same time.

The current challenge that IVHS faces is the translation of these ideas into applications that are practical, cost-effective, and user-friendly. IVHS will offer drivers much more data than previously available with paper maps and atlases, including data that can be updated on a continuous basis, such as status reports on traffic and environmental conditions.

Currently, freeways in the developed world can handle about 2,000 vehicles per lane per hour. More traffic than this per freeway lane causes stop-and-go traffic, which leads to accidents and gridlock. If the average vehicle is traveling 60 miles per hour on a freeway, then the average density of automobiles per lane is one every 135 feet. If the vehicle is 16 feet long, then 118 feet of freeway is going to waste. One concept behind IVHS is to increase the number of vehicles traveling at high speeds, packing together into a platoon, so that up to 7,000 vehicles per lane per hour could exist on a modern freeway—one car every half second. If this could be accomplished, the present freeway system could be used in a much more efficient way, preventing the addition of extra freeway lanes and the double-deckering of freeways through the most congested urbanized areas. Double-deckering in a variety of U.S. cities, including San Francisco, Seattle, and New York, has always been met with strong environmental opposition.

Automobile electronic components make up about 10% of the value of today's automobile. With smart streets and smart highways, this figure could jump to 25% by the year 2000. IVHS technologies range from real-time routing and congestion information being broadcast to the auto driver via radio to allowing the car to drive by itself on an automated roadway. New electronics associated with IVHS provided real-time information on accident, congestion, and roadway incidents. Traffic controllers, which have information, beam it to motorists who can select new routing strategies or use roadside services (Figure 4.37). Collision avoidance systems using radar, lane tracking technologies that platoon or stack vehicles at high density, and readout terminals on the dashboard that display a map of the city, as well as locations of accidents and the shortest route between two points based on real-time traffic flow information, will be given.

IVHS technology has progressed most in Europe and Japan. In Europe, a consortium of 12 European automakers, more than 75 research institutes and universities, and an additional 100 subsidiary electronics and computer firms have combined in a project called Program for European Traffic with Highest Efficiency and Unparalleled Safety (PROMETHEUS).

This $1 billion experiment began in 1986 and has borne other programs, including the Dedicated Road Infrastructure for Vehicle Safety in Europe (DRIVE) and the Leitund Information System Berlin (LISB). In Berlin, for example, the LISB experimental route guidance information system operates more than 2,500 miles of roadways, 5,000 intersections, and 1,500 traffic signals. Cars receive data from 250 infrared roadside signaling devices at the rate of 8,000 characters per second.

FIGURE 4.37
(top) The Global Positioning System (GPS). (bottom) The dashboard in a smart car. (Source: Trimble Navigation Limited)

In Japan, smart highways are operated under Advanced Mobile Traffic Information Communications Systems (AMTICS). The Japanese government and electronics firms operate mapping and navigation equipment built on the dashboard of the car with teleterminals connected by cellular radio to traffic control observation centers along the roadway.

The United States has gotten a late start in IVHS but has come on strongly in the last several years, pledging

$20 billion during the next 10-year period by the U.S. Department of Transportation. The California Department of Transportation (CALTRANS) presently operates experimental corridors on a 12-mile stretch of the Santa Monica freeway, a 7-mile stretch of I-15 in San Diego, and surrounding major arterial streets. The smart corridor broadcasts through radio, using changeable message signs and menu driven telephone information systems. Squeezing cars closer together on the freeway at high speeds is the main approach to IVHS by California. The technology has been developed, but user acceptance and cost-effectiveness hurdles remain. One of the biggest fears is that if something goes wrong, it could create a massive traffic accident involving more than several hundred cars. Hospitals may be overloaded with demand. While the United States can tolerate 50,000 traffic fatalities per year from drunk drivers and human error, the public tolerance for deaths resulting from computer-operated vehicles on smart streets is almost zero.

Automatic Vehicle Identification

Even before the IVHS technology becomes fully operational, the *Automatic Vehicle Identification* (AVI) system will be operating. The concept behind AVI is to allow drivers to use express lanes or HOV (*High Occupancy Vehicle*) lanes and thus avoid congestion and gridlock by paying a toll. Toll booths for freeways in Pennsylvania and New York or bridge toll booths throughout the West often cause traffic delays. To enter the express lane or the HOV lane, a car needs to have an AVI tag attached to its rearview mirror. The tag costs $30 and is being developed by Texas Instruments. The tag is approximately the size of a credit card, contains a lithium battery, a microchip, and an antenna. When the car approaches an overhead radio device, fiber-optic cables transmit its identification to a control center that charges a congestion toll against the prepaid account of the motorist. If a vehicle enters the express lane without an AVI tag or without a prepaid balance on the motorist's account, surveillance cameras will alert a waiting police car or the motorist may be fined by being issued a warrant in the mail. Operators in the control centers, which are equipped with video cameras along the entire length of the freeway, raise and lower tolls based on levels of congestion, adjusting the signs equipped to flash numbers so that the maximum amount of cars can be processed without a slowdown.

Presently, a 10-mile stretch of toll freeway operates in Orange County, California. The $100 million roadway operated by the California Private Transportation company has leased new lanes that run down the median of State Route 91 from Orange County to Los Angeles. It is expected that congestion pricing of $2 may save as much as 45 minutes in the commuting journey to work.

High Speed Trains and Magnetic Levitation

Magnetic levitation technology eliminates mechanical contact between a vehicle and the roadbed, thus eliminating wear, noise, and alignment problems. The vehicle floats on a cushion of air one-half foot above the guideway supported by magnetic forces. The Japanese and German governments have spent approximately $1 billion each on magnetic levitation research, and a magnetic levitation high speed train prototype is planned for the Orlando to Miami corridor. In the future, *maglevs* may transfer passengers between U.S. cities separated by up to 300 miles, at over 300 miles per hour, using far less energy and time than automobiles, Amtrak, or even air carriers. One could shuttle between Boston and Washington, or Chicago and Minneapolis, or Los Angeles and San Francisco in less than an hour, downtown to downtown. Maglevs presently are twice as fuel efficient as automobiles and four times as efficient as airliners, producing little or no air pollution. In the future, maglevs may be built alongside highways and will occupy far less room than airports (the Dallas–Fort Worth airport consumes as much land as a 65-foot-wide right-of-way coast to coast).

In the meantime, high speed conventional rail systems have been improved to include *tilting train technology* (TTT). An example of this technology is found in the Swedish X2000 train that can travel up to 150 miles per hour and give service in the northeast corridor of the United States. The passenger car carriage tilts inward on curves, allowing increased speeds on existing track curvatures. Amtrak, the national railroad passenger corporation train service, is presently making heavy investments in tilting train technology. In the Texas triangle—Dallas, Houston, and San Antonio—the first commercial phase will open between Dallas and Houston in 1998. ISTEA, the Intermodal Surface Transportation Efficiency Act of 1991 has identified five existing rail corridors selected for development of high speed trains. These include San Diego–Los Angeles–San Francisco, Dallas–Houston–San Antonio, Miami–Orlando–Jacksonville, Pittsburgh–Chicago–Minneapolis, Washington–New York–Boston.

Supersonic Aircraft

Supersonic aircraft flying at four times the speed of sound could travel from Los Angeles to Tokyo in only two hours instead of the 12 hours required at present. In the age of globalization and Pacific Rim development, such aircraft would revolutionize business and air travel. Twenty-five years ago, the United States put supersonic air transport on a back burner because it was too expensive and too noisy. The SST, which flies from Paris and London into New York, has had problems with high fares, low ridership, and high noise levels. New technology has allowed supersonic air travel to be reconsidered. NASA has spent $1 billion to develop environmental compatible technology with aircraft engine manufacturers.

As national and international air transportation demand continues to grow with increases of a trillion revenue passenger miles projected for each decade between the year 2000 and the year 2020, supersonic aircraft will be a mode of the future. Two trillion revenue passenger miles are expected by the year 2000, and 4 trillion by the year 2020. Most of this projected growth is business travel between international markets, in the wake of globalization of the world economy. The North Atlantic air passenger market is expected to double between 1995 and the year 2005. Accordingly, the Pacific Rim market, with its longer distances, is expected to quadruple during the same period. Such supersonic transports, carrying 600,000 passengers per day by the year 2015, are expected to be introduced by the year 2005, cruising at speeds of mach 1.5 to mach 2.5 with ranges between 5,000 and 7,000 miles, each carrying 300 passengers.

Economic growth of the eastern Pacific Rim, especially of China, Japan, and the Four Tigers—Singapore, Hong Kong, Taiwan, and South Korea—are causing estimates for Asian air travel to grow twice as fast in the next two decades as air travel between North America and Europe.

Transportation Software

The transportation sector of the world economy, from the planning of traffic flows on city streets to the international flow of commodities via tanker and air freight, is encountering challenges of increasing magnitude. Some of these challenges include increasing construction costs, greater fuel costs, greater environmental sensitivity, and even opposition by local special interest groups that are limiting further expansion of local to international transportation systems from highway to port to freight and rail. At the same time, traffic volume at all scales continues to grow steadily, resulting in lengthy traffic delays and increased total costs. At the same time, throughout the developed and developing world, the transportation infrastructure, at all scales, is in desperate need of improvement. The number of issues that transportation logistic managers and engineers must face is increasing rapidly.

Transportation planners, engineers, geographers, and managers must work hard to get more mileage out of existing facilities and rolling stock. Decisions at the same time need to be made more quickly and efficiently. Yet the amount of data on which decisions have to be made has been increasing exponentially since the 1950s. Many transportation organizations, from the private to the public sector, still rely on data analysis and hand mapping methods that have been used for decades. These methods now hinder, rather than enhance, transportation companies' and public agencies' abilities to meet transportation challenges.

For example, in the case of public transportation planning and management at the city and state level alone, endless hours have been spent by manually combining, calculating, and transferring data from spreadsheet to map. These analyses include: pavement characteristics, address matching of facilities, residence locations and intersections, base mapping, right-of-way mapping, cut-and-fill calculations required to build a new right of way, environmental impact assessment, accident factors analysis, hazardous material routing, and facility sighting and vehicle routing. Endless hours have been spent totalling property values by hand along proposed route alignments for which property needs to be purchased.

With the help of an integrated GIS (discussed in Chapter 5), transportation planners can eliminate these and the many other problems that they are faced with and produce plans quickly in a coordinated and efficient manner. Numerous levels of data are georeferenced to the same scale and stored in the computer for easy manipulation and combination, which had to be done previously by hand.

Spatial Decision Support Systems

Prescriptive decision-making principles are designed to help decision makers reach decisions about spatial problems that confront them. Many geographic problems involve the organization of activities or resources spatially to reach some desired objective. For example, much of the world economy involves traders who buy in one set of locations and sell in others. What decision principles do they follow? Long before textbooks in management science were written, individual traders recognized that when the difference in prices of any commodity between any two markets is greater than the cost of transporting the good between the markets, then profit can be gained by buying in the market where the price is low and selling in the market where the price is high. In 1938, the great French statistician Cournot formulated the law that the difference between prices in all markets should, in the absence of restrictions on trade, equal the cost of transporting the product between them. It is only comparatively recently that econometricians have developed the methods to determine, from information on potential production possibilities in each region at given relative prices, what the equilibrium price of each commodity will be in each market and what the pattern of trade will be between all markets. As the geographic information base improves, it becomes easier to compute possible production levels in any region. At this point, producers and traders begin to ask for advice on actions that will bring them highest returns on their efforts.

These optimization methods increasingly are being adopted and applied to the daily operation of the world economy. We illustrate by describing a SDSS called TRAILMAN (Transportation and Inland Logistics Manager) developed by a research team at the University of Tennessee for the United States Agency for International Development (USAID) to assist in distributing food aid in Africa. TRAILMAN is a microcomputer-based software package for logistics planning in Africa. That is, the system can be used to choose routes and schedules for shipping commodities from supply points (often overseas) to demand points. In addition to its programs for modeling the movement of goods in Africa, the system contains graphics and editing software for maintaining geographic databases and building planning scenarios. The entire system is accessed through a series of menus (TRAILMAN, 1993).

The system addresses six questions:

1. What is the least expensive way to move food aid from supply points to demand points so that all demands are met?
2. What is the fastest way to move food aid from supply points to demand points so that all demands are met?
3. What is the best combination of time and cost minimization?
4. What is the maximum amount of food aid that can be moved from supply points to demand points in a fixed amount of time?
5. What is the shortest distance, cost, time, or cost and time path between any two nodes by a given set of modes?
6. What is the best strategy for moving food aid from major distribution centers to outlying demand nodes? (TRAILMAN, 1993, p. 1.4)

The mathematical optimization methods used, known as linear programming, are complex. However, the software system calls the appropriate models as it decides the appropriate method that should be used to answer the question posed. These answers are determined from the questions it presents in its menus to the user. These prescriptive decision principles are those that an expert in food aid processing and distribution would use after many years of effort and experience in the field. The TRAILMAN software is trying to capture the knowledge of experts and embed it so that decision makers with far less experience will make decisions as good as those of the expert.

Figure 4.38, taken from the TRAILMAN manual, shows the major road system of the countries in the southern half of Africa. The box brought up on the computer screen illustrates how the system is first used to modify the road network. The illustration shows how a particular road link from Lusaka to Chipata can be

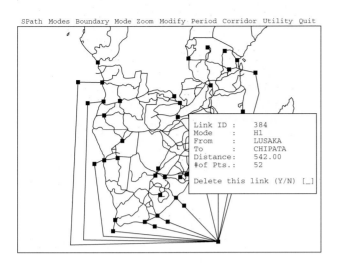

FIGURE 4.38
The delete link option. *(Source: TRAILMAN, 1993, p. 7.37)*

cilities exist, where they are needed, for the amounts of food aid that would arrive under each solution scenario.

Figure 4.40 shows the detailed distribution tours of food aid within one country in the region and shows the locations of the warehouses that would be used. This system is currently in use in southern Africa, where the confluence of several forces that have negatively affected the level of indigenous production of basic foodstuffs have left destitute large proportions of the populations of several countries in the region. These forces include a 5-year drought, attempts by governments in some areas of the region to control the production processes that have led to sharp reductions in production levels, fierce civil wars, and continued high rates of population growth. The distribution of food aid does not solve any of these fundamental problems; it merely averts catastrophe for the moment and gives these countries some breathing time to solve these problems.

deleted from the network for the purposes of a set of analyses. Users might want to delete this link if they have knowledge that the link is not available at the time of year when the food aid would be moving, or that it is broken by warfare and unsafe. Much of the work in using such decision support systems is the task of creating an accurate and relevant geographic database for subsequent analysis. Figure 4.39 illustrates a solution to a particular problem of identifying the ports and the routes into the interior of southern Africa where food aid is needed. Accompanying such a map are tables indicating the precise quantities of shipments that would move through each port and down each route. The program ensures that flows sent down any particular route are not more than the route can handle and that storage fa-

Typical GIS Applications for Transportation

GIS technology facilitates the coordination of many and varied functions of a transportation company. Traffic analysis; alignment studies; roadway design; rail, pavement, and river route tracking systems; right-of-way or corridor analysis along the network; parking studies; optimum location of service facilities; optimum location of schools, fire stations, and other public facilities; optimum routing through a system; and optimum allocation and flow modeling functions are accomplished through such geographic GIS software available today. Data can be pulled together from a variety of scales and sources, which before meant endless hours of matching by hand. The transportation industry uses

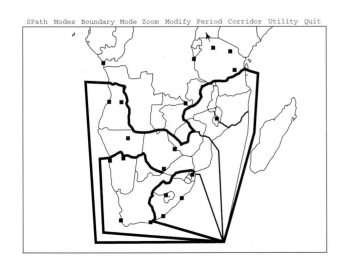

FIGURE 4.39
A linear programming flow map. *(Source: TRAILMAN, 1993, p. 7.43)*

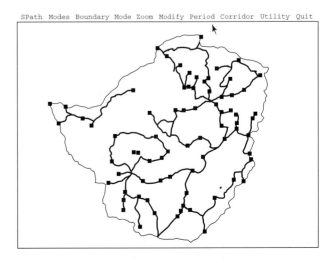

FIGURE 4.40
A warehouse routing flow map. *(Source: TRAILMAN, 1993, p. 7.46)*

◈

T a b l e 4.8

Typical GIS Applications for Transportation

Facility Inventory and Maintenance
Optimum Routing and Scheduling
Emergency Management and Evacuation
Highway and Transit Planning and Traffic Forecasting
Environment Impact Assessment
Traffic Operations and Flow Speed
Accident Analysis and Prevention
Pavement Management and Summaries
Optimal Facility Location Modeling
Network Design and Evaluation
Market Analysis and Competition
Remotely Sensed Data from Satellite Mapping
Photologging of Network Conditions

information in tabular form, such as data from the U.S. Census, pavement summaries, accident locations, computer-assisted design data (i.e., roadway or rail design), survey data and public opinion information, market information and competition, image and remotely sensed data from satellites, surface data on roadway or routeway elevation, line drawings, scan documents (i.e., permits and boring logs), video backup images (i.e., photologging of network conditions) at every mile, and scan photographs. All of these and more can be integrated, managed, and mapped at the same scale and georeferenced using GIS software (see Table 4.8).

The value of GIS software lies in the ability to draw together and analyze tabular and map data from the nation, state, or local level, effectively cutting across functional divisions of agencies each with their own scale or data format. A GIS can store an entire digital, georeferenced network of rail, pavement, canals, rivers, fiber-optic telephone lines, or pipelines as a part of an executive information system that can support decisions on maintenance and improvements systemwide. The U.S. Census Bureau has digitized the road and highway system of the United States. This *TIGER File*, as it is called, is now widely used for transportation planning and management. Transportation planners and economic managers can use the GIS to see the entire system, analyze traffic flow, and compare the results against land use, economic projections, and demographic data in order to predict future levels of service that will be demanded. Engineers can use GIS to analyze simultaneously geologic, hydrologic, topographic, climatic, and terrain-data features for transportation route structure design and maintenance. In addition, a GIS is well suited for assessment of environmental, economic, political, and cultural impacts of

new transportation projects. The major producers of such GIS software include the Environmental System Research Institute of Redlands, California, which produces ARC/INFO; Tydac Corporation, which produces SPANS; Caliper Corporation, which produces GIS PLUS and TRANSCAD; and Urban Systems Analysis Group, which produces TRANPLAN.

COMMUNICATION IMPROVEMENTS

Modernization of transportation has integrated the economic world, but equally important are technical developments in communication. Traditional forms of communication such as the letter post have been joined by telecommunications and computer-based methods of information transfer. In many ways, communication is the invisible layer of transport supplementing the physical transport links between cities, regions, and countries (see Chapter 7: Smart Cities).

In the 19th century, a major development in communication was that of the telegraph, which made pos-

In North America today, 30% of the workers telecommute at least one day per week. Workers seem to be in a Darwinian struggle, with the computer literate atop the food chain. The Cold War's demise has helped reduce the U.S. federal deficit from $290 billion in 1992 to $60 billion today. Thus, interest rates are down, and dollars have been shifted away from military spending. At the same time, countries embracing capitalism for the first time have craved American products, enabling U.S. companies to ring up rising sales from countries such as Russia, China, and Chile. U.S. exports reached a record $830 billion in 1996. (*Photo: Ron Chapple, FPG International*)

sible the worldwide transmission of information concerning commodity needs, supplies, prices, and shipments—information that was essential if international commerce was to be conducted on an efficient basis. By 1886, a skeleton of international telegraph links was completed when a cable along the west coast of Africa was laid. Today, an international telecommunications network based on fiber optics and digital technology has replaced the telegraph. Breakthroughs in information technology (IT) not only increase the productivity of business, but also make it feasible to perform service tasks at a much wider range of locations than in the past (Figure 4.41). The "wired home" can now become

the workplace as well as the base for many interactions that do not require travel. The *global office* is also becoming a reality. For example, in 1989, Texas Instruments opened a software development facility in Bangalore, India, that is linked by satellite to its headquarters in Dallas.

Computers Automate Business

Firms are starting to use computers to revolutionize the world economy, instead of merely improving old established mass production practices of the industrial

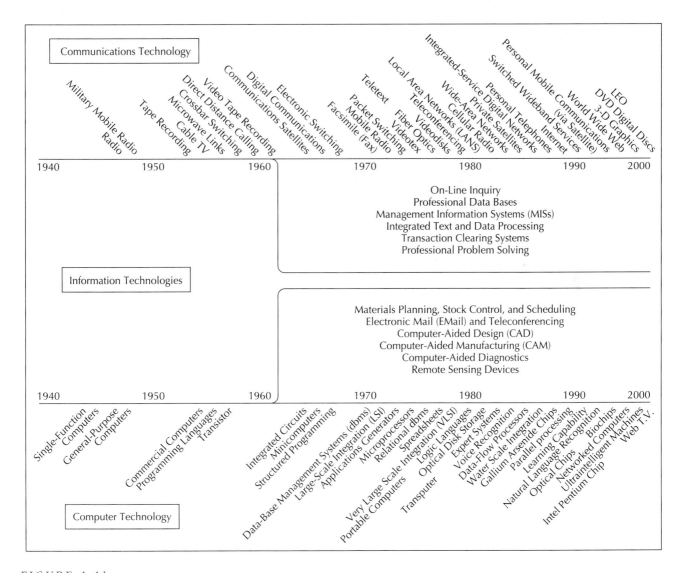

FIGURE 4.41

The convergence of computer technology and communications technology yields information technologies (IT). Two initially distinct technologies have now combined to impact the world economy. Communications technology is concerned with the transmission of information, while computer technology is concerned with the processing of information. The two technologies started to merge in the 1960s. Without information technologies, today's complex world economic system could not exist and develop. *(Source: Expanded from Freeman, C., 1987. The Challenge of New Technologies NOECD, Interdependence and Cooperation in Tomorrow's World, OECD, Paris, pp. 123–156)*

age. The workplace of the past has been filled with older generations used to working in hierarchial command and control systems doing repetitive tasks with very little thought. The first computers allowed the bureaucracies of the giant organizations to shuffle papers more rapidly. The new high speed printers produced even more reports, more analyses, and more financial statements. Computers in the past permitted bureaucracies to get larger, without requiring them to become more effective. Workers became only a little more efficient at secondary work but were not generally used for innovative work. Programmers, technicians, system analysts, and other specialists were added to the bureaucracy, and work was highly centralized in large mainframes (see Chapter 8: Business Reengineering).

Today, computers are finally being adapted to truly automate the operations of a bureaucracy—repetitive, technical, clerical and professional work, including redundant calculations, allowing the labor force to organize, process, and make decisions concerning data that allows business to change and adapt to the new marketplace. The adoption rates of new technologies is accelerating (Figures 4.42 and 4.43). Firms are starting to realize that conducting business in the information age is not simply making over old hierarchies or improving old methods little by little. It is about changing old methodologies and starting over. In 1993, firms began to embrace ideas of the latest best seller, *Reengineering the Corporation*, by Michael Hammer and Robert Champy, 1993. The theme was to start from scratch and reengineer—organize around the customer and around results-oriented processes, not activity-based functions.

Computers are at the brink of automating white collar work. The computer revolution replaced bureaucrats and middle managers in the 1990s instead of making them more efficient. This is what is propelling job restructuring. The power of mainframe computers has been put into desktop computers (Figure 4.44). By the year 2005, we will see information technologies projected so that 10 Cray supercomputers' power will fit on a single microchip that costs less than $100. Intel, the world's largest chip manufacturer, introduced its powerful *pentium chip* for the

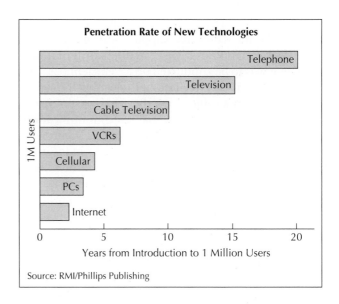

Source: RMI/Phillips Publishing

FIGURE 4.42
Penetration rates of new technologies. Penetration rates are accelerating while S-curves are getting shorter with increasing new technology.

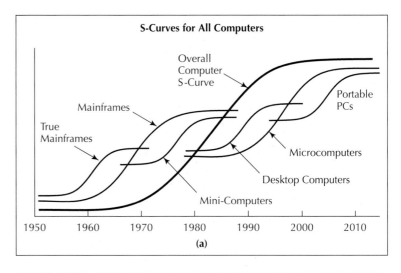

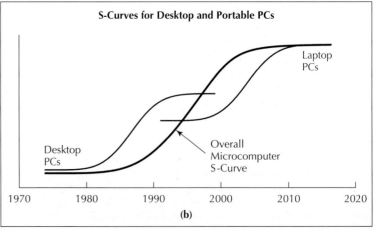

FIGURE 4.43
Overall product evolution of computers shows mainframes giving way to minicomputers. Minicomputers have given way to desktop computers, followed by portable PCs that are now networked together. *(Source: See Fig. 2.36, Dent and Smith, 1993)*

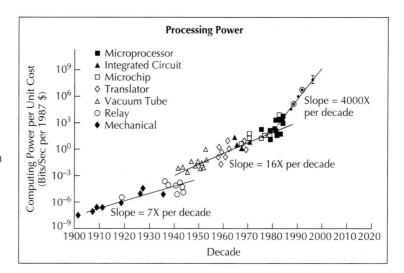

FIGURE 4.44
Computer processing power per unit cost has been increasing exponentially. The application of new software that links manufacturing, inventory, and other functions is dramatically increasing industry efficiency. Management practices are finally catching up to the technology base of computers. The bottom line on what a factory can produce is increasing.

world of IBM computers and IBM-compatibles in 1995. Then, Apple introduced its line of power PCs with a chip between 60% to 100% more powerful than the pentium chip. Such powerful computers have revolutionized the ordinary business of the firm; they can process and deliver inexpensive information, through the guidance of *expert systems* (rule-based instruction for decision-making), to satisfy customers' needs without waiting on bureaucrats to intervene at various levels of authority (Figure 4.44).

The information revolution is occurring first at the level of the firm and the university, and then filtering into homes, schools, and government agencies. A major trend today is the smart cellular phone, cable transmitting voice, data, faxes, messages, calendars, spreadsheets, schedules, rollodexes, and more. In 1996, Motorola launched its first low orbiting satellite, allowing smart phone users to access global wireless networks of information. These powerful, handheld computers are not only communication devices, capable of voice interaction, but capable of accepting data input from any location on the World Wide Web. The smart cellular phones are being networked to home and office databases of information accessible anywhere at anytime. By the year 2001, Micrcosoft and Sprint plan to have an even better network of such satellites in orbit that will allow everyone access to the wireless network.

Digital television-computers, both in the workplace and at home, are being used as much as the mini computer is used today on the Inter-

net. Video graphics, data sets, and telephone audio linkages through fiber-optic cables are coming online. Such "web TV" or television computers (Toshiba calls it PC cable) will become the prime centers of work in the firm. At home (1) learning, (2) entertainment, and (3) communications will be performed by these new *smart televisions* (Figure 4.45).

Using increased computer power means firms will be able to access managers, consultants, accountants, and lawyers without having to have them on staff.

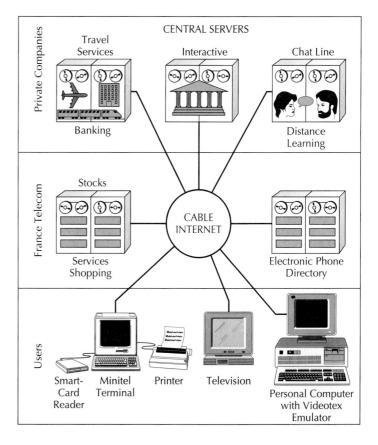

FIGURE 4.45
France's Transpac videotex home communication system. *(Source:* Scientific American, *March 1990, p. 88)*

American companies and their workers went through a restructuring frenzy in the early 1990s. However, as a result, these companies and workers are now punishingly competitive on the world scene. Adding to the U.S. advantage is high-tech know-how. Americans are far ahead of their European and Japanese counterparts in embracing computers and communication systems in homes and offices. With companies increasingly able to generate more productivity from their workers and from their capital investments, corporate profits, and hence stock prices, have risen sharply. The Standard and Poor 500 stocks, for example, have tripled since 1991. (*Photo: Bill Bachmann, Photo Researchers, Inc.*)

computer libraries, and law classes. This is the new educational system, featuring interactive formats for learning based on an individual's needs, not on a general, broad, liberal education. These educational approaches will be the fast-track to running a business, to proceeding on a particular career path, or to gain access in the information age.

Networked computers will allow for tremendous information power to be downloaded at each computer work station in the firm (Figure 4.46). Huge international databases from the Internet linked with expert systems will allow salespeople or service representatives to pursue higher level tasks. Mobile executives are those telecommuting from home or who are already using portable computers to respond quickly for rapid changes in the stock market, world events, customer needs, and international pricing. These mobile executives and salespeople bypass the typical accounting function and order processing of the large business hierarchy, approving orders directly from one computer to another until the shipment arrives at the plant. *Flexible factories* (manufacturing) will produce small runs of niche products, using compact, efficient teams that require little or no inventory and no typical supervision or coordination.

Outsourcing (subcontracting) through computer technology is bringing down costs. Computers with *expert system* software will make inquiries of bureaucrats in the office a thing of the past. Firms will pay for renting time or pay royalties to access such expert systems, but these costs will be much less expensive than having a professional on staff in each area of decision making.

The smart television will also increase the speed of learning for new training programs. The interactive computer will permit acceleration in learning the business practice of the firm for tomorrow. The information revolution is being felt in learning circles, and this learning is speeding the ability to perform one's job for increased productivity for individuals and organizations.

Labor will be able to experience a myriad of training exercises in today's information-loaded, communication-rich world. Any employee who wants to learn can attend classes or participate in live discussion at just about any university in the country (*distance learning*). However, beyond the formal degree, access is now available for specialty magazines, books, seminars, home study courses, videotapes, audiotapes, CD rom,

Flexible Factories

The Dell Computer Corporation has designed its newest factory without room for inventory storage. The Chrysler Corporation can increase vehicle production without building new factories. The General Electric Company expects to save millions of dollars by buying spare parts for its plants over the Internet.

Dell Computer, which doubled in size to $8 billion in annual revenue in 1995 and 1996, is perhaps the purest example of the efficiencies made possible by information technology. Unlike the rival Compaq Computer Corporation, which uses a vast network of resellers, Dell sells all its systems directly to its customers. The company waits until it has received an order before it begins to build a machine.

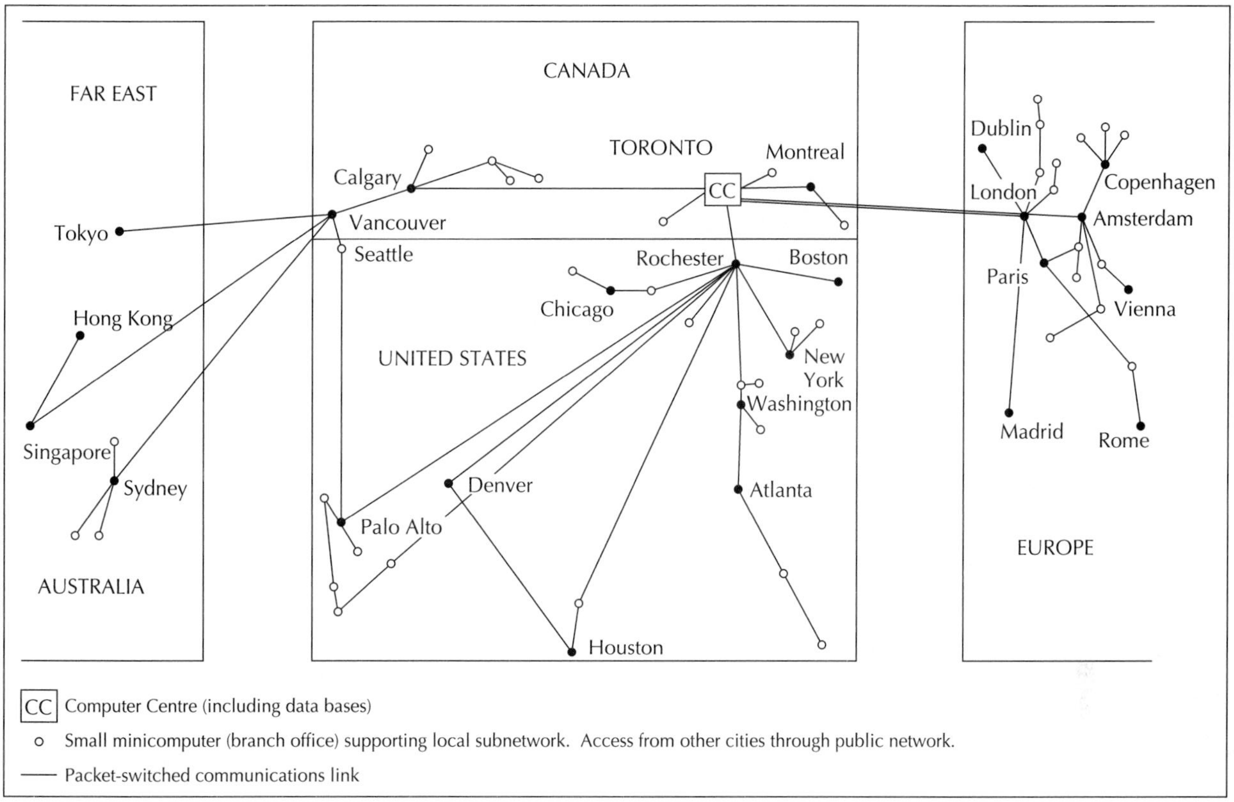

CANADA

FAR EAST

Calgary
TORONTO Montreal
CC

Tokyo
Vancouver
Seattle

Hong Kong
Rochester Boston
Chicago

UNITED STATES
New York
Washington

Singapore
Denver
Atlanta

Sydney
Palo Alto

AUSTRALIA

Houston

Dublin
London Copenhagen
Amsterdam
Paris Vienna

Madrid Rome

EUROPE

CC	Computer Centre (including data bases)
o	Small minicomputer (branch office) supporting local subnetwork. Access from other cities through public network.
——	Packet-switched communications link

FIGURE 4.46
The Canadian-based multinational I. P. Sharp network. *(Source: Hepworth, 1990)*

Dell recently redesigned its computers so that each model incorporates as many of the same component parts as possible. The company minimizes the number of parts it holds at any one time by relying on its suppliers to stock its factories in much the same way as packaged goods companies stock their products on supermarkets shelves.

By January 1997, the company had 12 days of inventory, meaning that it replenishes its parts stocks some 30 times a year. That allows Dell to respond quickly to new technology, shifts in customer demand, and changing prices. The company's new factory in Austin, Texas, is designed to drive inventory levels even lower. The new plant has no storage space. If businesses no longer overproduce during booms, they won't be forced to cut production and lay off workers when demand slackens.

On the other side of the equation, information technology has also allowed companies to produce more with less. When Chrysler decided to increase car production, it used computers and overtime to squeeze more cars out of its existing operations rather than build a new plant.

Indeed, even high-technology companies have not managed to immunize themselves from traditional inventory bloat. Another computer maker, the Digital

Equipment Corporation, stumbled badly in 1996, when it was caught with thousands of unsold personal computers that were built before a plunge in memory chip prices. When PC prices followed chip prices lower, Digital sustained heavy losses, forcing a revamping that cost 7,000 jobs.

Nevertheless, the world is in the beginning of a process that will accelerate with the rapid adoption of the Internet as the primary means of commercial communications. General Electric (GE), the diversified industrial and financial giant, is converting its entire supply chain from paper to the Internet, a shift executives say will save hundreds of millions of dollars a year. GE's lighting division has already begun using the Internet to solicit bids for spare parts for its factories overnight instead of over two weeks. The solicitations go out over the global network to a selected group of suppliers, who reply with their bids the next day.

Fiber-optic and Satellite Systems

Fiber-optic cable and satellite systems are the communication links of the future. Today, four dozen fiber-optic laying ships are operating throughout the world. AT&T's PAT9 cable, now in service under the Atlantic Ocean,

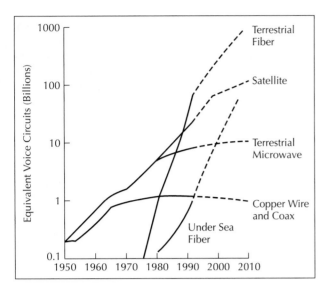

FIGURE 4.47

Global transition capacity. Fiber-optic and satellite transmission have surpassed copper wire and the coaxial cable. The increase has been an exponential one regarding voice circuitry. The U.S. government's breakup of AT&T and deregulation of telecommunications have helped turn the telephone into an important growth industry and changed the economic geography of service, allowing Sioux Falls, South Dakota, and Omaha, Nebraska, to become major processing centers for credit cards and telemarketing.

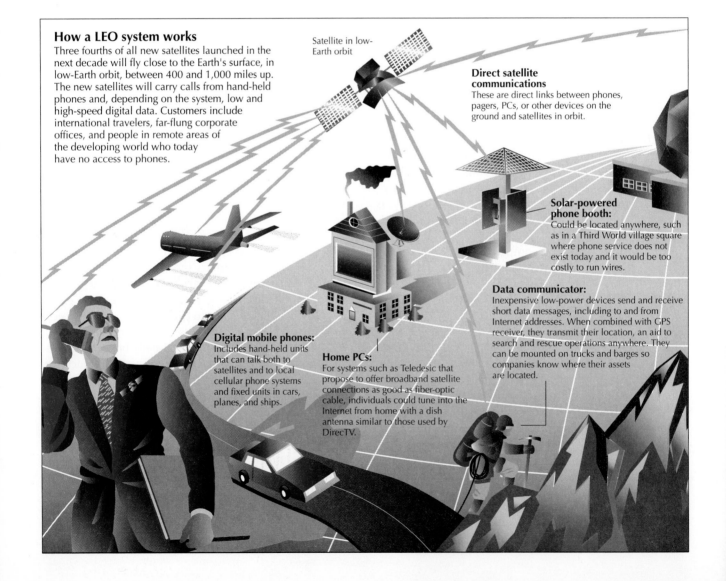

How a LEO system works

Three fourths of all new satellites launched in the next decade will fly close to the Earth's surface, in low-Earth orbit, between 400 and 1,000 miles up. The new satellites will carry calls from hand-held phones and, depending on the system, low and high-speed digital data. Customers include international travelers, far-flung corporate offices, and people in remote areas of the developing world who today have no access to phones.

Satellite in low-Earth orbit

Direct satellite communications
These are direct links between phones, pagers, PCs, or other devices on the ground and satellites in orbit.

Solar-powered phone booth:
Could be located anywhere, such as in a Third World village square where phone service does not exist today and it would be too costly to run wires.

Data communicator:
Inexpensive low-power devices send and receive short data messages, including to and from Internet addresses. When combined with GPS receiver, they transmit their location, an aid to search and rescue operations anywhere. They can be mounted on trucks and barges so companies know where their assets are located.

Digital mobile phones:
Includes hand-held units that can talk both to satellites and to local cellular phone systems and fixed units in cars, planes, and ships.

Home PCs:
For systems such as Teledesic that propose to offer broadband satellite connections as good as fiber-optic cable, individuals could tune into the Internet from home with a dish antenna similar to those used by DirecTV.

processes 100,000 transmissions at one time and has the capacity to carry in one day what the first cable laid in 1956 carried in 20 years. AT&T's transpacific cable can, as of 1996, handle 600,000 simultaneous calls between North America and Asia. By 1997, there were 1.1 million voice circuits between the United States and Europe. New techniques of data transmission, such as the so-called frame delay format, raises speeds of transmission nearly 30 fold over the 1990 technology (Figure 4.47).

Large scale, *low earth orbit* (LEO) satellite systems, like Motorola's 66 Satellite Iridium System, are private global satellite constellations that are now coming online (Figure 4.48). They provide telephone communications to and from any place on earth. These private, global satellite constellations will transmit television, radio, fax, computer, and voice images.

Teleworking, or *telepresence*, will aid productivity in the future and is being touted as the answer to reduced transportation costs, easing demands on energy and reducing environmental impacts. There is a growing trend toward wireless terminals because the wiring of computers and peripherals to networks is such a costly undertaking. Wireless terminals include computers and other devices that can communicate with machines with infrared or electromagnetic signals (like television remote control devices). This allows computers to be moved easily and yet continue to function within company or international networks. The first generation of

FIGURE 4.48

By the year 2010, the planet will be encased in a delicate web of fiber-optic cable and circled by a galaxy of communication satellites as networks are capable of transmitting symphony-quality audio, and snapshot quality video images to wall-size panels mounted in homes and exurban mini-office nucleations. The social, economic, and technological impact of this new communication networking will be profound, driving the *globalization of the economy* and decentralizing population clusters from major cities. (*Source: U.S. News & World Report*)

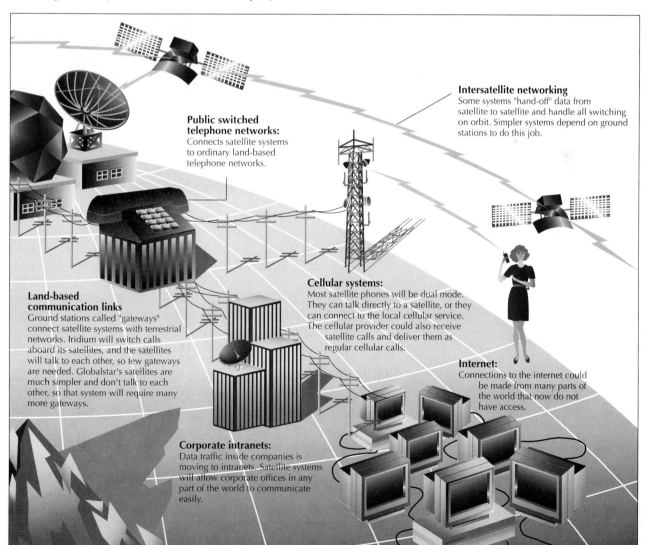

Intersatellite networking
Some systems "hand-off" data from satellite to satellite and handle all switching on orbit. Simpler systems depend on ground stations to do this job.

Public switched telephone networks:
Connects satellite systems to ordinary land-based telephone networks.

Land-based communication links
Ground stations called "gateways" connect satellite systems with terrestrial networks. Iridium will switch calls aboard its satellites, and the satellites will talk to each other, so few gateways are needed. Globalstar's satellites are much simpler and don't talk to each other, so that system will require many more gateways.

Cellular systems:
Most satellite phones will be dual mode. They can talk directly to a satellite, or they can connect to the local cellular service. The cellular provider could also receive satellite calls and deliver them as regular cellular calls.

Internet:
Connections to the internet could be made from many parts of the world that now do not have access.

Corporate intranets:
Data traffic inside companies is moving to intranets. Satellite systems will allow corporate offices in any part of the world to communicate easily.

The most important improvement in global communications has been the development of satellite technology. The first communication satellites date from 1965, with the launching of Early Bird, able to carry 250 telephone conversations or two television channels simultaneously. Since then, more advanced communication satellites carry 100,000 circuits of simultaneous telephone communications or television channeling. This has allowed for the age of personal, mobile telephones, and fax machines that can reach every corner of the globe instantly. In the commercial space, it appears that the planet will be completely covered by satellite-based telephone and message capability before the year 2000. The Russians have entered the commercial launch marketplace and have driven launch prices down. This has encouraged an expansion of the commercial use of space. Personal phones have also become miniaturized. (*Photo: MI Sinibaldi, The Stock Market*)

wireless terminals is now popular in the form of desktop and laptop PCs. Some palm-sized units can trade files between one another by pointing and clicking in the same direction. Palmtop units can send large data files or e-mails, using satellite communications technology. This will eventually eliminate or replace pagers and cellular phones. The trend toward wireless terminals is significant because it allows more portability and eliminates the need to be connected and disconnected from local area networks. Cellular networks also use satellite communications technology. One can be in constant cellular contact in any North American populated area.

The Internet/Information Superhighway

The manner in which we will work and live in the new century will be determined by a vast web of electronic networks called the *information superhighway*. This system delivers large amounts of services to our homes, offices, and factories, including telephone calls, TV programs and other video images, text, and music. The system enables students at rural schools to use computers to tap resources at Stanford University or researchers in a small college in Arkansas to use the power of the supercomputer at the University of Illi-

nois. The superhighway allows doctors to check patients from their homes, and it permits doctors in several remote cities to collaborate on a patient's care by immediately sharing multimedia computer screens.

Like the railroad system of the 19th century and the interstate highway system of the 20th century, the information highway of fiber-optic cables, satellites, and wireless grids links millions of computers, telephones, faxes, and other electronic products all over the country and all over the world.

The term electronic or information superhighway is a catchall name for several different architectures. America's information highway is a network of networks, controlled by many companies. Almost 45 million people trade information via the Internet, which uses existing telephone lines. The superhighway has two groups of customers. The first group is individuals at home who shop, bank, work, and entertain themselves without leaving their homes. These customers pay for services they get, in a way similar to the way we pay for the telephone or cable TV services. The second group of customers are organizations and businesses. These customers use the networks for hundreds of applications.

Companies conduct important electronic meetings and train employees in many locations simultaneously,

Normally a million people used wireless phones a decade ago when they were viewed mainly as toys or for rich people or executives. With falling prices and the ability to call from anywhere in an emergency, wireless telephone has become increasingly popular. Today, 20% of all North American households have wireless phones. The driving force behind such growth is the new digital service called PCS (personal communication system) that was created by the Federal Communications Commission. PCS uses digital signals that enable PCS to offer such features as e-mail, caller ID, paging, and links to the Internet, as well as compact disc quality sound and greater security from wireless thieves. (*Photo: Paul Barton, The Stock Market*)

using advanced multimedia tools and intelligent computer-aided tutoring. Using the information superhighways, people are able to see places before they travel to them and even "feel" scenarios.

The Internet is the world's largest computer network; it is actually a network of networks. It was composed of over 25,000 connected networks with over 2.5 million computers and 50 million users in 1997 (Figure 4.49). The largest network is Internet, and it includes over 2,000 universities and government repositories. The core countries of the world, including Europe, North America, and Japan, are wiring homes, businesses, and schools into the network. Semi-periphery and periphery nations will move from remote locations and from the medieval age to the communication information age almost overnight.

The growth of the Internet has been phenomenal. The Internet was started by the Defense Department's Advance Research Projects agency in 1968. By 1985, 200 agencies, schools, and research labs were connected to the Internet. Between 1990 and 1996, the number of Internet users connected went from 500,000 to 50 million, many of these international. Transmission bun-

dles of information swelled from 50 million a month in 1988 to 40 billion a month in 1997. Due to commercialization and advertising, software, data, and information are offered over the Internet free. In addition to the network growth, Internet's transmission speeds are increasing as well (now measured in gigabites per minute). The use of the Internet has diffused from large

FIGURE 4.49
Internet adoptions have grown exponentially. In 1993, there were 1 million Internet hosts; there were 20 million in 1998. Web use for education and training puts it at the leading edge of a trend. Businesses are just beginning to leverage Web browser power and Internet technologies for internal information dissemination. In 10 years, this will be the standard way to do business in a large organization. The days of having most of your workforce based in a 40-story tower with a centralized library and training room are over. People now work where it makes the most sense. With Web technology, they can take the library and training room with them, wherever they are.

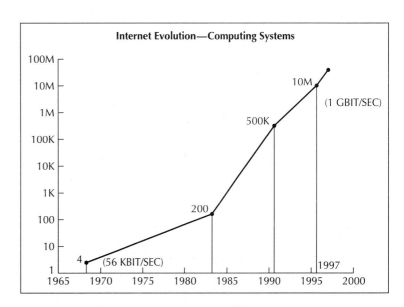

U.S. government agencies to universities, institutes, corporations to small businesses, libraries, grade schools and high schools, and homes (Figure 4.50).

The availability of e-mail, the Internet, and public network services provides opportunities to conduct business electronically. A large number of potential customers can be reached at a low cost via the Internet. Therefore, advertising on the Internet is becoming popular. In addition, major publishers have put their books on the Internet.

Societal Implications of Information Technology (IT)

The social implications of information technology (IT) are far reaching. Use of IT already has had many direct beneficial effects on complicated human and social problems such as medical diagnosis, computer-assisted instruction, government program planning, environmental quality control, and law enforcement. Problems in these areas could not have been solved economically—or at all—without IT.

Opportunities for People With Disabilities

Integration of some AI (artificial intelligence) technologies such as speech recognition and vision recognition into a CBIS is creating new employment opportunities for people with disabilities. Adaptive equipment permits people with disabilities to perform ordinary tasks with computers. Visually impaired California geographer Reg Gollege has invented an audio guide system for Santa Barbara, California, using a hand-held computer with a TIGER File and a satellite CPS.

Changing Role of Women

IT is changing the "traditional" role of women in the workplace. The opportunity to work at home helps women with young children assume more responsible managerial positions in organizations. This could lead to better pay for women who can devote more attention to business while they still carry on duties at home.

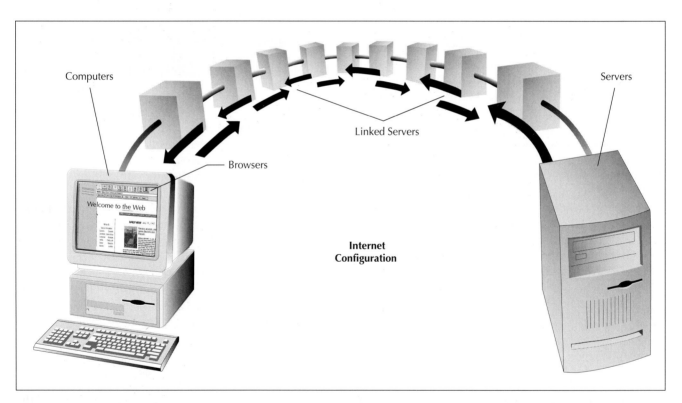

FIGURE 4.50

The Internet comprises computers, linked servers, browsers, and users. Computers are now more powerful than ever because they are linked to virtually any source of stored information through modems that convert data such as photos and text into electronic signals. The "Web" refers to all those interconnected data sources one can access via the Internet. Observing the World Wide Web requires a browser program, a remote controlling apparatus that assembles photos, text, and audio on Web pages that appear on your computer screen. The Internet backbone is like a phone system that ties callers from around the world together via a web of data transmission lines on the Net. Large computers called servers are linked together via telephone lines and fiber-optic cable. Most of the work on the Net is done by special server computers that dish out data. Web browsers type in an address like those http:// commands that connect them to the servers.

HOW THE INTERNET WORKS

The personal computer is the key to the Internet, whether it be in a home, a business, school, or in a briefcase. Computers are more powerful than ever because they can be linked to virtually any source of stored information throughout the World Wide Web by a modem that converts data, like text, statistics, and photos, into electronic signals. The *World Wide Web* (or simply "the Web") refers to all those interconnected data sources that one can access via the Internet.

Browsers are remote controllers that assemble text, facts, figures, and photos on the Web pages to appear on computer screens. Seeing content on the Web requires a browser program. Internet Explorer is Microsoft's latest browsing tool. Netscape Navigator still dominates the market and has tremendous loyalty among early users.

The Internet is a worldwide integrated "phone system" that ties callers from almost any location around the world together via a web of data transmission lines. On the Net, giant computers called *servers* are networked together via telephone lines and fiber-optic cable.

The real work on the network is done by *servers*—special computers that distribute data to computer users. Web browsers type in addresses (they start with http://) that connect them to the servers. Companies are now putting browser software on anything with a screen and a modem. The first step is the Internet TV, followed by network computer online video gaming machines and net *surfing* cell phones. Organized around a powerhouse electronics alliance that includes Sony, Nintendo, NEC, and IBM, learning is to make the Internet usable to just about anyone, anywhere.

Improvements in Health Care

IT has caused major improvements in health care delivery, ranging from better and faster diagnoses to expedited research and development of new drugs, to more accurate monitoring of critically ill patients. One technology that has made a special contribution is artificial intelligence. AI supports various tasks carried out by physician and other health care workers. Of special interest are expert systems that support diagnosis of diseases and the use of machine vision in enhancing the work of radiologists (see Chapter 7: Smart Cities).

Help for the Consumer

Several IT products are in place, and many more will be developed to help the layperson perform tasks that are skilled or undesirable. TaxCut is an expert system product that can help in tax preparation; Wilmaster is an ES that helps laypersons draft a simple will; and Wines on disk advises consumers how to select wines. Intelligent robots will clean the house and mow the lawn.

Quality of Life

On a broader scale, IT has implications for the quality of life. An increase in organizational efficiency may result in more leisure time for workers. The workplace can be expanded from the traditional 9-to-5 job at a central location to 24 hours a day at any location. This expansion provides flexibility that may improve the quality of leisure time, even if the total amount of leisure time is not increased. Health and safety could also be improved, because robots can work in uncomfortable or dangerous environments.

Information Technologies and Their Impact

EDI, Electronic Data Interchange, can be defined as the electronic movement of standard business documents between and within firms. EDI uses a structured machine-retrievable data format that permits data to be transferred between networked computers without rekeying. Like e-mail, EDI enables the sending and receiving of messages between computers connected by a communication link such as a telephone line (Figure 4.51).

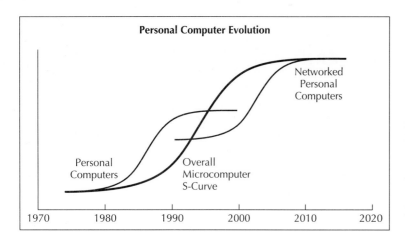

FIGURE 4.51
Logistic growth (S-curves) for network computers. In 1950, 60% of all jobs in North America were of the unskilled variety, but by 1990, that figure fell to 25%. By 2005, it is expected to fall still farther to 15%. Conversely, finding adequate supplies of "knowledge workers" who derive power and efficiency from networked computers will require companies to search the labor force and shift offices and factories more rapidly than in the past.(*Source: Dent, 1997*)

However, EDI has some special characteristics:

1. Business transaction messages. EDI is used primarily to transfer repetitive business transactions. These include purchase orders, invoices, approvals of credit, shipping notices, and confirmations. In contrast, e-mail is used mainly for non-standard correspondence.
2. Data formatting standards. Because EDI messages are repetitive, it is sensible to use some formatting standards. In contrast, there are no data formatting standards for e-mail because it is usually not formatted.
3. EDI translators: The conversation of data sent into standard formats is done by special EDI translators.
4. EDI uses VAN. In contrast to e-mail, which uses regular telephone lines, EDI uses a value-added network (VAN).

Although a company can lease dedicated lines from telephone companies for conducting communications in some situations, it is usually better to use the services of a commercial VAN. One company may use several vendors. This arrangement is cost-effective because VAN vendors work with large volumes, have expertise in maintenance, and can help customers develop the necessary interfaces. Many companies use EDI to change their business processes, whereas others use it to gain competitive advantage. The major advantages of EDI are:

1. EDI enables a company to send and receive large amounts of information around the world in real time.
2. Information can flow among several trading partners consistently and freely.
3. Companies have the ability to access partners' computers to retrieve and store standard transactions.

4. EDI fosters a true partnership because it involves a commitment to a long-term investment and refinement of the system over time.
5. EDI creates a paperless environment, saving money and increasing efficiency.
6. The time for collecting payments can be shorted by several weeks.
7. Data may be entered off-line, a batch mode, without typing up ports to the mainframe.
8. When an EDI document is received, the data may be used immediately.

Electronic data interchange has its limitations and costs. A company may have to use several EDI translators; the VAN services may be expensive; there could be security problems on the networks; and there could be communication problems with some of the business partners.

Multimedia

Multimedia refers to a pool of human-machine communication media, some of which can be combined in one application. In information technology, an interactive multimedia approach involves the use of computers to improve human-machine communication by using several items of the media pool with the computerized system as the center of the application. Such integration allows the combination of the strengths of voice, GUI, and other media. The construction of an application is called authoring. Multimedia can merge the capabilities of computers with TVs, VCRs, CD players, and other entertainment devices.

Hypertext and Hypermedia

Multimedia software is mainly used for making dynamic presentations—providing information or computer-based training in an interactive and exciting manner that makes use of video, sound, animation, and graphics as well as text. Multimedia software tools use a hypermedia navigational technique incorporated into a wide variety of products. Hypermedia itself is based on hypertext navigation, which was text-based but incorporated innovative mechanisms for interacting with a computer. Hypertext software is a dynamic approach to navigating, or moving through a database where information is linked together and is accessible in any form desired, regardless of its location. With hypertext, some links lead to other links, which further links to other topics.

Hypermedia improves on the hypertext concept by linking together not only text but also graphics, sound, animation, and video. The major benefit of computerized hypermedia is that readers do not need to know the physical location of data. They simply select the links they desire. Some hypermedia products allow you to interact with them.

ISDN

Integrated Services Digital Network (ISDN) is a communications protocol that provides functions and services of the Internet via cable. For geographic areas with access to these services, a single ISDN phone line provides each user with simultaneous phone and medium-speed communications. For years, providers of local telephone services have tried to sell ISDN to major corporations as a way of extending data communications to the desktop of any employee who already has a telephone. But it has been a hard sell because most organizations have already installed local area networks (LANs) that run at 10 Mbps or higher; a 64 Kbps channel does not impress them. ISDN is viewed as ideal for home offices, small business, or at-home workers who can use a single ISDN line for voice, facsimile, and data communications.

The Information Warehouse

Data or information can be viewed as a tangible asset that needs to be stored, tracked, and made readily available to those who need it, when they need it. Borrowed from the principles and practices used in the warehousing of physical assets, the field of study and practice within the IT area is known as data warehousing or *information warehousing*. Although definitions vary, an information warehouse is generally thought of as a decision-support tool, collecting information from multiple sources and making that information available to end users in a consolidated, consistent manner. The concept started in the 1970s, when corporations realized they had isolated "islands"

of information systems that could neither share information nor project an enterprise-wide picture of corporate business. Recently, there has been a resurgence of interest in this concept, as corporations seek distributed computing architectures while they leverage isolated legacy systems. Rather than trying to unite all the systems into one or linking all systems in terms of processing, why not just combine the data in one place and make it available to all systems?

In most cases, a data warehouse is a consolidated database maintained separately from an organization's production system databases. It is significantly different from a design standpoint. Production databases are organized around business functions or processes such as payroll and order processing. Many organizations have multiple production databases, often containing duplicate data. A data warehouse, in contrast, is organized around informational subjects rather than specific business processes. The data warehouse is used to store data fed to it from multiple production databases in a format that is more readily accessed and interpreted by end users.

Providing a consolidated view of corporate data is better than many smaller views. Another benefit, however, is that data warehousing allows information processing to be off-loaded from individual systems onto lower-cost servers. Once done, a significant number of

One of the satellites in the tracking and data relay satellite system in orbit around the earth is the TDRSS satellite. The system provides advanced tracking and telemetry services for a number of other satellites, as well as commercial telecommunications services. (*Photo: NASA Headquarters*)

end-user information requests can be handled by the end users themselves, using graphic interfaces and easy-to-use query and analysis tools. Accessing data from an updated information warehouse should be much easier than doing the same thing with older, separate systems. Furthermore, some production-system reporting requirements can be moved to decision support systems—thus freeing up production processing. Additionally, the performance of production processing will likely be optimized for its particular type of processing—typically for simple and frequent transactions and updating. The information warehouse, on the other hand, could also be optimized for ad hoc, complex queries, such as those needed by decision makers. It would likely be very difficult, if not impossible, to optimize existing databases for both operational production purposes and for decision-making needs.

Wal-Mart Stores, Inc., the country's largest retailer, records every sale in every one of its 2,268 (1997) stores in a giant information warehouse. The company uses the data to fashion targeted marketing strategies while distributing products to the stores where the most people are likely to buy them. Programs automate and seamlessly link all aspects of the company's operations. The system processes orders, schedules and tracks manufacturing, dispatches and tracks goods, and makes sure they are invoiced properly so that the bills sent to customers match the orders.

SUMMARY

For simplicity, industrial location theory assumes that transport costs are proportional to distance; however, in the real world, transport costs are much more complex. In our review, we consider some of the factors other than distance that play a role in determining transport costs—the nature of commodities, carrier and route variations, and the regimes governing transportation. Transport costs remain critical for material-oriented and market-oriented firms, but they are of less importance for firms that produce items for which transport costs are but a small proportion of total costs. For these firms, transit time is more crucial than cost. Modernized means of transport and reduced costs of shipping commodities have also made it possible for economic activities to decentralize. Multinationals have taken full advantage of transport developments to establish "offshore" branch-plant operations, especially in Asia's economic tigers (Taiwan, South Korea, Hong Kong, Singapore, and Thailand).

Movements of goods, people, and information take place over and through transport networks. We begin this chapter by discussing how geographers analyze these networks by means of graph theory, reducing them to a set of vertices and a set of edges. Graph the-

oretic measures may be used to determine nodal accessibility and network connectivity. Geographers also examine other properties of networks—their shape and density. Using models of network change, they further demonstrate how the growth of transport media is inextricably tied to the process of economic growth.

In a discussion of transport development, we explain how improvements over the centuries have resulted in time-space and cost-space convergence. They integrated isolated points of production into a national or a world economy. Although the friction of distance has diminished over time, transport remains an important locational factor. Only if transportation were instantaneous and free would economic activities respond solely to aspatial forces such as economies of scale.

The prediction of flow in transportation networks consists of the estimation of the attraction of places based on their size and their distance apart. The gravity model is a good approximation for the prediction of commodity flow, information flow, and person movement between anchor points. The law of retail gravitation helps establish rules for the breaking point between cities and the edges of their spheres of influence. New developments in flow systems include spatially separated computer networks linking facilities of multinational corporations. Computer innovation now ties the home environment to remote information sources, including business and banking services, travel and commercial services, and library and telephone directory services.

Innovation in urban transportation systems is necessary because of the tremendous increase in travel demand in large cities of the developed world. For example, in the United States, vehicle miles traveled, automobile ownership, and total vehicle trips are increasing rapidly. IVHS technologies are aimed at accommodating this tremendous increase in travel with present roadways, while TOD seeks to redesign cities in higher densities to foster the use of public transportation systems. Because transportation systems, planning, construction, and management have become so complex, GIS computer software is now a common tool used by transportation decision makers and analysts.

Communications and information technology (IT) are transforming the world economy at rates never before thought possible. Profound implications, even many that the world cannot yet measure, accompany this IT explosion. The amount of information is said to be doubling every 18 months. In some areas of the world economy, the growth of information is so rapid that by the time this book is published in 1997, portions will be outdated. At the center of this information explosion is the microprocessor, networked computers, and the Internet.

Chapter 4

STATISTICAL APPENDIX*

For those who have had no previous contact with the basic statistical methods (scatter diagrams, correlation, regression), this appendix provides a brief and much simplified introduction. It is not designed to make the reader operational in the methods of theory concerned, but it *is* designed to enable the reader to *interpret* various equations and coefficients as they have been used in transport study and to understand their conceptual implications.

One example is the interaction between a single city *(i)* and 15 other cities *(j* = 1, . . . , 15). the *actual* interaction (I_{ij}), which, in turn, will be based on a simple gravity model $(P_i P_j / d_{ij})$. Table 4A.1 lists both the actual traffic *(Y)* and the gravity model estimates.

The first step is to graph the two variables so as to get a better picture of their relationship. Figure 4A.1 is a *scatter diagram* with the actual passengers (I_{ij}) represented along the *y*-axis and the gravity model estimates (I^{ij}) along the *x*-axis. For each link between city A and one of the other 15 cities, a point is on the graph with the actual traffic as its *y*-value and the gravity model estimate as its *x*-value. For example, city 3 has actual traffic of 200 city with A, and its gravity model estimate is 300. Thus, the point on the graph for city 3 is located at the intersection of the *y*-value of 200 and the *x*-value of 300, as shown by the dotted lines on Figure 4A.1. By statistical convention, the *x*-axis variable is referred to as

◈

T a b l e 4A.1
Actual and Projected Air Passengers

To City	Projected Passengers from City A (P/D) X	Actual Passengers from City A Y
1	371	339
2	219	271
3	300	200
4	420	299
5	409	337
6	185	240
7	261	238
8	420	390
9	189	201
10	425	400
11	312	291
12	440	400
13	360	270
14	112	270
15	150	180

FIGURE 4A.1
Scatter diagram. This scatter diagram shows the relationship between gravity model expectations on the *x*-axis and actual traffic on the *y*-axis. The trend of the scatter indicates a positive relationship between the two.

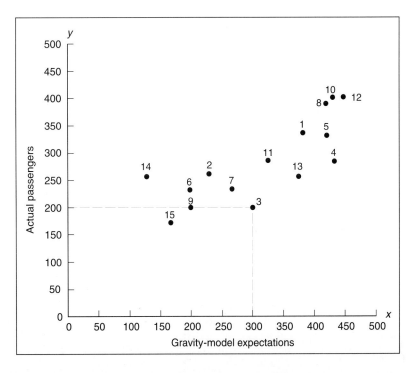

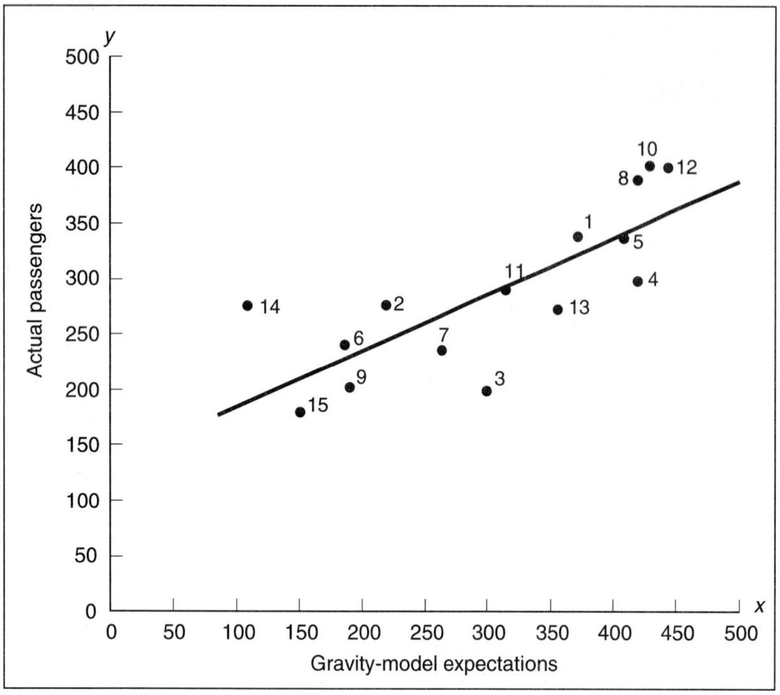

FIGURE 4A.2
Scatter diagram and regression line. A regression line has been fitted to the points in the scatter diagram to give a more precise measure of the relationship between the actual and expected traffic. The regression equation is in the form of the equation for a straight line, $y = A + bx$, or, in this case, $y = 133.7 + 0.5x$.

the *independent* variable; the *y*-axis variable is referred to as the *dependent* variable.

We are most interested, however, in the *general* relationship between *x* and *y* as shown by the entire scatter of dots. The dots are fairly well grouped together and they trend upward and to the right. This indicates a *positive* relation between the *x* and *y* variables. As gravity model estimates increase, actual traffic also increases.

Three useful measures may be derived from the scatter diagram: the regression line, the correlation coefficient, and the residual. The *regression line* is a line fitted to the scatter of points. This provides a clearer picture of the *average relationship* between the *x* and *y* variables. We can refer to this as a functional relationship and think of *y* as some function of $x, y = f(x)$. The straight line we have fitted in Figure 4A.2 provides a useful measure of that relationship in the form of the basic equation of a straight line. For our purposes, it is convenient to express this as $y = a + b(x)$. Here *y* and *x* are variables and can represent each of the 15 observations in turn. The *constants a* and *b* can be replaced with a single number that will be the same for all 15 observations. The constant *a* locates the *y*-intercept, the point at which the line would strike the *x*-axis if the lines were extended back to a value of $x = 0$. The constant *b* indicates the *slope* of the line or the number of units of *y* for every single-unit change in *x*. If *b* is positive, it indicates that *x* and *y* are positively related: as *x* increases, *y* also increases. If *b* is negative, it means that *x* and *y* are negatively related: as *x* increases, *y* decreases.

The equation for the straight line fitted to the scatter on Figure 4A.2 is $y = 133.7 + 0.5x$. These values indicate that the *y*-intercept is 133.7 and the slopes is 0.5. This tells us that *y* will equal 133.7 when *x* equals zero and that *y* will increase by 0.5 units for every unit in *x*. We can use the equation to give us a better description of the average relationship between *y* and *x*. The *a*-value of 133.7 and the *b*-value of 0.5 are the *parameters* of the equation. If we were just to use a *ratio* between *y* (passengers) and *x* (the gravity model estimate), we would have an average relationship based on a single number. This would be equivalent to the dashed line on Figure 4A.3, which starts at the *origin*, since $y = O$ when $x = O$. The slope indicates that, on the average, air passengers *(y)* increase 0.95 units for every unit increase in gravity model estimates *(x)*. The regression line obviously fits the scatter of dots much better and thus provides a better estimate of an average relationship by considering *both* slope and intercept.[1]

---

[1] Some caution must be exercised in interpreting the meaning of the intercept. The idea that there could be an average of 133.7 passengers *(y)* when the gravity model estimate *(x)* was zero is, of course, meaningless, since a zero value for *x* could only occur with zero population product or infinite distance. The descriptive utility of the regression equation is strictly confined to the range of empirical values on which it has been based. Also, a negative intercept obviously has no meaning in terms of expected passengers. It does, however, serve to position the regression line so as to provide a better statistical description of the relationship between *x* and *y*.

FIGURE 4A.3
Scatter diagram, regression line, and ratios. The dashed line represents the ratio between actual and expected air traffic. This ratio of 0.9 air passengers per unit of air passenger expectation is clearly less accurate a description of the relationship between the two than is the regression based on the equation $y = 133.7 + 0.5x$.

We can now return to city 3 and use the regression equation of $133.7 + 0.5x$ to calculate an average relationship between air passengers and gravity model estimates. We multiply the x-value (or gravity model estimate) for city 3 by 0.5 and add 133.7 to it. Since the gravity model estimate for city 3 was 300, the equation becomes $y = 133.7 + 0.5 (300)$. The resultant figure of 284 would, of course, fall right on the regression line and would represent an estimate of the passengers between city A and city 3 based not just on the gravity model estimate but on the *average* relationship between the gravity model estimate and the actual traffic between city A and all 15 destinations.

The specific parameters or constants of 133.7 and 0.5 for a and b, respectively, may be obtained in several ways. It would even be possible, although unnecessarily inexact, to draw in a subjectively estimated line by hand and then measure the slope and intercept on the scatter diagram. There are, however, a number of easy computational procedures for determining a line of best fit. The line shown in Figure 4A.2 is based on the minimizing of the squared departures of the plotted points from a calculated regression line. Most commonly, the procedures are carried out on a computer using any one of a wide variety of statistical software packages. There are also many small hand calculators that have regression and other basic statistical programs built in.

The second useful measure associated with the relation between an x and a y variable is the measure of closeness of fit. As shown in Figure 4A.4, the correlation coefficient r varies from 0.0 for no correlation to

FIGURE 4A.4
Scatter diagram and correlation. The correlation coefficient, r, may be regarded as an index of the closeness of the relationship between the scatter of points and a regression line. In (a), r would be close to 0.0 since no regression could provide an effective description of the nearly circular pattern of data. In (c), on the other hand, r would be close to 1.0 since most of the data would fall on or near a straight line. (b) would fall somewhere between these two extremes, with an r of 0.7 or 0.8, for example.

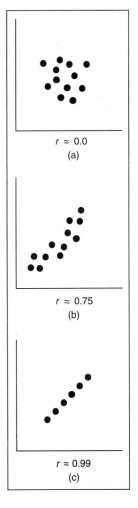

$r \approx 0.0$
(a)

$r \approx 0.75$
(b)

$r \approx 0.99$
(c)

+1.0 for a perfect fit of a positive regression line to all points on the scatter diagram. If there were a perfect fit to a negative slope for all points on the scatter diagram, the *r*-value would be −1.0. The value r^2, the coefficient of determination, indicates the percentage of variation that is associated with (statistically "-explained" by) the variation in the *x*-variable. The r^2 measure is frequently employed as a useful indicator of the closeness of the relationship between the two variables. In the case of our example, the *r*-value is 0.785, and the r^2-value is 0.617, indicating that 62 percent of the variation in city A's air passengers is associated with the variation in gravity model estimates.

A common misuse of the correlation coefficient and the r^2 is to confuse it with causation. A high value of *r* or r^2 simply means that the *x* and variables tend to *co-vary*. As one increases, the other tends either to increase or decrease consistently. Although the closeness of this covariation, as measured by r^2, is called statistical "explanation," it is not necessarily explanation at all. Only if the investigator is willing to advance a plausible explanation from other theoretically or empirically based evidence may a causal interpretation be inferred from a close numerical covariation between the *x* and *y* variables. The validity of that relationship, however, lies not in the closeness of fit, or the statistical covariation, but in the plausibility of the postulated relationship. The measure is frequently referred to in the text as a useful indicator of the closeness of the covariation of the variable concerned.[2]

The third useful measure associated with the scatter diagram relationship between the *x* and *y* variables is the *residual*, or the difference between the actual value of *y* and *y* calculated from the regression equation. These residuals may be calculated by subtracting the estimated from the actual values, *y* − *y*. A *positive residual* indicates that the actual value of the dependent variable is *greater* than would be expected from the average relation between it and the interdependent variable. A *negative* residual indicates that there is *less* than would be expected.

In the case of traffic between city A and city 3, for example, a computed *y* = value of 284 was obtained when the *x* = value of 300 (based on gravity model expectations) is put into the regression equation. The *actual* number of passengers to city 3 is 200. The residual

(*y* − *y*) is therefore −84. There were 84 fewer passengers between city A and city 3 than would be expected from the *average relationship* between actual passengers and gravity model expectations.

Examination of residuals suggests additional relationships between dependent and independent variables, and such examination serves as a corrective to the oversimplified view of complex relationships provided by a simple linear regression, where the only measure is the degree to which air passengers flows are adequately described by a straight-line relationship to a simple population-distance ratio. Large residuals, either positive or negative may indicate that an important explanatory variable is missing.

A particularly geographic representation of these residual deviations from a particular model is provided by Figure 4A.5, which shows how the regression equation *y* = 133.7 + 0.5*x* fails to describe the relation between actual air passenger flow and the gravity model estimates. The map shows some of highest and lowest residuals. The shaded cities are those with *positive* residuals. Traffic to these cities was *greater* than would be expected from the average relation between actual and estimated traffic as expressed by the regression line. The cities left blank have *negative* residuals, indicating *less* traffic than would be expected.

Several reasons for the residual deviations on Figure 4A.5 might be suggested. For example, the positive residual at city 14 might be associated with city function. The city might have a strong recreational function and attract more traffic than its size would suggest. Size of city itself seems to be understated as a factor in the gravity model estimates. The largest cities, such as cities 10 and 8, have positive residuals, indicating that their traffic may be less affected by distance than the smaller cities' traffic.

Examination of mapped residuals might also suggest certain consistent spatial relationships. For example, the large and medium-size long-haul cities seem more likely to have positive residuals than the large and medium size short-haul cities, suggesting that passenger traffic does not seem to fall off with distance as rapidly as expected from the model.

Of the many possible modifications, extensions, and variations on the simple linear regression model in this example, two have been used most frequently in this text. One has to do with the nature of the functional relationship itself. Seldom will the relationships as examined in their original form be clearly linear. One common practice has been to transform the variables to logarithms. Figures 4A.6 and 4A.7 show two different cases in which log transformations provide a better description of a linear relationship than does the original, unmodified arithmetic data.

In Figure 4A.6a, the data are clustered near the origin, and their variability increases greatly as the

---

[2] It should be noted that, in this brief discussion, we have avoided the topic of statistical inference and significance tests. These extremely important considerations form much of the basis for most introductory statistics texts. As applied to regression and correlation, however, such terms as *significant* or *not significant* still do not permit causal interpretations.

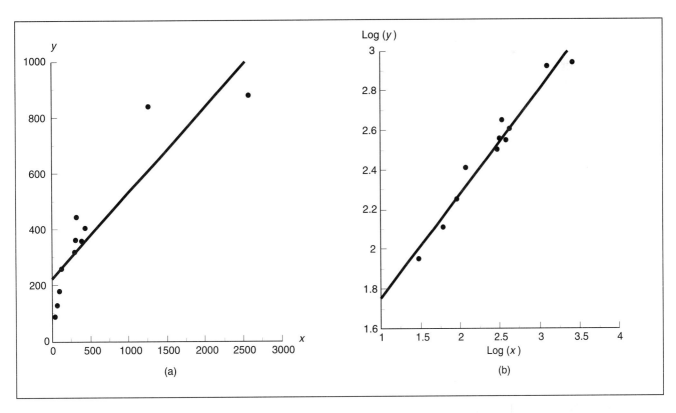

FIGURE 4A.5
Residuals map. On this map, the cities with positive residuals are shaded; the cities with negative residuals are left blank. Cities with positive residuals, $y - y$, are those at which the actual traffic is greater than would be expected from the average relationship between actual traffic and the gravity model relationships as represented by the regression line.

FIGURE 4A.6
Log transformation to correct data clustering. A straight-line fit to clustered data does not provide as effective a description of the raw data (a) as of the log-transformed data (b).

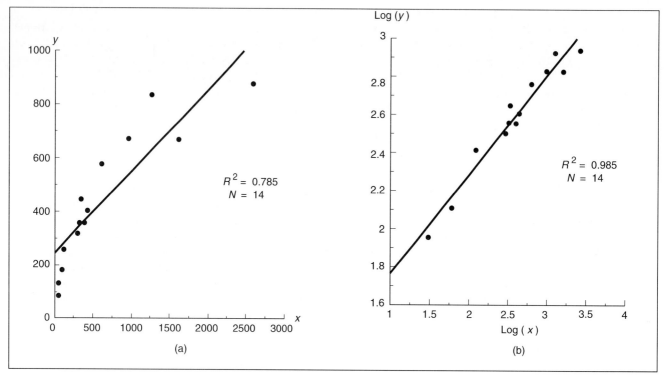

FIGURE 4A.7
Log transformation to correct for curvilinearity. A straight-line fit to curvilinear datas does not provide as good a fit to the raw data (a) as to the log-transformed data (b).

values of x and y increase. We could fit a straight line with ordinary regression procedures, but it would not provide a very good description of the relationship. In Figure 4A.6b, however, the scatter diagram based on the logs of the same variable shows a much improved fit. The log transformation has the effect of spreading out the cluster of low values and pulling in the higher values. The resulting regression equation is $\log y = a + b \log x$, which becomes $y = A(x)^b$ when taken out of logs. It provides a better description of the relation between y and x than did the original untransformed equation of $y = a + bx$.

A more common reason to use logs or some other transformation is shown in Figure 4A.7 where the relationship is obviously curvilinear rather than linear. Again, a straight line provides a poor fit, and a log transformation is one way to provide a better description.

Another way in which the log transform emerges is in the transformation of the probability model to a conventional regression format. In the chapters on discrete choice we used models for $P_i = e^{a+bxi}/(1 + e^{a+bxi})$ for the probability that the discrete choice variable takes a value of zero or one. In this model, we can vary x_i over a set of different values and obtain predictions for P_i, which range between zero and one. In order to regress P_i against observed values of x_i we need to take the

transformation $\log e(P_i/1 - P_i)$, which conveniently equals $a + bx_i$.[3]

Common logs tell us to what power 10 must be raised to get a given number. Thus, $100 - 10^2$ may be written as $\log_{10} 100 = 2$. Natural logs tell us to what power the number e (2.71828) must be raised. There are several different notations used for this text: $2.71828 = e^1$ or $\exp(1)$, $\log_e 2.71828 = 1$, or $\ln 2.71828 = 1$.[4]

---

[3] The number $e = 2.71828$ has many useful mathematical properties. The number itself is derived from the limit of the expression $(1 + 1/n)^n$ as n approaches infinity. Thus,

$$e = \lim (1 + 1/n)^n = 2.71828 \quad n \to \infty$$

[4] The transformation used is as follows:

$$P_i = \frac{e^{a+bxi}}{1 + e^{a+bxi}}$$

and so

$$1 - P_i = \frac{1 + e^{a+bxi}}{1 + e^{a+bxi}} - \frac{e^{a+bxi}}{1 + e^{a+bxi}} = \frac{1}{1 + e^{a+bxi}}$$

and if we calculate $P_i/(1 - P_i)$,

$$\frac{P_i}{1 - P_i} = \frac{e^{a+bxi}}{1 + e^{a+bxi}} \frac{1 + e^{a+bxi}}{1} = e^{a+bxi}$$

Therefore, when we take the natural logarithm of both sides we see that

$$\log_e[P_i/(1 - P_i)] = a + bx_i$$

In addition to the nature of the functions involved, the possibility of additional factors must be considered. Since most problems involve the effects of more than one factor, a *multiple regression* is commonly used in which the dependent variable y is considered to be dependent on more than just a single independent variable x. In the present example, we might include a second independent variable to account for the recreational function of cities. The equation $y = a + b_1 x_1 + b_2 x_2$ might then include the gravity model estimate as x_1 and the percent of employment in service industries as x_2. The resultant percentage of "explained" variation would then be expressed as R^2, which would take into account the combined effect of both independent variables. If four factors affecting traffic can be identified, the regression equation could be expressed as $y = a + b_1 x_1 + b_2 x_2 + b_3 x_3 + b_4 x_4$.[5]

We can combine multiple regression and log transformation. The gravity model estimate, for example, can be split into two parts, the population product $(P_i P_j)$ and the intervening distance (d_{ij}). It can then be transformed into logs to isolate the distance exponent. The equation $\log PASS_{ij} = a + b_1 (\log P_i P_j) + b_2 (\log d_{ij})$ could be taken out of logs to provide the equation $PASS_{ij} = A(P_i P_j)^b 1(d_{ij})^{b2}$. An example with numerical parameters might be $\log PASS_{ij} = 2 + 0.2(\log P_i P_j) - 0.3(\log d_{ij})$, which would become $PASS_{ij} = 100(P_i P_j)^{0.2}(d_{ij})^{0.3}$. The number 0.3 is, therefore, an empirically derived value of the average distance exponent for the city pairs examined.

A particularly useful type of regression, logistic regression, has been used to deal with categorical (either/or) variables and probabilistic models.

As an example of a probability model we can consider a commuter choosing between two modes, bus and auto, for the journey to work. The attractiveness of the bus (x_1) equals 1 and that of the auto (x_2) equals 2. We can first express the model as $x/\sum x$, which takes the attractiveness score of each alternative and expresses

it as a fraction of the combined attractiveness of all the alternatives (in this case just two alternatives). This expression can be regarded as a proportion of attraction associated with each alternative. As the expression varies from zero to one, we have a measure that is restricted in its range of values. For the bus, $x_1/(x_1 + x_2) = 1/(1 + 2) = 1/3$. The expression $x_1/(x_1 + x_2)$ may also be rewritten as $1/(1 + x_2/x_1)$ by dividing both the numerator and denominator by x_1. Substituting the numerical values we see the same result: $1/[1 + (2/1)] = 1/3$.

The previous example used x_1 and x_2 for simplicity. In fact, in many logit models, the attributes of the alternatives are measured in exponential form. Thus, we compare $\exp(x_1)$ with the sum of $\exp(x_1)$ and $\exp(x_2)$. (The rationale for this exponential format is complex and is derived from assumptions about the distribution of the error terms.) In the case of the small numerical example, we compute $\exp(1)/[\exp(1) + \exp(2)] = e^1/(e^1 + e^2) = 2.71828/(2.71828 + 7.3891) = 2.71828/10.10738 = 0.27$. Thus, 0.27 could be regarded as the probability of the individual choosing the bus.

In the exponential case also, the expression $e^{x1}/(e^{x1} + e^{x2})$ can be simplified to a more convenient form by dividing the numerator and denominator by e^{x1}: Note that dividing e^{x2} by e^{x1} is the same as subtracting exponents, or $e^{x2}/e^{x1} = e^{x2=x1}$. The result is that $e^{x1}/(e^{x1} + e^{x2}) = 1/(1 + e^{x2-x1})$. Numerically this would give $1/(1 + \exp(2 - 1)) = 1/(1 + e^1) = 1/(1 + 2.71828) = 1/3.71828 = 0.27$. The form of the equation $1/(1 + e^{x2-x1})$ is one example of the binary logistic model.

There are, of course, many pitfalls in the use of multiple regressions and transformations, although they have been widely utilized in a variety of transportation studies. Discussion of such pitfalls as multicolinearity, modifiable units, and spatial autocorrelation are found in many texts on statistics and statistical geography.[6]

---

[5] We will also use notation that recognizes the existence of a random error, Σ, which includes the variation (presumably random) not accounted for by the model, or $Y_c = -b_0 + b_1 X_1 + \ldots + b_n X_n + \Sigma$. In this case, we use b_0 instead of a for the intercept.

[6] These include A. D. Cliff and J. K. Ord, *Spatial Autocorrelation* (London: Pion, 1973); P. Taylor, *Quantitative Methods in Geography: An Introduction to Spatial Analysis* (Prospect Heights, IL: Waveland Press, Inc., 1977); N. Draper and H. Smith, *Applied Regression Analysis*, 2d ed. (New York: Wiley, 1981); N. Wrigley and R. J. Bennett, eds., *Quantitative Geography: A British View* (London: Routledge and Kegan Paul Ltd., 1981); W. A. V. Clark and P. L. Hosking, Statistical Methods for Geographers (New York: Wiley, 1986); G. M. Barber, *Elementary Statistics for Geographers* (New York: Guilford Press, 1988); D. A. Griffith and G. Amrhein, *Statistical Analysis for Geographers* (Englewood Cliffs, NJ: Prentice-Hall, 1991); and R. Earickson and J. Harlin, *Geographic Measurement and Quantitative Analysis* (New York: Macmillan, 1954).

KEY TERMS

accessibility index
areas of dominant influence (ADI)
artificial intelligence
automatic vehicle identification (AVI)
backhauling
beta index
break-of-bulk point
communication
computer network
connectivity
cost-insurance-freight (CIF) pricing
cost-space convergence
deregulation
distance decay
distance learning
elasticity of demand for transportation
electronic data interchange
expert systems
flexible factories
freight-on-board (FOB) pricing

friction of distance
global office
graphs
gravity model
hub-and-spoke networks
information warehouse
intelligent vehicle highway systems (IVHS)
Internet
ISDN
journey to work
law of retail gravitation
least builder's cost network
least user's cost network
line-haul costs
maglev
multimedia
Nationwide Personal Transportation Survey (NPTS)
network density

networked computers
outsourcing
Paul Revere network
privatization
route demand
Shimble index
smart cars
smart streets
spatial interaction
stepped freight rates
telepresence
terminal costs
TIGER File
time-space convergence
transit oriented development (TOD)
transit time
Transpac Network
transport costs
traveling salesman network

SUGGESTED READINGS

Brunn, S. D., and Leinbach, T. R., eds. 1991. *Collapsing Space and Time: Geographic Aspects of Communications and Information.* London: Harper Collins.

Dempsey, P. S., and Goetz, A. R. 1992. *Airline Deregulation and Laissez Faire Mythology.* New York: Quorum Books.

Desta, E., and Pigozzi, B. W. 1991. Further experiments with spatial structure measures in gravity models. *Tijdschrift voor Economische en Sociale Geographie,* 82(3):220–226.

Dimitriou, H. T. 1990. *Transport Planning for Third World Cities.* London: Routledge.

Gayle, D. J., and Goodrich, J. N., eds. 1990. *Privatization and Deregulation in Global Perspective.* London: Pinter.

Hall, D. R., ed. 1993. *Transport and Economic Development in the New Central and Eastern Europe.* London: Belhaven.

Hanson, Susan, ed. *The Geography of Urban Transportation.* 2nd ed. New York: Guilford, 1995.

Hartshorn, Truman A. *Interpreting the City: An Urban Geography.* 2nd rev. ed. New York: John Wiley & Sons, Inc., 1992.

Hayuth, Y. 1992. Multimodal freight transport. In B. S. Hoyle and R. D. Knowles (eds.), *Modern Transport Geography* (pp. 199–214). London: Belhaven.

Hepworth, Mark E. *Geography of the Information Economy.* York: Guilford Press, 1990.

Hilling, D., and Browne, M. 1992. Bulk freight transport. In B. S. Hoyle and R. D. Knowles (eds.), *Modern Transport Geography* (pp. 179–198). London: Belhaven.

Hoyle, B. S., and Knowles, R. D. 1992. *Modern Transport Geography.* London: Belhaven.

Office of Technology Assessment, Congress of the United States (1995). *The Technological Reshaping of Metropolitan America.* Washington, D.C.; U.S. Government Printing Office.

Parrott, R., and Stutz, F. P. 1992. Urban GIS Applications. In Maguire, D. J., Goodchild, M. F., and Rhind, D. W. (eds.), *Geographical Information Systems: Principles and Applications,* (pp. 247–260) (Vol. 2). London: Longman.

Stutz, F. P. 1994. Environment Impacts of Urban Transportation. In S. Hanson (ed.), *Urban Transportation Geography.* New York: John Wiley.

Stutz, F. P., and Parrott, R. 1992. Charting Urban Space-Time Population Shifts Using Trip Generation Models. *Urban Geography,* 13:468–475.

Taaffe, E. J., Gauthier, H. L., and O'Kelly, M. E. 1996. *Geography of Transportation.* Upper Saddle River, N.J.: Prentice Hall.

WORLD WIDE WEB SITES

THE TRIP
http://thetrip.com
This site allows travelers to check the status of aircraft in flight; in other words, enter a flight number and find that the plane is 20 miles southwest of Buffalo at 32,000 feet traveling at 600 mph.

TRAVELOCITY
http://www.travelocity.com
A great way to check prices on air travel and compare fares from different cities. Owned by American Airlines, but no obvious bias for American flights. Also allows booking.

INFOSPACE
http://www.infospaceinc.com
Search for addresses, phone numbers, e-mail addresses. Also a "My Town" profile that, using phone book data, gives you a personal profile of any town in the United States. Want to know the names of Chinese restaurants in Boise? Click. Want to see them on a map? Click. Want written directions on how to get there? Click.

MAPQUEST
http://www.mapquest.com
Type in an address and get a clickable, zoomable map of that location.

TRIPQUEST
http://www.tripquest.com
Type in a starting location and an ending location. The program provides door-to-door or city-to-city directions.

TRAFFIC FLOW
http://www.lohan.sdsu.edu/faculty/fstutz/
Check highway networks in real time for L.A., Orange, and San Diego counties, color coded by average speed and congestion levels, off Frederick Stutz' homepage. Start with "click map of San Diego." Then click on "San Diego Source," then "News and Reports," then "Weather and Traffic Information."

ETRADE
http://www.etrade.com
Flat rate broker allows buying or selling of stock for a flat $14.95 fee for 5,000 shares or less, a penny a share more above that.

ATM LOCATOR
http://visa.infonow.net/usa.html
Locate the nearest automatic teller machine anywhere in the United States.

EARTHCAM
http://www.earthcam.com
The locations of nearly every live camera on the Internet. Spy on a classroom, watch people work, check out a live skyline.

RAND MCNALLY
http://www.randmcnally.com/home/
Good travel site enhanced by a searchable index of major road construction projects all over the United States.

HOTBOT
http://www.hotbot.com
How to search the Internet without drowning in information. Those of you interested in qualitative research may want to check out QualPage—Resources for Qualitative Researchers at: **http://www.ualberta.ca/~jrnorris/qual.html**

GENERAL "ECONOMIC" SITES
http://www.progress.org/econolink
Econolink are the best web sites that have anything to do with economics. Their listings are selective with brief descriptions of site content; many referenced sites themselves have web links.
http://netec.wustl.edu/WebEc/WebEc.html
A much more extensive set of site listings in the *WebEc: World Wide Web Resources in Economics.*

U.S. TRANSPORTATION RESEARCH BOARD, STATE AND FEDERAL DOTS
http://www.prenhall.com/stutz

INTELLIGENT TRANSPORTATION SYSTEMS
http://www.prenhall.com/stutz

SO IS TELECOMMUTING FOR ME?
http://cybertime.com.sg/Features/telepart3.html

TELECOMMUTING: FAD OR FUTURE?
http://cybertime.com.sg/Features/telepart1.html

AMERICAN TELECOMMUTING ASSOCIATION
http://www.knowledgetree.com/ata.html

TIPS FOR TELECOMMUTING PRODUCTIVITY
http://www.smartbiz.com/sbs/arts/mos56.htm

TELECOMMUTING JOBS WEB PAGE
http://www.tjobs.com/

EXERCISES FOR EACH CHAPTER OF THIS BOOK CAN BE FOUND AT
http://www.prenhall.com/stutz

CHAPTER

5

WORLD AGRICULTURE AND RURAL LAND USE

OBJECTIVES

- To discuss the origin and diffusion of agriculture
- To help you appreciate the effects of agricultural practices on the land
- To acquaint you with world subsistence agricultural practices, associated crops, and regions
- To acquaint you with commercial agricultural practices, associated crops, and world regions
- To describe the agricultural policies of the United States and the former Soviet Union and the shortcomings of these policies
- To examine von Thünen's deductive model of agricultural land use

Peasants plowing rice fields prior to planting in central Luzon, Philippines. *(Photo: World Bank)*

Agriculture, the world's most space-consuming activity and humanity's leading occupation, is the science and art of cultivating crops and rearing livestock in order to produce food (and fiber) for sustenance or for economic gain. It is the basis for the development and betterment of humanity. Throughout most of our existence as a species, we were hunters and gatherers, subsisting on what nature chanced to provide. Agriculture made possible a nonnomadic existence; it paved the way for the rise of cities and fostered the development of new technologies. Until the 19th century, however, agriculture produced little food per worker, so most of the population worked full time or part time on the land. The small surplus released few people for other pursuits. Not until the agrarian revolution that occurred in European settlement areas during the last 200 years did large-scale employment in manufacturing and service activities become possible. The shift of labor from the agricultural sector to other sectors constitutes one of the most remarkable changes in the world economy in modern times. In the United States and the United Kingdom, less than 2% of the economically active population now work directly in agriculture. In contrast, about 90% of the population in a number of African and Asian countries are engaged in the agricultural sector.

Economic geographers are concerned with problems of agricultural development and change as well as with patterns of rural land use. Where was agriculture discovered? How did it diffuse? Why do farmers so often fail to prevent environmental problems? What are the characteristics of the main agricultural systems around the world, and what are their goals? What is the effect of industrialized agriculture on farmers and the traditional rural countryside? What principles can help us understand the spatial organization of rural land use? In this chapter, we seek answers to these questions.

Of critical importance to many of the issues addressed in this chapter is the decision-making environment of land users and land managers. Who makes decisions to manage land, how are they made, and what are their consequences? Frequently, individual farmers make direct land-use decisions, but they often must choose from a predetermined range of options. Farmers may be denied access to common property resources, such as water or grazing land. Landlords, multinationals, the state, or social or market demand may force them to grow certain crops. They may be faced with fluctuations in prices for export commodities. It is incumbent on land managers to devise strategies to cope with such pressures and apply these strategies to their land, which, itself, is subject to changes in nature. To appreciate the response of land managers to changes in their circumstances, we must recognize the significance of different scales. Patterns of

production and land use are the outcome of a series of forces operating at a series of scales.

TRANSFORMING ENVIRONMENTS THROUGH AGRICULTURE

The course toward a technological culture was marked by the rise of farms at the expense of the wilderness and by the rise of cities at the expense of the countryside. Agriculture was the first instance of human land use that significantly altered the natural environment. Before agriculture, landscapes evolved according to the laws of nature.

Revolutions in Agriculture

Most likely through a series of accidents and deliberate experiments, people eventually learned how to produce food and fiber plants by using the components of soil, moisture, and the atmosphere. They also learned how to herd animals and to control animal breeding.

Domestication of plants and animals probably emerged as an extension of food-gathering activities of preagricultural hunters and gatherers and as a response to a slow, sustained increase in population pressure.

Although scholars have been unable to determine exactly where and when the earliest experiments in food production occurred, they suspect that the first agricultural revolution began in the *Fertile Crescent* of the Middle East nearly 10,000 years ago (Figure 5.1 and Table 5.1). This was a well-watered area, extending from the highlands of the eastern Mediterranean through the foothills of the Taurus and Zagros mountains. Archaeological finds also indicate that domestication began early in parts of Central America and Southeast Asia.

A reliable food supply liberated people from food gathering. Increased security and leisure, resulting from the new way of life, allowed time for arts and crafts. Communities became involved in spinning, weaving, and dyeing cloth from vegetable fibers, cotton, silk, and wood and in manufacturing pottery and containers. Adequate food supplies also allowed for the exchange of specialized goods in markets. In addition, plant cultivation weakened the forces that scattered populations and strengthened the forces that concentrated them. The new way of life allowed people to live in villages and towns, which reached population densities far higher than those of preagricultural communities.

Farming practices that emerged during the Neolithic period changed little until the creation of a feudal hierarchy in medieval Europe. In this hierarchy, secular or religious overlords protected the serfs who farmed fields and paid taxes in *kind* (goods or commodities) or

FIGURE 5.1

The Fertile Crescent. This area stretches from the Persian Gulf in the southeast, north to the southern border of Turkey, and to the Mediterranean Sea in the southwest. The Nile Valley of Egypt, a rich agricultural region, and its Delta are sometimes included in the Fertile Crescent. The territory was one of the most important in terms of successions of great empires throughout world history, including the Assyrians, Babylonians, Medes, Persians, Egyptians, Israelites, Venetians, and Turks. Many important early developments in the domestication of agricultural plants originated in the region between the Tigris and Euphrates rivers, which comprise present-day Syria, Iraq, Turkey, and Iran.

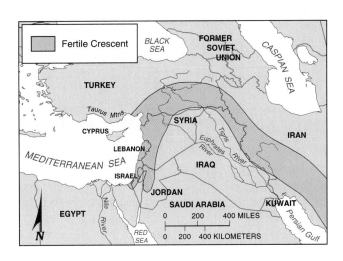

◫

T a b l e 5.1

Three Agricultural Revolutions

	I. Beginnings and Spread	*II. Subsistence to Market*	*III. Industrialization*
Time	Pre-10,000 B.P. to twentieth century	C. A.D. 1650 to present	1928 to present
Key Periods	Neolithic Medieval Europe	Eighteenth-century England Nineteenth-twentieth century in European settlement areas	Present
Key Areas	Europe South and East Asia	Western Europe and North America	Former Soviet Union and Eastern Europe North America and Western Europe
Major Goal	Domestic food supply and survival	Surplus production and financial return	Lower unit cost of production
Characteristics	Initial selection and domestication of key species Farming replaces hunting and gathering as way of life and basis of rural settlement and society Agrarian societies proliferate and support population growth Subsistence agriculture: labor intensive, low technology, communal tenure	Critical improvements, mercantilistic outlook, and food demands of industrial revolution replace subsistence with market orientation Agriculture part of sectorial division of labor: individual farm family becomes "ideal" way of life and for making a living Commercial agriculture develops growing reliance on technological inputs and infrastructure	Collective (socialist) and corporate (capitalist) ideologies and common agrotechnology favor integration of agriculture production into total food-industry system Emphasis on productivity and production for profit replace agrarian structure and farm way of life Collective/corporate production uses economies of scale, capital intensity, labor substitution, and specialized production on fewer, larger units

Source: Troughton, 1985, p. 256.

money according to the custom of the manor. English religious manors extracted the *tithe*—one-tenth of a farmer's annual production. Bishops and abbots put little into the farming business and took none of the risks but harassed farmers at harvest time.

The most important innovations associated with farming in medieval Europe were the heavy plow, the replacement of oxen with horses for plowing, and the development of the open-field system, consisting of two or three large fields on each side of a village. These advances increased agricultural production, intensified human concentration in villages and towns, generated commerce, and changed patterns of environmental exploitation. For example, the forested lowlands of western Europe were gradually cleared when the heavy plow was invented. Clayey lowland soils could not be cultivated with the old Mediterranean scratch plow, which was suitable only for light limestone soils.

Medieval farming methods prevailed in western Europe until capitalism invaded the rural manor. This resulted from a vast population increase in the new trading cities that depended on the countryside for food and raw materials. Another force that brought the market into the countryside was the alienation of the manorial holdings. Lords, who needed cash to exchange for manufactured goods and luxuries, began to rent their lands to peasants rather than having them farmed directly through labor-service obligations. Thus, they became landlords in the modern sense of the term.

A second agricultural revolution that began with the demise of the manorial system replaced subsistence agriculture with market-oriented agriculture. Open fields were enclosed by fences, hedges, and walls. Crop rotation replaced the medieval practice of fallowing fields. Seeds and breeding stock improved. New agricultural areas opened up in the Americas. Farm machines replaced or supplemented human or animal power. The family farm came to represent the core model of commercial agriculture.

Since the late 1920s, a third agricultural revolution has been taking place. This revolution, which some observers believe is the logical extension of the second agricultural revolution, points to the resolution or rationalization of the distinction between family and corporate models of agriculture. In other words, it signifies the elimination of distinct agrarian economies and communities. Although the third agricultural revolution is incomplete, industrial agriculture has become the dominant form in most developed countries, capitalist and socialist, and is being applied to export enclaves of developing countries. Key elements of industrial agriculture are capital intensity, technological inputs, high energy use, concentration of economic power, and a quest for lower unit costs of production. Although industrial agriculture has increased output

per unit of input, it has also depleted water and soil resources, polluted the environment, and destroyed a way of life for millions of farm families.

Industrial agriculture has drastically reduced the number of farmers in North America. In the United States, the number of farmers declined from 7 million in 1935 to around 2 million in 1990. In Canada, 600,000 farm operators existed in 1951 but only half that number were still in operation in the late 1970s. Europe has witnessed similar trends. In Great Britain, for example, an annual 1.5% decline in the number of farm workers occurred during the 1980s.

The Diffusion of Agriculture

The Fertile Crescent was one of several locales for plant and animal domestication (Figure 5.2). The spread of agriculture from these centers was slow. For example, archaeologists have calculated that it took from 6000 B.C. to 3000 B.C. for a form of shifting cultivation to spread along the Danube and Rhine corridors. (Shifting cultivation is a type of agriculture in which clearings are used for several years then abandoned and replaced by new ones.) Another 1,000 years elapsed before agriculture reached southern England.

By A.D. 1500, on the eve of European overseas expansion, agriculture had spread widely throughout the Old World and much of the New World. In Europe, the Middle East, North Africa, central Asia, China, and India, cereal farming and horticulture were common features of the rural economy. Nonagricultural areas of the Old World were restricted to the Arctic fringes of Europe and Asia and to parts of southern and central Africa. Agriculture had not spread beyond the eastern Indian islands into Australia.

By the time of the first European voyages across the Atlantic, the cultivation of maize, beans, and squash in the New World had spread throughout Central America and the humid environment of the eastern half of North America as far north as the Great Lakes. In South America, only parts of the Amazon Basin, the uplands of northeastern Brazil, and the dry temperate south did not have an agricultural economy.

These patterns of agriculture persisted until the era of European overseas settlements. From the Age of Discovery to the mid-17th century, Europeans did not attempt to establish large overseas settlements. Eventually, European settlement assumed two forms: (1) farm-family colonies in the middle latitudes of North America, Australia, New Zealand, and South Africa; and (2) plantation colonies in the tropical regions of Africa, Asia, and Latin America. These two types of agricultural settlements differed considerably.

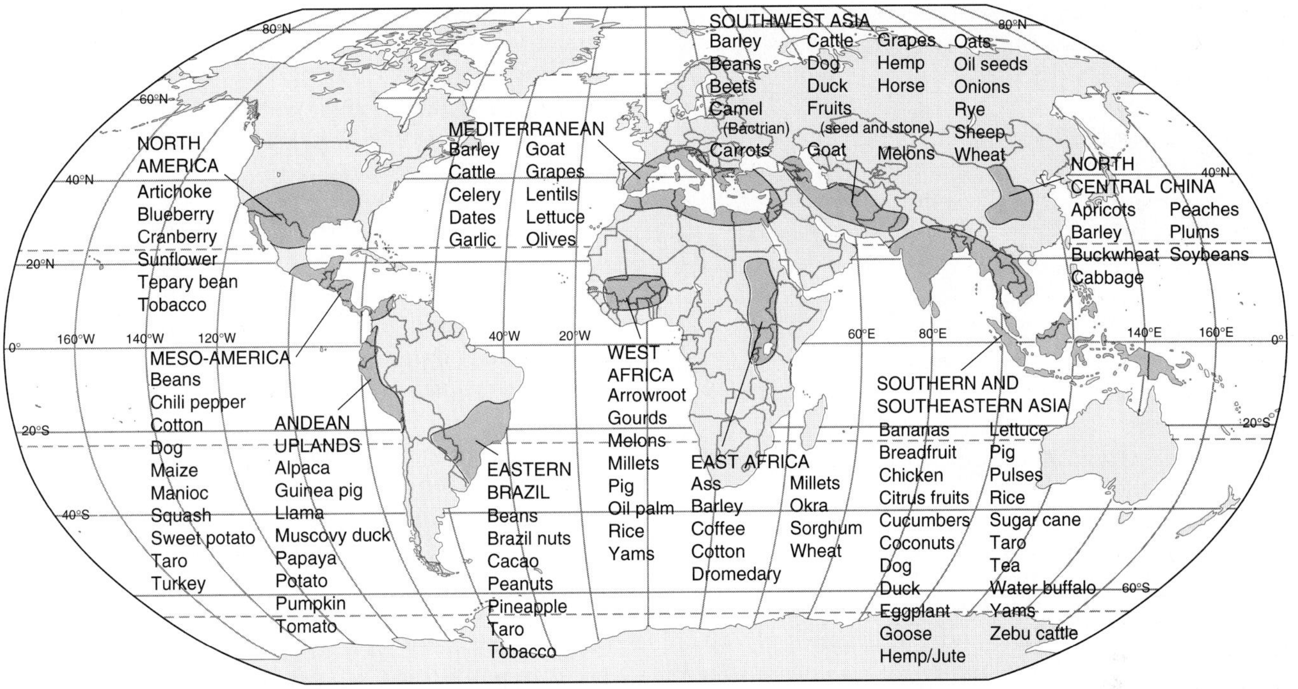

SOUTHWEST ASIA
Barley · Cattle · Grapes · Oats
Beans · Dog · Hemp · Oil seeds
Beets · Duck · Horse · Onions
Camel · Fruits · · Rye
(Bactrian) · (seed and stone) · Sheep
Carrots · Goat · Melons · Wheat

MEDITERRANEAN
Barley · Goat
Cattle · Grapes
Celery · Lentils
Dates · Lettuce
Garlic · Olives

NORTH AMERICA
Artichoke
Blueberry
Cranberry
Sunflower
Tepary bean
Tobacco

NORTH CENTRAL CHINA
Apricots · Peaches
Barley · Plums
Buckwheat · Soybeans
Cabbage

MESO-AMERICA
Beans
Chili pepper
Cotton
Dog
Maize
Manioc
Squash
Sweet potato
Taro
Turkey
Pumpkin
Tomato

ANDEAN UPLANDS
Alpaca
Guinea pig
Llama
Muscovy duck
Papaya
Potato
Pineapple
Taro
Tobacco

EASTERN BRAZIL
Beans
Brazil nuts
Cacao
Peanuts

WEST AFRICA
Arrowroot
Gourds
Melons
Millets
Pig
Oil palm
Rice
Yams

EAST AFRICA
Ass · Millets
Barley · Okra
Coffee · Sorghum
Cotton · Wheat
Dromedary

SOUTHERN AND SOUTHEASTERN ASIA
Bananas · Lettuce
Breadfruit · Pig
Chicken · Pulses
Citrus fruits · Rice
Cucumbers · Sugar cane
Coconuts · Taro
Dog · Tea
Duck · Water buffalo
Eggplant · Yams
Goose · Zebu cattle
Hemp/Jute

FIGURE 5.2

Origins of plant and animal domestication. Animals were probably first domesticated as household pets by prehistoric peoples. The domestication of plants came from two sources. The first was *vegetated planting*, which was the planting of pieces of existing plants, such as stems and roots. Plants first found growing wild that were useful to the household were cut up and transplanted. The second source was *seed agriculture*, which involved the planting of seeds and was the direct result of natural, annual fertilization of plants.

For example, farm colonization in North America depended on a large influx of European settlers whose agricultural products were initially for a local market rather than an export market. Europeans introduced the farm techniques, field patterns, and types of housing characteristic of their homelands, yet they often modified their customs to meet the challenge of organizing the new territory. For example, the checkerboard pattern of farms and fields that characterizes much of the country west of the Ohio River resulted from a federal system of land allocation (the *Township and Range System*). It involved surveying a baseline and a principal meridian, the intersection of which served as a point of origin for dividing the land into 6-by-6-mile townships, then into square-mile sections, and still further into quarter sections a half-mile long. This orderly system of land allocation prevented many boundary disputes as settlement moved into the interior of the United States.

In tropical areas, Europeans, and later Americans and Japanese, imposed a plantation agricultural system that did not require substantial settlement by expatriates. *Plantations* are large-scale agricultural enterprises devoted to the specialized production of one tropical product raised for the market. It is believed that they were first developed in the 1400s by the Portuguese on islands off the tropical West African coast. Plantations produced luxury foodstuffs, such as spices, tea, cocoa, coffee, and sugarcane, and industrial raw materials, such as cotton, sisal, jute, and hemp. These crops were selected for their market value in international trade, and they were grown near the seacoast to facilitate shipment to Europe. The creation of plantations sometimes involved expropriating land used for local food crops. Sometimes, by irrigation or by clearing forests, new lands were brought into cultivation.

Europeans managed plantations; they did no manual labor. The plantation system relied on forced or poorly paid indigenous labor. Very little machinery was used. Instead of substituting machinery for laborers when local labor supplies were exhausted, plantation managers went farther afield to bring in additional laborers. This practice was especially convenient because world demand fluctuated. During periods of increased demand, production could be accelerated by importing additional laborers. This practice made the need for installing machinery during booms unnecessary and minimized the financial problems of idle capital during slumps.

The effect of centuries of European overseas expansion was to reorganize agricultural land use worldwide. Commercial agricultural systems have become a feature of much of the habitable world. Hunting and

On the world's largest rubber plantation, at Harbel, Liberia, more than 36,000 hectares, or 30% of the total land area of Liberia, are cultivated by Firestone. The company has also established plantations in Brazil, Ghana, Guatemala, and the Philippines. How do plantations benefit host societies? *(Photo: Bridgestone/Firestone, Inc.)*

vegetation zones show signs of clearing, burning, and the browsing of domestic animals. The impoverishment of vegetation has led to the creation of successful agricultural and pastoral landscapes, but it has also led to *land degradation* or a reduction of land capability.

Hunters and gatherers hardly disturb vegetation, but farmers must displace vegetation to grow their crops and to tend their livestock. Farmers are land managers; they upset an equilibrium established by nature and substitute their own. If they apply their agrotechnology with care, the agricultural system may last indefinitely and remain productive. On the other hand, if they apply their agrotechnology carelessly, the environmental base may deteriorate rapidly. How farmers actually manage land depends not only on their knowledge and perception of the environment, but also on their relations with groups in the wider society—in the state and the world economy.

As agriculture intensifies, environmental alteration increases. Ester Boserup (1965) proposed a simple five-stage model of agricultural systems based on frequency of land use. Stage 1, forest-fallow cultivation, involves cultivation for 1 to 3 years followed by 20 to 25 years of fallow. In Stage 2, bush-fallow cultivation, the land is cultivated for 2 to 8 years, followed by 6 to 10 years of fallow. In Stage 3, short-fallow cultivation, the land is fallow for only 1 to 2 years. In Stages 4 and 5, annual cropping and multicropping, fallow periods are either very short—a few months—or nonexistent. Boserup noted that the transition from one form of agriculture to another was accompanied by an increasing population density, improved tools, the increasing integration of livestock, improved transportation, a more complex social infrastructure, more permanent settlement and land tenure, and more labor specialization.

Forest-fallow, or *shifting cultivation*, survives in areas of the humid tropics that have low-potential environmental productivity and low population pressure. Under ideal conditions, this form of agriculture leaves much of the original vegetation intact. Farmers make

gathering, the oldest means of survival, has virtually disappeared, although it still sustains groups such as the Bushmen of the Kalahari and the Pygmies of Zaire. Pastoralists, such as the Masai of Kenya and Tanzania who drive cattle in a never-ending search for pasture and water, have declined in numbers. Subsistence farming still exists, but only in areas where impoverished farmers, especially in developing countries, barely make a living from tiny plots of land. Few completely self-sufficient farms exist; most farmers, even in remote areas of Africa and Asia, trade with their neighbors at local markets.

Human Impact on the Land

The emergence of agriculture and its subsequent spread throughout the world has meant that little if any land still can be considered natural or untouched. Vegetation has been most noticeably changed. Virtually all

A tapper on the Firestone plantation in Liberia makes an incision in a rubber tree. The latex will flow down the incision through a spout and into a cup attached to the tree. Some of the latex is carried in pails by women to collecting stations. What would this tapper do if he were not working for Firestone? *(Photo: Bridgestone/Firestone, Inc.)*

A scene showing the deforestation in the rainforest in Acro, Western Brazil. Forests have been burnt to the ground to create temporary pastures for cattle. The nation's rainforests are being cut down at a rate 50% faster today than they were 10 years ago. Rainforest loss creates or contributes to a number of intractable problems: it contributes to greenhouse warming, eliminates the cleansing of the atmosphere, creates new semideserts, increases large-scale flooding, and threatens wildlife habitats. *(Photo: P. Sudhakaran, United Nations)*

small, discontinuous clearings in forests. They girdle some trees and cut down others, burn the debris, and prepare the soil by digging holes in a pattern of points for a variety of crops—groundnuts, rice, taro, sweet potato. Because no fertilizer is used, soil nutrients are quickly depleted. Thus, farmers abandon their plots and establish new gardens every few years, but they rarely move their residences. Except on steep slopes, where soil erosion can be a serious problem, shifting cultivation can be a sustainable system of agriculture. It allows previous plots to regenerate natural growth. In Papua, New Guinea, for example, about one-quarter of the country's forested area is well-developed secondary forest created and maintained by shifting cultivation.

However, shifting cultivation can lead to degradation when an increasing population demands too much of the land or when new forces intrude into the farming system. In one sequence, common in Latin America, shifting cultivation agriculturalists follow loggers and oil prospectors into an area. After cropping is finished, the land is seeded with grass and sold to commercial ranchers who produce beef for the North American and European markets. Heavily grazed, the land quickly declines in carrying capacity and is then abandoned as the ranchers move to new areas.

In contrast, *permanent agriculture* (annual cropping and multicropping) usually occurs in areas of high-potential environmental productivity and of high population pressure. Under permanent cultivation, the land becomes totally transformed. Yet the beauty, fertility, and endurance of the land may not be impaired. Soils of the Paris Basin have been intensively cultivated for hundreds of years, and still they remain highly productive. In many parts of the Orient, carefully terraced hillsides have maintained the productivity of valuable soil resources after thousands of years.

In general, modern farming practices pose the main danger to land. Clean tillage on large fields, monoculture (the cultivation or growth of a single crop), and the breaking down of soil structure by huge machines are a few factors that may destroy the topsoil. Droughts and dust storms during the 1930s, 1970s, 1980s, and 1990s in the Great Plains of the United States gave testimony of how nature and industrial agriculture can combine to destroy the health of a steppe landscape and transform it into a desert.

Whether farmers achieve a harmonious relationship with nature does not necessarily depend on either their technologies or their political philosophies. Farmers with simple tools and technologies can destroy the long-run food-producing capacity of the land. Mechanical agriculture in both capitalist and socialist countries alike can degrade the land.

Agriculture threatens ecological balances when people begin to believe that they have freed themselves from dependence on land resources. In developed countries, there is an inherent tendency to exploit the land as a result of pressure to reproduce economic conditions of production or to maximize profits. Household and corporate producers want to make land use more efficient and land more productive; thus, farming is often viewed as just another industry. However, we must remember that land is more than a means to an end; it is finite, spatially fixed, and ecologically fragile.

FACTORS AFFECTING RURAL LAND USE

Rural land-use patterns, which are arrangements of fields and larger land-use areas at the farm, regional, or global level are difficult to understand. Worldwide, hundreds of farm types exist. The most interesting aspect of the world's agricultural land-use areas or regions is not their extent, but the uniformity of land-use decisions farmers make within them. Given any farming region, why do farmers make similar land-use decisions? For example, why does one farmer on the slopes of Mount Kilimanjaro decide to mix coffee bushes with banana stands, and likewise all other neighbors? The land-use pattern on Kilimanjaro, as elsewhere, reflects a host of factors. Geographers identify at least four groups of variables that determine land use: (1) site characteristics, (2) cultural preference and perception, (3) systems of production, and (4) relative location. We discuss these four groups of variables in more detail next.

SITE CHARACTERISTICS

Variations in rural land use depend partly on site characteristics, such as soil type and fertility, slope, drainage, exposure to sun and wind, and the amount of rainfall and average annual temperature. As an example, consider the climate milieux in which crops grow. Plants require particular combinations of temperature and moisture. An optima-and-limits schema shows the range for a hypothetical crop. Increasing rainfall is plotted on the horizontal axis and increasing temperature on the vertical axis. Absolute physical limits of the crop are "too wet," "too dry," "too cold," and "too hot." A series of isopleths, which connect points of equal dollar yield per hectare, mark optimum conditions. The diagram emphasizes that a particular combination of temperature and moisture conditions characterizes each site. Absolute climatic limits are wide for some crops, such as maize and wheat, but narrow for others, such as pineapples, cocoa, bananas, and certain wine grapes.

WORLD CULTURAL PREFERENCE AND PERCEPTION

Food preferences and prejudices are one of the most important variables in determining the type of agricultural activity at a given site. Some cultural groups would rather starve than eat edible but taboo food. Many Africans avoid protein-rich chickens and their eggs. Certain Hindus abstain from eating all meat, but particularly beef. Muslims do not eat pork; hence, pig raising is absent from the Muslim world, which stretches from Mauritania and Morocco to Pakistan and Bangladesh and to parts of Indonesia (Figure 5.3). The Chinese and some other people of East and Southeast Asia abstain from drinking milk or eating milk products. In the United States, a consumer preference for meat leads American farmers to put a greater proportion of their land in forage crops than European farmers, who grow more food crops.

People interpret the environment through different cultural lenses. Their agricultural experiences in one area influence their perceptions of environmental conditions in other areas. Consider the settlement of North America. The first European settlers were Anglo-Saxons accustomed to moist conditions and a tree-covered landscape. They equated trees with fertility. If land was to be suitable for farming, it should, in its natural state,

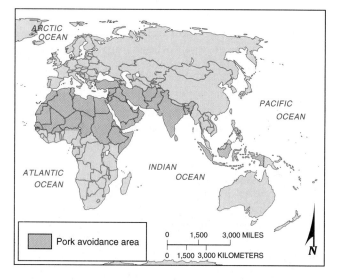

FIGURE 5.3

Pork avoidance areas of the world. Cultural preferences and perceptions against certain foods by a particular culture are called *food taboos*. The pork avoidance, or pork taboo, for North Africa and the Middle East occurred in early biblical times. The Semitic peoples were nomadic herders; pigs were more conducive to sedentary agricultural regions. Pigs were considered unclean because of their environmental setting. Today, the Muslim areas of the world, which subscribe to some Old Testament biblical laws, are the largest areas of pork avoidance. The Muslim faith spread throughout the Middle East to India and Southeast Asia, areas that are not nomadic herding regions; nonetheless, they have adopted the pork taboo because of religious practices. India is only 15% Muslim and therefore does not have a taboo against pork by the majority of the population; however, a beef taboo is practiced in this world region by the majority.

have a cover of trees. Thus, the settlers of New England and the east coast realized their expectations of a fertile farming region. When Anglo-Saxons edged onto the prairies and high plains west of the Mississippi River, they encountered a treeless, grass-covered area. They underestimated the richness of the prairie soils, in particular, and the area became known as the "great American desert." In the late 19th century, a new wave of migrants from the steppe grasslands of eastern Europe appraised the fertility of the grass-covered area more accurately than the Anglo-Saxons who preceded them did. The settlers from eastern Europe, together with technological inventions such as barbed-wire fencing and the moldboard plow, helped to change the perception of the prairies from the "great American desert" to the *great American breadbasket.*

Ignorance of land degradation and its perception are a function of the rate and accumulated degree of degradation as well as the intelligence of land managers. In the mountains of Ethiopia, where cultivation has been occurring for 2,000 years with a fairly low rate of soil loss, the cumulative erosion of good soil has resulted in a serious decline in the capability of the land. In comparison, in the hills of northern Thailand, where rates of soil loss are much higher, the local land management system has compensated for soil erosion, and the capability of the land has been maintained.

However, land is sometimes devastated by land managers—not because of ignorance or stupidity but because of calculated human agency. A strong market imperative, a need to occupy new land for cropping, or a belief that humans can and must master nature can ruin land in sensitive environments. Much land in the former Soviet Union has been degraded as a result of attempts by the state to "transform nature."

SYSTEMS OF PRODUCTION

Systems of agricultural production set their imprint on rural land use. Like manufacturing, the agricultural endeavor is carried out according to three systems of production: (1) peasant, (2) capitalist, and (3) socialist. The major distinction among these systems is the labor commitment of the enterprise. In the peasant system, production comes from small units worked entirely, or almost entirely, by family labor. In the capitalist system, family farming is still widespread; but, as in the socialist system, labor is a commodity to be hired and dismissed by the enterprise according to changes in the scale of organization, the degree of mechanization, and the level of market demand for products.

In any geographic region, one system of production dominates the others. For example, capitalist agriculture dominates parts of South America, whereas peasant agriculture dominates other parts of the continent. Capitalist agriculture finds expression in a vast cattle ranching zone extending southwest from northeastern Brazil to Patagonia; in Argentina's wheat-raising Pampa, which is similar to the U.S. Great Plains; in a mixed livestock and crop zone in Uruguay, southern Brazil, and south central Chile, which is comparable to the U.S. Corn Belt; in a Mediterranean agriculture zone in middle Chile; and in a number of seaboard tropical plantations in Brazil, the Guianas, Venezuela, Colombia, and Peru. Peasant agriculture dominates the rest of the continent. There is shifting cultivation in the Amazon Basin rain forest, rudimental sedentary cultivation in the Andean plateau country from Colombia in the north to the Bolivian Altiplano in the south, and a wide strip of crop and livestock farming in eastern Brazil between the coastal plantations and livestock ranching zones.

SUBSISTENCE, OR PEASANT MODE, OF PRODUCTION

Subsistence agriculture, also called *peasant agriculture*, is associated with developing countries, and it is labor intensive (labor centered). Farmers are small-scale producers who invest little in mechanical equipment or chemicals. They are interested mainly in using what they produce rather than in exchanging it to buy things that they need. Food and fiber are exchanged, particularly through interaction with capitalist agriculture at global, national, and local scales, but farm families consume much of what they produce. This is called *use value* rather than *exchange value*. To obtain the outputs required to be self-supporting, peasant farmers are frequently willing to raise inputs of labor to very high levels, especially in crowded areas where land is rarely available. Highly intensive peasant agriculture occurs in the rice fields of South, East, and Southeast Asia. Most of the paddies are prepared by ox-drawn plow, and the rice is planted and harvested by hand—millions of hands. The term agriculture involution refers to the ability of the agricultural system in the densely populated parts of Asia, including Japan, to absorb increasing numbers of people and still provide minimal subsistence levels for all in rural communities.

An example of the peasant mode of production exists in the semiarid zone of East Africa. This zone includes the interior of Tanzania, northeast Uganda, and the area surrounding the moist high-potential heartland of Kenya. As in most parts of the developing world, peasant agriculture in this region has been complicated by the colonial and post-colonial experience.

People in the semiarid area of East Africa earn a living by combining several activities. They eat their crops and livestock and sell or exchange agricultural surpluses at markets. They grow cash, or export, crops such as cotton. They maintain beehives in the bush and sell part of the honey and wax. They brew and sell beer. They hunt, fish, and collect wild fruits. They earn income by cutting firewood, making charcoal, delivering water, and carrying sand for use in construction. Some of them have small shops or are tailors. Most important, people sell their labor, both short term and long term, nearby and far away (Figure 5.4).

To farm and herd successfully in the semiarid zone, land managers must meet certain requirements set by the environment and by the nature of crops and animals. Livestock require water, graze, salt, and protection from disease and predators. To meet these needs day after day, year after year, land managers must have considerable skill and knowledge. They must know a great deal not only about the ability of animals to withstand physiological stress, but also about environmental management—which grass to save for late grazing,

where and when to establish dry-season wells to enable the stock to withstand the rigor of the daily journey between water and graze. With respect to crops, land managers must know about plant-moisture and nutrient needs. They must also be sensitive to the variability of rainfall.

Most of the time, this system of agriculture in East Africa provides peasants with an adequate and varied food supply. In bad times, there are mechanisms for sharing hardship and loss so that those farmers who are hardest hit can usually rebuild their livelihoods after bad times end. However, the peasant mode of production has been forced to adjust to pressures from governments during colonial and post-colonial periods.

Subsistence Agriculture: Crops and Regions

Most of the world's farmers, including the people of Latin America, Africa, and Asia, practice subsistence agriculture. These regions have several characteristics in common:

1. The majority of workers are engaged in agriculture instead of manufacturing, services, and processing information.
2. Agricultural methods and practices are primitive. Farms and plots are small in comparison with those of the developed world, labor is used intensively, and mechanization and fertilization are used only infrequently.
3. Agricultural produce that is harvested on the farms is used primarily for direct consumption. The family, or the extended family, subsists on the agricultural products from the farm. Although in certain years surpluses may be produced, this is rarely the case.

James Rubenstein (1995) divides subsistence agriculture into three categories: shifting cultivation in the tropics, pastoral nomadism in North Africa and the Middle East, and intensive subsistence agriculture in South and East Asia, where rice is grown (Figure 5.5).

SHIFTING CULTIVATION
Shifting cultivation is a type of agriculture practiced in three main tropical rain-forest areas of the world: (1) the South American Amazon region, (2) the central African Congo region, and (3) Southeast Asia, Indonesia, and New Guinea (Figure 5.5). Rainfall is heavy in these regions, vegetation is thick, and soils are relatively poor quality.

When shifting cultivation is practiced, the people of a permanent village clear a field adjacent to their settlement by slashing vegetation. After the field is cleared with axes, knives, and machetes, the remaining stumps

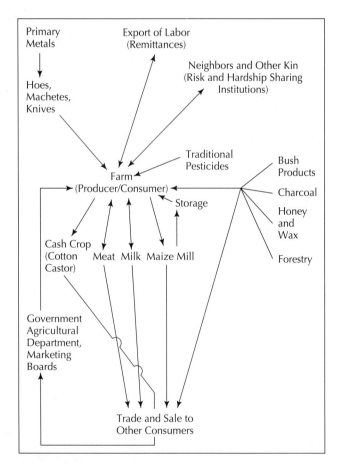

FIGURE 5.4
The structure of the semiarid peasant agricultural system in East Africa. *(Source: Porter, 1979, p. 33)*

are burned. Daily rain returns the ash and nutrients to the soil, temporarily fertilizing it. (Because of this clearing technique, shifting cultivation is sometimes called *slash-and-burn* agriculture.) The field is used for several years. At the end of this time, the soil is depleted, and the village turns elsewhere to clear another field. Eventually, the forest vegetation again takes over, and the area is refoliated. The soil is thus allowed to replenish itself.

Using hoes or knives, farmers plant the fields by hand with tubers or seeds. An indentation is made in the soil. A stem of a plant is submerged or a seed is dropped into a hole, and soil is pushed over the opening by hand. Mechanization and animals are not used for plowing or for harvesting. The most productive farming occurs in the second or third year after burning. Following this, surrounding vegetation rapidly regenerates, weeds grow, and soil productivity dwindles. The plot, sometimes called a *swidden* or *milpa*, is abandoned. Then, a new site is selected nearby. Usually, the village does not permanently relocate. The villagers commonly return to the abandoned field after 6 to 12 years, by which time the soil has regained enough nutrients to grow crops again.

The predominant crops grown in shifting agricultural areas include corn and manioc (cassava) in South America, rice in Southeast Asia, and sorghum and millet in Africa. In some regions, yams, sugarcane, and other vegetables are also grown.

The patchwork of a swidden is quite complex and seemingly chaotic. On one swidden, a variety of crops can be grown, including those just mentioned, as well as potatoes, rice, corn, yams, mangoes, cotton, beans, bananas, pineapples, and others, each in a clump or small area within the swidden.

Only 5% of the world's population engages in shifting cultivation. This low percentage is not surprising because tropical rain forests are not highly populated areas. However, shifting cultivation occupies approximately 25% of the world's land surface and therefore is an important type of agriculture. The amount of land devoted to this type of agriculture is decreasing because governments in these regions deem shifting cultivation to be economically unimportant. Consequently, governments in developing countries are selling and leasing land to commercial interests that destroy the tropical hardwoods and rain forests.

These forests are important for oxygen production and a global, ecologically balanced environment. Many scientists believe that maintenance of the tropical rain forests contributes to a healthy atmosphere and guards against *global warming*. The fewer rain forests, the more carbon dioxide builds up, enhancing the *greenhouse effect* by trapping solar radiation in the atmosphere. Species extinction from rain forest depletion is also a serious problem. Therefore, even though shifting cul-

tivation is an inefficient form of agriculture for large populations, it helps maintain tropical forests.

One principal agreement of the 1992 Rio de Janeiro United Nations Conference on Environmental Development was the preservation of tropical hardwoods. Presently, the World Bank and the International Monetary Fund (IMF) will loan monies to developing nations only if those nations practice certain environmental protection measures. One such measure is maintenance of tropical hardwoods.

PASTORAL NOMADISM

Shifting cultivation and *pastoral nomadism* can be classified as extensive, or nonintensive, subsistence agriculture. Areas in which pastoral nomadism is practiced include North Africa and the Middle East, the eastern plateau areas of China and Central Asia, and eastern Africa's Kenya and Tanzania (Figure 5.5). According to Rubenstein (1995), only 15 million people are pastoral nomads, but they occupy 20% of the earth's land area. Combining this area with that in which shifting cultivation is practiced, we can see that almost 50% of the world's area is included. The pastoral nomads occupy areas that are climatically opposite those of shifting cultivators. The lands occupied by pastoral nomads are dry, usually less than 10 inches of rain accumulate per year, and typical agriculture is normally impossible, except in oases areas.

Instead of depending on crops as most other farmers do, nomads depend on animal herds for their sustenance. Everything that they need and use is carried with them from one forage area to another. Tents are constructed of goats' hair, and milk, clothing, shoes, and implements are produced from the animals. Pastoral nomads consume mostly meat and grain. Sometimes, in exchange for the meat, other needed goods are obtained from sedentary farmers in marginal lands near the nomads' herding regions. It is common for pastoral nomads to farm areas near oases or within floodplains that they occupy for a short period of the year. Nomadic parties usually include 6 to 10 families who travel in a group, carrying bags of grain for sustenance during the drier portions of the year.

A cyclical pattern of migration is entrenched in the nomadic way of life, and it lasts for generations. Pastoral nomads are not wandering tribesmen; they follow a 12-month cycle in which lands most available with forage are cyclically revisited in a pattern that exhibits strong territoriality and observance of the rights of adjacent tribes. The exact migration pattern of today's pastoral nomads have developed from a precise geographic knowledge of the region's physical landforms and environmental provision.

Nomads must select animals for their herds that can withstand drought and provide the basic necessities of

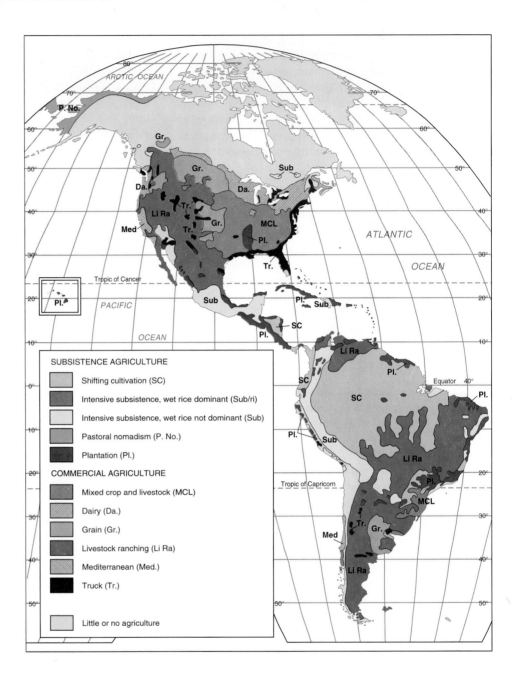

FIGURE 5.5

World agricultural regions. Geographers and agricultural economists divide the world into three agricultural regions. The first is subsistence, or peasant, agriculture. The second is commercial, or capitalist, agriculture. The third is the socialist, or command economy, agricultural system. On this map, commercial and socialist systems are shown as commercial agriculture. Subsistence agriculture includes nomadic herding, slash-and-burn agriculture, and intensive sedentary subsistence agriculture. (See Color Insert 2 for more illustrative map.) Africa, Southeast Asia, and the Amazon Basin are the principal regions of shifting agriculture. North Africa, Southwest Asia, and central East Asia are the principal areas of nomadic herding, or pastoral nomadism. Intensive subsistence agriculture includes rice-dominant cultivation and rice mixed with other crops. These regions include East Asia, South Asia, and Southeast Asia. Plantation agriculture is primarily in tropical and subtropical regions of Latin America, Asia, and Africa.

the herdsmen. The camel is the quintessential animal of the nomad because it is strong, can travel for weeks without water, and can move rapidly while carrying a large load. The goat is the favorite small animal because it requires little water, is tough, and can forage off the least green plants. Sheep are slow moving and require more water, but they provide other necessities: wool and mutton. Small tribes need between 25 and 60 goats and sheep and between 10 and 25 camels to sustain themselves.

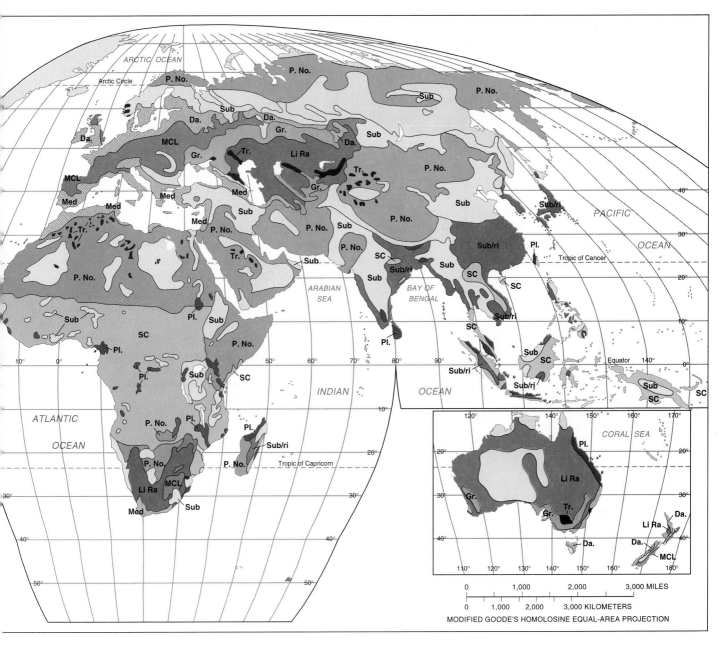

MODIFIED GOODE'S HOMOLOSINE EQUAL-AREA PROJECTION

Commercial regions, on the other hand, include mixed crop and livestock farming, which exists primarily in the northern Unites States, southern Canada, and central Europe. It also includes dairy farming near the population centers of the northeastern United States and northwestern Europe. Grain cultivation exists in Argentina, the Great Plains of the United States, the central Asian Plains of the former Soviet Union, and Australia. Livestock ranching includes areas too dry for plant cultivation, including the western mountain and plain regions of Canada and the United States; southeastern South America; southwestern regions of the former Soviet Union, centered on Kazakhstan; and large portions of Australia. Mediterranean agriculture specializes in horticulture—fruit and vegetable regions—and includes areas surrounding the Mediterranean Sea, regions of the southwestern United States, central Chile, and the southern tip of Africa. Finally, truck farming—commercial gardening and fruit farming—is found in the southeastern United States (to supply the northeastern population center of America) and in southeastern Australia.

In early times, before the railroad and telegraph, pastoral nomads were the communication agents of the desert regions, carrying with them innovations and information. This is no longer the case, now that their host governments have joined the information and technol- ogy age. Many of these governments have modern weapons and can control vast territories. However, no- madic herding has been allowed to remain because these vast dry areas of the world cannot be used for other eco- nomic activity. Furthermore, government attempts to settle pastoral nomads have met with little success, although the former Soviet Union and China have settled large numbers of them. In the future, pastoral nomads will be allowed only on lands that do not have energy resources or precious metals beneath the surface, or on lands that cannot be easily irrigated from nearby rivers, lakes, or groundwater aquifers. In any case, the number of pastoral nomads is declining because of increasing population pressures on the countries in which they live.

INTENSIVE SUBSISTENCE AGRICULTURE

Intensive subsistence agriculture is practiced by large populations living in East, South, and Southeast Asia, Central America, and South America (Figure 5.5). Whereas shifting cultivation and pastoral nomadism are extensive low-density, marginal operations, intensive subsistence agriculture, as the term implies, is a higher intensity type of agriculture in the majority of the densely populated developing areas of the world. Rice is the predominant crop because of its high levels of carbohydrates and protein. Most farmers using inten-

sive subsistence rice agriculture use every available piece of land, however fragmented, around their villages. Most often, a farm encompasses only a few acres.

Intensive subsistence agriculture is characterized by several features:

1. Most of the work is done by hand, with all family members involved. Occasionally farm animals are used, such as water buffalo or oxen. Almost no mechanization is used because of lack of capital to purchase such equipment and because plots are tiny.
2. Plots of land are extremely small by Western standards. Almost no piece of land is wasted. Even roads through agricultural regions of intensive subsistence are made narrow so that all cultivatable areas can be used.
3. The physiological density (that is, the number of people that each acre of land can support) is very high.
4. Principal regions that are cultivated are river valleys and irrigated fields in low-lying, moist regions in the middle latitudes.

Because rice is a crop that has a high yield per acre and is rich in nutrients, it is a favorite in intensive subsistence agricultural regions (Figure 5.6). First, the field is plowed with a sharpened wooden pole that is pulled by oxen. Next, the field is flooded with water and planted with rice seedlings by hand. Another method is to spread dry seeds over a large area by hand. When the rice is mature, having developed for three-fourths of its life underwater, it is harvested from the rice *paddy*, or *sawah*, as it is called in Indonesia. To separate the husks from the rice itself, the farmers thrash the rice by beating it on a hard surface or by trampling it underfoot. Sometimes it is even poured on heavily traveled roads. The chaff is thus removed from the seeds, and sometimes the wind blows the lightweight material far from the pile of rice itself. This process is known as *winnowing*.

Some year-round, tropical, moist areas of the world allow *double cropping*. This means that more than one crop can be produced from the same plot throughout the year. Occasionally in wet regions, two rice crops are grown, but more frequently a rice crop and a different crop, which requires less water, are produced. The field crop is produced in the drier season on nonirrigated land.

In the higher-latitudes of East Asia, rice is mixed with other crops and may not be the dominant crop. In western India and the northern China plain, wheat and barley are the dominant crops, with oats, millet, corn, sorghum, soybeans, cotton, flax, hemp, and tobacco also produced.

In Indonesia, harvesting rice is an example of labor-intensive peasant culture. *(Photo: World Bank)*

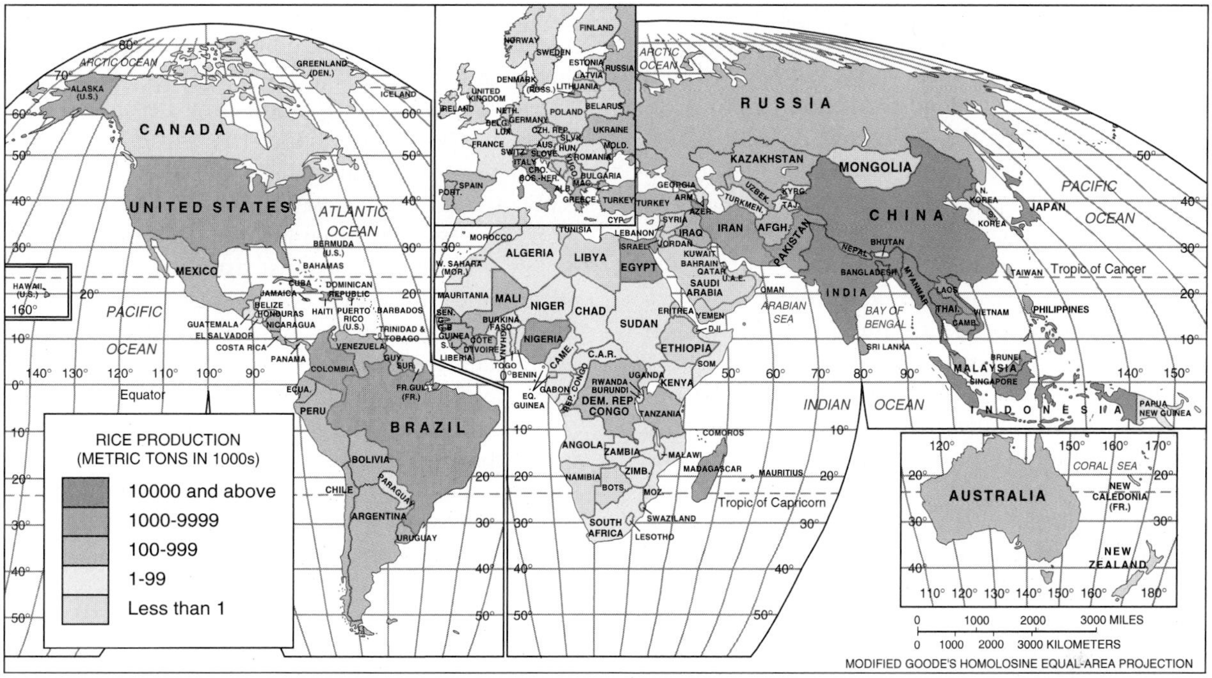

FIGURE 5.6

World rice production. Anthropologists tell us that rice was domesticated in East Asia more than 7,000 years ago. Unlike corn in the United States, rice is almost exclusively used for human consumption. Almost 2 billion people worldwide are fed chiefly by rice. It is the most important crop cultivated in the most densely settled areas of the world, including China in East Asia, India in South Asia, and Southeast Asia. These areas produce more than 90% of the world's rice. Populations in these regions use rice for between one-third and one-half their total food intake. Rice can be grown in only very humid and tropical regions because it depends on water to develop. Rice requires a large amount of labor and tedious work, but the returns are bountiful. Rice produces more food per unit of land than any other crop; thus, it is suitable for the most densely populated areas of the world, such as China and India. Outside Asia, Brazil is now the leading rice producer, followed by the United States. Each of these two countries provides less than 2% of the world's rice supply. (See Color Insert 2 for more illustrative map.)

Problems Faced by Subsistence Agriculturalists

Subsistence agriculture is subjected to variations in soil quality, availability of rain from year to year, and, in general, environmental conditions that can harm crop-production levels and endanger life. In addition, subsistence agriculturalists lack tools, implements, hybrid seeds, fertilizer, and mechanization that developed nations have had for nearly 100 years. With such drawbacks, subsistence agriculturalists can barely provide for their families, and net yields have not increased substantially for many generations. These families and countries do not have enough capital to purchase the necessary state-of-the-art equipment to improve their standard of living.

Finally, all too often, developing countries turn to their only source of revenue to generate the cash flow needed for infrastructure, growth, and military equipment. They must produce something that they can sell in the world market. Occasionally, they sell mineral resources and nonmineral energy resources or fuels. Most frequently, however, developing countries generate funds from the agricultural sector, which also needs further development. Often, these countries sell cash crops on the world market to generate foreign revenue; thus, the food is not used to sustain its own population. Fruits and vegetables are examples.

Another category of agricultural products that can generate the needed revenue is nonfood or not-nourishing crops, such as sugar, hemp, jute, rubber, tea, tobacco, coffee, and a growing harvest of cotton to satisfy the world's need for fabric and denim. How can starving nations feed themselves when a large proportion of their agricultural productivity and acreage is devoted to nonfood crops? This is the plight of many African, South American, Central American, and Asian countries today. As a result, sometimes alternative sources of income are inviting, even if they are illegal.

ILLEGAL DRUG TRADE

The peoples of South America and Asia sometimes produce crops that are sold on the world market and that bring much higher profits than conventional crops do. These high-profit crops are the coca leaf, from which cocaine and crack cocaine are produced; marijuana; opium; and hashish. Even though they are illegally grown, these crops provide farmers with hundreds to thousands of times the monetary return of the same acreage planted with conventional crops.

There is a geographic setting and contiguity to most major drug-producing areas of the world (Figure 5.7). For example, in South America, the countries of Bolivia, Peru, Ecuador, and Colombia produce the coca leaf. Half the supply comes from Peru alone, and the next largest amount comes from Colombia. However, Colombia is the target of U.S. DEA (Drug Enforcement Agency) activity because it is the refining center and distribution point of 85% of the cocaine from South America that enters the United States. Marijuana, the illegal drug used most widely in the United States, comes primarily from Mexico. Smaller amounts come from Colombia, Jamaica, and Belize.

Some of the most remote and hostile regions of the world produce opium from the juice of opium poppies. Iran, Afghanistan, Pakistan, Myanmar, Thailand, and Laos are the principal supply countries. Pakistan, Afghanistan, Lebanon, and Morocco are principal producers of hashish.

Developing countries are now beset with a great problem. On the one hand, the world frowns on the production of illegal drugs, and the IMF, the World Bank, and the United Nations attach conditions to their loans and bail-out programs according to the amount of law enforcement against drug production. On the other hand, such governments face tremendous international debt and a lack of products that generate foreign exchange. Consequently, although most of these countries' official policies disallow drug production, in reality enforcement against illegal drug production is impossible in certain regions of each country because of the potential for insurrection and the political power of drug money. For example, in 1996 alone, 16 mayors and 5 judges were assassinated in Colombia.

COMMERCIAL MODE OF PRODUCTION

Agriculture in the United States epitomizes the commercial system (see Figure 5.8). Modern American farming is quick to respond to new developments, such

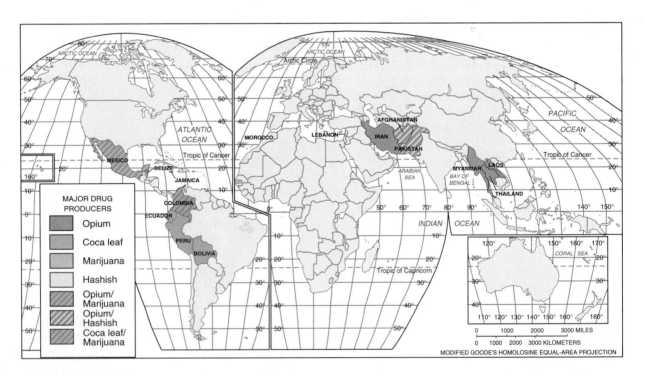

FIGURE 5.7

Sources of drugs worldwide. Drugs flow from the impoverished Third World countries to the demand centers in the developed world of North America, Europe, and cities throughout the remainder of the world. The flow of marijuana from South America has decreased, whereas the flow of hard drugs from South America, primarily the coca leaf, from which cocaine and crack cocaine are made, has increased substantially. Sources of the coca leaf include Colombia, Ecuador, Peru, and Bolivia. Morocco, Afghanistan, and Pakistan are the chief sources of hashish, whereas Iran, Afghanistan, Pakistan, and Southeast Asia are the principal supply regions of opium.

Harvesting cocaine in Peru. This peasant gathers coca leaves four or five times per year. One hectare furnishes 2 tons of leaves per year, which makes it possible to manufacture 40 kilos of pâte. This pâte is smoked in a form of a stock, or transformed into cocaine. Most of the production occurs in Colombia, South America. The traffic in coca leaves is one of the major problems in Peru. Under American pressure, the government set up a plan for tearing up the plants, calling for the destruction of 600 hectares per year. Even so, this is a small amount, considering there are more than 200,000 hectares now devoted to producing coca leaves in Peru. *(Photo: J. C. Criton/Sygma)*

as new production techniques. Consequently, farmers with sizable investments of money, materials, and energy can create drastic changes in land-use patterns. For example, farmers in the low-rainfall areas of the western United States have converted large areas of grazing land to forage and grass production with the use of center-pivot irrigation systems. Other farmers grow sugar beets and potatoes in western oases through federally subsidized water projects.

U.S. farmers are more vulnerable to catastrophic events than their peasant counterparts are. For the most part, peasant farmers can provide their families with food, clothing, and shelter. Most U.S. farmers are completely tied to an elaborate marketing system. If their communication lines with the wider space economy (Figure 5.8) were cut, they would quickly run out of the essentials: fuel, spare parts, fertilizer and seeds, and store-bought food and clothing.

At the frontier of American farming is *agribusiness*, which is associated with the trend on the part of such giant food companies as Ralston Purina, General Mills, General Foods, Hunt-Wesson, and United Brands to control the whole food chain from "seedling to supermarket." The concept that describes the food companies' control of production, processing, and marketing is *vertical integration*. The promise of high profits and a favorable tax structure has also attracted nonfood companies to move into food production. These companies include tractor firms, fertilizer and pesticide manufacturers, oil companies, and aircraft companies.

At the farm level, agribusiness is *capital intensive* and *energy intensive*. The very high per capita productivity results in rural farm depopulation. Although the fam-

ily farm remains the basic unit in the American agricultural system, the direct role of agribusiness is increasing. The importance of corporate farming is growing in *market gardening*, which is sometimes called *truck farming*. Modern food-production truck farms

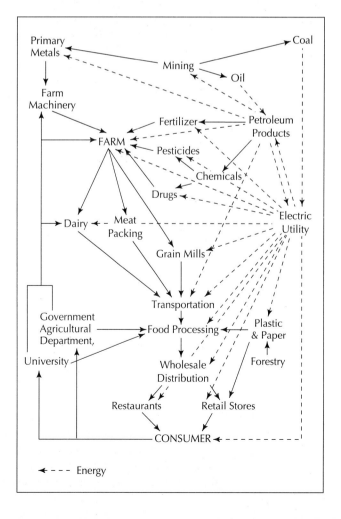

FIGURE 5.8
The structure of the U.S. agricultural system. *(Source: Knight and Wilcox, 1976, p. 23)*

Feed lots for beef cattle at Grosford, California. According to the "Code of the West," cattle ranchers owned little land, only cattle, and grazed open land wherever they pleased. New cattle breeds introduced from Europe, such as the Hereford, offered superior meat, but were not adapted to the old ranching system of surviving the winter by open grazing. In moist areas, crop growing supplanted ranching because it generated a higher income per acre. Some cattle are still raised on ranches, but most are sent for fattening to feed lots along major interstate highways or railroad routes. Many feed lots are owned by agribusiness and meat processing companies, rather than by individuals. *(Photo: Joe Munroe, Photo Researchers, Inc.)*

specialize in intensively cultivated fruits, vegetables, and vines, and they depend on migratory seasonal farm laborers to harvest their crops. Other examples of modern food production include poultry ranches and egg factories. Agribusiness has also extended livestock farming immensely.

At one time, livestock farming (for example, in western Iowa) was associated with a combination of crop and animal raising on the same farm. In recent years, livestock farming has become highly specialized. An important aspect of this specialization has been the growth of factory-like feedlots, where companies raise thousands of cattle and hogs on purchased feed. Feedlots are common in the western and southern states, in part because winters there are mild. These feedlots raise more than 60% of the beef cattle in the United States.

American corporate farming is also extending overseas to become a worldwide food-system model. Family farming is still dominant in western Europe, but beef feedlots are found in the Italian Piedmont. Poultry raising operations in Argentina, Pakistan, Thailand, and Taiwan are like those in Alabama or Maryland. Enterprises such as United Brands, Del Monte, Unilever, and Brooke Bond Oxo are diverting more and more food production in developing countries toward consumers in developed countries.

U.S. Commercial Agriculture: Crops and Regions

The main characteristic of *commercial agriculture* is that much of it is produced for sale off the farm, at the market. Commercial agricultural areas include the United States and Canada, Argentina and portions of Brazil, Chile, Europe, the former Soviet Union countries, South Africa, Australia, New Zealand, and portions of China.

Following are some of the characteristics of commercial agriculture:

1. Populations fed by commercial agriculture are normally nonfarm populations who are living in cities and engaged in other types of economic activity, such as manufacturing, the services, and information processing.
2. Only a small proportion of the countries' population is engaged in agriculture.
3. Machinery, fertilizers, and high-yielding seeds are used extensively.
4. Farms are extremely large, and the trend is toward even larger farms.
5. Agricultural produce from commercial agriculture is integrated with other agribusiness, and a vertical integration exists that stretches from the farm to the table.

Let us look more closely at the last four points.

COMMERCIAL AGRICULTURE AND THE NUMBER OF FARMERS
The percentage of laborers in developed countries working in commercial agriculture is less than 5% overall. In contrast, in some portions of the developing world where intensive subsistence agriculture is practiced, 90% of the population is directly engaged in farming, and the average is 60% overall. Today, U.S.

The development of the center pivot irrigation system in the 1950s enabled large-farm operators to transform huge tracts of land in sandy or dry regions of the United States into profitable cropland. Here, alfalfa is being irrigated in Montana. *(Photo: Tim McCabe/USDA-SCS)*

farmers can produce enough food for their family and 70 other families.

In 1990, the United States had approximately 2 million farms, compared with 5.5 million in 1950 (Figure 5.9). This reduction in the number of farm families as a percentage of the population is a result of push factors and pull factors. *Push factors* are economic factors (such as the high cost of equipment) that push families off the farms. *Pull factors* include the advantages of urban life in the United States, Europe, and other developed countries. The opportunity for college education, specialized occupations in the services, finance, insurance, real estate, and telecommunications has lured farm children off the land. The attraction of these alternatives is the promise of a higher standard of living and a shorter working day. However, according to the 1990 census, for the first time in American history more people have moved from the cities to the countryside. But most of these urban-to-rural migrants have moved to enjoy a relaxed pace of life and retirement, not the arduous work of the farm. One serious problem in the United States, and a push factor, is the encroachment of the metropolitan area onto the best farmland, directly adjacent to the urban area. Suburban sprawl, brought by interstate highways that reduce the commute and penetrate into the flat countryside, has usurped viable topsoil and farmland around Los Angeles, Chicago, Amsterdam, Paris, and Buenos Aires.

MACHINERY AND OTHER RESOURCES IN FARMING

The second aspect unique to commercial agriculture, besides the small percentage of farmers in the population, is the heavy reliance on expensive machinery, tractors, combines, trucks, diesel pumps, and heavy farm equip-

ment, all amply fueled by petroleum and gasoline resources, to produce the large output of farm products. To this has been added miracle seeds that are hardier than their predecessors and that produce more impressive tonnages. Commercial agriculture is also fertilizer intense.

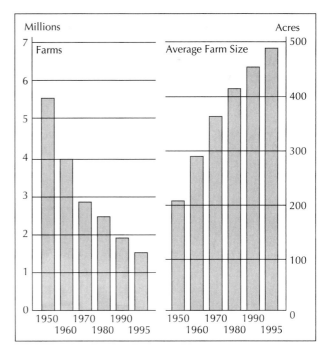

FIGURE 5.9
The number of U.S. farms compared with the average farm size. For the last 40 years, the number of U.S. farms has decreased by almost two-thirds, to approximately 2 million. At the same time, the average farm size has increased two and one-half times from 200 acres to almost 500 acres. *(Source: U.S. Bureau of the Census, 1994, p. 241)*

Improvements in farm-to-city transportation have resulted in less spoilage. Products arrive at the canning and food-processing centers more rapidly than they did in previous generations. By 1850, the farm was well connected to the city by rail transportation. More recently, the motor truck has supplanted rail transportation, and the advent of the refrigerator car and the refrigerator truck means that freshness is preserved. Cattle also arrive at packing houses by motor truck as fat as when they left the farm, unlike 100 years earlier, when long cattle drives were the order of the day, connecting cattle fattening areas in Texas, Oklahoma, and Colorado with the Union-Pacific rail line, stretching from St. Louis to Kansas City to Denver.

Agricultural experiment stations are now part of every state and are usually connected with land-grant universities. These stations have made great strides and improvements in agricultural techniques, not only in improved fertilizers and hybrid plant seeds, but also in hardier animal breeds and new and better insecticides and herbicides, which have reduced pestilence. In addition, local and state government farm advisors can provide information about the latest techniques, innovations, and prices so that the farmer can make wise decisions concerning what should be produced, when it should be produced, and how much should be produced.

FARMS AND THEIR RELATIONSHIPS WITH OTHER BUSINESSES

Farming in the developed world is an agribusiness venture. Large food-processing companies, such as General Mills, General Foods, Ralston Purina, and Campbell, purchase most of the farm products directly from the farmer. Therefore, the farmer is integrated into a large variety of economic activity, which features interindustry cooperation to provide food to the tables of not only Americans, but also other citizens of the world. These activities include food transportation, processing, canning, packaging, inspection, sorting, storing, wholesaling, and retailing. However, agribusiness includes capital goods production as well, including fertilizer producers, seed producers and distributors, tractor and combine manufacturers, and energy companies.

Types of Commercial Agriculture

James Rubenstein (1996) divides commercial agriculture into six main categories: mixed crop and livestock farming, dairy farming, grain farming, cattle ranching, Mediterranean cropping, and horticulture and fruit farming (see Figure 5.5).

MIXED CROP AND LIVESTOCK FARMING

Mixed crop and livestock farming is the principal type of commercial agriculture, and it is found in Europe, Ireland, Russia, Ukraine, North America (west of the Appalachian Mountains and east of 100° west longitude), South Africa, central Argentina, eastern Australia, and New Zealand.

The primary characteristic of mixed crop and livestock farming is that the main source of revenue is livestock, especially beef cattle and hogs. In addition, income is produced from milk, eggs, veal, and poultry (see Figure 5.10). Although the majority of the farmlands are devoted to the production of crops such as corn, most of the crops are fed to the cattle. Cattle fattening is a way of intensifying the value of agricultural products and reducing bulk. Because of the developed world's dependence on meat as a major food source, mixed crop and livestock farmers have fared well during the last 100 years. However, widespread acceptance of the healthfulness of the low-fat diet may affect this type of farming in the future.

In developed nations, the livestock farmer maintains soil fertility by using a system of crop rotation in

Corporate farming in the United States. An employee watches a television monitor to see when a truck is in position to receive its computer-calculated load. *(Photo: Michael Lawton, USDA)*

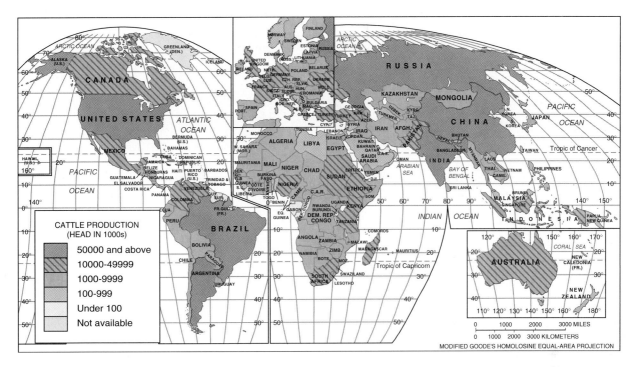

FIGURE 5.10
World beef and veal production. The developed countries produce the most beef and veal products because a large amount of the grain crops can be fed to cattle to fatten them. Poorer nations must consume all available food supplies directly or use them as revenue-producing exports. The United States, Western Europe, the former Soviet Union, Brazil, Argentina, and Australia are the leading producers of beef and veal worldwide. (See Color Insert 2 for more illustrative map.)

which different crops are planted in successive years. Each type of crop adds different nutrients to the soil. The fields become more efficient and naturally replenish themselves with these nutrients. Farmers today use the *four-field rotation system*, wherein one field grows a cereal, the second field grows a root crop, the third field grows clover as forage for animals, and the fourth field is fallow, more or less resting the soil for that year.

Most cropping systems in the United States rely on corn (Figure 5.11) because it is the most efficient for fattening cattle. Some corn is consumed by the general population in the form of corn on the cob, corn oil, or margarine, but most is fed to cattle or hogs. The second most important crop in mixed crop and livestock farming regions of central North America and the eastern Great Plains is the soybean. The soybean has more than 100 uses, but it is used mainly for animal feed. In China and Japan, tofu is made from soybean milk and is used as a major food source. Soybean oil is also used. Children with galactosemia must drink soy milk to avoid the lactic acid in cow's milk.

DAIRY FARMING
Dairy farming accounts for the most farm acreage in the northeastern United States and northwestern Europe. It also accounts for 20% of the total value of agricultural products from commercial agriculture.

Ninety percent of the world's milk supply is produced in these few areas of the world. Most milk is consumed locally because of its weight and perishability. Figure 5.12 shows the distribution of milk production in the world.

Dairy farming is an intensive land-use activity, as the von Thünen model, discussed later in this chapter, shows. Because milk is heavy and highly perishable, dairy farms are near cities, and the milk is trucked into the cities daily. The area around the city from which milk can be shipped in daily without spoilage is known as the city's *milkshed* region. Most farms in Australia, New Zealand, Europe, and North America are within the milkshed of a consuming city because of the rapid transportation of refrigerated trucks on interstate highways and railroad cars on railroad lines.

Some dairy farms produce butter and cheese as well as milk. In general, the farther the farm is from an urban area, the more expensive the transportation of fluid milk, and the greater proportion of production in more *high value-added* commodities, such as cheese and butter. For example, the Swiss have discovered ways of concentrating their milk products into high value-added chocolates, cheeses, and spreads that are distributed worldwide. These processed products are not only lighter, but also less perishable. On the other hand, in the United States, the proximity of farms to Boston,

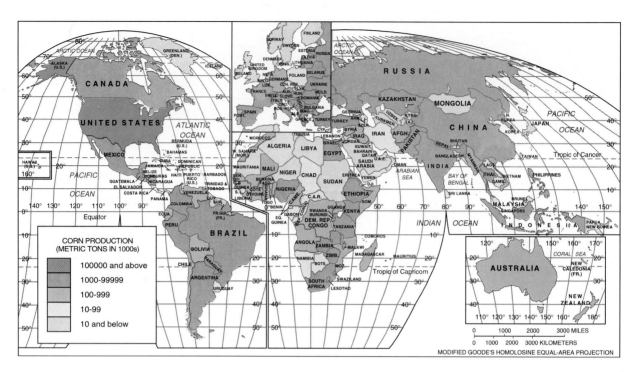

FIGURE 5.11

World corn (maize) production. Corn was domesticated in Central America more than 5,000 years ago and exported to Europe in the 15th century. The United States accounts for more than 30% of the world's corn production, and 90% of this corn crop is fed to cattle and livestock for fattening and meat production. Outside the United States, corn is called maize. China is the second-leading producer of corn, but in this case, the majority of the crop is consumed by humans. Because meat produces a greater market value per pound than selling corn does, U.S. farmers convert corn into meat by feeding it to livestock on farms and feedlots. In the United States, the Corn Belt is also the livestock region of North America, and it is located in the western Midwest and the eastern Great Plains, centered on the state of Iowa. Argentina and Brazil, as well as Europe, also have sizable corn-production areas. (See Color Insert 2 for more illustrative map.)

New York, Philadelphia, Baltimore, and Washington, D.C., on the East Coast, and to Chicago and Los Angeles in the Midwest and West, means that these farms primarily produce fluid milk. Farms throughout the remaining areas of the United States primarily produce butter and cheese. For example, only 10% of dairy farms in Wisconsin produce fluid milk because of their distance from the market, whereas only 10% of New York and Pennsylvania farms produce butter and cheese because of their proximity to Philadelphia, New York and the northeastern seaboard, Pittsburgh, and Detroit. Worldwide, remote locations such as New Zealand, for example, devote three-fourths of their dairy farms to cheese and butter production, whereas three-fourths of the farms in Great Britain, with a much higher population density at close proximities, produce fluid milk.

The cow is an amazing conversion machine that produces large amounts of consumable food products that are vitamin and protein rich. Certain types of cows, such as the holstein, produce enormous quantities of milk twice daily. Other types of cows, such as the jersey,

produce lower but richer quantities of milk with more cream. Most dairy farms have a variety of cows that produce both high-quality milk and large quantities. Frequently the milk is mixed to produce the best results.

Dairy farms are labor intensive because cows must be milked twice a day. Most of this milking is done with automatic milking machines. However, the cows still must be herded into the barn and washed, the milking machines must be attached and disassembled, and the cows must be herded back out and fed. The difficulty for the dairy farmer is to keep the cows milked and fed during winter, when forage is not readily available and must be stored.

GRAIN FARMING

Commercial grain farms are usually in drier territories that are not conducive to dairy farming or mixed crop and livestock farming. Most of the grain, unlike the products of *mixed crop and livestock* farms, is produced for sale directly to consumers. Only a few places in the world can support large grain-farming operations. These areas include China, the United States, Canada,

FIGURE 5.12

World milk production. A high correspondence exists between levels of economic development and the amount of per capita fluid milk production. Dairy farms are expensive to operate; thus, milk is a luxury in most areas of the world. Note Africa's paucity of milk production, except for South Africa. Most large cities of developed countries have milk production close by because of its heavy weight and perishability. Dairy farms exist further away from large cities, but in these cases most milk products are converted to less perishable and less heavy commodities, such as cheese and butter. For example, almost all the milk produced in Wisconsin is converted to cheese and butter, while 99% of the milk produced in Pennsylvania and New York is for direct fluid milk consumption. Countries such as New Zealand that are noted for specialized milk commodities but are a distance from the world's market must also convert the majority of their milk products into less perishable forms in order for them to be shipped to the world markets. New Zealand, for example, produces only 5% liquid milk from its dairy products, compared to 70% for the United Kingdom. (See Color Insert 2 for more illustrative map.)

Harvesting wheat by combine in the United States exemplifies capital-intensive and energy-intensive agriculture. Grain, such as the wheat shown here, is often a major crop on most farms. Commercial grain agriculture is different from mixed crop and livestock farming in that the grain is grown primarily for consumption by humans, rather than by livestock. In developing countries, the grain is directly consumed by the farm family or village, whereas in commercial grain farming, output is sold to manufacturers of food products. *(Photo: Doug Wilson, USDA)*

the former Soviet Union, Argentina, and Australia (Figure 5.13). Wheat is the primary crop and is used to make flour and bread. Other grains include barley, oats, rye, and sorghum. These grains are less perishable than fluid milk or corn and can be shipped long distances. Of the group, wheat is the most highly valued grain per unit area and is the most important for world food production. Figure 5.14a shows that grain yield and production have increased markedly in developing countries between 1970 and 1990, while cropland area has increased only slightly. However, production per capita is much more disappointing in developing countries (Figure 5.14b). Africa is declining.

In North America, the Spring *Wheat Belt* is west of the mixed crop and livestock farming area of the Midwest and is centered in Minnesota, North Dakota, South Dakota, and Saskatchewan (Figure 5.15). Another region, just south, near the 100th parallel of longitude, is the Winter Wheat Belt, which is centered in Kansas, Colorado, and Oklahoma. Because winters are harsh in the Spring Wheat Belt, the seeds would freeze in the ground, so instead *spring wheat* is planted in the spring and harvested in the fall; the fields are fallow in the winter. *Winter wheat*, however, is planted in the fall and moisture

accumulation from snow helps fertilize the seed. It sprouts in the spring and is harvested in early summer.

Like corn-producing regions, wheat-producing areas are heavily mechanized and require high inputs of energy resources. Today the most important machine in wheat-producing regions is the combine, which not only reaps but also threshes and cleans the wheat.

Large storage devices called *grain elevators* are a prominent landscape feature as one traverses the Great Plains of the United States and Canada. Major cities in the Great Plains, such as Minneapolis and St. Paul, built their fortunes off flour milling and distribution.

Wheat is the international commodity because it leads the list of world food products that are transported between nations. The United States and Canada are the leading export nations for grains and together account for 50% of wheat exports worldwide (see Table 5.2). Although the large grain trade agreements between the United States and the Soviet Union in the early 1970s are no longer in effect, the North American wheat-producing areas have been appropriately labeled the *world's breadbasket* because they still provide the major source of food to many deficit areas, including the starving nations of Africa (Figure 5.16).

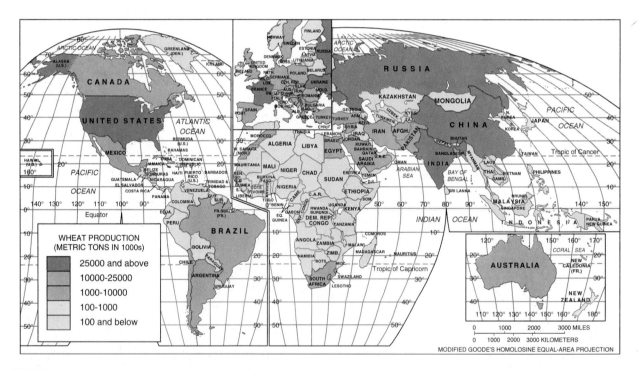

FIGURE 5.13

World wheat production. The United States, Canada, Australia, and Argentina are the primary wheat exporters, whereas the former Soviet Union countries, especially Kazakhstan, and India and China import the most wheat. China is the world's leading wheat producer, followed by the former Soviet Union and the United States. Wheat can be stored in grain elevators. Therefore, current wheat prices worldwide reflect not only growing conditions for that year, but also supplies from commercial and subsistence operations that have been stored throughout the world. (See Color Insert 2 for more illustrative map.)

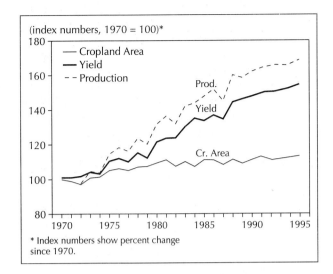
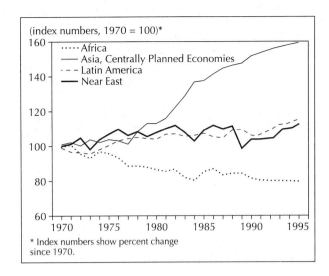

FIGURE 5.14

(a) The world's output of major food crops has increased dramatically during the last 20 years. The most dramatic increase has been in cereals, from 1.3 billion metric tons in 1970 to almost 2 billion metric tons in 1990. Milk, meat, fish, fruit, and vegetables have also made gains in production worldwide. In the last 20 years, every region of the world has increased production. Most notable has been the Asian centrally planned economies, which have doubled their production, whereas the Near East and Latin America have produced a 60% to 80% increase. Production in Africa has increased more than 40% during the last 20 years, even though the per capita production has decreased. The figure shows that the index of grain production has increased substantially during the last 20 years primarily as a result of an increase in yields, instead of an ever-increasing cropland. (Source: *World Resources*, 1997, p. 95.) (b) Index of per capita food production in developing regions, 1970–1995. While the index of grain production and yields in developing countries showed notable increases between 1970 and 1990, the prognosis is much poorer when one includes the effect of population growth. As this figure indicates, African countries showed a per capita decline in food production, with the worst declines occurring in Angola, Botswana, Gabon, Mozambique, Rwanda, and Somalia. Sudan also fell dramatically. Many factors contribute to these fluctuations, including inadequate rainfall, plunging world prices, ethnic unrest, and civil war. *(Source:* World Resources, *1997, p. 568)*

T a b l e 5.2

World Grain Trade since 1935: U.S. + Canada = World's Bread Basket

	Amount Exported or Imported (million metric tons)[1]					
Region	*1935*	*1950*	*1960*	*1970*	*1980*	*1990*
North America	5	23	39	56	131	123
Latin America	9	1	0	4	−10	−12
Western Europe	−24	−22	−25	−30	−16	28
Eastern Europe and former USSR	5	0	0	0	−46	−38
Africa	1	0	−2	−5	−15	−31
Asia	2	−6	−17	−37	−63	−82
Australia and New Zealand	3	3	6	12	19	15

[1]No sign in front of a figure indicates net export; a minus sign in front of a figure indicates a net import.

Sources: 1935 through 1980: U.N. Food and Agriculture Organization, FAO Production Yearbook *(Rome: various years); U.S. Department of Agriculture, Foreign Agricultural Service,* World Rice Reference Tables *(unpublished printout) (Washington, D.C.: June 1988); USDA, FAS,* World Wheat and Coarse Grains Reference Tables *(unpublished printout) (Washington, D.C.: June 1988), 1990;* World Resources, *1994–95 (New York, 1994).*

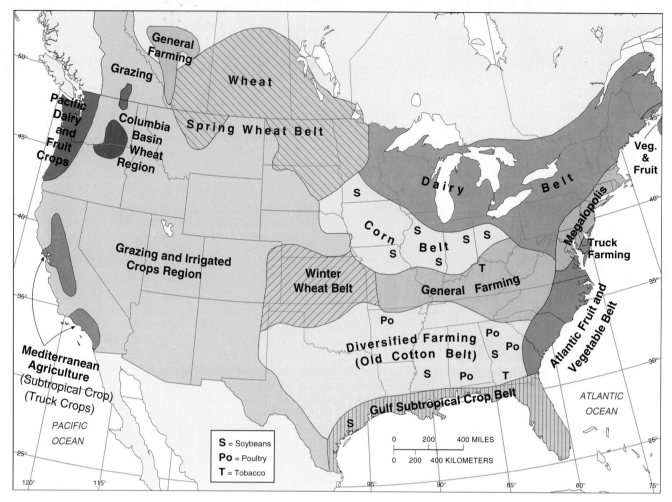

FIGURE 5.15
Major agricultural regions of the United States and Canada.

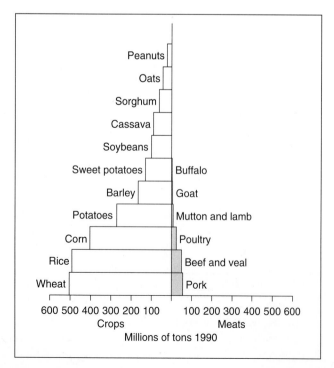

FIGURE 5.16
A diversity of crops offers protection against catastrophe in case any one crop should fail. Modern commercial agriculture is increasingly specialized, however. Today, most of humankind is dependent upon a handful of plants and animal species for food. Three-quarters of the human diet is based on eight crops alone.

CATTLE RANCHING

Cattle ranching is practiced in developed areas of the world where crop farming is inappropriate because of aridity and lack of rainfall. Cattle ranching is an extensive agricultural pursuit because many acres are needed to fatten cattle. In some instances, cattle are penned near cities, and forage is trucked to these cattle-fattening pens, which are called *feedlots*.

Amerindians and Native Americans were agriculturalists and hunters, not cattle ranchers. America received its first domesticated cattle from the Iberian Peninsula as a result of the second voyage of Christopher Columbus and the succeeding voyages of other explorers. Animal husbandry was part of these explorers' cultures, so cattle ranching was introduced to the New World and was successful because of the hardiness of the animals and the large, extensive areas of rangeland.

Major cities grew up across the western United States because of the services provided by their slaughterhouses and stockyards. Denver, Dallas, Kansas City, and St. Louis are examples. If a cattleman could get a steer to one of these cities, it was worth ten times as much as the $5 it was worth on the range. Early American ranchers were not as concerned about owning territory as they were about owning heads of cattle. Consequently, the range was open and the herds grazed as they went toward market. Later, farmers bought up the land and established their perimeters with barbed-wire fences. Until about 1887, the ranchers cut the barbed-wire fences and continued to move their herds about wherever they pleased. However, after that point, the farmers seemed to win the battle, and ranchers were forced to switch from long cattle drives and wide territories of rangeland to stationary ranching. The land was divided according to the availability of water and the amount of rainfall. Farmers used the land that was productive for farm crops, such as grains and wheat, whereas the ranchers received the land that was too dry for farming. Rubenstein (1996) points out that, ironically, today 60% of cattle grazing occurs on land leased from the U.S. government. (Beef and veal production are shown in Figure 5.10.)

The most popular variety of beef cattle today is the Hereford. It was introduced from England, was not as hardy, and could not withstand the long, cold winters of the U.S. Great Plains as well as the longhorn cattle that were first raised. However, the Hereford had a greater proportion of beef and, if properly cared for with stockpiled forage in the winter, could produce a good profit to the rancher.

Today, ranches in Texas and the West cover thousands of acres because the semiarid conditions mean that several acres alone are required to fatten one head of cattle. In addition, extremely large ranches are owned by meat-packing companies that can fatten the cattle, slaughter them, and package the meat all on the same ranch.

South America, Argentina, Uruguay, and Brazil each have a major cattle-ranching industry (see Figure 5.5). These regions, as well as Australia and New Zealand, followed a similar pattern of cattle-ranching development. First, cattle were grazed on large, open, government tracks with little regard for ranch boundaries. Later, when a conflict with farming interests occurred, cattle ranches moved to drier areas. When irrigation first began to be used in the 1930s and 1940s, farms expanded their territory, and ranchers moved to even drier areas and centered much of their herds on feedlots near railroads or highways directed toward the markets. Today, ranching worldwide has become part of a vertically integrated agribusiness meat-processing industry.

MEDITERRANEAN CROPPING

Mediterranean regions of the world grow specialized crops, depending on soil and moisture conditions. These regions include the lands around the Mediterranean Sea, coastal southern California, central Chile, South Africa, and southern Australia. In these regions, summers are dry and hot, and winters are mild and wet.

The Mediterranean Sea countries produce olives and grapes. Two-thirds of the world's wine is produced from the Mediterranean countries of Europe, especially Spain, France, and Italy. In addition, these countries and Greece produce the world's largest supply of olive oil for cooking. They also grow wheat for the production of staple bread products and pasta.

In California, the crop mix is slightly different because of consumer demand and preference. Most of the land devoted to *Mediterranean agriculture* is taken up by citrus crops, principally oranges, lemons, and grapefruit. Nut trees, tomatoes, avocados, and even flowers are produced in the balmy Mediterranean climates of California. San Diego county alone produces 80% of the North American supply of poinsettias, for example (Stutz, 1993). Because of its high value-added Mediterranean crops, San Diego county ranked fifth in the United States in 1992 terms of the value of agricultural products. Similar products are grown in Chile, South Africa, and Australia. Australia also produces kiwi fruit and the jojoba nut.

Unfortunately for Mediterranean farmers, these areas of the world are some of the most prized for their wonderful climates. Northern Europeans have discovered the Mediterranean lands and turned them into a series of rivieras. The closest, and most expensive, is the French Riviera, centered on Nice. The next closest, and next most expensive, is the Italian Riviera, centered on Portofino. Europeans have also discovered the less expensive Costa del Sol and Costa Brava in Spain and the Algarve in southern Portugal. No discussion is necessary to describe the population pressure of tourism and the burgeoning growth of southern California. Ninety percent of the Chilean population lives in the Mediterranean lands in the central one-third of the country, centered on Santiago. Condominium projects, time-shares, and burgeoning suburban developments for major cities, especially Los Angeles, are rapidly dwindling our Mediterranean agricultural lands.

HORTICULTURE AND FRUIT FARMING

Because of taste and preferences and a severe winter season, there is a tremendous demand in U.S. East Coast cities for fruits and vegetables not grown locally. Shoppers in Philadelphia, New York, Washington, D.C., Baltimore, and Boston are willing to pay dearly for *truck farm* fruits and vegetables, such as apples, asparagus, cabbage, cherries, lettuce, mushrooms, peppers, and tomatoes. Consequently, a horticulture and fruit-farming industry exists as close as possible to this portion of the United States, as temperature and soil conditions allow. Stretching from southern Virginia through the eastern half of North Carolina and South Carolina to coastal Georgia and Florida is the *Atlantic Fruit and Vegetable Belt* (Figure 5.15). This is an intensively developed agricultural region with a high value per acre. The products are shipped daily to the northeastern cities for direct consumption or for fruit and vegetable packing and freezing.

As in the case of Mediterranean agriculture and subtropical cropping in southern California, the Atlantic Fruit and Vegetable Belt of horticulture and truck farms relies on inexpensive labor. In California, the laborers are primarily from Mexico and Central America and often enter the United States illegally. On the Atlantic Coast, the laborers are primarily from the Caribbean and Puerto Rico, and their immigration status is also often questionable. Inexpensive labor is one way for specialized agriculturalists to maintain production in areas under pressure for urban growth and expansion. In southern California, the retirement and entertainment industries apply pressure, whereas on the Atlantic Coast, pressure results from relocated manufacturing plants from the Northeast and the incipient laborers and populations that they generate.

U.S. Agricultural Policy

In the early days of America, farms were family owned, were small, and served local markets. In those days, farm prices were stable and predictable. Nature accounted for sudden weather changes and soil conditions, which led to price fluctuations. From time to time, a certain region of the country's farms enjoyed bumper crops; other times, it suffered through droughts.

More recently, farming has been highly mechanized, and technological improvements have revolutionized agriculture. By the beginning of the 20th century, farms had become much larger. An individual farm family could manage as many as a thousand acres with new mechanized equipment. More acres were cultivated, and high yields were produced. With improved transportation to the markets and between countries, the U.S. farmer now served a much wider market area. The early years of the 20th century were fairly prosperous for U.S. farmers, especially during World War I when they provided a large amount of food for Allied troops. However, many farmers lost their fortunes during the 1920s and 1930s with the Great Depression, and many farms ceased to operate. World War II created another upswing for agricultural pursuits as farmers once again provided food for a much larger army, the Western allies, and a hungry nation.

After the war, the farmers' fortunes dwindled in the 1950s and 1960s until the U.S. government agreed to major grain trade agreements with the Soviets. As a result of these agreements, farms prospered even more until the early 1970s. Since then, the national economy has slowed, and farms have felt the effect. At the same time, world markets for U.S. grain have dwindled as foreign countries have become better able to produce more of their own food. For example, as a result of Green Revolution technology, India, formerly a net food importer, is now a net food exporter. At the same time, the prices of farm operation—including machinery, fertilizer, land, and transportation—have increased drastically. These less profitable times for farmers, in which costs have outrun income, seem to be continuing into the 1990s. Compared with other sectors of the U.S. economy, the farm sector has faced higher operating costs and lower revenues and is a problem industry.

The Farm Problem in North America

One reason that agricultural markets are currently in such desperate straits is that the demand for farm products is inelastic. Consumers do not demand more food when farm prices are low, so the reduction in price does not lead to a substantial increase in the quantity demanded. This phenomenon is coupled with the fact that the yield from U.S. agriculture has increased manyfold during the last 100 years. Technological and mechanical improvements and hybrid seeds have increased yields more than anyone expected. These innovations make U.S. farm productivity the highest in the world. Finally, the quantity of farm products has increased much more rapidly than the demand has. These three factors have pushed down prices, as shown in Figure 5.17.

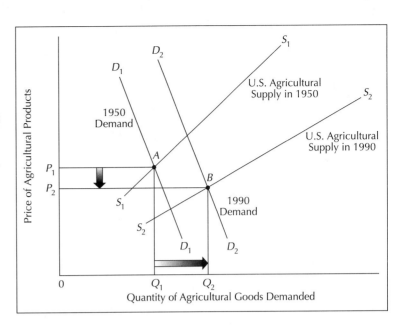

FIGURE 5.17
Falling U.S. agricultural prices. During the last 100 years, U.S. farm production has burgeoned remarkably because of increased productivity, increased mechanization, and improved fertilizers, pesticides, and hybrid seeds. The U.S. farm supply in 1994 was substantially more than it was in 1950, despite the Soil Bank program and other methods used to keep land out of production. At the same time, because food is an inelastic commodity, demand has increased, but not as much as supply. The result has been increased quantities and reduced prices. There are more than enough agricultural commodities to supply the United States and export quotas.

By looking at Figure 5.17, we see the three tendencies of American farming during the last 100 years: drastically increased supply, moderately increased demand, and falling prices. The result has been marginal returns to farm families and has spelled disaster for many farmers. With lower prices and increased quantities, more and more farmers cannot afford the rapidly rising costs for machinery, fertilizer, transportation, and labor. World farm prices have likewise fallen (Figure 5.18). The result to U.S. farmers has been a continuing reduction in return for their investment. Considering our production possibilities curve (see Chapter 1, Figures 1.21 and 1.22), farmers should now move their productivity to another, more profitable industry. However, unlike a store that can change hands and change function, or a high-tech manufacturing plant that can change products, a farm is difficult to adapt to a new economic use.

There has been a large movement of farm families and farm labor away from the farm (Table 5.3). In 1910, 35% of the U.S. population lived and worked on farms. By 1995, this figure had dropped to less than 1.5% of the total population. However, considering present production, prices, and consumption, we still have too many farmers in America today. In a normal market situation, resources would have shifted away from agriculture into other categories of economic use. However, because of U.S. government price supports for farms, this has not been the case, and there has been no movement along the production-possibilities curve (Table 5.4).

FIGURE 5.18
Trends in agricultural commodity prices. The trends show declines worldwide during the past two decades. Most farm prices have dropped to 60% or their 1970 levels. The decline in agricultural commodity prices has negatively affected many developing countries that base their economies on farm exports. This worldwide decline in farm commodity prices is a result of the inelasticity of food demand in developed nations, increased food supplies from Green Revolution techniques, and increases in food productivity, subsidies to producers in industrialized economies, and low income growth of farmers in developing nations. In the United States, for example, export subsidies designed to use farm overproduction abroad has kept world prices down by increasing the supply. *(Source:* World Resources, *1992, p. 97)*

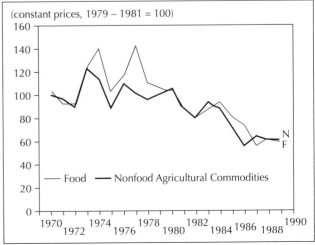

Note: *Food commodities* include beverages (coffee, cocoa, tea), cereals (maize, rice, wheat, grain sorghum), fats and oils (palm oil, coconut oil, groundnut oil, soybeans, copra, groundnut meal, soybean meal), and other food (sugar, beef, bananas, oranges). *Nonfood agricultural commodities* include cotton, jute, rubber, and tobacco.

Table 5.3
The Decline of the American Farm Population

Year	Farm Population (millions)	Farm Population as a Percentage of the Total Population
1910	32.1	35
1920	31.9	30
1930	30.5	25
1935	32.2	25
1940	30.5	23
1945	24.4	18
1950	23.0	15
1955	19.1	12
1960	15.6	9
1965	12.4	6
1970	9.7	5
1975	8.9	4
1980	7.2	3
1985	5.4	2
1987	5.0	2
1990	4.4	1.7

Source: Statistical Abstract of the United States and Economic Report of the President 1992, U.S. G.P.O. p. 69.

The U.S. Farm Subsidy Program

As early as 1933, Congress passed the Agricultural Adjustment Act to aid American farmers. This act was designed to help a large proportion of the population (as much as 33%). The main foundation of this act was to artificially raise farm prices so that farmers could enjoy a "fair price," or *parity price*, for their products. A parity price was defined as "equality between the prices farmers could sell their products for, and the price they would spend on goods and services to run the farm." The parity period was from 1910 to 1914, when farm prices were relatively high in comparison with other products. Let us illustrate the concept of parity price: If in 1914 a farmer could sell a bushel of corn and with the income from it buy 5 gallons of gas, the ratio between the price of farm products and the consumer price index would be maintained into the future.

Since 1933, however, the parity ratio declined for farmers until 1990, when it was approximately 50% of the original 1914 parity established in 1933. In other words, without parity, farmers could sell products and purchase only 50% of what they could in the earlier period. Real farm income, even in the short run, has been on a downward trend since the high prices of the early 1970s during the Soviet grain trade agreements.

The federal government stepped in to establish a subsidy program, or a price floor. The government established price supports for key agricultural commodities. These supports were minimum prices that the government could assure farmers. The *price floor* is a guaranteed price above the market price. For example, the government bought all corn and wheat from farmers and sold it at what the market would bear. It stored many of these commodities in its own storage facilities. In 1994, the U.S. government offered farmers *target pricing*, which is similar to the price supports of the 1950s through the 1980s. With target pricing, the government pays directly to the farmer the difference between the market selling price and the target price that the government has set. The government no longer takes control of the product.

Table 5.4
Parallels between Plant and Animal Farming

	Plant	Animal
Major products	Grains, fruits, and vegetables for food	Meat, dairy products, and eggs for food
Other important products	Oils, fabrics, rubber, specialty crops (spices, nuts, etc.)	Labor, leather, wool, manure, lanolin
Modern practices	Industrialized agriculture on former grasslands and forests	Ranching, dairy farming, and stall feeding
Traditional practices	Subsistence agriculture on marginally productive lands	Pastoral herding on nonagricultural land
Current global land use	3.7 billion acres (1.5 billion ha) (11% of land surface)	7.6 billion acres (3.1 billion ha) (25% of land surface)

Nebraska farm auction. Because of drastically increased supply, moderately increased demand, and falling world farm prices, margined returns to farm families has spelled disaster for many farmers. Auctions, such as this one in Nebraska, allow farmers to liquidate in hopes of raising enough money to cover mortgages on machinery and land. There has been a large movement of farm families and farm labor away from the farm. *(Photo: Andy Levin, Photo Researchers, Inc.)*

Figure 5.19 shows the effect of price supports on agricultural products. The subsidy process and its results are outlined next:

1. The market cannot arrive at an equilibrium price through its normal market mechanism.
2. Farmers produce a larger amount of surplus goods.
3. Buyers pay more than they would if market conditions prevailed.
4. Farmers' incomes are artificially raised by government subsidies.
5. Resources are poorly allocated and some are wasted.

As shown in Figure 5.19, the parity price is established above the market price. With the parity price artificially high, farmers will supply the intersection of the price line and the supply curve at K for a total quantity of q_1. However, with higher prices, the consumers will demand q_2. The difference between q_1 and q_2 is the surplus that the government would purchase under the old price-support plan. Regardless of whether the subsidization is in the form of price supports or target pricing, the result is an extra cost to the taxpayer.

These price-support and target-price programs relegated artificially high agricultural prices to U.S. agriculture for the last 40 years. The hope, of course, was

FIGURE 5.19
U.S. agricultural price supports. The U.S. government has supported farmers by establishing a parity price above the equilibrium price, according to supply and demand factors. For the last 60 years, the effect of the price support has been to establish prices higher than they would normally be, thus producing a surplus that the government was required to purchase with tax dollars. The market cannot obtain an equilibrium at E because surplus goods are produced and too often wasted. The farmers' incomes are artificially raised, but buyers in the marketplace must pay more than the goods would warrant under normal conditions. Unfortunately, resources are artificially allocated and therefore misallocated as price shifts from P_1 to P_2, demand drops back to point L, while supply moves up to point K.

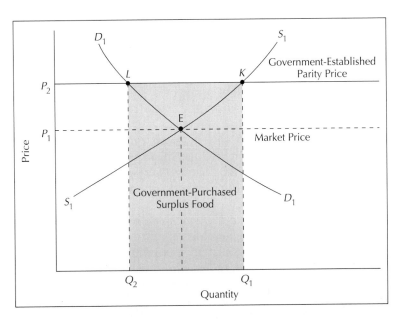

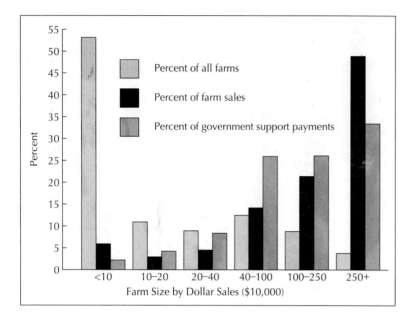

FIGURE 5.20
The proportion of farms, the proportion of farm sales, and the percentage of government support payments by farm size and dollar sales. Most U.S. farms are small and produce small proportions of total farm sales. However, most government support payments go to large farms. With regard to farming and farm policy, the rich appear to be getting richer, whereas the poor or small farms, who were the original focus of price supports, are getting relatively poorer and scarcer.

that market prices would rise to parity, and they did during World War II and during the Soviet grain trade agreements of the early 1970s. However, most of the time, the price of farm products was much less than the parity price. The government also attempted to reduce production with the Soil Bank program, which paid farmers to keep acreage out of production. Initially, this approach worked, but the per-acre yields increased amazingly and completely overshadowed the lost acreage in terms of total yield.

Unfortunately, the small American farmer as a cultural institution is an endangered species. For years, the government price-support programs kept inefficient farmers in business. Now, an even more terrible conclusion to the farm problem is resulting. It appears that the subsidy program benefits the larger farmers, the agribusinesses, and not the small family farms as originally expected (Figure 5.20). In essence, the large corporate agribusiness farms have become richer and, with their production and the lion's share of U.S. government subsidy, forced food prices even lower. This has, in effect, continued to force the small farmers off the land.

The obvious solution to America's farm problems is to design new uses for farm products that are not currently demanded in the United States. One example is an attempt to generate gasohol from corn and other agricultural products to fuel automobiles. A second use of the food surplus is *Food-for-Peace programs*, which allow agricultural surpluses to be distributed to starving nations instead of being liquidated or exterminated by dumping or destroying storehouses of food. Locally, the *food-stamp program* for America's poor operates off the large agricultural surplus.

SOCIALIST MODE OF PRODUCTION

The former Soviet Union is one example of a country with a socialist system of agriculture. This mode of production is based on the labor theory of value, in which the state, representing peasants and workers, distributes wealth according to need rather than ownership of land, factories, or stores.

Before the Bolshevik Revolution, agricultural land consisted of a mixture of small peasant holdings and estates of the rich. The Communists subsequently organized the land into collective farms and state-owned enterprises in response to the poorly organized peasant holdings that they inherited from the Bolsheviks. State-owned enterprises are generally larger and better equipped than collective farms and tend to specialize. In the 1970s, cultivated land was fairly evenly divided between state-owned enterprises and collective farms. However, by 1980, state-owned enterprises represented nearly 70% of the total agricultural area. The government of the former Soviet Union intends eventually to convert the remaining collective into state farms that will be cultivated not by peasants, but by workers who will receive the same regard as their industrial counterparts.

Tiny private plots, or in official parlance, "personal subsidiary holdings," exist alongside the giant socialized farms. These plots consist of small gardens in which a typical collective farmer or state farm worker may keep a cow, a few pigs, and some chickens. The private plot helps compensate for the deficiencies of socialized farming in the labor-intensive operations of animal husbandry and fruit and vegetable production.

Compared with the transformation of Chinese agriculture, the organization of socialist agriculture in the former Soviet Union was not strikingly successful. Soviet agricultural achievements also were less impressive than their industrial accomplishments. Between 1930 and 1970, agricultural production increased by only 70% to 80%, whereas industrial output increased more than tenfold.

The former Soviet Union attempted to increase agricultural production in three ways: first, by opening up new, but mainly marginal, lands on the cold and dry fringes; second, by improving farming methods (e.g., irrigation) and crops (e.g., drought-resistant varieties); and third, by mechanizing, especially its wheat and other grain lands. None of these methods was totally successful. Aside from recurring weather problems, the food-production fiasco was attributed to a low level of past agricultural investments, to poor management, and to a lack of incentives for farm workers to increase their output.

Agricultural Policy in the Former Soviet Union

In the former Soviet Union, another type of agricultural policy was in effect. Prices were set artificially low so that all consumers would have enough money to purchase the food that they needed. The government fixed input and output prices, and these prices held for many years, bearing no relationship either to their economic value or to the prices that a free and competitive market would dictate. Such low prices meant that the demand was much greater than the supply. Consider, for example, Figure 5.21. The perfectly inelastic supply curve, S_1, is vertical in a command economy. In this case, agricultural products are provided at a particular price and quantity. Prices and supplies are set by government quotas. A normal demand curve slopes down to the right, as D_1 does in this figure.

Given the inelastic supply S_1, the equilibrium price is p_3 at Point B. But the new price is established by the government, below the equilibrium price, at p_1. Food is available to everyone, even the poor. The problem is that demand now shifts far to the right, and not everyone who wants to purchase food at the artificially low price can do so because of the shortage. At p_1, the quantity demanded, q_1 at Point E, is much greater than the amount supplied. This explains the long lines and empty shelves for food items and many consumer products in the former Soviet Union.

COMPARISON OF THE THREE SYSTEMS OF PRODUCTION

We can make some general observations about peasant, capitalist, and socialist agriculture. The peasant system of production is the most efficient in terms of value of output per hectare. Capitalist agriculture, epitomized by U.S. agribusiness, is the most productive, but it uses costly inputs. Finally, socialist agriculture, as in the former Soviet Union, is less efficient than peasant agriculture and less productive than capitalist agriculture.

RELATIVE LOCATION OF FARMS

Despite the growing importance of public companies and corporations, farming is still, for the most part, a family business. An important factor that shapes individual farmers' land-use decisions is the relative location or situation of a place in terms of its access to other places. Worldwide, the importance of situational components in agriculture increased as market exchange economies grew and developed. At one time, before commercial agriculture, a farmer's site relations—links

FIGURE 5.21
Soviet food price-fixing dilemma. The former Soviet Union's centrally planned economy tampered with food prices. An inelastic supply was established, shown by supply curve S_1. The government provided food at a particular price and quantity well below market equilibrium at point C. The price was established at P_1 and, consequently, the demand increased to point E. The amount supplied was shown at the intersection of S_2 and Q_3. A massive shortage developed and resulted in long food lines and disgruntled populations. Not everyone who wanted to purchase food at the artificially low price could do so because of the shortage. Since the Soviet Union's collapse, prices have floated up to the equilibrium position at C. These prices are much higher than they were under the old system. The shortage has been reduced substantially, but prices have increased dramatically.

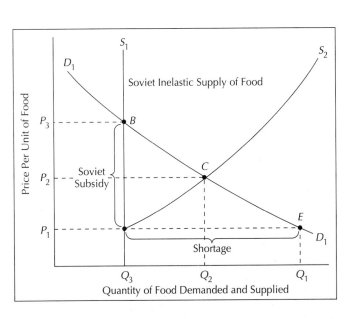

with soil, sun, rain, and crops—were overwhelmingly important considerations in earning a living. Today, site relations have not ceased to be important; farmers still depend on the weather. However, site relations have weakened as farmers have been drawn increasingly into situational relationships, with transport lines between farms and the market linking them ever more strongly to a wider spatial economy.

Von Thünen's Model

Relative location determines agricultural land use in several dimensions of space. The importance of relative location in rural land use was first discussed by Johann Heinrich von Thünen, a north German estate owner interested in economic theory and local agricultural conditions. From his experiences as an estate manager, he observed that identical plots of land (sites) would be used for different purposes, depending on their accessibility to the market (situation). The meticulous records that he kept provided a framework for his book, *The Isolated State*, which was published in 1826. Von Thünen's aim was to uncover laws that govern the interaction of agricultural prices, distance, and land uses as landlords seek to maximize their income. His methods in many ways constitute the first economic model of spatial organization; his conclusions continue to be discussed and debated by modern economic geographers.

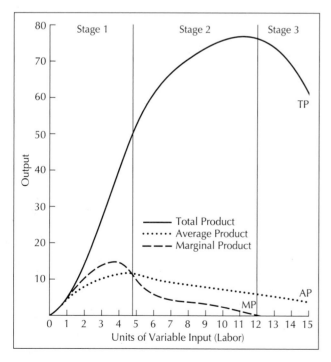

FIGURE 5.22
The stages of production.

The Law of Diminishing Returns

The *law of diminishing returns* relates to the situation that confronts farmers in the short run. It considers existing possibilities of managing land, labor, and capital inputs. It considers the state of technical knowledge as given and assumes no fundamental cost-reducing production changes. This law is as follows: As successive units of a variable input (say, labor) are applied to a fixed input (say, land), total product (output) passes through three stages. First, the total product increases at an increasing rate; second, it increases at a declining rate; third, it declines.

A standard example illustrates the principle of diminishing returns (Figure 5.22). Assume one fixed input (land) and one variable input (labor). Average productivity (AP) of labor is the total product (TP) divided by the number of labor units. Marginal productivity (MP) of labor is the addition to total output attributable to the last labor input employed. Throughout Stage 1, marginal product exceeds average product. The intersection of the marginal-product curve with the average-product curve marks the end of Stage 1 and the beginning of Stage 2. During Stage 2, the marginal-product curve declines until it finally becomes zero. This marks the end of Stage 2 and the beginning of Stage 3.

Knowledge of total, average, and marginal productivity establishes some general boundaries for rational zones of agricultural production. If farmers are trying to obtain a maximum return for their investments, they will never operate in Stage 1. The level of intensity is too low; that is, the amount of variable inputs (labor) per unit of land area is too small. Land is used too extensively. Farmers would want to take advantage of increasing returns to scale and add more variable inputs to intensify their operations. The boundary between Stages 1 and 2 is termed the *extensive margin of cultivation*.

Farmers who are trying to maximize their returns will also never operate in Stage 3. Obviously, no rational farmer will operate in the range in which additional units of labor decrease total production, causing negative marginal-product values. The boundary between Stages 3 and 2 is regarded as the *intensive margin of cultivation*. This leaves Stage 2 as the *zone of rational production*. In the real world, however, many enterprises—particularly large ones—operate successfully in Stage 3 because of government regulations, subsidies, and lack of true economic competition.

Some scholars find fault with the law of diminishing returns. Given the law's assumption of profit maximization, they claim it to be tautological in that the conclusions are concealed in the definitions. Farmers, for example, hire workers as long as they produce a surplus above their wages. When the additional profit falls to zero (i.e., when the marginal product from the last unit of labor added equals zero), farmers stop adding workers.

Economic Rent

The concept of economic rent is central to von Thünen's discussion of agricultural land use. *Economic rent* is a relative measure of the advantage of one parcel of land over another. More precisely, it is the difference in net profits between two units of land. Net profit per unit area of land is equal to the total value of production minus the total costs involved in bringing forth the product. Differential rents may result from variances in productivity of different parcels of land and/or variances in the distance from the market.

At the beginning of the 19th century, British economist David Ricardo (1912) presented the idea that rent variations result from the effect of physical factors on productivity. We can illustrate Ricardo's ideas of rent variations attributable to fertility conditions in a productivity schema for a spatially restricted area (Figure 5.23). As we move away from a crop's optimum physical conditions, costs per hectare increase and rents decrease. A cross section is drawn through the line *A-A'*. The side view is a space-cost curve, which graphs changes in cost across distance. Assume that the market price of a crop is $80 for 1 hectare of production. In this imaginary case, limits of production are determined by the intersection of the market-price line and the space-cost curve. No production occurs outside the $80 isoline, but rent increases toward the optimum.

What happens if the market price for a crop rises to $100 because of increasing demand? (see Figure 5.24). Spatial margins to profitability (SMP) spread out. Previously submarginal land is brought into production, and higher-rent land is used more intensively. On the other hand, if the market price falls to $60, spatial margins to profitability draw back. Lower-quality land is abandoned, and superior-quality land is used less intensively.

Von Thünen provided an alternative view of economic rent. Holding land quality constant, he demonstrated that rent is the price of accessibility to the market. In other words, rents decline with the distance from a market center. Geographers often use the term *location rent* as opposed to *economic rent* to express the concept of decline in rents with an increase in distance from the market.

The Isolated State

To explain agricultural land use, von Thünen described an idealized agricultural region about which he made certain assumptions. He envisioned an isolated state with a large central city serving as the only marketplace. A uniform plain surrounded the city. The farmer used a single mode of transport—the horse and cart—to supply the market with produce. The farmers were price takers, who attempted to maximize their profits. There were no extraneous disturbances in this ideal landscape; social

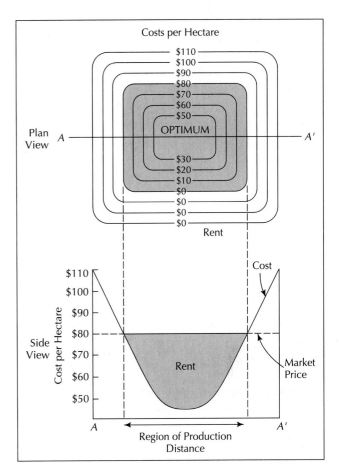

FIGURE 5.23
Optimal and marginal limits: the space-cost curve.

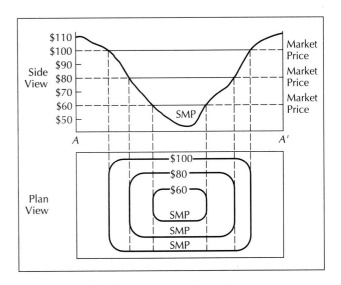

FIGURE 5.24
Spatial margins to profitability.

classes and government intervention were absent. In addition to these constraints, von Thünen introduced one variable: transport to the central town—its costs increased at a rate proportional to distance.

Von Thünen's conditions are not representative of actual conditions in the early 19th century or in this century. Indeed, von Thünen regarded the Isolated State as an Ideal State—the ultimate stage in the development of bourgeois society. In his view, this Ideal State represented a goal that humankind should strive toward. When it is attained, no further change is necessary. People live in a harmonious society free of exploitation.

THE PROBLEM

After stating his assumptions in *The Isolated State*, von Thünen posed the problem that he wanted to investigate. Paraphrasing von Thünen, the problem we want to solve is this: What pattern of cultivation will take shape in these conditions, and how will the farming system of the various districts be affected by their distance from the town? We assume throughout that farming is conducted absolutely rationally.

It is on the whole obvious that near the town will be grown those products that are heavy or bulky in relation to their value and that are consequently so expensive to transport that the remoter districts are unable to supply them. Here also we find the highly perishable products, which must be used very quickly. With increasing distance from the town, the land will progressively be given up to products cheap to transport in relation to their value. For this reason alone, fairly sharply differentiated

concentric rings or belts will form around the town, each with its own particular staple product.

From ring to ring, the staple product, and with it the entire farming system, will change; and in the various rings, we will find completely different farming systems.

Thus, Von Thünen suggested that in a landscape free from all complicating factors, locational differences were sufficient to produce a varied pattern of land use. After he observed the role of transport costs, von Thünen relaxed his rigid assumptions and introduced other variables to see how they modified his ideal pattern of land use.

LOCATION RENT FOR A SINGLE CROP GROWN AT ONE INTENSITY

To illustrate von Thünen's concept of differential rent, let us assume an isolated state producing one commodity (say, wheat) grown at a single intensity on yield per acre per year. Let us further assume that the market price of wheat is $100 per hectare per year, that it costs every farmer in the state $40 to produce a hectare of wheat, and that transport costs are $5 for each hectare of wheat (Figure 5.25). Under these conditions, what would be the net profit per hectare for farmers located 0, 1, 6, and 12 kilometers from the market? Farmers adjacent to the market pay no transport costs; therefore, their net profits would simply be market price ($100) minus production costs ($40)—or $60. Farmers 1 kilometer from the market pay $5 in transport charges; thus, their net profits would be $55. At 6 kilometers from the market, farmers would earn a net profit of $30, and at 12 kilometers, net profits would be zero. Beyond 12 kilometers, it would

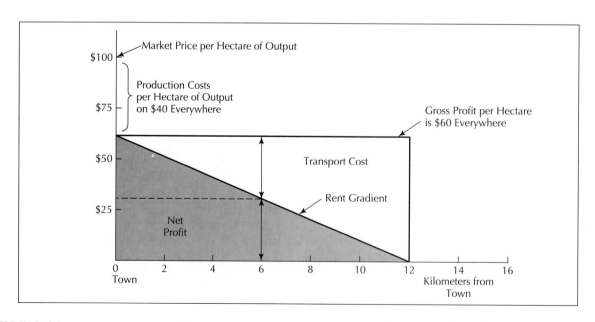

FIGURE 5.25

Net income from wheat production. Von Thünen suggested that the net income would decline with increasing distance from the town where the product had to be sold. Transportation costs would eat into the net income as one moved farther away from the town. In this example, production costs are $40 per hectare at every location, and the margin of profitability is 12 kilometers. *(Source: Based on Wheeler and Muller, 1981, p. 314)*

be unprofitable to grow a crop for the market. In this outer area, only subsistence cultivation could be pursued, and cash would have to be earned by migrant labor.

Our example has shown that farmers near the central market pay lower transport costs than farmers at the margin of production. Clearly, the net profits of the closer farmers are greater, and the difference is known as *economic rent*. Farmers recognize this condition, and they know that it is in their best interest to bid up the amount that they will pay for agricultural land closer to the market. Bidding continues until bid rent equals location rent. At that price, farmers recover production and transport costs, and landowners receive location rents as payments for their land. Competitive bidding for desirable locations cancels income differentials attributable to accessibility. The *bid rent*, or the trade-off of rent levels with transport costs, produces a spatial-equilibrium situation. It declines just far enough from the market to cover additional transport costs; hence, farmers are indifferent to their distances from the market.

We can simplify Figure 5.25 by including production costs in a single expression with market price. This is illustrated in Figure 5.26a, which shows a *rent gradient* sloping downward with increasing distance from the central market. When the *rent gradient* is located around the market town, it becomes a rent cone, the base of which indicates the extensive margin of cultivation for a single crop grown at a single intensity (Figure 5.26b).

LOCATION RENT: AN EXAMPLE
Location rent for any crop can be calculated by using the following formula:

$$R = E(p - a) - Efk$$

where:

R = location rent per unit of land
E = output per unit of land
k = distance to the market
p = market price per unit of output
a = production cost per unit of land (including labor)
f = transport rate per unit of distance per unit of output

Thus, if we assume that a wheat farmer 20 kilometers from the market obtains a yield of 1,000 metric tons/km², has production expenses of $50/ton/km² to transport grain to the market, and receives a market price of $100/ton at the central market, the location rent accruing to 1 square kilometer of the farmer's land can be calculated as follows:

$$R = 1000(\$100\backslash\$50)\backslash1000(\$120) = \$50,000\backslash\$20,000$$
$$= \$30,000$$

At 50 kilometers from the market, the location rent per square kilometer of land in wheat is $0. Obviously, no

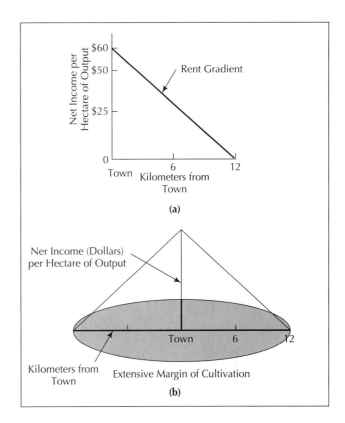

FIGURE 5.26
From a rent gradient to a rent cone.

rational farmer in a competitive market economy would grow wheat beyond 50 kilometers from the market.

LOCATION RENT FOR A SINGLE CROP GROWN AT DIFFERENT INTENSITY LEVELS
Now let us suppose that wheat is grown at two intensity levels, reflecting two farming systems (Figure 5.27). The more intensive farming system has a steeper rent curve and is the most profitable system as far as 36 kilometers from the market. The less-intensive system is most profitable from 36 kilometers to the limits of wheat farming, at 70 kilometers. At the margin of transference, the location rent for the two farming systems is the same. Separation between more-intensive and less-intensive systems illustrates the principle of *highest and best use*. According to this principle, land is used for the purpose that earns the highest location rent for its owner but not necessarily the highest wage for the workers.

LOCATION-RENT GRADIENTS FOR COMPETING CROPS
In von Thünen's analysis, patterns of agricultural land use form according to the principles of highest and best use as measured by the location rent at each distance from the market. Consider location-rent gradients for an isolated state in which farmers have three

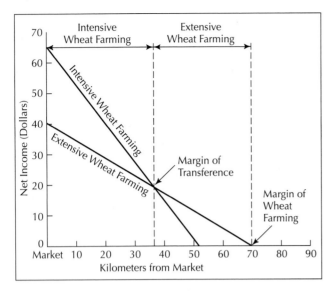

FIGURE 5.27
Rent gradient for a single crop grown at different intensities.

land-use choices: vegetable production, dairying, and beef production (Figure 5.28). A farmer close to the market could profitably carry on any one of the three activities, but which activity would maximize the farmer's income? Vegetable production has the highest rent-paying capability. All farmers seeking to maximize their incomes make the same decision: They grow vegetables between 0 and 10 kilometers from the market. Dairying is the choice between 10 and 25 kilometers, and beef production is the choice between 25 and 50 kilometers from the market. Beyond 50 kilometers, no commercial land use is feasible.

VON THÜNEN'S ORIGINAL CROP SYSTEM
In his theoretical Isolated State, von Thünen described six farming systems arranged in a series of concentric circles around the central city. The innermost zones produced perishable products (fluid milk and fresh vegetables) and heavy, bulky commodities in proportion to their value (fuelwood and lumber). On land most distant from the market, where transport costs were highest, land was used only for animal husbandry, which requires little investment but large amounts of space (livestock ranching). Between these inner and outer rings, agriculture consisted of intensive and extensive arable farming.

At first, it seems odd that von Thünen put a forestry zone close to the market. This arrangement does not fit with our image of reality. However, timber and fuel were in great demand in early 19th-century Germany. Consumers were not willing to pay high prices for items that were expensive to haul over long distances. The fact that patterns of agricultural land use in developed parts of the world in the late 20th century differ from those of the early 19th century does not undermine von Thünen's methodology; they simply reflect higher levels of transport efficiency and time-space convergence (see Chapter 4).

MODIFIED PATTERNS OF AGRICULTURAL LAND USE
Von Thünen was acutely aware that many conflicting factors—physical, technical, cultural, historical, and political—would modify the concentric patterns of agricultural land use. He modified some of his initial assumptions—the transportation assumption, for example—to approximate actual conditions more closely. Although he retracted the condition of a single-market town, he did not elaborate on the effects of several competing markets and a system of radiating highways. We can presume, however, that the tributary areas of competing markets would have had a variety of crop zones enveloped by those of the principal market town and that a radiating highway system would

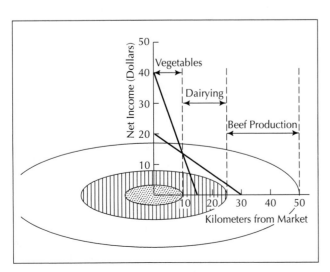

FIGURE 5.28
Location rent gradients for competing crops. Because of the different sloping bid rent curves based on market prices and transportation costs for various crops, the highest and best use and the highest level of net income will result from a combination of crops being grown at different distances from the market. Vegetables have the highest rent-paying ability because of their relatively high yield per hectare. Due to their weight, however, the bid rent curve drops off dramatically, as transportation costs eat into profits. Vegetables are the best choice up to 10 kilometers from the market, beyond which dairying becomes a better choice. At approximately 24 kilometers from the market, dairying and its associated relatively heavy and perishable products become more expensive to ship than the less expensive beef production. From this point out to the margin of production at 50 kilometers, beef is the wisest choice. Beyond that point, no commercial land use is feasible.

have produced a "starfish" pattern. Von Thünen retracted other conditions as well, such as uniform physical characteristics, and considered other complicating factors such as the effects of foreign trade, taxes, and subsidies. He also emphasized the effect of distance on agricultural land use at all scales.

R. F. Dodson (1991) has developed this model further in an interactive computer program so that students can alter the geographic characteristics of a hypothetical area and then observe the effects on the land-use pattern generated by the decision-making principles. The program permits the user to locate many markets, set different prices at the markets, construct a railroad network, set transport rates differently for road and rail, and describe yields at various locations for different crops. The program is constructed on top of a GIS program (IDRISI, 1992). Such GIS programs are able to evaluate and display earth resources data from satellites. It is also possible to enter socioeconomic data such as observed yields of crops and market prices and to simulate how real world economic systems might change if natural environmental change occurs, such as might occur with global warming, or if national policies of crop price guarantees were changed. A student who wanted to see how such systems are now beginning to be developed could see it in the contents of recent issues of *Earth Observation Magazine, GIS World*, or *GeoInfosystems*. The Food and Agricultural Organization (FAO) has developed an Agro-Ecological Zone model to assess Africa's agricultural suitability for many potential food crops. The model uses the Advanced Very High Resolution Radiometer (AVHRR) sensor on U.S. National Oceanic and Atmospheric Association (NOAA) satellites. From these data, computer maps supply a continuous overview, in time and space, of the entire continent of Africa's ecological conditions (van der Laan, 1992).

VON THÜNEN'S MODEL AND REALITY

Does the intensity of agricultural land use and the price of land increase toward the market as von Thünen's model suggests? We can answer this question by examining agricultural locations at local, national, and international scales.

Thünen effects at the local scale can be observed in the developing world, where localized circulation systems resemble those of early 19th-century Europe. Ronald Horvath (1969) found such a pattern of land-use around Addis Ababa, Ethiopia. Von Thünen's original farming system placed forestry in the second land-use ring, from which the city drew wood for building and fuel. Horvath described an inner wood-producing zone of eucalyptus forest that surrounded the Ethiopian capital (Figure 5.29). The zone was wedge-shaped rather than a ring, reflecting the greater accessibility to the city along major roads. Horvath also showed the expansion of the eucalyptus zone between 1957 and 1964, indicating transport improvements in the Addis Ababa area. The improvements permitted wood to be shipped to the

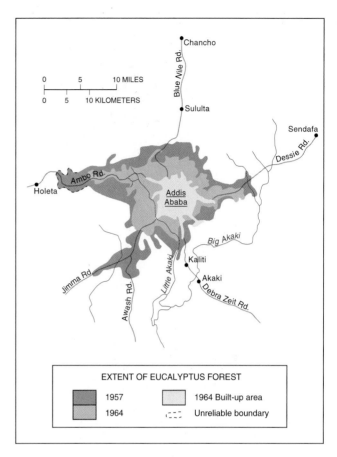

FIGURE 5.29
The forest zones surrounding Addis Ababa, Ethiopia.
(Source: Based on Horvath, 1969)

city over increasingly greater distances and released more land near the city to be used for vegetable production. Horticulture was a major activity of the innermost ring in von Thünen's ideal schema.

Additional studies indicated distance-related adjustments in land use in the Third World. Piers Blaikie (1971) observed that small farmers in north India adjust land use to distance from their villages in order to reduce the total amount of work to be completed. Farmers living in these villages must walk to the land that is under cultivation; therefore, the greatest effort is applied to land near the village, and farming becomes less intensive in the outlying fields (Figure 5.30).

Agricultural land use at the national level represents another change of scale. Suppose that we consider U.S. agricultural production going from the hypothetical to the real. Figure 5.31a is a map of hypothetical land-use rings. It assumes that the United States is a homogeneous plain, that New York City is the only national market, that transport costs are uniform in all directions from New York City, and that crops are ranked by rent-paying ability. These land-use zones, of course, are not consistent with reality. Other assumptions, such as

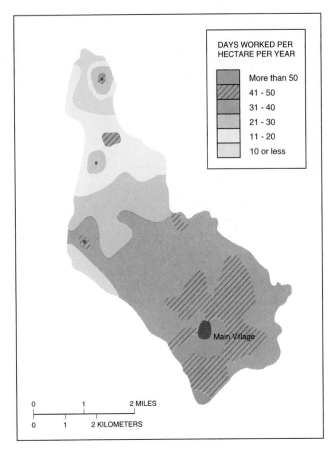

FIGURE 5.30
Intensity of farming around a village near Jaipur in Rajasthan, India. *(Source: Based on Blaikie, 1971)*

framework of von Thünen's model. By 1900, early 19th-century crop zones had expanded with improving transport technology from local isolated states to the entire country. As the agricultural structure changed, the enlarged original Thünian zones were modified: (1) the first ring developed a distinct horticultural zone and a surrounding Dairy Belt, (2) the forestry ring was displaced to the marginal areas of the system because the railroad could haul wood cheaply, (3) the crop ring subdivided into an inner mixed-crop/livestock belt that produced meat (the Corn Belt) and an outer cereal-producing area, and (4) the ranching area remained a peripheral grazing zone that supplied young animals to be fattened in the Corn Belt. This super-Thünian regional system was anchored by a supercity—the northeastern Megalopolis (Figure 5.15).

Although the map of North American agricultural regions does not exhibit the classic rings, certain regularities are apparent. Most important, the intensity of farming declines with distance from the national market. Thus, the Atlantic Fruit and Vegetable Belt, Dairy Belt, Corn Belt, Wheat Belts, and Grazing Belt conform to the model's structure. Deviations from the schema are the result of environmental variations, and special circumstances. Central Appalachia supports only isolated valley farming (general farming). Areas of the dry western mountains that have been reclaimed through hugh federal investments in irrigation projects support oasis farming. California and the Gulf Coast-Florida region have mild winters and, with the help of irrigation

a north-south temperature gradient, result in a more complex and realistic pattern (Figure 5.31b).

Now consider the map depicting the actual regionalization of North American agricultural production (see Figure 5.15). The major agricultural regions, established for more than a century, developed largely within the

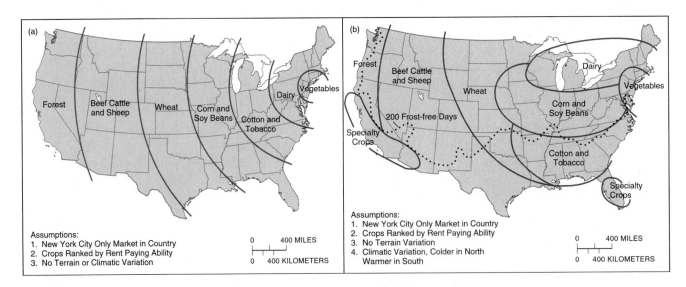

FIGURE 5.31
Theoretical land-use rings in the United States. *(Source: Kolars and Nystuen, 1974, p. 258)*

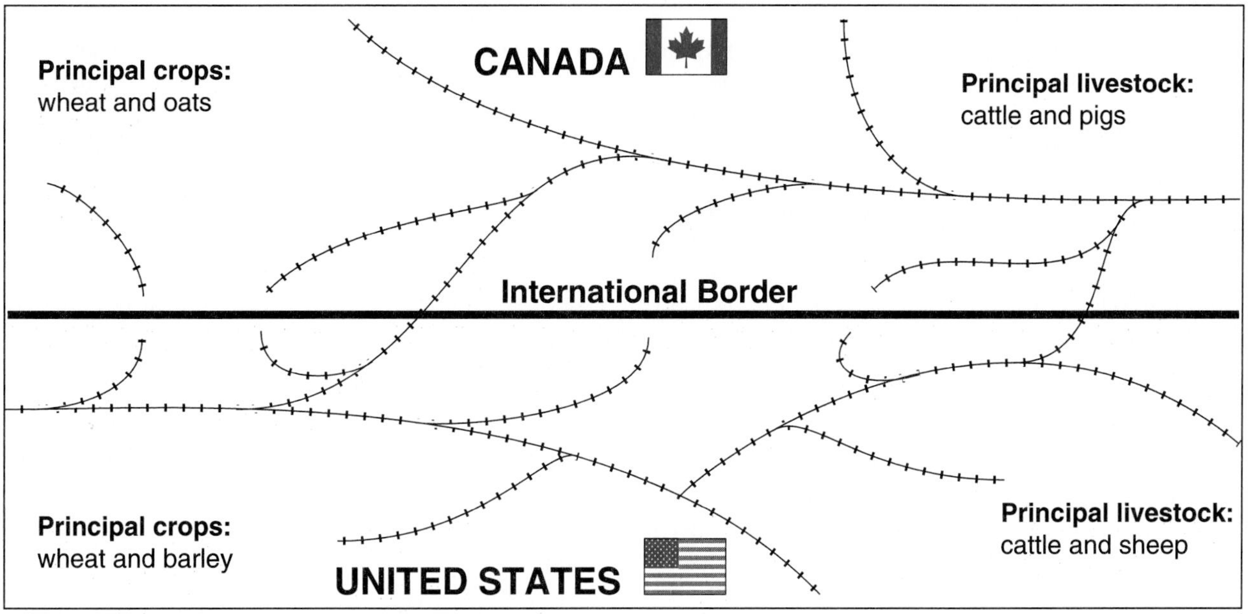

FIGURE 5.32
Different crops and livestock predominate on different sides of the U.S.–Canadian border. While the principal crops in Canada are wheat and oats, the principal crops in the United States are wheat and barley. Cattle and pigs predominate in Canada, while cattle and sheep predominate in the United States. The differences aren't so much environmental or market driven but result from the two countries' agricultural support policies. A second feature illustrated by the map is that transportation networks frequently approach borders, but seldom cross them. Fewer crossings mean an easier time of managing trade flow and immigration.

and refrigerated transport, produce large crops of fruits and vegetables. The farms of the Pacific Northwest serve the local population—the nation—with potatoes, apples, and other specialty crops, and the Pacific Basin countries with wheat. The old Cotton Belt, once poorly integrated into the national economy, specializes in the production of beef, poultry, soybeans, and timber.

Finally, we can increase our scale of observation to the international level. Canadian agriculture patterns also generally follow a von Thünen scheme, but differ somewhat from that of the U.S. due to different agriculture support policies (Figure 5.32). In Europe, the intensive cash-cropping areas that extended only a few miles from the market have expanded far beyond von Thünen's estate, southeast of Rostock, in Mecklenburg, Germany. We can now visualize Europe as a set of von Thünen rings. Production is most intense in the area centered on the Low Countries, Denmark, north Germany, northern France, and southeastern England.

The world itself may also be viewed as a set of von Thünen rings. At this scale, the world is the Isolated State, and Europe–North America is the Thünian "Town."

SUMMARY

For many thousands of years, a complex relationship has evolved between people and the natural environment.

All people depend on agriculture for their well-being, but in our agricultural pursuits, we necessarily modify the land. Good land managers give back to nature what they take; poor land managers degrade the environment.

Collections of plants and animals define fields that are organized around farms. Collections of farms make up farming regions. When geographers speak of agricultural regions, they refer to the artificial division of the world into homogeneous farming types.

The most intriguing aspect of farming regions is not their number or extent but the similarity of land-use decisions farmers make within them. After reviewing agricultural origins and dispersals and the effects of food production on the land, we identify four basic factors that influence agricultural land-use patterns. These factors are site characteristics, cultural preference and perception, systems of production, and relative location.

Agriculture in developing countries is primarily of the subsistence variety. Subsistence agriculture includes shifting cultivation, intensive farming, and pastoral nomadism. Developing countries have usually more than 50% of their labor force engaged in agriculture and use few pieces of mechanized equipment.

By contrast, agriculture in developed countries employs a small percentage of the labor force and uses large-scale mechanized equipment and large inputs of energy and fertilizers. While outputs per hectare are comparable to intensive farming and even somewhat less, outputs per worker are as much as 50 times greater. Farmers who raise crops and livestock on huge farms are part of a vast agricultural system that includes machinery manufacturers; fertilizer, pesticide, and energy supplies; grain mills and slaughterhouses; food processing and wholesale and retail distribution.

Although mixed crops and livestock farms are the most common in developed countries, other types of commercial agriculture also exist, including dairy farming, commercial grain farming, usually centered on wheat, cattle ranching, Mediterranean horticulture, and tree farms.

One theory that helps us to understand the distribution and location of agriculture was formalized by Johann Heinrich von Thünen in the early 19th century. We describe the normative model that von Thünen developed to explain patterns of land use around the town in north Germany where he lived. We then present some of the conclusions that can be drawn from this model: (1) There is an inverse relationship between lo-

cation rent and transport costs; (2) there is a limit to commercial farming on a homogeneous plain with an isolated market town at its center; (3) land values and intensity of land use increase toward the market; and (4) crop types compete with one another and are ordered according to the principle of the highest economic rent.

In the last section of the chapter, we see how the basic Thünian principles can be applied to agricultural land-use patterns at scales ranging from the village to the world. Contemporary Thünian effects at the microscale are best observed in the developing world, where localized circulation systems provide a transport setting similar to that of early 19th-century Europe. Improvements in transport technology and the development of refrigeration have permitted the Isolated State to expand from the micro- to the macroscale. Thus, in the United States and Europe, the model is no longer centered on a single city but on a vast urbanized region. At the global level, core nations are the market around which production zones of peripheral countries develop.

KEY TERMS

agribusiness
Atlantic Fruit and Vegetable Belt
capital intensive
commercial agriculture
environmental perception
extensive margin of cultivation
four-field rotation system
highest and best use
high value added
inelastic supply curve
intensive margin of cultivation
intensive subsistence agriculture

Isolated State
labor intensive
location rent
marginal product
margin of cultivation
Mediterranean cropping
mixed crop and livestock farming
parity price
pastoral nomadism
peasant agriculture
price supports
rent

shifting cultivation
slash and burn
space-cost curve
stages of production
subsistence agriculture
target pricing
use value
vertical integration
virgin and idle lands program
von Thünen's Isolated State
wheat belt
world's breadbasket

SUGGESTED READINGS

Bebbington, A., and Carney, J. 1990. Geography in the international agricultural research centers: Theoretical and practical concerns. Annals of the Association of American Geographers, 80:34–48.

Bowler, Ian. *The Geography of Agriculture in Developed Market Economies*. New York: John Wiley & Sons, 1993.

Boycko, Maxim, Andrei Shleifer, and Robert Vishny. *Privatising Russia*. Washington, D.C.: Brookings Institution, 1994.

Brown, L. R., and Young, J. E. 1990. Feeding the world in the nineties. In *State of the World* (pp. 101–141). New York: Norton & Company.

Christopherson, R. W. 1996. *Geosystems*. Upper Saddle River, N.J.: Prentice Hall.

Clawson/Fisher. 1998. *World Regional Geography: A Development Approach, Sixth Edition*. Upper Saddle River, N.J.: Prentice Hall.

Grigg, David. *An Introduction to Agricultural Geography*. New York: Routledge, 1995.

Nebel, B. J., and R. T. Wright. 1996. *Environmental Science, the Way the World Wood Works*. Upper Saddle River, N.J.: Prentice Hall.

Raven, P. H., Berg, L. R., and Johnson, G. B. 1995. *Environment*. Philadelphia: Saunders College Publishing.

Reganold, I. P., Papendick, R. I., and Parr, J. E. 1990. Sustainable agriculture. *Scientific American*, June, 88–95.

Rhoades, R. E. 1991. The world's food supply at risk. *National Geographic*, 179(4): 81–94.

Rubenstein, J. M. 1996. *The Cultural Landscape: An Introduction to Human Geography*. New York: Macmillan.

Turner, B. L. II, and Brush, S. B., eds. 1987. *Comparative Farming Systems*. New York: Guilford.

World Resources Institute. 1996. *World Resources: A Guide to the Global Environment, 1996–97*. New York: Oxford University Press.

WORLD WIDE WEB SITES

AG AGENT HANDBOOK
http://hammock.ifas.ufl.edu/txt/fairs/aa/31541.html
This online handbook from the University of Florida has links to infomration on: soils, field and forage crops, turf-grass, vegetable crops, sustainable production practices, forestry, dairy cattle, beef cattle, poultry, rabbit production, swine, control of small mammals and birds, beekeeping, agricultural engineering, farm and resource economics, and pest control.

AGRICULTURE
http://point.lycos.com/reviews/database/stag.htm
This site has links to online agricultural resources and also rates them according to content, presentation, and experience.

AGRICULTURAL GENOME INFORMATION SERVER
http://probe. nalusda.gov:8000/index.html
This site, sponsored by the U.S. Department of Agriculture and the University of Maryland, has links to information on plants, livestock animals, and other organisims, plus newsletters, journals, and other publications. You will also find links to FTP and Gopher resources.

AGRICULTURAL OUTLOOK
http://www.econ.ag.gov/epubs/pdf/agout/ao.htm

AGRICULTURE RESOURCES
gopher://gopher.mountain.net/11/Agriculture
This Gopher site is a comprehensive guide to agricultural resources on the Internet.

FOOD REVIEW
http://www.econ.ag.gov/epubs/pdf/foodrevw/
   foodrevw.htm

HORTICULTURE IN VIRTUAL PERSPECTIVE
http://hortwww-2.ag.ohio-state.edu/hvp/HVP1.html
This award-winning site is maintained by the Department of Horticulture and Crop Science at Ohio State University. Stop by to find links to valuable online horticultural resources on the Net, including a plant dictionary and student internships.

INFORMATION SERVICES FOR AGRICULTURE
http://www.aginfo.com/agsearch.html
This page has a search tool to locate agricultural information from a database of over 10,000 documents.

INVERIZON INTERNATIONAL
http://www.inverizon.com
Inverizon International Inc. is involved with the discovery, development, and delivery of innovative technology for improved food, feed, and fiber production. Their well-done site is an excellent place to visit if you want to know about a business in crop production. You will find links to national and international agricultural companies, as well as to agri-biotechnology information.

LIFE ON THE FARM
http://web2.airmail.net/bealke
What is it like to live on a farm? Visit Chuck Bealke's virtual farm and get his bi-weekly remembrance or comment on farm life—much of it based on farming (soybeans, wheat, corn, hay & polled herefords) west of St. Louis. Links to many agricultural online resources.

MAPQUEST
http://www.mapquest.com
As you travel in agri-cyberspace, you may find a virtual world atlas helpful. At this site you will find an interactive street guide with access to maps anywhere in the world. This feature also provides city-to-city driving directions for cities in the continental United States, Canada, and Mexico.

NATIONAL AGRICULTURAL LIBRARY (NAL)
http://www.nalusda.gov
The NAL is part of the Agricultural Research Service of the U.S. Department of Agriculture and is one of four national libraries in the United States. NAL is a major international source for agriculture and related information.

RURAL CONDITIONS AND TRENDS
http://www.econ.ag.gov/epubs/pdf/rcat/rcat/htm

RURAL DEVELOPMENT PERSPECTIVES
http://www.econ.ag.gov/epubs/pdf/rdp/rdp.htm
The Economic Research Service of the United States Department of Agriculture has recently made these publications available (Adobe Acrobat) [.pdf] format only). Agricultural Outlook, the monthly short and long term commodity outlook publication is now available graphics and charts. Food Review, issued three times a year, deals with "trends in [US] food consumption, food assistance, nutrition, food development, food safety, and food product trade.

**UNITED STATES DEPARTMENT OF AGRICULTURE
   HOME PAGE**
http://www.usda.gov

URBAN LAND USE: THEORY AND PRACTICE

OBJECTIVES

- To explore the relationship between urban growth and development
- To explain how the process of city growth operates under free-market conditions
- To extend von Thünen's model to urban land-use configurations
- To introduce land-use models that describe the spatial dispersion of activities in cities
- To help show that the free market for space in the metropolis has produced a pattern of sprawl and social problems

The heart of Amsterdam, the Netherlands. *(Photo: Guido Alberto Rossi, The Image Bank)*

A city is a built environment—a tangible expression of religious, political, economic, and social forces that houses a host of activities in proximity to one another. Cities, the foundation of modern life, represent humanity's largest and most durable artifact (Vance, 1977). They are living systems—made, transformed, and experienced by people.

Although the world pattern of cities is primarily the result of events triggered by the 19th-century Industrial Revolution, the city originated thousands of years ago. The first cities emerged in the Mesopotamian area of the Middle East about 7,000 years ago. Cities also developed early in the Nile Valley (about 3000 B.C.), in the Indus Valley (by 2500 B.C.), in the Yellow River Valley of China (by 2000 B.C.), and in Mexico and Peru (by A.D. 500). The raison d'être of cities from the start was to exchange goods and services with surrounding communities. As urbanization spread from its ancient hearths, it was incorporated into the cultures of various regions.

The manifestations of the urban process display dazzling diversity. Because the historical antecedents of modern patterns of daily living differ from one part of the world to another, the structure of the city differs from region to region. For example, North American cities contrast strikingly with those of Europe or Asia. North American cities are largely the creation of the last 200 years, a period of free-market capitalist activity politically based on the concepts of democracy and a relatively egalitarian spirit. Most European cities have grown since the start of the 19th century, but the tradition of privatism has not been the only structural influence on their urban growth. For the majority of European cities, other socioeconomic structures (e.g., feudalism, absolutism, mercantilism), never significantly present in North America, have also played a role.

In Europe, urban life began more than 2,000 years ago. Few European cities were created on virgin territory; most evolved from rural settlements. Some European cities existed before the growth of the Roman Empire. Apart from their own city-states, the Ancient Greeks were responsible for the foundation of other Mediterranean cities such as Naples, Marseilles, and Seville. By the end of the Roman Empire, many of Europe's largest present-day cities had been established.

From the fall of the Roman Empire to the early modern period, cities in Europe grew slowly or not at all. They ceased to be important during the period loosely referred to as the Dark Ages, a time when long-distance trade and rural-urban interaction drastically declined. Cities revived from the 16th century onward with the pursuit of profit in a period of incipient and later burgeoning capitalist economic activity. In commerce, a new middle class developed, and the revolution in the countryside reduced the peasant class and helped establish the working class. The accumulation of capital, the growth of new social classes, the use of inexpensive labor in the colonies, and scientific and technological breakthroughs destroyed the feudal fetter on production and created a new function for the city—industrialization. The emergence of the *industrial city*, a product of capitalism, resulted in lower transportation and communication costs for entrepreneurs who needed to interact with one another; hence, most commercial and industrial enterprises concentrated in and around the most accessible part of the city—the *central business district* (CBD).

During the period of urban decline and rebirth in Europe, the urban process of the rest of the world exhibited different patterns. Before A.D. 1500, Europe was a mere upstart in a world system that included major interlocking subsystems of central places stretching from the Mediterranean region to China. These subsystems were dominated by cities such as Constantinople, Baghdad, Samarkand, Calcutta, and Hangchow that had greater continuity and played a more permanent role in the world economy than their European counterparts did. Not until the commencement of European colonization were the urban civilizations of Asia, Africa, and the Americas threatened. Centuries of European penetration and occupation resulted in the growth of many cities that owe their origins to colonial foundations or to trading requirements.

Eventually, *colonial cities* dominated the urban patterns of Africa, Asia, and Latin America. Political independence and the development of the new international division of labor allowed underdeveloped countries to experience a transformation of the urban process as profound as that in 19th-century Europe and North America. Indeed, urban growth is now occurring more rapidly in the underdeveloped world than it did in Europe during its period of fastest growth in the late 19th century. During the second half of the 20th century, increasing affluence and the technologies of mass transportation and modern communications began a trend toward urban decentralization in Europe and North America.

Waves of change have washed through cities, remodeling and redefining their shapes and details, but rarely have the traces of their historical legacies been completely obliterated. The legacy of history is of immense importance to economic geographers who study cities and attempt to find solutions to urban problems and crises. Questions of interest to the economic geographer include the following: What types of societies and associated modes of economic exchange give rise to cities? What economic factors account for cities? What are the most vital influences on urban structure? What are the issues at the core of the urban process? Answers to these questions are the concern of this chapter.

CITIES AND SOCIETIES

Basic Forms of Society

Cities require the existence of a particular type of society in order to grow and develop. In the context of our discussion, a *society* refers to a group of people organized around a self-sufficient operating system that outlives any individual member. To maintain conditions of self-sufficiency, human groups must have forms of social organization capable of producing and distributing goods and services.

We can identify three main types of societies with associated forms of economic exchange (Fried, 1967). First, *egalitarian societies* are established through voluntary cooperative behavior. The economies of these societies are dominated by *reciprocity*, of which market barter is an example. Reciprocity involves trading without the use of money in a mutually beneficial exchange of goods.

Second, there are *rank societies*, examples of which are tribal and feudal societies. The economy of a rank society is dominated by *redistribution.* For example, African chiefs would exact gifts in kind (goods or commodities) or money from their tribes according to custom. Under feudalism in Europe, serfs, who were bound to the land of some estate, owed the overlord (by tradition, or by force if necessary) food, labor, or military allegiance.

Third, there are *stratified societies*, in which members do not have equal access to the resources that sustain life. The economies of these societies are dominated by *market exchange.* A market-exchange society adapts to scarcity by selling goods and services at a price. Pricing is the mechanism that connects the economic activity of large numbers of individuals and controls many decentralized decisions. Market exchange facilitates division of labor, specialization of production, and technological and organizational advances. It produces wealth for society out of scarcity, but often at the cost of even greater scarcity for the already poor. Socially created scarcities cannot be eliminated in market-exchange economies.

Cities do not evolve in egalitarian societies dominated by reciprocity. However, reciprocal forms of interaction, such as the mutually beneficial exchange of goods and services, do occur in cities. Everyday examples of reciprocity in an urban setting are exchanges among neighbors: lending a snowblower, gossiping, helping out when life crises occur.

Cities evolve in societies that can organize the exploitation of a surplus product. Rank and stratified societies have an especially suitable hierarchical structure to extract, appropriate, and redistribute a socially derived surplus product. A *social surplus product* is the part of the annual product of any society that is neither consumed by the direct producers nor used for the reproduction of the stock or the means of production. In rank and stratified societies, the social surplus product is appropriated by the ruling group. Surpluses are extracted from outside the confines of a city, as in the case of agriculture, and from inside a city, as in the case of manufacturing.

Stratified societies provide the most favorable conditions for the growth of cities. These conditions include unequal access to resources that sustain life, socially created resource scarcities, and institutions of market exchange. Except in countries that claim to have socialist economies (e.g., North Korea, China, and Cuba), contemporary cities exist in stratified societies.

Transformation of Market Exchange

Prior to the Industrial Revolution, market exchange was an appendage to the redistributive economy of the rank society. Under feudalism, European cities were usually extensions of the personalities of those who governed them. For example, Venice, Italy, was the city of the doges. Located at the seaward margins of the marshy Po Delta, Venice became one of Europe's most important

A canal in the merchant city of Bruges, Belgium. This Hanse town was the north European counterpart to Venice during the mercantile period. *(Photo: Marvullo, Belgian National Tourist Office)*

centers of manufacturing and long-distance trade during the Middle Ages and early modern period. The dominant economic institutions of Venice and other European towns were the *guilds*—craft, professional, and trade associations. In Europe, before the Industrial Revolution, anyone who wanted to produce or sell any good or service was required to join a guild, which regulated members' conduct in all their personal and business activities. With the Industrial Revolution came a steady penetration of market exchange through the fabric of society. Cities ceased to be reflections of individual rule; they became instruments of industrial growth.

Individual capitalism was the hallmark of the early stages of the Industrial Revolution. Indeed, the period from the mid-1840s to 1873 has been called the golden age of individual or competitive capitalism. Under competitive capitalism, cities registered high rates of industrial innovation and prodigious increases in productive power. Their standards of achievement were based on industry and technology. If both prospered, the city was considered to be good. The best city was the busiest one—the one that was growing most quickly and recording the largest increases in bank clearings. However, these producer cities were ugly creations and horrifying environments for the laboring poor. The British coined the term *Black Country* to refer to the grimy industrial cities of the English Midlands. Interestingly, the industrial city, which functioned as a workshop for production and capital accumulation, does not attract 20th-century tourists in search of urban beauty. Tourists avoid rich industrial cities such as Toledo, Ohio, and overrun poor preindustrial cities such as Toledo, Spain.

In the late 19th century, capitalism took on a different form. There was a drift from competitive capitalism to monopoly, or corporate, capitalism. Through the elimination or absorption of small competitors, large industrial and financial corporations emerged. Their emergence diminished the community of competition. In today's developed countries (and in many underdeveloped countries) important areas of manufacturing and strategic industries are dominated by a relatively small number of multinational corporations.

Large corporations have had a major influence on the 20th-century Western city. Corporate administrative buildings and home plants dominate skylines and extensive land areas. For example, the organizational headquarters of such corporations as Standard Oil of Indiana and Sears Roebuck have helped to shape the image of Chicago.

The geography of contemporary Western cities has also been affected by the need of corporate enterprises to find ways to absorb their surpluses. An American example is the corporate penchant for disposing of their surpluses through urban renewal projects. Funds are sometimes used in projects sponsored by local governments to replace run-down, low-income housing with luxury office and residential buildings. Corporations and the federal government have poured resources into urban renewal projects in cities including Atlanta, Boston, Dallas, Houston, Minneapolis, New York, and Philadelphia.

In addition, U.S. cities have been affected by the need for corporate enterprises to increase demand for their products and services. *Need creation*, a process in which luxuries are marketed as necessities, is exploited by corporations through daily appeal to potential customers. Need creation also operates as a consequence of urban spatial organization. For example, low-density metropolises, such as Kansas City, Dallas, Houston, and Los Angeles, make a car, or two cars, a necessity. Residents are literally forced to drive the cars that the automobile industry produces.

Relative Importance of Different Modes of Exchange

All three modes of economic exchange (*reciprocity, redistribution,* and *market exchange*) operate in most cities, but the emphases have varied with time. The cities of medieval Europe reflected the dominant influence of redistribution, but market exchange also operated. In general, cities in the zone between the North Sea and Italy were more supportive of the market than those on the margin of the region were. In the commercial area, as the old feudal ordering of society declined, the social and political influence of merchants grew. When commerce was permitted to operate freely, as in Venice and Florence, the market became a notable feature of city structure. Yet the disposal of wealth through the construction of massive cathedrals, public buildings, and universities emphasized the preeminence of cultural values over worldly economic concerns. The imprint of cultural values is also unmistakable in other places with a long history of urbanization. For example, Lahore, the cultural focus of Islamic Pakistan, founded in the first or second century A.D., is adorned with palaces and mosques.

From the late medieval period onward, the importance of market exchange increased. Large commercial cities such as London, Amsterdam, and Antwerp boasted the triumph of the market over redistribution. By the 19th century, market exchange dominated life in western Europe and in its overseas progeny.

In North America, where the medieval order did not exist, the 19th-century city was an expression of economic influence. The power of the American city did not rest in the nominal government; rather, it was based in the dominant economic institutions, which were usually industrial establishments. Nonetheless, city government did play a redistribution role that grew more important as the Industrial Revolution pro-

Typical Moscow apartments at night. There is a light burning almost everywhere because entertainment possibilities at night are very rare. In this city of 10 million inhabitants, only hard currency facilities are open after 11 P.M. Apartments are typically small and noisy. Single-family dwellings are very rare. *(Photo: Wim Van Cappellen, Impact Visuals Photo & Graphics, Inc.)*

gressed. Urban bureaucracies collected taxes and provided a range of public services. Although much less prominent than the other modes of integration, reciprocity existed at every level of society, especially in working-class neighborhoods, where it provided residents with a degree of social solidarity and security.

In the modern Western city, the role of reciprocity is abridged. Market exchange prevails, but it is challenged by redistribution. Because of the growth of big business and government, redistribution is more important now than it was in the 19th century. The hierarchical status of employees in modern institutions is reminiscent of the structure of rank societies. Corporate and government bureaucracies have become major agents of redistribution in urban areas. For example, governments appropriate resources and return them to the populace at large in the form of various public services, welfare programs, public projects, and subsidies. The provision of funds for elementary schools and high schools is a good example of government redistribution at the local level. Most of the money for schools, which usually represent the largest expense for local government, is generated from property taxes.

THE PROCESS OF CITY BUILDING

What general forces attract business and industry to cities? In a purely competitive market, we can account for the attraction of firms to cities in terms of two opposing forces. These forces are *scale economies* and *transportation costs*. The influence that they exert on city building becomes apparent as we survey the classical economic principles that relate to the production and cost behavior of the single firm.

Cost Behavior of the Single Firm

Firms have short-run and long-run planning periods relative to the production function. The *production function* is a mathematical statement of the relationship between the inputs (land, labor, and machines) used by a firm and the flow of output (goods and services) that results in a particular time period (week, month, year). By definition, *long run* is a period long enough to permit a firm to vary the quantity of all the inputs in its production function; *short run* is a period short enough for at least one input to be fixed in amount and invariable.

Suppose that we have a simplified production function consisting of two inputs—machines and human labor. How would we illustrate the firm's production cost? A common device in economic analysis is to use the relationship between the level of output and the cost *per unit of output* (*average cost*). Some basic data are presented in Table 6.1 for a hypothetical firm manufacturing golf balls.

◈

T a b l e 6.1
Cost Relationships for a Firm

Labor (1)	Machines (2)	Output (3)	Total Cost (4)	Cost per Unit of Output (5)
10	5	6,000	$1200	$0.20
20	10	12,000	2400	0.20
40	20	24,000	4800	0.20
80	40	48,000	9600	0.20

The relationship between the inputs (labor and machines, or Columns 1 and 2) and the output (golf balls, or Column 3) is given by the firm's production function. If 10 units of labor are employed (10 people for 1 day each) and 5 machines are used (1 day each), the output is 6,000 golf balls. If each day of labor costs the firm $100 per employee and each day's use of one machine costs $40, the firm's total cost of producing 6,000 golf balls is $1,200 (Column 4). Cost per unit of output, or average cost, is $0.20 (Column 5).

In this case, if the firm decides to expand its output, its production function tells it that the optimal way to do so is to increase both inputs by the same proportion (e.g., doubling both, tripling both, increasing both by 50%, etc.). Suppose that the firm wants to double its output (to 12,000) and does this by doubling its use of both labor and machines (to 20 and 10, respectively). If the price of each input remains the same, the firm's total costs double to $2,400. However, the cost per unit of output (average cost) remains constant at $0.20. In this case, continued increases in output have the same effect. As output doubles, total costs double and cost per unit of output remains unchanged at $0.20. This phenomenon is called *constant returns to scale*. Figure 6.1a shows the relationship between average cost and the level of output. Economists refer to this graph as the *long-run average-cost* (*LRAC*) curve.

Constant returns to scale is not the only possibility. For some firms in some industries, output rises by a greater proportion than inputs are increased (*increasing returns to scale*, or *economies of scale*), and average cost decreases as output increases. In other cases, output rises by a smaller proportion than the increase in inputs (*decreasing returns to scale*, or *diseconomies of scale*), and average cost rises as output rises. These cases are presented and discussed in the next section of this chapter and in more detail in Chapter 10.

Two types of agglomeration economies (described in more detail in the next section of this chapter) can have constant returns to scale: *Localization economies* refer to declining average costs for firms as the output of the industries of which they are a part increases (Figure 6.1b). *Urbanization economies* refer to declining average costs for firms as cities increase their scales of activity (Figure 6.1c).

Scale Economies and Diseconomies of the Single Firm

SCALE ECONOMIES

Scale economies are a key for understanding why economic activities concentrate in cities. The concept refers to a set of conditions in which average costs of a firm decrease as the scale of production increases. Changing the scale of production means altering proportionately all inputs used in production. Costs of production

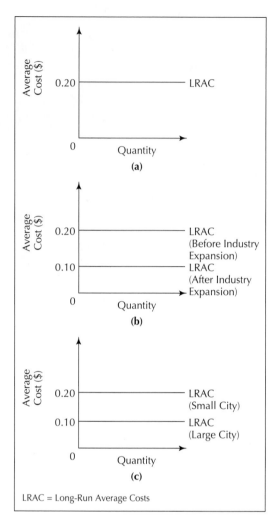

FIGURE 6.1
Constant returns to scale for a firm: (a) no agglomeration or internal scale economies; (b) localization economies; (c) urbanization economies.

under these conditions are called long-run costs. Dividing long-run costs by output yields the long-run average costs of a firm.

Two sets of forces influence a firm's long-run costs: internal scale economies and external scale economies.

INTERNAL SCALE ECONOMIES

Internal scale economies are subject to direct management control. Managers may be able to take advantage of labor economies (more efficient labor specialization), technical economies (larger, more efficient machines), market economies, and managerial economies. Internal economies come into play because many production factors (inputs) are indivisible and can be used more efficiently at larger scales of output. Thus, the concept of internal scale economies refers to cost-reducing changes that tend to lower the average costs of firms as they grow in size (Figure 6.2a).

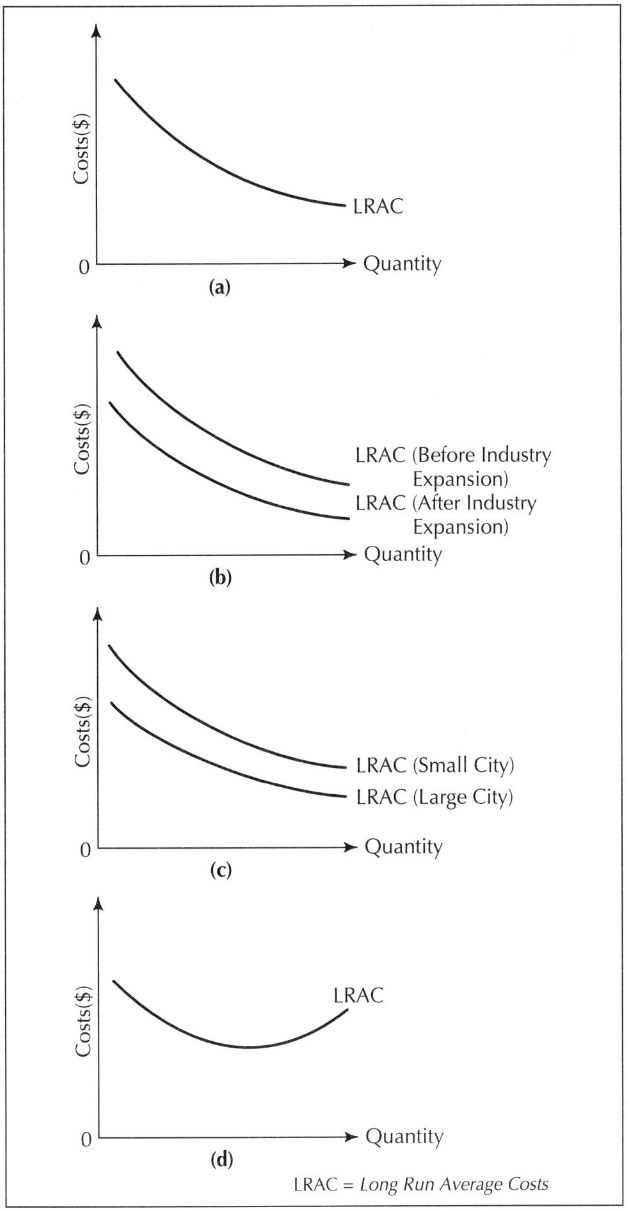

FIGURE 6.2
Scale economies and diseconomies for a firm: (a) internal economies; (b) localization economies; (c) urbanization economies; (d) internal economies and diseconomies.

EXTERNAL SCALE ECONOMIES

External scale economies represent two forms of *agglomeration economies*: (1) localization, or industry, economies and (2) urbanization economies. *Localization economies* refer to declining average costs for firms as the output of the industries of which they are a part increases (Figure 6.2b). These economies stem from benefits that industries derive within restricted geographic areas, such as the development of a large labor pool with skills needed by the industry.

Urbanization economies refer to declining average costs for firms as cities increase their scales of activity (Figure 6.2c). Cost reductions stem mainly from technologies that stimulate production on a scale that can be achieved only with firm specialization; that is, when plants perform only one or a few functions in the overall production process. As a result of transportation costs, firm specialization leads to geographic clustering, which in turn promotes more geographic specialization and concentration. The garment industry of New York and London and the metal trades of Ohio and the English Midlands are outstanding examples of geographic clustering of specialized firms.

J. Vernon Henderson (1988) extensively investigated localization and urbanization economies. He used sophisticated statistical analyses to estimate the relationships between output per worker (labor productivity) in particular industries in urban areas and two variables: (1) the total output of the same industry in urban areas and (2) the total population of urban areas.

The first relationship provides evidence of localization economies. If higher labor productivity in, say, garment manufacture occurs in urban areas having a high rate of garment output, then, other things being equal, garment manufacturers in urban areas near other firms in the same industry will have lower costs per unit of output. On the other hand, a positive (direct) relationship between labor productivity in an industry and total urban population suggests that firms in that industry reap lower unit costs from being located in larger urban areas, regardless of the types of firms located there—an urbanization economy.

Table 6.2 shows that localization economies are evident in many industries in each country. However, it is clear that the results are not identical in all three countries. Table 6.3 shows the significant cases of urbanization economies. This second category of external economies is far less common than localization economies.

Results must be interpreted with caution, however. Data limitations and significant statistical problems exist with any studies of this sort. Furthermore, as pointed out by Henderson, these studies do not deal with the possibility of external economies being realized from the scale of *related* industries as opposed to the industry itself (a localization economy) or the scale of the whole urban economy (an urbanization economy). Nonetheless, the finding of limited urbanization economies is significant. It suggests that the productivity-enhancing (cost-saving) advantages in urban areas arise from concentrations of firms in the same industry rather than from large urban size.

SCALE DISECONOMIES

Internal and external economies accrue only up to a point. Consider internal diseconomies. At first, firms

◼

T a b l e 6.2

Localization Economies

United States/Canada	Brazil	Japan
Auto manufacturing	Iron and steel	Iron and steel
Aircraft manufacturing	Nonelectrical machinery	Nonferrous metal
Apparel manufacturing	Transport equipment	Nonelectrical machinery
Computers	Chemicals (including petrochemicals)	Electric machinery
Primary metals	Textiles	Transport equipment
Electrical machinery		Precision machinery
Machinery		Textile products
Petroleum		Apparel
Pulp and paper		Pulp and paper products
Food products		Metal products
Wood products		Rubber and plastics
		Chemicals

◼

T a b l e 6.3

Urbanization Economies

United States/ Canada	Brazil	Japan
Nonmetallic minerals	Printing and publishing	Textile products
Brewing		Food products
Printing and publishing		Lumber products
Food catering		Furniture
		Printing and publishing

experience cost-reducing internal economies, but after a certain scale of production is attained, it becomes impossible to vary proportionately all inputs used in the production process (Figure 6.2d). Management, for example, does not grow proportionately as firms expand. Managers are forced to spread themselves ever more thinly over wider areas of decision making. Decreasing returns set in, with an eventual decrease in efficiency.

External diseconomies must also be accounted for. For example, urbanization diseconomies—rising average costs accompanying an increasing scale of activity within a city—may arise for at least three reasons. First, firms may experience higher costs as a result of land scarcity. Second, competition for labor and high living costs may force firms to pay workers higher wages. Third, firms may encounter transportation congestion, parking problems, pollution, crime, and financial difficulties.

Transportation Costs of the Single Firm

There would be no need to worry about transportation costs if resources were ubiquitous, if production technology were the same everywhere, and, of course, if movement were instantaneous and free. But resources are rarely ubiquitous, technical aspects of production are highly variable, and movement over geographic space always encounters resistance. Transportation costs represent the alternative output that we relinquish when we commit inputs to the movement of people, goods, information, and ideas over geographic space. They are the swimming pools and libraries that must be surrendered for roads and railways.

What effect does the cost of overcoming the friction of distance have on the location of economic activities? Because transportation costs are directly correlated with distance, profit-maximizing firms select sites close to their inputs, to other firms, or to consumers who buy their products. Workers decide to live near their places of employment. Accessibility pays off in the form of transportation-cost savings that concentrate firms and workers. Moreover, incentives to concentrate activities are intensified by existing transportation systems, which provide a high degree of access to only a limited number of geographic areas.

Economic Costs and City Building

Having explored the meaning of scale economies and transportation costs, we are now in a position to see how these forces influence city building. First, let us consider an area in which people are dispersed geographically and a production technology in which economies of scale are absent. As the sizes of firms increase, they experience only constant returns to scale; that is, their output increases proportionately as the amount of all inputs increases. Firms could choose to locate in proximity to one another, but if there were no scale economies, such a choice would result in rising unit costs and falling profits. Geographically concentrated production would increase the distance, and hence the cost of transporting goods and services, from producers to consumers. With scale economies absent, there would be no incentive to profit-maximizing firms to concentrate production.

Now let us alter the situation. Suppose that a new technology is developed for the garment industry that allows a specialized machine to make buttonholes in coats cheaply and in large quantities (Hoover, 1984, p. 110). Suppose also that no single coat manufacturer can keep the machine busy all the time, but if the coat manufacturer were located near other coat manufacturers, their total demand for buttonholes would make it profitable for an entrepreneur to invest in such a machine and rent time on it to each manufacturer. Geographic concentration of garment firms would reduce their cost per unit (a localization economy).

If consumers remained dispersed geographically, such concentration would again increase transportation costs. But this time, transportation-cost increases could be offset by a reduction in another cost (e.g., skilled labor). If so, profit-maximizing garment firms would locate near one another and their employees would live nearby. The buttonhole firm and other suppliers of goods and services to these households would also locate nearby. A city would be created. In this way, city building is cumulative; concentration leads to more concentration.

Cities grow when they gain firms and population. To illustrate this process, geographers sometimes use the principle of *cumulative causation*. This principle states that increases in urban economic activity create an increase in population. In the case of urban decline, the reverse holds true. Cities decline when they lose firms and population, conditions that create negative cumulative causation.

INTRAURBAN SPATIAL ORGANIZATION

We have seen that activities locate in cities for sound economic reasons. But what factors influence where the various activities will locate in a city? Why are some parts of a city zoned for commercial land use, others for industrial use, and still others for mixed single- and multifamily housing? Classical urban location theory provides one answer. This theory concerns the private land-use decision process; it is based on von Thünen's concept of location rent. Assumptions embodied in urban location rent theory are these: First, the central business district (CBD) is the most economically productive location because of its concentration of transportation facilities. Second, rent falls to zero at the fringe of the city (it falls to a level of agricultural value). Third, firms are competitive price takers, not price makers. Finally, cities exist in competitive market-exchange economies without government and social classes.

Although the concept of differential rent is a useful arranger of land uses, a word of warning is in order. In capitalist societies, the pursuit of the most profitable use of urban land is inhibited, to varying degrees, by monopoly, societal class divisions, racial discrimination, and public authority. However, despite its deviations from real-world conditions, urban location theory is still used by many economic geographers to indicate and interpret problems of land use in cities.

The Competitive-Bidding Process

Classical location theory states that activities locate in cities according to the outcome of the *competitive-bidding process*. People willing and able to pay the highest price for a particular site win the competition and put the land to the economically highest and best use. Highest and best use, of course, can change as external market forces change. These forces include effective demand, public tastes and standards, and land-use regulations.

Why do users bid for particular parcels of land? In some instances, a user may value the inherent characteristics of a site. For a residential user, a scenic vista may be desirable; for a commercial user, nearness to potential customers may be crucial. In other instances, site attributes such as natural hazards may detract from the value of a tract to all potential developers or bidders.

Although part of every tract's value depends on site characteristics, relative location is usually more important. For particular activities, some land parcels may be more desirably located than others because they are more accessible; that is, they reduce users' transportation costs. For example, if accessibility to work is important to residential users, bid-rent curves reflect higher bids near places of employment.

Ceiling Rents

According to classical economic theory, the competition for the use of available locations results in the occupation of each site by the user able to derive the most utility or profit from it, and therefore able to pay the highest rent. The maximum rental that a particular user pays for a site is called the *ceiling rent*.

Consider a hilly tract of land, commanding excellent views, on the outskirts of a city. Ceiling rents for this parcel of land might be $9,000 for residential, $6,000 for retailing, and $5,000 for manufacturing. Clearly, this tract of land is likely to be sold or leased for residential land use. But if the tract owner thinks that the site is worth more than $9,000, the land will remain vacant.

Another tract is for sale a few kilometers away from the parcel put to residential use. It is relatively flat and located adjacent to a major highway. Ceiling rents might be $3,000 for residential, $10,000 for retailing, and $7,000 for manufacturing. In this case, the tract is sold or leased for retailing—say, a shopping center.

The Residential Location Decision

An important criterion for most people in selecting a home is accessibility to where they work in the city. The choice of a residential location depends in part on how much money a family can afford to spend on overcoming relative distance. For purposes of our discussion, we assume a single-centered city; that is, a city with only one center of employment—the CBD (central business district).

First, let's consider patterns of residential land use and the cost of commuting to work. A family's budget must account for living costs, housing costs, and transportation costs. Assume that people are either rich or poor. Poor families, who have little money to spend on commuting after living and housing expenses are deducted from their income, have sharply negative bid-rent curves. They attach much importance to living close to where they work. The only way the inner-city poor can afford to live on high-rent land is to consume less space. On the other hand, rich families have plenty of money to spend on transportation. Proximity of residential sites to places of employment is of little conse-

quence to them. They can trade access for agreeable lots away from the center of the city.

But now let's consider the effect of time costs on the residential decision. Time spent commuting is time that could be devoted to earning income. From this standpoint, distance is more critical for the rich, whose time is more valuable than that of the poor. Rich households have steep bid-rent curves, and they are located near the center of the city. Meanwhile, the poor, with shallow bid-rent curves, live farther away. This land-use pattern is characteristic of many Third World cities; the poor often live in peripheral squatter settlements, whereas the rich occupy high-rent city housing.

To this point, a single mode of travel to work has been assumed. A more realistic situation would involve two modes of travel—walking and driving. The steepest bid-rent curve is for the rich walking to work, followed by the poor walking, the rich driving, and the poor driving.

Although cost, time, and mode of travel to work have important implications for residential location, many other dimensions to accessibility, such as nearness to services, and general supply and demand factors must be considered. These and other factors influence bid prices. In every situation, however, the rich can outbid the poor. The results of the competitive-bidding process in market societies are always relatively advantageous to the rich and relatively disadvantageous to the poor.

HOUSING PRICE DIFFERENTIALS

Frederick Stutz and Art Kartman provide a method of explaining the basis of housing price differentials using the supply-demand framework (Figure 6.3). The approach is based on the idea that the actual price at which any item is sold is the outcome of a complex set of factors that determine the behavior of buyers and sellers. Demand factors are those that determine the number of potential buyers, the amount each wants to

Stamford Town Center, Connecticut. Beyond the "zone of better residents" is a broad commuter zone, an incompletely built-up area of small satellite towns and middle-class and upper-class residences, with major malls and shopping centers at the intersection of beltways. *(Photo: The Taubman Company)*

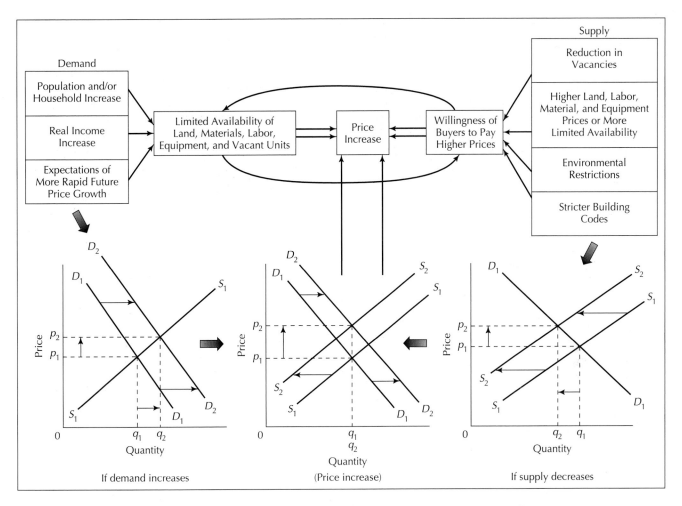

FIGURE 6.3
The economic geography of housing price increases in the United States. Supply and demand factors raise or lower the price.
(*Source: Stutz, 1994, p.329*)

buy, and the maximum price each is willing to pay. Supply factors determine the number of sellers, how much each will offer for sale, and the minimum price necessary to induce each seller to offer one item for sale. Except in special cases, neither set of factors is the sole determinant of the price at which an item is sold. For example, even if there is only one seller, a monopolist, the ability of the monopolist to set the price is limited by how much money or income buyers have ("what the market will bear").

Key *demand factors* in the housing market are population (both the number of people and the household size); inflation-adjusted, or "real," income per person or per household; and, in the case of owner-occupied housing, the home buyers' expectations of the future change in home prices (*speculation*). Important *supply factors* are the availability of items needed in building new housing units (land, labor, and construction equipment and materials) and construction standards or building codes. Availability of financing may be viewed either as a demand factor (long-term mortgages) or as a supply factor (construction loans).

As an example of the effect of a demand factor, consider an increase in the number of households in an area as a result of in-migration. If vacant housing of the desired kind is limited, the price of existing housing will be bid up. This will induce new construction and create new demand for land, labor, and raw materials. If these items are abundant, their prices should remain constant. However, if supplies are limited, their prices will be bid up. This increase in builder cost will lead to even further increases in housing prices, with the amount depending on how willing home buyers are to pay the higher prices. Increases in income and a heightened expectation of future increases in home prices will have similar effects.

On the supply side, consider a case of vacant land where the local authorities suddenly change the zoning from residential to nonresidential. The reduction in available supply will raise land prices and, hence, builder costs. The ability of builders to pass on the higher costs depends on the willingness of buyers to pay them. The more willing the buyers are to pay, the higher the demand pressures remain on land, labor, and so forth, and the more limited these items are relative to the demand, the greater the upward pressure on home prices. Similar responses occur for reduced vacancies; higher labor, equipment, and material prices; and stricter building codes.

Better access to amenities—the CBD, jobs, shopping, schools, and high-rent districts adds to the willingness of buyers to pay higher prices and to overall demand. The three graphs at the bottom of Figure 6.3 describe how supply and demand curves shift upward, thus raising the price of housing. At the bottom left, the demand curve shifts to the right, raising the price of housing, whereas the supply curve remains fixed. The graph at the lower right describes a situation in which the supply curve shifts upward to the left, raising the price of housing, whereas the demand curve stays fixed. When these processes occur simultaneously, the center graph describes an even larger increase in price. The diagram shows average U.S. home prices in 1996 (Figure 6.4).

Site Demands of Firms

Firms also compete for urban space, but because of the nature of their activities, the criteria that they use in making locational decisions are not the same as those for households. Given a rational market-exchange economy, firms want to maximize profits, and households want to maximize satisfactions. If intraurban accessibility is important to sales, firms should be willing to make higher bids for locations that are central to all potential customers. Increasing distance from the more productive locations in the urban area should increase costs to customers and therefore reduce sales. This would mean decreasing revenues, and, hence, lower profits. For *nonbasic firms* this is the case, but for basic firms it is not. *Basic firms* export what they produce to surrounding areas; nonbasic firms sell goods or provide services to city residents and businesses.

Revenues of nonbasic firms decline as the firms move away from downtown or from other central lo-

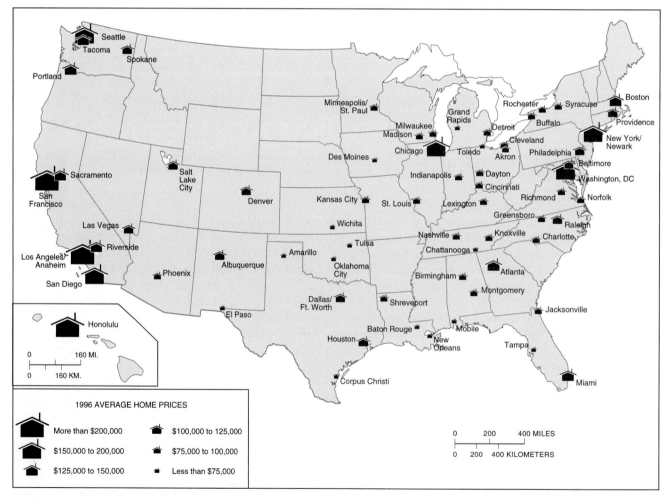

FIGURE 6.4
Home prices in the United States, 1997. The 71 largest cities are shown. Cities of megalopolis and subtropical Pacific Coasts in California and Hawaii show the highest prices. *(Source: Modified from Stutz and Kartman)*

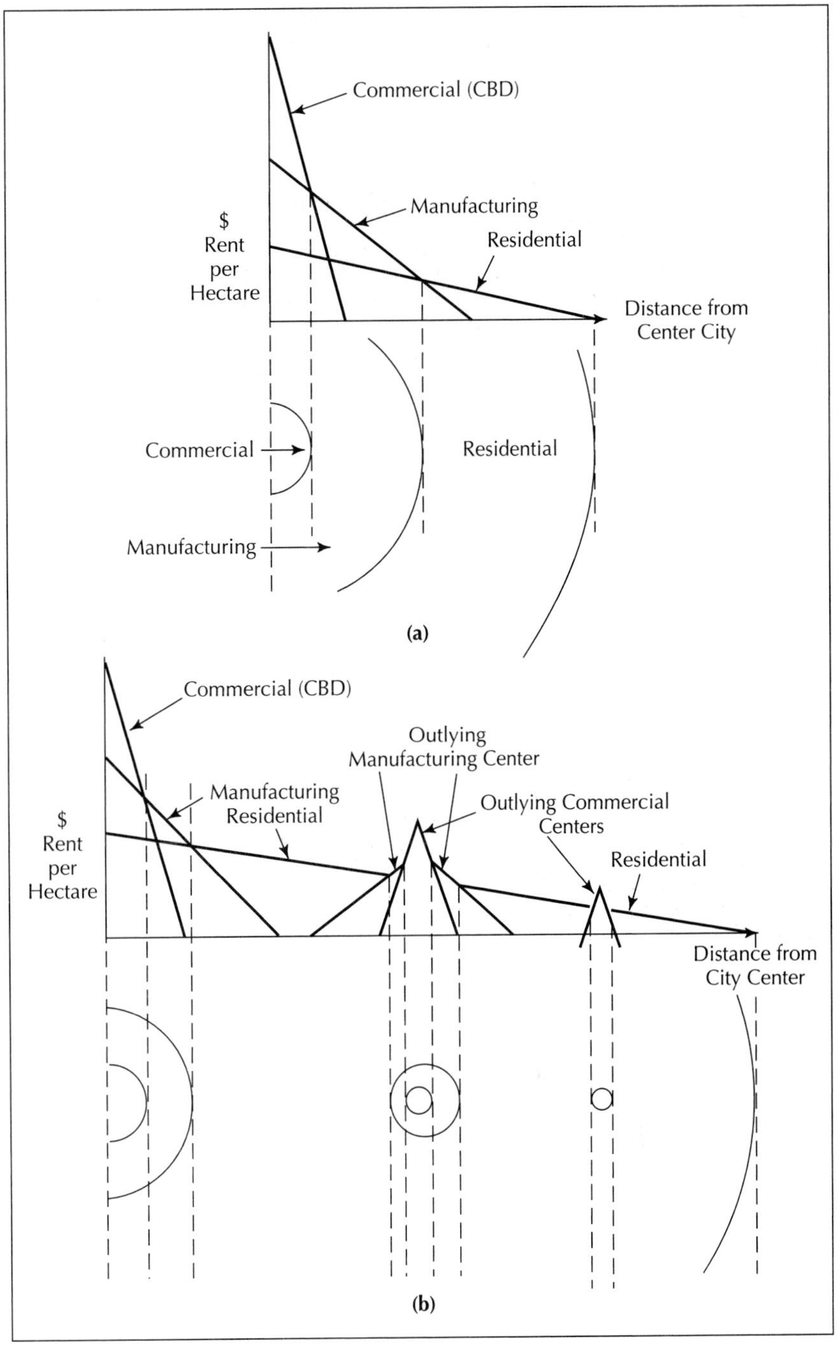

FIGURE 6.5

Multiple land-use patterns: (a) Single-centered city. Commercial space, such as that found in the CBD, has a higher intensity of land use because of the potential mark-up of its goods and services and the volume of business that it can produce. Mark-up times volume, called turnover in the business community, results in a relatively high ceiling rent for commercial land uses. The bid rent curve for manufacturing is less steep because manufacturing is not able to generate as much revenue off each square unit of land. It can produce more off each unit of land than can residential; therefore, it has a higher ceiling rent. The residential bid rent curve is even less steep because of a lower ceiling rent at the center of the city, as well as a lower use intensity on a low sensitivity to travel distances. (b) Multicentered city. In a multicentered city, more than one location affords the user high levels of accessibility to the market. While the commercial CBD retains the highest level of accessibility and, therefore, the highest ceiling rents, outlying shopping centers or manufacturing centers, perhaps at the intersection of beltways in suburbia, also provide relatively steep bid rent curves. Because exposure to the regional market increases at such accessible locations, the opportunity for volume and turnover increases, increasing the maximum utilization of the land by commercial and manufacturing entrepreneurs.

cations. For example, department stores have high revenue requirements for profitable scales of operation. Traditionally, these firms required access to all parts of a city. They were willing to pay high rents for downtown sites where intraurban transportation lines converge. Many nonbasic firms, such as grocery stores and beauty parlors, have much lower revenue requirements. Their revenue conditions allow smaller geographic scales of operation in the city.

Location within the city has little effect on the revenues of basic firms, but more of an effect on some of their costs. Firms requiring a lot of space might purchase sites at marginal locations where land costs are lower. Those drawing labor from residential areas throughout the city might be willing to pay high rents for central locations. Movement away from central lo-

cations could result in higher wage bills. To attract necessary labor, firms might be forced to increase wages to compensate for higher journey-to-work costs.

Market Outcomes

We have looked separately at locational decisions of households and firms as they deal with distance frictions. Now let us fashion a model, analogous to von Thünen's crop model, that shows how a multitude of individual decisions combine to produce a pattern of urban land use in which rents are maximized and all activities are optimally located.

Consider three land-use categories: manufacturing, commercial, and residential. Figure 6.5a shows distinctive bid-rent curves for the three types of land use in our hypothetical single-centered city. Commercial activities that require the most productive central sites have steeply sloping rent gradients. Manufacturing firms have shallower bid-rent curves. They cannot afford to pay the high costs of a central location. Residences have gently angled bid-rent curves and are relegated to the outer ring, where land prices are lower. We can complicate matters by considering a land-rent profile in a multicentered city (Figure 6.5b). Apart from secondary peaks, perhaps at intersections of main traffic routes, the rent gradient still shows price bids declining outward from the CBD.

CLASSICAL MODELS OF URBAN LAND USE

So far, our study of urban structure has been static. Now let us turn to an urban land-use model that helps explain city-building processes in North America. It is the concentric-zone model of Ernest Burgess (1925), a sociologist. This is a historical generalization about the layout of the city; therefore, no particular city fits this type exactly.

The Concentric-Zone Model

Burgess's *concentric-zone model* emphasizes centripetal forces that focus economic activity on the CBD, which was the dominant center of urban spatial organization in the industrial city. For simplicity, the model assumes a uniform land surface, universal accessibility, and free competition for space. Under these assumptions, cities expand symmetrically in all directions. Burgess suggested a sequence of five zones from center to periphery (Figure 6.6; see also color insert 2).

1. *The central business district.* This zone is the focus of commerce, transportation, and social and civic activity. It encompasses department stores, specialty shops, office buildings, banks, headquarters of organizations, law courts, hotels, theaters, and museums. Encircling the downtown retail district is a mixture of wholesaling and light-manufacturing operations and truck and retail depots.
2. *The zone of transition.* This area reflects residential deterioration. Older private homes have been subdivided into rooming houses. Mansions have been taken over for offices and light manufacturing (functional change). Abandoned dwellings have been torn down to provide space for urban renewal (morphological change). According to Burgess, this is the zone of slums, with their attendant disease, poverty, illiteracy, unemployment, and underworld vice.

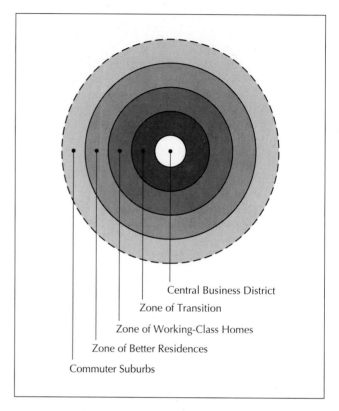

FIGURE 6.6
Burgess's concentric-zone model of urban structure. (See Modeled City Structure in color insert.)

3. *The zone of independent workers' homes.* This ring is characterized by decreasing residential density and increasing quality and cost of homes. It is inhabited by blue-collar workers who have "escaped" the zone of residential deterioration but who need to live close to work. It is regarded as an area of second-generation immigrants who have had enough time to save the money to buy homes of their own.
4. *The zone of better residences.* Still farther from the CBD, working-class residences give way to newer, more spacious single-family dwellings and high-rise apartment buildings occupied by middle-class families.
5. *The commuter suburbs.* Beyond the zone of better residences is a broad commuter area. It is an incompletely built-up area of small satellite towns and middle- and upper-class residences along railway lines and major highways.

The Burgess model suggests that as cities grow, resistances are encountered. Characteristic land uses of one zone exert pressure on future land uses of the next outer zone. This process is called *invasion and succession.* Residents of one zone try to improve their situation by moving outward into a zone of better housing units. New housing constructed at the edge of the city triggers a complex chain of moves. Dwellings vacated by the out-migration of middle- and high-income families are filled by lower-income families moving from the next inner zone. At the end of the chain, the working poor move out of the zone of transition, leaving behind the least fortunate families and abandoned housing units. The result is an inner-city slum. This *filtering process*, which exerts downward pressure on rents and prices of existing housing, enables lower-income families to obtain better housing. The major reason the filter-down process occurs is that the poor, with the strongest latent demand for housing, are the least able to afford new housing. In contrast, the rich can most easily afford to move into new housing and leave their old homes to others. The demand for high-income housing is generally elastic—a new demand generates a quick response from the private housing industry.

The Sector Model

The *sector model* takes into account differences in accessibility and, therefore, in land values along transport lines radiating outward from the city center. According to Hoyt's model, a city grows largely in wedges that radiate from the central business district. One wedge may contain high-rent residential; another, low-rent residential; and still another, industrial. Hoyt believed the contrasts in activity along various sectors usually became apparent early in the city's history, and continued to be marked as the city grew. (See color insert 2.4.)

The Multiple-Nuclei Model

The concentric and sector models describe single-centered cities. However, most modern cities have *multiple nuclei*: a downtown with satellite centers on the periphery. In 1945, Harris and Ullman described a model city that develops zones of land use around discrete centers. Their model of urban structure encompasses five areas: (1) the central business district, (2) a wholesaling and light-manufacturing area near interurban transport facilities, (3) a heavy industrial district near the present or past edge of the city, (4) residential districts, and (5) outlying dormitory suburbs.

Harris and Ullman recognized that the number and location of differentiated districts depends on the size of the city and its overall structure and peculiarities of historical evolution. They also gave reasons for the development of separate land-use cells. The pattern of multiple cells might result from specialized requirements of particular activities, repulsion of some activities by others, differential rent-paying ability of activities, and the tendency of certain activities to group together to increase profit from cohesion. (See color insert.)

MODELS OF CITY STRUCTURE IN DEVELOPING COUNTRIES

The classical models are based on North American experience and are not universally applicable. They are tied to a particular culture. Although the forces of urban change may eventually result in closer similarity of city structure in the non-communist world, attitudes toward density, land-use arrangements, open spaces, and architectural preferences will continue to vary. In addition, institutional factors—zoning laws, building codes, the role of government in the housing market—will vary with the culture and the level of technology. Models of the structural elements of cities in Latin

Rio de Janeiro—Favela de Roanhia. Here in Brazil, many favelas are on the outskirts of town. Open sewers line the streets, and residents live in tiny, illegally built huts, lacking sanitation and fresh water. The worst of these poverty-stricken areas include people so poor that they live on the streets, often in doorways or in cardboard boxes they carry with them. Because of total denuding of the landscape in search of firewood for cooking and warmth, when rains come, mud slides generally take lives. *(Photo: Alain Keler, Sygma)*

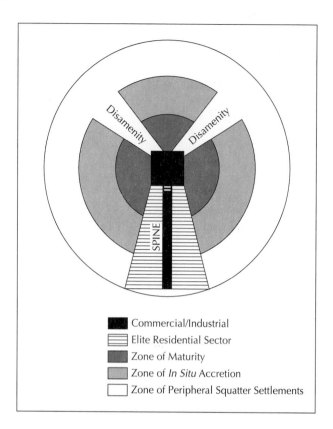

FIGURE 6.7
Model of Latin American city structure. *(Source: Griffin and Ford, 1980)*

America and Southeast Asia exemplify different patterns of land use in different regions.

Latin American City Structure

Ernest Griffin and Larry Ford (1980) proposed a model of the Latin American city. Blending traditional ele-ments of Latin American culture with modernizing processes, the framework of the idealized city is a composite of sectors and rings. The heart of the city consists of a vibrant and thriving CBD. A reliance on public transportation and nearby concentrations of the affluent ensure the dominance of the inner city, with a landscape that increasingly exhibits skyscraper offices and condominium towers.

Outward from the inner city is a *commercial/industrial spine* surrounded by a widening *elite residential sector.* The spine/sector is an extension of the CBD, characterized by offices, shops, high-quality housing, restaurants, theaters, parks, zoos, and golf courses, which eventually gives way to wealthy suburbs. Within the spine/sector, the centrifugal forces are similar to those operating in North American cities (Figure 6.7).

Three zones, which reflect traditional Latin American characteristics, are home to the less fortunate, who account for the majority of city residents. The *zone of maturity*, attractive to the middle classes, features the best housing beyond the spine/sector. Filtered-down colonial homes and improved self-built dwellings are common. In the *zone of in situ accretion,* the housing is much more modest, interspersed with hovels and unkempt areas. It is a transition zone between inner-ring affluence and outer-ring poverty. The outermost *zone of peripheral squatter settlements* houses the impoverished. Although this teeming, high-density ring looks wretched to a middle-class

Cut off from the city's mainstream and many of its services, the urban poor live in makeshift shelters. In many Third World cities, more than half the residents live in slums such as this squatter settlement on the urban fringe of Djakarta, Indonesia. *(Photo: World Bank)*

North American, the residents perceive that neighborhood improvement is possible; they hope, in time, to transform their communities.

The final structural element of the Latin American city is the *disamenity sector*. In this sector, we find slums, known as *favelas* in Brazil. Open sewers line streets; residents live in tiny huts built illegally and lacking sanitation. The worst of these poverty-stricken areas include people so poor that they live on the streets, often in doorways or in cardboard boxes that they carry with them.

Southeast Asian City Structure

If the Griffin-Ford model provides an interpretation of the organization of a city in the developing world, then Terry McGee's (1967) generalization about the Southeast Asian city provides a departure that occurs in colonial port cities that have continued to grow rapidly in the post-independence era (Figure 6.8). McGee's land-use diagram illustrates the old port zone, which is the city's focus, together with a surrounding commercial district. A formal CBD is absent, but its elements occur as separate clusters: a sector of government buildings; a European commercial area; a crowded alien commercial zone, where the bulk of the Chinese merchants live and work; and a mixed land-use strip of land along a railway line for various economic activities, including light industry. Other nonresidential zones include a peripheral market-gardening ring and, still farther from the city, a new industrial park. The residential zones in McGee's model are reminiscent of those in the Griffin-Ford model: a new high-class suburban residential area, an inner-city zone of comfortable middle-class housing, and peripheral areas of low-income squatter settlements with substandard sanitation and inadequate water supplies.

SPRAWLING METROPOLIS: PATTERNS AND PROBLEMS

The Spread City

The classical models of land use fitted earlier patterns of North American city growth better than present-day patterns. The concentric-zone model, developed in the 1920s, emphasized centripetal forces that concentrated economic activity in the downtown of the inner city. Subsequently, the sector model and then, even more so, the multiple-nuclei model stressed centrifugal forces that have decentralizing influences. In the second half of the 20th century, centrifugal forces have gained the ascendancy. As a result of automobile-based intraurban dispersal, the city has evolved into a restructured form called *multicentered metropolis*. The classical models cannot easily accommodate this new urban reality.

The rapid outward spread of urban North America owes much to the completion of the radial and circumferential freeway network, which resulted in near-equal levels of time convergence across the metropolitan area. In effect, the freeway system destroyed the regionwide advantage of the CBD, making most places along the expressway network just as accessible to the metropolis as the downtown was before 1970. No longer on the cutting edge, the downtown gave way in the 1970s and 1980s to an ever-widening suburban city that was being transformed—new neighborhoods, new business centers, and new shopping malls. Many Americans now live, work, play, shop, and dine within the confines of this freeway culture.

The spreading out of the American city has captured the imagination of geographers. In the early 1960s, an extreme form of a spread city was described by Jean Gottmann (1964), who coined the term *megalopolis* to

FIGURE 6.8
Model of Southeast Asian city structure. *(Source: McGee, 1967)*

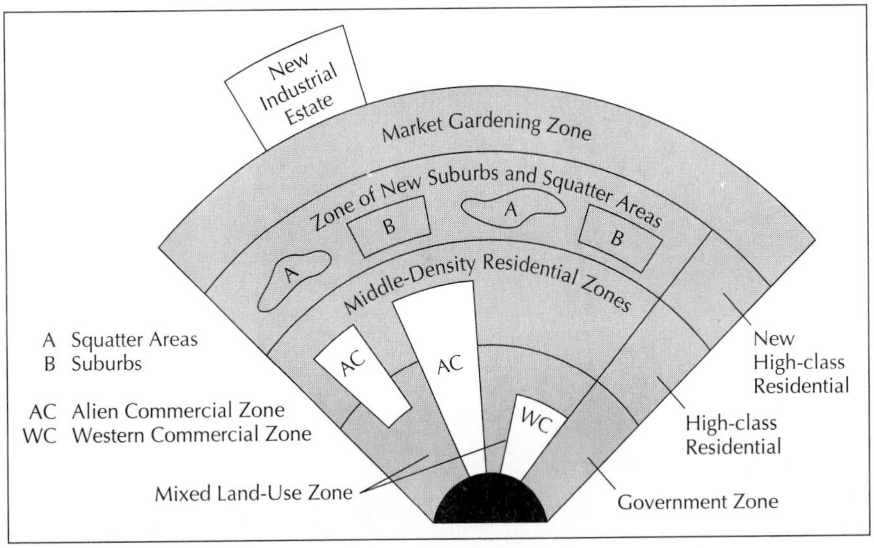

describe the coalesced metropolitan areas on the Atlantic Seaboard. This superurban region stretches from Boston to Washington; hence, it is sometimes referred to as *BoWash*. It includes a network of cities fused by expressways, tunnels, bridges, and shuttle jets. Another evolving supercity is the loosely knit lower Great Lakes urban region, also known as *ChiPitts*, centered on Chicago, Detroit, Cleveland, and Pittsburgh.

By the year 2000, trend projections indicate that the United States will consist of four superurban regions:

1. *BoWash*, extending along the Atlantic Seaboard
2. *ChiPitts*, stretching from Chicago to Pittsburgh, and merging with BoWash via the "Mohawk Bridge" to form the Metropolitan Belt
3. *SanSan*, a belt from San Francisco to San Diego
4. *JaMi*, a strip from Jacksonville to Miami

These megalopolitan networks will be supplemented with about 22 other major urban regions, each containing at least a million people (Figure 6.9). According to Jerome Pickard (1972), five-sixths of the population will be concentrated on one-sixth of the country's land area.

Outside the United States, *conurbations*, or "systems of contiguous cities," are also evident. In Canada, there is the Windsor-Quebec city axis, which geographer Maurice Yeates (1975) called *Main Street* (Figure 6.10). More than half of all Canadians live in this multicultural megalopolis, many of whom make their homes in Toronto and Montreal.

In Great Britain, megalopolitan England developed as an axial belt from London through Birmingham to Liverpool, Manchester, and Leeds (Figure 6.11). The five conurbations of megalopolitan England—Greater London, West Midlands, Southeast Lancashire, West Yorkshire, and Merseyside—as well as those of Tyneside and Central Clydeside, are home to one-third of the British population of 57 million. In the western Netherlands, the *Randstad* (or Ring City) runs in a horseshoe-shaped line approximately 170 kilometers long (Figure 6.12). It centers on major conurbations grouped around the four cities of Rotterdam, The Hague, Amsterdam, and Utrecht. From this complex, the E36 motorway joins the Netherlands to the heart of Europe, linking the Randstad with another vast urban agglomeration of continental Europe—the Rhine-Ruhr. The Dutch Randstad and the cities of the Rhine-Ruhr

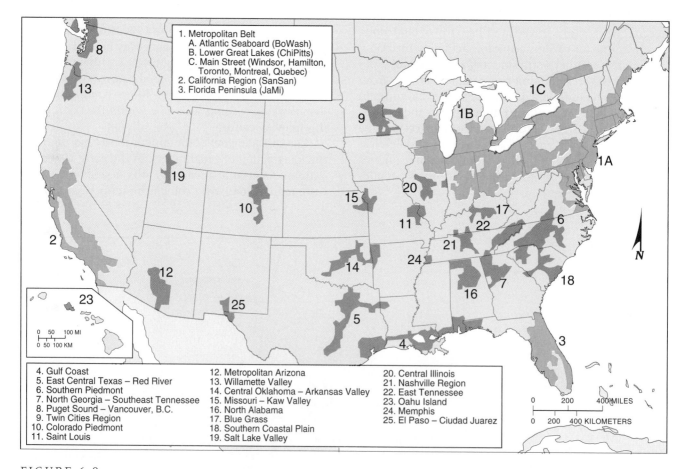

FIGURE 6.9

Projected growth of U.S. urban regions with populations of 1 million or more by the year 2000. The four superurban regions are shown in a darker shade. *(Source: Updated from Pickard, 1972, p. 143)*

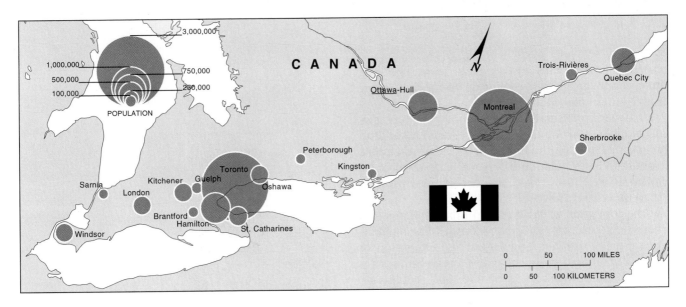

FIGURE 6.10
Canada's Main Street. *(Source: Yeates, 1984, p. 242)*

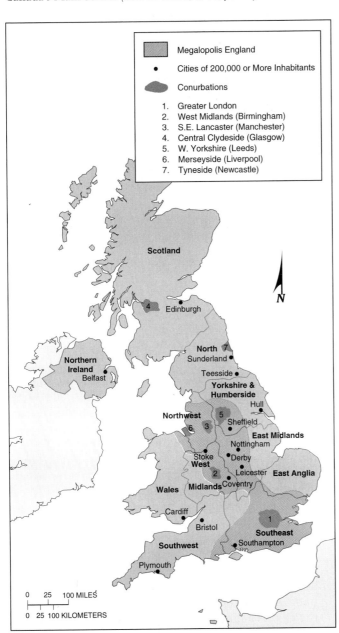

Legend:

Megalopolis England

• Cities of 200,000 or More Inhabitants

Conurbations

1. Greater London
2. West Midlands (Birmingham)
3. S.E. Lancaster (Manchester)
4. Central Clydeside (Glasgow)
5. W. Yorkshire (Leeds)
6. Merseyside (Liverpool)
7. Tyneside (Newcastle)

FIGURE 6.11
Megalopolitan Great Britain. *(Source: Based on Herbert, 1982, p. 234)*

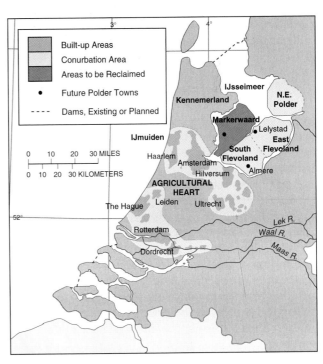

Legend:

Built-up Areas

Conurbation Area

Areas to be Reclaimed

• Future Polder Towns

- - - Dams, Existing or Planned

FIGURE 6.12
Randstad, the Netherlands: a horseshoe-shaped ring of cities, each performing specialized functions—government in The Hague, commerce in Rotterdam, shopping and culture in Amsterdam—with a central agricultural, or green, heart. *(Source: Based on Hall, 1982, p. 236)*

coalesced in the 1980s into a gigantic urban region: a European megalopolis stretching down the river Rhine from Bonn to the Hook of Holland.

In Japan, too, there is an enormous, high-density megalopolis. Half of Japan's more than 120 million people are crowded into three areas—the conurbations around Tokyo, Osaka, and Nagoya. One-third of all Japanese live within 150 kilometers of their emperor's palace in central Tokyo.

The formation of regional and urban systems also characterizes Third World urban processes. The rapid increase in the number, population, and expansion of urban agglomerations in some developing countries is leading to the growth of megalopolises. Such systems are developing in Brazil, where the major nodes are Río de Janeiro–Belo Horizonte–Sao Paulo; in Mexico, where Mexico City–Puebla–Vera Cruz form the major centers; and in Egypt, where Alexandria and Cairo are the major cities. Elsewhere, urban growth is concentrated around one or two cities, as in Lagos and Ibadan, Nigeria; in Djakarta and Surabaja, Indonesia; and in Seoul, South Korea. Nearly 40% of South Korea's 42 million people live in or near Seoul, the nerve center of the nation.

The geographer Peirce Lewis (1983) provided a provocative description of the outward spread of cities. For the United States, he coined the term *galactic metropolis* to refer to a vast continuum of urbanization stretching from coast to coast. In his national vision, huge urban concentrations are interspersed with small towns and cities, as well as with loosely separated clusters of houses and businesses around freeway interchanges.

Causes of Urban Spread

What is at the heart of this spreading out of the American city? A number of economic and noneconomic factors are involved. Let us now look at one or two of them with reference to the locational decisions of households and firms.

Suburbanization of households is closely associated with intraurban transportation improvements (Figure 6.13). With each revolution in transportation, travel costs were lowered, and families become less willing to pay high rents for central locations. Since 1945, the desire for a single-family home in the suburbs has resulted in rapid suburban expansion. Factors that have reinforced this trend are (1) low mortgage interest rates, (2) loan guarantees provided under federal housing and veterans' benefit programs, (3) property-tax reductions for owner-occupied homes, (4) cheap transportation, (5) massive highway subsidies, and most of all (6) cheap land.

The freeway-dominated automobile era has removed virtually all restrictions on intraurban population mobility so that residential land use is feasible

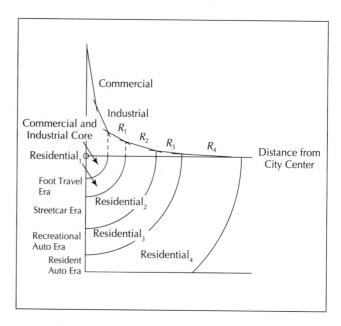

FIGURE 6.13
Passenger transportation improvements and housing density. With each innovation in transportation, travel costs were lowered, and families became less willing to pay high rents for central locations. *(Source: Abler, Adams, and Gould, 1971, p. 358)*

almost anywhere in a metropolitan area, especially with automobile companies buying and closing down intraurban streetcar lines across the country (Figure 6.14).

THE ECONOMICS OF COMMUTING
The economics of residential land and commuting time are attributed to William Alonso (1964) and Richard Muth (1969). According to these theorists, residential location choice is a utility-maximization process. In the simple model, the assumptions include these: (1) each household has one laborer, (2) housing is a homogeneous factor and only location varies, (3) transportation costs are based solely on the commuting distance to work, and (4) all employment is fixed at the center of the city. Each householder chooses the location where the best combination of land, rent, and transportation costs (as well as a set of other factors) occurs.

In the residential land theory, costs of housing and land decline farther from the center of the city, as land becomes less in demand. All commerce and business occur in the center of the city, where the price of land is bid up. With increasing distance from the city, there is less opportunity for commerce and less demand for space as a function of area, or Πr^2 (pi times radius squared). For example, if all housing were the same price—a uniform price for land at every location—all business owners and householders would choose the same location, the center of the city. On the other hand, transportation costs increase with distance from the city, as a direct function of time in movement.

FIGURE 6.14
The development cycle ushered in by the Highway Trust Fund. This legislation placed a tax on gasoline and ear-marked the revenues to be used exclusively for building new roads. (The fund still exists, but it was modified in 1991 to allow a portion of the money to be used for public transit in order to reduce congestion.) The Trust Fund actually fostered the development of open land and commuting by more drivers from more distant locations. Average commuting distances have doubled since 1960, but average commuting time has remained about the same. The increase in commuting distance requires more fuel but generates more money for the Highway Trust Fund, thus continuing the cycle.

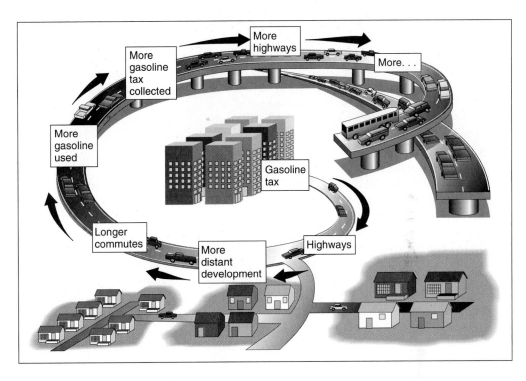

The optimal location for a householder is the point where the marginal savings in housing are equal to the marginal increase in transportation costs. That is, householders will, according to their preferences for transportation and housing space, try to maximize both by finding the place away from the center of the city where the decrease in housing costs is just offset by an increase in the commuting costs. Studies have shown that the housing preference is the stronger of the two. That is, people are willing to offset extra transportation costs to consume a larger *market basket* of housing, even if doing so means a longer commute. The theory continues by postulating that, with increased incomes, more housing will be consumed farther from the city, where the unit cost of consumption is less.

What if transportation costs decline because highway improvements or a mass-transit line reduces travel time? According to Figure 6.15, as transportation costs decline from T_{D1} to T_{D2}, the optimum, or equilibrium, location for housing moves from D_1 to D_2, outward toward the suburbs, as a larger market basket of housing can be consumed for the same transportation cost.

ΔH_D is the householders' savings on housing costs by moving away from the center of the city. The original equilibrium location is D_1, where the housing-cost decline curve intersects the transportation-cost decline curve. The new equilibrium location is D_2, where the housing-cost decline curve, ΔH_D, intersects the new transportation-cost curve, T_{D2}, according to Guliano.

Since World War II, gasoline prices in the United States have been extremely low compared with those of other highly developed nations around the world. For example, in the 1950s, a gallon of gasoline averaged $0.20. In the 1960s, it averaged $0.35. During the 1970s, a gallon of gasoline remained at $0.35, only creeping up to $0.65 by 1978. During the 1950s through the 1980s, America saw its most dramatic suburbanization push, associated with

FIGURE 6.15
The effect of reduced transportation costs (T_{D2}) on residential location (D_2). *(Source: Guliano, 1986, p. 251)*

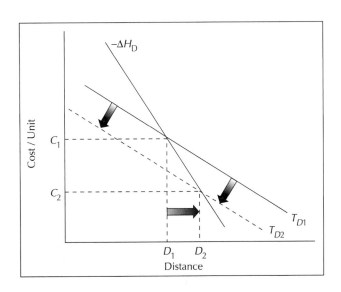

freeway building and inexpensive gasoline, relative to other commodities and high demand.

So far, two types of costs have been discussed: out-of-pocket gasoline costs and time costs. Each affects social classes differently. From a business standpoint, an increasingly rapid movement toward decreased travel-time costs benefits the wealthy more than the low-income groups because their time is worth more. Keeping the price of gasoline low also benefits the wealthy more because, compared with other family needs, gasoline is a lower overall proportion of the household budget. Other things being equal, travel-time costs and the cost of gasoline have more of an effect on higher-income groups.

Now, according to residential rent theory, as the demand for suburban residential locations increases, the relative price of properties closer to the city will decline. Regardless of lower prices, residential rent theory predicts declining population densities in the central and nearby areas, and increasing population densities farther away, which has been the case in American cities. Improvements in transportation that reduce commuting time have a strong decentralizing effect on a city, especially for the higher-income groups.

The flight of households to the suburbs, the inhabitants of which outnumbered those of the central city by 1970, may also be a consequence of rising real incomes, population growth, and the postwar movement of rural blacks to central cities (Table 6.4). However, although it is true that upper-income households are concentrated in the suburbs, there is an unresolved debate as to whether this is the result of income per se or some other variable. In the 1960s, Richard Muth (1969) argued that income is the cause. He argued that higher income led to a higher demand for housing (a positive-income elasticity), and this higher demand for housing led to a utility-maximizing move away from the city. Although higher income also meant an increased commuting-time cost (the time cost is estimated as a fraction of a person's explicit or implicit hourly wage) and would tend to make the house-

holder more interested in remaining near the job site, Muth's statistical studies led him to conclude that the added push away from the CBD exceeded the added pull toward it. In 1987, William Wheaton published another study on the same question. His results showed that the push effects and the pull effects were almost the same. Thus, his work suggests that higher income per se is not the reason for the suburban concentration of upper-income individuals.

GENTRIFICATION OR SUBURBANIZATION

Between 1940 and 1970, many southern blacks were displaced from mechanized farms and moved to northern cities. Black migration from the rural South to the urban North peaked in the late 1950s. As blacks moved into city neighborhoods, white middle-class families experienced property devaluation, often spurred by the *blockbusting* of real-estate brokers. Meanwhile, valuations of suburban locations increased. The number of whites who moved to the suburbs in order to maintain a *social distance* from the immigrant blacks of the inner city has not been documented, however.

During the 1970s and 1980s, reverse migration occurred; middle-class families moved from the suburbs to the inner cities. The number of families involved was relatively small, however. For every family that moved back to the inner city, eight moved to the suburbs. Residential revitalization in and around the CBD clearly did not spawn a return-to-the-city movement by suburbanites. In fact, most reinvestment was undertaken by those already living in the central city. Such inner-city neighborhood redevelopment involves *gentrification*—property upgrading by high-income settlers—which frequently results in the displacement of lower-income residents.

Suburbanization of retailing was a response to the residential flight to the suburbs, new merchandising techniques, and technical obsolescence of older retailing areas. The automobile provided customers with a convenient mode of transportation to shopping places,

◈

T a b l e 6.4
U.S. Population Change, 1980–1990

	U.S. Population		
	1980	*1990*	*Change (%)*
Metropolitan areas	170,540,000	189,877,000	+11.3
Inside center cities			
Outside center cities			
Nonmetropolitan areas	57,217,000	61,646,000	+7.7
Total	227,757,000	251,523,000	+10.4

Shopping in Tysons Corner, Virginia, the archetypal edge city on the outskirts of Washington, D.C. It is the largest retail concentration on the East Coast after Manhattan, New York.

but downtown parking facilities were scarce and expensive. A need to improve the parking situation and to increase profits impelled retailers to the suburbs.

The decentralization trend began in the 1920s as stores began spreading out from the downtown along main thoroughfares. Yet it was not until the postwar years that retailers moved to the suburbs in large numbers. First came the strip center, or neighborhood shopping center, consisting of a string of 10 to 30 shops, usually anchored by a supermarket. Then came the larger community center with a small department store or a variety store as the principal tenant. The success of these early centers, which catered to a limited trade area, depended on a main-road location, free parking, and the persuasiveness of "super" everything, "drive-in" everything, self-service stores, and discount outlets.

Neighborhood and, later, larger community shopping centers became vulnerable to more attractive regional shopping centers that appeared after 1955. The newest and biggest of these centers in distant suburbs have several floors, three or more department stores, and scores of specialty shops. Surrounded by huge parking lots, these shopping complexes are usually enclosed so that customers can shop in climate-controlled comfort. Unlike early suburban shopping centers, the giant regional shopping centers are catalysts attracting a variety of activities to the area (Table 6.5).

Decentralization of manufacturing began before the turn of the century. Technical advances, such as the development of continuous-material flow systems, induced many manufacturers, especially those engaged in large-scale production of industrial goods, to spread out along suburban railway corridors where land was relatively cheap and abundant. Nonetheless, most manufacturers, despite truck transportation, decided

◈

Table 6.5

Decline in Population of American Cities 1950–1994

City	Population (thousands)					Percent change
	1950	1960	1970	1980	1992	1950–1994
Baltimore, MD	950	939	905	787	726	−23
Boston, MA	801	697	641	563	552	−31
Buffalo, NY	580	533	463	358	323	−44
Cleveland, OH	915	876	751	574	503	−45
Detroit, MI	1850	1670	1514	1203	1012	−45
Louisville, KY	369	391	362	298	271	−26
Minneapolis, MN	522	483	434	371	363	−30
New York (Bronx), NY	1451	1425	1472	1169	1195	−18
Oakland, CA	385	365	362	339	373	−03
Philadelphia, PA	2072	2003	1949	1688	1553	−25
St. Louis, MO	857	705	662	453	384	−55
Washington, D.C.	802	764	757	638	585	−27

to remain in or near the central city until the 1960s, when two technological breakthroughs occurred. These innovations involved (1) the completion of the urban expressway system and (2) the scale economies in local trucking operations. Completing the freeway network helped to neutralize the transportation cost differential between inner city and suburb. As the locational pull of central city water and rail terminals declined, most of the remaining urbanization economies of downtown were nullified (Muller, 1976, p. 33).

In the freeway metropolis, the economic advantages of a central-city location have disappeared. Consequently, the spatial organization of the manufacturing industry is responding increasingly to noncost factors. Manufacturers are relatively free to select the most prestigious sites that they can find in the outer city.

Expansion of offices into the suburbs began in the early postwar years, when large corporations began looking for new office headquarters. For example, General Foods, IBM, Reader's Digest, Union Carbide, and ESSO-Standard Oil left New York for the suburban countryside. This trend of the large corporations prompted an avalanche of similar moves by a host of small office firms. In that they were unable to create their own environments in the manner of the large corporations, the small firms began to rent or lease space in office parks. Tenants of these office buildings were attracted by the convenience, amenity, and prestige of an office-in-a-park address. The major factor in office-site selection is accessibility to an expressway.

Initially, suburban business and commerce located at any convenient highway intersection or at a site near a freeway. Today, geographers can compute the relative potential for a business at several locations to determine the optimal location (Figure 6.16; see Target Marketing, Chapter 7).

Suburban economic activities have a growing locational affinity for one another. Without doubt, the focal point of the *edge city* is the huge regional shopping center. Super shopping malls are catalysts for other commercial, industrial, recreational, and cultural facilities. The result is the emergence of miniature downtowns called edge cities. In many metropolitan areas, edge cities are unplanned, loosely organized, multifunctional nodes, and they are strongly shaping the geography of suburbia.

Urban Realms Model

James Vance (1977) put forward the *urban realms model*, which he constructed from his observations of the San Francisco Bay area and its sprawling metropolis. Peter Muller (1981, 1989), and others, applied this model to New York City and the Los Angeles metropolitan area. The urban realms model identifies suburban regions that have independent suburban downtowns as their foci, and yet are within the sphere of influence of the central city and its metropolitan CBD. Each suburban downtown coexists with other suburban downtowns, as well as the principal CBD, each being self-sufficient in form and function. Hartshorn (1992) calls this model the "culmination of the impact of the automobile on urban form in the 1990s" (p. 234).

According to Vance, the existence and form of each urban realm in modern Anglo-American cities today depends on four factors: (1) the overall size of the metropolitan region, (2) the amount of economic activity in each urban realm, (3) the topography and major land features, which can help to identify urban realms, and (4) internal accessibility of each realm with regard to

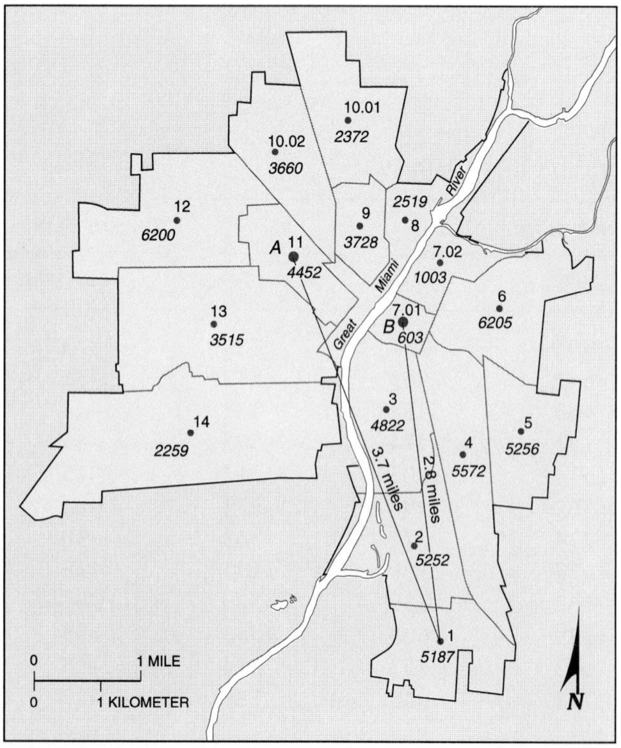

FIGURE 6.16

Geographers can compute the relative potential for a service at various locations. For example, to determine whether a service should be located in Tract 7.01 or Tract 11 in Hamilton, Ohio, follow these steps: (1) Determine the potential population of patrons in each of the other neighborhoods; these figures are shown in italics. (2) Measure the distance from each neighborhood to one of the alternative locations; for example, the distance from Tract 1 to Tract 7.01 (location *B*) is approximately 4.5 kilometers (2.8 miles). (3) Divide the population in each neighborhood by the distance from that neighborhood to the proposed location of the service. (4) Sum all the populations and divide by the distance; this figure is the relative potential of a service at a particular location. (5) Repeat steps 2 through 4 for alternative location *A*. The optimal location for a service is the tract with the highest relative potential.

daily economic functioning and travel patterns. Also according to Vance, a region has a high probability of becoming a self-sufficient urban realm if the following are true: (1) the overall size of the metropolis is large, (2) there is a large amount of decentralized economic activity in the region, (3) the topography and barriers isolate the suburban region, and (4) it has good internal accessibility for self-sufficient daily commerce and functioning (especially if it is tied to an airport or metropolises).

Somewhat later than in North America, European cities expanded outward from the inner cities to the suburbs and beyond. In the 1950s, for the first time, the population of major British cities started to decline. Often inner-city depopulation was a result of an official policy to relieve overcrowding. By the 1960s, these major cities began to lose business. Old industries around which the cities had grown up, and that supported the economy of the inner cities, were in decline. The newer growth industries could not make up the job losses and were often developed on greenfield sites far from the central cities, where building and land were less expensive. Although large European cities are beginning to resemble expanding doughnuts as they decentralize, their CBDs remain more vibrant than many in North America.

JAPANESE SPRAWL

Urban sprawl is also a feature of Japanese cities. Tokyoites who want to purchase their own homes have been pushed farther and farther from the middle of the city by high land prices. Of Tokyo's 5 million or more daily commuters, 2.75 million live outside the city, itself an area larger than Luxembourg. Even though land prices in central Tokyo are astonishingly high by New York and London standards, Japanese companies still want their headquarters near the ministries and banks of the capital. Eventually, Tokyo—a megalopolis with the profile of a pancake—may become so big that it decentralizes. Technology is the key to decentralization. Telecommunications and high-speed, magnetically levitated trains would make Japan's second city, Osaka, a 60-minute ride from Tokyo against today's 3.5-hour ride in a "bullet train." The line would pass through Nagoya and create an urban corridor along central Japan's Pacific coast. Away from this spine would lie suburban centers linked by advanced telecommunications networks. Tokyo would become a vast, wired megalopolis of more than 60 million people stretched over 500 kilometers. This vision is less odd to the Japanese than to Europeans. Japanese cities do not have centers such as Paris or London, but a series of subcenters, more like a grand-scale, high-density Phoenix or Tucson.

By European or American standards, Japanese cities are overcrowded to Dickensian proportions, but so too are the large cities of developing countries. In many a developing country, the largest city is growing uncontrollably. Rapid growth is linked to what economic

geographers call the *economies of agglomeration.* Firms locate in the large city because of the existence of modern infrastructure. When a plant locates in the city, it brings with it new jobs. However, industry is not the only source of work in the metropolis. The growing bureaucracy needed to administer public investment in schools, transportation, communications, water, and sanitation provides many jobs, as does the semilicit service economy. Work in the industrial plant, bureaucratic office, or black market offers a higher standard of living than that available in the provincial town or village. The superior standard of living has a powerful attraction for the people of peripheral regions, who flock to the large, crowded cities.

Problems of the City

The free market for space in the metropolis has produced a pattern of suburban sprawl in advanced capitalist societies. From a rare social entity at the beginning of the 20th century, suburbs have evolved into major growth centers for industrial and commercial investment, and a suburban way of life has been adopted by millions of people. The decentralized metropolis fostered large-scale consumption and prosperity in the past, but it is causing real problems now. In some instances, the urban fringe has pushed out farther than workers are willing or able to commute. Urban sprawl has generated externalities such as uneven development, pollution, and the irrational use of space, which increasingly impinge on the life of urban residents. Furthermore, recurrent fiscal crises threaten to bankrupt central cities.

In the United States, the rationalization of the metropolis for the purpose of planned development is blocked by the political independence of the suburbs. The suburbs have resulted from the differential ability of various groups to organize and protect their advantages. They are not willing to abdicate clear-cut, short-run benefits for less certain long-run gains. Thus, metropoliswide planning in the face of a bewildering multitude of rigid and outdated municipal boundaries—1,200 of them in New York—is extremely difficult, if not impossible, to implement. Yet without planning, without redrawing areas of municipal authority, the continued profitability and stability of the metropolis and capitalist society are threatened.

Most cities in North America and Western Europe are now faced with a roughly similar situation—they are dying in the middle. Jobs are moving out with modern technology and communications, and so is shopping and entertainment. Violence is moving in. The need for the city, as far as the middle class is concerned, has diminished.

To be sure, there have been attempts to revitalize inner cities. For example, in England there are tax-incentive plans for the rebirth of blighted areas. The

A homeless person on the streets of New York. *(Photo: Thomas Hoepker/Magnum Photos)*

intent is to create jobs out of urban wastelands and to transform old warehousing and waterfront districts. In London, dock-land revival involves some light manufacturing, but it is dominated by office and commercial development as well as the construction or refurbishment of housing units (gentrification). The residential units are for the rich up-and-comers; they cost too much for the majority of the original East-Enders. Small schemes aimed at revitalizing inner-city areas are narrow technical solutions to a broad problem created by development.

Most governments of Western Europe and North America have concentrated, often for decades, on relieving congestion and welfare pressures by demolishing block upon block of old housing, factories, and other buildings. A major problem for local governments has been where to rehouse the people affected by clearance. One solution has been to replace crowded terraced streets with blocks of tall apartments. But besides being more expensive to build than houses, highrise housing developments are generally regarded as dehumanizing environments.

Most redevelopment schemes can rehouse only about half the displaced population. The so-called overspill must move out. Governments have tried a variety of methods to relieve the overspill problem. For London in the 1940s, it was proposed that a green belt of open space be preserved around the build-up area, and that a number of new towns to house people from central London be developed beyond the green belt. The eight new towns that were built in the 1950s and 1960s were fashioned after the Garden City model, advocated in Ebenezer Howard's (1946) book, *Garden Cities of Tomorrow*, first published in 1902. This model called for self-contained, medium-sized cities with large areas of public open space separating urban functions.

New towns have been built in other parts of Great Britain as well, especially northern industrial areas (Figure 6.17). The British new-town model has also been adopted in the United States, Japan, and other European countries. In the United States, examples include Columbia, Maryland; Reston, Virginia; and Irvine, California.

New towns and other population redistribution schemes, such as planned public-housing estates, have not solved big-city problems. For example, a deep-seated polarization between inner and outer city remains. Increasingly, the center city is home to the metropolitan disadvantaged and includes a few specialized services, whereas suburb and satellite cities house the affluent and support a range of activities that were previously city bound. This pattern of spatial organization favors rich people, who collect a disproportionate share of the surplus product with respect to the location of services and job opportunities. Examples of how the rich tend to be favored in U.S. metropolitan areas follow.

RETAIL STORES

Shop location is a reflection of the economic behavior of entrepreneurs. However, their decisions are usually

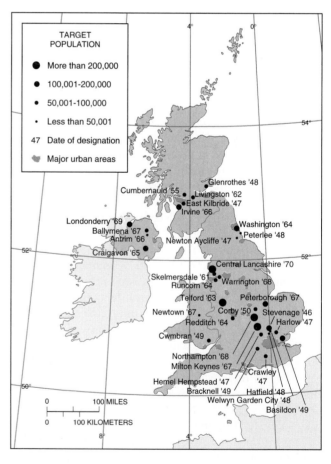

FIGURE 6.17
New towns in Great Britain, 1946–1980. *(Source: Based on Hall, 1982, p. 167)*

subject to some public control through zoning. Decision makers select locations with a high demand potential. Because demand is income related, it is natural for entrepreneurs to choose affluent suburban areas first. For example, many large supermarkets locate in the suburbs and sell their produce at lower prices than inner-city neighborhood stores do. Differential prices and accessibility to retail outlets contributed to the urban riots of the 1960s and 1970s.

MEDICAL CARE

Health care services provided by general practitioners, internists, and pediatricians are located mainly by private action. Because they, too, are sensitive to demand potential, they tend to locate in affluent areas even though the need for medical care is likely to be higher in low-income neighborhoods. In the inner city, the needy sick are taken care of in public hospitals, which are often poorly equipped and understaffed.

PUBLIC UTILITIES

Provision of water, electricity, sewage and sanitation, and transportation services provides additional examples of the inequitable appropriation of the surplus. For instance, consider water. In some metropolitan areas, water prices are not seasonally adjusted, yet the peak demand for water in summer comes mainly from suburban families, who use it to sprinkle lawns and fill swimming pools. Inner-city residents have neither yards to sprinkle nor pools to fill. When pricing systems for water fail to reflect seasonal demands, they effectively subsidize the already wealthy.

NOXIOUS ATTRIBUTES

The value of a dwelling varies according to proximity to noxious attributes such as smoke, dust, noise and water pollution, and traffic congestion. The affluent, who wield political clout, usually manage to exclude noxious facilities from their neighborhoods. But these facilities must go somewhere. Traditionally, facilities such as power stations have been located in inner-city communities and in rural areas.

JOBS/HOUSING BALANCE

A relatively new solution to the problem of increased suburban traffic congestion caused by explosive employment growth gained credibility in the late 1980s in many areas of the United States. This problem resulted from the increased commuting times to these work centers from low-density residential subdivisions that shifted farther into the suburban fringe. Some observers (Porter, 1985; Porter and Lasser, 1988) have noted that if nearby affordable housing development had accompanied job center growth, perhaps less auto traffic would have resulted (Porter, 1985). This led to a ground swell of support for the concept of *jobs/housing balance*.

Local and regional planners increasingly support the notion of expanding the supply of housing in job-rich areas and the quantity of jobs in housing-rich areas. Ideally, a ratio of one job per dwelling (a ratio of 1:1) would represent a balance. However, in many households more than one person is employed, so more jobs than housing units would provide a better balance.

The Southern California Association of Governments (SCAGS) has studied this issue for nearly 20 years and projected in the mid-1980s that, between 1984 and 2010, a growth of nearly 6 million people would occur in a particular region, along with the addition of 3 million jobs (Porter and Lasser, 1988). Moreover, the projections indicated that most of the job growth would occur in Los Angeles and Orange County, whereas housing growth would occur in the fringe counties of San Bernardino, Riverside, and Ventura. The projections indicated that this imbalanced growth would further overload freeways and average speeds on them would drop 50% to 19 mph. The SCAGS growth management plan, adopted in 1989, recommended a jobs/housing ratio of 1.22 to 1 within each subregion of the area. The ratio represents the projected regionwide balance between jobs and housing in 2010. Now the question becomes how can one achieve such a balance? What is the appropriate geographic area in which to achieve balance? What price ranges should be encouraged? As answers to these and other questions are worked out in California, and elsewhere in the coming years, evidence mounts that the jobs/housing crisis is worsening.

URBAN DECAY

Affluent residents moving to the suburbs begins the process of *exurb migration* and *urban blight*. Local governments' major source of revenue to pay for municipal services—public schools, road maintenance, police and fire protection, garbage collection and disposal, fresh water and sewers, welfare services, libraries and local parks—comes from local property taxes. The local property tax is the yearly proportional assessment according to real estate market value. If property values are increased or decreased throughout the year, the property taxes are adjusted accordingly. Most central city governments are separate entities from the governments of the surrounding suburbs.

Therefore, exurban migration causes, through the laws of supply and demand, escalating suburban property values because of the influx of affluent people. Suburban governments enjoy increasing tax revenue with which they can expand local services. However, property values in the central city decline because of declining demand leading to an eroding tax base. Those who are unable to move to the suburbs are usually lower income and disadvantaged people who are non-property owners requiring public assistance; thus, cities must carry a disproportionate burden of welfare obligations, as well as increasing crime and deterioration of infrastructure due to declining tax revenues (Figure 6.18).

The eroding tax base forces local governments to cut local services and raise the property tax rate. Therefore, property taxes on a home in the central city are often several times as high, proportional to assessed valuation, as a home in the suburbs. This difference could be $5,000 to $8,000 per year in many cities based on an average size home. At the same time, parks, schools, streets, libraries, and garbage collection services deteri-

orate from neglect, leading to the vicious cycle of exurban migration and urban decay. This downward spiral of conditions is referred to as *urban decay*.

OUT IN THE EXURBS

Affordable housing for most workers is not being provided in adequate quantities near growing suburban employment centers of edge cities, commuting trips are becoming longer, and pollution levels from automobile emissions continue to increase. The problem is graphi-

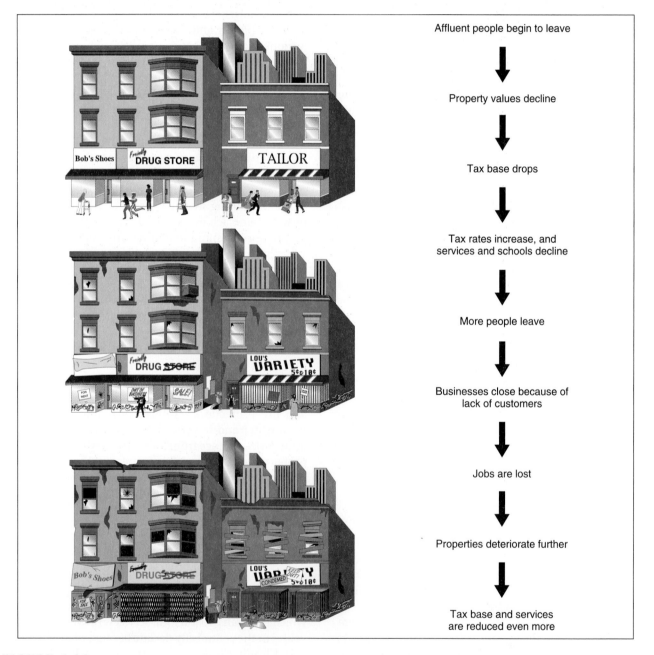

FIGURE 6.18

The cycle of suburban migration and urban decay. People moving to the suburbs from the city create the cycle of urban blight, which continues in most U.S. cities to this day. Most cities are responsible for providing public schools, maintenance of local roads and facilities, police and fire protection, waste collection, public water and sewer, welfare services, libraries, and local parks. The major source of revenue to pay for these services is property taxes.

cally portrayed by the experience of exurbanites from Victor Valley, in San Bernardino County, the fastest-growing county in California. In 1990, only 50,000 people lived there; now the population is 186,000. The nearest major employment centers are in Orange County, 70 miles to the south, and in Los Angeles, 90 miles to the southwest. Figure 6.19 illustrates population shift between cities and rural areas over a 50-year period.

Rapidly growing communities like these, the *exurbs*, lack urban amenities and are "boomtowns" in remote boondock areas for only one reason—access to affordable housing. The tradeoff for people living in the exurbs is the stress of the long commute and the strain it places on family relations. This problem has now spread to most metropolitan areas of the country, including New York, Chicago, and San Francisco, and to many areas in Florida. Long work commutes have fueled the clamor for growth management in which the jobs/housing balance issue and traffic congestion play an important role.

The information revolution and the customized flexible economy will allow businesses, consumers, and entrepreneurs to move to the exurbs, lower-cost areas of the country, thereby increasing their standard of living and quality of life (Figure 6.20). Technologies and infrastructures open up new low-cost land, which happened in the move to the cities in the late 1800s and, after the Great Depression, the massive move to the suburbs. The move to the suburbs was made possible by automobiles, highways, telephones, and electrical energy. Cheaper land meant larger properties and cheaper prices. Eventually the era of suburban growth matured into high-cost places to live. By 1990, families were looking to small cities beyond the metropolitan area to escape high costs of land, houses, and utilities, and high taxes, congestion, and crime. By 1990, home values in some suburbs started to sink.

The introduction of basic innovation is usually initiated in major business relocation and population shifts. The steam engine and the invention of iron smelting brought on by the first Industrial Revolution allowed people to move to the city in record numbers. Next, in the 1870s, 55 years later, railroads and steel making began to emerge, and a substantial portion of the U.S. population migrated from the eastern portions of the country westward. Migration continued from the farms to the cities. After the innovations of electricity, telephone, and automobiles on the 1920s, and after cities had become increasingly saturated and costly places to live, from approximately 1935 to 1965, the growth into the mainstream meant population shift to the suburbs, 60 years after the migration to the cities led by railroads. (The move to the suburbs continued with the fourth long wave cycle, or *kondratiev cycle*, through the 1970s; see Chapter 8.)

However, the microelectronics revolution, which is the basis of the information superhighway, will allow in

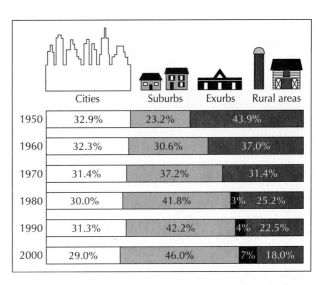

FIGURE 6.19
Migration from rural areas to cities, to the suburbs, and finally to exurbs. By 1970, more of the U.S. population lived in suburban portions of the metropolitan statistical areas than lived in the central cities or in rural areas. While this trend continues, an increasing number of city dwellers and suburbanites have moved back to small towns in exurbia, 100 miles beyond the city limit of large central cities.

the coming decades a massive population shift from the now expensive suburbs to the exurbs. Typically, exurbs are 50 to 150 miles beyond the metropolitan fringe. They include small resort towns, recreationally oriented towns, which have lower costs and high appeal, perhaps with a university or with potential for business infrastructure. Businesses will relocate to the exurbs or small towns to reduce their costs and to become more competitive, providing they can still reach their markets. Such businesses that move to smaller towns will receive much higher average growth and less competition at the outset. Retirees, of course, as well as baby boomers fed up with the suburbs, will flock to the exurbs. The exurbs provide great opportunity in the coming decades for entrepreneurial job and career advancement with improved lifestyle at lower costs (see Chapter 2: Exurbs). The effect of this shift on central cities and suburbs could be disastrous unless cities start planning now (see Smart Cities; Chapter 7).

EMPLOYMENT MISMATCHED FOR MINORITIES
Private and public discrimination (e.g., redlining by lending institutions, actions of realtors, screening devices adopted by subdivision developers) and exclusionary zoning practices give the poor, especially minorities, little recourse but to locate in inner-city areas. Meanwhile, most new employment opportunities matched to the work skills of these people are created in the suburbs. This is known as the *spatial mismatch theory*.

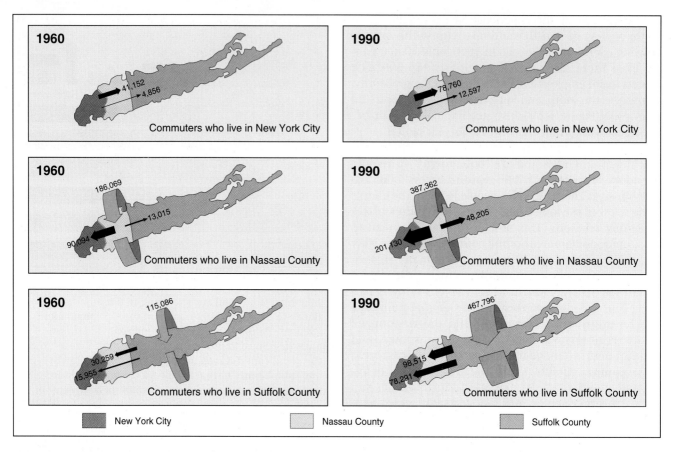

FIGURE 6.20

These maps of commuter work trips to New York's Long Island show that more and more people work within their own home counties between 1960 and 1990. Individual automobile transportation necessarily replaces mass transit into the city. Job movement has been to the exurbs.

Thus, the poor are faced with the problem of either finding work in stagnating industrial areas of the inner city or commuting longer distances to keep up with the dispersing job market. Although *reverse commuting* has increased, barriers abound. These barriers include *transportation constraints*—such as increased time and cost of the daily journey to work, and inadequate public transportation for those without cars—and *communication constraints*—such as difficulty in obtaining timely information concerning new job opportunities. Other serious obstacles to suburban employment faced by the inner-city poor include few and substandard work skills and biased hiring practices. In the face of these problems, many otherwise employable persons give up job hunting altogether and contribute instead to the growing number of unemployed in inner-city neighborhoods.

The city of Philadelphia suggests that the spatial mismatch theory is an important factor in the high unemployment levels of black youths. Overall, unemployment levels for both black and white youths increase with greater distances to job opportunities. The principal reason for higher black unemployment rates is that the commuting time is higher for blacks than for whites. However, the mismatch hypothesis accounts for only approximately half the unemployment, leaving another 50% to be explained by labor-market discrimination, motivation, differences in education, and so forth.

In a study that compared 50 metropolitan areas in 1991, Ihlanfeldt and Sjoquist (1990) conclude that (1) the mismatch hypothesis explains 25% of the gap between black and white employment rates and 35% of the gap between hispanic and white employment rates, and (2) the spatial mismatch theory is more strongly operative in larger urban areas. The smaller the city, the lower the gap between black and white unemployment rates explained by poor access. Ihlanfeldt found in the 50 metropolitan areas only a 3% gap in small cities but a 14% gap, which resulted from inferior access, in medium-sized cities.

The research just mentioned suggests that some questions still need to be answered to explain unemployment in the ghetto and in poor sections of the city. The spatial mismatch theory cannot answer these questions.

However, any government can promote the suburbanization of minorities in a number of ways and thus reduce access problems and the long commutes to available job opportunities. Governments need to enforce fair housing laws that reduce housing discrimination and poor treatment of minorities in suburban housing markets. Zoning laws also play a key role in blocking from suburbia multifamily housing that minorities can afford. Consequently, local governments must consider providing a full range of government and social services to suburban communities, reducing exclusionary zoning, and providing for a full range of housing types in suburbia.

Visions of Future Metropolitan Life

According to Anthony Downs, Senior Fellow at the Brookings Institution, the current vision of a desirable metropolitan area includes four elements:

1. Ownership of single-family homes on spacious lots.
2. Ownership and use of personal motor vehicles.
3. Low-density suburban office or industrial workplaces and shopping centers.
4. Residence in small communities with self-governance.

However, inconsistencies in these key elements result in four corresponding flaws in the vision:

1. Low-density residential and work patterns generate a tremendous amount of vehicular travel and traffic congestion.
2. There is no provision for housing for low- and moderate-income families.
3. There is no consensus regarding how to finance the required infrastructure.
4. No mechanism exists to resolve disagreements about facility locations.

According to Downs, a new vision for suburban America can be crafted from key elements of existing visions, if residents begin to think and behave more in terms of community and the collective impacts of individual decisions, rather than the way they currently think and behave. Key elements of the new vision follow:

1. Sizable areas must be set aside for moderately high-density residential and workplace uses.
2. Residents must live closer to work.
3. The governance structure must include provisions for local authority administered in the context of area-wide needs.

Oxford Street is a major shopping district in London's West End. London is not only the seat of government; it houses the headquarters of 80% of major British transnational corporations. It is the center of banking, insurance, publishing, fashions, advertising, and the legal system for Britain. London's influence stretches far beyond the city to all of the British Isles and even northwest Europe. (*Photo: Alan Becker, The Image Bank*)

4. Incentive arrangements are required to encourage individuals and households to take into account more realistically the collective costs of their decisions. (Examples include peak-hour pricing on highways, and linkage zoning requiring the simultaneous development of housing and workplaces.)
5. Equitable strategies to finance the infrastructure must be developed.

Any serious effort to eliminate the polarized or dualistic metropolis must involve planning on a metropoliswide basis. Whether such an effort can succeed remains an open question; certainly, it will be made more difficult by governments reactive to the needs of dominant economic interests.

If the dynamics of capitalist development have created urban problems in developed countries, they have created even more severe urban problems in the noncommunist countries of the underdeveloped world. The rapid growth of Third World cities creates not only

social divisions and tensions within them, but also a po-larization of metropolis and periphery. For Third World countries, this polarization can lead to political and en-vironmental disaster.

Policies to curb the expansion of major cities, which is largely a product of economic growth, are not easily established. It is more efficient from an economic standpoint to invest in the large city than to invest in the periphery. Thus, public and private capital tend to concentrate in the major city. The cost of providing public services—schools, roads, transportation, com-munications systems, water, and sanitation—is high in the metropolis, but the government of an undeveloped country is under considerable pressure to invest its cap-ital there. Part of the pressure is political. The stability of the country often depends on preventing unrest in the major city. So that all runs smoothly, public services are often provided in the metropolis before they are provided in the periphery.

The relatively high standard of living in the major city is the force that drives the concentration of popu-lation. To reduce the flow of people toward major cities, the differential in living standards between city and countryside must be reduced. This can be accom-plished by slowing the pace of national economic growth and/or raising the standard of living in the countryside.

Migration toward major urban areas has been lim-ited by economic depression, as in the case of Peru and Chile. It has also been curbed in socialist countries: Stringent policies have reduced the growth of Havana, Cuba, and Ho Chi Minh City, Vietnam. More interest-ing to policymakers are countries in which population flow to the bright city lights has been checked by im-proving the quality of life of those who live in the pe-riphery. Examples include Argentina and Venezuela, countries that have achieved a high level of economic development. The case of Sri Lanka illustrates that even a poor country can upgrade living conditions in the countryside. However, in this case, national economic growth and social cohesion were sacrificed.

In the foreseeable future, most public investment will take place in the major cities of underdeveloped countries. Without this investment, the stability of many of these countries may be threatened. However, some of the adverse effects of rapid concentration can be countered by governments who also pay attention to areas beyond the traffic-choked, smog-wrapped cities.

Summary

Cities exist in societies that create the conditions neces-sary for the appropriation of the *surplus product*. These conditions are met in *stratified market-exchange* societies.

In the 19th century, the stratified societies of Europe and North America experienced an urban transforma-tion. During this period of widespread innovation, cities, especially large manufacturing ones, were ugly cre-ations and horrifying environments for the poor. Denied access to the fruits of rapid economic growth, the worker bore the social costs of urban industrialization. The early 19th-century industrial city was character-ized by many small, relatively powerless enterprises. Toward the end of the 19th century, however, the mar-ket mode of economic integration took on a different appearance. There was a drift from individual to mo-nopoly capitalism. As a result, control of the most im-portant industries became more and more concentrated. Today, large corporations have a pervasive influence on cities throughout the capitalist world.

To explain how certain general forces tend to con-centrate activities in cities, we consider the model of *pure competition*. This model, which approximates 19th-century capitalism, shows that profit maximizing deci-sions lead to concentrated clusters of firms at nodes where production and assembly costs are minimized. In addition, the location of many firms in proximity to one another helps to reduce the transportation costs of shifting secondary inputs and outputs among them. Workers live close to their places of employment, in dense residential districts scattered around the indus-trial and commercial heart of the city.

In a capitalist society, urban land-use arrangements are structured by a *rent-maximizing land market*. We use classical urban location theory to illustrate why land-using activities are located where they are. Although the private appropriation, exchange, and use of urban land are steadily being eroded by the progressive so-cialization of space through planning, urban land use in contemporary North America and Western Europe is governed primarily by market exchange.

From the operation of the private land market emerge characteristic patterns of land use: a commercial core, a scattering of industry, and socially segregated neighborhoods. We describe three widely accepted models that capture the essence of the urban land-use system in North America before the advent of suburbia: the *concentric-zone*, *sector*, and *multiple-nuclei* models. De-partures from these patterns appear in cities in different cultural realms, as exemplified by the models of Latin American and Southeast Asian cities.

The focus of the last part of the chapter is on the patterns and problems of urban sprawl in developed and underdeveloped countries. The growth of cities, which is the inevitable concomitant of economic growth, has witnessed a host of deleterious break-downs and conflicts. In developed and developing countries, the predicament-laden course of urban ex-pansion and land-use development highlights the need for social control and management.

KEY TERMS

agglomeration economies	exurbs	morphological change	scale diseconomies
average total costs	filtering process	multicentered metropolis	scale economies
basic firms	fixed costs	multiple-nuclei model	sector model
ceiling rent	functional change	need creation marketing	social surplus product
central business district (CBD)	gentrification	new-town model	spatial mismatch theory
colonial city	income elasticity	nonbasic firms	spread city
competitive-bidding process	invasion and succession	production function	squatter settlement
concentric-zone model	jobs/housing balance	pure competition	stratified societies
constant returns to scale	localization economies	Randstad	urban land-use models
conurbation	LRAC curve	rank societies	urban realm
cumulative causation	market basket	redistribution	zone in transition
ecumenopolis	megalopolis	rent-maximizing land market	zone of working-class homes

SUGGESTED READINGS

Alonso, W. 1964. *Location and Land Use.* Cambridge, Mass.: Harvard University Press.

Benevelo, L. 1991. *The History of the City,* 2nd ed. Cambridge, Mass.: MIT Press.

Bourne, L.S., and Ley, D.F., eds. 1993. *The Changing Social Geography of Canadian Cities.* Montreal: McGill-Queen's University Press.

Brunn, S.D., and Williams, J.F. 1993. *Cities of the World: World Regional Development.* 2nd ed. New York: Harper & Row.

Cadwallader, M. 1985. *Analytical Urban Geography.* Englewood Cliffs, N.J.: Prentice Hall.

Carter, H. 1995. *The Study of Urban Geography.* 4th ed. London: Edward Arnold.

Clark, D. 1996. *Urban World/Global City.* London: Routledge.

Clark, W.A.V., and Dieleman, F. M. 1996. *Households and Housing.* New Brunswick, N.J.: Center for Urban Policy Research.

Drakakis-Smith, D. 1987. *The Third World City.* New York: Methuen.

Dutt, A.K., et al., eds. 1994. *The Asian City: Processes of Development, Characteristic and Planning.* Dordrecht/Boston/London: Kluwer Academic Publishers.

Friedan, B. J., and Sagalyn, L. B. 1989. *Downtown Inc.: How America Rebuilds Cities.* Cambridge, Mass.: MIT Press.

Garreau, J. 1991. *Edge City: Life on the New Frontier.* New York: Doubleday.

Geyer, H.S., and Kontuly, T.M. 1996. *Differential Urbanization.* New York: John Wiley & Sons.

Gilbert, A., and Gugler, J. 1992. *Cities, Poverty and Development: Urbanization in the Third World.* 2nd rev. ed. New York: Oxford University Press.

Gottmann, J. 1964. *Megalopolis.* Cambridge, Mass.: MIT Press.

Hart, J. F., ed. 1991. *Our Changing Cities.* Baltimore: Johns Hopkins University Press.

Knox, P. L. 1994. *Urbanization: An Introduction to Urban Geography.* Englewood Cliffs, N.J.: Prentice Hall.

Miles, E. S., and McDonald, J. F., eds. 1992. *Sources of Metropolitan Growth.* New Brunswick, N.J.: Center for Urban Policy Research.

Parrott, R., and Stutz, F. P. (1992). Urban GIS applications. In Maguire, Goodchild, and Rhind, eds., *GIS: Principles and Applications* (pp. 247–260). London: Longman.

Stutz, F. P. 1992. Maquiladoras branch plants: Transportation—labor cost substitution along the U.S./Mexican border. In Janelle, D., ed., *Snapshots of North America* (pp. 62-64). New York: Guilford Press.

Stutz, F. P. 1992. Urban and regional planning, Chapter 19. In T. Hartshorn, ed., *Interpreting the City: An Urban Geography Textbook* (pp. 445–476). New York: John Wiley & Sons.

Stutz, F. P., Parrott, R., and Kavanaugh, P. 1992. Charting urban space-time population shifts using trip generation models. *Urban Geography,* 13(5):468–475.

WORLD WIDE WEB SITES

STATE OF THE NATION'S CITIES
http://www.policy.rutgers.edu/cupr/

UNITED NATIONS POPULATION INFORMATION NETWORK (POPIN)
http://www.undp.org/popin/popin.html
Access to a wealth of data on the world, regional, and country levels.

USA CITYLINK PROJECT
http://usacitylink.com//
Comprehensive listing of WWW pages featuring American cities.

URBAN GEOGRAPHY ON THE WEB
http://www.staff.uiuc.edu/dgrammen/ugsg.html
The official home page of the Urban Geography Specialty Group of the Association of American Geographers. It has facts on the subdiscipline as well as links to web sites of interest.

EVOLUTION OF THE URBAN SYSTEM
http://sparky.sscl.uwo.ca/Demo2.html
This site, at the University of Western Ontario, Canada, combines the notion of urban evolution with a short icon sequence.

THE CALIFORNIA GEOGRAPHICAL SURVEY
http://130.166.124.2/CApage1.html
A digital atlas of California which contains more than 300 maps at the moment and many more to come.

C H A P T E R

7

CITIES AS RETAIL AND SERVICE CENTERS

O B J E C T I V E S

- To explain the concepts of marketing threshold, range, and hierarchy in the world economy
- To introduce central-place theory and the mercantile model of settlement
- To help you appreciate empirical regularities of the central-place concept
- To explore the market arrangements of different cultures of the world
- To acquaint you with some practical applications of the central-place theory
- To introduce you to the computerized smart city

Main Street, Schenectady, New York, in the 1940s.
(Photo: Library of Congress)

In Chapter 6, we concentrated on individual cities and their patterns of urban land use. However, urban society reveals itself not as one city but as many cities linked in an integrated *hierarchy of centers* of different functions and sizes. Why should there be urban hierarchies? What mechanisms control the size and spacing of cities? This chapter answers these questions by developing the concept of an urban hierarchy and by exploring patterns of service centers that have emerged to satisfy economic demands at local, regional, and international levels.

Although many factors determine locational patterns of cities, one classical location theory, *central place theory*, provides insights into the urban hierarchy. Central-place theory considers the locational pattern of market-oriented retail and service firms and the hierarchy of urban places insofar as they are market centers. It deals with relationships between market centers and consumers *within* regions. Other types of relationships also link regions to one another. For example, wholesale trade is conducted primarily *between* large regional centers; consequently, the locational pattern of these metropolises is determined by external trade linkages.

The metropolis-and-region network of retail and wholesale centers is a fairly accurate reflection of a domestic urban hierarchy. However, industrial restructuring throughout the world has led to the emergence of a global urban hierarchy. Today, a few international cities serve as centers of business decision making and corporate strategy formulation. To understand the urban hierarchy, these new business and corporate functions must also be considered.

One new strategy is the smart community or smart city: a geographical area ranging in size from a neighborhood to a multi-county region whose residents, organizations, and governing institutions are using information technology to transform their region in significant ways. Cooperation among government, industry, educators, and the citizenry, instead of individual groups acting in isolation, is preferred. The technological enhancements undertaken as part of this effort should result in fundamental, rather than incremental, change.

Central places and their hinterlands

Locational Patterns of Cities

The concentration of large populations in cities is essentially the result of the spatial organization of secondary and tertiary activities. These activities can be conducted more profitably when they are clustered rather than dispersed. Locational patterns of cities typ-

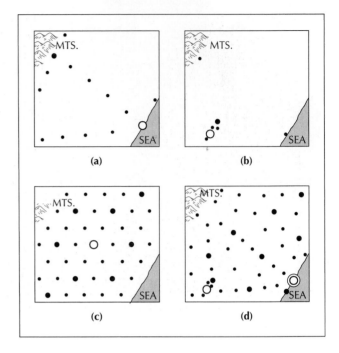

FIGURE 7.1
Different patterns of urban location: (a) transportation centers, aligned along railroads or at the coast; (b) specialized-function centers; (c) theoretical distribution of central places; (d) a theoretical composite grouping of different types of centers.

ically consist of three elements, which often work in concert (Harris and Ullman, 1945) (Figure 7.1):

1. A lined pattern of *transportation centers* that perform break-of-bulk and related services organized in relation to communication routes. These centers grow along transportation routes and at the junctions of different types of transportation, such as at road and railway junctions and at the head of sea, lake, or river navigation. Most of the largest U.S. cities originated along the seacoast, major rivers, and the Great Lakes, where there was a necessary break in transportation.

2. A pattern of *specialized-function centers*, which develop, either singly or in clusters, around a localized physical resource. Each center or cluster of centers is usually dominated by one activity, such as mining, manufacturing, or recreation. Examples are steel-making and metal-finishing cities in Pennsylvania, in proximity to the coal resources of the Allegheny Plateau, and resort towns along the California and Florida coasts.

3. A uniform pattern of centers that exchange goods and services with their hinterlands. Centers for the local exchange of goods and services are referred to as *central places*. Every settlement—large or small—is a central place, even though most

BELOW: Geographers and agricultural economists divide the world into three agricultural regions. The first is subsistence, or peasant agriculture. The second is commercial, or capitalist, agriculture, and the third is the socialist, or command economy, agricultural system. On this map, commercial and socialist systems are shown as commercial agriculture. Subsistence agriculture includes nomadic herding, slash and burn agriculture, and intensive sedentary subsistence agriculture. Southeast Asia and the Amazon Basin are the principle regions of shifting agriculture. Southwest Asia, and central East Asia are the principle areas of nomadic herding, or pastoral nomadism. Intensive subsistence agriculture includes rice-dominant cultivation and rice mixed with other crops. These regions include East Asia, South Asia, and Southeast Asia. Plantation agriculture occurs primarily in tropical and subtropical regions of Latin America, Asia, and Africa.

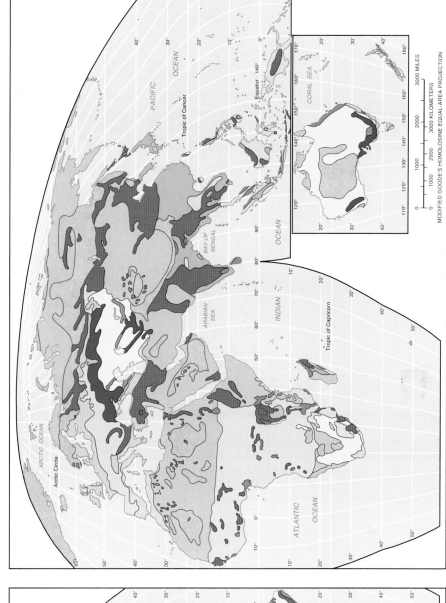

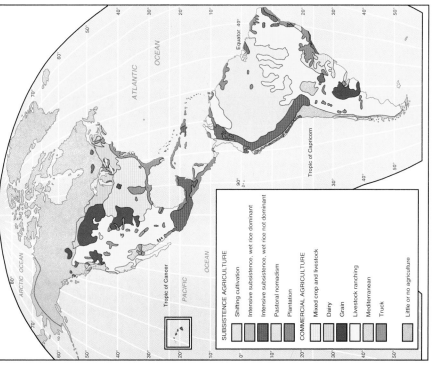

SUBSISTENCE AGRICULTURE

Shifting cultivation

Intensive subsistence, wet rice dominant

Intensive subsistence, wet rice not dominant

Pastoral nomadism

Plantation

COMMERCIAL AGRICULTURE

Mixed crop and livestock

Dairy

Grain

Livestock ranching

Mediterranean

Truck

Little or no agriculture

MODIFIED GOODE'S HOMOLOSINE EQUAL-AREA PROJECTION

2.1

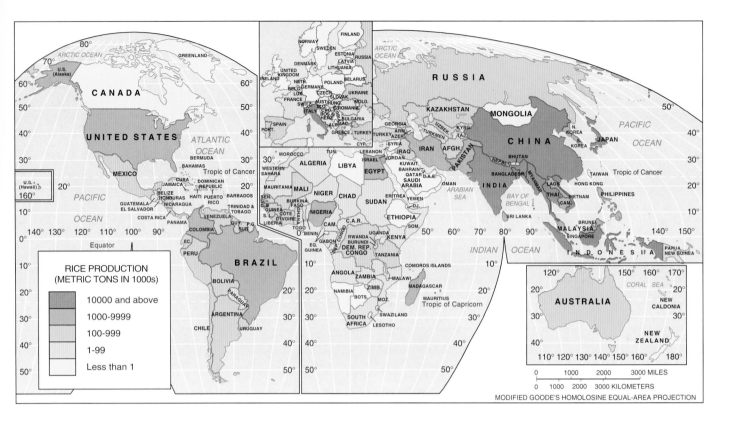

ABOVE: Anthropologists tell us that rice was domesticated in East Asia more than 7000 years ago. Unlike corn in the United States, rice is almost exclusively used for human consumption. Almost 2 billion people worldwide are fed chiefly by rice. It is the most important crop cultivated in the most densely settled areas of the world, including China in East Asia, India in South Asia, and Southeast Asia. These areas produce more than 90% of the world's rice.

BELOW: The developed countries produce the most beef products because a large amount of the grain crops can be fed to cattle to fatten them for slaughter. Poorer nations must consume all available food supplies directly or use them as revenue-producing exports. The United States, Western Europe, the former Soviet Union, Brazil, Argentina, and Australia are the leading producers of beef worldwide.

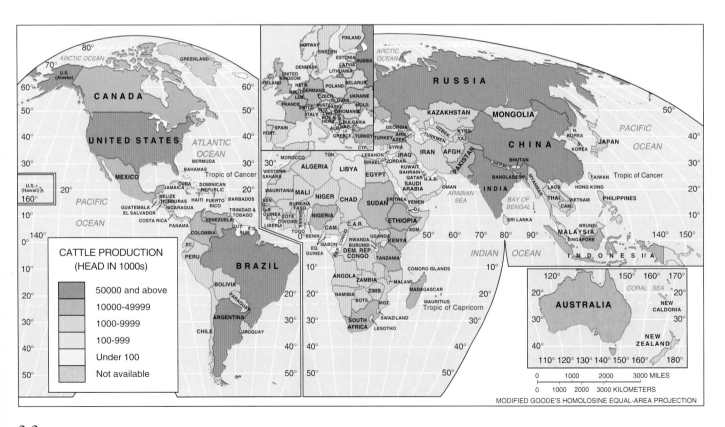

2.2

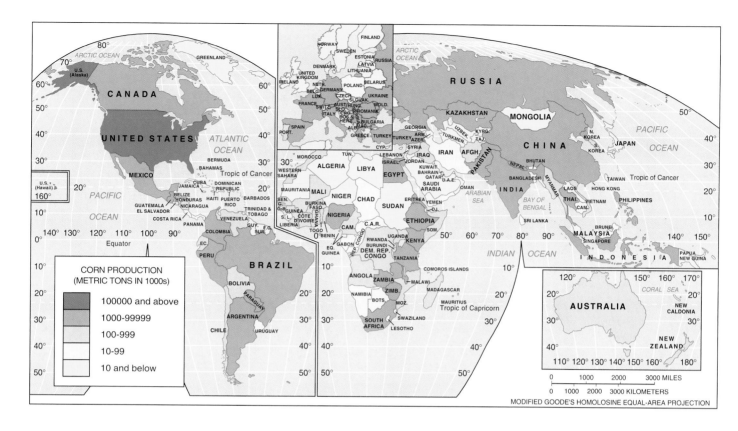

ABOVE: Corn was domesticated in Central America more than 5000 years ago and exported to Europe in the fifteenth century. The United States accounts for more than 30% of the world's corn production. Ninety percent of this corn crop is fed to cattle and livestock for fattening and meat production. Outside of the United States, corn is called maize.

BELOW: There is a high correspondence between the levels of economic development and the amount of per capita fluid milk production. Dairy farms are expensive to operate; thus, milk is a luxury in most areas of the world. Note Africa's paucity of milk production, except for South Africa. Most large cities of developed countries have milk production in facilities nearby due to milk's heavy weight and perishability.

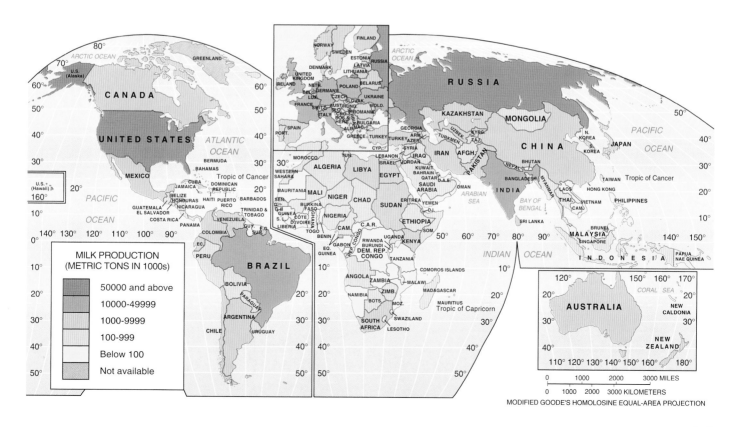

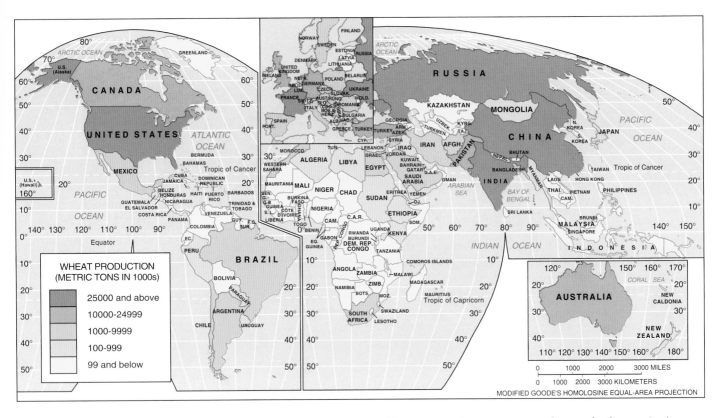

ABOVE: The United States, Canada, Australia, and Argentina are the world's primary wheat exporters; whereas the former Soviet Union countries(especially Kazakhstan), India, and China import the most wheat. China is the world's leading wheat producer, followed by the former Soviet Union and the United States. Wheat can be cultivated in a great variety of environments—from sea level to 10,000 feet and where levels of rainfall range from 12 to 70 inches a year. The few countries that have surplus grain to export have power and influence over those countries that must rely on food imports.

BELOW: According to the concentric zone model of E.W. Burgess, a city grows outward from the central business district (CBD) in a series of roughly concentric rings, each with a different land use. The zone in transition contains industry and low income housing. As with the concentric zone model, industries are confined to a district and have different housing submarkets.

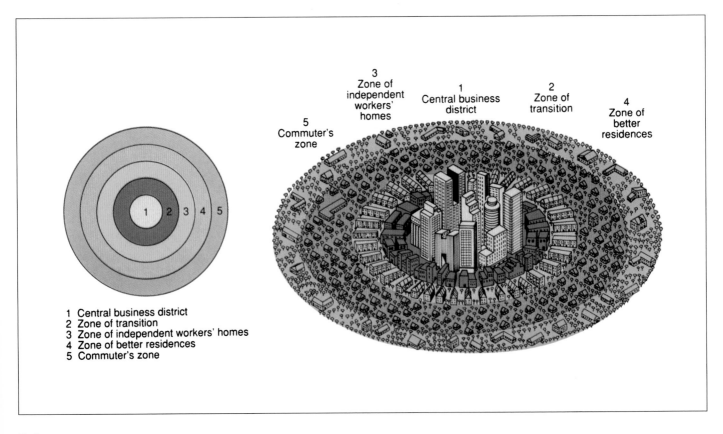

2.4

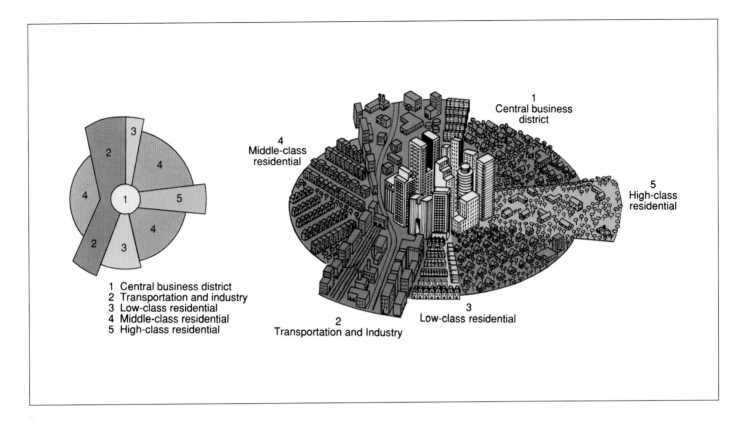

1 Central business district
2 Transportation and industry
3 Low-class residential
4 Middle-class residential
5 High-class residential

ABOVE: According to the sector model of H. Hoyt, a city grows outward from the central business district (CBD) in a series of wedges or sectors centered on routes of transportation.

BELOW: According to the multiple nuclei model of Harris and Ullman, cities develop around a series of centroids or nuclei. The usual land uses are identifiable in areas around various nucleations in downtown, but especially in the suburbs. No one city represents a pure form of any model. Most cities have visible aspects of all three models and are, therefore, composite.

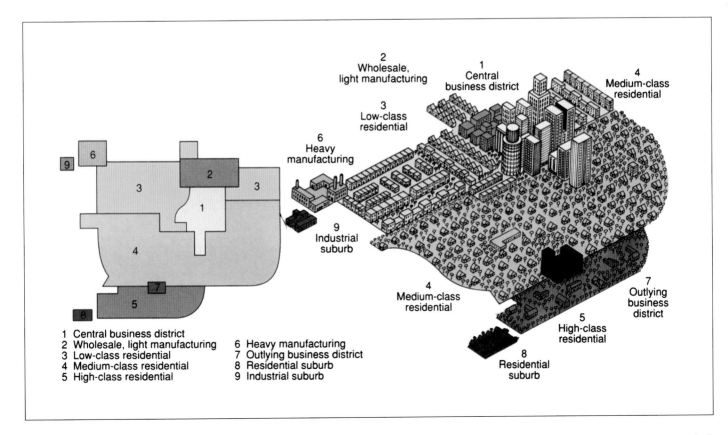

1 Central business district
2 Wholesale, light manufacturing
3 Low-class residential
4 Medium-class residential
5 High-class residential
6 Heavy manufacturing
7 Outlying business district
8 Residential suburb
9 Industrial suburb

2.5

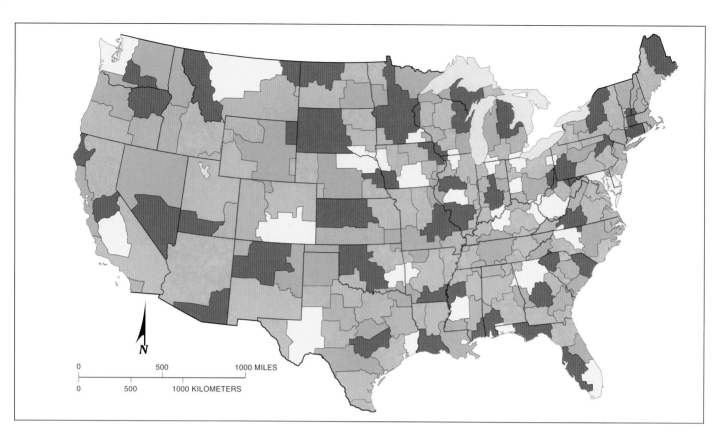

ABOVE: 1997 Daily Urban Systems. The metropolitan trade areas of the United States, based on commuting to the central city, on newspaper circulation, and on a number of other patronization/marketing indexes, are shown in this map. Dividing the country into such regions shows that all households have access, even if distant access, as in the case of Helena, Montana, to jobs, schools, shops, and cultural activities. How many daily urban systems can you name?

BELOW: An example of weight-losing, or bulk-reducing, raw material is copper ore. Copper has a high material index, meaning that 99% of the ore is wasted, and only 1% can be reduced to pure copper. Manufacturing is concentrated near the copper mines. This orientation to the source of raw material eliminates moving wasted ore tailings. Refining smelters are also located near the mines. In the United States, Arizona and New Mexico are the most important foundry areas.

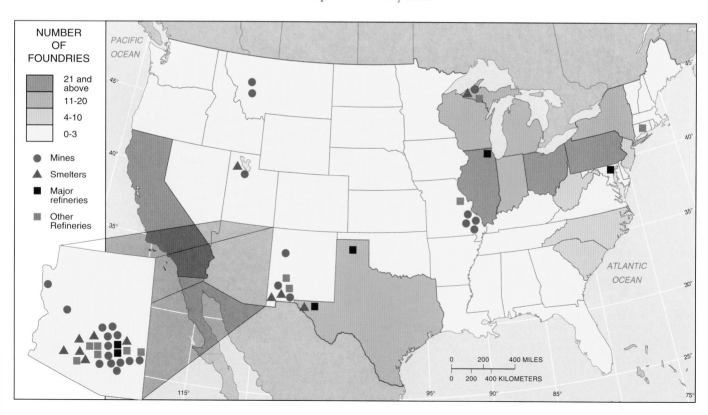

2.6

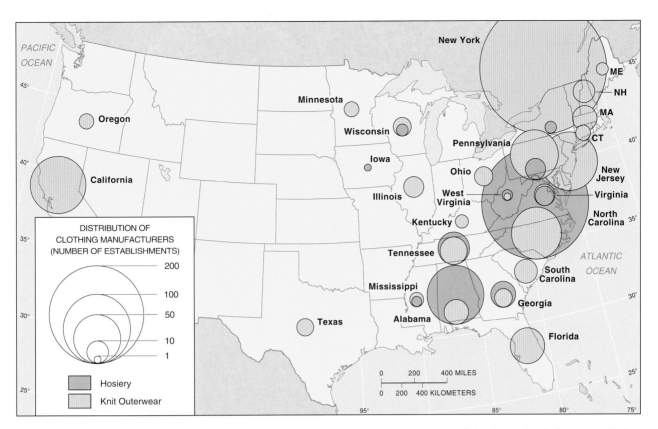

ABOVE: Hosiery and knit outerwear manufacturing establishments. Hosiery manufacture is a labor-intensive industry, and it has moved into the southeastern United States where the labor rate is lower. Knit outerwear manufacturers have remained primarily in New England and the Northeast because of the industrial inertia and the availability of more skilled workers, especially in the New York City area.

BELOW: The U.S. distribution of computer equipment manufacturers. California has the largest number of plants followed by clusters in the Northeast. Electronics manufacturing plants gravitate to highly skilled labor. The finished products are valuable, and higher wages can be paid. There is a concentration of these manufacturers in California because early electronics production, as part of the aircraft manufacturing industry, was centered on the West Coast.

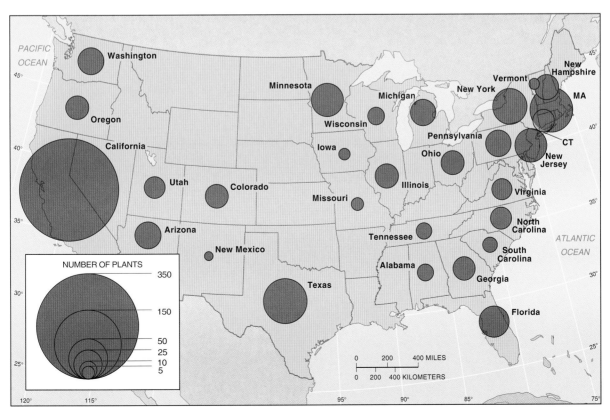

2.7

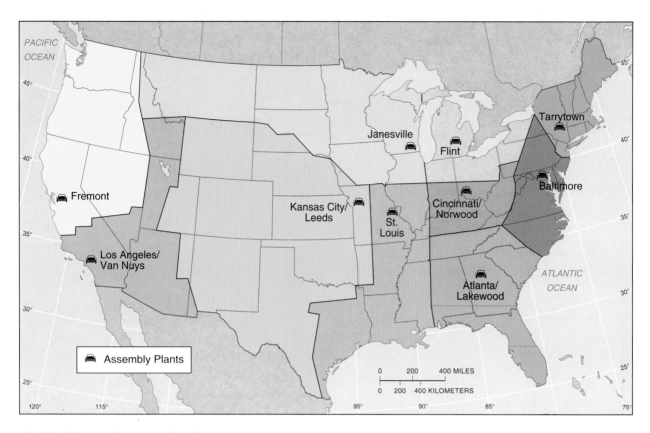

ABOVE: The distribution of 10 assembly plants across the United States in 1983 helped Chevrolet minimize the cost of distributing bulky products to consumers in each region. Recently, Chevrolet and other automobile manufacturers have reconcentrated their plants in the Midwest. Interior locations near Detroit are a result of the large increase in the variety of automobile models produced in North America. Automobile companies now operate a single assembly plant specializing in the production of one model for distribution throughout the United States and Canada, and therefore take advantage of scale economies.

BELOW: There has been a marked national shift from a disbursed coastal pattern during the last 20 years to a concentrated pattern near the original automobile core. Japanese-owned automobile parts manufacturing plants have followed U.S. motor vehicle assembly plants and clustered in the Midwest.

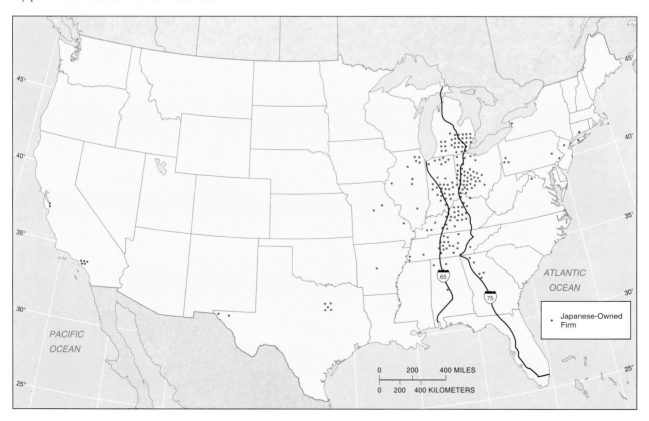

2.8

cities do not depend exclusively on central-place functions. Central places provide retailing and wholesaling services; banking, insurance, and real-estate services; governmental and administrative services; and recreational, medical, educational, religious, and cultural facilities.

Cities and Trade

No single locational theory applies to all three components of a settlement pattern, but the arrangement and distribution of central places have been the topics of much research. Central-place theory emphasizes that cities perform extensive services for their hinterlands. Business conducted totally within the hinterland is called *settlement-forming trade*. A true central place is based exclusively on these activities and can never support a population that transcends its hinterland. The number of jobs and, therefore, population size is a direct function of the demand generated in the hinterland. However, most settlements also have other functions that do not depend on hinterland size. For example, a manufacturing plant that sells to a national or an international market does not depend directly on the local retail and service hinterland. Such activity is called *settlement-building trade*. Each settlement also does some internal business—sales of goods and services to the residents of the center. This business is called *settlement-serving trade*.

Hinterlands

Central places serve areas larger than themselves. These areas are called *hinterlands*, tributary areas, trade or *market areas*, or *urban fields*. A trade area may be theoretically continuous. For instance, consider the circulation of a

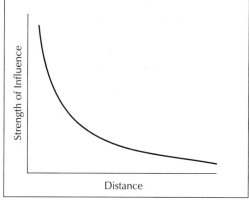

FIGURE 7.2
Cross section of an idealized trade area.

city's newspaper. Most papers are purchased by people who live in or near the city, but some may be purchased by people living thousands of kilometers away. In a graph describing this theoretical relationship, the curve approaches the distance axis infinitely closely but never reaches zero (Figure 7.2). For practical purposes, however, a city's trade area ends much closer to the origin.

Geographers have often used a *median*, or *line-of-indifference*, boundary to delimit trade areas. For example, the median boundary in newspaper circulation is the line between two cities along which 50% of the purchased newspapers are from one city. To further illustrate the concept, we shall consider the line-of-indifference boundary for six goods and services provided by Mobile, Alabama (Figure 7.3). The outermost lines—the isopleths for business in wholesale meat, wholesale produce, and wholesale drugs—mark the approximate boundary of Mobile's influence. From this map, we could delimit the

FIGURE 7.3
Areas served and influenced by Mobile, Alabama. Where would you draw the hinterland of Mobile? *(Source: Ullman, 1943, p. 58)*

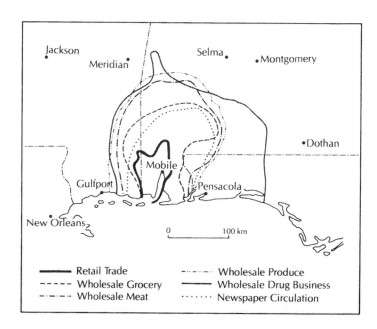

territory over which the city exerts more or less total dominance.

The map of areas influenced by Mobile is the result of fieldwork. To adopt the same method for a number of cities would be tedious. Alternatively, a shortcut means of determining a city's trade area is to measure one activity that is particularly expressive of the trade area. In the United States, for example, metropolitan trade areas can be determined by the extent and intensity of long-distance telephone calls, the length of the journey to work, or newspaper circulation. Newspaper circulation and commuting are conceptually good indicators of the social, economic, and cultural ties between a city and its tributary region. This is especially evident in metropolitan areas. People in the tributary area look to the regional newspaper for information on sales and social events, or to the city as an employment location. Areas of dominant influence are called *daily urban systems* (Figure 7.4).

Areas that focus on central places through circulation networks are known as *functional* or *nodal regions*. Every city has a nodal region. The size and shape of this region depends on the size of the city, the influence or competition of dominant neighboring centers, and the ease of travel.

The Law of Retail Gravitation

When satisfactory data cannot be obtained to determine urban trade areas, a modification of the *law of retail gravitation*—as described in Chapter 4—can be used to provide estimates. Although the break-point model provides a shortcut technique for determining trade-area boundaries, it has some deficiencies. City populations are assumed to be homogeneous masses; that is, the formula does not take into account cultural, economic, and other differences among people. Multipurpose trips and ease-of-transportation variables are not considered. The model also ignores the fact that each service has its own threshold and range characteristics.

The break-point model is therefore too rigid, a difficulty partially overcome in the probability model suggested by geographer David Huff (1963). This model still uses gravitation-model principles, but assuming that consumers have several centers from which to choose, it specifies that probability of choosing each center. The results can be mapped to produce probability surfaces for consumers choosing to shop in each center. Probability models have been used by applied economic geographers to assess the likely effect of adding to existing shopping centers, or more important,

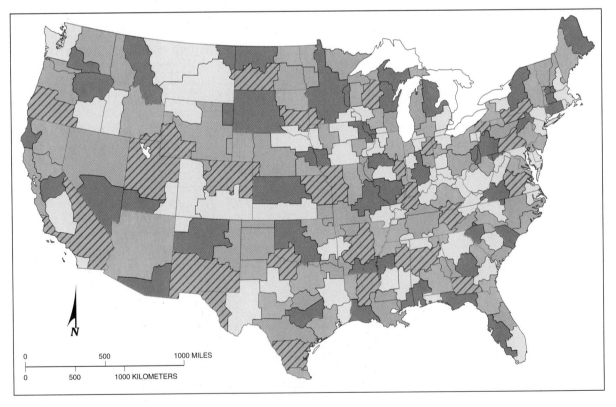

FIGURE 7.4
Daily urban systems. The metropolitan trade areas of the United States, based on commuting to the central city, on newspaper circulation, and on a number of other patronization/marketing entices, are shown in this map. Dividing the country into such regions shows that all households have access, even if distant access, as in the case of Helena, Montana, to jobs, schools, shops, and cultural activities. How many daily urban systems can you name?

the consequences of introducing a new center into an existing system of centers. The equation is:

$$I_{ij} = O_i \frac{A_j f_{(dij)}}{\Sigma A_j f_{(dij)}}$$

I_{ij} is the number of travelers going from place i to j.
O_i is the number of travelers leaving residential zone i.
A_j is the attraction of the shopping center in zone j.
$f_{(dij)}$ is the friction between i and j.

Gravitation models are not as useful for understanding the processes underlying retail and social behavior as they are for providing descriptions of the behavior of large populations. Moreover, they do not help us to understand the process underlying the formation of trade areas. Central-place theory does, however, provide a good basis for understanding the formation of trade areas.

The Questions of Central-Place Theory

Central-place theory attempts to answer four questions about cities in a regional economy:

1. How many cities (or central places) will develop?
2. Why are some cities larger than others?
3. Where will the cities locate?
4. What will be the size of each city's trade area?

An Elementary Central-Place Model

To create a general theory of central places, geographers commonly begin with a normative model that assumes the following:

1. An *isotropic surface* (uniform transportation costs in all directions).
2. A given level and uniform distribution per capita of demand and population.
3. Equal ease of transportation in all directions.
4. Settlements depending totally on hinterland trade.
5. Optimizing producers and consumers.
6. A steady-state economy free of government or social classes.
7. Ubiquitous (available at every place) production inputs at the same price.
8. No shopping externalities. Shopping externalities exist with complimentary goods (one-stop shopping) and imperfect substitutes (comparison shopping).

The model also assumes a *linear market*, with consumers evenly spaced along a road that extends across the isolated plain (Figure 7.5). Given these nine constraints, we next investigate the number of central places

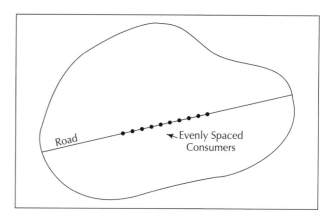

FIGURE 7.5
A linear market.

required to meet consumer demand, the size of trade areas, and the most efficient spacing of central places.

Threshold and Range

Threshold and range are key concepts in central-place theory. For a firm to offer a good—a *central function*—at a point along the road, it must sell enough to meet operating costs. The minimum level of effective demand that will allow a firm to stay in business is called the *threshold of a good*, or its scale economy. But given the assumption of an evenly distributed population and purchasing power, we can also speak of the minimum number of people necessary to support a central function.

The *range of a good* is the maximum distance that people are willing to travel to obtain the good at market price. A consumer who lives next door to the shop pays the store price ($0.50) for a loaf of bread. However, a consumer who lives at some distance from the shop must pay the store price plus the cost of travel to the central place. If the travel cost is $0.20 per kilometer, a consumer who lives 5 kilometers from the shop pays $1.50 for the bread (Figure 7.6a). Clearly, the price that a consumer pays is a direct function of distance. If price increases with distance from a central place or distribution point, demand should decline with distance. We can find a distance from the central place at which demand is zero. That point is the range of a good (Figure 7.6b). The range of a good is the distance, R, in both directions from a distribution point in a linear market (Figure 7.7).

Order of a Good and a Center

Different goods have different thresholds. Inexpensive, frequently purchased, everyday necessities have

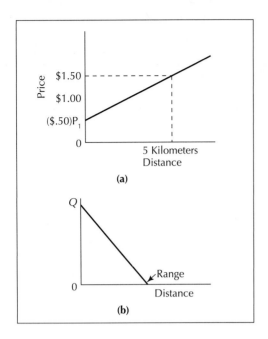

FIGURE 7.6
(a) Price and (b) range of a good.

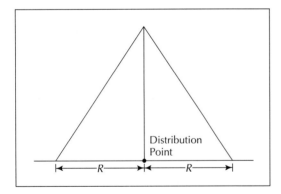

FIGURE 7.7
Range as distance.

low thresholds. Goods that are costly and purchased infrequently have higher thresholds. Items with low thresholds, such as eggs purchased at a supermarket, are called *low-order goods*. Goods with higher thresholds, such as furniture, are called *higher-order goods*. Thus, central functions can be ordered on the basis of their threshold size. The highest-order good has the highest threshold, and the lowest-order good has the lowest threshold.

Just as we can order goods, we can order centers. The order of a center is determined by the highest-order good offered by the center. Low-order centers offer only low-order goods; high-order centers offer high-order goods.

Emergence of a Central-Place Hierarchy

We are now in a position to derive a *hierarchy* of central places. Assume that highest-order places, or *A*-level places, offer all goods from 1,000 to 1 (Table 7.1). The market area of each *A*-level place must include at least 1,000 people. If 10,000 consumers live along our imaginary linear market, 10 *A*-level centers can be established. Two *A*-level centers and their market areas are shown in Figure 7.8. We assume a population density of 10 people per kilometer, so the minimum market area of each *A*-level center is 100 kilometers, or 50 kilometers on either side of each *A*-level center. We also assume that competition forces market areas to be as small as possible—that is, 100 kilometers equals 1,000 consumers. This minimizes the travel cost that the most distant consumer must pay. A consumer on the dividing line between two *A*-level centers will purchase goods from both centers in equal measure.

The good that defines the *A* level of the hierarchy has a threshold of 1,000, but there is also a good with a threshold of *A* divided by two, or 500. The threshold market area for that good is 50 kilometers long, or 25 kilometers on either side of a distribution point. A market area of 500 centered on A_1 and A_2 allows an addi-

◙

Table 7.1
Goods and Threshold Size

	Goods														
Centers	*1000,*	*999,*	*998,*	*...,*	*502,*	*501,*	*500,*	*...,*	*252,*	*251,*	*250,*	*...,*	*3,*	*2,*	*1*
A	X	X	X	X	X	X	X	X	X	X	X	X	X	X	X
B					X	X	X	X	X	X	X	X	X	X	X
C											X	X	X	X	X

FIGURE 7.8
A *K* = 2 hierarchy.

50 km	A_1	50 km	*B*	50 km	A_1	50 km

1000 People
A-Level   1000 People
A-Level

500 People
B-Level   500 People
B-Level   500 People
B-Level

tional 500-person market area centered on the midpoint between the two *A*-level centers. A central place locating there is at the *B* level; it can offer all goods with a threshold of 500 or less. The good that defines the *B* level has a threshold of 500 and is called a *hierarchical marginal good*. A hierarchical marginal good is the highest-order good offered by a given level of the central-place hierarchy. What threshold size will define *C*-level centers? Their threshold size is *B* divided by two, or 250. They locate midway between higher-order centers; thus, they occur every 25 kilometers along our imaginary road.

Our linear hierarchy follows the "rule of twos": Each successive level is defined by a function with a threshold one-half the size of the next-highest hierarchical marginal good. The rule of twos also applies to market-area sizes and the spacing of centers. In *central-place theory*, this type of hierarchy is known as a *K*-equals-two hierarchy, because two is the constant parameter of the system. The letter *K* stands for the German word *Konstant*.

So far in our discussion, we have constructed a hierarchy of central places based on the concepts of threshold and range. We have seen that the number of required centers is minimized, the number of consumers served is maximized, and the distance that consumers must travel for a given set of central functions is minimized. We have seen that higher-order centers, which offer more functions and therefore employ more people, have larger populations than lower-order centers. Higher-order centers are also more widely spaced, serve larger market areas, and occur less frequently. In summary, *our central-place hierarchy can be regarded as a multiple system of nested centers and market areas. Lower-order centers and their market areas nest under the market areas of higher-order centers.* Next, let us look at the more interesting two-dimensional hierarchy.

A Rectangular Central-Place Model

Let us examine a small region of the midwestern United States, Jay Hawk County, Kansas, which has 100,000 people and an economy based on only three

Madison, Indiana, from across the Ohio River, a low-order central place. *(Photo: Piet van Lier, Impact Visuals Photo & Graphics, Inc.)*

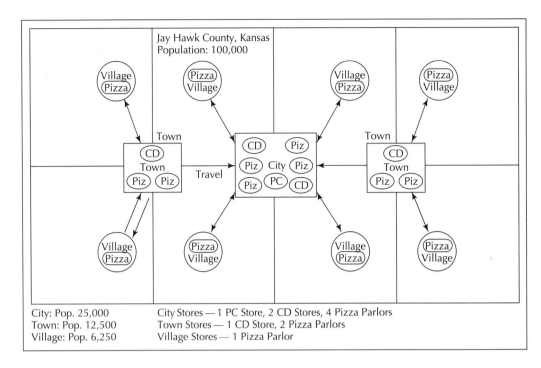

Jay Hawk County, Kansas
Population: 100,000

City: Pop. 25,000
Town: Pop. 12,500
Village: Pop. 6,250

City Stores — 1 PC Store, 2 CD Stores, 4 Pizza Parlors
Town Stores — 1 CD Store, 2 Pizza Parlors
Village Stores — 1 Pizza Parlor

FIGURE 7.9
Jay Hawk County,
Kansas: a rectangular
central-place system.
*(Source: Stutz, 1994,
p. 370)*

products: pizza, CDs, and PCs (Figure 7.9). These three products have different per capita demands and thresholds (or scale economies):

1. PCs: Thresholds are high, relative to demand, and each PC store requires a population of 100,000, so only one store is possible. It will serve the whole region by locating in the center of Jay Hawk County.
2. *CDs:* The threshold is medium, compared with PCs and per capita demand. Each CD store requires 25,000 people to support it. Because no excess profits are allowed, four CD stores emerge.
3. Pizza: Pizza is demanded more often than PCs and CDs because it is cheaper and because it has a small scale economy. Thresholds, therefore, are low compared with PCs, CDs, and per capita demand. Each pizza parlor requires a population of 6,250 people, so 16 pizza parlors emerge.

Remember that the central-place model is concerned only with retailing and service firms, and their markets, not manufacturing. Therefore, stores locate according to access to customers. This is why the PC store will locate in the center of Jay Hawk County, where the cost of traveling to the store is minimized. The raw materials, or inputs, cost the same everywhere, and the population is initially spread evenly. As the PC store locates in the center of Jay Hawk County, other types of stores with different thresholds also locate there, and a higher population emerges than in the rural parts of the county. Some of these people in the

city are employees of the PC store and the other stores because they want to minimize their commuting costs.

The demand for CDs is greater than the demand for PCs. Likewise, the threshold for a CD store is one-quarter that of a PC store. Consequently, two CD stores will locate in the city, and two additional CD stores will locate outside the city, equidistant to the edge of Jay Hawk County. These locations will minimize the customers' travel costs. Employees of the CD stores, and others, will locate near the stores. As a result, these areas will also have higher densities than the rural areas of the county. The region can support only four CD stores: two in the city and one in each new smaller city, or town.

Next, the pizza parlors enter the Jay Hawk County economy. Because the thresholds of a pizza parlor are one-sixteenth that of a PC store, and one-fourth that of a CD store, four pizza parlors can locate in the city, two additional ones in each of the towns, and eight more elsewhere in Jay Hawk County. As a result, the rural area is divided into eight equal trade areas. Again, these latter eight places are more densely populated than the rural parts of the county, and they develop as small towns, or villages.

Jay Hawk County has a total of 11 central places. Everyone in the county travels to the city to buy a PC. The city is in the center of the county and has a population of 25,000. This means that it is large enough to support two CD stores and four pizza parlors. (It will have other functions or types of stores as well, but for simplicity's sake we chose to use a three-function economy.)

Each of the two towns has a population of 12,500 and a market large enough to support one CD store and two pizza parlors. People who live in villages and want to purchase CDs will come to these towns—if the towns are closer than the city. The four villages on the outskirts of the county are closer to the two towns, so these people will shop for CDs at the towns. The four center villages are as close to the city as to the towns, so these people will shop for CDs at the city. Each village has a population of 6,250 and can support only one pizza parlor.

Figure 7.10 shows the frequency distribution of central places in Jay Hawk County by size and by range, from large (rank=1) to small (rank=4–11). There is only one city, and it has high thresholds for its goods (PCs). It can support many goods and services. The small central places have low populations and can support only low-threshold goods (pizza). Goods and services flow down the hierarchy from large places to smaller places. PCs are purchased in the city and flow down to the towns and villages. CDs are purchased in the towns and flow down to the villages as people come to the towns to shop. Two central places of the same order, or size, do not trade with each other because neither has anything that the other needs.

This example includes three levels of central places, each with a distinct trade-area size and city spacing. Each has a set population. Because the city requires many people to support its high-order functions, in this case a PC store, there can be few big cities.

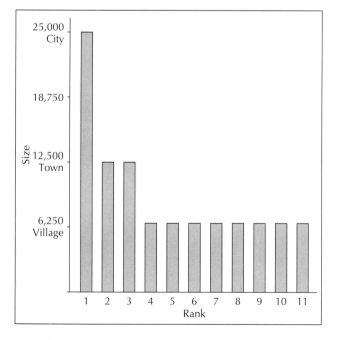

FIGURE 7.10
Rank-size urban hierarchy in Jay Hawk County, Kansas.
(Source: Stutz, 1994, p. 371)

A Hexagonal Central-Place Model

In the early 1930s, the German geographer Walter Christaller (1966) established the foundations of central-place theory. On the basis of the simplifying assumptions that we introduced earlier and the concept of a range of a good, Christaller constructed a deductive theory to explain the size, number, and distribution of clusters of urban trade and institutions. To demonstrate hierarchical interrelations between places in a competitive market society, he built three geometric models. His central places are arranged according to marketing, transportation, and administrative principles. According to Christaller, these three principles underlie the most efficient system of central places.

THE MARKETING PRINCIPLE

The *marketing principle* assumes the largest provision of central-place goods and services from the minimum number of central places (Figure 7.11). Each *B*-level central place is midway between three neighboring centers of the next-highest order. Midway points are corners of hexagonal market areas of the next-highest order. Each higher-order place is surrounded by six places of the next-lowest order.

Figure 7.12 illustrates the progression of central places and market areas for a three-level hierarchy. The market area of each *A*-level center passes through six lower-order *B*-level central places. *A*-level market areas are three times larger than *B*-level market areas: Each *A*-level market area includes the *B*-level market area centered on the *A*-level area, plus one-third of the six surrounding *B*-level market areas [1 + (1/3)(6) = 1 + 2 = 3B = level areas]. Distances separating places at the same level of the hierarchy are the same. If lower-order places are 1 unit apart, rival higher-order places, dominating three times the area and three times the population, are K-equals-three networks in which the number of trade areas with successively less specialized levels progresses by a "rule of threes"; that is, the number of central places increases geometrically: 1, 3, 9, 27. For example, one *A*-level area contains the equivalent of three *B*-level areas or nine *C*-level areas.

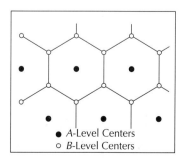

FIGURE 7.11
Location of centers: a
K = 3 hierarchy.

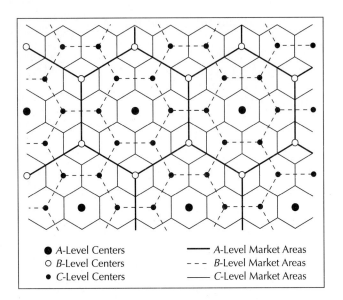

FIGURE 7.12
Market areas: a *K* = 3 hierarchy.

Christaller argued that the system of central places developed on the basis of the range of a good is rational and efficient from an economic viewpoint. However, he did note that real-world conditions produce deviations from this system. Actual conditions include historical circumstance, government interference, social stratification, income differences, and topographical variations not accounted for in the normative model.

THE TRAFFIC PRINCIPLE

If an optimum road network is superimposed on a marketing hierarchy, *B*-level centers do not lie on the *A*-level road network (Figure 7.13a). More roads must be constructed if *B*-level centers are to be connected. Christaller rejected traffic routes in the marketing system and asked this: How can connectivity between places be maximized and network length minimized? By shifting *B*-level centers to a point midway between each pair of *A*-level centers, he found the answer (Figure 7.13b). The pattern of centers arranged according to the *traffic principle* results in a *K*-equals-four hierarchy. The *B*-level market areas are one-fourth the size of the *A*-level market areas. Each *A*-level market area dominates the *B*-level area centered on it, plus one-half the surrounding six (equals three), for a total of four. The number of market areas at successively less specialized levels of the hierarchy progresses by the "rule of fours"; that is, the number of central places increases geometrically: 1, 4, 16, 64.

Compared with the marketing principle, the traffic principle requires more centers at each level of the hierarchy if the entire hexagonal landscape is to be adequately provided with central places for the distribution of goods and services. The advantage of a more efficient transportation system for moving goods cheaply is counterbalanced by the additional distance that consumers must travel to reach a center at a given level of the hierarchy.

THE ADMINISTRATIVE PRINCIPLE

The *administrative principle* requires sociopolitical separation of market areas. This is achieved when each central place controls six dependent centers (Figure 7.14). Hinterlands nest according to the "rule of sevens" (*K*-equals-seven). They are larger than those in either *K*-equals-three or *K*-equals-four systems, which means that consumers must travel farther to reach a center of a given level in the system.

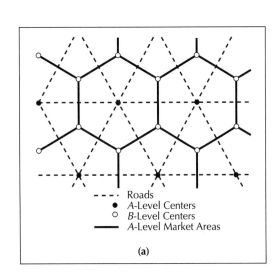

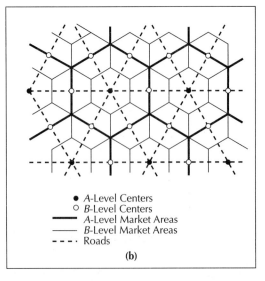

FIGURE 7.13
The traffic principle: (a) optimum transportation network: maximum connectivity for *A*-level centers (*K* = 3); (b) a *K* = 4 hierarchy.

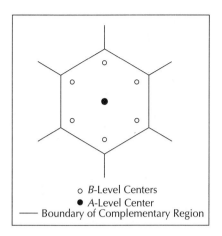

FIGURE 7.14
Two-level hierarchy of central places under the administrative principle.

Southern Germany

Christaller tested his central-place model in southern Germany and bordering areas of France, Switzerland, and Austria (Table 7.2). He recognized a seven-level hierarchy ranging from market hamlets to regional capital cities in which centers at each level dominate approximately three times the area (Column 4) and three times the population (Column 5). The distance between similar centers increases by three over the preceding smaller category (Column 2). The smallest centers are about 7 kilometers apart because 4 or 5 kilometers, roughly a 1-hour walking distance, corresponded to the market area for the smallest centers. Centers of the next order of specialization, township centers, are 12 kilometers apart.

Networks of central places nest within the market areas of regional capital cities. The cities are Munich,

Frankfurt, Stuttgart, and Nuremburg, together with the border cities of Strasbourg in France and Zurich in Switzerland.

Christaller found that the expected pattern was approached most closely in poor, thinly settled farm districts that were virtually self-contained. Thus, deviations from the rational pattern were uncommon in many parts of southern Germany. On the other hand, the theoretical ideal was not evident in the highly industrialized Rhine-Ruhr region.

WHOLESALING AND THE MERCANTILE MODEL OF SETTLEMENT

Geographer James Vance (1970) introduced the argument that central-place theory is too parochial, sufficing as an explanation only for the trading-settlement structure of an area. Central-place theory deals with relationships between customers and sellers of goods *within* regions. It does not account for the wholesale trade that links regions. Wholesaling, which involves the sale of goods from one entrepreneur to another, is conducted primarily *between* higher-order centers, and these external linkages influence their locational patterns. In addition, local settlement hierarchies are influenced by long-distance external ties.

Wholesalers subject regions to external change and stimulate growth of wholesaling centers. Vance thought of these centers as "unraveling points" in the geography of trade. They link production areas, mediate trade flows, and determine the metropolitan centers from which central-place patterns develop to meet the consumers' demands.

Vance doubted whether any region in North America was economically isolated enough to have begun in a

🔷

Table 7.2
The Urban Hierarchy in Southern Germany

Central Place (1)	Towns		Tributary Areas	
	Distance Apart (2)	Population (3)	Size (Sq Km) (4)	Population (5)
Market hamlet (Marktort)	7	800	45	2700
Township center (Amtsort)	12	1500	135	8100
County seat (Kreisstadt)	21	3500	400	24,000
District city (Bezirksstadt)	36	9000	1200	75,000
Small state capital (Gaustadt)	62	27,000	3600	225,000
Provincial head city (Provinzhaupstadt)	108	90,000	10,800	675,000
Regional capital city (Landeshaupstadt)	186	300,000	32,400	2,025,000

Source: Ullman, 1940–1941, p. 857.

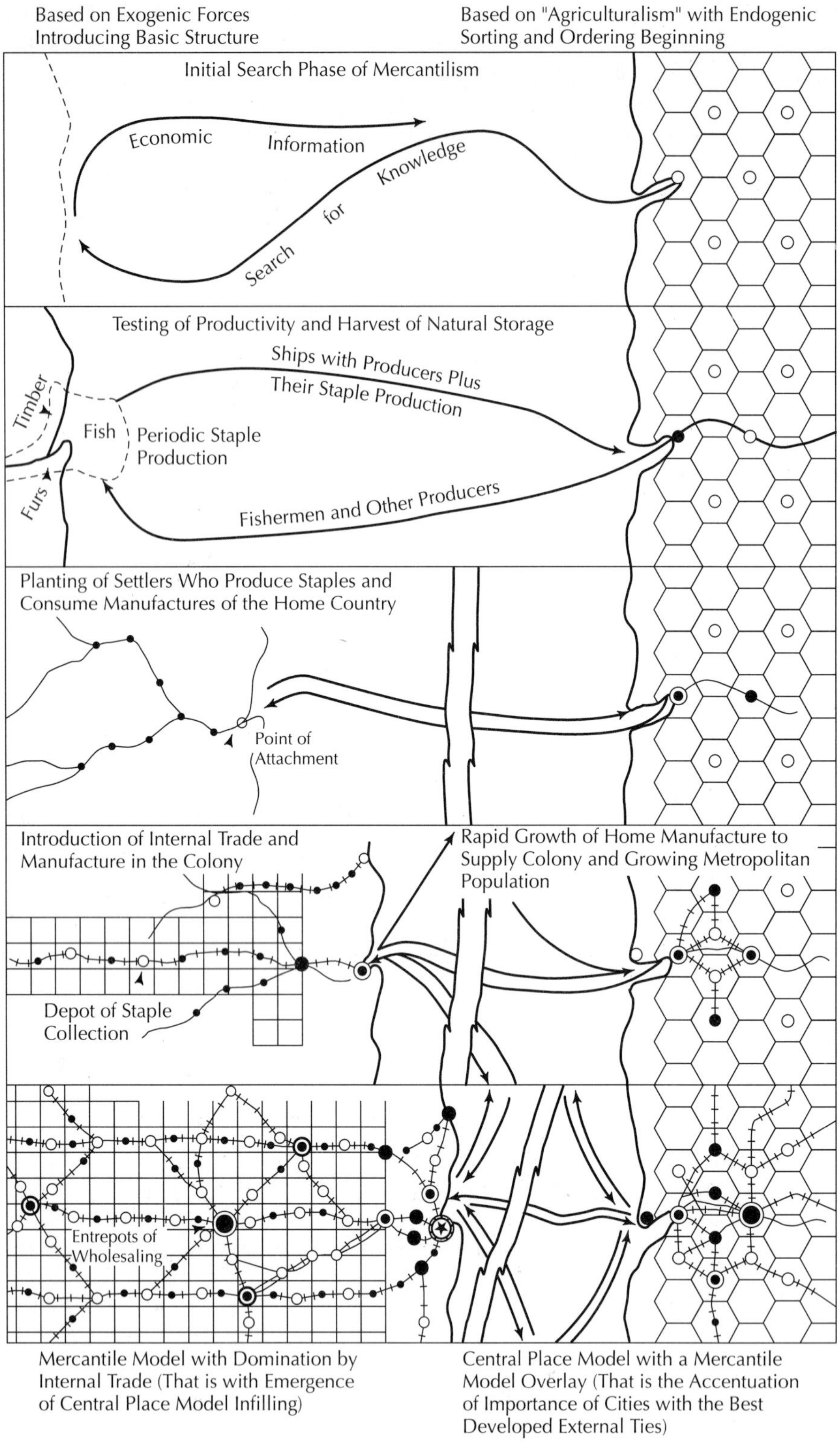

FIGURE 7.15
Urban evolution in mercantile and central-place models. *(Source: Vance, 1970, p. 151)*

Based on Exogenic Forces Introducing Basic Structure

Based on "Agriculturalism" with Endogenic Sorting and Ordering Beginning

Initial Search Phase of Mercantilism

Economic Information Knowledge Search for

Testing of Productivity and Harvest of Natural Storage

Ships with Producers Plus Their Staple Production

Timber Fish Periodic Staple Production Furs Fishermen and Other Producers

Planting of Settlers Who Produce Staples and Consume Manufactures of the Home Country

Point of Attachment

Introduction of Internal Trade and Manufacture in the Colony

Rapid Growth of Home Manufacture to Supply Colony and Growing Metropolitan Population

Depot of Staple Collection

Mercantile Model with Domination by Internal Trade (That is with Emergence of Central Place Model Infilling)

Entrepots of Wholesaling

Central Place Model with a Mercantile Model Overlay (That is the Accentuation of Importance of Cities with the Best Developed External Ties)

New York, Philadelphia, Baltimore, and Charleston—were traders' towns. Created before their hinterlands expanded, they were unraveling points for the distribution and collection of goods. Subsequently, interior mercantile cities developed to serve as primary collecting points for resources shipped back to the East Coast. Cities such as Chicago, Cincinnati, Memphis, Minneapolis, St. Louis, Kansas City, and Omaha owe much of their early growth to their wholesaling function. Although Vance regarded only cities of more than 50,000 people as true distribution centers, many small towns also developed on the basis of long-distance trading connections. Eau Claire, Wisconsin, for example, was a workplace town of about 10,000 people in the 1880s; however, the products of its many lumber companies were shipped to places as far away as St. Louis.

Vance described the development of distant trade between Europe and North America and the subsequent evolution of the American urban hierarchy. After ascertaining that an area had sufficient economic potential, Europeans established mercantile centers within it. These centers linked European countries and their sources of raw materials, and the centers grew as the size of the trading system increased. Eventually, the central-place model began to characterize American settlement, with a subsequent parallel growth of settlement in accordance with both central-place and mercantile models.

Vance suggested that central-place dynamics may have sketched the European settlement pattern (Figure 7.15). For example, in areas such as Christaller's southern Ger-

closed local region. He preferred to view the history of settlement patterns not only in terms of local trading patterns (central-place model) but also in terms of long-distance trading connections. He argued that broad-scale settlement of North America must be considered in the context of long-distance trade. Colonial towns—Boston,

many, economies were fairly isolated from one another in the feudal Middle Ages. However, by the early modern period the feudal economy was giving way to mercantilism, and central places with developing external ties grew rapidly. Long-distance trade accentuated the importance of Bristol, St.-Malo, Seville, Cadiz, and Lübeck, to name only a few merchant cities.

EVIDENCE IN SUPPORT OF CENTRAL-PLACE THEORY

Upper Midwest Trade Centers

We begin our examination by looking at the trade centers of America's northern heartland (Borchert, 1987), or the Upper Midwest. The region coincides with the Ninth Federal Reserve District, which extends 2,500 kilometers along the Canadian border, from Montana in the West to the Upper Peninsula of Michigan in the East. In this rather homogeneous region, we might anticipate the regular pattern of hexagonal market areas suggested by Christaller. But the underlying density of the farm population varies considerably. Rural population densities are greatest in the southern and eastern regions of Minnesota, which represent the northern and western margins of the Midwest agricultural heartland. Densities decline toward the west and the north. In the west, aridity reduces the carrying capacity of the land. In the north, infertile soils and short growing seasons produce lower yields per unit of farmland. Compared with southern parts of the region, the amount of land in farms is small. More than 90% of the land in southern Minnesota is in farms, but this figure falls to less than 10% in northern Minnesota, where forests dominate. A hexagonal lattice that conforms to the distribution of the farm population supports Isard's theory—cells are smaller in areas that are more densely populated and larger in areas of sparse settlement.

Hierarchy of Business Centers

John Borchert (1963), a geographer at the University of Minnesota, demonstrated that a hierarchy of central-place functions exists in the Upper Midwest. He selected 46 functions and determined those that were typical of various orders in the hierarchy of business types. He grouped central functions into convenience, specialty, and wholesale categories for eight types of trade centers (Figure 7.16).

Hamlets, the lowest-order central place recognized by Borchert, have only gasoline service stations and eating and drinking establishments. The next two levels, *minimum-convenience* and *full-convenience centers*, provide everyday necessities. Minimum-convenience centers

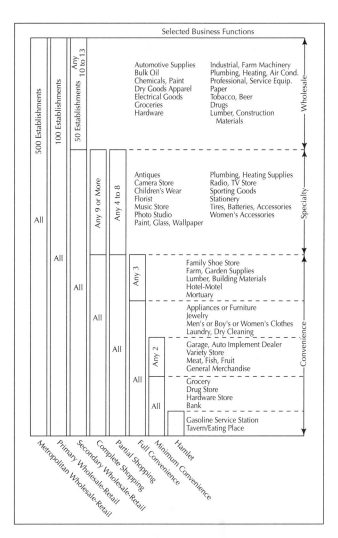

FIGURE 7.16
Trade-center types in the Upper Midwest. *(Source: Borchert, 1963, p. 12)*

have hamlet-level functions plus a hardware store, a drugstore, a bank, a grocery store, and two other convenience functions, such as a variety store.

Full-convenience centers have all hamlet-level and minimum-convenience functions, as well as stores dealing in laundry or dry cleaning, jewelry, appliances or furniture, clothing, lumber, building materials, shoes, and garden supplies. In addition to shops, most full-convenience centers have a hotel or a motel. Still higher in the hierarchy of business types are *partial shopping* and *complete-shopping centers*, offering specialty goods and services. *Secondary wholesale-retail*, *primary wholesale-retail*, and *metropolitan retail centers* are the highest-order places. With regard to the frequency of trade-center types, as the hierarchical level increases, the number of trade centers decreases. There is also a strong relationship between trade-center types and population size (Table 7.3).

▣

Table 7.3

Frequency and Median Size of Trade-Center
Types in the Upper Midwest

Type of Center	Number of Centers	Median Population (Thousands)
Wholesale-Retail Centers		
Metropolitan	1	1440.0
Primary	7	55.4
Secondary	10	32.2
Shopping Centers		
Complete	78	9.5
Partial	127	2.5
Convenience Centers		
Full	111	1.5
Minimum	379	0.8
Hamlets	1539	0.2

Source: Borchert, 1963, p. 11.

Minneapolis-St. Paul is the largest and only metropolitan wholesale-retail center in the Upper Midwest. Besides convenience, specialty, and wholesale functions, the Twin Cities provides other services for its massive trade area, such as regional head offices of insurance companies, and specialized medical, educational, and administrative facilities. People living as far as 1,600 kilometers away may never visit the Twin Cities. Instead, they obtain goods and services from lower-order trade centers. The highest-level centers that many people living beyond the Twin Cities need to reach are primary and secondary wholesale-retail centers such as Eau Claire, Fargo-Moorhead, and Duluth. Nonetheless, the Twin Cities is the controlling center of the Upper Midwest economy. Trade-area residents feel the influence of the metropolis through communications, banking, agricultural marketing, and retail-wholesale relationships.

Central-Place Pattern

The geographic distribution of trade centers in the Upper Midwest conforms roughly to *central-place theory* (Figure 7.17). Wholesale-retail centers are widely spaced, and hamlets are the most numerous centers. Wider spacing of all classes of trade centers to the north and west is a striking feature on the map.

It is generally agreed that there are too many trade centers in the Upper Midwest. Most of the 2,500 settlements that dot the Upper Midwest map were established under conditions quite different from those of today. After the railroads opened up the region in the late 19th century, immigrants established small farms. To meet the needs of farm families, low-order central places developed. These places were closely spaced in response to slow, difficult travel conditions. In recent decades, two changes have produced stress within the original framework of settlements: migration and transportation.

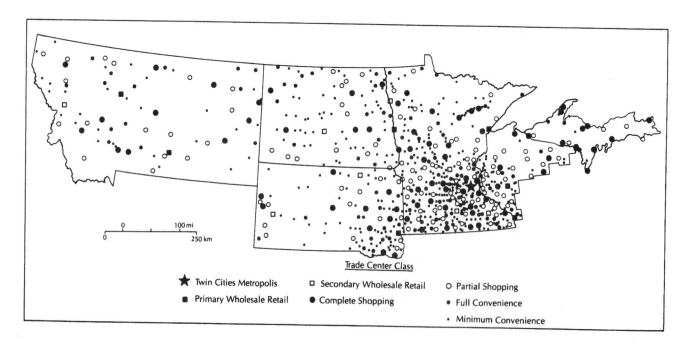

FIGURE 7.17
Distribution of trade centers in the Upper Midwest. *(Source: Borchert, 1963, pp. 13–14)*

Migration has had the most influence on the structure of trade centers. In the past 60 years, the farm population has declined sharply as farms increased in size and farmland was abandoned. In addition to a declining farm population, the region has experienced net out-migration. Within the region itself, the population has shifted progressively from farms and small trade centers into larger urban areas, and from central cities into the suburbs and the countryside. Examples of trade centers that have grown rapidly are the Twin Cities, Rochester, and Fargo-Moorhead.

Modern highways crisscrossing the region illustrate a second influence on the structure of trade centers. Improvements in transportation after 1914, such as paved roads and the widespread use of automobiles, enabled consumers to bypass smaller centers and to patronize larger ones. Increased consumer mobility meant that small trade centers could not compete with larger towns that offered a wider variety of services in larger quantities.

Dispersed Cities

In the years to come, most of the Upper Midwest population will be concentrated in Minneapolis-St. Paul (metro cluster) and other low-density metropolises (urban clusters). These urban clusters, or dispersed cities, are products of modern transportation and communication networks and are formed by linkages of complete and wholesale-retail centers (Figure 7.18). They all share the following features:

1. The length of any link or corridor can be traveled in less than 60 minutes or 30 minutes, respectively.
2. Each cluster has multiple shopping and service centers, which are more complementary than competitive.
3. Each cluster has many low-order retail and service centers.
4. Each cluster has industrial and wholesale zones, public higher-education facilities, public hospital facilities, and newspapers and broadcasting stations.
5. Each city within a cluster functions independently to a large extent. However, there is also considerable interdependence—travel in every direction for business, shopping, education, health care, and social and recreational purposes.

FIGURE 7.19
Higher-order trade centers and trade areas in the United States. *(Source: Based on Borchert, 1967)*

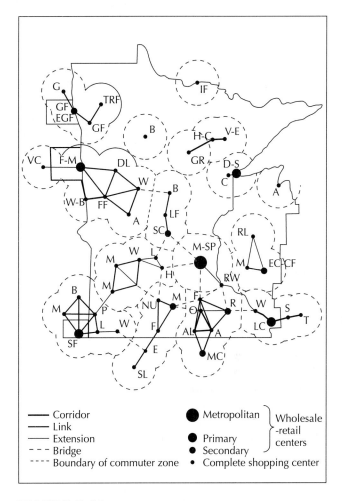

FIGURE 7.18
Dispersed cities. *(Source: Borchert and Carroll, 1971, p. 14)*

High-Order Central Places

Minneapolis-St. Paul is one of 24 high-order central places in the United States (Figure 7.19). Borchert (1967) divided these important trade centers into three

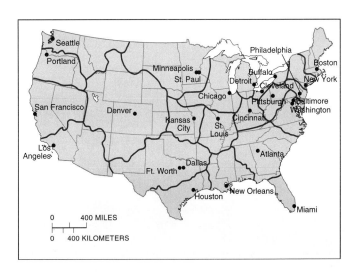

orders according to their size and functional complexity. The first-order center is New York City, the national metropolis providing the widest range of specialized activities. There are 6 second-order centers in addition to New York City (which is also a second-order center). These cities are Chicago, Boston, Philadelphia, Detroit, San Francisco, and Los Angeles. The second-order centers are regional metropolises for much of the U.S. market. Added to the 7 second-order centers are 17 metropolitan centers, for a total of 24 high-order central places. These cities are Baltimore, Washington, D.C., Atlanta, Miami, Buffalo, Cleveland, Cincinnati, Pittsburgh, New Orleans, St. Louis, Kansas City, Minneapolis-St. Paul, Dallas-Fort Worth, Houston, Denver, Seattle, and Portland.

Rank-Size Rule

In his study of the Upper Midwest, Borchert noted that the hierarchical structure of trade centers is reflected not only in their functional complexity, but also in their relative size. We can obtain an urban-size hierarchy by ranking centers according to their population sizes. The most well-known representation of this hierarchy is the *rank-size rule*, an empirical finding popularized by G. K. Zipf (1949). The rank-size rule states that if all settlements in an area are ranked in descending order of population size, the population of the rth city is $1/r$ the size of the largest city's population. When plotted on double logarithmic graph paper, the relationship produces a straight, downward-sloping line with a gradient of 45 degrees. The hypothetical rank-size distribution describes an urban system containing a few large metropolises, a large number of medium-sized cities, and a still larger number of smaller towns.

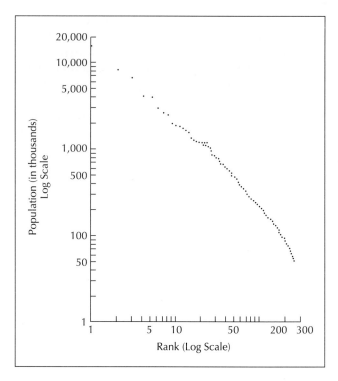

FIGURE 7.20
Rank-size distribution of urbanized areas in the United States, 1980. *(Source:* U.S. Bureau of the Census Statistical Abstract of the United States, *1980)*

The urban-size hierarchy of the United States conforms closely to the rank-size rule (Figure 7.20). For example, New York, the first-ranking city, is nearly twice as large as Los Angeles, the second-ranking city, and nearly three times larger than Chicago, the third-ranking city. However, in many countries the popula-

The downtown Minneapolis skyline. Minneapolis–St. Paul represents the highest-order central place in the hierarchy of Upper Midwest trade centers. *(Photo: Greater Minneapolis Convention and Visitors Association)*

tion of the first or largest city is much greater than would be expected from the rank-size distribution, so a condition of *primacy* exists.

Brian Berry (1961), an urban geographer, interpreted city-size distributions through a comparative study of 37 countries (Table 7.4). Berry said that as countries become politically, economically, and socially more complex, they tend to develop straight-line rank-size distributions (Figure 7.21). In the early stages of national development, a simple pattern of primacy prevails. This pattern gradually transforms into a rank-size distribution, which is the steady state of an urban-growth process.

To what extent is there a correspondence between central-place theory and the rank-size rule? The central-place hierarchy is based on the *functional size* of centers. Functional size is determined by the number and order of central functions offered by a place and is tied to the role of settlement-forming functions. The total population of a place is a function of both settlement-forming and settlement-building functions. Two centers at the same level of the hierarchy (that is, equivalent in functional size) may differ somewhat with respect to population. The rank-size rule is based on the population of centers—not their functional size. Population size is reflected by a smooth rank-size curve; functional size produces a stepped hierarchy. The discrepancy between the stepped and continuous curves may also be a function of scale. Rank-size distributions apply to large economic areas, such as the United States, and the central-place model to their smaller subsystems, such as the Upper Midwest.

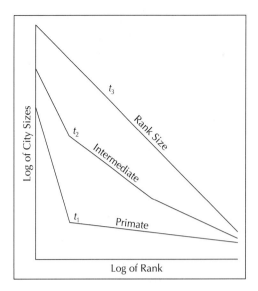

FIGURE 7.21
Idealized evolution of the city: distributions under increasing size through three time periods.

Structural Elements of the Central-Place Hierarchy

Much empirical research has focused on the structural elements of the central-place hierarchy (Berry, 1967). All the elements of central-place theory are structured or tied together logically and proportionately. To illustrate, let us graphically summarize four relationships: (1) the relationship between population size and functional units, (2) the relationship between population size and central functions, (3) the relationship between establishments and population size, and (4) the relationship between trade-area size and population density.

POPULATION SIZE AND FUNCTIONAL UNITS
The term *functional unit* refers to the provision of a central function each time it is offered. A graph of the relationship between population and functional units in southwestern Iowa shows that most centers fall close to the regression line, which is the straight line fitting the data (Figure 7.22). Two centers, Red Oak and Glenwood, however, have larger populations than one would expect, given the number of functional units in each. These discrepancies can be explained by the relatively large settlement-building functions of the centers. When the population of each town that is supported by these noncentral functions is subtracted from the centers' total populations, the number of people supported by settlement-forming functions can be estimated. These estimates fit the regression line well.

POPULATION SIZE AND CENTRAL FUNCTIONS
The term *central function* describes a good or a service offered by a central place. There is a curvilinear or log-linear relationship between population size and the number

Table 7.4
City-size Distributions

Countries with Rank-Size Pattern	Countries with Pattern of Primacy	Countries with Intermediate Patterns
Belgium	Sri Lanka	Australia
Brazil	Denmark	Canada
China	Dominican	Ecuador
El Salvador	Republic	England and Wales
Finland	Greece	Malaya
India	Guatemala	New Zealand
Italy	Japan	Nicaragua
Korea	Mexico	Norway
Poland	Peru	Pakistan
South Africa	Portugal	
Switzerland	Sweden	
United States	Thailand	
West Germany	Uruguay	

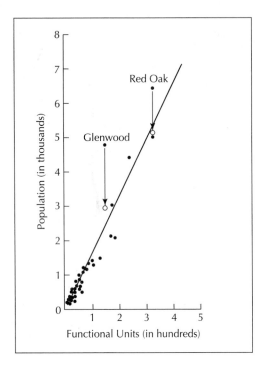

FIGURE 7.22
Relationship between population and functional units in southwestern Iowa. *(Source: Based on Berry and Meyer, 1962)*

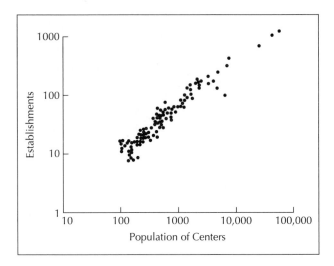

FIGURE 7.23
Relationship between establishments and population of centers.

of central functions performed by centers. This relationship indicates that the population of a central place is a function of the total number of business types offered. Again, there are deviations from the norm. Some settlements that have more central functions than expected may be tourist centers, such as Las Vegas or Reno, Nevada, where excess central functions are supported by transient populations. Other settlements have larger populations than expected, which is usually indicative of the settlement-building functions of the centers.

ESTABLISHMENTS AND POPULATION SIZE
The term *establishment* connotes ownership and control. Increases in the number of establishments are not proportional to increases in center sizes (Figure 7.23). The population/functional-unit and population/central-function relationships are not proportional—an observation that raises the question of how centers meet increases in demand brought about by population changes. Existing establishments may expand, new establishments of the same functional type may be added, or a combination of the two responses may occur. Empirical evidence as revealed in the graph indicates that increasing the number of establishments tends to occur more often—particularly in the case of lower-order functions and centers.

TRADE-AREA SIZE AND POPULATION DENSITY
There is a strong tendency for trade-area size to adjust to variations in population density (Berry and Meyer,

1962) (Figure 7.24). The 45-degree line in the figure indicates constant population density. Each level of center is arrayed along a line with a slope of more than 45 degrees. Thus, variations in trade-area size are greater than variations in total population influenced by central places. This difference suggests density adjustment.

The *central-place hierarchy* is sensitive to local variations in population density, but even more so to re-

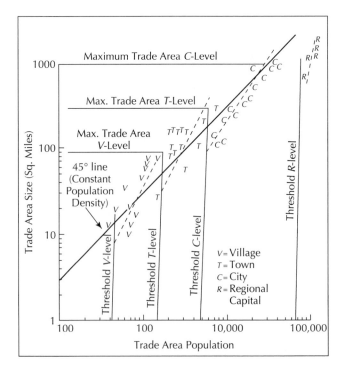

FIGURE 7.24
Density adjustment (southwestern Iowa). *(Source: Based on Berry and Meyer, 1962)*

Copenhagen, Denmark. A primate city in northern Europe. *(Photo: S. Cordier, Photo Researchers, Inc.)*

gional variations. If we examine the relationship between trade-area sizes and total population serviced by central places along a traverse from the densely settled area of Chicago, through the corn and dairy lands of Illinois, Wisconsin, Iowa, and Minnesota, on to the wheat lands of the Dakotas, and into the rangelands of Montana, we discover remarkable regularities. Instead of what appears as a random scattering of settlements on the map, we find a highly regular set of relationships (Figure 7.25). In the urban areas near Chicago, population densities are high, trade areas are small, and numbers of people served, large. As we move through the suburban areas and across the Corn Belt to the wheat lands and rangelands, population densities decline, but the regular relationship between the trade area and the population served is maintained.

Consumer Travel as a Mirror of the Hierarchy

A central-place system depends on consistent consumer behavior. Christaller's theory of central places postulates that consumers behave predictably in that they will always obtain goods and services from the nearest possible center. However, in reality, this may not be the case.

Figure 7.26 shows the primary purchase movements of rural and urban consumers for clothing in southwestern Iowa. Clothing is the highest-level good (city level) followed by dry cleaning (town level). Groceries are the lowest-level function (village level). Consumers must consider the order of a good because it limits the number of possible purchase points. Within the framework of where goods are actually of-

fered, however, consumers have some flexibility as to where they make their purchases. They will generally go to the nearest place to obtain *convenience goods*. For *shopping goods*, however, they will often travel to higher-level centers, even if these goods can be obtained locally.

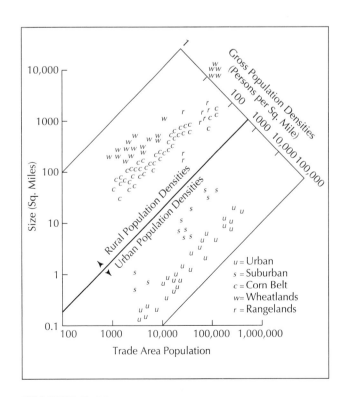

FIGURE 7.25

Density adjustment: large regional variations in population density. *(Source: Berry and Meyer, 1962)*

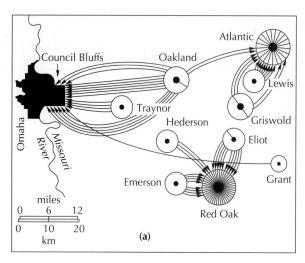
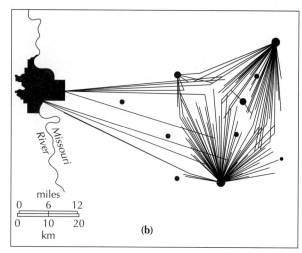

FIGURE 7.26
Purchase movements of (a) urban and (b) rural consumers for clothing in southwestern Iowa. *(Source: Berry and Meyer, 1962)*

The maps of travel patterns in southwestern Iowa do not replicate the geometrical precision of the central-place model. This is because the normative constraints of the model do not actually exist. Central-place theory can help us better understand and predict spatial patterns of consumer behavior. But we must bear in mind that people are not economic optimizers, although they do adhere to the principle of distance minimization to a limited extent.

Space-Preference Structures

Gerard Rushton (1969) attempted to add reality to the central-place system of central Iowa. In the 1970s, he observed hundreds of locational choices by rural residents as they traveled to the market to consume goods and services. He was interested in how closely the trade areas of a hierarchy of central places and relatively flat, rural Iowa would be described by the Christaller central-place system. He found that shopping behavior was a result of a mental screening process in which travelers sometimes bypass the closest town that offers the good or service that they need. Many times they travel to a farther destination if it offers a better variety of goods and services from which to choose.

Rural residents often minimize their travel time and costs by purchasing needed goods and services at the next-larger central place. The supplies for which they bypass the lower-order center, or the closest central place, are usually low-order convenience goods and services that a larger place can provide as well (e.g., groceries, gasoline, and basic agricultural services for their farms). This type of shopping behavior is some-

One of the most common central functions is that of a grocery store. Items are in high demand and are relatively inexpensive. As one climbs the urban hierarchy, increases in the number of establishments are less than proportional to city size because large companies absorb the smaller retailer and offer scale economies to the shopper. *(Photo: Frederick P. Stutz)*

times called *multipurpose travel*. Because multipurpose travel is the norm in Iowa and other modern semi-rural regions, Rushton noticed that the smallest towns have a significantly smaller number of places than is normally predicted by the central-place theory.

Rushton developed an approach called *space preference structures*, which measures the trade-off between the journey's distance to shop for goods and services, and the town size. In other words, the larger town size is a positive mental stimulus, and the increased distance to the town is a negative mental stimulus. When a town offered a good or a service that the rural resident needed, and the town was within the range of the good but was not patronized, Rushton assumed that the consumer consciously decided not to shop there, but at the next larger place. Rushton found that rural residents receive a certain payoff to go a certain distance, and the payoff can be measured by town size as a surrogate for the *market basket* of shopping possibilities that exist there.

All else being equal, including distance, rural residents prefer to shop in a town with a population of 5,000 rather than one with a population of 500 because the town of 5,000 offers more opportunities. These findings differed somewhat from the original central-place theory, which included only distance as an important variable.

Rushton also found that as rural residents shop for increasingly high-threshold and high scale economy goods, they are more willing to forgo closer places for a larger place that has a better selection. Rushton found that the space-preference structures of rural Iowans were relatively stable for different townships as well as for various ages. However, the pattern of trading off extra distance to get to a larger town with more choices existed only to a point. Rural Iowans actually preferred somewhat smaller places over very large towns, say those with populations of 200,000, perhaps because of a lack of public parking, increased traffic congestion, and resentment of large, high-density cities.

Rushton represented his space-preference structure findings as a series of curves on a graph shown in Figure 7.27. He called these curves *indifference curves*. Any place along an indifference curve is equally acceptable to the traveler. Because the curve slopes upward to the right, individuals trade extra distance to get to a larger town without being dissatisfied. Individuals along the indifference curves farther to the right must substitute greater differences than individuals along the indifference curves to the left must substitute. The curves farther to the right suggest lower population densities and more widely spaced towns.

If Rushton's space-preference structures are applied, the resulting trade areas of Iowa are shown in Figure 7.28. The *Thiessen polygons* at the left show the trade areas of central places in Iowa, centered on the Des Moines hierarchy. Areas within each trade-area

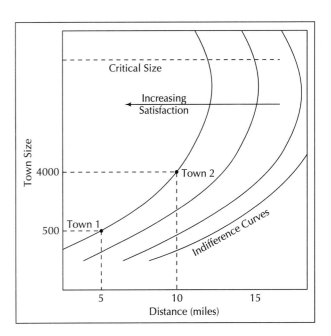

FIGURE 7.27
Consumer space-preference structure. *(Source: Adapted from Rushton, 1969, pp. 391–400)*

boundary must be closer to the midpoint than to any other point. That is, perpendicular bisectors of lines connecting points are trade-area edges. This is the assumption of central-place theory. Although the trade areas do not result in a perfect hexagonal arrangement in Figure 7.28a, they are relatively close, if we consider that all assumptions of the model are not being met (principally, rural population density spread evenly).

Figure 7.28b shows the assignment of trade-area boundaries on the basis of space-preference structures. The comparison of the two central-place systems—Thiessen polygons in Figure 7.28a, and the space-preference structure delimitations in Figure 7.28b—shows that many of the small central places in the Thiessen polygon delimitation would yield their trade areas entirely to the next-larger place. In this case, Des Moines and three satellite cities have much larger trade areas than they had in the original delimitation. Small hamlets and villages that are close to larger central places, and within the range of the good of these large places, are at greatest risk of losing their entire trade areas (retail function) while they maintain their residential function.

On the other hand, small central places farther from the main center retained their small trade areas. Consequently, we are left with central places of the same size in the hierarchy, with radically different levels of centrality or different central-place functions. With the advent of telecommunications and improved highway access, many hamlets and villages around large cities, throughout not only Iowa, but also the rest of the

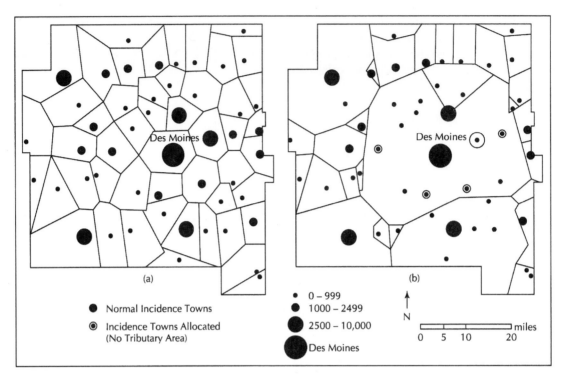

FIGURE 7.28
Alternate trade-area delimitations for grocery stores in Iowa: (a) Thiessen polygon delimitation and (b) space-preference structure delimitation. *(Source: Adapted from Bell, 1973, p. 293)*

world, function as important exurb bedroom communities, where the noise and problems associated with cities are far less. Their more distant rural counterparts function as true rural settlement sites with a variety of central-place functions. Each has its importance in the economic landscape.

A central-place hierarchy of goods and services also exists inside the city. This hierarchy is the *retail hierarchy*. At the lowest level are convenience centers, such as gas stations, minimarkets, and perhaps restaurants. This retail mix is comparable to the central functions of a hamlet. At the next level is a *neighborhood center*, which has more central functions at a somewhat higher threshold level. The rural counterpart of the neighborhood center is the village, which has the same central functions.

The hierarchy continues upward with *community centers*, *regional shopping centers*, and even the downtown central business district (CBD), with its wide variety of goods, services, and establishments. CBDs have been compared with the central functions and centrality of a town, a city, and a regional metropolis, respectively.

As in the rural area, when city travelers shop for convenience items with low threshold and scale economies, the closest places are frequently patronized—the convenience centers or the neighborhood centers. But for higher-order shopping, for goods and services

that have high thresholds and high scale economies, the closest center offering the goods is frequently bypassed. When several locations can satisfy the demand, comparison shopping is the rule, and the potential locations are selected and sequentially reviewed. While a shopper is on one of these higher-order shopping trips, he or she can purchase lower-order products as well.

Frequently, urban travelers pass the closest place that offers a good or a service and choose instead a new center, with higher reward levels because of variety, price, or shopping comfort and convenience. Thus, in the urban case, shoppers evaluate not only distance and the size of the center, as in Rushton's rural Iowa case, but also a variety of other mental stimuli, including newness of the center, the store mix, parking availability, and design characteristics, such as whether the shopping center is enclosed or open or has a variety of entertainment and restaurants associated with it (e.g., the newer shopping centers in America).

The space-preference structures originally designed by Rushton were also identified in urban travel to various churches in the San Diego region. One would expect that a parishioner would travel to the closest church of his or her denomination. However, this was far from the case in the San Diego Church of Christ study by Frederick Stutz and Phil Hinshaw (1976). The closest church was frequently bypassed in favor of a church

FIGURE 7.29
Churchgoing travel patterns reveal space-preference structures. *(Source: Stutz and Hinshaw, 1996)*

that was larger and offered more social and personal services. The existence of Sunday schools, counseling programs, Bible classes, Girl Scout troops, a gymnasium, and a day school—although these services were not offered on Sunday morning—were enough of an attraction to draw urban residents from far across town (Figure 7.29).

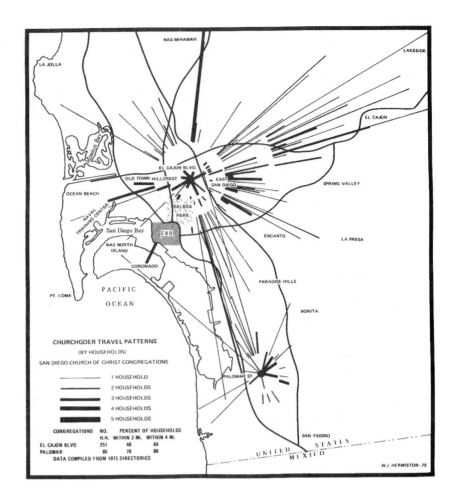

CHURCHGOER TRAVEL PATTERNS
(BY HOUSEHOLDS)
SAN DIEGO CHURCH OF CHRIST CONGREGATIONS

————	1 HOUSEHOLD
————	2 HOUSEHOLDS
▬▬▬▬	3 HOUSEHOLDS
▬▬▬▬	4 HOUSEHOLDS
▬▬▬▬	5 HOUSEHOLDS

CONGREGATIONS	NO. H.H.	PERCENT OF HOUSEHOLDS WITHIN 2 Mi.	WITHIN 4 Mi.
EL CAJON BLVD.	251	48	64
PALOMAR	65	78	88

DATA COMPILED FROM 1972 DIRECTORIES

W.J. HERMISTON–75

CROSS-CULTURAL PATTERNS

So far, our discussion has focused on cities as service centers in specialized societies; however, we have not yet considered how cultural differences affect local trade. Also, what of the situation in societies where specialization is less advanced? How do peasant societies organize their market activities? In seeking answers to these questions, we come to appreciate the rich variety of marketing and distribution systems throughout the world.

Cultural Differences in Consumer Travel

A study by Robert Murdie (1965) of an area in Ontario inhabited by both Old Order Mennonites and non-Mennonite Canadians provides an example of how cultural factors affect the geography of retailing and services in a dominantly specialized society. Mennonites use modern methods to manage their farms' businesses. But in dress, domestic consumption, and travel, they cling to a lifestyle that existed 200 years ago. Homemade clothes and the horse and buggy for transportation prevail. Few goods are demanded.

When Mennonites have needs similar to those of non-Mennonite Canadians, both groups use the central places in the area in much the same way; for example, Mennonites and non-Mennonite Canadians exhibit similar consumer behavior in their banking transactions. However, when the traditional beliefs of the Mennonites come into play, two distinct types of behavior with regard to central places become evident. Consider for example clothing purchased by non-Men-

nonite Canadians and yard goods purchased by Mennonites. The difference in mode of transportation is one factor crucial to the two types of consumer behavior. Non-Mennonite Canadians demand variety and go for it; hence, the maximum distance that they travel to purchase clothing is related to center size. On the other hand, the Mennonites purchase only a limited variety of yard goods and are restricted by the use of the horse and buggy; hence, the maximum distance that they travel does not vary with center size.

Periodic Markets

The majority of people in developing countries do not participate fully in the network of enterprises enjoined in the modern urban hierarchy. Rather than being involved in the production of goods for world markets, for the most part they subsist on what they can grow and trade. Their local transactions often take place at small rural markets, which often operate on a periodic rather than a permanent basis. The market is likely to be open only every few days on a regular basis because its size is limited by the level of transportation technology, and the aggregate demand for goods is insufficient to support

Fresh fruit and vegetable market, Guadalajara, Mexico. *(Photo: Joel Gordon)*

permanent shops. In this system, several places in a region profit from a day's trade, and each benefits from its participation in the wider network of interaction. People come to these *periodic markets* on foot, on bicycles, on the backs of their animals, or by whatever other means are available. Periodic markets are a feature of Southeast Asia, China, India, Central and South America, and large parts of Africa.

G. William Skinner described the periodic marketing systems of traditional rural China. In one system based on the lunar month (*hsün*), merchants moved between the central market and a pair of standard markets in a 10-day cycle divided into units of three: the central market (day 1), first standard market (day 2), second standard market (day 3), central (day 4), first standard (day 5), second standard (day 6), central (day 7), first standard (day 8), second standard (day 9), and central on day 10, when no business was transacted. The three-per-*hsün* cycle illustrates the periodicities of many market centers of different levels, and it resembles a Christaller-type *K*-equals-four hierarchy.

Elsewhere in the world, other cycles predominate: two-per-*hsün* cycles are common in Korea. Seven-day cycles occur in Andean Colombia (Symanski, 1971). In Africa, the market week varies from a 3-day to a 7-day week. The 3-day, 4-day, 5-day, and 6-day weeks stem from ethnic differences, and the 7-day week is a consequence of calendar changes introduced by Islam into Africa.

Periodic markets, then, form an interlocking network of exchange places. As each market in the network takes its turn, it is close enough to one part of the area so that people in the vicinity can walk to it, carrying what they want to sell or trade. The staggered pattern of local markets permits a small volume of produce to move through the market chain to larger, regional wholesaling markets where shipments are collected for interregional or perhaps even international trade. What is traded depends on the market's location. For example, in West Africa's savanna zone, sorghum, millet, and shea butter predominate. Further south, in the forest zone, yams, cassava, corn, and palm oil change hands. In the south, too, there are some imported manufactured goods, especially in markets near the relatively prosperous cocoa, coffee, rubber, and palm oil areas. But wherever the market, the quantities traded are small—a bowl of rice, a bundle of firewood—and their value is low.

The total amount of goods traded through the periodic marketplace, however, is tremendous; nobody

At Po in southern Burkina Faso, the periodic market occurs on a 4-day cycle. *(Photo: Alan Becker, The Image Bank)*

knows the volume or the value of goods that are actually distributed (Bohannan and Curtin, 1971). The task of examining indigenous marketing systems has been largely neglected, primarily in the belief that marketing systems to serve cultivators, distributors, and consumers are economically inefficient and exploitative. According to *The Marketing Challenge* (USDA, 1970), this belief is erroneous, an opinion shared by Earl Scott (1972), whose research in Nigeria indicated that indigenous marketing systems can promote development. In contrast, Linda Greenow and Vicky Muiz (1988) argued that stimulating market trade as it is currently organized is unlikely to encourage development. In their Peruvian study, they showed that market selling for most traders is a means of survival. According to Greenow and Muiz, "Traders, most of whom are women, rarely develop stable or growing businesses that support them and their families over a long run."

In general, periodic markets mainly serve people who live near the subsistence level. Market cycles are determined either by *natural* events, using the motions of the heavenly bodies, or they are *artificial*, without reference to natural cycles. Artificial cycles dominate today, with the periodicity of the markets influenced by population density. The more people in an area, the greater the aggregate demand, and the greater the frequency with which a market can operate. Eventually, the demand may be sufficient for a market to become continuous and permanent. Periodic markets are, therefore, logically related to the patterns of market centers observed in complex economies.

Just as the periodicity and types of commodity traded vary culturally, so do the locations of indigenous markets. In West Africa, rural populations live away from market sites, and the hierarchy of rural settlements is unrelated to the hierarchy of markets. In other regions, settlements and periodic markets may coincide. An understanding of periodic market sites requires an appreciation of local culture.

PLANNING USES OF CENTRAL-PLACE THEORY

Central-place concepts, confirmed by the empirical evidence, have been used for planning purposes. They were used to establish a hierarchy of market centers on the Dutch polders and to design a new system of settlements on the Lakhish plain in Israel. They were also used to guide the development of the planned suburban community of Park Forest, Illinois. Residential areas were organized into neighborhoods, each served by a local business center, and one large shopping center was established for the whole community with adjacent land set aside for a village hall and police and fire departments. In England, analysis of the distribution of

shopping centers and their service areas was an important component of the administrative reorganization of Greater London. These examples, cited by Berry (1967), are "symbolic of the practical uses of the central-place idea by regional and city planners for locating retail business, business centers, and market towns, or regionalizing an area" (p. 132).

Since the widespread adoption of the spatial organization theme in the 1960s, geographers have devoted much attention to the practical uses of the central-place concept. Making use of the complex variable, accessibility, they have analyzed and planned public and private service facilities. Geographers have also been called upon to give advice on questions of planning spatial patterns and structures in developed and underdeveloped countries.

New Central Places in the Netherlands

The Dutch are a classic example of a hard-working people who have reclaimed portions of their country from the North Sea by building dikes across inlets and pumping water off the land. In the *Zuider Zee reclamation scheme*, no settlement pattern existed, only marshy land and high points that swelled above the marshy level. Settlements and transportation systems were planned for this region. By 1930, the water was drained, which freed the polder of Wieringermeer. Four other polders were pumped and ready for development by 1968. Land was primarily pasture and agriculture, with the towns providing rural services and basic consumer goods.

Central-place theory was used to establish the original towns, but after World War II, transportation accessibility improved substantially. It became clear that too many small towns were too closely spaced for each to prosper economically. In later years, larger farm units and better town spacing, such as the case in East Flevoland and in South Flevoland, became common.

In the case of Lelystad, in the polder of East Flevoland, the rural setting became a destination for residents moving out of impacted Amsterdam. By 1992, Lelystad had a population of 70,000, and more than half the workers commuted to Amsterdam daily. The same became true of Almere in South Flevoland. About 70% of its residents were relocatees from Amsterdam, and the population in 1992 was approximately 60,000.

To conclude, central-place theory has been used not only in Israel, but also in the Netherlands, to plan new, undeveloped land. Because of improved transportation and the history of commuting in the Netherlands, town spacing is much wider than it was before World War II. Large settlements also exist within commuting distance of Amsterdam. These towns are not only bedroom communities to Randstad, the urban circular complex including Utrecht and Rotterdam, but also higher-order central places to the small, planned towns and market areas of the Zuider Zee reclamation project.

Market Area Analysis

Under the title "Being Close to Things and People," Peter Gould (1985) emphasized that often in geographic planning and consulting, questions of accessibility are close to the surface. He illustrated the idea of accessibility with the practical application of where to locate a new service facility (see Chapter 2: Target Marketing). Suppose that you are on a team of geographic consultants working in a poor country where there are only enough resources to build a new school to serve a rather large rural area. (You can substitute a clinic, family-planning center, or any other service because the nature of the problem remains the same.) Obviously, you need to know where the villages are and how many schoolchildren (or people to be served) are in each village. Villages *A* and *D* are close to one another, but *E*, with 52 school-age children, is more remote. Where will you build the school? What is the most accessible location for the new school?

To find out, you can build a simple analog computer, the *Varignon frame*. Glue a map of the area to a sheet of plywood. Then, at each village, drill a hole

and thread a smooth string through it. Next, tie the strings together on top of the map, and underneath attach weights proportional to the number of children in each village. If the knot represents the school, we can think of each village tugging the school toward it with a force equal to the number of children that it will serve. Where the knot locates is the best location; that is, the one that minimizes the aggregate distance that children must travel to school and back. If the children travel to school by bus, this location could save the school district a large amount of money in fuel costs.

The most efficient location for a new facility, however, is unlikely to be the most fair location from the standpoint of social justice. In our school example, the children in villages *A* to *D* would have nothing to complain about. But what about those living at *E*? They would have a long journey to school. Should we sacrifice efficiency for equity and build the school a bit closer to *E*? The question of equity, which arises whenever we deal with solutions to problems involving people, derives from the ideas of social justice (Harvey, 1973). Viewed in its geographic context, social justice is equated with territorial justice, which expresses a concern for a fair allocation of resources.

The simple analog computer can solve the problem of the best location for a *single* facility, but when you are searching for and evaluating all the possibilities in locating *multiple* facilities, high-speed computers are necessary. Geographers used computers as early as 1960 to find, for example, the best locations for hospitals in Sweden (Godlund, 1961) and administrative centers in southern Ontario (Goodchild and Massam, 1969). Literally hundreds of geographic applications for computer programs were produced in the 1970s and 1980s since these early studies. More recently, computer algorithms were used in a GIS in San Diego, California, to locate fire stations and areas for new development (Parrot and Stutz, 1992). (See Chapter 2.)

Accessibility is a key to measuring the welfare function of a consumer service. If there are variations in the degree to which a service is available to consumers, the welfare function of the service is not being fully used. In this view, the quality of life for consumers is affected to a degree by access to a basket of social goods— a doctor, a dentist, a public library (Smith, 1977).

No matter how hard geographers try, exact territorial equality in the provision of services can never occur. One reason is that there will always be a lag between demand and supply. There is considerable geographic and historical inertia in the locational distribution of services that leads to overprovision in some places and underprovision in other places.

Most facility location studies from the local scale to the world scale now take advantage of high-speed computers and very large data sets.

WORLD CITIES

To this point, we have considered cities as service centers at the national and regional levels. Now let us shift our attention to the international level, where cities function as centers of international business. The new international division of labor is forging the cities of the world into a composite system.

Industrial Restructuring and the Urban Hierarchy

The *industrial restructuring* that started to take place in the 1960s is changing the global urban hierarchy. The process of restructuring involves the movement of industrial plants from developed to developing areas within or between countries; the closure of plants in older, industrialized centers, as in the American Rust Belt; and the technological improvement of industry to increase productivity. Forces behind restructuring include the need for multinationals to develop strategies to locate new markets and to organize world-scale production more profitably, the national policies of developed countries to improve their future international

competitive position, and the national policies of developing countries to attract subsidiaries of multinationals. These multinational strategies and governmental policies have contributed to major shifts in employment and trade. The greatest effects have been felt in the urban centers of developed countries and in the larger cities of underdeveloped countries.

The U.S. Urban Hierarchy

Before industrial restructuring affected the organization of the labor process and the location of industry, the metropolis-and-region pattern of the United States was a fairly accurate reflection of the urban hierarchy. Each regional center—a Minneapolis–St. Paul, a Miami, or a Cleveland—was an important center of corporate services. For example, a large number of corporate law firms in Cleveland complemented that city's corporations. Major accounting firms were based in New York City and Chicago. Important regional banks were a feature of nearly all centers of corporate head offices (Figure 7.30).

By the 1970s, the metropolis-and-region network of corporate head offices and corporate services had

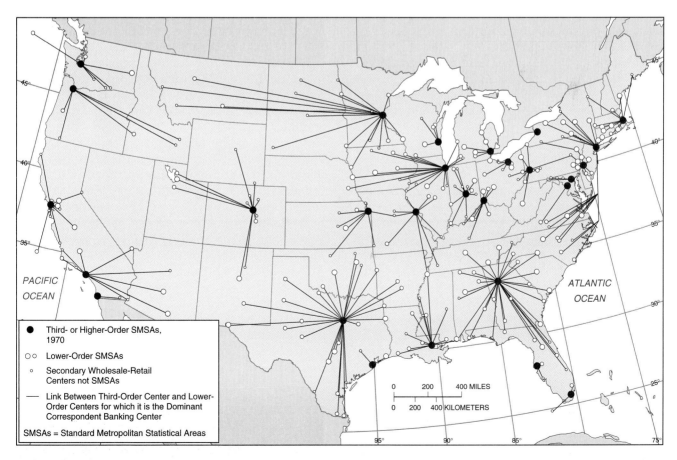

FIGURE 7.30
Banking linkages of third-order centers, 1970. *(Source: Borchert, 1972, p. 253)*

At the southern tip of Manhattan soar the skyscrapers of the financial district, symbolic of New York City's role as a global city. New York is the nation's preeminent international center for business decision making and corporate services. Firms headquartered in New York account for about 40% of foreign sales of Fortune 500 companies. More than 50% of foreign bank deposits in the United States are in New York banks. *(Photo: New York Convention and Visitors Bureau)*

expanded, but international business activity had become much more important. International decision making by major firms was concentrated in two cities, New York and San Francisco (Cohen, 1981). Cities that were important in the earlier, national-oriented phase of the economy began to lose ground to these *global cities.* Jobs connected with international operations did not develop as extensively in places such as St. Louis and Boston as they did in New York City and San Francisco.

Along with the growth of international operations, advanced corporate services—banks, law firms, accounting firms, and management firms—expanded their international presence in the 1970s. Similarly, these service activities developed strongly only in a handful of centers. Few banks with international expertise could be found in places other than New York City, San Francisco, and Chicago. Firms with international legal expertise were mostly confined to New York City, Los Angeles, and Washington, D.C. (Cohen, 1981).

In the 1980s, international activities remained concentrated. Los Angeles and Chicago joined New York City and San Francisco as top-level international business and banking centers. International activities of firms headquartered outside these cities became increasingly tied to financial institutions and corporate services located within them. In the next few years, it is expected that only a few additional cities will achieve interna-

tional status and that New York City and Los Angeles will remain the predominant American global cities.

The rise of Los Angeles from a regional metropolis in the 1960s to a global center of corporate headquarters, financial management, and trade in the 1980s has been remarkable. The Pacific Rim city has become an epicenter of global capital. The transformation of Los Angeles was accompanied by selective deindustrialization and reindustrialization. A growing cluster of technologically skilled and specialized occupations has been complemented by a rapid expansion of low-skill workers fed from the recycling of labor out of declining heavy industry and by a massive influence of Third World immigrants and part-time workers (Soja, Morales, and Wolff, 1987). Sprawling, low-density Los Angeles symbolizes the process of urban restructuring: It combines elements of Sunbelt expansion, Detroit-like decline, and free-trade zone exploitation. In sharp contrast to Los Angeles, arguably one of the world's most successful cities of reconstituted capitalism, Cleveland, which lost many blue-collar jobs because of plant closures, layoffs, and capital mobility, has been unable to attract many international business operations to cushion its economic decline. Although Cleveland may never become a global city, it may boom like Buffalo and experience an economic revival in the next few years because of the 1988 free-trade agreement between the United States and Canada.

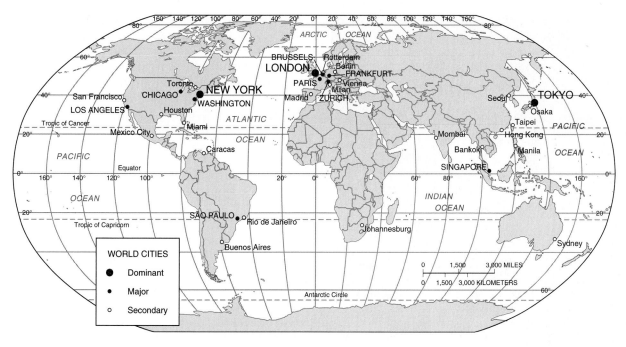

FIGURE 7.31
The system of world cities. It is possible to identify dominant, major, and secondary world cities on the basis of the location and transaction flow of financial centers, finance, major corporate headquarters, international institutions, communication nodes, and concentrations of business services.

The World Economy's World Cities

Singular and economically preeminent world cities are "basing points" for global capital flows (Friedmann, 1986). These *world cities* are usually the largest and most important cities in the international system that measures capital, population, employment and output. These cities are control points of the world economy where the critical mass of capital and articulation of production and marketing create dramatic world economic development. High levels of economic development have been primarily in developed countries; the globalization of economic activity has little touched developing countries. Therefore, the world cities—dominant centers of transnational corporations, business, international finance, and capital flows—are found in just a few locations. Friedmann identifies a hierarchy of dominant, major, and secondary world cities on the basis of their importance in the world economy, their hosting a major corporate headquarters, their possession of international banks and financial institutions, and their wealth of communications and business services (Figure 7.31). On this basis, only Sao Paulo and Singapore classified as world cities that are not in the *core countries.*

Three distinctive subsystems in world cities are identifiable from the figure, along an east/west axis—a western European regional system focused on Lon-

don, and to a lesser extent, the Rhine Valley, from Randstad to Paris to Zurich. The southern hemisphere is linked to this system via Johannesburg and Sao Paulo. Moving to the west, a North American subsystem is based on the four core cities of New York, Chicago, Los Angeles, and Toronto. Caracas, Sao Paulo, and Mexico City have direct links to the North American system of world cities. Thirdly, an Asian regional system is centered on the Tokyo–Singapore spine, with Tokyo being the dominant world city of East Asia and linked directly to Seoul, Osaka, Taipei, Hong Kong, and Bangkok. Singapore is the regional metropolis of Southeast Asia to which Sydney is directly linked.

Smart Cities

At least 65 of America's 100 largest cities are dying, in large part because cities in themselves are losing their relevance as governing units. People are moving to the suburbs and exurbs. Linking parts and functions of a city through networked computers, however, may be the answer to city decline. Such computer-networked cities are known as *smart cities*.[1] In the past, cities pros-

---

[1] Paraphrased from *Building Smart Communities,* Caltrans, 1997.

The tall, slender Canadian National Tower looms over Toronto at sunset, with the Toronto sky dome beside it. The calm waters of Lake Ontario stretch in front of this most dynamic Canadian city. (*Photo: John Mitchell, Photo Researchers, Inc.*)

pered as geopolitical entities because of their importance as transportation crossroads or as centers of industrial production. But telecommunications developments like telephones, fax machines, and electronically linked computers weakened the once inextricable connections between transportation systems and mobility. The *Economist* magazine calls it "the death of distance" (1994).

The new telecommunications network made it possible for businesses to produce, consumers to purchase, and workers to interact with one another without the need for a common physical location, allowing corporate and human citizens to choose where to reside based on a wide array of factors beyond mere physical convenience. Worse, communities suddenly found themselves competing for these residents not just with neighboring municipalities, but with cities across the country—or around the globe.

At the same time, telecommunications advances are transforming a world of discrete local and national economic spheres into a single global economy, placing local businesses and workers alike into direct competition with their counterparts on distant shores. Government and business leaders watched helplessly as

they began to lose control over their communities' economic destinies even as shifting business and population patterns eroded their economic base and their place in the urban hierarchy. As a result, economic and social institutions ranging from governments to economic development agencies to local schools, whose increasingly time-worn practices already were proving insufficient to the rapid technological changes and complex social and economic challenges of the late 20th century, were left with neither adequate resources nor adequate responses for solving the very difficult problems within their midst, further loosening their hold on the people and businesses who remain.

The Coming Digital Revolution

For all that telecommunications technology has provided, the biggest changes are still ahead. The coming digital revolution will remake lives. Work, education, commerce, and leisure will be redefined. Whole industries will vanish, and new ones will spring up overnight. The restraints of time and space, long the adhesive that has held many communities together, will virtually disappear; where technology once allowed a city's businesses and residents to strain their ties to the urban core, new technologies may break these links completely; where telecommunications once blurred national borders, soon it will wash away all but the most trivial vestiges of time and place, defining the concept of the "range of the good or service."

History offers a valuable lesson in this regard. In the early days of the American West, many once booming communities became literal ghost towns as the new railroad system passed them by. A century later, many of America's great central cities began to wither as suburban expressways and commuter rail lines ushered in the post World War II urban exodus.

Now, at the dawning of a new millennium, the "virtual transportation network" of computers and cables promises to remake the American city yet again. Communities unprepared for these changes risk being consigned to geopolitical obsolescence before they even know what hit them. In the process, they are likely to suffer a fate similar to that of many of the great cities of the past, becoming "electronic ghost towns" on the virtual frontier, abandoned by corporate and human citizens seeking a more enlightened leadership and a more electronically hospitable environment.

The New Shape of Competition

Smart communities of the coming century are the basis on which communities around the globe will be competing for businesses, jobs, and residents in the years ahead. They also are likely to be a primary way in which financially strapped local governments will be able to reduce the cost and burden of government while simultaneously increasing the quality and level of government service. A smart community is simply that: a community in which government, business, and residents understand the potential of information technology, and make a conscious decision to use that technology to transform life and work in their region in significant and positive ways.

Smart Community Architecture

The technological foundation of a typical smart community is an information network that links various users in a significant, common purpose. The network may range from a small, dedicated series of connections among a limited number of sites to a community-wide, broadband telecommunications grid. What matters is that its purpose be transformative—that is, that it dramatically change the way in which community members conduct an important activity or activities.

Such an information network generally consists of three primary elements: *infrastructure*, *access points*, and *applications*. The infrastructure, loosely, is the medium over which the information travels—telephone wires, copper or fiber-optic cables, even wireless or satellite pathways. The access points are the terminals at which users can interact with the network—usually personal computers, televisions, or kiosks. The applications are the uses to which the network's information and resources can be put.

Community-Based Applications

Once the infrastructure and access points are in place, communities can turn to the development of online applications—the core of most smart community projects. As every smart community advocate knows, the Internet itself provides users with access to a rich and growing base of information. No smart community has tried to replicate this vast data resource. In fact, the attractiveness and cost-effectiveness of an online government presence has become so obvious that, by the turn of the century, every major U.S. city, county, and state government is likely to have an interactive home page on the Internet.

Government

Placing government and social service information (office locations, departmental telephone numbers, city council minutes, government documents, and so on) online is a valuable first step toward building a smart community—but only a first step. Increasing people's access to such information serves the important function of creating a more well-informed citizenry, but it does little to improve the efficiency of government operations (except perhaps by reducing the number of telephone requests for information).

Ultimately, however, information technology's greatest potential in the government arena may lie in transforming the very nature of local government, making it possible to reconfigure traditionally monolithic, downtown City Halls into a network of small, neighborhood-based "branches" linked electronically to a slimmed-down city "headquarters." Under this scenario, almost all government and social services would be dispensed in the neighborhood—either from kiosks or in small, multifunction neighborhood service centers. Such structural reengineering could further reduce government staffing, operational, and office costs, minimize traffic congestion and pollution, and increase people's access to government officials and services, while creating a government more institutionally sensitive to neighborhood concerns.

More important, decentralizing government and staffing in this way is likely to distribute jobs and other economic resources (e.g., service contracts and employee spending) throughout a community rather than concentrating them in a downtown business district, thereby strengthening all of the community's neighborhoods, reducing neighborhood social problems, and improving the community's overall quality of life. This, of course, is one of the main objectives of the *sustainable communities* movement that recently has become popular in government circles. Local leaders intrigued by the promises of this concept, however, are likely to find that making their communities "smart" is an essential first step to making them "sustainable."

Business

Tens of thousands of companies, from small start-ups to nearly all of the Fortune 500, are already using the Internet to promote their businesses, and, thanks to secure payment mechanisms like First Virtual's Internet Payment System, a growing number are beginning actually to sell products and services online. Many smart community initiatives, like the Davis Community Network and the Blacksburg, Virginia, Electronic Village, are almost aimed at speeding this process along by helping local businesses to establish their own online presence.

Web sites, of course, are a passive form of advertising, and potential customers often do not encounter a particular company's site unless they happen to be looking for it—turning the World Wide Web into something of a high-tech Yellow Pages. In an effort to reach a captive Web audience, therefore, many companies have taken a cue from the television advertising model and have begun to sponsor high-profile, Web-based news, information, and entertainment sites where they can gain exposure to customers that they otherwise might never reach. Such efforts, however, are unlikely to end there. It is a short step from sponsoring a program to creating one. Smart communities (or coalitions of companies) of the future may find it profitable to develop high-profile Web-based informational, educational, or entertainment programming on their own for the primary purpose of attracting interest in their communities, products, or services.

For all their promotional potential, however, information networks are more than effective advertising and marketing vehicles. They can, in fact, change the very nature of work. In much the same way that government telecommunications networks may allow government agencies to distribute their workforces among communities, large businesses can establish dedicated neighborhood *telework centers* that give them access to potential employees, such as rural residents, homemakers, people with disabilities, or individuals without transportation, whom they otherwise might not be able to attract. By moving operations out of congested and high-priced central cities, telework centers simultaneously can increase workers' productivity while reducing companies' rent, transportation, and labor costs.

Telecommuting, of course, is an idea that has been around for decades, and the practice has not yet taken off in the way many of its advocates forecast. This has been due in part to business's lack of interest in telecommuting and to its reliance on traditional, fixed-site management practices. But the growing number of home-office, temporary, and contract workers makes telework not only feasible, but increasingly necessary. With workers demanding more flexible schedules, shorter commutes, and relief from traffic congestion, companies may find telecenters are a powerful tool for retaining or attracting high-quality workers. What's more, with videoconferencing, e-mail, and high-speed computer networks, telecommuting finally has become technologically practical in a wide range of job categories and work situations.

Education

Information technology's most obvious potential in education lies in giving students access to computers and the Internet. That is, in fact, the substance of most of the smart community educational initiatives now taking place at local and federal levels. Bringing computers into the classroom will increase young people's access to information, help to equalize educational resources among poorer and wealthier school districts, and better prepare today's students for tomorrow's workforce.

Not just schools are ripe for technological revamping. Library systems, like the San Francisco Public Library, are remaking themselves into comprehensive information and research centers, while others, such as San Diego's and Chula Vista's new libraries, are being planned from the start as "smart libraries," relying on multimedia, virtual reality, and global networks. A few educational technology vendors—for example, San Diego's Creative Learning Systems—are building entire turnkey educational environments around computer-based, collaborative, multimedia learning environments.

Indeed, technology is forcing educational planners to reevaluate the entire concept of mass-produced, discipline-based education, in which students are herded around schools to assigned classrooms according to fixed schedules and one-size-fits-all curricula. In the smart communities, students—adults as well as young people—will learn when they want to, how they want to, and at their own pace. Already, predominantly rural states like Montana and Iowa are using telecommunications to link schools in sparsely populated areas into a statewide educational network, allowing instructors and students to interact from around the state within the confines of a single "virtual classroom."

Health Care

Escalating health care costs have severely strained the nation's health care system. Much of this cost increase is due to an enormous volume of record-keeping and data transfer—a problem well-suited to a technological solution. In the United States alone, $40 billion to $80 billion per year could be saved by using advanced communications technology for the routine transfer of laboratory tests and more orderly collection, storage, and retrieval of patient information.

Technology also could be used to counteract the growing shortage of primary care physicians, particularly in traditionally underserved areas. Telecommunications-enabled *distance care*, like that conducted by the University of California at Davis for the rural Sacramento Valley and Foothill health systems, is a model that easily could be transplanted to urban areas, bringing quality care to currently poor served neighborhoods at a fraction of the cost of building new clinics or hiring new medical staffs.

Networked Communities

With fewer than 40,000 residents, the university town of *Blacksburg, Virginia,* is small by traditional population standards. But in the world of cyberspace, it is a giant. In fact, *Reader's Digest* recently labeled it "The Most Wired Town in America." In January 1992, Blacksburg town officials joined with the local university, Virginia Tech, and private industry to explore how they could work together on a project none of them could accomplish on their own—to create an "electronic village." Five years later, more than 40% of the town is on the Internet, and 62% use electronic mail. Through cooperative efforts with the public schools and the public library, all school children who desire it have free e-mail accounts and free, direct access to the World Wide Web. Local business is on board as well. More than two-thirds of Blacksburg companies advertise on the Internet with Web sites and online services, and they're prospering. One local real estate agency, for instance, received more online visits in a single month than it had had at its physical office for the entire previous year. Corporate partner Bell Atlantic deems the project a success—a judgment that, at this point, certainly seems warranted.

Touchscreen kiosks are an increasingly important part of smart community projects. They attract public attention, require only limited space and wiring, and are an effective and exciting way of presenting information to the public. Unfortunately, they also can come with a steep price tag: up to $30,000 per kiosk for proprietary software, hardware, interface development, and maintenance costs. With an annual budget of less than $300,000 for its entire community network, *Charlotte's Web*—a smart community initiative sponsored by the Charlotte and Mecklenburg County, North Carolina, public library system—obviously could not afford such a heavy investment in a single element of its network. Administrators therefore turned to an off-the-shelf solution: free or inexpensive software and a minimal hardware configuration that kept costs low while still producing a high-quality kiosk system. Web administrators relied on an inexpensive, standard World Wide Web browser (Netscape) for their kiosk interface, and on Linux—a free, Unix-like development tool—for their operating system. They also chose a basic hardware set-up, centered around a standard 486 PC, that held their total hardware and software costs to less than $3,000 per kiosk.

Like Blacksburg, Virginia, *Davis, California,* is a small university town in which the local university—the University of California at Davis—took the lead in launching a highly successful smart community initiative. The Davis Community Network (DCN) is a public/private partnership aimed in part at bringing Davis's small business community online. Working in conjunction with Dynasoft, an Internet consulting firm, DCN provides free Internet accounts, technical support, World Wide Web hosting, and—through Dynasoft's new Davis Virtual Market (DVM) service branch—personalized, one-on-one assistance to local businesses. As a consequence of this business-friendly environment, more than 956 small businesses have come online. DVM currently hosts more than 80 business Web sites, and these num-

bers are growing by an average of two sites per week. At the same time, DVM has donated hundreds of hours to developing locally based content in such diverse areas as pet care, the Girl Scouts, and education. DCN's online business presence has not been limited to the usual information-only postings. There are several teleshopping experiments that support ordering everything from pizzas to stereos online. In fact, one local comic book dealer has become an international distributor of collectibles thanks to its DVM Web page, while a local computer seller has grown from a small mail-order firm to a storefront with its own Internet service providership.

The Los Angeles Freenet (LAFN) provides low-cost ($15/year) Internet access, including e-mail, to more than 10,000 subscribers in the Greater Los Angeles area. Unlike commercial Internet service providers, the Freenet's goal is to offer access to all users. To that end, LAFN waives even its nominal fee for those users registering from one of the many Freenet-supported public terminals. The LAFN is run primarily by a base of more than 200 volunteers. In addition to its original goal of providing online medical information, the LAFN offers a multitude of other services, including electronic mail, World Wide Web access, and voter information. The LAFN subscriber base is 30,000.

The *Net of Two Rivers* (N2R), a nonprofit organization, is one of the first regional community networks in the nation. Headquartered in California's capital city of Sacramento, the project stretches from the center of the state to its eastern border and into Nevada. N2R's reach extends outward in a more important way, however—to more than two dozen collaborating organizations. Funded initially by the U.S. Department of Commerce, N2R has since brought business, government agencies, hospitals, libraries, schools, and nonprofit and volunteer groups into its thriving partnership. United Way, for instance, donated office space; Sacramento computer user groups provide technical support. California State University at Sacramento coordinates Internet training, and Apple Corporation made a sizable equipment donation. To date, N2R has established 50 public access sites throughout its 15-county area, helping to alleviate the significant disparities in computer access that exist in this largely rural region. In addition to its core focus on literacy training, N2R offers free Internet instruction in exchange for volunteer hours, and provides localized information to the public on such topics as wellness, disaster preparedness, job-seeking skills, voting, child-abuse prevention, and crime prevention.

Relying on off-the-shelf technology and a spirit of sharing, the San Francisco Public Library and its numerous public, private, and nonprofit partners are building bridges in cyberspace to link together neighborhoods, organizations, and individuals. Called "CityLink/Bridge," the new program is designed to transform libraries from their traditional role as book repositories into comprehensive community resources. *CityLink/Bridge* developers plan to make cultural, educational, health care, and information services available free to the public over a variety of telecommunications media, including the Internet and public broadcasting television stations. In addition, Citywatch 54, San Francisco's government access channel, is offering the City's first interactive television service, "Response TV." Viewers can use an ordinary telephone keypad to bring information to the television screen, and soon will be able to participate in interactive community polls. Central to CityLink/Bridge's approach is an active outreach program. The first step: a survey to identify the telecommunications resources available to public school students. In order to encourage public involvement, the library further has established free community access centers at its major new central facility, and free Internet terminals at each of its 30-plus branches.

The *Seattle Public Access Network* (PAN) consists of a free public World Wide Web site, an FTP and e-mail server, and a bulletin board system developed, operated, and funded since February 1995 by the City of Seattle. PAN's primary purpose is to serve as an electronic City Hall, allowing Seattle citizens to communicate with city officials and to obtain city information and services electronically. Registering more than half a million "hits" per month, PAN is designed to reach the widest possible audience by serving a user population with diverse electronic access capabilities. PAN is accessible not only by personal or business computers connected to the Internet but by more than 300 public library terminals and more than two dozen public access workstations in community centers throughout the Seattle region. PAN advertises its site as being unsurpassed in the amount of Seattle information it contains. It achieves this volume by allocating content-creation responsibilities among departments and organizations both inside and outside city government. In the future, PAN hopes to make all city publications available online, and to provide online access to all transactions, like licensing and permitting, that the city routinely conducts with its citizens and businesses.

Summary

In this chapter, we examine the assumptions, content, limitations, and extensions of classical central-place theory, concentrating first on cities as service centers at the national or regional level and ending with examples of urban hierarchies at the international level.

Classical central-place theory relies on three concepts to explain spatial equilibrium: *threshold, range,* and *hierarchy.* Central-place activities are arranged in a hierarchy according to the functions that they perform. High-order places bind regions of a national market-exchange econ-

omy. Within each metropolitan trade region we find a chain of low-order centers—smaller cities, towns, villages, and hamlets. These centers function as markets for the distribution of goods and services.

The assumptions on which classical central-place theory is based are unrealistic. Nowhere do we find an even *population distribution.* Resources are never the same everywhere, and landscapes are always in a state of disequilibrium. We discussed how actual conditions distort and transform theoretical networks of central places. Nonetheless, one of the strengths of a good scientific theory is that it can be modified to more closely fit reality by relaxing its assumptions. For example, we pointed to several empirical studies conducted by geographers in different areas that reveal systematic relationships between the sizes of towns, their distances from one another, the trade areas that they serve, and the densities of their surrounding populations. These empirical regularities make the central-place concept valuable for planning purposes. It has proven useful in deciding where to locate educational and medical facilities, as well as shopping centers. It has been used to recommend change in the spatial pattern of settlements.

To some geographers, however, the application of central-place theory, as well as other theories of spatial organization, is troublesome. British geographer David Harvey (1972) feared that our theories of spatial organization may well be tools for the frustration of social change. Edward Relph (1976) pointed out that much physical and social planning is divorced from places as we know and experience them in our daily lives. He argued that when we reduce places to points or areas, with their most important quality being (profitable) development potential, we are taking an approach that rationalizes draining wetlands for the construction of new regional shopping centers or displacing single-family residences with high-rise office buildings.

A major attribute of central-place theory is that it helps us understand the emergence of an *integrated hierarchy of cities* of different functions and sizes. A weakness is that it fails to explain the underlying forces that control the development of such hierarchies. The urban hierarchy is an important concept that geographers are trying to understand better, particularly because it is instrumental in the development of international capitalism. The consequences of the emergence of a few *global cities*—London, Tokyo, New York, Los Angeles—for decision making and corporate strategy formulation are far-reaching for people in developed and underdeveloped countries.

Smart communities are geographic regions ranging in size from neighborhood to a multi-county region whose people, organizations, and governments are using IT to transform their region into efficient "whales" of fundamental interdependence and cooperation.

KEY TERMS

administrative principle
break-point model
central function
central place
central-place theory
complete shopping center
convenience good
daily urban system
full convenience center
functional size
functional unit
global city

gravity model
hierarchical marginal good
hierarchy
hierarchy of central places
higher order goods
hinterland
industrial restructuring
isotropic surface
law of retail gravitation
linear market
lower order goods
market basket

market segments
marketing principle
mercantile model
minimum convenience
 centers
multipurpose travel
order of good and center
periodic market
range of a good
rank-size rule
settlement-building
 function

settlement-forming
 function
settlement-serving
 function
shopping good
smart cities
space-preference structure
Thiessen polygon
threshold of a good
trade area
tributary area
Varignon frame

SUGGESTED READINGS

Benevelo, L. 1991. *The History of the City*, 2nd ed. Cambridge, Mass.: MIT Press.

Bourne, L.S., and Ley, D.F., eds. 1993. *The Changing Social Geography of Canadian Cities*. Montreal: McGill-Queen's University Press.

Brunn, S.D., and Williams, J.F. 1993. *Cities of the World: World Regional Development*. 2nd ed. New York: Harper & Row.

Cadwallader, M. 1985. *Analytical Urban Geography*. Englewood Cliffs, N.J.: Prentice Hall.

Carter, H. 1995. *The Study of Urban Geography*. 4th ed. London: Edward Arnold.

Christaller, W. 1966. *The Central Places of Southern Germany*. Trans. C. W. Baskin. Englewood Cliffs, N.J.: Prentice Hall.

Clark, D. 1996. *Urban World/Global City*. London: Routledge.

Daniels, P. W. 1985. *Service Industries: A Geographical Appraisal*. New York: Methuen.

Drakakis-Smith, D. 1987. *The Third World City*. New York: Methuen.

Dutt, A.K., et al., eds. 1994. *The Asian City: Processes of Development, Characteristic and Planning*. Dordrecht/Boston/London: Kluwer Academic Publishers.

Friedan, B. J., and Sagalyn, L. B. 1989. *Downtown Inc.: How America Rebuilds Cities*. Cambridge, Mass.: MIT Press.

Garreau, J. 1991. *Edge City: Life on the New Frontier*. New York: Doubleday.

Geyer, H.S., and Kontuly, T.M. 1996. *Differential Urbanization*. New York: John Wiley & Sons.

Gilbert, A., and Gugler, J. 1992. *Cities, Poverty and Development: Urbanization in the Third World*. 2nd rev. ed. New York: Oxford University Press.

Hart, J. F., ed. 1991. *Our Changing Cities*. Baltimore: Johns Hopkins University Press.

Jones, K. G., and Simmons, J. W. 1990. *The Retail Environment*. London and New York: Routledge.

Knox, P. L. 1994. *Urbanization: An Introduction to Urban Geography*. Englewood Cliffs, N.J.: Prentice Hall.

Lowder, S. 1988. *Inside Third World Cities*. London: Routledge.

Stutz, F. P. 1992. Urban and regional planning, chapter 19. In *Interpreting the City*, T. Hartshorn, ed. New York: John Wiley.

Stutz, F. P., and Parrott, R. 1992. Urban GIS applications. In *GIS: Principles and Procedures*, D. Maguire, M. Goodchild, and D. Rhind, eds. (pp. 247–260). London: Longman.

Whyte, W. H. 1988. *City: Rediscovering the Center*. New York: Doubleday.

WORLD WIDE WEB SITES

SOURCEBOOK OF CRIMINAL JUSTICE STATISTICS 1995
http://www.cs.wisc.edu/scout/report/archive/scout-961129.html#3

UNIFORM CRIME REPORTS 1990–1993
Abstract: http://www.cs/wisc.edu/scout/report/archive/scout-961129.html#3
Site: http://www.lib.virginia.edu/socsci/crime/index.html
http://govinfo.kerr.orst.edu/usaco-stateis.html
From Census Bureau, USA Counties compiles 'useful demographic, economic, and governmental information spanning several years.'

MAPPING URBAN SPRAWL IN THE SAN FRANCISCO BAY REGION
http://geo.arc.nasa.gov/esdstaff/william/urban.html
A database of urbanization for the San Francisco Bay area.

THE BEST PRACTICES DATABASE
http://www.bestpractices.org/
A searchable database with solutions to common urban problems.

STATE OF THE NATION'S CITIES
http://www.policy.rutgers.edu/cupr/SoNC.html
The Center for Urban Policy Research at Rutgers University.

THE CITY IN PICTURES
http://hubcap.clemson.edu/mwsmith/citypix.html

ESPROMUD: URBANIZATION MODEL
http://cchp3.unican.es/ESPROMUD/espromud53.html
Explores the impact of various actions on earth surface processes.

INDUSTRIAL LOCATION: FIRMS

O B J E C T I V E S

- To consider industrial location in terms of business and management decision making
- To present the basic elements of industrial location theory
- To trace the rise of multinational corporations
- To show the trend toward flexible manufacture, flexible labor, and the flexible economy
- To describe business process, reengineering, and downsizing
- To describe Kondratiev long-wave models of industrial evolution
- To show the impact of information technology and the location of producer services

A General Motors robotic welding line. *(Photo: General Motors Media Archives)*

This chapter and the next deal with a crucial activity— manufacturing. To manufacture is to make things—to transform raw materials, under humanly created conditions and in controlled environments, into goods that satisfy our needs and wants. Manufacturing considerations include (1) what will be produced, (2) how it will be produced, (3) where it will be produced, and (4) for whom it will be produced. Manufacturing is important because not only does it produce goods that sustain human life, but it also improves our standard of living, provides employment, and generates economic growth. It has played this developmental role since the Industrial Revolution in England in the late 18th century.

Geographers approach the study of manufacturing from a viewpoint that emphasizes either firms or places. When firms are of primary significance, interest focuses on the locational choices that firms make. When areas are emphasized, attention centers on the nature of industries in a city, a region, or a country. In this chapter, we concentrate on firms; in Chapter 9, we examine the changing geography of industrial areas.

Whether they are considering firms or areas, geographers can adopt a variety of theoretical frameworks for interpreting industrial location. These frameworks include normative industrial location theory; the behavioral approach; and the Marxist, or structural, perspective. Normative industrial location theory derives from and shares the conservative ideology of classical economics. This theory uses abstract models to search for best, or optimal, locations. The behavioral approach focuses on the decision-making process. Rather than considering how decisions *should* be made, it examines how decisions *are* made. This liberal and more practical approach recognizes the possibility of suboptimal behavior. The Marxist, or structural, perspective challenges the ideology of the normative and behavioral industrial theories, which approach the question of location from a strictly managerial perspective. The structural approach calls for "a greater awareness of the social implications of shifts in industrial activity" (Marshall, 1979, pp. 675–676).

The three frameworks not only reflect the views of the geographers who use them, but also reflect changes in the nature of manufacturing itself. Normative industrial location theory, which prevailed until the 1960s, was formulated in the early 20th century, when most manufacturing businesses were single-plant firms and when basic heavy industry was in the vanguard of industrial progress. The behavioral approach came to the fore in the 1960s, when rapid economic growth in developed countries provoked an increased academic and political interest in the decision-making process. At first, attention centered on the decision-making behavior of single-plant firms; however, this focus became too narrow with the rise of large enterprises. The late 1960s and 1970s witnessed an upsurge of interest in the ge-

ography of large corporations. Meanwhile, critical world economic issues, affecting the location of industry, could no longer be ignored. Geographers devoted attention to the role of manufacturing in regional development theory and planning (Pred, 1977) and to structuralist interpretations of industrial location change (Massey, 1984). By the late 1980s, few geographers were involved in the development of normative industrial location theory, but many were involved in analyses of industrial restructuring—selective deindustrialization and reindustrialization in developed countries and industrial revolution in parts of the developing world.

This chapter, which emphasizes normative and behavioral approaches, begins with a discussion of the general circumstances that influence industrial locations of single-plant firms. Although most locational factors apply whether or not a firm is a single plant, today individual plants are more often part of larger enterprises. In fact, one of the fundamental revolutions in the global structure of manufacturing is the role played by multiproduct, multiplant, multinational operations. Accordingly, a section of the chapter is devoted to the spatial behavior of large industrial enterprises. Just as companies evolve, so do industries. The trend in the world economy today is away from the standardized economy and towards flexible manufacturing, flexible labor, and the flexible economy. The flexible economy is being birthed through a process called business process reengineering (BPR), downsizing, and information technology (IT). BPR and IT are having an impact on the location of producer services.

THE NATURE OF MANUFACTURING

Manufacturing involves four distinct phases: deciding what is to be produced (what?), gathering together the raw materials at a plant (where?), reworking and combining the raw materials to produce a finished product (how?), and marketing the finished product (for whom?). These phases are called *selection, assembly, production,* and *distribution* (see Chapter 1: Four Questions of the World Economy). The assembly and distribution phases require transportation of raw materials and finished products, respectively. Normative industrial location theory attempts to identify the plant locations that will minimize these transportation costs. The production phase—changing the form of a raw material—involves land, labor, capital, and management, factors that vary widely in cost from place to place. Thus, each of the three steps of the manufacturing process has a spatial or a locational dimension.

Changing the form of a raw material increases its use or value. Flour milled from wheat is more valuable than raw grain. Bread, in turn, is worth more than flour. This increase in labor power is termed *value added by manufacturing*. The value added by manufacturing as a percentage of the total value of a shipment is quite low in an industry engaged in the initial processing of a raw material. For example, turning sugar beets into sugar yields an added value of about 30%. In contrast, changing a few ounces of steel and glass into a watch yields a high added value—more than 60%. The cost of labor, or the availability of skills, plays an important role in high-value-added manufacturing; the cost of raw materials is the key variable in other industries. The relative importance of production factors is called *orientation*. Geographers frequently use a term such as *raw-material-oriented* or *market-oriented* to specify the key variable for a given industry. The orientations of industries affect their patterns of geographic locations and concentrations.

UNEVEN DISTRIBUTION OF RAW MATERIALS

Von Thünen's model of agricultural production assumed an even distribution of resources. Points of manufacturing (cities) would develop even if all resources were ubiquitous. Manufacturing would operate at selected points and incur only two kinds of costs: *production costs*, which would arise from interrelationships among other factors of production and demand, and *distribution costs* that would result from transporting the finished product to dispersed markets.

However, in reality, resources are not evenly distributed—especially the raw materials required for basic heavy manufacturing. Even the industries that use manufactured goods as their raw materials face an uneven distribution of inputs. Therefore, manufacturing involves a third kind of cost: *assembly costs*—the price that must be paid to bring raw materials from diverse locations to one plant. Assembly costs are the main concern of classical industrial location theory.

The Simple Weberian Model: Assembly Costs

Classical industrial location theory is founded on the work of Alfred Weber (1929), a German economist. Weber was, of course, influenced by the period in which he wrote. Therefore, when we evaluate his model, we must take into account the considerable changes in the manufacturing industry since the early 20th century. Nonetheless, Weber taught geographers to think about the distinction between material- and market-oriented industries.

Weber attempted to determine the manufacturing patterns that would develop in a world of numerous, competitive, single-plant firms, given a set of normative constraints. He began by assuming that transportation costs are a linear function of distance. He required that

producers, who face neither risk nor uncertainty, choose optimal locations. He also implied that the demand for a product is infinite at a given price. Producers could sell as many units as they produced at a fixed price. They could sell none at a higher price, and charging a lower price would not affect the total demand for the product. The producer's strategy was therefore to assemble the product at the lowest possible cost in order to maximize revenue. Weber's system is often called a *least-cost approach* because he assumed that such locations are optimal.

Transportation sets the general regional pattern of manufacturing. This pattern is in turn distorted by spatial variations in the cost of labor. The final determinant is the local factors. In Weber's approach, as each set of forces is considered in sequence, the complexity of analysis increases.

Raw-Material Classes

The first cost faced in the manufacturing process is that of assembling raw materials. In Weber's simple classification system (Table 8.1), raw materials are first classified by their frequency of occurrence. *Ubiquitous raw materials*, such as air, are universally distributed; they always have a transportation cost of zero. *Localized raw materials*, such as coal, are found only at specific locations; their transportation costs are a function of the distance that they must be moved. Second, Weber classified raw materials on the basis of how much weight they lose during processing. *Pure raw materials*, such as an automobile transmission, lose no weight in processing, whereas *gross raw materials*, such as fuels, do lose weight in processing. A weight-losing raw material is assigned a *material index*, which indicates the ratio of raw-material weight to finished-product weight. Pure raw materials have a material index of one. Weight-losing raw materials have material indexes of more than one. Fuels have the highest material indexes because none of their weight affects the weight of the finished product.

We are now ready to discuss transportation costs as they relate to different kinds of raw materials. For each case we must consider (1) the costs of assembling raw materials (*RM*), (2) the costs of distributing the finished product (*FP*), and (3) the total transportation costs (*TTC*). In all cases, we assume the existence of a single market point. The best location for a manufacturing plant is the point at which the total transportation costs are minimized.

Ubiquities/Localized Raw Materials

Only localized raw materials attract production. Ubiquities merely add to the pull of the market. Ubiquitous raw materials occur everywhere, so their assembly

▨

Table 8.1
Solutions to Weber's Locational Problems

Raw-material Classes	Plant Location
Ubiquities Only	Market
Localized and Pure	
One pure	Anywhere between source of raw material and market
One pure and ubiquities	Market
More than one pure	Market
More than one pure and ubiquities	Market
Localized and Weight-Losing (Gross)	
One weight-losing	Source
One weight-losing and ubiquities	Source or market, depending on relative size of input
More than one weight-losing	Indeterminate (mathematical solution)
More than one weight-losing and ubiquities	Indeterminate (mathematical solution)

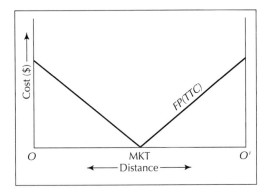

FIGURE 8.1
Weber's model: ubiquities only. Because ubiquities are found everywhere, including at the market, finished product (FP) and total transportation cost (TTC) are lowest at the market.

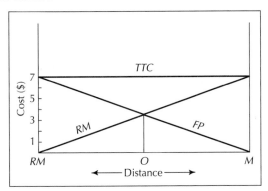

FIGURE 8.2
Weber's model: one localized pure raw material, located at *RM*. The line *RM* represents assembly costs, while the line *FP* represents finished product distribution costs, which are lowest at the market (*M*). Since the raw material does not lose weight, total transportation costs (*TTC*) are the same everywhere along a straight line between the raw material and the market.

costs are always zero. Only finished-product costs are important, and they are reduced to zero when the plant location is at the market point, as illustrated by the graph in Figure 8.1. Raw-material costs (*RM*) are the line *O-O′*. Finished-product costs rise steadily with increasing distance (in either direction) from the market. The cost line *FP* also marks the total transportation costs, which are minimized at the market.

ONE LOCALIZED PURE RAW MATERIAL
Figure 8.2 indicates the costs of a product that requires one localized pure raw material. The raw material is localized at Point *RM*, and the market is at *M*. The line *RM* represents the assembly costs, which increase as a function of distance from the source of the raw material. Similarly, the line *FP* represents the distribution costs for the finished product. Total transportation costs (*TTC*) are the sum of *RM* and *FP*. At *RM*, *TTC* equal $7.00 (*RM* = $0, *FP* = $7.00, *TTC* = $7.00). At *O*, *RM* = $3.50 and *FP* = $3.50, so *TTC* = $7.00. Because the total transportation costs are exactly $7.00 everywhere along a straight line between the source and the market, a manufacturing plant located anywhere along this line can minimize costs.

ONE LOCALIZED PURE RAW MATERIAL PLUS UBIQUITIES
Figure 8.3 shows the costs of a product requiring one localized pure raw material plus ubiquities. The assembly costs for the localized raw material (*RM*) are minimized at Point *RM*. Ubiquitous assembly costs are zero everywhere, and finished-product distribution costs are minimized at *M*. Ubiquitous raw materials, once processed, add to the weight of the finished prod-

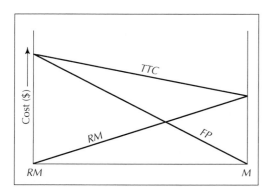

FIGURE 8.3
Weber's model: one localized pure raw material plus ubiquities. Ubiquities add to the weight of the finished product. Ubiquities are available everywhere; therefore, total transportation costs (*TTC*) are lowest at the market.

uct, so the total transportation costs (*TTC*) are minimized at the market (*M*). In other words, a plant located at the market eliminates the need to move the ubiquitous material in its processed form.

Bottled and canned soft-drink manufacturing exemplifies the use of one pure raw material (syrup concentrate) and water (in Weberian analysis, a ubiquitous raw material). The strong association between soft-drink manufacturing and population indicates that the industry is market-oriented. If the plant locates at the market, the ubiquitous raw material (water), which makes the largest contribution to the weight of the finished product, does not need to be moved (Figure 8.4).

SEVERAL LOCALIZED PURE RAW MATERIALS
Figure 8.5 indicates costs for two localized pure raw materials, but the outcome would be the same for more than two. Once again, the single market is at *M*; one pure raw material is localized at Point RM_1, and the other is localized at Point RM_2. The transportation costs for each raw material are given by the lines RM_1 and RM_2. We assume that the raw materials are used in equal amounts (1 ton each) and that transportation costs are \$1 per ton-kilometer. At Point RM_1, the cost of RM_1 is zero, and the cost of RM_2 is \$6, for a total of \$6. But the finished product weighs 2 tons; hence, an additional \$6 in transportation costs is required to ship the finished product back to *M* (2 tons shipped 3 kilometers at \$1 per ton-kilometer equals \$6). Total transportation costs (*TTC*) at Point RM_1 are therefore \$12, and they are the same at Point RM_2. At *M*, however, total transportation costs equal only \$6. Locating the plant at the market eliminates the need to

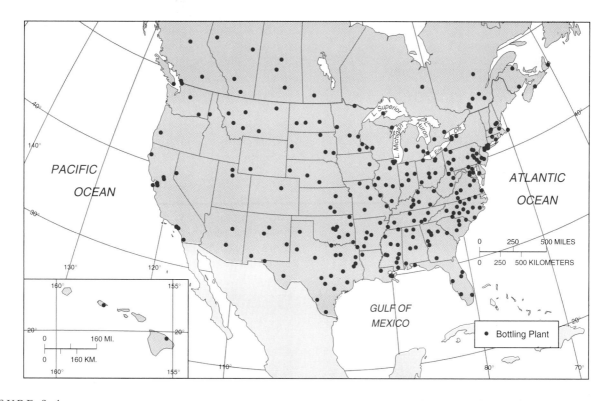

FIGURE 8.4
Bottled and canned soft drink manufacturing exemplifies the use of one pure raw material, syrup concentrate, and one ubiquity, water. Bottling plants must be located near the market because of the relatively high cost of transportation in this bulk-gaining industry. Consequently, there are several hundred bottling plants throughout North America, each located near a large urban center.

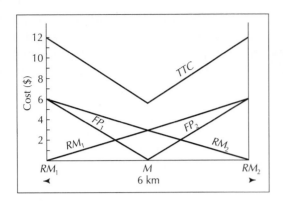

FIGURE 8.5
Weber's model: two localized pure raw materials.

backhaul a raw material, so total transportation costs are minimized.

SEVERAL LOCALIZED PURE RAW MATERIALS PLUS UBIQUITIES

Remember that ubiquities always add to the pull of the market. In the graph in Figure 8.6, we assume that 1 ton each of RM_1, RM_2, and the ubiquitous raw material are used; thus, the finished product weighs 3 tons. Finished-product distribution costs at Points RM_1 and RM_2 are $9 (3 tons shipped 3 kilometers at $1 per ton-kilometer). Localized raw-material costs, RM_2, equal

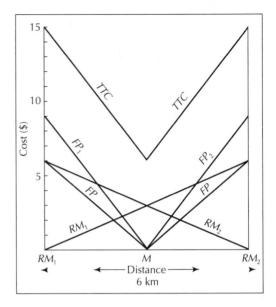

FIGURE 8.6
Weber's model: several localized pure raw materials plus ubiquities.

$6, and finished-product distribution costs, FP_2, equal $9. Total transportation costs equal $15 at Points RM_1 and RM_2, but equal only $6 at M. The pull of the market is considerably strengthened by the addition of ubiquitous raw materials.

Ready-mixed concrete production is a good example of an industry that uses more than one pure raw material plus ubiquities. The industry uses portland cement (pure), aggregate (pure and often ubiquitous), and water (considered to be ubiquitous). Weberian theory indicates that costs are minimized at the market. In the United States, the high correlation between industry distribution and population indicates a market orientation.

PERISHABLE GOODS

Perishable or time-urgent goods and industries are also an example of weight- or bulk-gaining raw materials. Examples of perishable goods include fresh-baked foods, fresh milk, and other types of manufactured food that require rapid delivery to preserve them. Although butter and milk, which have longer shelf lives,

The Hull Rust iron-ore pit near Hibbing, Minnesota. From the late 19th century to shortly after World War II, high-grade ore from Minnesota's iron ranges was shipped via the Great Lakes to iron and steel plants. When much of the accessible high-grade ore was exhausted, attention shifted to the exploitation of low-grade ore, known as taconite. Before shipment, taconite is beneficiated and pelletized in plants near the mines. Thus, it falls into the category of a localized weight-losing raw material. *(Photo: Owen Franken, Stock Boston)*

can be produced near the source of the raw materials—at rural dairies—milk and cream production is fixed closer to the markets to minimize the time and cost of delivery to places where these products are consumed.

Communications media, especially the printed medium, are an example of a perishable industry. Newspapers must be printed near the location where they are consumed to minimize the distance that these bulky items must be transported. Consequently, producing a nationwide newspaper is a difficult logistic problem. The *New York Times*, the *Wall Street Journal*, and *U.S.A. Today* are not printed and published in New York City or Washington, D.C., and distributed to the entire United States. Instead, these newspaper companies transmit news and columns via satellite to scattered locations such as Atlanta, Chicago, and Los Angeles, where the papers are printed and then distributed to the nation.

Other specialized manufacturers, such as designer-clothing manufacturers, are attracted to large markets as well. Being near accessible locations where buyers can view the merchandise is important. High-fashion clothing distribution is also a perishable industry because of the speed with which clothing styles change. As a result of the decisions by national and international buyers, merchandising, production, and sales occur at the market. New York City's garment and manufacturing district is a good example of this type of specialized manufacturer. In addition, cloth manufacturers and suppliers of shoulder pads, clasps, pins, zippers, elastic, and thread cluster near the principal garment manufacturers in New York City.

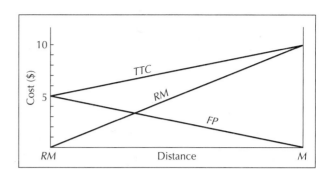

FIGURE 8.7
Weber's model: one localized weight-losing raw material. Because the raw material loses weight, the total transportation costs (*TTC*) are minimized at the raw material source (*RM*) because locating the processing plant at the market (*M*) would incur transportation costs to partially move leftover waste.

Weight- or Bulk-Losing Raw Materials

ONE LOCALIZED WEIGHT-LOSING RAW MATERIAL

Figure 8.7 illustrates the costs of a product requiring one localized weight-losing raw material. Assume that the raw material, which is localized at Point *RM*, loses half its weight in processing. The material index is therefore 2. The transportation costs for each point between the source and the market are indicated by the line *RM*. Each unit of the raw material shipped to the market at *M* costs \$10. Each unit of the finished product shipped from the source, however, costs only

The classical principles of industrial location theory are evidenced in the river valley and railroad site of Bethlehem Steel Corporation's plant at Bethlehem, Pennsylvania. This huge plant, which extends for nearly eight kilometers along the south bank of the Lehigh River, converts raw materials—Appalachian coking coal and Minnesota iron ore—into structural shapes, large open-die and closed-die forgings, forged steel rolls, cast steel and iron rolls, ingot molds, and steel, iron, and brass castings. The main market for the steel products is the American Manufacturing Belt with its abundance of metal-using industries. *(Photo: Bethlehem Steel Corporation, IN)*

$5. Total transportation costs (*TTC*) are minimized at the raw-material source. The distribution of the copper industry illustrates this situation. Copper ore has a high material index (99), and manufacturing is concentrated near copper mines. A raw-material orientation eliminates the transportation costs of moving waste (Figure 8.8).

ONE LOCALIZED WEIGHT-LOSING RAW MATERIAL PLUS UBIQUITIES

Of importance when a product requires one localized weight-losing raw material plus ubiquities is the ratio of the weight lost through processing to the weight of the ubiquitous material. Two extreme cases are illustrated in Figure 8.9. In the first case (Figure 8.9a), the weight-losing raw material is a fuel, and all of its weight is lost in the manufacturing process. Assume

that 1 ton of fuel is localized at Point *RM* and that 1,000 kilograms of the ubiquitous pure raw material are required to produce 1,000 kilograms of the finished product. Total transportation costs are minimized at the source of the weight-losing raw material.

In the second case (Figure 8.9b), assume that a weight-losing raw material has a material index of 2. Half the weight of the localized raw material is lost in processing. In this case, however, we assume a 3-to-1 ratio of the ubiquitous raw material to the localized raw material. Two tons of the localized raw material plus 3 tons of the ubiquitous raw material are processed into a finished product weighing 4 tons. At Point *RM*, total transportation costs are $4, but at *M*, total transportation costs are only $2, which is the transportation cost for the amount of the localized weight-losing raw material required for 1 unit of the finished product. Therefore,

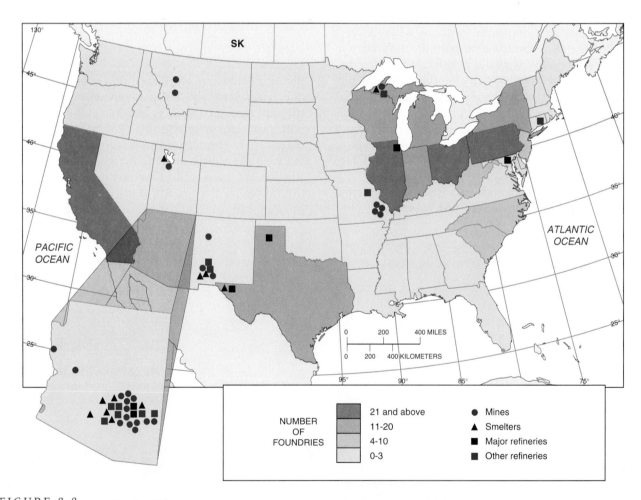

FIGURE 8.8
An example of a weight-losing, or bulk-reducing, raw material is copper ore. Copper has a high material index, meaning that 99% of the ore is wasted, and only 1% can be reduced to pure copper. Manufacturing is concentrated near the copper mines. This orientation to the source of the raw material eliminates moving wasted ore tailings. Refining smelters are also located near the mines. In the case of the United States, Arizona and New Mexico are most important. (See Color Insert 2 for more illustrative map.)

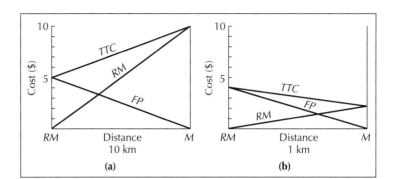

FIGURE 8.9
Weber's model: one localized weight-losing raw material plus ubiquities. (a) Best location at raw material source, and (b) best location at market.

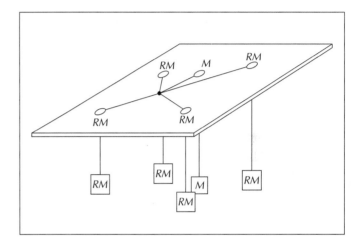

FIGURE 8.10
The Varignon frame, a mechanical device of weights, strings, and pulleys formerly used to find the best location for a plant requiring several localized weight-losing materials. Today, high-speed computers can juggle numerous materials, markets, and other factors (e.g., labor, capital, tax structure) to find the best location.

total transportation costs are minimized at the market (*M*). This case typifies commercial brewing. Barley and hops are the localized raw materials that lose weight in processing, but the major ingredient by weight—water—is ubiquitous in Weberian analysis. Brewers are market- rather than raw-material-oriented.

SEVERAL LOCALIZED WEIGHT-LOSING RAW MATERIALS

The situation becomes more complex when several weight-losing raw materials are necessary. The Varignon frame simplifies the problem. The localized raw-material sites are located on a map that is glued onto a board (Figure 8.10). Holes are drilled in the board at each site. A pulley, which reduces friction, is located at each hole. A raw material is simulated by a weight proportional to the total weight required to produce 1 unit of the finished product. Finished-product distribution costs are simulated by a weight equal to the finished-product weight at the market point (*M*). Ubiquitous raw materials are simulated by adding to this weight. Cords are run from the weights through the pulleys and tied into a single knot. When the weights are released, the final location of the knot indicates the optimal location. This type of analysis has been applied to the steel industry, in which several weight-losing materials are processed.

EXTENSIONS OF WEBER'S MODEL

Space-Cost Curves

Weber's basic system can be extended by using the space-cost curves. Assume that equal amounts of two localized weight-losing raw materials are required to produce 1 unit of a finished product. The material index of each raw material is 2, so 1 ton of RM_1 and 1 ton of RM_2 yield a 1-ton finished product. In Figure 8.11a, concentric circles have been drawn around each raw-material source and the market point. Weber called these isocost lines for each point isotims.

Total transportation costs are the sum of all costs to all three points. At Point *X* (Figure 8.11b), for example, RM_1 costs $3, RM_2 $2, and the finished product $2 to be transported. We can find total transportation costs for as many points as we want and connect points of equal value to produce total-cost isopleths, which Weber called isodapanes (Figure 8.11c). If we visualize Figure 8.11c in three dimensions, we map a depression (Figure 8.11d). Smith assumed that the market price for the finished product is a spatial constant, the line *MP-MP*′. The intersections of the space-cost curve and the market-price line delimit the *spatial*

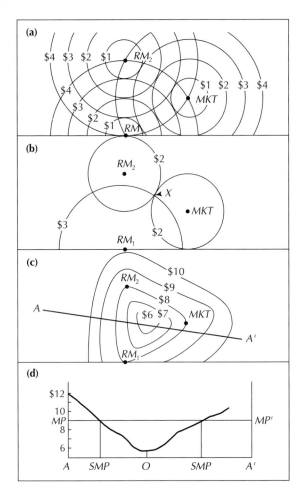

FIGURE 8.11
The development of a space-cost curve: (a) isotims, (b) total cost to Point *X*, (c) isodapanes, (d) space cost along the line *A–A'*.

margins to profitability. The best location (*O*) is at the lowest point on the space-cost curve.

David Smith's extension of Weber's ideas brings us a step closer to reality. It acknowledges that a least-cost location is not essential for economic survival. Profits can be realized by entrepreneurs who locate farther from the optimal location at Point *O*.

Distortions of the Isotropic Surface

Weber's assumption of a completely isotropic surface is modified when we account for the effects of localized resources. Thus, we begin a transition from a normative model to a descriptive model. The regular patterns implied by central-place theory are distorted to minimize the pull of nonubiquitous resources. The role of the natural environment (localized resources) distorts the ideal patterns of an isotropic surface.

It becomes apparent, then, that the forces controlling the location of manufacturing may be quite different from those that control the location of central places. The tertiary sector (i.e., retail and services) is market-oriented, but the orientation of the secondary sector (i.e., manufacturing) varies from industry to industry. Some types of manufacturing, as we have seen in the Weberian model, are market-oriented. If all industries were of this type, the manufacturing pattern would match the central-place pattern. Some industries, however, have cost structures so dominated by localized input costs that they are material-oriented, which distorts the pattern.

This distortion is illustrated in Figure 8.12. Point *I* is the ideal location for a central place. The shaded area is a major resource deposit. City *A* was established to exploit this resource and became a manufacturing center. City *A* also supplies central functions to the surrounding area. Because of its initial establishment and growth, it superseded the establishment of service functions as the theoretical ideal location (*I*). The purely spatial pattern was distorted by the uneven distribution of resources.

Real-world patterns are evolutionary; they are not the result of decisions made by optimizers. Most real-world decisions do not result in best (most profitable) locations. Locational decisions, once made, lead to inertia, in this case, *industrial inertia.* This tendency to continue investing in a nonoptimal site may be strong enough to perpetuate the distorted pattern, even if more optimal locations are discovered in the future. For example, we now have the analytic skills to locate state capitals at the centroids of state or national populations, but the investment that we already have in most state or national capitals precludes us from doing so. Tension develops between ideal spatial patterns and the patterns produced by localized resources. As technology (especially transportation) improves, ideal spatial patterns (from the entrepreneur's viewpoint) become more feasible, but the inertia resulting from past actions exerts a constant deterrent on actualizing these patterns.

We inherited one aspect of inertia from the 19th century. Locational patterns were then dominated by

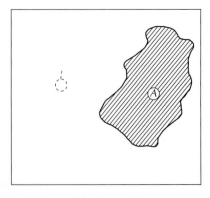

FIGURE 8.12
Distortion produced by localized resources.

coal, which was used as a source of carbon in the iron and steel industry, as a source of fuel for steam engines, and as a source of heating for homes and businesses. As a result, businesses and populations gravitated to the coalfields in the mid and late 19th century. Although other forms of energy gradually replaced coal, the population distribution has changed more slowly. Many of the depressed industrial regions in developed countries are found on coalfields.

Weber Diminished in Today's World

Let us conclude this discussion of Weber's basic system and its extensions with a brief appraisal. His model distinguishes between material- and market-oriented industries. Gross localized materials encourage material orientation, so bulk-reducing industries—mineral processing, metal smelting, timber processing, fruit and vegetable packing—are frequently found near material sources. But manufacturing is more complex than it was in the early 20th century. Many plants begin with semifinished items and components rather than with raw materials. Producers' goods seldom lose large amounts of weight; therefore, there is not much tendency toward material orientation.

Ubiquities encourage material orientation. However, few materials can be classified as ubiquities without qualification. For example, water is scarce in many areas. Water may be ubiquitous for firms that use a little (e.g., a bakery), but not for firms that use a lot (e.g., a steel mill).

Weber's model has been criticized for its unrealistic view of transportation costs as a linear function of distance. Because of fixed costs, especially terminal costs, long hauls cost less per unit of weight than short hauls do. Plants tend to locate at material or market points rather than at intermediate points, unless there is an enforced change in the transportation mode, such as at a port (Hoyle and Pinder, 1981). However, with the expansion of the modern trucking industry and its flexibility in short hauls, the disadvantages of intermediate locations have been reduced.

Three other transportation developments have a bearing on industrial location in the late 20th century:

(1) *Freight rates have risen more sharply on finished items than on raw materials.* As a result, raw materials are often shipped nearer to the market to reduce the distance that finished goods must be shipped.

(2) *Transportation costs have been declining.* This decline increases the importance of other locational factors. Labor now is the most important industrial location determinant. This is most obvious in firms producing high-value and high-tech products. For these firms, transportation costs are relatively unimportant. Yet for firms that distribute consumer goods (e.g., soft

Table 8.2

Value Added in Manufacturing Products

Product	Added value ($/pound) rough 1990 estimates
Satellite	20,000
Jet fighter	2,500
Supercomputer	1,700
Airplane engine	900
Jumbo jet	350
Video camera	280
Mainframe computer	160
Semiconductor	100
Submarine	45
Color television	16
Numerically-controlled machine tool	11
Luxury car	10
Standard car	5
Cargo ship	1

drinks) to dispersed markets, transportation costs remain a significant factor.

(3) *Brainpower is producing muscle and machine power* and transforming natural resources (Chapter 3). Natural resources are no longer as important in the growth of economies. Instead, there is *transmaterialization* of resources as smaller, lighter, "smarter" products are manufacturing from resources to which high technology and brainpower have been added (Table 8.2).

PRODUCTION COSTS AT THE SITE

So far, we have not considered the costs of the actual manufacturing process. After materials are assembled at a point, they must be reworked and combined to produce a finished product. Production costs include land, labor, capital, and managerial and technical skills. All these are necessary for production, and all exhibit spatial variations in both quantity and quality.

The Cost of Land

Since World War II, to minimize distribution costs, industry has been moving closer to the market. Within the urban area, however, there has been a *centrifugal drift* to suburban industrial properties since the 1970s. A large contemporary factory needs a large amount of land, preferably in a one-story building. Large parcels of industrialized land are more likely to be available in the suburbs than in central-city locations. Land in the suburbs is also much less costly than it is in the center of the city, as we learned in Chapter 6. The cost might be a million dollars per acre in the city, but only a few thousand dollars in a distant suburb near a beltway.

More reasons why industrial properties have expanded into the suburbs include locations that are easily accessible to motor freight by interstate highway and beltway, and serviced properties, which have ample sewer, water, parking, and electricity. Industries may also be attracted to the suburbs because of nearness to amenities and residential neighborhoods. Suburban locations minimize the laborers' journey to work. In addition, the U.S. regions of the South and the West have been attractive recently because of the demand for recreational resources, a mild climate, and opportunities for enjoying the outdoors. Some firms might base their decisions partially on other land factors, including accessibility to educational facilities, cultural facilities, or even major-league sports franchises.

Truman Hartshorn, in his book *Interpreting the City* (1992), identified eight factors that have accelerated the development of suburban industrial parks:

1. the expansion and dispersion of light assembly and the distribution of its facilities
2. lack of suitable industrially zoned land in older central cities
3. blight, traffic congestion, and cramped conditions of older industrial areas
4. a change in plant design from multistory, mill-type buildings to single-story plants permitting horizontal line production, which requires larger sites
5. more dependence on the automobile for commuting
6. increased transportation of industrialized products by trucks, which requires unloading space
7. a preference by institutional investors for financing construction and planned districts where investor security is more certain
8. the convenience and economy of not having to take care of the details that development and management organizations now handle

The Cost of Labor

Labor inputs are required for all forms of economic production, but the relative contribution of labor to the value added by manufacturing varies considerably among industries. For example, the contribution of labor costs is high in the automobile industry but low in the petroleum-products industry. The supply and demand for labor vary, but those industries in which labor costs play a major role are much more sensitive

◈

Table 8.3
Variations in Labor Productivity

Plant	Total Hourly Wages and Fringe Benefits (per Worker)	Output per Hours in Units (per Worker)	True Labor Cost per Unit Productivity
A	$ 5.50	100	5.5¢
B	$ 7.80	200	3.9¢
C	$14.20	400	3.5¢

to local variations in the cost of this input. Under capitalism, the real cost of labor is determined by the relative productivity of labor rather than the dollar cost of wages and fringe benefits. The hypothetical data in Table 8.3 illustrate this point. The labor cost per unit produced is lowest in Plant *B* even though the hourly total of wages and fringe benefits is highest.

Weber considered the cost of labor to be a regional factor controlling manufacturing patterns. The initial manufacturing pattern is set by transportation costs and then distorted by variations in the cost of labor. Weber's model assumed an infinite amount of available labor at different points, but the cost of labor varies from point to point.

Weber's analysis of this problem is illustrated in Figure 8.13. A product requires two localized raw materials and is distributed to a single market at *M. Isodapanes*, which are isopleths of total transportation costs, are indicated in the diagram. These costs are minimized at Point *T*, $4. As we move away from *T*, total transportation costs increase. At Point *L*, labor costs are $2 per unit less than at *T*. This unit labor savings is

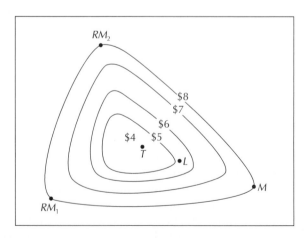

FIGURE 8.13
Critical isodapane for labor.

used to determine the value of a *critical isodapane*. No savings in total unit cost will result if the plant is moved to another point where the increased transportation costs exceed the labor savings at that point. In other words, the cost of the move must not exceed $2 per unit. Transportation costs at T ($4) plus labor savings at L ($2) determine the amount that cannot be exceeded ($6). Point L lies inside this critical isodapane and is clearly an economic move. Moving to a point on the critical isodapane would result in zero savings; moving to a point outside the line would result in higher total costs. Thus, within the limits established by critical isodapanes, variations in labor costs distort the pattern established by general transportation costs.

In theory, equilibrium conditions should balance regional differences in the supply and demand for labor, which is a mobile production factor. A high demand for labor in one place and an excess supply in another should be brought into equilibrium by labor migration. Such migration has certainly occurred. Examples are the 19th-century migration of people from rural areas to the city in countries such as Great Britain, and the 20th-century movement of labor from the Third World and other peripheral countries to the core of industrial Europe: from the Caribbean and South Asia to Great Britain, from North Africa to France, and from Turkey to Germany.

As in the case of all production factors, the response of labor is not instantaneous. Skilled labor, in particular, has a relatively high degree of inertia, especially in the short run. People are reluctant to leave familiar places, even if jobs are plentiful in other areas. They will sit out short periods of unemployment or accept a smaller net income than could be earned elsewhere. Liberal welfare policies, such as unemployment payments and workers' compensation, have reduced the plight of the unemployed and underemployed in advanced industrial countries in this century.

Cotton textile manufacturing in the United States was in New England, but is now located in the lower labor cost Middle Atlantic states of Virginia, North Carolina, South Carolina, Georgia, Alabama, and Tennessee. No particular agglomeration exists, but many medium-sized and small towns have textile plants. For example, Mt. Vernon Mills and Regal Textile each have 25 plants in this belt. However, the wool industry remains in the Northeast because of larger numbers of skilled laborers and because of inertia. Workers must use complex equipment to accomplish precise cutting and weaving. Another example in the United States of high-skill worker demand is the semiconductor and computer manufacturing industry concentrated in California and the Northeast, especially Massachusetts and New York City.

The lack of instantaneous adjustment in labor demand and supply has resulted in variations in the cost of labor within and among countries. In the United States, wages are often higher in the more industrial states, in densely settled areas, and in highly urbanized environments (Table 8.4).

Table 8.4
Hourly Compensation Costs in U.S. Dollars for Factory Workers

	1985	1995		1985	1995
Germany	$ 9.60	$31.88	Australia	8.20	14.40
Switzerland	9.66	29.28	Ireland	5.92	13.83
Belgium	8.97	26.88	United Kingdom	6.27	13.77
Austria	7.58	25.38	Spain	4.66	12.70
Finland	8.16	24.78	Israel	4.06	10.59
Norway	10.37	24.38	New Zealand	4.47	10.11
Denmark	8.13	24.19	Greece	3.66	8.95
Netherlands	8.75	24.18	Korea	1.23	7.40
Japan	6.34	23.66	Singapore	2.47	7.28
Sweden	9.66	21.36	Taiwan	1.50	5.82
Luxembourg	7.72	20.06	Portugal	1.53	5.35
France	7.52	19.34	Hong Kong	1.73	4.82
UNITED STATES	**$13.01**	**$17.20**	Mexico	1.59	1.51
Italy	7.63	16.48	Sri Lanka	.28	.45
CANADA	**10.94**	**16.03**			

Note: Latest figures for Luxembourg and Sri Lanka are for 1994.

At the world scale, developed countries have higher wage rates than newly industrializing countries. One factor responsible for differential wage rates is the level of worker organization. Higher rates of unionization are associated with higher wages. Unionization is generally more prevalent in the older, established industrial countries than in the newly industrializing countries. Thus, considerable advantages can be gained by companies that relocate to, or purchase from, newly industrializing countries, especially if these countries are characterized by low levels of capital-labor conflict. The capital-labor conflict, manifested in industrial disputes, is a powerful force propelling the drift of industrial production outward from the center to the periphery of the world system.

The Cost of Capital

Capital, another necessary production factor, takes two forms: *fixed capital* and *liquid, or variable, capital*. Fixed capital includes equipment and plant buildings. Liquid capital is used to pay wages and meet other operating costs. Liquid capital is theoretically the most mobile production factor. The cost of transporting liquid capital is almost zero, and it can be transmitted almost instantaneously in our "wired world." About $9 trillion in electronic funds transfers are completed annually over international communications lines, an amount equal to two-thirds of the global gross national product (GNP). Fixed capital is much less mobile than liquid capital; for example, capital invested in buildings and equipment is obviously immobile and is a primary reason for industrial inertia.

Any type of manufacturing that is profitable has an assured supply of liquid capital from revenues or borrowing. Most types of manufacturing, however, initially require large amounts of fixed capital to establish the operation—or, periodically, to expand, retool, or replace outdated equipment or to branch out into new products. The cost of this capital, which is interest, must be paid from future revenues. Investment capital is not uniformly distributed and does not display great mobility.

Investment capital has a variety of sources: personal funds; family and friends; lending institutions, such as banks and savings and loan associations; and the sale of stocks and bonds. Most capital in advanced industrial countries is raised from the last two sources. The total supply of investment capital is a function of total national wealth and the proportion of total income that is saved. Savings become the investment capital for future expansion.

Whether a particular type of manufacturing, or a given entrepreneur, can secure an adequate amount of capital depends on several factors. One factor is the supply of and demand for capital, which varies from place to place and from time to time. Of course, capital can always be obtained if users are willing to pay high enough interest rates. Beyond supply-and-demand considerations, investor confidence is the prime determinant of whether capital can be obtained at an acceptable rate. Investor confidence in a particular industry may exist in one area but be lacking in another. For example, Henry Ford, of Ford Motor Company, failed to raise investment capital in one area of the United States but was able to secure it in his hometown of Detroit.

The Ford Motor Company steel mill and auto assembly plant at River Rouge near Detroit, Michigan, is a vertically integrated system. Basic raw materials are used to produce steel, which is incorporated into the engines, frames, bodies, and parts of finished automobiles. It is a transnational corporation with 360,000 employees located in 30 countries. *(Photo: Ford Motor Company, Detroit)*

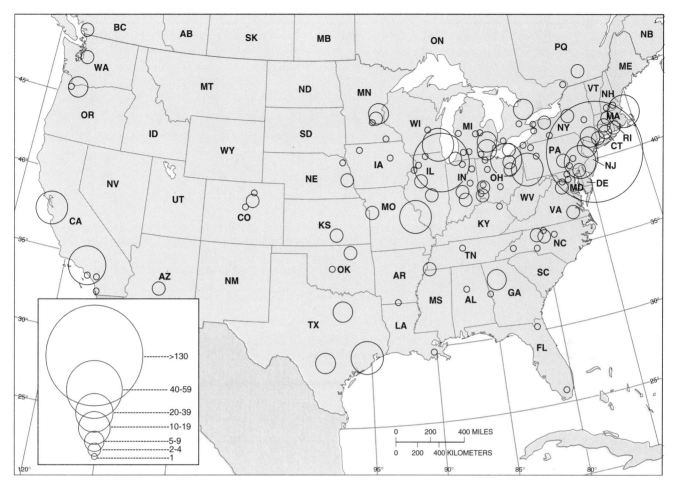

FIGURE 8.14
Corporate headquarters: 500 largest corporations.

Managerial and Technical Skills

Managerial and technical skills are also required for any type of production operation. Corporations are the primary agents responsible for specific industrial patterns. The general pattern is, of course, determined by the localization of other production factors, subsequent transportation costs, and spatial margins to profitability.

Theoretically, management should be much more mobile than labor because fewer people are involved, and the higher income of managers should make moving less of a financial burden. But managerial skills, like labor skills, are often highly concentrated. Where are the organizational headquarters of the 500 largest American corporations? Most are in the largest urbanized areas of the country (Figure 8.14). The top 10 metropolitan headquarter areas are New York City (157 head offices), Chicago (41), Los Angeles (28), San Francisco (25), Philadelphia (18), Minneapolis–St. Paul (15), Detroit (13), Boston (12), Pittsburgh (12), and Houston (Rand McNally, 1996).

Technical skills are the skills necessary for the continued innovation of new products and processes. These skills are generally categorized as research and development (R&D). In the early phases of industrialization in developed countries, product development was usually carried out in tandem with production by small firms, many of which, together with their innovations, failed to survive. Today, the R&D required for new products is a large and expensive process, involving long lead times between invention and production; therefore, it is beyond the scope of small firms. The cutting edge of advanced industrial economies, R&D is concentrated in a few major research-university clusters and established areas of innovation (Figure 8.15). Three of these in the United States are Silicon Valley, the region south of San Francisco Bay in the vicinity of Stanford University; Boston and Route 128; and the Research Triangle of North Carolina, so called because of three universities located there—the University of North Carolina at Chapel Hill, Duke University in Durham, and North Carolina State University at Raleigh. Roughly

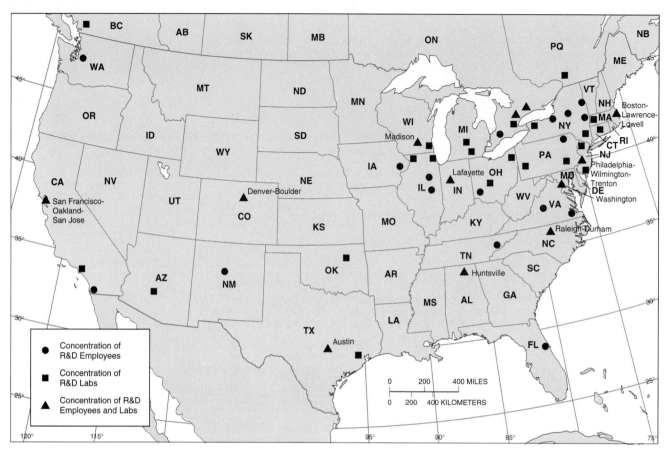

FIGURE 8.15
Major concentrations of research and development activity. (*Source: Based on personal communication with Malecki, 1997*)

equidistant from the three main cities, Research Triangle Park is home to the laboratories of IBM, Burroughs Wellcome, Northern Telecom, and other major companies conducting R&D.

For most of the 20th century, the United States has been the world's leader in technological innovation. In the 1970s, however, the number of new inventions and granted patents declined. Factors that slowed U.S. innovation included a decline in federal support for basic research, an increase in government red tape, an increased managerial interest in improving existing products to yield quick returns, and difficulties in obtaining risk capital. Basically, the United States was not spending enough on R&D. Although outlays for R&D began to rise in the 1980s as foreign competition forced firms to automate in order to reduce costs and increase productivity, U.S. expenditures for R&D represented only 2.7% of the GNP (United Nations, 1985). This figure compares favorably, however, with the expenditures of other developed countries—Japan (2.6%), Germany (2.5%), Great Britain (2.4%)—which, as a group, dominate the world in the number of R&D scientists and engineers and in the amount of R&D expenditures.

TECHNOLOGICAL CHANGE

As we learned in Chapter 4, the time involved in and the cost of transporting materials, products, and information have decreased substantially in the last 50 years as a result of technological innovations and transportation engineering breakthroughs. The ability to transmit information at faster and faster speeds led to the growth of transnational corporations. In terms of transportation and communications technology, New York City and Tokyo are much closer today than Philadelphia and New York City were during World War I. "The new telecommunications technologies are the electronic highways of the information age, equivalent to the role played by the railway systems in the process of industrialization" (Henderson and Castells, 1987, p. 6). Communications technology is valuable to all types of economic activity, but especially to industries associated with development processing and information transmission—business services, including finance, insurance, and real estate (FIRE).

Satellites, transoceanic cables, and facsimiles (faxes) have revolutionized world communications and international business (which shows the tremendous change in information technology during the last 40 years). The number of communication satellites alone grew astronomically in the mid to late 1960s after the Early Bird (Intelsat 1) satellite was launched. In 1980, satellites could carry 150,000 bidirectional circuits for international communications, and by the year 2000, they will carry a million circuits each.

Fiber-optic cable, a new technology in telecommunications, allows a high carrying capacity and signal strength at extremely high speeds across continents and oceans alike. There has been an increase in satellite and cable capacity during the last 20 years. Fiber-optic links are under construction in the Pacific Rim to produce the global digital highway around the world. More than 100,000 simultaneous messages can be carried among North America, Western Europe, and Japan through the new fiber-optic cables that allow people or facsimile machines to converse with one another.

Texas Instruments, the large transnational American computer and electronics company, operates more than 50 plants in 19 countries, and each plant is linked by satellite communications. Through satellite communications and fiber-optic cables, production planning, cost accounting, financial planning, marketing, customer service, and personnel management are conducted from more than 300 remote planning terminals, 8,000 inquiry terminals, and 140 distributed computers around the world. A similar computer network was presented for I. P. Sharp Canadian Financial Planning Company in Chapter 4.

FLEXIBLE MANUFACTURING

Management has attempted to recognize and control the nature, speed, and quality of work through five developmental stages of the labor and production process:

1. *Manufacture:* assembling labor workshops and dividing labor into specific jobs and tasks (circa 1800)
2. *Machinofacture:* further dividing labor by the use of mechanical power through machines in factories (circa 1860)
3. *Scientific management (Taylorism):* enhancing the degree of the division of labor after scientific study, together with increased supervision and control of the manufacturing process (circa 1920)
4. *Fordism:* managing assembly-line work units to control and improve the pace and quality of production (circa 1920)
5. *Flexible Production System* (circa 1980)

Fordism in the past, most notable in the automobile industry, produced assembly-line products at rapid speeds using standardized techniques for a mass-consumption market. The mass production in the Fordist approach is being replaced by a lean production, which is necessary in competitive world markets today. The most important aspect of this new, or lean, system is flexibility of the production process itself, including the organization and management within the factory, and the flexibility of relationships among customers, supplier firms, and the assembly plant.

The key to production flexibility lies in the use of information technologies in machines and operations. These permit more sophisticated control over the process. With the increasing sophistication of automated processes and, especially, the new flexibility of the new electronically controlled technology, far reaching changes in the process of production need not be associated with increase scale of production. Indeed, one of the major results of the new electronic computer-aided production technology is that it permits rapid switching from one process to another and allows the tailoring of production to the requirements of individual customers. Traditional automation is geared to high volume standardized production; the newer flexible manufacturing systems are quite different (see Chapter 4).

Flexible manufacturing allows goods to be manufactured cheaply, but in small volumes as well as large volumes. A flexible automation system can turn out a small batch, or even a single item, of a product as efficiently as a production-line, mass-assembled product.

The minimum-change approach to industrial mass production is not necessarily the most cost effective since the advent of flexible manufacturing. Rapid technological change of new products becomes much less costly and risky. The new approach achieves profitability through targeting segmented markets (discussed in Chapter 2) and being able to adapt production systems to local conditions, needs, and demands.

JUST-IN-TIME MANUFACTURING

The Japanese developed *just-in-time manufacturing* systems shortly after World War II to adapt U.S. practices to car manufacturing. Just-in-time refers to a method of organizing immediate manufacturing and supply relationships among companies to reduce inefficiency and increase time economy. Stages of the manufacturing process are completed exactly when needed, according to the market, not before and not later, and parts required in the manufacturing process are supplied with little storage or warehousing time. This system reduces idle capital and allows minimal investment so that capital can be used elsewhere.

Occasionally machines are idle because they run only fast enough to meet output. If machines run more quickly than the market requires, they must be shut off and manufactured items warehoused. The manufacturing run proceeds only as far as the market demands, and no faster. Thus, suppliers and producers of raw materials must warehouse their inventories. Buffer stocks are very small and are only replenished to replace parts removed downstream. Workers at the end of the line are given output instructions on the basis of short-term order forecasts. They instruct workers immediately upstream to produce the part they will need just-in-time, and those workers in turn instruct workers upstream to produce just-in-time, and so on. In practice this means that buffers between workers are extremely small. In short, it is a pull rather than a push system (Sayer and Walker, 1993).

Flexible Economy

The major impacts of the information and telecommunications revolution is still coming. It will occur from 1996 well into 2020s. Companies will reverse their old principles of hierarchial, bureaucratic assembly-line (Fordist) processes as they switch to customized, flexible, consumer-focused processes that can deliver personal service through niche markets at increasingly lower costs and faster speeds. Because of the telecommunications revolution, products and services once thought to be confined to premium and niche markets will move rapidly into the mainstream. No technological innovation of the past, not automobiles, railroads, plastics, iron or steel, ever came close to the power of the change being brought on by the information technologies. The power of 16 Cray supercomputers on a single chip is simply unprecedented in the history of the world, which will result in rapid changes in how computers produce business productivity. The baby boom spending cycle connected to the technological revolution of the microcomputer will create social and organizational trends resulting in a brand-new, customized *flexible economy* (see Chapter 2).

The customized flexible economy will provide an even greater innovation than the assembly line of the past Fordist productivity, which in itself allowed an amazing array of standardized products to move into massive affordability and a 10-fold average wage labor increase for the American worker. In the coming decades, the computer will help to make customized, flexible products and services increasingly affordable. The microcomputer revolution and the innovative market demand by baby boomers will dictate the growth for decades to come and will be based on these principles: flexibility and customization to individual needs and wants; higher quality and higher value added; rapid response and delivery just in time; and improved personal service and follow-up.

Customization and flexibility are the watch words of business in the immediate future. Companies that can adjust to the customized flexible economy are the ones that will lead the boom of the immediate future (Figure 8.16). The customized flexible economy means

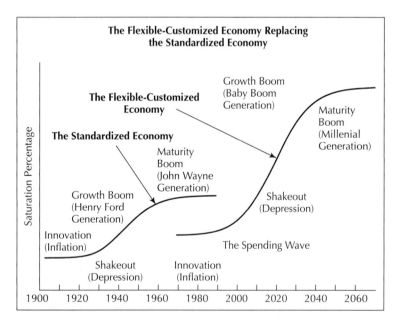

FIGURE 8.16
Integrating the standardized economy and the flexible customized economy: S curves. The flexible customized economy is replacing the standardized economy. The Henry Ford generation, and the Fordist assembly-line approach, led off the innovation phase of the standardized economy. The late 1960s, early 1980s, has been a comparable innovation period, launched by the baby boomers. During such a time, new technologies supplant the old. More and more new technologies go into new products and services to create new benefits. This period always is characterized by inflation, the economy's way of financing the transition from old technologies to the new, from one economy to another. Low productivity, through restructuring, usually results from the combination of new technologies being in their early learning phases, and old technologies beginning to slow and falter. This characterized the United States from the 1970s through the 1990s. By the year 2000, today's microcomputer technology, comparable to the automobile of the early part of the century in innovation, is multiplied in its potential by powerful, flexible software. This software and hardware can revolutionize any industry, just as the assembly line revolutionized industries throughout the Fordist period of the last century. The growth boom and the spending wave create a time when innovations begin moving out of their niches into the mainstream, driven by the power of the individualist. New technologies put downward pressure on prices, but rising consumer demand from the spending wave exerts an upward pressure. The United States is entering such a baby boomer growth period at the year 2000. Shakeout is a time when the profit is less, and there exists a pool of excess capacity. A shakeout occurs, leading to survival of only the fittest companies. *(Source: Dent and Smith, 1993:See Figure 2.26)*

custom design of products and services around individual needs and customers. Instead of greater capital investments in infrastructure and machines, greater capital investments will come in software, allowing marketing databases to estimate needs of customers and identify niche markets. Such software will allow the production and service machinery to make "short runs" for individual and market niche needs, quickly changing markets without setup costs and delays of the old standard assembly-line systems. The business trend of the customized flexible economy will mean that premium niche products and services of the past will move into mainstream affordability as the computer and software forces down the price and as bureaucracies become restructured and more efficient, causing a flexible labor force (see Business Process Reengineering later in this chapter).

LOCATIONAL COSTS

Now that we discussed production costs, we can examine their influence on location in more detail. Smith (1986) pointed out that the establishment of any manufacturing plant in a market economy involves three interdependent decision-making criteria: (1) *scale*—the size of the operation that will determine the volume of total output, (2) *technique*—the particular combination of inputs that are used to produce an output, and (3) *location*. In this section, we concern ourselves with location as a function of input costs and consider technique and scale in subsequent sections.

Let us assume that technique and scale are constant and that variations in demand, if they exist, are solely a function of price. These assumptions allow us to portray three general industrial location cases (Figure 8.17). In Case (a), market price (revenue) is a spatial constant, and costs vary across space. The optimum location is then the lowest point on the space-cost curve, and the spatial margins are where costs equal revenue. Total revenue (demand) exhibits spatial change in Case (b), and costs are a spatial constant. The optimum location is the highest point on the revenue curve, and the spatial margins to profitability are, again, where costs equal revenue. Variations in both cost and revenue across space are shown in Case (c). The optimum location (*O*) is the place where revenue exceeds costs by the greatest amount. For all three cases we can show that *both* curves determine the spatial margins to profitability, that the variable with the *steepest* gradient determines the best location, and that the *slope* of the curve indicates the relative importance of locational costs.

The concept of spatial margins to profitability is noteworthy because it incorporates *suboptimal* behavior. Profits are possible anywhere within the defined limits. The graphs in Figure 8.18 represent the most

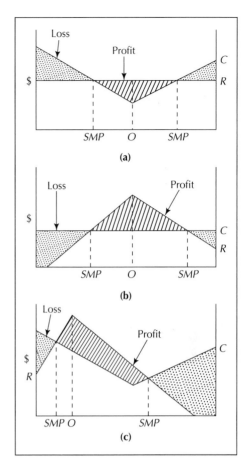

FIGURE 8.17
Spatial margins to profitability: (a) cost variable, revenue constant; (b) revenue variable, cost constant; (c) revenue and cost variable. *(Source: Based on Smith, 1981, p. 113)*

general statement that can be made about locational viability. Defining these margins in reality and determining specific locations to be occupied within them,

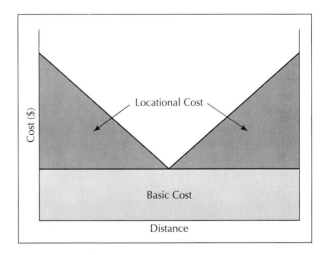

FIGURE 8.18
Spatial margins to profitability and clustering.

however, is a much more difficult problem. We can still make one generalization about real-world patterns within spatial margins to profitability—industries that are clustered must face limited spatial margins to profitability and high location costs. This situation is illustrated in Figure 8.19. Remember that any kind of cost and/or revenue may be a critical factor in determining the spatial margins to profitability.

Weber's theory is preoccupied both with transportation costs and with finding least-cost locations. A more general theory should consider total cost and the possibility of suboptimal decisions. Walter Isard (1956) suggested that locational factors can be divided into three general classes on the basis of their geographic occurrence: (1) *transfer charges*—costs that can be portrayed as a regular function of distance; (2) *spatially variable costs*—costs that vary across space (labor, power, capital, managerial skills) but do not vary systematically with distance; and (3) *aspatial costs*—factors that can influence costs but that are independent of location, such as scale changes.

Smith (1981) extended Weber's analysis to all types of costs. He assumed that each input has a least-cost point. Least-cost points for materials may be mines or the factories of parts producers. There is a least-cost point for the particular kind of labor required, and there is a point at which finished-product distribution costs are minimized. Each of these points exerts a certain pull on location (recall the Varignon frame). The relative weight of all these pulls determines the least-cost location.

Smith also acknowledged the distinction between *basic* and *locational costs*. Basic costs are the minimum costs that must be paid regardless of location; they represent the lowest point on the cost surface of a particular input. For example, the basic labor cost is the minimum wage. Locational costs are all costs exceeding the basic cost. They vary with location and may rise as a function of the distance from the least-cost location. Figure 8.19 illustrates the two kinds of costs. We assume that some workers will accept the minimum wage.

Their location represents the lowest point on the labor-cost surface, or the basic cost. Away from this point, workers demand more than the minimum wage, but the additional amount takes the form of locational costs.

We see then that the total cost of any input is the sum of locational and basic costs. The relative pull of a given input therefore depends on the slope of the (locational) space-cost curve and the percentage contribution of an input to the total cost of output. An input accounting for a large proportion of the total basic cost or varying widely in locational costs should have the most influence on plant location.

Smith (1981) examined this proposition with variable-cost models. He demonstrated that locational costs can occur either because transportation costs per unit of a particular input are high, or because a large quantity (basic cost) of the input is required, even though the unit cost may be low. The latter point can be illustrated by coal, which had a profound effect on past industrial locational patterns for heavy industry. Coal can be moved relatively cheaply on a per ton basis, but the large quantity required in some types of manufacturing has resulted in high locational costs and a pull on industries to locate near coal deposits. This example illustrates the difference between transportation costs per unit and locational costs.

THE LOCATIONAL EFFECTS OF TECHNIQUE

Technique, or the particular combination of inputs used to produce a given finished product, can have an important effect on a firm's locational decision. A certain amount of land (resources), labor, and capital is needed to produce any finished product, but, within limits, capital may be substituted for labor, resources may be substituted for labor, and so forth.

The most evident trend in modern manufacturing has been substitution of capital in the form of machinery and robotics for labor. More and more autonomous manufacturing systems, which apply sophisticated technology to improve the quality and efficiency of production, are replacing certain kinds of labor. Whether or not substitution between production factors occurs depends on the relative cost of the two inputs and the scale and locational decisions already made by the firm. If, for example, labor costs rise at a given location, the firm may choose to substitute capital for labor at that location, or it may opt to change locations to take advantage of lower labor costs and thus maintain the same labor-to-capital ratio.

The limits set on substitution vary considerably from industry to industry. Petroleum refining, for example, can be readily automated, whereas garment manufacture cannot. The textile industry, therefore, is

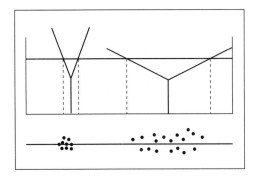

FIGURE 8.19
Basic and locational costs.

much more sensitive to changes in labor costs than the refining industry is. In the late 19th and early 20th centuries, the U.S. textile industry shifted from old multistoried New England mills to new mills in the Southern Piedmont as labor costs rose in the Northeast. This is an example of the influence that options in technique exert to determine the locational decision. The increased labor costs outweighed the costs of moving the industry. Of course, the wage advantage of the South did not persist; as new industry moved south, wages there rose. This has forced textile industries to move into other areas of the United States where pools of cheap labor are available (e.g., the depressed coal-mining towns of eastern Pennsylvania) or to migrate farther afield—to Mexico, Brazil, South Korea, and Singapore. If capital substitution had been a viable option, the textile industry might not have moved.

Many times a firm may want to change its scale to increase output and to earn extra profits. A change in scale may also require a change in location and/or technique.

SCALE CONSIDERATIONS IN INDUSTRIAL LOCATION

Scale, along with location and technique, is important because it is one of the three interdependent production criteria that drive decision-making (Figure 8.20). Smith (1981) stated:

> The choice of location cannot be considered in isolation from scale and technique. Different scales of operation may require different locations to give access to markets of different sizes. . . . Different techniques will favor different locations, as firms tend to gravitate toward cheap

sources of the inputs required in the largest quantities, and location itself can influence the combination of inputs and hence, the technique adopted. (pp. 23–24)

But scale is also important because producers are concerned with the unit cost of production—and adjustments in scale can produce considerable variations in unit cost. Scale is the means by which production is "tuned" to meet demand. In some economies, this tuning may be done by the state; in others, by private entrepreneurs.

Principles of Scale Economies

DIVISION OF LAND AND CAPITAL
Along with standardization of parts, *division of labor* is a primary component of mass production. Workers who perform one simple operation in the production process are much more efficient than those who are responsible for all phases. Division of labor not only speeds up production, but also facilitates the use of relatively unskilled labor. A worker can learn one simple task in a short time, whereas the skills required to master the entire operation may take years to learn. Division of labor, however, requires a relatively large scale because a large pool of workers is generally necessary. A common way to measure the size of a firm is by the number of employees. Capital, once invested in machinery and buildings, becomes fixed capital and produces income only when in operation. A three-shift firm makes much more efficient use of its fixed capital than a single-shift firm does. The three-shift firm is three times larger in scale, measured by employment, yet its fixed capital investment may be no more than that of the single-shift firm.

FIGURE 8.20
The four questions of economic geography and the industrial production process.
(Source: Stutz, 1994, p. 432)

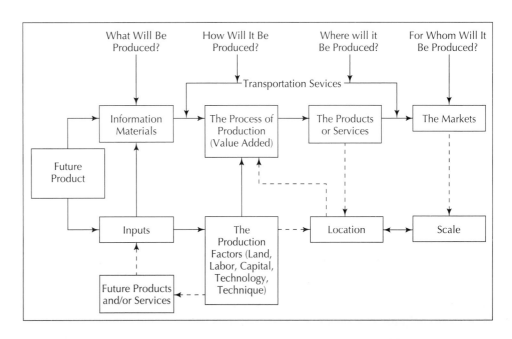

IMPERFECTLY DIVISIBLE MULTIPLES

Scale economies operate when inputs do not allow one-to-one increases in scale. Let's say that we are manufacturing hammers. The machine that produces the hammerheads can output 300 per hour, whereas the hourly output of the handle machine is 400. If we run both machines for an hour, we produce 300 complete hammers and 100 extra handles. Hammer production involves *imperfectly divisible multiples*. The scale of operation can be increased, however, until we find a unitary combination of head and handle machines. Suppose that we increase the number of head machines to four and the number of handle machines to three. Now we produce 1,200 complete hammers per hour. It is easy to envision the kinds of savings that result from this principle in an industry such as automobile manufacturing, in which perfectly divisible multiples may be reached only in very large operations.

VOLUME PURCHASES

Large firms generally pay much less for material inputs than small firms do. For example, Ford Motor Company can obtain tires for a much lower unit price than an individual dealing with the same tire company can, because Ford buys millions of tires a year. Increasing scale, in other words, generally lowers the unit cost of inputs.

Possible Scale Economies

Economists portray scale economies as a curve of long-run average costs (LRACs), which graphs the unit costs as a function of scale. Several possible LRAC curves are indicated in Figure 8.21. Notice that unit costs decrease, reach an optimum point, and then began to increase. The rise in the curve is termed *diseconomies of scale* (diminishing marginal returns to scale) and occurs when a firm becomes too large to manage and operate efficiently. The optimum scale of operation is very small in Industry *A*, very large in industry *C*, and fairly wideranging in industry *B*. Firms in Industry *A* should be small, they should be large in Industry *C*, and they can range from small to large in Industry *B*.

Possible scale economies also indicate how firms in an industry can expand production. A firm in Industry *A*, for example, can build a branch plant; increasing the size of operations on the original site would produce diseconomies of scale. Firms in Industry *B*, however, can increase production either by expanding existing plants or by building new ones.

Implications of Scale Economies

To explore the implications of scale economies, let us consider a company that operates two small breweries,

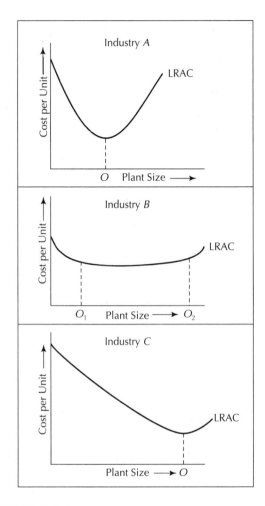

FIGURE 8.21
Variations in long-run average cost (LRAC).

each in a different town. The entire output of the firm is sold in the two towns. The LRACs (per barrel) for the firms are illustrated in Figure 8.22a. The firm operates the breweries at scale, S_1, and costs are $5 per barrel. The firm can reduce its cost per barrel by consolidating its operations into a single plant at S_2 ($2.50 per barrel). One plant can be closed and the remaining plant doubled in size, or both plants closed and a new, larger brewery built. However, the firm minimizes its finished-product distribution costs by manufacturing beer in each town. Additional distribution costs will be incurred if a single brewery is used. The question is whether the savings from scale economies outweigh the increased transportation costs. These two items balance at a transportation cost of $2.50 per barrel. Figure 8.22b shows the isocost line of $2.50 per barrel for each town. Notice that the two lines intersect. This intersection indicates that the scale savings outweigh the increased transportation costs, so the larger plant can operate in either city or at an intermediate point with a lower cost per barrel than is possible with the two small-scale plants.

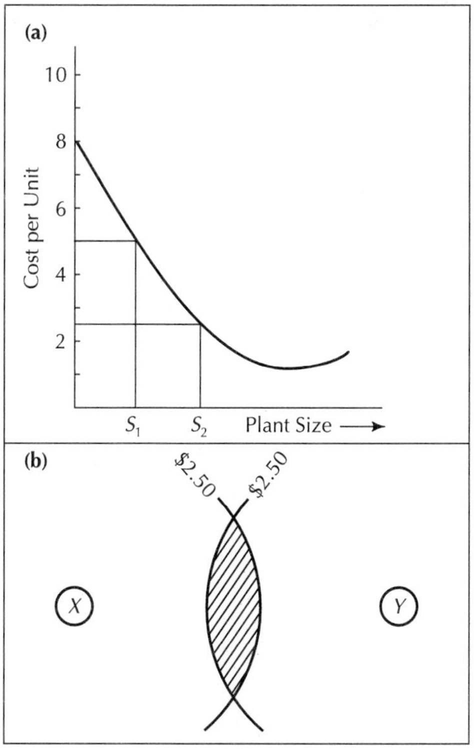

FIGURE 8.22
Spatial implications of scale economies: (a) long-run average costs and (b) transportation cost isolines for break-even total cost with plant size S_2.

Integration and Diversification

Scale refers to anything that changes the volume of a firm's total output. Besides simply increasing plant size, two other means are commonly employed for effecting scale changes. Some firms purchase raw material sources or distribution facilities. This is called *vertical integration* (or vertical merger) in that the firm controls more "up and down" in the total production process. Some large automobile manufacturing firms, for example, own iron and coal mines and produce their own steel ("down" in the process). They may also own dealerships and do their own transporting and marketing ("up" in the process). Large oil companies are also often vertically integrated; they control exploration, drilling, refining, and retailing. In contrast, *horizontal integration* (or horizontal merger) occurs when a firm gains an increasing market share of a given niche of a particular industry.

Vertical and horizontal integration generally refers to a single finished product. The vertical integration of Ford Motor Company, for example, focuses on controlling the inputs and marketing required for automobile production. However, the trend among corporations in the United States, Japan, and Western Europe has been

a strategy of *diversification*. Many large corporations, through conglomerate merger, control the production and marketing of diverse products. A company may produce many unrelated products, each with elements of horizontal and vertical integration. Diversification spreads risk and increases profits. Diminishing demand for the products of one division may be offset by rising sales in another.

Most industrial location theory is based on the firm, which implies a small, single-plant operation producing a single product. Large corporations are much more complex, but they deal with all the variables of location theory and must still make locational decisions. Although large enterprises may seem to be more concerned with technique and scale decisions, each locational decision has an effect on scale and technique. We should consider two points: First, large firms may be able to operate in less than optimal locations and still have a significant effect on the market through the control that they exert over government policies and the prices and sources of raw materials. Conversely, large firms may be able to make optimal locational choices through their employment of the scientists and technical personnel who help top management make more profitable decisions.

Interfirm Scale Economies: Agglomeration

To this point, we have been concerned with intrafirm scale economies. However, scale economies also apply to clusters of firms in the same or related industries—for example, the computer firms localized in California's Silicon Valley and the metal trades concentrated in the West Midlands of England. By clustering and increasing the spatial scale, unit costs can be lowered for all firms. These economies, often called *externalities, agglomeration economies,* or *linkages,* take several forms. *Production linkages* accrue to firms locating near other producers that manufacture their basic raw materials. By clustering, distribution and assembly costs are reduced. Close physical links between related businesses were more common a century ago than they are now. Today, component supplies are often far apart. The Boeing 767 is a case in point. Boeing manufactures its airplane frame in Seattle with parts from Japan and Italy. The engine is assembled in Ohio with parts from Sweden, France, Germany, and Italy. Advances in technology, communications, and transportation have given momentum to the economies of globalization (see Chapter 1: Toy Story).

Service linkages occur when enough small firms locate in one area to support specialized services. The garment industry in New York City provides an example of service linkages. Firms in the garment district are small, but they require specialized service and maintenance activities,

IBM designed the Yamato Laboratory with the aim of making it the focal point of development of high-tech products for IBM Japan. The lab integrates all development groups at a single location and uses IBM's systems to create products suited to local customer needs. The Yamato Laboratory is one of more than 30 IBM basic research institutes and product development labs around the world. *(Photo: IBM Corporation)*

such as the repair of sewing and cutting machines. The clustering of the garment industry in Manhattan has also provided the impetus for increased numbers of investment specialists who deal almost exclusively with loans to the garment industry. They understand the special needs of the industry and are much more likely to advance risk capital than other investors are.

In addition to production and service linkages, there are *marketing linkages,* for which the garment industry again is an example. These linkages occur when a cluster is large enough to attract specialized distribution services. The small firms of the garment industry in New York City have collectively attracted advertising agencies, showrooms, buyer listings, and other aspects of finished-product distribution that deal exclusively with the garment trade. Firms within the cluster have a cost advantage over isolated firms that must provide these specialized services for themselves or that must deal with New York firms at a considerable distance and cost.

Some economies are not the result of interfirm linkages, per se, but occur from locating in large cities or industrial complexes. Firms in these locations have an advantage, within limits, over similar firms in more rural areas. Cities provide markets, specialized labor forces and services, utilities, and transportation connections required by manufacturing. *Urbanization* or *industrial-complex economies,* therefore, are a combination of production, service, and marketing linkages concentrated at a particular location.

OBSTACLES TO OPTIMAL LOCATION

The final location chosen by an industry is not always determined by the cost of raw-material assembly and distribution, as was Weber's principal question, nor by

production costs at the site, including land, labor, capital, and managerial and technical skills. James Rubenstein (1996) listed six factors that complicate locational decisions:

1. A firm may have more than one critical site or situation factor, each of which suggests a different location.
2. Even if a firm clearly identifies its critical factor, more than one critical location may emerge.
3. A firm cannot always precisely calculate costs of situation and site factors within the company or at various locations because of unknown information.
4. A firm may select its location on the basis of inertia and company history. Once a firm is in a particular community, expansion in the same place is likely to be cheaper than moving operations to a new location.
5. The calculations of an optimal location can be altered by a government grant, loans, or tax breaks.
6. Individual choice and whim play important factors in *foot-loose industries* that have gravitated to coastal areas in the Sunbelt of the United States because of recreational opportunities, availability of amenities, and lifestyle factors.

EVALUATION OF INDUSTRIAL LOCATION THEORY

Industrial location theory helps us gain insight into how individual manufacturing establishments are located with reference to the production factors and the distribution of customers, suppliers, and competitors. It also helps us to appreciate plant-location decisions. Are industrial location patterns rational? Do firms search for optimal locations? These questions were partially addressed in Chapter 5. Here, we distinguished between two types of decision making leading to locational patterns: *adoptive* and *adaptive behavior.*

During the 19th and early 20th centuries, when capitalism came into its own in Europe and North America, decision making was of the adoptive kind. Decisions were made arbitrarily and left behind a pattern of survivors lucky enough to have selected good locations for their plants—potteries, textile factories, and ironworks and steelworks. In the competitive economic environment, location mattered. Weber's theory therefore helps us understand the success stories of economic history. It is a framework for understanding *what is.*

In contrast, adaptive behavior focuses on rational decision making. It involves a systematic analysis of alternative locations that leads to the development of rational industrial landscapes. This type of decision making is expected of multinationals in the emerging one-world economy of the late 20th century. Location theory can guide these enterprises in selecting optimal locations for their manufacturing plants, development centers, and research laboratories. It is a normative framework suggesting *what should be.*

From the perspective of the behavioral geographer, the main defect of normative industrial location theory, based on recognition of spatial limits to profitability, is that it fails to say what decision makers actually do (Pred, 1967). How do enterprises actually select profitable sites for branch plants? John Rees (1972, 1974), a British geographer, provided an answer to this question. In his studies of the investment location decisions of large British and American firms, he reported interviews with executives confirming the validity of a framework of lo-cational search, learning, and choice evaluation. The first phase of the decision-making process is the recognition that a growth problem (stress threshold) exists with respect to demand satisfaction. (Rees pointed out that the question of demand—whether or not there exists a potential market area of sufficient size to consume the output of a plant—is the prime variable in the locational choice of a large, modern manufacturing firm.) The alternative responses to in situ expansion are relocation, acquisition, or a new plant. A new plant involves a three-stage search procedure, the outcome of which leads to a decision and, finally, to the allocation of resources. It also generates feedback into learning behavior and into the decision-making environment. A major virtue of Rees's model is the distinction between short- and long-term responses of the organization. Classical industrial location theory tends to ignore the timeframe within which profit maximization is sought. The behavioral approach is much more realistic in recognizing that the environment in which the enterprise operates is in a constant state of flux.

Industrial location theory has been criticized by structuralists, primarily because it focuses on firms as abstract entities, without effective structural relationships to the rest of the economy. For structuralists, locational analysis must begin from the top, with the world's capitalist system, not from the bottom, with individual firms. The actual behavior of the individual firm takes on its meaning in the broader economic context that the structural approach seeks to reveal. Working up from the bottom can explain neither the individual elements nor the system as a whole. According to structuralists, then, industrial location theory is idealistic because it abstracts elements that form only a small part of reality. This criticism extends to the new approaches in which the simple conception of the single-plant firm has been replaced

Small firms most often use adaptive behavior. This small-scale bike shop has chosen a location near a very large West Coast university. However, because land rents are expensive next to the school, this firm is located 10 blocks away. (*Photo: Walter Hodges, Tony Stone Images*)

with a model of the firm as a complex organizational structure.

Global Manufacturing Location Today

Since 1962, worldwide exports have increased from 12% to more than 31% of world GNP. Worldwide exports totalled 4 trillion in 1994. The MIT Sloan School of Management estimates that 70% of goods now operate in an international marketplace. Sloan has developed new dynamics of global manufacturing site location for the year 2000. Global corporations of the future according to Sloan Management School will develop a manufacturing network of decentralized plants based in large, sophisticated regional markets. Each plant will be smaller and more flexible than is typically found in today's manufacturing environment. The location of such plants will be based more on regional infrastructure, local skill levels, and government policies than on purely weberian cost-based factors. Consider the following (based on the report by MacCormac, et al., 1994):

1. The development of large and sophisticated overseas markets dictates a global presence of leading manufacturers.
2. Increasing levels of nontariff barriers are forcing firms to localize production resources.
3. The evolution of a world trade system based on regional trading blocs creates incentives for firms to follow direct investment strategies that give them a manufacturing presence in each region of significant demand.
4. Regionalization of trading economies is increasing the benefits to decentralized manufacturing organizations.
5. Exchange rates and other aspects of risks are forcing firms to be flexible in terms of capacity and location and to view their global networks in a holistic way.
6. The emergence of manufacturing technologies and methods, such as flexible manufacturing systems, just-in-time manufacturing, and total quality control have reduced scale, increased the importance of worker education and skilled development, and placed demands on local infrastructure.
7. Large, centralized manufacturing facilities in low-cost countries with poorly skilled workers are generally not sustainable.

According to Sloan, traditional approaches to production location no longer apply. Large, centralized manufacturing facilities have given way to decentralized manufacturing structures, with smaller, lower scale plants serving demands and regional markets. Location depends increasingly on educational and institutional infrastructure. These factors can be synthesized into a new framework for industrial location strategy for the third millennium. Sloan proposes a four-phase procedure to aid decision-makers:

Phase 1: Establish the critical success factors of a business, the degree of global orientation necessary, and the required manufacturing support role.

Phase 2: Access options for regional manufacturing configuration, considering market access, risk management, consumer demand characteristics, and the impact of production technologies on plant scale.

Phase 3: Design a set of potentially usable sites, based on infrastructure, that adequately support the business and manufacturing strategies of the parent corporation.

Phase 4: Rank the most cost-effective solutions using a quantitative analysis of remaining location possibilities and define the manner of development.

To conclude, advances in technology, changes in management approach, and shifting market requirements are the new dynamics that shape the firm's facility location decision today. These trends suggest that global corporations of the future will move to a manufacturing network of decentralized plants, based in large sophisticated regional markets, but linked with information technology to one another. Specific locations will be based more on local infrastructure, such as workforce capability, training programs, and government policies than on traditional cost base considerations. Plants will be smaller than current ones, more flexible, yet have significantly more ability to produce multiple products.

THE LARGE INDUSTRIAL ENTERPRISE

Although small, single-plant operations remain the most common type of firm, we live in an era in which giant corporations with transnational bases control a large share of the world economy. In 1988, the 600 largest companies in the world—the "billion dollar club" (their annual sales exceed $1 billion each)—created 20% of the world's total value added in manufacturing and agriculture (*The Economist*, 1988a). The effect of big companies on the global economy, which is out of all proportion to their numerical significance, is steadily increasing. What are the trends in industrial organization? What is the relationship of large enterprises to small firms? Why do firms grow? How do they grow? How are corporate systems geographically organized? Answers to these questions help us to appreciate the role played by multiplant, multiproduct, multinational enterprises in the world economy.

Trends in Industrial Organization

One accessible measure of business size is annual sales. Table 8.5 lists the rank order of 20 of the largest 500 U.S. industrial corporations in terms of this measure. The majority of these enterprises also appear in the combined top-20 ranking of the largest U.S. and non–U.S. corporations. These huge companies have annual sales that exceed the GNPs of many countries.

Industry concentration and *aggregate concentration ratios* are frequently used to measure the economic power of large companies. The industry concentration ratio indicates the percentage of total sales accounted for by the largest enterprises (typically between three and eight) in a particular market. However, a business that exerts little influence in a sector can have a great influence on the economy as a whole. This situation can be represented by calculating an aggregate concentration ratio, which indicates the percentage share of national manufacturing sales accounted for by the largest (typically the top 100) companies.

Most large corporations owe their growth and size to diversification. An example of a *multiproduct diversified enterprise* is Tenneco (Table 8.6). For multiproduct enterprises, aggregate concentration is a better measure of corporate power than industry concentration is. As with industry concentration, aggregate concentration in manufacturing increased in the 1950s and 1960s and stabilized or even declined in the 1970s. Does the reversal signify a change in the scale of economic organization? No, it does not. Many of the largest manufacturing companies have expanded into nonmanufacturing activities (Hughes and Kumar, 1984).

Large multiproduct companies are usually *multiplant enterprises*. Their geographic bases are as broad as their product ranges. With factories and offices in other countries, these area-organizing institutions are also *multinational enterprises* (Table 8.7).

The emergence of a global production system, having at its heart the multinational corporation, is a recent phenomenon. As late as 1950, most large corporations were barely multinational. But by 1970, the situation had changed dramatically. Large corporations, which for years had exported goods to foreign markets, had set up foreign subsidiaries in numerous countries.

In the 1960s and 1970s, both American and worldwide perceptions of multinational corporations were

◈

T a b l e 8.5
The Largest U.S. Industrial Corporations, 1990

Rank	Company	Headquarters	Sales ($000)
1	General Motors	Detroit	102,813,700
2	Exxon	New York	69,888,000
3	Ford Motor	Dearborn, MI	62,715,800
4	International Business Machines	Armonk, NY	51,250,000
5	Mobil	New York	44,866,000
6	General Electric	Fairfield, CT	35,211,000
7	American Tel. & Tel.	New York	34,087,000
8	Texaco	White Plains, NY	31,613,000
9	E. I. du Pont de Nemours	Wilmington, DE	27,148,000
10	Chevron	San Francisco	24,351,000
11	Chrysler	Highland Park, MI	22,513,500
12	Philip Morris	New York	20,681,000
13	Amoco	Chicago	18,281,000
14	RJR Nabisco	Winston-Salem, NC	16,998,000
15	Shell Oil	Houston	16,833,000
16	Boeing	Seattle	16,341,000
17	United Technologies	Hartford	15,669,157
18	Procter & Gamble	Cincinnati	15,439,000
19	Occidental Petroleum	Los Angeles	15,344,100
20	Atlantic Richfield	Los Angeles	14,585,802

Source: Rand McNally, 1995.

Table 8.6
Companies and Products Controlled by Tenneco

Division	Products
Tenneco Oil	Crude oil, natural gas, refining, service stations
Tennessee Gas Transmission	Natural gas pipelines
J. I. Case	Two- and four-wheel-drive agricultural tractors and implements, loader/backhoes, crawler and wheel loaders, excavators, trenchers, industrial and materials handling cranes, skid steer loaders, forklift and compaction equipment
Tenneco Automotive	Automotive exhaust systems, shock absorbers and ride-control products, jacks and lifting equipment, filters, wheel oil seals, fans, pulleys, manifolds
Tenneco Chemicals	Fine, intermediate, and hydrocarbon chemicals; plastic resins, stabilizers, plasticizers; paint colorants and dispersions; chemical foam products and fabricated plastic materials; synthetic and organic chemicals; paper and specialty chemicals
Newport News Shipbuilding	Naval and merchant ship construction and repair, nuclear vessel refueling, components and services for the nuclear power industry, heavy castings and sheet-metal products for industrial use
Packaging Corporation of America	Corrugated containers, paperboard, folding cartons, molded pulp products
Tenneco West	Agricultural products (fresh fruits, vegetables, almonds, pistachios, dates, raisins); commercial, recreational, and residential real estate

generally negative. They were widely labeled as exploitative giants—and they still are by radical scholars who view their actions as socially disruptive and likely to promote a general tendency toward world economic stagnation. In contrast, traditional scholars and policymakers in the market-conscious 1980s and 1990s view multinationals as sources of employment and revenue rather than inherent exploiters.

Multinationals increase employment in their host countries. It is estimated that direct employment of multinationals is 65 million, or 3% of the world's labor force. Add indirect employment, and these companies may generate 6% of the world's employment. In 1984, U.S. multinationals employed about 6.5 million people abroad, 32% of these in developing countries, 42% in Europe, 5% in Japan, and 14% in Canada.

Multinationals also increase a host country's output and exports. This is especially important for LDCs in need of fast growth and foreign exchange to service bank debt. Foreign-owned companies accounted for 45% of Singapore's employment in the manufacturing industry in 1997, 63% of its manufacturing output, and 90% of its exports of manufactured goods. They produced 71% of Zimbabwe's industrial output. In 1997, nearly 35% of Argentina's manufacturing output and exports came from multinationals.

American, west European, and Japanese enterprises own most of the world's multinational assets, but new sources of capital are emerging. Some LDCs—Argentina, Brazil, Hong Kong, India, Mexico, Singapore, and Taiwan—have firms that have been establishing foreign direct investment.

As multinationals have spread across the world, there has been an increasing interpenetration of capital. For example, much Japanese private foreign investment goes to the United States. Japanese direct U.S. investment, however, lags far behind that of the Europeans. At the end of 1987, the Dutch invested $47 billion and the British $74.9 billion. Most of the new capital from abroad favors the traditional Manufacturing Belt of the Northeast and Great Lakes and the newer Sunbelt areas of the West and the South (McConnell, 1983) (see Map, Chapter 1). The interpenetration of flexible multinational capital means that countries virtually everywhere are facing increased competition from foreign suppliers. It is estimated, for example, that approximately 74% of all U.S. goods produced by domestic corporations face stiff competition inside the United States from foreign suppliers. This level of competition has had a significant effect on the dynamics of U.S. manufacturing and has registered dramatically as a loss of jobs in Rust Belt industries.

❖

T a b l e 8.7
Locational Determinants for Manufacturing

Factors Considered by Alfred Weber	Additional Factors that Are Important Today
Raw materials	Capital Transmaterialization
Labor force	Technology Information
Market	Governmental Regulations Infrastructure
Transport Costs	Political Stability Info Technology
Agglomeration	Inertia Decentralized MNC Behavior

Product	Added Value ($/pound) 1990 Estimates
Satellite	20,000
Jet fighter	2,500
Supercomputer	1,700
Airplane engine	900
Jumbo jet	350
Video camera	280
Mainframe computer	160
Semiconductor	100
Submarine	45

In addition to foreign direct investment, which increased fivefold between the mid-1970s and 1990, other forms of international business are open to multinationals. For example, international industrial firms engage in turnkey projects, arrangements in which the contractor not only plans and builds the project, but also trains the buyer's personnel and initiates operation of the project before "turning the key" over to the buyer. Corporations also engage in *licensing ventures* such as *franchising*. For example, Pepsi-Cola licenses the use of its name and the right to manufacture and sell its drink abroad. Part of the contract, however, requires that the foreign licensee buy the syrup from Pepsi; thus, the company enjoys royalty and export advantages.

Another form of international business engaged in by multinationals is the *joint venture*. In this situation, a subsidiary is owned jointly by two or more parties. The joint-venture partners may be either from the private sector of the investing company's home country, from a third country, or from the host country. Corporations also engage in *international subcontracting*, sometimes

called *offshore assembly* or *foreign sourcing*. An important form of international subcontracting, especially in textiles, is an arrangement whereby firms based in the advanced industrial countries provide design specifications to producers in underdeveloped countries, purchase the finished products, then sell them at home and abroad.

Multinational corporations

MNC Organization Structure

Other countries' multinational corporations (MNCs) have stepped up their involvement in the world economy and, not surprisingly, have developed a characteristic organizational form. Invariably, the parent company is headquartered in the country in principal ownership, but this is not always the case. For example, Royal Dutch Shell is 60% Dutch owned and 40% British, yet it maintains headquarters in both The Hague and in London. A MNC's stock is publicly held almost always, and is made accessible to individual investors of any nationality. It is true, however, that a number of outstanding multinationals are still privately held. An MNC's headquarters represents a control center responsible for decisions made worldwide. Early on, an MNC's headquarters, primary management, and staff are generally residents of the home country. Over time, the headquarters often welcome talented individuals from its overseas associates. Decisions regarding issues such as company-wide planning, goods to be manufactured, location of potential markets, and derivation of raw materials are made by the management group.

It is common for MNCs to follow the Japanese example by forming their overseas affiliates into integrated systems where exchange of materials, capital, and goods occurs on a daily basis. East Asian affiliates of QualComm, a San Diego headquartered leader in global digital wireless communications, exchange parts, subassemblies, and finished phones, and ground systems from one country to another in an elaborate intracompany network. It is also characteristic to find an MNC maintaining connections with other foreign enterprises in the form of joint ventures, licensing agreements, and distributorships.

Role in Global Redistribution of the Factors of Production

The role an MNC symbolizes is that of an effective agent for *transferring capital*, managerial skills, technology, product design, and commodities among countries. An MNC represents a major agent of technological change by transferring innovations among nations working to achieve a global equalization of factors. This can be done

Japanese cars are lined up on a pier at Boston's waterfront ready for import. Japanese automobiles have made major penetration into U.S. markets, led by Toyota, Nissan, and Honda. Sales dropped off because of the expensive yen on world markets, raising the price of Japanese cars.*(Photo: Spencer Grant, Photo Researchers, Inc.)*

by introducing the scarcities of one country to a country with surpluses.

A key task the MNC is responsible for includes the transfer of managerial skills often from one country to another. The MNC is able to execute this transfer much better than a local company because of its great size and financial strength. The second key task an MNC performs, once the decision to invest in a new foreign undertaking has been made, is the transfer of capital. Financial capital or real capital (equipment and machinery) changes hands at this stage. It is possible for the MNC to borrow funds from the host country or elsewhere, or generate its own capital from within.

Often a new affiliate of a host company is the recipient of an infusion of technology from the headquarter company. This is the third major function of the MNC, to create technology and transfer it throughout the system. Research and development should not only be taking place at the headquarters, but at the site of the overseas affiliated companies as well.

As the primary generator of international trade, the MNC fulfills its fourth and final function. This consists of the transfer of raw materials, components, and finished products that take place among the company's many branches, hence, promoting integrated production and marketing between the parent and its overseas affiliates. Ultimately, the company strives to integrate its international operations for the purpose of serving every national market with a full line of its products and to do so at the lowest unit costs.

Theories of the Multinational Enterprise

Although there is a lack of substantial literature on the multinational enterprise because of its relatively recent

frame, the growing body of literature offers answers to two sets of questions: (1) What causes a firm to go abroad and succeed in a foreign environment against competition from both domestic firms and other multinationals? (2) What happens to the character of a firm during the course of internationalization?

An Outline of Leading Theories

A common trait that characterizes firms entering into multinational production is the expectation of greater profits from foreign ventures than those received by local competitors in the same regions. MNCs demand greater returns because they must conquer problems not borne by domestic firms. Because the MNC almost always maintains superior knowledge and a larger size and scope of operations, a greater return is almost always guaranteed.

Superior knowledge manifests itself in the form of leadership innovation which is often monopolized by the MNC. This innovativeness is an important element in the technology-intensive industries such as pharmaceuticals and electronics. MNCs spend more capital on research and development (R&D) than other affiliated firms, and are able to transfer this technology with little additional cost. To acquire its superior knowledge, the MNC must create a home environment that offers increased levels of technical and managerial skills and has a well-developed market.

Another advantage for MNCs is their size and ability to produce in many countries. As a result, the MNCs achieve the most economical scales of operations. Often local markets are too small to absorb all of their output; hence, the surpluses are sent abroad to their other branches. Because of their large size, MNCs may assume a greater range of functions that contribute to the secrecy of their technology while minimizing effects of governmental restrictions.

Because MNCs pay close attention to their competitors' actions, they are highly interdependent in their decision making. When one firm decides to enter a foreign market, it is not surprising to find rivals following suit to minimize the risk to their market shares. This behavior is entitled "bandwagon effect" or "reverse investment," in which MNCs compete in each other's home markets.

Competing successfully in an alien environment is something an MNC must be able to do. The first step is to choose which countries to enter carefully. Country characteristics that come into play are: location, resource, endowments, size and nature of market, and political environment. It is not a surprise that U.S. multinationals prefer to invest in countries with familiar cultures with large markets, most notably English-speaking Canada, Britain, and Australia. These markets offer economies of scale. MNCs generally prefer foreign direct investment when operating in those countries with stable governments and large, prosperous markets. As poorer, less developed countries tend to involve higher levels of political risk, MNCs rely mainly on exporting, together with some licensing of local production, but with only a minimum of direct investing.

globalization, to manufacture in other countries. This decision is a result of discovering that exports to a given national market have reached a level sufficient to justify building a factory there. Once the decision is made to set up production in the overseas market, the firm must then decide whether to build a wholly new factory or to acquire an existing firm in the foreign area. Once production is underway in multiple overseas locations, the firm may integrate its foreign subsidiaries in a "mesh," either vertically or horizontally, to gain economies of specialization (Figure 8.23).

(3) *Globalization of outlook and organization.* This third phase occurs as a result of foreign operations generating such a large percentage of the company's total revenue that the headquarters management views the firm's business on a global level. Escorting this change of perspective is the globalization of the headquarters staff through worldwide recruitment of cosmopolitan executives who no longer identify with the firm's home country but with the company as a global entity. These prospective cosmopolitans are rich in three assets, according to Rosabeth Kanter, former Harvard Business School dean: concepts— the best knowledge and ideas; competence—the ability to operate at the highest standards of any place in the world anywhere; and connectedness—the best relationships, which provide

Globalization of the Firm

The second phase of MNC theory grapples with the question of how, when, and why a firm becomes multinational and what happens to its organizational arrangements during this process. Globalization of a company most typically occurs in three stages.

(1) *Move to exports.* It is here that a parent begins to export some of its output, as opposed to one that serves the domestic market only. At this juncture, the company has neither the specialized facilities to manufacture shipments for export nor personnel skilled in export procedures; hence, outside specialists are hired to perform these functions. As orders from abroad increase, it will become more economical to establish an export department or foreign division, constituting a handful of clerks and an experienced manager.

(2) *Globalization of production.* The company decides, at this phase of

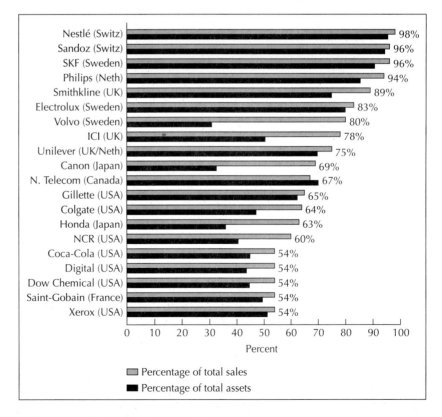

FIGURE 8.23
Global companies with over 50% of total sales and assets outside their home countries.

access to the resources of other people and organizations throughout the world (Kanter, 1995). While the headquarters still makes vital decisions affecting the system as a whole, leadership by cosmopolitans allows greater autonomy of overseas branches, and decentralized networks may evolve, ultimately resulting in a fully globalized firm.

A "New" Theory of International Trade and Transactions

Evidence that MNCs are responsible for a major share of world trade, and for a variety of other kinds of international economic transactions that do not fit the conventional mold, are central issues fueling the recent dissatisfaction with the explanatory powers of traditional theory. This dissatisfaction generated alternative approaches from the 1970s onward, each offered as a replacement for existing theory.

The appearance of a theory of industrial structure arose, seemingly establishing a relationship between trade and industrial organization—more specifically, the multinational enterprise. With this new theory of foreign direct investment, the new group of writers set about simultaneously explaining both the overseas production of multinationals and the international trade conducted by these enterprises (Dunning, 1995).

A distinct set of assumptions has been the backbone for what these theorists have relied upon, straying from those rooted in conventional trade theory. They have concluded that:

1. Individual firms possess unique competitive advantages.
2. Firm assets are mobile among branches of the enterprise.
3. The economic functions performed by various branches of a firm are decided in accordance with the spatial distribution of firm assets.

The Dual Economy

The dominant position of large enterprises in the modern world economy owes much to their relationships with smaller firms. Robert Averitt's (1968) notion of a *dual economy* captures the essence of these relationships and puts them in perspective. According to Averitt, there are two distinct types of business enterprises. On the one hand, there are a few *core* or *center firms*—large, complex organizations that represent the nucleus of the economy and that account for a high proportion of its production and profit-making potential. On the other hand, there are numerous *satellite* or *periphery firms*—small, straightforward organizations that manage to survive in the market by minimizing labor costs and maximizing labor exploitation. There is no precise boundary between center and periphery firms, but the leading 200 or so industrial corporations form the heart of the U.S. center economy. At the world level, the international center economy consists of around 500 to 700 firms.

For Averitt, small is not so beautiful from the perspective of industrial relations. The relationship between center and periphery firms is one of dominance and dependence, as reflected in purchasing policies, franchising agreements, and advisory and management contracts. The terms of these arrangements are dictated by the core firms. For example, IBM purchases components for its computers from many smaller firms. However, IBM can, through its purchasing policies, make or break its business partners and dictate their location within a specified distance of IBM's manufacturing facilities (Susman and Schutz, 1983). This relationship emphasizes that the power of core enterprises is often much greater than that implied by industry and aggregate concentration ratios, which fail to take account of unequal relationships between business organizations.

What are the prospects for the survival of small firms in the advanced economies? This question is the subject of considerable debate (Curran and Stanworth, 1986; Storey, 1986). One argument is that small firms will continue to flourish because of their competitiveness, flexibility, dynamism, and innovativeness. The counter-argument is that small firms can play only a restricted role in a world in which the terms of production and competition are set by large firms. Indeed, the survival of the small firm is not related to internal advantages so much as to protection from market forces arising from ties with large firms. Moreover, the characteristics of competitiveness, flexibility, dynamism, and innovativeness may be just as applicable to large firms as to small ones.

Why Firms Grow

The notion of the dual economy conflicts with traditional economic thinking about absolute limits on firm size imposed by diseconomies of scale. Most large companies operate at a scale far beyond the initial point on the LRAC curve. In fact, evidence of increasing returns to scale has led to a reappraisal of the theory of the firm.

To explain why firms increase their scales of operation, economists make two distinctions. First, they separate *economies of size* (scale) from *economies of growth*. Economies of growth may exist independently of economies of size. For the "enterprising firm," unused productive services are "a challenge to innovate, an incentive to expand, and a source of competitive advantage. They facilitate the introduction of new combinations of resources—innovation—within the firm" (Penrose, 1959, p. 85). Second, economists separate *actual* from *perceived* scale economies. The actual relationship between unit cost and size is irrelevant. The relationship that is relevant is that which executives believe holds true.

The tendency toward increasing scales of operation is therefore based on the motivating force of growth. Firms expand for two reasons: survival and growth. Both goals are promoted by horizontal and vertical expansion and by diversification.

The view that corporate growth is part of a natural progression is deterministic, however. It flies in the face of reality. The majority of firms in an economy remain small and peripheral. Only some firms, especially those that manufacture capital goods, have the potential to develop into large corporations. Financial barriers prevent most firms from making successive transitions from a small regional base to larger national organizations and then to multinational operations. Access to finance—banking capital, venture capital, and international bond and currency markets—has become increasingly uneven, favoring some firms and not others. Because these finance gaps have become wider, a small firm has a much less chance of evolving into a corporate giant today than it did a hundred years ago.

How Firms Grow

How a firm grows depends on the *strategy* that it follows and the *methods* that it selects to implement its strategy. As we discussed earlier in the chapter, growth strategies are *integration* or *diversification*. In the United States, horizontal integration predominated from the 1890s to the early 1900s, vertical integration came to the fore in the 1920s, and diversification has been the principal goal since the 1950s. This three-stage sequence provides a framework for understanding the interrelationship of the various strategies. The early growth of large enterprises involves the removal of competition by absorption leading to oligopoly. This is followed by a period in which the oligopoly protects its sources of supply and markets by vertical integration. Once a dominant position is achieved, rapid corporate growth can proceed only with diversification.

Methods for achieving growth are *internal* or *external* to the firm. Growth can be financed internally by the retention of funds or new share issues. Or it can be generated externally by acquiring the assets of other firms through mergers. Most large firms employ both means, but external growth is particularly important for the largest and fastest-growing enterprises.

Whatever strategy and method are adopted, corporate growth typically involves the addition of new factories and, thus, a change in geography. Initially, much of the employment and productive capacity of a firm concentrates in the area in which it was founded. An example is Ford Motor Company. For a long time, most of its operating plants were in the Midwest. As enterprises grow, they become more widely dispersed multiplant operations, which is sometimes accompanied by decreasing dominance of the home region. Exceptions tend to be companies confined to one broad product area and based in a region where there is a historical specialization within that product area.

The choice of growth strategy affects corporate geography. Horizontal integration frequently involves setting up plants over a wider and wider area. The geographic consequences of vertical integration vary according to whether the move is backward ("down" in the production process) or forward ("up" in the production process). *Backward integration*, in which a firm takes over operations previously the responsibility of its suppliers, can lead a firm into resource-frontier areas. An example is the development of iron-ore deposits by U.S. and Japanese companies in Venezuela and Australia. Conversely, *forward integration*, in which a firm begins to control the outlets for its products, can lead a resource-based organization to set up plants in market locations. Diversification does not have such predictable consequences for the geography of large enterprises.

The method of growth also affects the geography of multiplant firms. When growth is achieved internally, enterprises can carefully plan the location of new branch plants. When growth is achieved externally, enterprises inherit facilities from acquired firms; hence, there is less control over their locations. Moreover, the attractiveness of new facilities often lies in their economic, financial, and technical characteristics. Nonetheless, geography does play a role in the decision process. Firms typically confront the uncertainty and risk of expansion by investing first in geographically adjacent or culturally similar environments.

Geographers have developed models of how firms grow. Most of these models postulate a single development path beginning with a small, single-plant operation and culminating with the multinational enterprise. L. Håkanson, for example, proposed a five-stage model that incorporates the transition from home country to overseas operations. The top, left diagram of Figure 8.24 illustrates the firm's action space, divided into a core area where it was founded, the remainder of the home country, and an outer circle representing the rest of the world. In Stage 1, a single-plant firm is tied to the immediate environment. In Stage 2, the firm penetrates the home market through sales offices, the expansion of central management, and new production capacity away from the original plant. Stage 3 sees the first incursions into foreign markets through a network of sales agents; at home, production capacity may be expanded outside the original core area. In Stage 4, sales offices replace some of the overseas agents. Finally, production plants

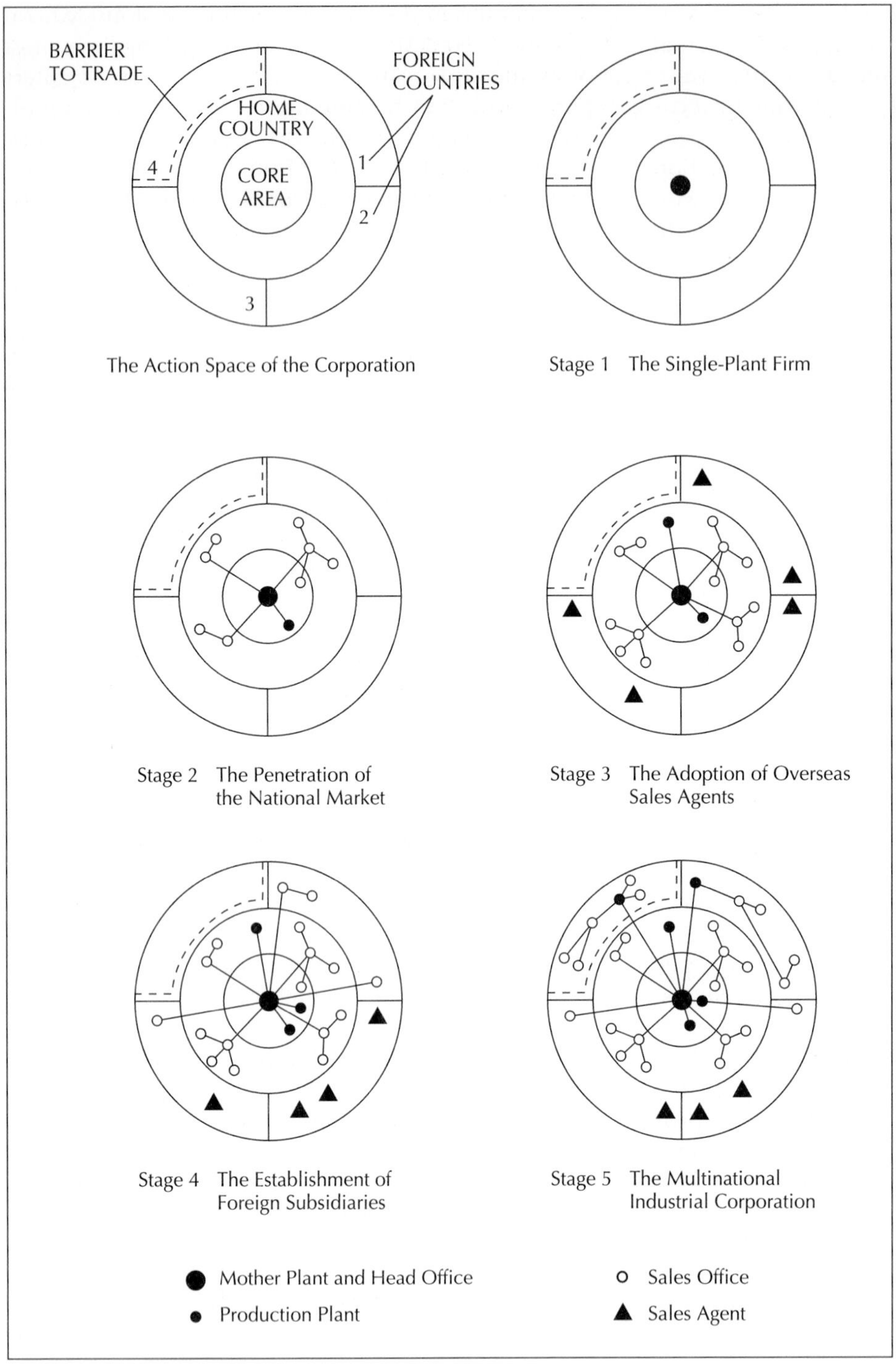

The Action Space of the Corporation

Stage 1   The Single-Plant Firm

Stage 2   The Penetration of the National Market

Stage 3   The Adoption of Overseas Sales Agents

Stage 4   The Establishment of Foreign Subsidiaries

Stage 5   The Multinational Industrial Corporation

● Mother Plant and Head Office

• Production Plant

○ Sales Office

▲ Sales Agent

FIGURE 8.24
A model of stages of growth and geographic expansion of the industrial corporation. Stage 1 is the single plant firm. A major step is taken in Stage 2 as a second plant is located within the same agglomeration as the initial plant. The firm sets up sales offices and expands to the regional scale. Stage 3 represents the penetration of overseas markets, which are developed via a network of sales agents and sales offices. Stage 4 is characterized by company-owned offices established in overseas markets, followed by manufacturing plants. Because of trade barriers and long distances, multinationals extend their geographic range by opening up plants in foreign countries to save on transportation costs, labor costs, and high tariffs. *(Source: Based on Håkanson, 1979, pp. 131–135)*

appear in foreign markets as acquisitions or subsidiaries. Stage 5, then, marks the fully fledged multinational.

This kind of evolution along a path from a local to a national and then to an international company is exceptional. Unequal access to finance makes it difficult, if not impossible, for many firms to expand beyond the subnational scale. In the late 20th century,

the size distribution of firms resembles a broad-based pyramid in which fewer and fewer firms can move from one level to another. Rather than the single development sequence that may have existed in the 19th century, today multinationals follow a distinctive path through a series of discrete development sequences.

Geographic Organization of Corporate Systems

Multifacility corporate systems, which include manufacturing plants, research laboratories, education centers, offices, warehouses, and distribution terminals, have their own distinctive geographies. To appreciate the internal geography of these systems, four issues must be considered: (1) the ways in which corporations are organized to maximize efficiency, (2) the influence of hierarchical management structures on the location of employment, (3) the effect of technology-based hierarchies on corporate spatial organization, and (4) the implications of locational shifts in the productive base of large companies.

Organizational Structure

Companies organize themselves hierarchically in a variety of ways to administer and coordinate their activities. The basic formats are (1) functional orientation, (2) product orientation, (3) geographic orientation, and (4) customer orientation. A fifth format, which is a combination of at least two of the basic formats, is called a *matrix structure*. Different companies may select different formats, but all formats are always subject to review and modification.

The organizational format that is based on various corporate functions—manufacturing, marketing, finance, and research and development—is illustrated in Figure 8.25a. With this framework, all the company's functional operations are concentrated in one sector of the enterprise. An example of a company with this type of organizational structure is Ford Motor Company. This form of organization works well for companies with relatively confined product bases.

Figure 8.25b illustrates the product-orientation organizational structure. Product groups can be cars, trucks, buses, and farm equipment for a major motor vehicle manufacturer. Although a corporate central staff is needed to provide company-wide expertise and to provide some degree of assistance to each product group, each group also has its own functional staff. Thus, a fairly high degree of managerial decentralization is required. The product-orientation format works well for companies with diverse product lines. Pan

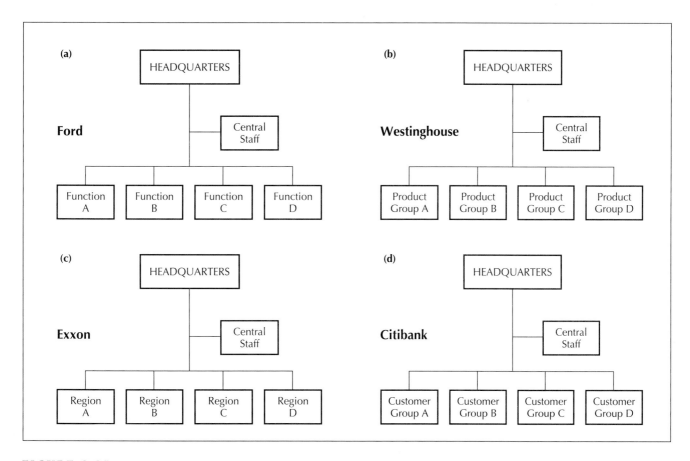

FIGURE 8.25
Organizational structures: (a) functional organization, such as the Ford Motor Company; (b) product orientation, such as Westinghouse; (c) geographic orientation, such as Exxon; (d) customer orientation, such as Citibank.

American Airways and Westinghouse are examples of companies organized in accordance with this format.

A third organizational format is based on geographic orientation—either the geographic location of customers or of the company's productive facilities. As shown in Figure 8.25c, the company is organized around regions rather than functions or products. Under this form of organization, most or all the corporation's activities relating to any good or service that is bought, sold, or produced within a region are under the control of the regional group head. Each geographic region is under a separate profit center. This organizational format is best suited for companies with a narrow range of products, markets, and distribution channels. It is popular among oil companies and major money-center banks.

Some companies organize according to the types of customers that they want to serve rather than the locations of customers (Figure 8.25d). For example, commercial banks are commonly organized around groups such as the personal, corporate, mortgage, and trust departments. Alternatively, manufacturing corporations might be structured around industrial, commercial, and governmental divisions according to the prevalent type of customer for each group.

The various organizational structures all have advantages, but none is ideal for all companies. Indeed, it is safe to say that these formats have drawbacks for most or all the companies that have adopted them. Nonetheless, a company usually chooses one basic format as the most satisfactory structure for its needs at a particular time in its evolution—or it creates a combination of two or more types.

H. A. Simon's (1960) analysis of these organizational structures identifies three tiers of activities. At the bottom are manufacturing and routine administrative activities. In the middle are coordinating functions that bind the various elements of the enterprise. At the top are strategic decision-making functions, which control the relationships of economic ownership (i.e., control the overall investment and accumulation process) and the relationships of possession (i.e., the means of production and labor power).

Simon's conceptualization has geographic implications. For the small, single-plant firm, strategy and production functions are not geographically separated; hence, there is no need for an intermediate tier of coordinating activities. As firms grow to become multilocational companies, more complex functional and spatial divisions of labor develop. One of the best-known forms of spatial organizations draws on the characteristics of large electronics companies. Strategy functions are performed at the headquarters. Coordinating functions are dispersed to regional offices that control a number of interdependent production facilities. For simplicity, suppose that two production facili-

ties exist: one branch plant manufactures complex components, the other assembles finished products. This organizational structure represents different degrees of removal of job control. It also represents a clear-cut distinction between the functions of conception on the one hand and execution on the other hand, with the parallel distinction between nonproduction and production employment.

Administrative Hierarchies

A major proportion of the employees of large corporations, even those primarily in manufacturing, are involved in nonproduction activities. The proportion is increasing because of the substitution of capital for labor and because of the growth of R&D activities. The ratio of nonproduction to production employment is less important from a geographic perspective than is the relative location of these activities.

Strategic head-office functions tend to cluster in a relatively few large metropolitan areas, especially in the case of huge manufacturing firms with a financial orientation rather than a production orientation. The concentration of corporate white-collar jobs in or around major metropolitan centers is further reinforced by the distribution of a company's R&D facilities. Corporate R&D establishments often locate close to headquarters. To be sure, there are exceptions. The labor factor has pulled R&D establishments to other locations—in France to the Côte d'Azur on the strength of its glorious climate, and in the United States to Lincoln, Nebraska, and to Austin, Texas, on the strength of their university research environments.

The contribution of head-office and R&D establishments to nonproduction employment within corporations has more strategic significance than numerical significance. Important administrative jobs are concentrated at the corporate core. But the majority of nonproduction jobs are dispersed among regional offices, branch plants, and depots. Similarly, a high proportion of R&D staff are not involved in basic research, but in the development of existing products and processes. These jobs are often dispersed to industrial manufacturing sites.

BUSINESS PROCESS REENGINEERING (FLEXIBLE LABOR)

Business process reengineering (BPR) refers to a major innovation in the manner in which organizations conduct their business. Such changes may be needed for increased profitability or mere survival. Business process reengineering involves changes in structure and in process. The entire technological, human, and organizational dimensions may be changed in BPR. Over 70% of large U.S. companies claim to be reengineering. As part of BPR, there are management realignments, mergers, consolidations, operational integration, and reoriented distribution practices.

According to Hammer and Champy (1993), certain common characteristics of BPR are:

1. Several jobs are combined into one.
2. Employees make decisions (empowerment of employees). Decision making becomes part of the job.
3. Steps in the business process are performed in a natural order, and several jobs get done simultaneously.
4. Processed batches have multiple versions. This enables the economies of scale that result from mass production, yet allows customization of products and services.
5. Work is performed where it makes the most sense, including at the customers' or suppliers' sites. Thus, work is shifted across organizational and international boundaries.
6. Controls and checks and other non–value added work are minimized.
7. Reconciliation is minimized by cutting back the number of external contact points and by creating business alliances.
8. A single point of contact is provided to customers or a deal structure.
9. A hybrid centralized/decentralized operation is used.

DOWNSIZING

Downsizing (sometimes euphemistically called "right sizing") is the most critical factor inhibiting wage inflation. Once seen mainly as a component of corporate reorganization aimed at achieving a healthier bottom line, the chilling effect of downsizing on workforce wage demands has effectively kept wage inflation in check, too.

From 1989 through 1997 in the United States, more than 3.3 million job cuts were announced, and employees who remain are reluctant to push for pay increases. They know management has placed a ceiling on the amount allotted for wages in order to avoid the inflationary move of charging higher prices and the risk of becoming noncompetitive.

The workforce is aware that a request for a pay increase could be met with dismissal, as management turns the downsizing valve, eliminating employees deemed too costly to keep. Employees also know that any vacated position could quickly be filled by someone willing to accept the present wage. Uncertainty over when management will turn the downsizing valve has insecure employees scrambling for jobs in the most productive, high-priority areas of a company.

No one wants to be targeted for dismissal because he or she stayed too long in an area perceived by management as too bureaucratic or obsolete. Even when employees manage to land a job in an area reasonably safe from the threat of downsizing, few are willing to stick their necks out and ask for a raise. Paradoxically,

Unloading automobile seats. Chrysler's *just-in-time system* supplies procurement in Detroit, Michigan. Just-in-time systems are more than simply a quick, efficient delivery of needed parts and supplies. They are part of a broader system of organization and production adopted by firms. Work is done only when needed and in the necessary quantity at a given time. Little waste occurs and little stock is kept on hand, which keeps prices low. Proximity among suppliers and manufacturers is essential because orders and deliveries may be made several times a day. Just-in-time operation systems tend toward localized agglomeration economies. A manufacturer can choose a single sourcing agent and, thus, buy supplies and products in large quantity, achieving scale economies. Other plants might choose multiple sourcing and spread the subcontracting network more widely, reducing the risk of procurement interruptions. *(Photo: Michael L. Abramson, Woodfin Camp & Associates)*

the price of using downsizing to control wage inflation has been high.

Downsizing has erased the traditional bond of employer-employee loyalty, has depleted corporate memory, and in many cases has eroded the work ethic, productivity, and morale of employees. The number of executives and managers who have been laid off two or three times is growing. With each new, unexpected twist in their careers, they become more uncertain and insecure.

Although most companies value creative thinkers who are willing to take calculated risks, many executives and managers who have felt the sting of a downsizing are inclined simply to recite the company line rather than gamble their careers on "out of the box" decision making.

In many companies, these more intangible, negative effects of downsizing are just beginning to be felt. Downsizing has proven effective as a means of keeping inflation in check and is here to stay as a means of making companies leaner and more competitive. But how long can companies remain competitive with a disloyal, insecure workforce? Management now must ask itself if there are alternatives to turning the valve of downsizing in order to keep wages stable and inflation low. In the long run, this method of controlling inflation may not be worth the price.

Restructuring and downsizing means layoffs and hurting people at the basic survival level. Downsizing challenges people's ability to survive, to provide shelter, food, and the services needed by their families. Today white-collar as well as blue-collar jobs are being lost. As shocking as this news is, it barely begins to describe the whole story of economic restructuring and the demographics of the firm. Today, downsizing is barely a ripple of the change in the *division of labor* for the firm that is to come. Yet beyond this job shock is a tidal wave of opportunity and progress. What is behind the change is the information technology (IT) revolution. It will dictate how firms operate and how they conduct business. It will dictate how labor works and the range of occupations available in the future. The primary tool of this revolution is the microcomputer and information technology (IT).

There can be an overall organizational transformation from traditional hierarchical to network-type organizations. There are also more isolated, specific reengineering efforts pertaining to one (or a few) cross-functional organizational process.

REAL TIME INFORMATION SYSTEMS (RTIS)

RTIS can segment customers into the smallest segments based on how their needs and service, quality selection, delivery information needs, and decision-making differ. It can then ask what people, skills, information, training, systems, and machinery would be required to focus on each market segment of customers' needs. RTIS can know those customers better than anyone else and serve them with a minimum of intervention from backline staff and management.

In the BPR frame of the future, every individual or team acts as a business. Self-managing teams are accountable to the customer. Every individual or team must have an understanding of what drives their productivity and must have clear productivity measures. Teams and individuals cannot be truly accountable or self-managing if they are not responsible for their revenues, costs, consumer satisfaction, and productivity. The firm of the future will teach each employee, each team, how to be managers and businesspeople instead of making decisions for them from the top down and telling them what to do.

Real time information systems are an important part of BPR. The whole point of the customized economy is to get self-managing individuals and teams to make more and more decisions with less and less bureaucracy to create higher service and flexibility to respond to customers at lower costs. Real time information systems hold individual entrepreneurs together. People need feedback as soon as possible to make faster and better decisions. They need to get information and approval on demand where possible so they can meet needs of their customers and then go on to the next opportunity. They need information now, not in a month or when the accounting department closes the books. Real time information systems link vertically all small teams and individuals and outside vendors into a larger whole and result in processes that are naturally a part of and to the customers directly. Real time information systems must link peers horizontally with other teams in geographic areas that share similar experiences and challenges so they can learn from each other. This also naturally requires human contact, such as frequent or occasional social functions or networking conferences. BPR real time information systems must link teams to experts and expert systems that can allow them to make faster decisions.

In the customized economy, firms and their strategic partners must operate like a school of minnows instead of like a cumbersome whale, able to tap critical expertise in economies of scale when necessary, but also able to, with real time information systems, turn on a dime and stay focused and competitive in the world of rapid change and customization.

IT AND STRATEGIC/COMPETITIVE ADVANTAGE

A study conducted by *Datamation* magazine in 1994 concluded that the use of IT to increase the competitive advantage of organizations is the most important issue faced by MNCs. Such systems are known as strategic information systems (SIS). Strategic information systems can be defined as systems that support or sharpen a business unit's competitive strategy. An SIS is characterized by the system's ability to change how business is done. This occurs through its contribution to the strategic goals of an organization and its ability to significantly increase performance and productivity. Neumann (1994) maintained that some conventional information systems that are used in innovative ways can become strategic.

Information technology (IT) contributes to strategic management in many ways. Consider these three:

1. IT creates applications that provide direct strategic advantage to organizations.
2. IT supports strategic changes such as reengineering. For example, information technology allows efficient decentralization by providing speedy communication lines, and it streamlines and shortens product design time by using computer-aided engineering tools.
3. IT provides business intelligence by collecting and analyzing information about innovations, markets, competitors, and environmental changes. Such information provides strategic advantage because, if a company knows something important before its competitors, or if it can make the correct interpretation of the information before its competitors, then it can introduce changes and benefit from them.

Competition, according to Michael Porter (1990), is at the core of a firm's success or failure. Competitive strategy is the search for a competitive advantage in an industry, either by controlling the market or by enjoying larger than average profits. Such a strategy aims to establish a profitable and sustainable position against the forces that determine industry competition. SIS are designed to provide or support competitive strategy. Only when SIS are combined with structural changes in the organization can they be very beneficial in providing a strategic advantage. The shift of corporate operations from a competitive to a strategic orientation (of which competition is only one aspect) is fundamental. IT has a significant impact on the profitability of an organization and even on its survival. The use of an SIS can be important to a business in many ways. The most obvious is that it can give a firm a competitive advantage in the marketplace.

Technological Hierarchies

In addition to administrative hierarchies, there are technology-based hierarchies. Product cycles and production systems help us to appreciate the importance of technological considerations in corporate spatial organization. The *product life cycle*, which begins with a product's development and ends when it is replaced with something better, is important geographically because products at different stages of production tend to be manufactured at different places within corporate systems. Moreover, at any given stage of the cycle, the various operations involved in the manufacture of a product such as a camera are not necessarily concentrated at a single factory. Production of a camera's complex components occurs at a different place from where the final product is assembled (see Chapter 4).

Economist Simon Kuznets (1930) developed the concept of the three-stage product cycle (Figure 8.26). In Stage 1, innovators discover, develop, and commercially launch a product. They also benefit from a temporary monopoly and all the special privileges—high profits—that result from it. In Stage 2, competitors buy or steal the new idea, which forces an emphasis on low-cost, standardized, mass-production technologies. Sales of the product increase for a while, but the initial high returns diminish. By Stage 3, the product begins

FIGURE 8.26
A typical product life cycle. Stage 1 is the monopolistic phase in which initial discovery and development are followed by the commercial launching of the product. Rapid sales ensue. The company may enjoy a monopoly during this period, at which time they attempt to improve the products. Stage 2 is characterized by the entry of competition. Emphasis is now on mass produced, inexpensive items that are standardized and directed toward expansions of the market. Competition begins to erode a large share of the innovating firm's sales. In Stage 3, a large share of the market has been lost to new products and other companies. Overall sales of the product declines as alternative products and manufacturing processes are introduced. *(Source: Chapman and Walker, 1990, p. 28)*

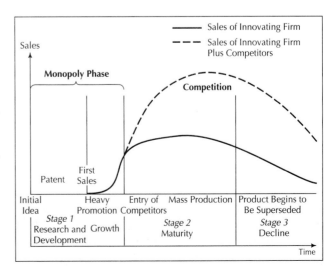

to be superseded. Markets are lost to new products, and manufacturing capacity is reduced.

Innovation begins in an advanced industrial country. These countries have the science, the technology, and the internal market to justify R&D. As a result, they also have an international advantage, and they export their product around the world. But as the technology becomes routinized, other producers appear on the scene, first in the other advanced countries, then on the periphery (Figure 8.27). Meanwhile, back in the rich country, investment in the newest generation of sophisticated technology is the cutting edge of the economy.

There is no doubt that developed countries are the innovators of the world economy and that LDCs increasingly specialize in the laborious task of transforming raw materials into commodities. But developed countries are also engaged in activities associated with the second and third stages of the product cycle. Indeed, Great Britain and Canada have expressed concern about their recipient status. This concern has also been voiced in the United States.

The product life cycle is not the only way that production technology affects corporate spatial organiza-

tion. Corporations frequently establish fragmented production systems or part-process structures in which the division of labor principle is taken a step further. As described by K. Chapman and D. Walker (1987), "The various tasks are geographically separated so that the motorway networks and shipping routes which link them effectively become integral parts of the assembly line" (p. 114). This type of system, established for a long time on a regional scale, now operates on a world scale.

Not all manufacturing operations are fragmented. Corporate branch plants are often *clones*, supplying identical products to their market areas. Examples abound. Medium-sized firms in the clothing industry often have this structure, as do many multiplant companies manufacturing final consumer products. Part-process structures tend to be associated with certain industrial sectors, such as electronics and motor vehicles, characterized by complex finished products comprising many individual components.

Labor is an important variable in the location of facilities making components. Manufacturers seek locations where the level of worker organization, the degree of conflict, and the power of labor to affect the actions

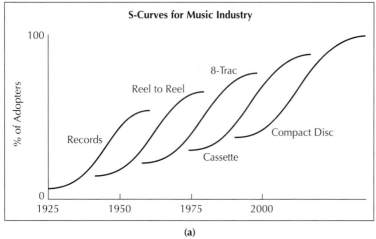

(a)

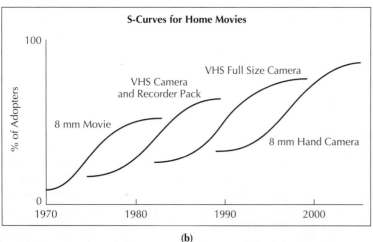

(b)

FIGURE 8.27

S curves for (a) the recording industry and (b) home movies. Since 1980, the U.S. economy lost 40 million manual and manufacturing "sunset jobs" through downsizing and structural changes, but it gained 70 million "sunrise jobs" in the service and technology sectors. *(Source: After Stutz, 1998)*

of capital are more limited than in long-established production centers such as Detroit, Coventry, and Turin. Starting in the early 1970s, Fiat began to decentralize part of the company's production away from its traditional base in Turin to the south of Italy. Compared with the workers of Turin, who were relatively strong and well organized, the workers of the south were new to modern industry and had little experience of union organization. At the international level, Ford adopted a similar tactic when it invested in Spain and Portugal in the 1970s. Ford management perceived that it could operate trouble-free plants in a region of low labor costs. The labor factor is further emphasized by the practice of *dual sourcing*. To avoid total dependence on a single workforce that could disrupt an interdependent production system, companies such as Ford and Fiat are willing to sacrifice economies of scale for the security afforded by duplicate facilities in different locations.

Locational Adjustment

Corporate production systems undergo continuous locational adjustment. Shifts may be inspired by technical and organizational developments internal to an industry or by changes in the external environment in which they operate, such as the oil-price hikes of 1973. Particularly significant from a geographic viewpoint are adjustments in response to major shocks or stresses placed on an enterprise. For example, when faced with the challenge of competition from lower-cost regions and with a falling rate of expansion of global markets, an enterprise can adopt a number of strategies—rationalization, capital substitution, outright closure, reorganization of productive capacity associated with the closure of older plants—which all in one way or another result in losses of employment. The recent industrial experience of Great Britain provides many illustrations of painful corporate restructuring programs. The 10 largest manufacturing employers in the West Midlands reduced their British employment by 25% between 1978 and 1981 while increasing their overseas workforce by 9% (Gaffikin and Nickson, 1984). This shift in the productive base of these companies abroad undermined the economic well-being of this area. Such employment withdrawals are an aspect of the growing international integration of production and mobility of capital.

One of capital's crucial advantages over labor is geographic mobility; it can make positive use of distance and differentiation in a way that labor cannot. Corporations take advantage of such flexibility by shifting production to low-wage regions, setting up plants in areas with low levels of worker organization, or establishing plants in areas that offer incentive policies. Many LDCs offer tax relief and capital subsidies for new industries.

INDUSTRIAL EVOLUTION

Just as firms evolve, so do industries—groupings of firms that have common elements. Industries evolve according to a sequence of developmental stages analogous to youth, maturity, and old age. Models that capture this evolutionary process are industry and Kondratiev cycles.

The Industry Life-Cycle Model

Industries tend to experience a period of experimentation, a period of rapid growth, a period of diminished growth, and a period of stability or decline. Each period may be related to the technology of an industry. Because of the link between the technology of an industry and its locational requirements, these similarities find geographic expression in characteristic distributions associated with industries in their youth, maturity, and old age.

YOUTH
During this stage of development, an industry is preoccupied with the design and commercialization of a new product. The industry consists of a number of firms, mainly new start-ups, which pursue the adaption of an innovation. The preoccupation with product design and commercialization leads to the geographic concentration of the industry in relatively limited areas. For example, the electronics and computer-related industry began to concentrate in an area to the south of San Francisco in the 1950s. Many of the most significant advances in the electronics industry resulted from the work of scientist-entrepreneurs operating in converted garages and workshops in the area.

MATURITY
During this phase, growth rates in output rise and then slacken as the industry shifts toward mass production and market penetration. Firm size increases, and the number of firms decreases. Geographically, this phase is associated with production decentralization at interregional and international levels. Cheaper labor costs, better business climates, and proximity to markets pull the more routinized parts of the production process away from the innovating centers of the industry.

OLD AGE
The final stage of the industry life cycle is characterized by market saturation and rationalization. A good example of market saturation is the iron and steel industry in developed countries. Major steel-using industries are relatively less important in developed countries now than they were in the 19th century. Consequently, their steel output has stabilized or even declined. In the United States and the United Kingdom, iron and steel plants have closed down. The industry has tended to

reconcentrate production in a few places, especially at coastal lower-cost production sites.

Kondratiev Cycles

A common criticism of the life-cycle model, which treats industrial history as a natural rather than a social process, is that it neglects relationships between industries. These relationships, of significance to scholars, are interpreted in terms of innovation cycles in the process of economic growth. The cycles are called *Kondratiev cycles*, after the Soviet economist Nikolai Kondratiev (1935), who first identified them in the 1920s.

Kondratiev hypothesized that industrial countries of the world have experienced successive waves of growth and decline since the beginning of the Industrial Revolution with a periodicity of 50 to 60 years' duration. But it was left to Joseph Schumpeter (1939), a German economist, to explain Kondratiev's observation in terms of technical and organizational innovation. Schumpeter suggested that long waves of economic development are based on the diffusion of major technologies, such as railways and electric power. More recently, another German economist, Gerhard Mensch (1979), argued that throughout capitalist history, inno-

vations have significantly bunched at certain points in time—around 1764, 1825, 1881, and 1935—just when the model of long waves would demand. According to Mensch, innovations come in clusters in response to social needs; they coincide with periods of depression that accompany world economic crises.

Kuznets (1954) described the Kondratiev cycles in terms of successive periods of recovery, prosperity, recession, and depression. The upswing of the first cycle was inspired by the technologies of water transportation and the use of wind and captive water power; the second by the use of coal for steam power in water and railroad transportation, and in factory industry; the third by the development of the internal combustion engine, the application of electricity, and advances in organic chemistry; and the fourth by the rise of chemical, plastic, and electronics industries. In the present period of world economic crisis, with higher energy costs, lower profit margins, and growth of the old basic industries exhausted, scholars are asking whether a *fifth wave* is emerging (Figure 8.28).

A new technoeconomic paradigm does seem to be emerging based on the extraordinarily low costs of storing, processing, and communicating information. In this perspective, the structural crisis of the 1980s and 1990s is a prolonged period of social adaptation to the growth

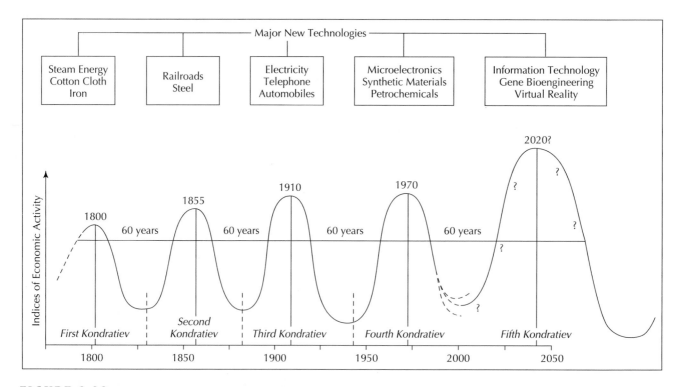

FIGURE 8.28

Kondratiev long waves of economic activity. Kondratiev, or K-waves, last approximately 50 years each and have four phases of activity, including boom, recession, depression, and recovery. Each period of economic activity has its associated major technological breakthroughs that power economic growth and employment. The world economy is presently in a recession as it creeps closer to the year 2000. On the horizon, there appears to be a new boom cycle based on information technology, biotechnology, space technology, energy technology, and materials technology. *(Source: Expanded from Dicken, 1992, p. 99)*

of this new technological system, which is affecting virtually every part of the economy, not only in terms of its present and future employment and skill requirements, but also in terms of its future market prospects.

INFORMATION TECHNOLOGY: THE FIFTH WAVE?

Some scholars, especially Christopher Freeman (1988), argue that a fifth Kondratiev cycle appears to have begun in the 1990s, and it is associated with *information technology*. Information technology production in the future is based on microelectronic technologies, including microprocessors, computers, robotics, satellites, fiber-optic cables, and information handling and production equipment, including office machinery and fax machines. Information technology, production, and use has a strong production pattern in Japan, in Far Eastern newly developing countries, in the United States and Canada, and in Western Europe, notably Germany, Sweden, and France. Freeman (1988) argues that the importance of information technology results from the convergence of communications technology and computer technology (Chapter 4; Figure 4.4). Communications technology involves the transmission of data and information, whereas computer technology is concerned primarily with the processing, analysis, and reporting of information.

Long Waves of Sector Shifts

The economic sector composition of employment in North America between 1900 and 1997 reveals a five-part subdivision: primary, secondary, tertiary, quaternary, and quinary. The proportional share of employment in the tertiary share (providing services directly to the consumer, such as retail, wholesale) has remained relatively constant throughout this century (Figure 2.2). The principal economic shift in labor composition after the 1910 Kondratiev long wave peak was the quaternary sector job growth—the sector accounting for the transformation of mass production. The most significant changes since the 1970 Kondratiev long wave peak have involved rapid expansion of the quinary sector—the sector responsible for reengineering the new social and economic progression of the flexible economy.

With each of these peaks in the Kondratiev long wave cycle a new geography of the spatial division of labor has developed. The tertiary sector that provides goods and services directly to consumers was described in Chapter 7 as central-place theory, consisting of urban hierarchies and market areas. The geography of the quaternary sector providing services to producers

emerged after 1970, produced by the scale economies from the mass production technologies. Because of internal scale economies, the quaternary long wave produced a shift of managerial emphasis from the individual firm to that of the multi-establishment and multinational corporation. Such business services as advertising, purchasing, auditing, inventory control, and company financing (*producer services*) were shifted from the local level to the regional head office, restricting the function of individual firms to the delivery of the main product or service to the local region. As a result, the new geography of world cities emerged; a command and control economy centered on nodal cities encompassing high thresholds of the first-order central places. Stanback, et al., (1981) suggests that these high-level quaternary service nodes include: (1) National nodal centers, metropolitan complexes in which there is a massive concentration of corporation headquarters, banking activity, and great intensity and diversity of specialist producer services, including advertising, consulting, and investment banking; (2) Regional nodal centers, small metropolitan bases from which large corporations can administer their industrial and commercial operations in a country's main regional markets.

With the peak of the 1970 Kondratiev long wave has come a host of new services of a specialized technical and business nature. The defining characteristic of these new services is that they create and manipulate knowledge products in almost the exact same way as the manufacturing industries that peaked in 1800, 1855, and 1910 transformed raw materials into physical products. These services have become the salient forces of the new post-industrial society and the information economy. These are the leading forces that are restructuring the geography of manufacturing, because they are the basis of productivity increases—technological innovation, better resource allocation through expert systems, increased training, and education.

According to Neusbaumer (1987), the principal functions of these advanced producers services are:

1. Knowledge carrying, including education and training, producer services, management consultant, and advisory services.
2. Linkage forming, including the construction of channels of communication and the capacity for carrying information, as well as the creation of markets.
3. Communication, the sending of information.
4. Information itself, the substantive material that is transmitted.

These producer services allow businesses to improve in efficiency, while at the same time reducing costs. Such services include research and development activities, management and consultant support, risk reducing procedures, improvement of health and safety to the corporation, and increased quality and life of products. Such information services allow companies to react to market conditions rapidly as a set of linked individuals, rather than one large bureaucratic, slow-moving organization (minnows versus whale).

SUMMARY

In this chapter, we discussed manufacturing from the standpoint of firms rather than areas, with emphasis on normative and behavioral approaches to industrial location theory. Classical location theory stresses that manufacturing patterns are caused by geographic characteristics—*locational factors*—rather than by underlying social relations. According to least-cost location theory, assembly costs are incurred because the raw materials required for a particular kind of manufacturing are not evenly distributed. Production costs vary because of the areal differences in the costs of labor, capital, and technical skills. Finished-product distribution costs are incurred when producers must sell to dispersed or widely scattered markets. All these costs are collectively called *locational costs*. Classical location theory provides a rationale to help find the points of production at which locational costs are minimized.

Once a point of minimum locational costs is determined, however, other decisions must be made. These pertain to the *scale* at which the firm will operate and the particular combination of inputs (*technique*) that will be used. A producer must also be concerned with the actions of competitors. We can see that the location problem is complex, but by applying the concept of *spatial margins to profitability*, we can reduce the complexity. Locational costs, scale, technique, and locational interdependence together determine spatial margins to profitability. By definition, all viable manufacturing must take place within these margins. How these limits are empirically determined and how locations within them are chosen, however, are usually discussed by geographers in a behavioral or decision-making context.

Most geographers now question the empirical usefulness of industrial location theory in light of the revolutionary role played by *multiproduct, multiplant, multinational operations* in the global structure of manufacturing. Accordingly, we devoted a major portion of the chapter to the spatial behavior of large industrial enterprises. Attention was given to trends in industrial organization, the relationship of large firms to small ones, the reasons for corporate growth, and the internal geography of corporate systems. In the last section, we looked briefly at models of *industrial evolution*—the industry life-cycle model and the Kondratiev long-wave model. The Kondratiev model reminds us that the present world economic crisis may be the beginning of a fifth upswing, this one based on a cluster of microelectronics and information technologies.

KEY TERMS

assembly costs
backward integration
basic cost
business process reengineering (BPR)
centrifical drift
cloning spatial structure
conglomerate merger
demand-potential surface
diseconomies of scale
distribution costs
diversification
dual economy
dual sourcing
economies of scale
externalities
fifth wave
flexible labor
foreign sourcing
forward integration
franchising
horizontal integration
industrial complex economies
industrial inertia

industry concentration
industrial park
industry life cycle
information technology
integration
international subcontracting
isodapane
joint venture
Kondratiev cycles
least cost approach
licensing venture
localized raw material
locational costs
locational interdependence
machinofacture
marketing linkage
market-oriented
massing of reserves
material index
multinational corporation
multiplant enterprises
offshore assembly
oligopoly

orientation
product life cycle
production linkages
pure raw material
raw-material–oriented
real time information system
satellite firms
service linkages
space-cost curve
spatial margins to profitability
spatial monopoly
spatial oligopoly
Taylorism
transportation-cost surface
ubiquitous raw material
urbanization economies
value added by manufacturing
Varignon frame
vertical integration
weight-gaining raw material
weight-losing raw material

SUGGESTED READINGS

Britton, J.N.H., ed. 1996. *Canada and the Global Economy.* Montreal: McGill-Queens University Press.

Chapman, K., and Walker, D. 1993. *Industrial Location.* New York: Basil Blackwell.

Chapman, K., and Walker, D. 1991. *Industrial Location.* 2nd ed. Cambridge, Mass.: Basil Blackwell.

Cohen, B. J. 1990. The political economy of international trade. *International Organization,* 44:261–281.

Corbridge, S., ed. 1993. *World Economy.* The Illustrated Encyclopedia of World Geography. New York: Oxford University Press.

Dicken, P. 1992. *Global Shift: Industrial Change in a Turbulent World,* 2nd ed. London: Harper & Row.

Ettlinger, N. 1991. The roots of competitive advantage in California and Japan. *Annals of the Association of American Geographers,* 81:391–407.

Fransman, M. 1990. *The Market and Beyond: Co-operation and Competition in Information Technology in the Japanese System.* Cambridge: University Press.

Glasmeier, A. K. 1991. *The High-Tech Potential: Economic Development in Rural America.* New Brunswick, N.J.: Center for Urban Policy Research.

Harrington, J.W., and Warf, B. 1995. *Industrial Location.* New York: Routledge.

Henderson, J. 1989. *The Globalization of High Technology Production.* London: Routledge.

Hoare, A. G. 1993. *The Location of Industry in Britain.* New York: Cambridge University Press.

Hugas, J. W., and Seneca, J. J. 1996. *Regional Economic Long Waves.* New Brunswick, N.J.: Center for Urban Policy Research.

Knox, P.L., and Agnew, J.A. 1994. *The Geography of the Economy.* 2nd ed. London: Edward Arnold.

Krugman, P. 1991. *Geography and Trade.* Cambridge, Mass.: University Press.

Law, C. M., ed. 1991. *Restructuring the Global Automobile Industry.* London: Routledge.

Massey, D. 1984. *Spatial Divisions of Labor: Social Structures and the Geography of Production.* New York: Methuen.

Pitelis, C., and Suugden, R., eds. 1991. *The Nature of the Transnational Firm.* London: Routledge.

Porter, M. E. 1990. *The Competitive Advantage of Nations.* New York: Free Press.

Rosenfeld, S. A. 1992. *Competitive Manufacturing.* New Brunswick, N.J.: Center for Urban Policy Research.

Sayer, A., and Walker, R. 1992. *The New Social Economy.* Cambridge, Mass.: Blackwell.

Schmenner, R. 1992. *Making Business Location Decisions.* Englewood Cliffs, N.J.: Prentice Hall.

Stutz, F. P. 1992. Maquiladoras branch plants: Transportation—labor cost, substitution along the U.S./Mexican border. In 27 Congress of the International Geographical Union, ed., *Snapshots of North America.* Washington, D.C.: I.G.U.

WORLD WIDE WEB SITES

TOP TEN COUNTRIES WITH WHICH THE U.S. TRADES
http://www.census.gov/foreign-trade/www/balance.html

U.S. INTERNATIONAL TRADE IN GOODS AND SERVICES HIGHLIGHTS
http://www.census.gov/indicator/www/ustrade.html

U.S. TRADE WORLD COUNTRIES IN 1997
http://www.census.gov/foreign-trade/sitc1/1997/c5170.htm

THE WORLD TRADE ORGANIZATION
http://www.wto.org/
The principal agency of the world's multilateral trading system. Its home page includes access to documents discussing international conferences and agreements, reviewing its publications, and summarizing the current state of world trade.

THE WORLD BANK
http://www.worldbank.org/
A leading source for country studies, research, and statistics covering all aspects of economic development and world trade. Its home page provides access to the contents of its publications, to its research areas, and to related websites.

U.S. DEPARTMENT OF COMMERCE
http://www.doc.gov/
Charged with promoting American business, manufacturing, and trade. Its home page connects with the web sites of its constituent agencies.

BUREAU OF LABOR STATISTICS WEBSITE
http://stats.bls.gov/
Contains economic data, including unemployment rates, worker productivity, employment surveys and statistical summaries.

ECONOLINK
http://www.progress.org/econolink
Econolink has "the best web sites that have anything to do with economics." Its listings are selective with descriptions of side content; many referenced sites themselves have web links.

WEBEC: WORLD WIDE WEB RESOURCES
http://netec.wustl.edu/WebEc/WebEc.html
A more extensive set of site listings—though more purely "economic" than "economic geographic."

INDUSTRIAL LOCATION: WORLD REGIONS

O B J E C T I V E S

- To acquaint you with the major manufacturing regions of North America and the world
- To describe recent global shifts in the globalization of world manufacturing
- To examine the relocation of the U.S. manufacturing industry
- To explain how Japan became a tower of industrial strength
- To document the "East Asian Miracle"
- To present the problems of industrialization in developed and developing countries

Manufacturing Reeboks at the Lotus Plant, the Philippines. (*Photo: Susan Meiselas, Magnum Photos, Inc.*)

Having examined the locational choices of firms, let us now consider the effects of these choices on areas. This chapter explores the implications of industrial location for communities, regions, and countries. It seeks answers to the following questions: How do industrial areas develop and change? Why is there industrial growth in some areas, while others exhibit industrial decline? What are the major manufacturing trends in world regions? What is the recent history of the geography of manufacturing in advanced industrial countries and in newly industrializing countries? Geographers are intensely interested in these questions in this period of economic crisis. It is a period marked by the internationalization of production. Rationalization of this program of industrial restructuring has involved fragmentation of the labor process; closures, openings, and new production technology; and the increasing penetration and control of markets by giant corporations.

Machines and computers have released people from labor drudgery. On the other hand, new technologies have been applied in ways that have displaced millions of workers. Release from drudgery has, to some extent, resulted in the collapse of work.

What are the forces that drive this situation? From the traditional viewpoint, technology develops according to its own imperative, with techniques applied to achieve increased productivity with fewer employees. The alternative view sees technologies and techniques as social products, which are developed according to social relations. The alternative perspective alleges that the framework for analysis is society, not the technologies that it produces. The investigative problem lies not with technology, which has the potential for doing social good; rather, it lies with society, which has the power to misuse this potential.

A similar argument applies to the changing geography of industry and employment. Industrial restructuring can improve human welfare. What is wrong with the selective deindustrialization and reindustrialization of the polluted environments of the old manufacturing areas? What is wrong with industrial revolution in the developing world? Nothing is wrong with industrial restructuring, so long as it does not result in social inefficiency.

Unfortunately, however, industrial restructuring has been a painful process. The movement of British capital to the LDCs has had serious effects at home—high levels of unemployment, the destruction of communities, and the loss of valuable skills, plants, and equipment. In LDCs, the new manufacturing regions have attracted only selected industries or parts of industries, which makes them vulnerable to outside control by multinationals. In addition, the new industries rarely meet the urgent consumption needs of poor people, yet they can pollute environments, affect local cultures, and exploit labor—especially female labor. What drives such corporate behavior? Corporations relocate their operations in order to survive in a highly com-

petitive world. In their never-ending quest for profits, they must seek new production frontiers.

To appreciate how the industrial geography of a region changes, we must first learn about the economic structure of society. Thus, the opening section of this chapter provides an introduction to a theory of society, its economy, and the relations that influence locational decision making. Following this section, we look briefly at changes in the geography of world manufacturing, and then take a detailed look at industrial decline and growth in advanced industrial countries and at export-oriented industrialization in parts of the developing world. We offer a special look at the "East Asian Miracle." Almost all of the Southeast Asian countries have risen from poverty and colonialism to create huge industrialized economies. Why has East Asia emerged as the model for success, while Africa has seen mostly poverty, hunger, and economies propped up by foreign aid? Old industrial landscapes need not die; they can be rebuilt to fit new technologies. Whether or not rebuilding occurs does not depend on the physical age of the industrial region but on its social conditions, especially those relevant to profit making, and in particular the social relations between capital and labor. Social conditions exert a powerful influence on the geography of manufacturing.

FORCES OF PRODUCTION AND SOCIAL RELATIONS

People must produce objects to satisfy their physical needs. The production of material goods is, therefore, central to an analysis of society. According to this viewpoint, the basic elements of society are the *forces of production* and the *social relations of production*. Together these elements are known as the *mode of production*, and they constitute the *economic structure of society*.

The forces of production include (1) laborers, (2) natural resources, and (3) capital equipment. In the early stages of economic development, labor is the chief productive force. The ability to transform nature is limited, and the lives of people revolve around natural forces beyond their control. As the number of workers increases, and as the legacy of capital equipment grows, more and more of nature is harnessed. With more control over nature, people are better able to raise their living standards.

The crucial social relation of production is between owners of the means of production and the workers employed to operate these means. Under capitalism, the means of production are privately owned. Owners of capital control the labor process and the course of economic and social development. Private ownership has two dimensions: on the one hand, competitive relations exist among owners; on the other, cooperative and antagonistic relations emerge between owners (capital) and workers (labor).

Relations Among Owners

Capitalists make independent production decisions under competitive conditions. A raw competitive struggle for survival is fundamental to an appreciation of capitalist development. Competition in the market focuses on price, price depends on cost, and cost hinges on the productive forces used. Competition requires producers to apply a minimum of resources to achieve the highest output. It forces companies to minimize labor costs, which means extreme labor specialization and subordination of workers to machine automation. It demands large-scale production to lower costs and, if possible, to control a segment of the market. It also entails the acquisition of linked or competing companies and the investment of capital in new technology and in research and development (R&D).

Competition is the source of capitalism's immense success as a mode of production. But the tension between opposing elements cannot be solved without fundamental change. Consider, for example, the critical environmental issues generated by the contradiction between capitalism and the natural environment. For productive forces to continue to expand without a reduction in living standards, new values must be built into the production system. These values are already evidenced by the use of renewable energy sources and the imposition of pollution controls.

Relations Between Capital and Labor

Capital–labor relations are both cooperative and confrontational. Without a cooperative workforce, production is impossible. However, cooperative relations are often subordinate to antagonistic relations.

Because producers make decisions according to their desire to make profits, they try in every way possible to pay workers only part of the value produced by their labor. Value produced by workers in excess of their wages—called *surplus value*—is the basis for profit. On the other hand, workers try to increase their wages in order to enjoy a higher standard of living. They sometimes organize into unions and, if necessary, strike to demand higher wages. If management agrees to meet labor demands, cooperative relations may exist for a time before antagonistic relations resume.

Competition forces management to invest as much as possible in technology and research to increase productivity. As production increases, the struggle between employers and employees puts higher wages into the workers' hands. Increasing purchasing power means expanding markets that absorb a growing supply of commodities, and production development continues in a process of cumulative causation.

Machines and low-wage labor can replace high-wage labor. Low-wage peripheral regions can sell products to high-wage center regions. Industrial migration to the periphery removes jobs in the center, which disciplines organized labor. Pressures to increase wages slacken, and mass demand decreases. A problem of underconsumption develops. Thus, in capitalism, the solution to one problem may be the breeding ground for new problems.

Competition and Survival in Space

Relations among owners and between capital and labor are sources of change in the geography of production. Competition among owners may cause a company to relocate all or parts of its operation to a place where it can secure low-wage labor. From the company's perspective, this strategy is mandatory for survival; if other companies lower their costs and it does not, it will inevitably lose in the competitive struggle. Capitalists must expand to survive, and the struggle for existence leads to the survival of the biggest. In their search for profits, giant corporations have extended their reach so that few places in the world remain untouched.

The incessant struggle of companies to compete successfully is especially evident in the entrepreneurial response to differential levels of capital–labor conflict. Old industrial regions of the core—Europe, North America—have high conflict levels. In contrast, peripheral regions have various combinations of lower conflict levels and lower wages. Organized labor in the old industrial areas induced the owners of capital to switch production and investment to countries that were not yet industrialized or to newly industrializing countries. The reason that mobile capital could avoid the demands of organized labor was the development of productive forces—an increased ability to traverse space and conquer the technical problems of production—and the emergence of a huge alternative labor force in the developing world following the colonial revolution in Asia, Africa, and the Caribbean.

These dramatic changes in the 1960s and 1970s ended the original international division of labor that was formalized in the 19th century. Under the old imperial system, the advanced powers were the industrialists, and the colonies were the agriculturalists and producers of raw materials. After decolonization, light industry and even some heavy industry began to emerge in the former colonies. The advanced economies assisted this process. The increasing globalization of production was accompanied by a new international division of labor. The world became a "global factory," in which the developed countries produced the sophisticated technology, and the developing countries were left with the bulk of the low-skill manufacturing jobs.

The emergence of this new international division of labor, mainly a consequence of the activities of the foot-loose multinational corporations, resulted in deindustrialization in the old industrial regions of advanced economies and a precarious export-led industrial revolution in parts of the developing world.

WHERE INDUSTRY IS LOCATED

Four major areas account for approximately 80% of the world's industrial production (Figure 9.1). These areas are northeastern North America, northwestern Europe, western Russia and the Ukraine, and Japan. In this section, we examine each of the first three regions as well as several nearby subregions. Japan is discussed in detail later in this chapter.

North America

North American manufacturing is still centered in the northeastern United States and southeastern Canada (Figure 9.2). This region is called the North American Manufacturing Belt. James Rubenstein (1994) stated that this area encompasses only 5% of the land in North America, but accounts for one-third of the North American population and nearly two-thirds of North American manufacturing.

This area was the first region settled by Europeans in the 17th and 18th centuries. It was tied to the European markets and possessed the raw materials, iron ore, coal, and limestone necessary to produce the heavy machinery and manufactured items on which the industrialization of America was based. In addition, this region had many markets and a large labor pool.

The transportation system included the St. Lawrence River and the Great Lakes, which were connected to the East Coast and the Atlantic Ocean by the Mohawk and Hudson rivers. This transportation system allowed the easy movement of bulky and heavy materials. Later, the river and lake system was supplemented by canals and railroads.

NEW ENGLAND DISTRICT

Within the North American Manufacturing Belt, there are several districts. The oldest is southern New England, centered on Massachusetts and the greater Boston metropolitan area. Historically, this area was the textile and clothing manufacturing center of the early 19th century. Cotton was brought from the southern states to be manufactured into garments, many of which were consumed locally and some of

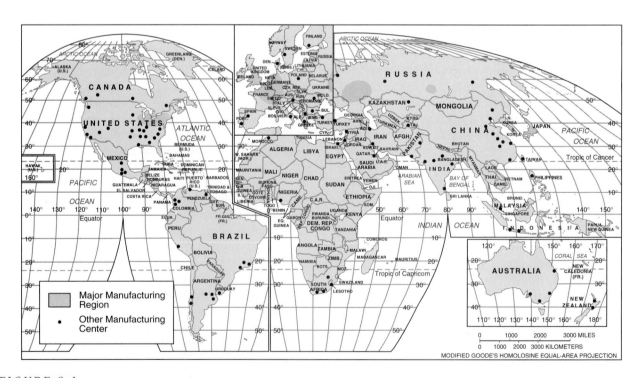

FIGURE 9.1

Worldwide distribution of manufacturing. The four main manufacturing regions include the northeastern United States and southern Great Lakes region, the northwestern European region, the eastern Soviet Union and Ukraine region, and the Japan/South Korea region. See the text for a detailed elaboration of districts within each of these regions, as well as other manufacturing regions shown as dots.

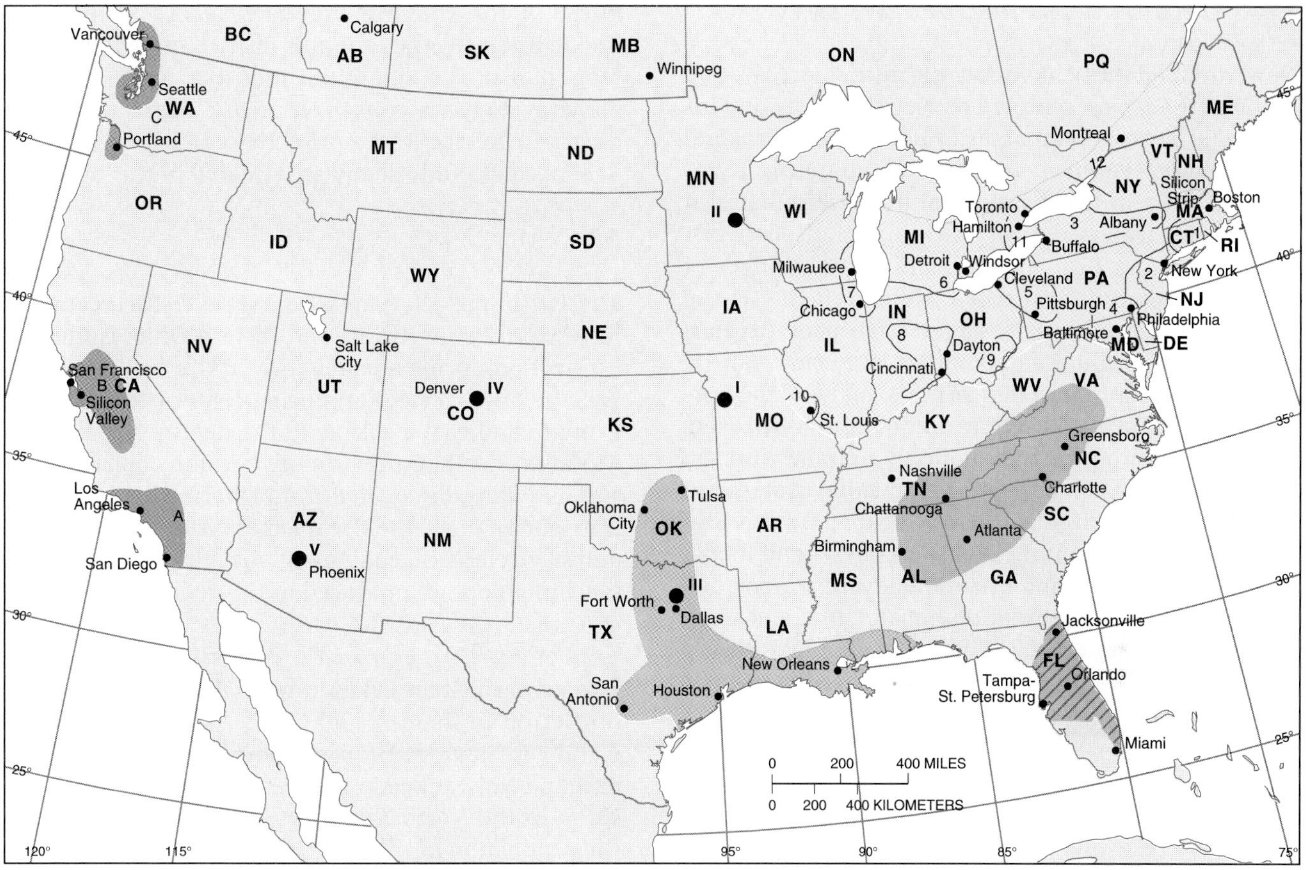

FIGURE 9.2
Manufacturing regions and districts throughout the United States and Canada. *(Source: Fisher, 1992, p. 155)*

Anglo-American Manufacturing Region	6 Southeast Michigan district	Southeastern Manufacturing Region	C Pacific Northwest district (Portland-Seattle, Vancouver)
1 New England district	Automobiles	Textiles	Aircraft
Electrical machinery	Iron and steel	Apparel	Lumber products
Machinery	7 Lake Michigan district	Transportation equipment	Food processing
Fabricated metals	Iron and steel	Furniture	● Other Centers of Manufacturing
Textiles	Fabricated metals	Food processing	I Kansas City
Electronic products	Machinery	Lumber	Food processing
Apparel	Printing and publishing	Primary metals	Automobile assembly
2 Greater New York district	Electrical machinery	Gulf Coast Manufacturing Region	II Minneapolis-St. Paul
Apparel	8 Southwest Ohio-Eastern Indiana district	Petroleum refining	Food processing
Printing and publishing	Iron and steel	Chemicals	Machinery
Machinery	Fabricated metals	Primary metals (aluminum)	Fabricated metals
Food processing	Machinery	Electrical machinery	III Dallas-Ft. Worth
Fabricated metals	Electrical machinery	Electronic products	Transportation equipment
Chemicals	Paper manufacturing	Central Florida Manufacturing Region	Food processing
3 Central New York district	9 Great Kanawha and Middle Ohio Valley district	Food processing	IV Denver-Pueblo
Electrical machinery	Chemicals	Electrical machinery	Food processing
Chemicals	Primary metals	Electronic products	Chemicals
Optical machinery	Glass	West Coast Manufacturing Region	Iron and steel
Iron and steel	10 St. Louis district	A Los Angeles-San Diego district	V Phoenix
4 Mid-Atlantic district	Transportation equipment	Aircraft	Electrical machinery
Apparel	Iron and steel	Electrical equipment	Electronic products
Iron and steel	Fabricated metals	Automobile assembly	
Chemicals	Food processing	Apparel	
Food processing	11 Ontario Peninsula district (Canada)	Petroleum refining	
Machinery	Iron and steel	B San Francisco district	
5 Pittsburgh-Cleveland district	Machinery	Electronic products	
Iron and steel	Chemicals	Food processing	
Machinery	Food processing	Shipbuilding	
Electrical equipment	12 St. Lawrence Valley district (Canada)	Machinery	
Rubber	Pulp and paper		
Machine tools	Primary metals (aluminum)		

which were exported to Europe. As the low-wage European immigrant laborers settled and achieved a higher standard of living and unionization, wages became higher, and the textile industry moved to the South. Today, although it still produces some high-value-added textiles and clothing, the New England district manufactures electrical machinery, fabricated metals, and electronic products. The region is noted for highly skilled labor and ingenuity, with nearby universities—including Boston College, Boston University, Massachusetts Institute of Technology, and Harvard University—providing the chief supply of both. Boston is now called *Silicon Valley Northeast* (see Figure 9.2).

NEW YORK AND THE MIDDLE ATLANTIC DISTRICT

New York and the Middle Atlantic district (areas 2 and 4 in Figure 9.2) are centered on New York City and include the metropolitan areas of Baltimore, Maryland; Philadelphia, Pennsylvania; and Wilmington, Delaware. The Great Lakes industrial traffic terminates in New York City via the Mohawk and Hudson rivers. From New York City, foreign markets and sources of raw materials can be reached. New York City is the largest market and has the largest labor pool. Because of its central location among other large cities on the Eastern Seaboard, as well as its ports, many of the Fortune 500 firm headquarters are in this district. The New York district is in proximity not only to trade with the rest of the world, but also to the population centers and manufacturing hubs of America (Figure 9.2). It is also near financial, communications, and news and media industries, which are important for advertising and distribution. This region produces apparel, iron and steel, chemicals, machinery, fabricated metals, and a variety of processed foods. In addition, it is the headquarters of the North American publishing industry. Many major book-publishing companies are found in this region.

CENTRAL NEW YORK AND THE MOHAWK RIVER VALLEY DISTRICT

Another major industrial district within the North American manufacturing region is the central New York and Mohawk River valley district. In this region, electrical machinery, chemicals, optical machinery, and iron and steel are produced. These industries agglomerate along the Erie Canal and the Hudson River, which is the only waterway connecting the Great Lakes to the U.S. East Coast. Abundant electrical power produced by the kinetic energy of Niagara Falls provides inexpensive electricity to this district and explains the attraction of the aluminum industry, which requires above average amounts of electricity. The New York industrial cities of Buffalo, Rochester, Syracuse, Utica, Schenectady, and Albany are situated in this district.

PITTSBURGH–CLEVELAND–LAKE ERIE DISTRICT

The Pittsburgh–Cleveland–Lake Erie district, centered in western Pennsylvania and eastern Ohio, is the oldest steel-producing region in North America. Pittsburgh was the original steel-producing center because of the iron ore and coal available in the nearby Appalachian Mountains. When the iron ore became depleted, new supplies were discovered in northern Minnesota and transported in via the Great Lakes system. Besides iron and steel, electrical equipment, machinery, rubber, and machine tools are produced in this region.

WESTERN GREAT LAKES DISTRICT

The western Great Lakes industrial region is centered on Detroit in the east and Chicago in the west (areas 6 and 7 in Figure 9.2). In addition, Toledo, Ohio, in the east and Milwaukee in the west have a dominant position in the North American Manufacturing Belt for the production of transportation equipment, iron and steel, automobiles, fabricated metals, machinery, and printing and publishing. Detroit and surrounding cities, of course, have the preeminent position of automobile manufacture, and Chicago has produced more railroad cars, farm tractors and implements, and food products than any other city in the United States. The convergence of railroad and highway transportation routes in this area makes it readily accessible to

New England became the first and foremost textile manufacturing region in the United States in the late 18th century. The mills pictured here are in Lawrence, Massachusetts. By the 1940s, the textile region of southern New England had been in decline for more than 20 years. Firms left the region in search of more profitable operating conditions, and workers were forced to seek other employment. The region has experienced a revival as new industries, notably electrical engineering, have replaced the older, declining ones. *(Photo: Lionel Delevingne, Stock Boston)*

the rest of the country, and a good distribution point to a national market.

ST. LAWRENCE VALLEY–ONTARIO DISTRICT

Canada's most important industrial region by far stretches along the St. Lawrence River Valley, on the north shore of the eastern Great Lakes (areas 11 and 12 in Figure 9.2). This area has access to the St. Lawrence River–Great Lakes transportation system, is near the largest Canadian markets, has skilled and plentiful labor, and is supplied with inexpensive electricity from Niagara Falls. Iron and steel, machinery, chemicals, processed foods, pulp and paper, and primary metals, especially aluminum, are produced in this district. For example, Toronto is a leading automobile-assembly location in Canada, whereas Hamilton is Canada's leading iron and steel producer.

OTHER DISTRICTS WITHIN THE NORTH AMERICAN MANUFACTURING BELT

Three other districts are important. One is in southwest Ohio and in eastern Indiana and is centered on the cities of Columbus, Dayton, and Cincinnati, Ohio, and Indianapolis, Indiana. This district is noted for iron and steel, fabricated metals, machinery, electrical machinery, and paper manufacture. A second is Great Kanawha and middle Ohio Valley district, which specializes in chemicals, metals, and glass. The last district centers on the greater St. Louis metropolitan area. In this region, automobiles, transportation equipment, iron and steel, fabricated metals, and processed foods are produced. Brewing and beverage manufacture are also popular.

SOUTHEASTERN MANUFACTURING REGION

The *southeastern manufacturing region* of the United States, sometimes called the *Piedmont region*, stretches south from central Virginia through North Carolina, western South Carolina, northern Georgia, northeast-ern Alabama, and northeastern Tennessee (Figure 9.2). It centers on the towns of Greensboro and Charlotte, North Carolina; Greenville and Columbia, South Carolina; Atlanta, Georgia; Birmingham, Alabama; and Chattanooga and Knoxville, Tennessee. The district stretches around the southern flank of the Appalachian Mountains because of poor transportation connections across the mountains. Textiles are the main product, the industry having moved from the Northeast to the South to take advantage of less expensive, nonunion labor. Transportation equipment, furniture, processed foods, and lumber are also produced. Aluminum manufacturers moved to this region because of the inexpensive electricity produced by the more than 20 dams built by the Tennessee Valley Authority, and Birmingham has long been the iron and steel center of the southeastern United States because of the plentiful iron-ore and coal supplies nearby.

Figure 9.3 shows the distribution of hosiery and knit outerwear manufacturers. The hosiery industry is centered on areas from North Carolina southward, which have traditionally supplied low-cost labor. Knit outerwear requires more skill to manufacture and has not decentralized from the Northeast. The New York City region and New England still have the most knit outerwear manufacturers.

GULF COAST MANUFACTURING REGION

The *Gulf Coast manufacturing region* stretches from southeastern Texas through southern Louisiana, Mississippi, and Alabama, to the tip of the Florida panhandle. Principal cities include Houston, Texas; Baton Rouge and New Orleans, Louisiana; Mobile, Alabama; and Pensacola, Florida. Because of nearby oil and gas fields, petroleum refining and chemical production are important. The region also produces primary metals, including aluminum, and electrical machinery and electronic products.

San Jose, California. A $1 billion downtown renaissance has helped establish San Jose, capital of Silicon Valley, as California's third and the United States' eleventh largest city. *(Photo: San Jose Convention and Visitors Bureau)*

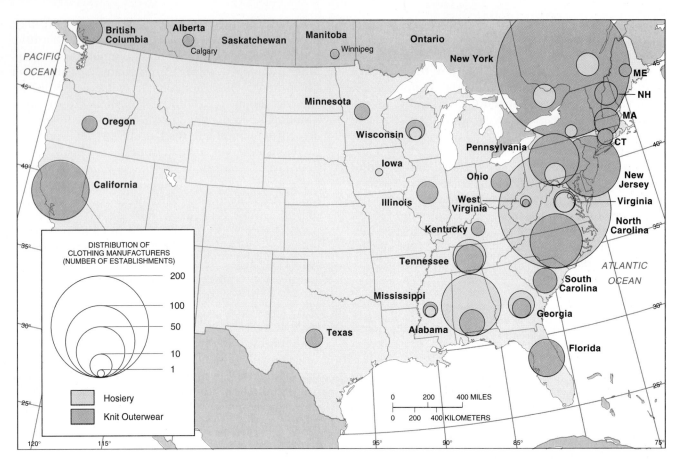

FIGURE 9.3

The distribution of hosiery and knit outerwear manufacturing establishments. Hosiery manufacture is a labor-intensive industry, and it has moved into the southeastern United States, where it can obtain cheap labor. Knit outerwear manufacturers have remained primarily in New England and the Northeast because of industrial inertia and because of the availability of more skilled workers, especially in the New York City region. (See Color Insert 2 for more illustrative map.)

CENTRAL FLORIDA REGION

The *central Florida manufacturing region* includes the cities of Jacksonville, Tampa, St. Petersburg, Orlando, and Miami. Processed foods are the most important product, but electrical machinery and electronic parts are also produced.

WEST COAST MANUFACTURING REGIONS

The Los Angeles and San Diego district in southern California specializes in aircraft and aerospace manufacture and electrical equipment. In the 1930s, the airline industry chose this location because favorable weather 330 days of the year meant unimpeded test flights and savings on heating and cooling the large aircraft plants. Because of the myriad electronic parts and equipment and the associated high-tech sensing and navigational devices required in aircraft manufacture, the electronics industry was also attracted to this region and was anchored there 30 years later (Figure 9.4). Today, aircraft, apparel manufacture, and petroleum refining are important in Los Angeles,

whereas San Diego also specializes in pharmaceutical production and in national defense manufacturing and transport equipment industries.

The San Francisco district is the second most important *West Coast manufacturing region*. Electronic products, processed foods, ships, and machinery are produced in the district. *Silicon Valley*, the world's largest manufacturing area for semiconductors, microprocessors, and computer equipment, is located just south of San Francisco in the Santa Clara Valley (Figure 9.2).

The Pacific Northwest district includes the cities of Seattle, Washington, and Portland, Oregon. Boeing Aircraft is the single largest employer, followed closely by the paper, lumber, and food-processing industries.

OTHER U.S. MANUFACTURING REGIONS

Other manufacturing centers include Kansas City, Missouri, with its food-processing and automobile assembly industries; Minneapolis and St. Paul, Minnesota, which produce wheat, processed foods, machinery, and fabricated metals; and Dallas–Fort Worth, Texas, which pro-

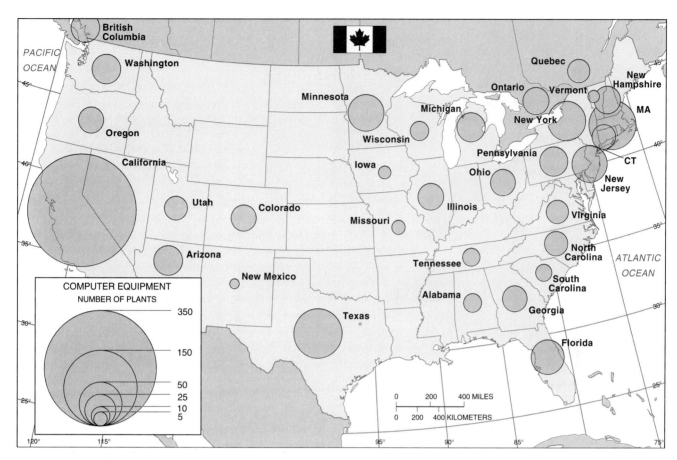

FIGURE 9.4

The distribution of computer equipment manufacturers. California has the largest number of plants, followed by a cluster in the Northeast. Electronics manufacturing plants gravitate to highly skilled labor. The finished products are more valuable than are items of clothing, and higher wages can be paid. There is a concentration of these manufacturers in California because early electronics production as a part of the aircraft manufacturing industry was centered on the West Coast. (See Color Insert 2 for more illustrative map.)

duce transportation equipment and processed foods. In addition, Denver and Pueblo, Colorado, have food-processing, chemical, and steel manufacturing industries, and the Phoenix metropolitan region manufactures electrical machinery and electronic products.

The electronics and computer industries require high-skilled workers, so, unlike clothing manufacturers, who have gravitated to low-wage districts, manufacturers of electronic and computer equipment have located in relatively high-wage districts to attract skilled workers. The largest high-tech clusters include New England, centered in Boston; California, centered in Silicon Valley; and Texas, centered in Austin.

Europe

Europe has developed some of the world's most important industrial regions (Figure 9.5). They are located in a north–south linear pattern, starting from Scotland and extending through southern England, continuing from the mouth of the Rhine River valley in the Netherlands, through Germany and France, to northern Italy. Good supplies of iron ore and coal provide fuel to the countries in these industrial regions. In addition, competition among countries has resulted in several subareas within Europe, each near large markets of consumers.

THE UNITED KINGDOM

The Industrial Revolution started in the United Kingdom in 1750. It had its basis in iron and steel production and textile and woolen manufacture. Because many dependent nations have since learned to produce their own iron, steel, and textiles, the world currently has an oversupply of these items, and the market for British goods has decreased substantially.

Great Britain's outmoded factories and deteriorating infrastructures have also reduced its overall global competitiveness for products. In contrast, Germany and Japan, with U.S. assistance, rebuilt after World War II,

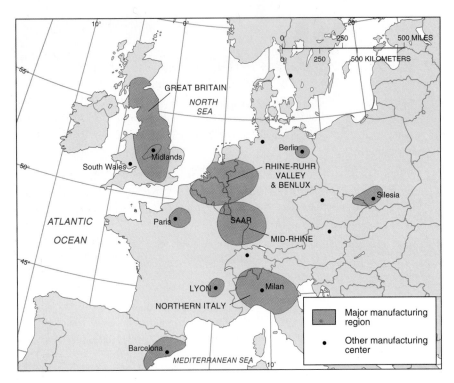

FIGURE 9.5

European manufacturing regions. Much European manufacturing exists in a linear belt from Scotland to the Midlands of Great Britain, to the South, including the London area. This belt continues onward from the low countries of Belgium, Luxembourg, and the Netherlands, south along the Rhine River, including portions of France and Germany, and into northern Italy. These areas became major manufacturing regions not only because of the concentration of skilled laborers, but also due to the availability of raw materials, principally coal and iron ore. In addition, good river transportation was available, as well as large consuming markets for finished products.

modernizing their plants and industrial processes at the same time. As a result, Germany and Japan are industrial successes in the world today, whereas Great Britain, more so than any other modern industrialized country, has suffered an industrial depression.

RHINE–RUHR RIVER VALLEY REGION

The most important European production region today is the northern European lowland countries of Belgium and the Netherlands, northwestern Germany, and northeastern France. In this region, the Rhine River and the Ruhr River meet. Although no single city dominates the region in Germany, Dortmund, Düsseldorf, Essen, and Duisburg are the centers of this district.

This region's backbone has been the iron and steel industry because of its proximity to coal and iron-ore fields. Production of transportation equipment, machinery, and chemicals helped lead Western Europe into the industrial age long before the rest of the world.

The Rhine River, which is the main waterway of European commerce, empties into the North Sea in the Dutch city of Rotterdam. Consequently, because of its excellent location, Rotterdam has become the world's largest port.

While exports of iron and steel are down from what they were 30 years ago, the region has been better able to avoid the depression of the U.K. because of its greater internal conversion of steel into high-quality finished products, which are in demand worldwide.

UPPER RHINE–ALSACE-LORRAINE REGION

Straddling the chief transportation artery for western Europe, the Upper *Rhine–Alsace-Lorraine Region* (called Mid-Rhine in Figure 9.5) is in southwestern Germany and eastern France. It is the second most important European industrial district, after the Rhine–Ruhr River valley.

Because of its central location, this area is well situated for distribution to population centers throughout western Europe. The main cities on the German side include Frankfurt, Stuttgart, and Mannheim. Frankfurt is almost perfectly located in the center of what was West Germany and hence became the financial and commercial center of its railway, air, and road networks. Stuttgart, on the other hand, is a center for precision goods and high-value, volume, manufactured goods, including the Mercedes Benz, Porsche, and Audi automobiles. Mannheim, located along the Rhine River, is noted for its chemicals, pharmaceuticals, and inland port facilities.

The western side of this district, in France, is dominated by the industrial cities of Metz, Nancy, Strasbourg, and Mulhouse. This area, known as Alsace-Lorraine, produces a large portion of the district's iron and steel. Germany and France have fought numerous wars over the occupation of this district, because of its ethnic French- and German-speaking peoples, as well as the rich iron-ore fields that extend northward into Luxembourg. Consequently, this region has changed hands many times. The present political arrangement is a result of World War II. The French were part of the victorious Allies, so they extended the borders of their country to the western bank of the Rhine River.

PO VALLEY OF NORTHERN ITALY

The remaining large industrial district of western Europe is the *Po River valley* in northern Italy. The principal industrial cities in this area include Turin, Milan, and Genoa, but other cities of industrial importance include Cremona, Verona, and mainland Venice. This area includes only one-fourth of Italy's land, but more than 70% of its industries and 50% of its population (Rubenstein, 1996).

This region specializes in iron and steel, transportation equipment (especially high-value automobiles), textile manufacture, and food processing. The district is bordered on the north by the Swiss Alps, Italy's breadbasket as well as its industrial backbone. The Alps, a barrier to the German and British industrial regions, give the Italian district a large share of the southern European markets. The mountains also provide Italian industries with cheap hydroelectricity and therefore reduced operating costs. Compared with workers in the American Manufacturing Belt, Italian laborers are willing to work for less because of their lower cost of living, thus this region attracts labor-intensive industries, such as textiles, from northern Europe.

Russia and the Ukraine

Five major industrial regions exist within the former Soviet Union. Four are primarily within Russia, and one is centered on the southern Ukraine (Figure 9.6). Since the demise of the Soviet Union in 1991, production statistics and information on the livelihoods of these regions are not as readily available as in the past.

MOSCOW, OR CENTRAL, REGION

The *Moscow, or central, industrial region* is near the population center of Russia and takes advantage of a large, skilled labor pool as well as a large market, even though natural resources are not plentiful. This region produces more than a quarter of the total industrial output of the former Soviet Union.

The largest single item produced is textiles: linen, cotton, wool, and silk fabrics. This manufacturing complex is set around the city of Ivanovo. Moscow industry also specializes in iron and steel, transportation equipment, chemicals, and automobiles and trucks. Novgorod, the automobile-producing Soviet Detroit, lies 100 miles north of Moscow.

UKRAINE REGION

The *Ukraine industrial region* relies on the rich coalfield deposits of the *Donets Basin*. The iron and steel industry

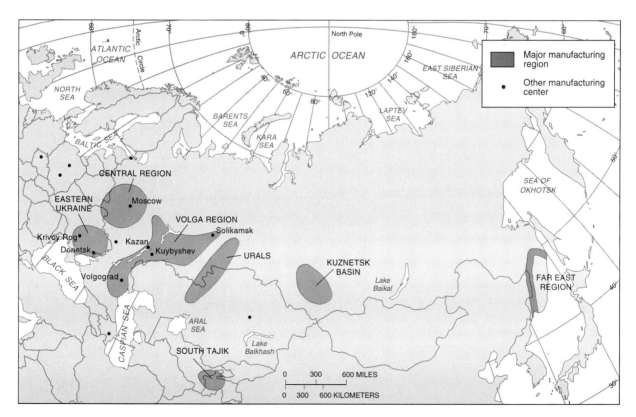

FIGURE 9.6
Manufacturing regions of Russia, Eastern Europe, and central Asia.

base is the city of Krivoy Rog, with nearby Odessa as the principal Black Sea all-weather port. The area is collectively known as *Donbass*. Like the German Ruhr area, Ukraine's industrial district is near iron-ore and coal mines, a dense population, and a large agricultural region, and is served by good transportation facilities.

The Ukraine industrial district, no longer a part of the Soviet Union, is severely hampering the Russian industrial effort. Ukrainian leaders have decided to market their goods on an international scale instead of supplying Russia's needs, even though these needs are more urgent than ever.

VOLGA REGION

East of the Moscow industrial region is the linear *Volga region*, extending from Volgograd (formerly Stalingrad) in the south, to Perm in the north, and astride the Volga and Kama rivers. The Volga River, a chief waterway of Russia, has been linked via canal to the Don River and thereby to the Black Sea. This industrial region developed during and after World War II because it was just out of reach of the invading German army that struck not only the Ukraine, but also Moscow. It is the principal location of substantial oil and gas production and refining. Recently, a larger oil and gas field was discovered in West Siberia. Nonetheless, the Volga district is Russia's chief supplier of oil and gas, chemicals, and related products. Recently, one of the largest automobile plants in the world opened in Toglaitti, producing Fiat automobiles.

URALS REGION

Just east of the Volga region are the low-lying Ural Mountains that separate European Russia from Asian Russia (Figure 9.6). The Ural Mountains have the largest deposits of industrial minerals found anywhere in the former Soviet Union. Mineral types include iron, copper, potassium, magnesium, salt, tungsten, and bauxite. The central-lying Ural district was important during World War II because it also was beyond the reach of the German army. Although coal must be imported from the nearby Volga district, the Urals district provides Russia with iron and steel, chemicals, machinery, and fabricated metal.

KUZNETSK BASIN REGION

The Kuznetsk Basin—also called *Kuzbass* and centered on the towns of Novosibirsk, along the trans-Siberian railroad, and Novokuznetsk—is the chief industrial region of Russia east of the Urals. It is also known as *Russian Asia*. Again, as in the case of the Ukraine and the Urals districts, the Kuznetsk industrial district relies on an abundant supply of iron ore and the largest supply of coal in the country. The Kuznetsk Basin is a result of the grand design of former Soviet city planners. These planners poured heavy investments into this re-

gion, hoping that it would become self-perpetuating and eventually the industrial supply region for Soviet Central Asia and the Soviet Far East. It has become the industrial supply region envisioned by the planners, but the planners probably did not foresee the iron and steel industry smokestacks and air pollution that are visible for hundreds of miles.

GLOBALIZATION OF WORLD MANUFACTURING

As the new *international division of labor* asserted itself, the rate of world economic growth declined. Lower growth rates coincided with the end of the long boom of the quarter-century period after World War II and the beginning of a prolonged period of economic crisis. The world economic crisis started with a deep recession in 1974–1975 following the first oil shock in 1973. One of its most visible effects in the advanced economies was deindustrialization—reflected by the loss of manufacturing jobs. As firms restructured or went out of business in a climate of intense competition, workers were laid off.

Change in the geography of manufacturing began in the postwar period of rapid growth, but accelerated in the crisis of the 1970s and 1980s. Although the advanced countries maintained a huge share in world manufacturing output, their output grew less quickly than that of the underdeveloped countries (Tables 9.1 and 9.2). Nonetheless, manufacturing output in the advanced countries increased in the 1960s, with most economies coasting along with annual growth rates of between 5% and 8%. Only Great Britain, with a growth rate of 3.3%, hinted at future trouble. The manufac-

❖

Table 9.1

Share in Manufacturing Production
by Country Group

Country Group	Share in Production			
	1965	*1973*	*1985*	*1995*
Industrial market economies	85.4	83.9	81.6	79.5
Underdeveloped countries	14.5	16.0	18.1	20.1
Low-income	7.5	7.0	6.9	7.5
Middle-income	7.0	9.0	11.2	13.5
High-income oil exporters	0.1	0.1	0.3	0.4

Source: World Bank, 1996, p. 47.

Table 9.2
Growth in Manufacturing Production by Country Group

Country Group	*Growth in Production* 1965–1973	1973–1985	1985–1995
Industrial market economies	5.3	3.0	2.5
Underdeveloped countries	9.0	6.0	11.1
Low-income	8.9	7.9	8.1
Middle-income	9.1	5.0	7.9
High-income oil exporters	10.6	7.5	9.1

Source: World Bank, 1987, p. 47.

turing output of the advanced economies slowed dramatically in the 1970s, and production actually declined in Great Britain. The number of workers in manufacturing in the advanced countries, which grew in the early 1970s, stabilized in the mid-1970s, and fell in the late 1970s and in the 1980s. Job losses were most numerous in Great Britain and Belgium; each country lost 28% of its manufacturing workers between 1974 and 1983. During the same period, West Germany recorded a loss of 16%; France, 14%; and the United States, 8%. The highest rate of manufacturing job loss in the United States was in the Midwest—more than 11% between 1975 and 1982. However, manufacturing employment increased in new industrial areas—parts of the Southwest, the Mountain states, the Dakotas, and Florida. It also increased in areas successful in restructuring their industrial bases, such as southern New England. With few exceptions, U.S. regions that registered a rapid growth in manufacturing output in the late 1970s and 1980s were low-conflict or low-wage states.

Since 1960, manufacturing output has increased sharply in lower-wage industrializing periphery countries. The output of manufactures in southern Europe and Latin America grew at annual rates of between 5% and 11% in the 1960s (except Argentina). Although this performance was achieved during a period of unprecedented real growth in world output, these regions sustained their progress in the 1970s and 1980s. Their high production rates were usually sufficient to secure increases in industrial employment.

The most rapid growth of manufacturing output occurred in East and South Asia. Japan, with quite high wages but relatively low labor militancy, recorded spectacular annual increases in manufacturing output in the 1960s. Its manufacturing output also expanded impressively in the 1970s, but with a smaller labor force. Several newly industrializing countries equaled or exceeded Japan's annual growth rate in manufacturing output from 1960 to 1993. They also registered prodigious increases in manufacturing employment: 47% in South Korea between 1974 and 1996, 93% in Taiwan, 83% in Hong Kong, and 135% in Malaysia and Singapore.

In Africa and South Asia, the growth of manufacturing has been slower. One or two countries, such as Bangladesh and the Ivory Coast, have achieved vigorous growth rates on small manufacturing bases. On

The city-state of Singapore, which lacks space, minerals, materials, food, and energy, depends on global demand conditions and trade for its economic growth. Its major trading partners are Japan, the United States, and the European Union. Since the mid-1980s, the dampened international economic environment adversely affected Singapore's manufacturing sector and trade. For the 1990s and beyond, the Singaporean government is depending on the growth of "brain industries"—on the ability of its 2.7 million people to sell their skills and services to the rest of their region and the world. *(Photo: Dale Boyer, Photo Researchers, Inc.)*

the other hand, several countries, such as Tanzania and Congo (formerly Zaire), have registered reductions in their manufacturing output.

What conclusions can be drawn from this review of changes in the distribution of manufacturing output and employment? Much depends on your perspective. The traditional view is that deindustrialization in some places and industrialization in others are mirror images of each other. Industrial growth and decline are offsetting tendencies, representing a zero-sum, or even a positive, global game. The shift of production processes from the industrial heartland to the periphery releases a skilled labor force for more sophisticated forms of production in developed countries and allows labor in the developing countries to move from relatively unproductive employment to more highly productive employment in industry. The shift may lead to some transitional unemployment, but job losses in the industrial heartland are of little significance compared with the enormous rewards attached to a global reallocation of production.

Between 1974 and 1995, the advanced industrial countries lost 20 million jobs, whereas the newly industrializing countries gained 16 million jobs. Jobs lost in the advanced industrial countries paid from $5 to $15 an hour, but those gained in the newly industrializing countries paid only $2.50 per hour or less (Table 8.4). The gains from expansion in the newly industrializing countries were more than offset by the losses in the advanced industrial countries. Indeed, the shift led to lower global wage shares that may contribute to stagnation.

Nomadic capital, although it may serve individual company interests, can be socially inefficient. Those who hire labor control the work process, and distribution is always in favor of those who control the production location. Corporate allocation of production and investment is guided primarily by profitability concerns, where profitability is determined by the price of labor and the amount of work that can be extracted at

this price. Nomadic capital can also be socially inefficient because giant corporations are rarely faced with the full social costs of their locational decisions. Shifting production from country to country means that the advanced industrial countries must absorb not only most of the social costs of communities that are now abandoned because they can no longer be industrially competitive, but also the costs of the social infrastructure required by the newly industrializing countries. Some geographers view locational change as socially inefficient in a world dominated by giant firms with the power to set the terms under which they operate.

Globalization Shifts in the Textile and Clothing Industries

TEXTILE MANUFACTURE

Textile employment in the United States in 1987 was 831,000 workers, and in Japan, 670,000 workers. Since the 1960s, however, the pattern of textile employment and production has shifted dramatically from the developed nations to the developing nations of Eastern Europe and Asia (Figure 9.7). Employment in textile manufacture is strongest in East Asia. China has the most employees, followed by Russia and India.

Peter Dicken (1992), in his book *Global Shift: The Internationalization of Economic Activity*, documented the broad production changes in textile manufacturing between 1972 and 1987. Textile production between 1980 and 1987 rose by 6% worldwide and even increased by 12% in North America. However, Europe's production decreased overall, with the notable exception of Italy, which increased its yearly productivity and exports in textiles 2.5% between 1978 and 1987. France, Germany, and the Netherlands all experienced reduced productivity, as did the United Kingdom and Japan. In the developing nations, especially in Asia, however, production during the same period increased 18%. Asia, India, South Korea, the Philippines, and Indonesia dramatically increased their production, and Latin America, Chile, Colombia, Peru, and Venezuela did likewise. Among the eastern European nations that improved textile production were Bulgaria, Czechoslovakia, and Romania (Figure 9.7).

The Jackie Kennedy Doll in preparation. As Mexico recovers from its peso crisis and devaluation, which overshadowed any gains from NAFTA, it is drawing billions of dollars worth of investment in appliances, electronics, and textiles that once would have gone to low-cost Asian sites. Economic growth progressed at 5% in 1997, largely on the basis of exports to the United States, but the recovery has not trickled down to most Mexicans. Wages are still falling in real terms, and the average Mexican feels pinched by the government's tight anti-inflation policy. (*Photo: Paul Fusco, Magnum Photos, Inc.*)

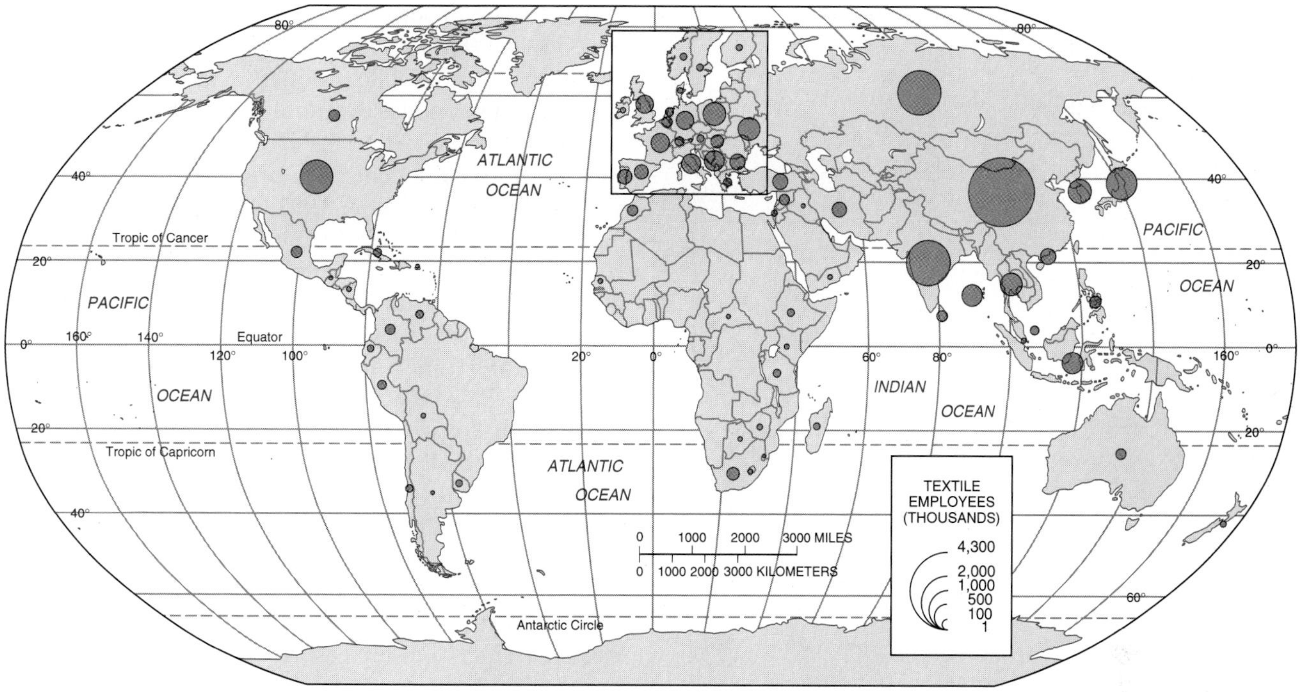

FIGURE 9.7

World distribution of textile manufacturing employment. China, the Soviet Union, and India, followed by the United States, have the largest employment in textile manufacturing. Textile employment in 1990 in the United States totaled 831,000 workers. Japan and Europe comprise the remaining centers of textile employment.

CLOTHING MANUFACTURE

Textile manufacture is the creation of cloth and fabric, whereas clothing manufacture uses textiles to produce wearing apparel. In 1987, the largest concentration of employment in clothing manufacturing was in the Soviet Union, and the next largest was in the United States. A third cluster, led by Japan, existed in East Asia. Clothing manufacture is still an important industry in western Europe, especially in Germany, France, the United Kingdom, Italy, and Poland; each has between 150,000 and 250,000 employees in the industry. For developing countries in 1987, Hong Kong and the Far East was the leading employment center of clothing manufacturers, with 282,000 employees in the industry. South Korea, the Philippines, Malaysia, and Singapore were also significant contributors to worldwide clothing manufacture.

Table 9.3 shows the changes in production in clothing manufacture between 1972 and 1987. During this period, the developed market economies fared rather poorly, while the developing market economies showed a much stronger increase, substantially stronger than the global shift in textile manufacture. For example, Germany's production fell to 77% of its 1980 production level, the Netherlands' production fell to 75%, and France's to 90%. North America, Canada, and the United Kingdom were somewhat more productive. However, the developing market economies—especially

of India, South Korea, Singapore, Malaysia, the Philippines, Venezuela, the Dominican Republic, Cyprus, Israel, and the formerly centrally planned economies of Eastern Europe—fared extremely well. The most rapid increase in clothing manufacture included Romania, Poland, and the Soviet Union, overall averaging 115% of the 1980 production figures.

Globalization Shifts in the Automobile Industry

Peter Dicken (1992) stated that more than 4 million workers are directly employed in car and motor truck manufacture throughout the world, and that twice as many are producing the materials and components that are used in the final assembly of automobiles and trucks. Although many people believe that the semiconductor and computer-chip industry has supplanted the automobile industry as the world's most important type of production, the automobile industry is more important to the world economy than it appears. "Twice in this century it has changed the most fundamental idea of how we make things. And how we make things dictates not only how we work, but what we buy and how we think and the way we live" (Womack, Jones, and Roos, 1990, p. 11). If we add the number of employees engaged in the direct manufacture of automobiles and components to those involved in sales and service, we

◈

T a b l e 9.3

Globalization of Clothing Manufacture, 1972–1992

	(1980 = 100)				
	1972	*1978*	*1983*	*1987*	*1992*
Developed Market					
Economies	101	104	94	94	92
North America	92	105	91	94	88
European community	112	105	96	94	73
Japan	—	102	99	98	97
Developing Market					
Economies	68	92	106	124	152
Asia	57	96	126	155	145
India	—	104	126	123	162
South Korea	—	102	139	206	225
Singapore	—	77	96	139	165
Malaysia	—	98	—	170	213
Philippines	—	54	—	367	380
Latin America					
Venezuela	—	87	—	214	210
Dominican Republic	—	70	—	156	160
Mediterranean					
Cyprus	—	89	—	136	140
Israel	—	96	—	129	131
Centrally Planned					
Economies	67	92	105	115	99
World	83	98	99	105	92.5

— No data available.

Source: United Nations, Industrial Statistics Yearbook, *1990, 1991, 1992, 1995, Dicken, 1992, p. 241.*

find a total world employment of 25 million in 1995, not far short of the entire population of Canada.

The automobile industry, more than any other, comprises giant transnational corporations. In no other industry in the world can so few companies dominate the world scene as can the automobile industry. For example, the world's 10 leading automobile manufacturers produce nearly 80% of the world's automobiles (Table 9.4). Each of these, from General Motors and Ford to the smallest automobile producer, has foreign assembly plants in other countries. Many have full-blown vertically integrated manufacturing operations, where all parts in the final assembly are foreign supplied. For example, the Ford Motor Company, the second largest automobile producer in the world, with an output of more than 4,500,000 cars in 1993, produces almost 60% of its cars outside the United States.

AUTOMOBILE MANUFACTURE

Worldwide automobile production has been rapidly increasing. Between 1960 and 1991, there was a world-wide increase of 358%, and 46,500,000 cars were produced in 1991. Figure 9.8 shows the 1991 world distribution of automobile production and assembly.

Three major nodes of automobile production exist—Japan, the United States, and Western Europe—with smaller production centers in Brazil, the former Soviet Union, and Australia. In 1991, Europe accounted for 32% of the world's automobiles; Japan, 23%; and the United States, 19%. Collectively, the other regions accounted for 21%. Table 9.5 shows the growth of automobile output by major producing countries between 1960 and 1991. In sum, the three developed regions of the world, East Asia, North America, and Europe, accounted for 86% of the automobiles produced. Unlike the trend in the textile and clothing industries, the developed countries clearly cornered the market in automobile manufacture. Only a few developing economies have shown a significant increase in automobile assembly, but not in their full-scale production.

The most dramatic shift in the automobile industry was the tremendous increase in Japan's productivity between 1960 and 1991. In 1960, Japan produced 165,000 cars; by 1991, this figure had increased to over 13 million, more than one-quarter of the world's total output, completely surpassing U.S. output. In 1960, the United States produced more than half the world's automobiles, but by 1991, only 18%. Great Britain produced more than 10% of the world's output in 1960 but less than 3.3% in 1991. Germany and France, however, continued to do well in automobile productivity, although their world share, 9.5% for France and 13% for Germany, declined 2% each during the 30-year span.

Globalization of Microelectronics

Microelectronic technology is the dominant technology of the later 20th century, transforming all branches of the economy and many aspects of society.

The radio was invented and produced as early as 1901, giving the first indications of an electronics industry, but the modern electronics industry was not born until the transistor was built in the United States in 1948 by Bell Telephone Laboratories. The transistor supplanted the vacuum tube, which was used in most radios, televisions, and other electronic instruments. The microelectronic transistor was a solid-state device made from silicon and acted as a semiconductor of electric current. By 1960, the *integrated circuit* was produced, which was a quantum improvement because transistors could be connected on a single small silicon chip. By the early 1970s, a computer so tiny that it could fit on a silicon chip the size of a fingernail came into production. Thus, the *microprocessor*, which could do the work of a roomful of vacuum tubes, was born. Increasing power and miniaturization progressed, and

❖

Table 9.4
World Auto Production—1992

Rank	1992 Passenger Cars	1992 Commercial Vehicles	1992 Total	World Share %	% Produced Outside Home Country
1. General Motors—U.S.	4,968,659	1,666,076	6,634,735	14.4	44
2. Ford—U.S.	3,452,039	1,686,321	5,138,360	11.3	46
3. Toyota—Japan	3,597,179	914,040	4,511,219	9.8	11
4. Volkswagen—W. Germany	2,921,481	166,952	3,088,433	6.5	30
5. Nissan—Japan	2,333,276	692,483	3,025,759	6.3	20
6. PSA—France	2,257,454	209,773	2,467,227	4.6	20
7. Renault—France	1,705,821	298,416	2,004,237	4.3	21
8. Honda—Japan	1,765,403	143,361	1,908,764	4.1	23
9. Fiat—Italy	1,636,838	261,717	1,898,555	4.1	7
10. Chrysler—U.S.	660,200	1,014,089	1,674,289	3.7	33
11. Mitsubishi—Japan	1,103,606	491,469	1,595,075	3.5	12
12. Mazda—Japan	1,250,714	300,541	1,551,255	3.3	7
13. Suzuki—Japan	542,128	370,650	912,778	2.0	7
14. Daimler-Benz—W. Germany	575,547	285,651	861,198	1.9	12
15. Hyundai—S. Korea	669,551	125,740	795,291	1.8	4
16. VAZ—C.I.S.	675,000	12,000	687,000	1.5	0
17. Daihatsu—Japan	420,313	250,168	670,481	1.5	0
18. Fuji—Japan	388,052	256,578	644,630	1.4	18
19. BMW—W. Germany	536,003	0	536,003	1.2	0
20. Isuzu—Japan	130,447	340,503	470,950	1.1	0
TOTAL 20 MANUFACTURERS	31,589,711	9,486,528	41,076,239		
OTHERS	698,879	1,462,882	2,161,761		
TOTAL PRODUCTION	32,288,590	10,949,410	43,238,000		

Source: Data compiled by AAMA from various sources. Information was obtained from published reports issued by various vehicle associations outside the U.S. and from a number of other sources considered reliable. Because of the numerous complex factors involved in determining this worldwide ranking, AAMA does not assume responsibility for the above classification. World Motor Vehicle Data, *American Automobile Manufacturing Association, Detroit, 1993.*

at the same time, new applications for the electronics industry were discovered, including calculators, electronic typewriters, computers, industrial robots, aircraft-guidance systems, and combat systems. New discoveries were applied to automobile construction for guidance, safety, speed, and fuel regulation. An entire new range of consumer electronics also became available for home and business use. The electronics industry, like textiles, steel, and automobiles before it, has come to be regarded as the touchstone of industrial success. Hence all governments in the developed market economies, as well as those in the more industrialized developing countries, operate substantial support programs for the electronics industry, particularly microprocessors and computers.

SEMICONDUCTOR MANUFACTURE

For nearly two decades, from the 1960s through the 1970s, the United States dominated the field of semiconductor manufacture. However, by 1990, Japan had taken over this role. World production of active electronic components, which includes semiconductors, integrated circuits, and microprocessors, is shown in Figure 9.9. The field is dominated by Japan and the United States, with other significant production in Western Europe and Southeast Asia. In 1995, Japan accounted for 42% of the world production of semiconductors; the United States, 25%; and Europe, 13%. Of the European share, Germany produced 31%; France, 19%; and the United Kingdom, 16%. In Southeast Asia, South Korea, the Philippines, Malaysia, Taiwan, Thailand, and Hong Kong were significant manufacturers.

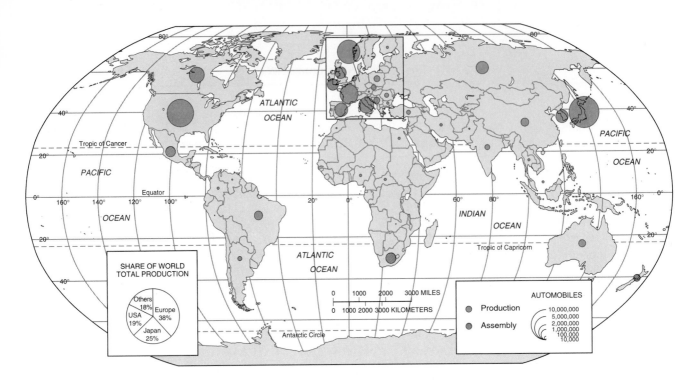

FIGURE 9.8

World distribution of automobile production and assembly. In 1991, Japan produced 13 million automobiles, which was 28% of the world's total output. In 1960, the United States produced half of the world's total automobiles, but by 1991, that proportion had dropped to 19%. The United Kingdom fell in production of automobiles during the same period, from 10.5% to 3.3% of the world total. In Europe, Germany and France comprise the two largest producers, followed by Italy. *(Source:* World Motor Vehicle Data, *American Automobile Manufacturing Association, Detroit, Michigan, 1993)*

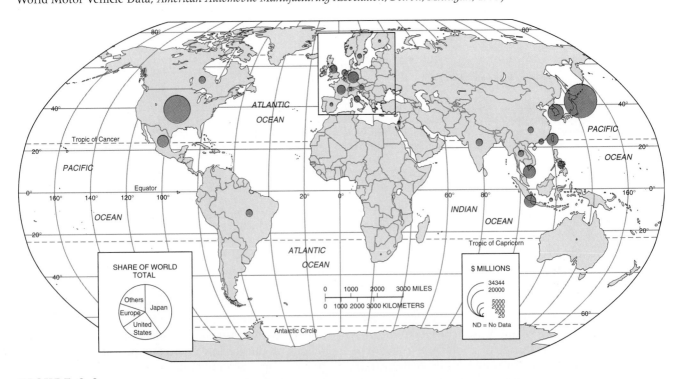

FIGURE 9.9

World production of electronic components, including semiconductors, integrated circuits, and microprocessors. In 1989, Japan produced 42% of the world's total electronic components, followed by the United States with 26%. Europe produced 12% of the world's total. Other leading producers included West Germany, France, and the United Kingdom, with 31%, 19%, and 16% of the European total, respectively. Global shifts in the electronics industry have been to the Pacific Rim with recent major centers of production developing in South Korea, Malaysia, Taiwan, Philippines, Thailand, and Hong Kong.

T a b l e 9.5
Growth of Automobile Output by Major Producing Countries, 1960–1991

Country	1960 Production (000 Units)	1960 World Share (%)	1991 Production (000 Units)	1991 World Share (%)	Average Annual % Change 1960–1991
Europe					
Belgium	194	1.5	337	0.7	+2.5
France	1175	9.0	3610	7.8	+7.2
West Germany	1817	14.0	5015	10.9	+6.1
Italy	596	4.6	1878	4.1	+7.5
United Kingdom	1353	10.4	1454	3.3	−0.3
Spain	43	0.3	2081	4.3	+160
Sweden	108	0.8	345	0.9	+8.4
North America					
United States	6675	51.4	8810	19.0	+1.1
Canada	323	2.5	1873	3.9	+16.5
Asia					
Japan	165	1.3	13,245	28	+273
South Korea	—	—	1497	3.2	—
Latin America					
Argentina	30	1.0	138	0.4	+12.6
Brazil	38	0.3	960	2.2	+83
Mexico	28	0.2	989	2.3	+118
Australia	—		293	1.0	—
USSR/CIS	139	1.1	2012	4.3	+46
Czechoslovakia	—		206	0.5	—
Poland	—		193	0.7	—
World	12,999	100.0	46,496	100.0	+8.9

—Data unavailable.
Source: World Motor Vehicle Data, *American Automobile Manufacturing Association, Detroit, 1993.*

The most important component of the semiconductor industry is computer memory, *RAM* (random access memory), production. Although the United States produced 100% of the world's total output in 1974, by 1988 it produced only approximately 20%, and Japan had claimed 75%. Similar to the shift in the automobile industry, there has been a tremendous global shift in the direction of East Asia, primarily to Japan, in RAM production.

CONSUMER ELECTRONICS MANUFACTURE
The world manufacture of consumer electronics is much more widely spread than that of the semiconductor industry. Although in the semiconductor industry, the United States, Japan, and Western Europe account for 80% of the total world production, in the consumer electronics industry, these three regions account for only 44% of television manufacture, for example. Third World countries, especially East Asian and Southeast Asian countries, are more involved in the consumer electronics industries. China emerges as the largest single television receiver manufacturer and produces 19% of the world's total output. In addition, South Korea and Japan each produce 14% of the world's televisions. Singapore and Malaysia are also significant producers. The United States' share of the world total was 11% in 1987, having fallen from 15% just 9 years earlier. As in the case of automobile and semiconductor manufacture, there has been a global shift from the developed nations to the Far East. Much of the television production that formerly occurred in the United States, Germany, and the United Kingdom now takes place in Japan. Outside Asia and North America, Brazil produces the most television receivers and manufactures 75% of the units used in Latin America.

This global shift is shown in Table 9.6. Between 1978 and 1990, world production increased by 108%, with a U.S increase of 40%, a European increase of 8.8%, and

✦

Table 9.6

Global Shift in Television Receiver Production by World Region, 1978–1990

Country/Region	Production (000 Units)		% Change 1978–1990	Share of World Production (%)	
	1978	1990		1978	1990
Asia (excluding former USSR)	21,175	74,802	+253	35.0	60
Japan	13,116	14,730	+12.6	21.7	11.7
South Korea	4826	17,503	+263	8.0	14
China	517	30,123	+5900	0.9	24
Malaysia	150	1421	+833	0.3	1.1
Singapore	725	3001	+313	1.2	2.4
Europe (excluding former USSR)	17,656	19,550	+8.8	29.1	18.9
West Germany	4391	3220	−22.5	7.3	3.3
United Kingdom	2417	2824	+21.0	4.0	2.8
Italy	2172	2321	+2.8	3.6	2.3
France	2101	2011	+3.9	3.5	2.0
United States	9309	13,203	+40	15.4	10.6
World total	60,592	125,161	+108	100.0	100.0

Source: United Nations, 1993, Industrial Statistics Yearbook, *1990, New York; United Nations.*

an Asian increase of 253%. Although Japan increased production only 12.6%, China, between 1978 and 1990, increased production a whopping 5,900%. Malaysia's output increased 833%, South Korea's, 263%, and Singapore's, 313%.

THE RELOCATION OF THE AMERICAN MANUFACTURING INDUSTRY

The United States experienced industrial devolution in the 1970s and 1980s, a period during which its share of world manufacturing output decreased. This points to a more rapid growth of manufacturing output in other countries, especially Japan. Can the relative decline of the American manufacturing economy be attributed to Japanese expansionism? Certainly Japan has increased its share of world industrial production and exports, but a different picture emerges if changes in world sales are classified by the nationality of the parent company. Although Japanese industrial capital made gains at the expense of U.S. capital in the 1960s, almost no further advances were achieved in the 1970s and 1980s. Thus, deindustrialization within the United States was occurring at a time when American capital was either increasing or maintaining its share of the world economy—at a time when American corporations were reacting to the prolonged economic crisis. Inside the United States, corporate profit rates were

declining. The advantages that promoted rapid capital accumulation in the old manufacturing areas were giving way to contradictions to high profits—rising real wages and obsolete infrastructure. As profit rates declined, corporations switched capital in space. The effect of locational change has been most severe in the American Manufacturing Belt.

The American Manufacturing Belt

North America has numerous manufacturing regions (Figure 9.2). By far the largest is the North American Manufacturing Belt, which accounts for about 53% of the manufacturing capacity of the United States and Canada. The U.S. portion of this region is called the *American Manufacturing Belt,* the historic heartland of the nation. The belt extends from Boston westward through upstate New York, southern Michigan, and southeastern Wisconsin. At Milwaukee it turns south to St. Louis, then extends eastward along the Ohio River valley to Washington, D.C. This great rectangle encompasses more than 10 districts, each with its own specialties that reflect the influences of markets, materials, labor, power, and historical forces.

The first major factories in the belt—the textile mills of the 1830s and 1840s—clustered along the rivers of southern New England. When coal replaced water as a power source between 1850 and 1870, and when railroads integrated the belt, factories were freed from the riverbanks. Industrialists began to pursue their profits in towns and cities.

Between 1850 and 1870, many urban areas of all sizes enjoyed rapidly expanding industrial production. But after 1870, manufacturing concentrated in a few large cities. Manufacturing employment in New York City, Philadelphia, and Chicago soared more than 200% between 1870 and 1900. The 10 largest industrial cities increased their share of national value added in manufacturing from less than 25% to almost 40% between 1860 and 1900 (Pred, 1966). Why did metropolitan complexes draw such a great proportion of manufacturing activity?

The orthodox view is that factories concentrated in cities such as Baltimore, Chicago, Cleveland, Cincinnati, Philadelphia, and Pittsburgh for a combination of the following reasons: (1) They could be near large labor pools; (2) they could secure easy railroad and waterway access to major resource deposits, such as the Appalachian coalfields and the Lake Superior–area iron mines; (3) they could be near industrial suppliers of machines and other intermediate products; and (4) they could be near major markets for finished goods. In other words, *agglomeration economies* accounted for the concentration of industrial activity in the belt.

The highly concentrated pattern of industrial production in the belt served the nation well for almost 100 years—roughly the century between 1870 and 1970. *Inertia*, the immobility of the investment forces and social relations, ensured considerable locational stability. Inertia was particularly pronounced in the capital-intensive steel industry.

However, cracks in the accumulation regime appeared as early as the late 19th century. Contradictions erupted. Labor unrest intensified. After 1885, the number of workers involved in strikes increased rapidly, with many of the most bitter strikes occurring in the largest manufacturing complexes. Gradually, owners lost power to labor—a power that enabled labor to negotiate higher wages than were paid previously or elsewhere, to organize high levels of unionization and extract good working conditions, and to command progressive welfare policies. By the early 20th century, the dense centralization of industrial workers in inner-city areas backfired on factory owners.

As labor control subsided, manufacturing started to move out of center cities to the suburbs. Between 1899 and 1909, center-city manufacturing employment increased by 40%, whereas outer-city manufacturing employment increased by nearly 100%. After the late 1920s, the use of the truck for freight transportation accelerated the movement of manufacturing to the suburbs. But as transportation costs equalized across the nation, even the suburbs of the older manufacturing belt cities were unable to compete with more agreeable labor environments in the South and the West. The 1960s marked the start of the steady gain of manufacturing employment in the South, parts of the Southwest, the Mountain states, and the Dakotas. Since the recession of the early 1980s, however, there has been evidence that these areas where class conflict is low are being bypassed in favor of even cheaper labor regions in Mexico and East Asia.

U.S Domestic Movement of Manufacturing

The locational change of manufacturing in the United States was particularly strong in the 1970s and 1980s. Virtually all states in the American Manufacturing Belt experienced manufacturing job loss, and virtually all states in the South and the West registered manufacturing job gains (Figure 9.10). The migration of employment from areas of high labor costs to areas where the labor costs were less saved U.S.–controlled companies billions of dollars. From 1960 to 1980, roughly 1.7 million jobs shifted from states with high labor costs to states with low labor costs.

The expansion of manufacturing in the West is not immediately explained by labor costs. California is characterized as a state of high labor costs, yet it has registered substantial increases in manufacturing employment. In California, class struggle loses its primary determining effect to physical and environmental factors, the role of the state (defense spending), and the dimension of consciousness (the "California image").

The West Coast manufacturing district does, however, represent an outstanding example of industrial restructuring in response to economic crisis and labor unrest. The Los Angeles–San Diego district has been extremely successful. Since the 1960s, the district has shed much of its traditional, highly unionized heavy industry, such as steel and rubber. At the same time, it has attracted a cluster of high-tech industries and associated services, centered around electronics and aerospace and tied strongly to the now depressed defense and military contracts from the U.S. government. Added to the combination of Frostbelt-like deindustrialization and Sunbelt industrialization has been the vigorous growth of "peripheralized" manufacturing, which resembles the industrialization of Hong Kong and Singapore and depends on a highly controllable supply of cheap, typically immigrant or female labor from Mexico and the Far East.

Meanwhile, industrial restructuring continues to be a painful process throughout much of the American Manufacturing Belt. The region contends with problems of obsolescence and reduced productivity, especially in such leading industries as steel, automobile manufacturing, and shipbuilding. It contends with inner-city areas littered with closed factories, bankrupt businesses, and struggling blue-collar neighborhoods.

The effect of disinvestment on workers and their communities has been devastating.

Victims of plant closings sometimes lose not only their current incomes, but their total accumulated assets as well. When savings run out, people lose their ability to respond to life crises. Although job loss respects neither educational attainment nor occupational status, some groups are more vulnerable than others. Because blacks are concentrated in areas where plant closings have been most pronounced, this group has been especially hard hit (see Chapter 8: Business Process Reengineering).

Although the widespread manufacturing decline has produced a lasting effect on people and communities in the manufacturing belt, all is not lost in the region. Already there are attempts to respond to the economic crisis. Old industrial cities such as Pittsburgh are building new bases for employment, and so too are the major urban complexes of Megalopolis. Southern New England stands out as a good example. This region, which suffered higher than average unemployment rates throughout much of the post–World War II period, now has the lowest unemployment rate in the belt. A new round of industrial expansion is taking place with a disciplined pool of highly skilled and unskilled workers. The industrial revival of southern New England emphasizes high-value products—electronic equipment, electrical machinery, firearms, and tools. A current worry, however, is the permanence of the revival. In the early 1990s, high-tech firms in Silicon Valley East were affected by the sluggish national economy and were forced to lay off workers, so doubt was cast on the "Massachusetts miracle."

Movement of the North American Automobile-Manufacturing Industry

In recent years, there has been a marked shift in the locations of automobile assembly in the United States. The most important reason for this shift has been to minimize the transportation cost of the finished product to the market. Another reason is the change in consumer preference, in that more models of each car type, such as Ford, are in demand, thus fewer plants are needed for each model (Figure 9.10).

In 1955, General Motors assembled identical Chevrolets at 10 cities to supply local consumers with cars (Rubenstein, 1996) (Figure 9.11). Because the cars were fairly perishable and expensive to transport, distribution cities were located across the country, evenly proportional to the population. Ford had a similar arrangement for its production of standardized, low-priced models. Top-of-the-line luxury cars, such as Lincoln and Cadillac, were still assembled in Detroit.

Beginning, however, in the 1960s, a variety of models of each particular automobile began to appear, ranging from subcompacts of 150 inches long to full-size vehicles of more than 220 inches long. Part of this diversity offered by U.S. manufacturers was a result of Japanese and European imports. Each time a successful import penetrated the market, the Detroit automobile manufacturers attempted to challenge the competition with another model in the same product line.

Soon consumers expected a wide choice in automobiles. With this increased demand for more models, major automobile producers began to build only one or two plants for each model, and to distribute the vehicles from these points to the rest of the country and Canada. Because fewer assembly plants were needed for each model, new assembly plants were located near the center of the country, again to minimize the distribution costs (Figure 9.12). Consequently, from 1970 to 1990, many coastal automobile assembly plants were closed.

James Rubenstein (1996) identified two types of automobile plants in existence today. Approximately 70 *automobile- and truck-assembly plants* exist in North America, as well as several thousand *component plants.* The component plants manufacture parts required in the assembly of the vehicle. Because of the increasing importance of *just-in-time inventory* and delivery, component plants locate close to assembly operations.

Just-in-time systems reduce not only labor and machine inefficiency, but also the amount of capital tied up in large inventories. Just-in-time systems deliver parts to the manufacturing plant only hours or days before they are used in the manufacturing process. Computerized automatic order and billing operate in real time and online between suppliers and final assembly plants so that parts needs can be filled immediately. Clustering of parts suppliers around automobile- and truck-assembly plants suggests a strong move to just-in-time delivery systems. In contrast, Japanese companies have located along interstates 65 and 75 from Tennessee, Kentucky, Indiana, and Ohio, to Michigan (Figure 9.13). Rubenstein (1996) determined that Japanese companies avoid large cities and those traditionally associated with automobile assembly because of the strong unions and high fixed operation costs in these areas.

The International Movement of American Manufacturing

The relocation of manufacturing within the United States is only one aspect of a wider dispersal of American manufacturing capital. Foreign direct investment by American enterprises was established as early as the end of the 19th century. But only since World

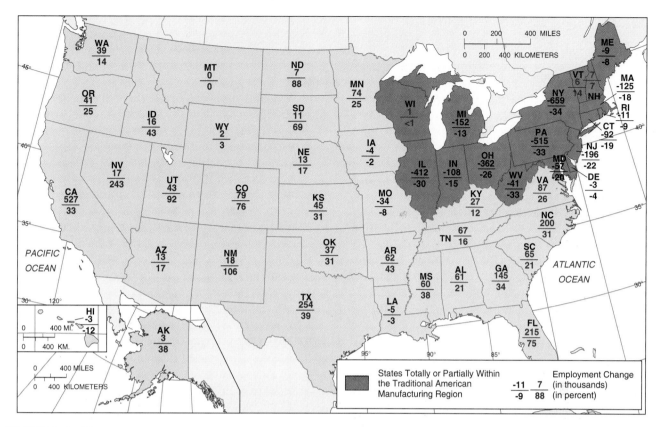

FIGURE 9.10

Employment shifts in manufacturing in the United States between 1967 and 1987. The top number shows employment change in thousands for the 20-year period, while the bottom number shows employment change as a percent for the 20-year period.

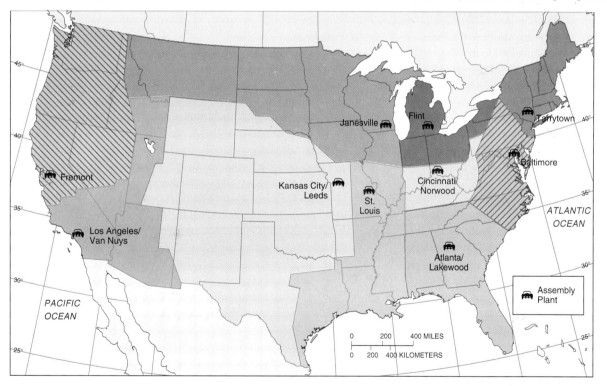

FIGURE 9.11

General Motors Chevrolet assembly plants in 1955. The distribution of 10 assembly plants across the United States helped Chevrolet minimize the cost of distributing the bulky products to consumers in each region. (See Color Insert 2 for more illustrative map.) Beginning in the early 1970s, automakers have located new assembly plants in the interior of the United States, rather than near coastal population concentrations, which had been the preferred locational pattern. Automakers now operate specialized assembly plants that build single models for distribution throughout the United States because of an increase in the variety of models produced in North America.

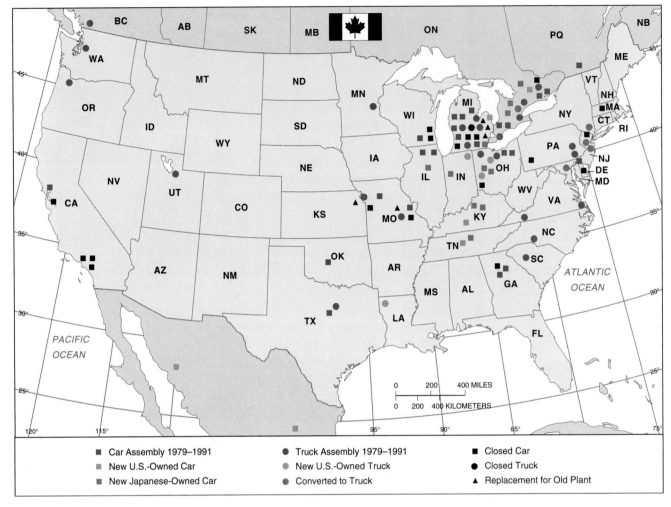

FIGURE 9.12

The distribution of auto and truck assembly plants in North America, 1991. There has been a marked national shift from a disbursed, coastal pattern during the last 20 years to a concentrated pattern near the original automobile core area in Michigan.

War II have American enterprises become major foreign investors. The 1940s saw heavy investment in Canada and Latin America; the 1950s in Western Europe; and the 1960s and 1970s in Western Europe, Japan, the Middle East, and South and East Asia. Overall, most of this investment occurred in advanced industrial countries rather than in developing countries (Figure 9.14).

Between 1945 and 1960, most U.S.–based companies were content to produce in the old industrial districts. But by 1960, the European Common Market and Japan had become competitors. Mounting international competition and falling profit rates at home coerced American companies to decentralize not only within the United States, but also abroad. By 1980, the 500 largest U.S.–based corporations employed an international labor force almost equivalent to the size of its national labor force. From 1960 to 1980, manufacturing employment outside the United States directly controlled by U.S. corporations increased from 8.7% to 17.5% of the total, a 169% increase compared with a 20% increase within the United States.

The dispersal of manufacturing investment to foreign lands resulted in a more competitive base and in enormous savings in wages for American firms. In 1980, the annual wage savings from 1.6 million jobs opened by U.S.–based corporations in developed countries between 1960 and 1980 was $8.4 billion, and the annual savings from 1.1 million jobs opened in developing countries between 1960 and 1980 was $14.5 billion. Even more savings were achieved in the 1980s as U.S.–based companies increasingly established manufacturing operations in low-wage regions of the developing world.

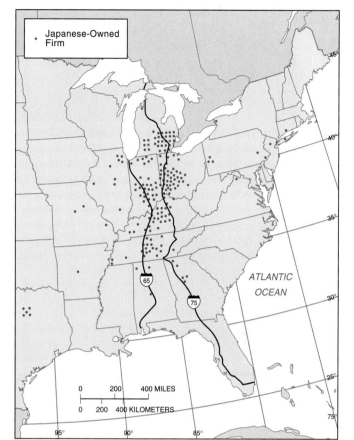

FIGURE 9.13

The distribution of Japanese-owned automobile parts manufacturing plants, 1991. This Midwestern concentration of parts plants facilitates rapid delivery to close-by automobile assembly plants. Japanese companies avoid large U.S. cities and those traditionally associated with auto assembly because of strong unions and high fixed costs of operating in these areas. Rising U.S. protectionist sentiment in the wake of growing trade deficits has obliged foreign MNEs to safeguard their shares in this crucial market by establishing production inside U.S. borders. Governments at all levels in the United States provide a generally hospitable investment climate for foreign companies. Not only does the federal government place few restrictions on foreign investors, but state and local governments compete vigorously for investments by offering all manner of tax and financial incentives to foreign firms.

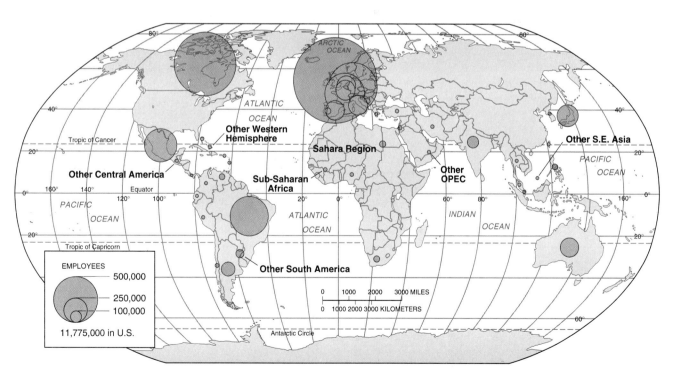

FIGURE 9.14

World employment in manufacturing of U.S.–based multinational corporations in 1992.

THE INDUSTRIALIZATION OF JAPAN

Japan's record of economic achievement, now tarnished by recession at home in Japan and worldwide, has no equal among advanced industrial countries in the post–World War II period. Between 1970 and 1984, when industrial production declined by 4% in Great Britain and increased by 48% in the United States, Japan's industrial output soared by 162% (World Bank, 1987). Unemployment in 1992, which registered 13.5% in Great Britain and 7.7% in the United States, was a modest 3.5% in Japan. When, in the 1980s, the U.S. trade deficit mushroomed, Japan's trade surplus mounted. In that decade, Japan took America's place as chief creditor, buying stocks, bonds, and real estate around the world. Given Japan's increasingly powerful economy, it is timely to answer the following questions: What is the basis for Japan's astonishing industrial record? What price is Japan paying in its bid to become the world's foremost industrial nation? Is Japan the working model for industrial success?

What gave Japan, an island nation that is slightly smaller than California, the opportunity to become an industrial juggernaut? What part of the answer lies in its raw-material base? Compared with the United States and Great Britain—countries with the physical resources to sustain an industrial revolution—Japan is much less well off. Except for coal deposits in Kyushu and Hokkaido, which provided a stimulus for 19th-

century industrial development, Japan is practically devoid of significant raw materials for major industry. It has depended in large measure on imported raw materials for its recent industrial growth.

Japan may lack physical resources, but human resources for industrial development are not scarce. There are more than 120 million Japanese, mostly crammed into an urban-industrial core that extends from Tokyo on the east to Shimonoseki at the western end of Honshu (Figure 9.15). The Japanese regard the homogeneity that characterizes their population as a major contributor to their nation's industrial success. Within this homogeneity, Confucianism and Buddhism, acquired from China and in place for more than a thousand years, have instilled such traits as self-denial, devotion to one's superiors, and the prevalence of group interests over individual interests. These social attributes lend Japan a high degree of *national consensus*. When a strong work ethic combines with a high level of collective commitment and incredible government support, it produces a work culture in which everyone, from the top to the bottom, knows his or her place and can pull in the same direction. A chilling degree of national consensus was apparent after World War II, when the Japanese set about the task of rebuilding their shattered economy.

Although permanent workers in large firms are well paid, especially when viewed in relation to part-time workers in small and medium-sized firms, the relative share of the product going to labor is lower in Japan than in any other advanced industrial country. Savings and profits are directed by the state toward whatever goals are set forth by a unique collaborative partnership between MITI—Japan's Ministry of International Trade and Industry—and private enterprise. This marriage between government and private enterprise, helping to make Japan the envy of the West, is sometimes called *Japan Incorporated.* The days of the successful relationship between MITI and industry, however, may be numbered. MITI has less controlling power than it once had. Japanese firms no longer need the watchful eye of MITI or much of its money. For example, Hitachi spent about $2.2 billion on R&D in 1988, or about 50% more than MITI's total R&D budget.

Substantial change may also be looming on Japan's labor front. The wealth generated by the Japanese system in the 1960s and 1970s has produced a workforce that is gingerly experimenting with traditions and customs. Japan's young people still adhere to the ideal of consensus, but they are unwilling to sacrifice their individuality. They are not satisfied with simply working hard. They prefer to work less and take all the vacations that their bosses offer. In the 1990s, more and more clashes will occur between the young and the old over the commitment of the individual to the ideals that propelled postwar rebuilding.

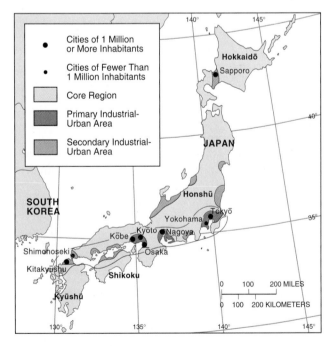

FIGURE 9.15

Japan's core region and selected cities. *(Source: Williams, 1989a, p. 331)*

The Problems of Japan Incorporated

In the postwar period, Japan set about the task of becoming a potent economic force. All other considerations were subordinated to that one overriding good. The drive for economic growth was at the expense of social welfare, the environment, and international relations. Japan is beginning to pay the price for its industrial achievement.

INTERNATIONAL RELATIONS

In the 1980s, the gap between Japan's imports and exports created tensions with its trading partners. As a result, Japan found its markets in key exports restricted. Protectionist tendencies have been sidestepped to some extent by foreign direct investment. Japan has invested heavily in big projects in the advanced industrial countries. It has also established numerous joint ventures that reduce the capital and political risks for Japanese companies. Examples include not only the link between American steel giant USX and Japan's Kobe Steel to manufacture tubular steel for automakers producing vehicles in the United States, but also the link between Boeing and Fuji Heavy Industries, Kawasaki, and Mitsubishi to build commercial aircraft.

THE DUAL ECONOMY

The most pressing Japanese problem is its dual economy. At the top are a handful of successful international corporations that thrive and change. Under MITI's guidance, Japan has relocated industries such as steelmaking and shipbuilding "offshore" in the newly industrializing countries (NICs), where labor costs are lower. Now, countries such as South Korea are developing their own higher technologies (e.g., consumer electronics). Meanwhile, Japan is challenging the United States in the newest generation of sophisticated technology. Even if they succeed, the Japanese worry about the transition to postindustrialism with less manufacturing and more stress on services. Manufacturing industry's share of both output and employment is predicted to decline by the turn of the century.

At the low end of the Japanese economy are thousands of tiny workshops. The large corporations job out parts of their production to the small firms with their underpaid and exploited workers. About 70% of Japan's labor force still works in these tiny firms. This mom-and-pop structure prevents Japan from playing its superpower role as the world's buyer of last resort, soaking up foreign goods to maintain international economic order and growth.

ENVIRONMENTAL POLLUTION

The pursuit of economic efficiency was the be-all and end-all of Japanese endeavor until the late 1960s, when the consequences of environmental pollution demanded immediate attention. The most serious aspects were air and water pollution, which killed hundreds of people who lived around the factories (Junkerman, 1987). In the 1970s, the government introduced pollution-control measures and tried to shift polluting industries out of congested areas. Still, Japan remains the most environmentally polluted country in the developed world.

REGIONAL IMBALANCE AND URBAN ILLS

Japan's uncontrolled and rapid industrial development created an unusually sharp economic divide between the core region and the rest of the country. In the late 1970s and 1980s, as efforts to overcome environmental pollution by relocating industry to less developed parts of the country started to take effect, the gap between the rich core region and the poor periphery began to narrow. Yet the regional contrast will persist throughout the 1990s.

The problems of Japan's large cities are another by-product of rapid economic growth. These problems include traffic congestion, noise, insufficient parking, accidents, air pollution, and, of course, land madness. When the Japanese government put 66 new apartments on the market in Tokyo in 1990, more than 28,000 people rushed to apply for what was considered a bargain: $300,000 for a small two-bedroom apartment. In one of the world's richest countries, most people cannot afford to buy their own homes. Housing is a great social problem, with young couples forced into 2-hour commutes into Tokyo just for the luxury of owning their own "rabbit hutches," as Japan's cramped quarters have come to be known. Although Japan has a higher gross domestic product (GDP) per capita than the United States, Canada, or most western European countries, because of scarcity and very high costs of living, its standard of living is actually less than the United States, Canada, and some European countries.

The Japanese Model

Japan's postwar industrialization, now slowed by world recession and an uncertain stock market, has nonetheless been remarkable. It has created a rich country but a poor people. Is Japan really the working model for industrial success? Even if it were, its model could not be readily adopted by other countries; it is the product of a unique culture and of a unique regional historical experience.

INDUSTRIALIZATION IN THE DEVELOPING WORLD

Deindustrialization in the Western Hemisphere in the 1970s and 1980s did not induce widespread industrialization in the developing world. In 1990, 40 countries accounted for 70% of manufacturing exports from developing countries; the top 15 alone accounted for about 60% (World Bank, 1992). Even more striking is that about one-third of all exports from the LDCs came from four Southeast Asian countries—Hong Kong, South Korea, Singapore, and Taiwan. Industrialization occurred, therefore, only in selected parts of the developing world.

Manufacturing was slowest to take hold in the poorest countries of the periphery, most of which are in Africa. It grew fastest in the NICs. These countries made a transition from an industrial strategy based on import substitution to one based on exports. The exporters can be divided into two groups. First, countries such as Mexico, Brazil, Argentina, and India have a relatively large domestic industrial base and established infrastructure. All four of these countries are primarily exporters of traditional manufactured goods—furniture, textiles, leather, and footwear—exports favored by natural-resource conditions. Second, countries such as Hong Kong, Taiwan, South Korea, and Singapore have few natural resources, small domestic markets, and little infrastructure. But by tailoring their industrial bases to world economic needs, they have become successful exporters to developed countries. These four peripheral countries emphasize exports in clothing, engineering, metal products, and light manufactures. The success of South Korea, for example, has encouraged other LDCs to adopt a similar program of export-led industrialization.

Why did import-substitution industrialization fail? What are the characteristics of export-led industrialization? What are the consequences of this strategy on economies and people? Can this form of industrialization in the periphery be sustained if the tendency toward slower economic growth in the center of the world economy continues? In the following sections, we attempt to answer these questions.

Import-Substitution Industrialization

In the post–World War II period, newly independent developing countries sought to break out of their domination by, and dependence on, developed countries. Their goal was to initiate self-expanding capitalist development through a strategy of *import-substitution industrialization*. This development strategy involved the production of domestic manufactured goods to replace imports. Only the middle classes could support a domestic market; thus, industrialization focused on luxuries and consumer durables. The small plants concentrated in existing cities, which increased regional inequalities. These "infant industries" developed behind tariff walls in order to reduce imports from developed countries, but local entrepreneurs had neither the capital nor the technology to begin their domestic industrialization. Foreign multinational corporations came to the rescue. Although projects were often joint ventures involving local capital, "indepen-

dent" development soon became *dependent industrialization* under the control of foreign capital. Many countries experienced an initial burst in manufacturing growth and a reduction in imports. But after a while, the need to purchase raw materials and capital goods and the heavy repatriation of profits to the home countries of the multinationals dissipated foreign-exchange savings.

Export-Led Industrialization

By the 1960s, it was apparent to LDC leaders that the import-substitution strategy had failed. Only countries that had made an early transition to *export-led industrialization* were able to sustain their rates of industrial growth. Once again, LDC development became strongly linked to the external market. In the past, export-oriented development had involved the export of primary commodities to developed countries. Now, export-oriented development was to be based on the production and export of manufactures.

The growth of export-led industrialization coincided with the international economic crisis of the 1970s and 1980s. It took place at a time when the demand for imports in the advanced industrial countries was growing despite the onset of a decline in their industrial bases. It was a response to the new international division of labor.

Export-oriented industrialization tends to concentrate in *export-processing zones*, where four conditions are usually met:

1. Import provisions are made for goods used in the production of items for duty-free export, and export duties are waived. There is no foreign exchange control, and there is freedom to repatriate profits.
2. Infrastructure, utilities, factory space, and warehousing are usually provided at subsidized rates.
3. Tax holidays, usually of 5 years, are offered.
4. Abundant, disciplined labor is provided at low wage rates.

The first export-processing zone was not established in the developing world; it was established in 1958 in Shannon, Ireland, with the local international airport at its core. In the late 1960s, a number of countries in East Asia began to develop export-processing zones, the first being Taiwan's Kaohsiung Export-Processing Zone, set up in 1965. By 1975, 31 zones existed in 18 countries. By the early 1980s, at least 68 zones were established in 40 principally developing countries. Most of them are in the Caribbean, in Central and Latin America, and in Asia (Figure 9.16).

Central to the growth of Third World manufacturing exports to the developed countries are multinational corporations, which establish operating systems

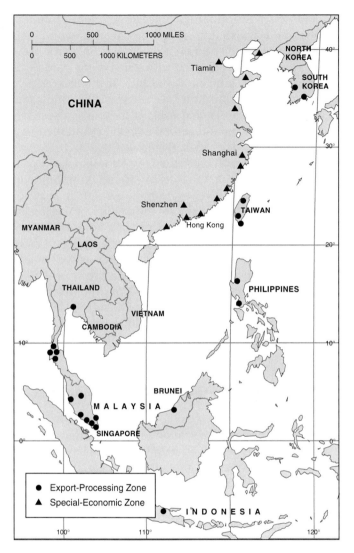

FIGURE 9.16
Export-processing zones in Asia. *(Source: Wong and Chu, 1984, p. 4)*

between locally owned companies and foreign-owned companies. The arrangement is known as *international subcontracting*, or offshore assembly and sourcing (*outsourcing*). Although numerous legal relationships exist between the multinational and the subcontractor, from wholly owned subsidiary to independent producer, the key point is that Third World exports to developed countries are part of a unified production process controlled by firms in the advanced industrial countries. For example, Sears Roebuck Company might contract with an independent firm in Hong Kong or Taiwan to produce shirts, yet Sears retains control over design specifications, advertising, and marketing. This is similar to the putting-out system in textiles that was developed in preindustrial England. With modern transport and communications, it is no more difficult for today's merchants to organize a putting-out system between New York and Hong Kong, or between Tokyo

and Seoul, than it was for the early English merchants to organize their putting-out system between London and the surrounding villages.

Consequences of Export-Led Industrialization on Women

Export-led industrialization moves work to the workers instead of workers to the work, which was the case during the long postwar boom. In some countries this form of industrialization has generated substantial employment. For example, since their establishment, export-processing zones have accounted for at least 60% of manufacturing employment expansion in Malaysia and Singapore. But, in general, the number of workers in the export-processing zones' labor forces is modest. It is unlikely that these zones employ more than a million workers worldwide.

Much of the employment in export-processing zones is in electronics and electrical assembly or in textiles. Young, unmarried women make up the largest part of the workforce in these industries—nearly 90% of zone employment in Sri Lanka, 85% in Malaysia and Taiwan, and 75% in the Philippines. Explanations for this dominance of women in the workforce vary; it is often attributed to sexual stereotyping, in which the docility, patience, manual dexterity, and visual acuity of female labor are presupposed. Of more significance is the fact that women are often paid much less than men are for the same job. Cheap labor is essential in the labor-intensive industries of the global assembly line.

According to A. Sivanandan (1987), export-led industrialization gives rise to *disorganic development*—an imposed economic system at odds with the cultural and political institutions of the people that it exploits. People produce things of no use to them. How they produce has no relation to how they formerly produced. Workers are often flung into an alien labor process that violates their customs and codes. For example, female factory workers often pay a high price for their escape from family and home production, especially in Asia, where women's family roles have been traditionally emphasized. Because of their relative independence, westernized dress, and changed lifestyles, women may be rejected by their clan and find it hard to reassimilate when they can no longer find employment on the assembly line.

Although export-oriented industrialization leads to growth in production and employment, as well as to increases in foreign exchange, it will not lead to the creation of an indigenous, self-expanding capitalist economy. The linkages between "export platforms" and local economies are minimal. Scholars who oppose Landsberg's view cite

South Korea as a shining example of a country that has completed a successful transition to industrial capitalism. But so far Korean industrial expansion has not taken place because of domestic demand. Rather, it has occurred because the Koreans have sought to increase exports and international competitiveness. This expansion is changing, partly because of the general global tendency to stagnation.

Economic stagnation in developed countries is a major concern of countries that have enjoyed tremendous success with the export-led industrialization strategy. For developed countries, where production and investment are moving out, purchasing power will be lost. The resultant spiraling down of general economic activity will choke off dependent industrialization and increase poverty and suffering for workers and peasants in LDCs.

East Asian "Miracle"

What then does it take, as Paul Kennedy (1993) asked, "to turn a 'have-not' nation into a 'have' nation? Does it simply require imitating economic techniques, or does it involve such intangibles as culture, social structure, and attitudes toward foreign practices?" (p. 33). Who is marching successfully forward to development?

Most successful have been the trading states of East Asia. South Korea, Taiwan, Singapore, and Hong Kong have followed the path pioneered by nearby Japan. Malaysia, Thailand, Indonesia, and the Pacific Rim of China are headed down the same road. After the devastation of World War II, Japan's economy was in ruins. By the mid-1990s, Japan's economy was the second largest in the world, two-thirds the size of the United States economy (Japan's population is only half as large). In the mid-1960s, South Korea was a land of traditional rice farmers who made up over 70% of the country's workforce. Its GNP per capita—$230—was the same as the West African country of Ghana. By 1993, Ghana's GNP per capita had risen to $430 while South Korea's had risen to $7,670, and over 70% of the people lived and worked in urban areas rather than on farms. South Korea's economy is now the eleventh most powerful in the world, ahead of such countries as Sweden, the Netherlands, and Australia. South Korea has become the world's largest shipbuilding nation and the world's fifth largest auto manufacturer. Its iron and steel and chemical industries are thriving, and with the largest number of Ph.D.s per capita in the world, South Korea has become a formidable competitor in research and development of semi-

conductors, information processing, telecommunications, and civilian nuclear energy. No other country has achieved as much economically in so short a time.

How has eastern Asia, with seven of the world's 12 largest ports, emerged as the hub of the increasingly prosperous Pacific Rim? Kennedy (1993) indicated basic characteristics that each of these Asian societies share, which, taken together, help to explain their sustained economic growth (Figure 9.17).

The first and probably the most important is the *commitment to education*. "This derives from Confucian traditions of competitive examinations and respect for learning, reinforced daily by the mother of the family, who complements what is taught in school" (Kennedy, 1993, p. 35). East Asian educational mores lead to social harmony and a well-trained workforce as well as encourage intense individual competitiveness. At the beginning of the decade, South Korea (population 43 million) had around 1.4 million students in higher education, compared with 145,000 in Iran (population 54 million) and 15,000 in Ethiopia (population 46 million).

The second factor, according to Kennedy (1994), is a *high level of national savings*. East Asian governments have encouraged personal savings by restricting the movement of capital abroad, maintaining low tax rates while keeping interest rates above the rate of inflation, and limiting the importation of foreign luxury goods. The result has been the accumulation of large amounts of low-interest capital that allowed Asian countries to finance education, infrastructure, manufacturing, and commerce. Singapore has the world's highest domestic savings rate—48% of gross domestic product (GDP) in 1993. Japan, Hong Kong, South Korea, China, Indonesia, Malaysia, and Thailand all save 30% or more

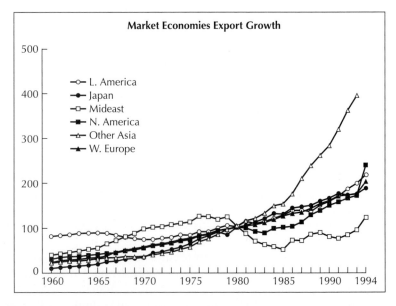

FIGURE 9.17
Export growth by market economies.

of their gross domestic product (GDP). In America, the domestic savings rate is about 13% of GDP.

Only recently, after economic "takeoff" was well underway, have Asian governments loosened financial policies to allow increased consumption and capital investment in consumer durables like new homes. Will such purchases, rather than savings, increasingly finance Asian prosperity? As Asian economies mature and their populations age, will savings rates decline and imported manufactures increase? Will industry give way to services (by 1996 nine of the world's 12 largest banks were headquartered in Tokyo)? Will East Asians tend more and more to consume rather than create wealth, as Americans tend to do?

The third feature marking East Asia's successful march to development has been a *strong political framework* within which economic growth is fostered. Industries targeted for growth were given a variety of supports—export subsidies, training grants, tariff protection from foreign competitors. Low taxes and energy subsidies assisted the business sector. Trade unions were restricted and democracy was constrained. This was accomplished by a variety of means, but the end was the same: an environment favorable to economic growth. In Japan, powerful government bureaucrats largely beyond public control promoted industrial expansion with little regard for the opinions or needs of Japanese consumers. Military governments in South Korea and Taiwan dealt harshly with industrial unrest and political dissidents. Authoritarianism ruled Hong Kong and Singapore. In no way did Asian governments follow a laissez-faire model. It is only lately that free elections and multiparty politics have been permitted outside Japan.

"Defenders of this system argued that it was necessary to restrain libertarian impulses while concentrating on economic growth, and that democratic reforms are a 'reward' for the people's patience" (Kennedy, 1993, p. 133). Thus, East Asian countries have not followed the consumer-driven economic policies of the United States, and they have successfully avoided the problems associated with American indebtedness and persistent deficits in trade and current accounts.

The fourth feature shared by East Asian NIC countries has been a *commitment to exports* (export led industrialization) in contrast to the import-substitution policies that characterized India, Africa, and Latin America until recently. Instead of encouraging industrialists with low labor costs to target foreign markets and compete there, governments in India, Latin America, and Africa decided to protect their economies rather than open them to international competition. They created their own food, steel, cement, paper, automobiles, and electronics-goods industries, which were given protective tariffs, government subsidies, and tax breaks to insulate them from international competition. As a result, their products became less competitive abroad. While it was relatively easy to create a basic iron and steel industry, it proved harder to establish high-tech industries like computers, aerospace, machine tools, and pharmaceuticals—so most of these states depend on imported manufactured goods, whereas exports still chiefly consist of raw materials such as oil, coffee, and soybeans.

A country that relies on the export of raw materials—mineral and agricultural commodities, "rocks and crops"—with little or no value added to the products

Pusan, South Korea, is a purpose-built city for shipbuilding and the export-import trade. Other towns in South Korea dedicated to specialized industrial functions include Changwan for machine tools and heavy plant, Ulsan for automobiles, and Yosu for petrochemicals. The experience of the rest of the world provides scant support for the notion that "protectionism" should be regarded as a vile epithet. The fastest growing economies this half-century has seen—China, South Korea, and Japan—have zealously protected their domestic markets. South Korea, which has come as close as almost any underdeveloped country in this century to approaching the income levels of the developed world, still imports less than 1 percent of its cars. (*Photo: World Bank*)

through finishing or manufacturing, and then has to import expensive (because of the immense value added) high-tech products, is not headed toward development unless the country commits its earnings to investment in quality exports and competitive high-tech manufacturing. Such countries need to get the fundamentals right: Keep inflation low and fiscal policies prudent, maintain high savings and investment rates, improve the education level of the population, trade with the outside world, and encourage foreign investment.

Throughout Latin America and Africa in the 1970s and 1980s that sort of bedrock financial responsibility was all too often missing. Governments poured money into state-owned enterprises, large bureaucracies, and oversized armed forces, paying for them by printing money and raising loans from Western banks and international agencies. Public spending soared, price inflation accelerated, domestic capital took flight to safe deposits in American and European banks, and indebtedness skyrocketed. By the mid-1980s, payments on loans consumed about half of Africa's export earnings. By 1989, Argentina owed industrial country banks and governments a staggering $1,800 for each man, woman, and child in the country. Zambia's debt rose to 334% of GNP. In more graphic terms, if the Zambian people were to give every penny they earned to their nation's foreign financiers beginning January 1, 1999, they would not eat again until May of 2002.

Just when Latin America and Africa needed capital for economic growth, countries there found themselves overwhelmed by debt, starved of foreign funds for investment, and with currencies made worthless by hyperinflation. By the end of the 1980s, poverty in developing countries outside East Asia had increased dramatically. Lands as well as people paid the price. Forests have been recklessly logged, mineral deposits carelessly mined, fragile lands put to the plow, and fisheries over exploited in a desperate effort to escape the poverty trap.

In the 1990s, "structural adjustment" policies—currency devaluation, export promotion, import reduction, sale of state-run industries, and government budget cuts—have been instituted all across Latin America and Africa. Hopefully this renewed economic growth will eventually raise living standards around the world. But as Latin America and Africa turn toward trade liberalization and export-led growth policies, they will notice just how far they have fallen behind countries of East Asia that set off down this road to development a generation or more earlier.

There are two other factors that favor development. First, because they enjoy a *geographical advantage*, some countries have a shorter road to travel to reach development. Singapore is not Zambia. Singapore thrives on the exchange of goods and services between the Malay Peninsula, Southeast Asia, Japan, the other economic powers on the Pacific Rim, and beyond. Zambia lies landlocked in the midst of the poorest continent, far removed from the world's trade routes. The trading states of East Asia have shown that it may not be necessary to be well-endowed with plentiful raw materials or farmlands in order to prosper, but their location on the Pacific Rim, adjacent to the powerhouse of Japan, is clearly an advantage as we move into the Pacific Century.

Second is the *demographic explosion*. Latin America's population was roughly equal to that of the United States and Canada in 1960. By 2025, Latin America's population will be more than double that of Anglo-America. From 1960 to 1990, Africa's population leaped from 281 million to 647 million; demographers forecast that the continent's population in 2025 could be nearly three times that of today. Such growth rates strain already from inadequate health care and education services and imply the crowding of millions of human beings into urban slums, pollution, and the degradation

Maquilladora textile operation in Ciudad Juarez. By January 1994 there were almost 3,000 maquilladora plants near the U.S. border in Mexico owned and operated by U.S. companies. Total employees number three-fourths of a million, and value-added approaches $4 billion. What effect will NAFTA have on the growth of American-owned plants in Mexico? *(Photo: Ted Soqui, Impact Visuals Photo & Graphics, Inc.)*

of forests, farmlands, grazing lands, and water supplies. These problems are not likely to help Latin America and Africa catch up with eastern Asia, where the commitment to education, manufacturing, and export-led economic growth has produced a steady rise in living standards and allowed those societies to make the demographic transition to smaller family sizes.

A small but growing number of countries is moving from "have-not" to "have" status, while many more remain behind. Such factors as cultural attitudes, education, political stability, capacity to carry out long-term plans, and locational advantage have influenced economic performance from one country to another. The race to development will as always surely have its winners and losers. Only this time, modern communications will remind us of the growing disparity.

WORLD INDUSTRIAL PROBLEMS

Global industrial problems center on a level, nonincreasing demand for many products that have traditionally increased in demand since the Industrial Revolution began. Dwindling growth rates of developed nations have reduced the increasing demand for many manufactured items. At the same time, as a result of technological innovation and because many other developing countries desire to generate their own industrial bases, the world is now saturated with excess output capacity, meaning that the supply is greater than the demand for industrial products.

Decreasing Demand for Industrial Products in the Mid-1990s

Since the mid-1970s, the industrial world has approached saturation for many consumer goods. Most of these countries have little population growth, and because of world recession, average personal disposable income, when adjusted for inflation, has not increased, but in most cases has decreased. Thus, the demand for industrial products has also decreased.

In addition, changing technology has, in part, reduced the demand for some products. For example, the global demand for steel is now less than it was in the 1970s. Part of the reason for this is that today's automobiles, which must be lighter weight in order to be energy efficient, use one-fourth less steel than they did 20 years ago.

At the same time, the quality of products has improved. In the early stages of manufacturing and demand for household goods, lower-quality products were mass produced. Now, emphasis is on high-quality products that last longer. The automobile industry is a prime example. Twenty years ago, the average American owned a car for 5 years. Today, the original buyer owns the average automobile for 11 years. Thus, again, with longer useful lives, fewer products are required.

Excess World Capacity

Although the worldwide demand is stagnant or decreasing for industrial products, capacity has increased. The higher industrial output now experienced is a result of three factors:

1. The diffusion from the developed world to the less developed world of the industrial revolution and basic levels of technology for manufacturing commonly demanded items, such as textiles, iron and steel, tools, motors, and clothing.
2. Increased output capacity by developed nations as a result of technological innovation, robotics, and flexible manufacturing systems.
3. The desire by developed, as well as developing, countries to maintain their own industrial bases because of positive effects on their overall economies, despite global overproduction.

For much of the 19th century, the United Kingdom dominated world industry. From the late 19th century onward, the United States, the then Soviet Union, Germany, and Japan, followed by China, Indonesia, and Mexico, all increased world industrial productivity. In addition, because most countries want to establish their own industrial bases as a hedge against world inflation and dependence on foreign imports, Asian countries such as South Korea, Singapore, Taiwan, and Hong Kong have recently contributed to the overproductivity. These countries can produce at cheaper rates than developed countries can, and thus have a place in the global competitive marketplace.

The steel industry is a prime example. Between 1970 and 1990, world steel demand did not increase. However, there was a global shift in this industry from Western Europe and North America to less developed regions of the world. During this period, one-fourth of the production in North America and Western Europe evaporated. This shift was partially to Eastern Europe and the former Soviet Union. But, output more than doubled in certain developing countries. In this short period of 20 years, the North American and Western European proportion of total global steel production declined from 67% to 50%, whereas developing countries' production levels increased from 10% to more than 20%.

Many steel-manufacturing firms have gone out of business as the global steel-production capacity exceeds global demand. Because of government subsidies, steel mills in some countries, especially in Europe, have remained open in the face of dwindling quotas. The U.S.

government, however, has been less willing to pay unemployment compensation to displaced workers and has allowed the U.S. steel industry to decline. Since the 1970s, U.S. production has decreased 33%, whereas employment in the steel industry has declined 66%.

Industrial Problems in Developed Countries

Developed countries are challenged to find new markets for their industrial output. At the same time, they are faced with reduced demand and increased unemployment levels in older industries. Their goal is to make their local industries competitive in an increasingly globally integrated world economy.

COMPETITION FROM MARKET BLOCS

Competition of markets has led to blocs of countries grouping to reduce trade barriers and to increase integration of supply and demand. One example is the cooperation of western European countries, called the *EU, or European Community.* In North America, the United States and Canada have a North American Free Trade Agreement, which reduces imports and exports of industrial goods. Cooperation in East Asia between Japan and other countries is also leading to trade liberalization.

Such trade agreements allow individual countries to take advantage of agglomeration economies or natural resources. Currently, the United States has passed the North American Free Trade Agreement (NAFTA) between Canada, the United States, and Mexico. With the addition of Mexico to the North American Free Trade

Zone, large markets are becoming available for Canadian and U.S. products. At the same time, it is expected that many North American labor-intensive industries will decentralize to Mexico to take advantage of labor rates that are 20% of American wage rates. American organized labor fears that many manufacturing jobs will be lost, whereas environmentalists fear that firms producing in Mexico, which has more lax environmental regulations, will increase the amount of pollution in an already very polluted environment. NAFTA proponents thus far have been unable to convince America that a free trade agreement that includes Mexico would benefit the American public. Time will tell.

MULTINATIONAL CORPORATIONS

Most giant industries in the highly developed countries are *transnational corporations,* meaning that they operate factories in countries other than their country of origin. The United States, Japan, Germany, Great Britain, and France each have numerous transnational corporations operating in many countries. Because of trade barriers and import limitations in desired foreign markets, transnational corporations operate in other countries to overcome restrictions placed on product imports when the corporation operates in its home country. In addition, it can increase its sales and decrease transportation costs by opening a factory in the country whose market it wants to penetrate. With regard to developing countries, transnational corporations have penetrated countries with extremely low site costs based on production factors. One site factor that varies from country to country for many industries is labor cost.

In this textile factory in Fortaleza, Brazil, women constitute the largest part of the workforce. Brazil is a major exporter of textiles to advanced industrial countries. Despite high rates of growth in many Third World countries, only a handful of economies—those of East Asia—actually have managed to narrow the gap with the northern, industrialized countries. Moreover, increased polarization between the richer and poorer countries has been accompanied by a rising trend in income inequality within countries such as in Brazil. The income share of the richest 20 percent has grown almost everywhere, while those in the lower ranks have experienced no rise in incomes to speak of. (Although they have benefited in kind through much improved educaitonal and medical provisions.) Even the middle classes have experienced little improvement. (*Photo: World Bank*)

However, the social problems created by industrial change and worldwide deindustrialization and reindustrialization in the global system have been substantial. Multinational corporations' *nomadic capital* and the switching of production from place to place in the 1970s through the 1990s, because of production innovation and transportation and communications savings, has disrupted some countries that have industrialized upon the transnational corporation's demand. In some cases, factory closures and job losses have been the result of shifting capital to increase profits and pay stockholders. In other cases, they have resulted from changing world demands and political postures. In the United States, up to a million manufacturing jobs per year were lost between 1978 and 1990, just by plant closures. Since 1991, the reduction of hostilities and the end of the Cold War has meant the loss of a million jobs in California alone because of reduced demand for aerospace industries, combat aircraft industries, and related electronics manufacturing and defense contracting. In some cases, the transnational corporation has no control over which plants it must close. The International Monetary Fund reported that economies of industrialized countries grew by slightly more than 1% in the mid-1990s.

RICH CITY–POOR CITY REGIONS

Within some countries and regions, disparities between relatively well off and relatively depressed economic areas exist. For example, unemployment in the United Kingdom is 50% higher in the northwest than in the south and in the east, and incomes average 30% higher in the south, near London. French industry is centrally distributed, with Paris being the area of wealth, whereas western France is relatively depressed and nets only two-thirds the per capita income of the Paris region. No regional disparities are greater than those of Italy. Italy has been described as two different countries, the north and the south, with per capita incomes three times higher in the northern industrial Po River valley compared with the areas south of Rome, in southern Italy, and Sicily. Sweden's regional disparities are likewise immense, with the prosperous southern region centered on Stockholm and Göteborg in sharp contrast with much less populated and poorer northern regions.

Germany presents a particularly interesting problem in its attempt to integrate former East Germany, with its depressed level of industry and high unemployment levels, with the much more prosperous west. In the United States, although the South has been the most depressed region historically since the 1960s, it has been the region of the most rapid industrial and per capita income gains. The original hearth area of the nation, New England, and the industrial heartland of the American Manufacturing Belt throughout the Midwest are now some of the more depressed areas of North America, but cities rich and poor vary by local region.

Industrial Problems in Developing Countries

Developing countries have a special set of problems in increasing their industrial output. As we discussed, many developing countries foster their own set of basic manufacturing industries, including iron, steel, and textiles, regardless of the glut on the world market for these products. Western countries built their power and wealth on industrialization, which developing countries see as an avenue to growth. Their problems center around accessibility to the distant world markets, lack of real investment capital, and lack of labor training capable of producing a management class. Other problems include an inadequate supporting infrastructure, which is critical for the development of industrial activity, and exploitation by transnational corporations, which aggressively control the developing countries' industrial potential.

Although the developing countries' own demand may be sufficient to warrant new industrialization, distant markets of the developed world compose a much larger source of revenue. But the United States, Western Europe, Japan, Russia, and some of the other former Soviet Union countries can produce enough industrial output to satisfy themselves. Thus, there is no excess demand to help the developing countries get started.

Frequently the education system of NICs cannot produce university-trained managers, accountants, and other white-collar professionals to support industrialization. In addition, capital is needed for basic demands such as clean water, transportation, and food production. Many times a country cannot afford to drain scarce capital from these basic necessities to aid industrialization.

The poor state of the transportation, communications, and service infrastructure in developing nations is often not at the level necessary to produce and support new factories. We have already discussed the consequences of export-led industrialization and import-substitution industrialization. Some African countries take advantage of proximity to raw materials, mineral resources, and the like. Most developing countries, however, have an abundant supply of cheap labor that is useful for certain types of industrialization.

However, transnational corporations involved in many developing countries take advantage of this low-cost labor supply and profit from selling the products produced by the labor. Consequently, often a developing country must compete against transnational corporations that have outside capital and large supplies of trained professionals to help in the industrialized management task.

Because of these reasons, industrialization of developing nations is extremely difficult, and no immediate future solution to the problem is on the horizon.

Sweatshops

Some Guess clothing is made by suppliers who use underpaid Latino immigrants in Los Angeles, sometimes in their own homes. Mattel makes tens of millions of Barbies a year in China, where young female Chinese workers who have migrated hundreds of miles from home are alleged to earn less than the minimum wage of $1.99 a day. Nike is criticized for manufacturing many of its shoes in tough labor conditions in Indonesia, and some of Disney's hottest seasonal products are being made by suppliers in Sri Lanka and Haiti—countries with unsavory reputations for labor and human rights. Soccer balls are sewn together by child laborers in Pakistan, and F.A.O. Schwarz's $150 Bernie St. Bernard was made in Indonesia. In an era when the world economy is global, it is impossible for consumers to avoid products made under less than ideal labor conditions. Moreover, what may appear to be horrific working environments to most citizens in the world's richest nation are not just acceptable but actually attractive to others who live overseas or even in "Third World pockets" of the United States. Anyone even casually familiar with how some Americans recompense their (usually immigrant) housekeepers understands the desire to work.

One icon of American culture whose manufacturing practices seem out of sync with its brand name is Disney, which is making a holiday merchandising blitz for its characters and movies. Disney maintains almost 4,000 contracts with other companies that assume the right to manufacture Disney paraphernalia, some of which are then sold in Disney stores. These licensees go to some of the world's lowest-cost-labor countries, including Sri Lanka and Indonesia, to produce stuffed animals and clothes. Disney itself rarely takes a direct hand in manufacturing.

Sears, which carries 200,000 products from manufacturers operating in virtually every country, is tightening up on buying goods from suppliers with dubious records. The Gap, after enduring criticism, also has become a model for manufacturing and sourcing products abroad.

In contrast to Disney, Mattel does most of its own manufacturing. It makes a staggering 100 million Barbie dolls a year in four factories, two in China and one each in Malaysia and Indonesia. The Barbie craze produced $1.4 billion in annual revenues for the El Segundo, California, company out of its total annual revenues of $3.6 billion in 1995.

Does a global economy mean consumers face no choice but to buy products made under conditions Americans don't want to think about? A number of U.S. companies say that intense global competition is no excuse for keeping working standards at the lowest possible level. Levi Strauss, for example, imposes its own "terms of engagement" on manufacturers who make its jeans products in 50 countries.

In many ways, what Americans buy is their most direct and intimate connection with the global economy. In a post–Cold War era in which governments seem to be losing their power to shape the lives of people, U.S. consumer spending can be an important tool in extending American values. The silver lining is that if Americans respond to even some of these concerns, they could enjoy their shopping and improve the conditions that millions of people around the world encounter in their daily lives.

SUMMARY

Four major regions of the world account for approximately 80% of the world's industrial production—northeastern North America, northwestern Europe, western Russia, and Japan. The textiles, clothing manufacture, and consumer electronics industries have shifted globally from the developed world to the developing world. Automobile production and semiconductor manufacture have also experienced global shifts, from North America and Europe to Japan and the Far East.

In this chapter, we explored the social relations that lead to industrial change, described worldwide manufacturing trends, and examined the recent history of industrial devolution in the developed world and of industrial revolution in the developing world. It was argued that the processes of *deindustrialization* and *industrialization* are not offsetting tendencies within the global system; rather, they constitute a negative-sum global game played by multinational nomadic capital. Multinationals switched production from place to place because of varying relations between capital and labor and new technological innovations in transportation and communications. With improved air freight, containerization, and telecommunications, multinational corporations can dispatch products faster, cheaper, and with fewer losses.

In the United States, 500,000 manufacturing jobs per year were lost between 1978 and 1992. These losses were hidden to some extent by selective reindustrialization and the migration of manufacturing from the American Manufacturing Belt to the South and the West. However, a worldwide recession and the end of the Cold War military buildup have also doomed many manufacturing companies. Overall, world industrial problems now center on overcapacity and the decreasing demand for industrial products.

Japan has been an industrial success story. However, its economic miracle has been achieved at the expense of social welfare, the environment, and international relations.

As we move into the third millennium, the developed countries (and now, the East Asian countries) seem to have all the trump cards—capital, technology, IT, plenty of food, and efficient and decisive MNCs. As technology erodes the value of labor and materials, LDCs fall farther behind.

KEY TERMS

American Manufacturing Belt
assembly plants
capital-labor relations
component plants
deindustrialization
dependent industrialization
disorganic development
East Asia Miracle
economic structure of society
export-led industrialization
export-processing zones

forces of production
import-substitution industrialization
industrial restructuring
industrialization
inertia
integrated circuit
international subcontracting
Japan Incorporated
just-in-time manufacturing
materialist science
microprocessor

mode of production
NAFTA
nomadic capital
outsourcing
post-industrial society
quaternary economic activities
social relations of production
surplus value
transnational corporations
value added

SUGGESTED READINGS

Britton, J.N.H., ed. 1996. *Canada and the Global Economy.* Montreal: McGill-Queens University Press.

Chapman, K., and Walker, D. 1991. *Industrial Location.* 2nd ed. Cambridge, Mass.: Basil Blackwell.

Corbridge, S., ed. 1993. *World Economy.* The Illustrated Encyclopedia of World Geography. New York: Oxford University Press.

Drakakis-Smith, D., ed. 1990. *Economic Growth and Urbanization in Developing Areas.* London: Routledge.

Fransman, M. 1990. *The Market and Beyond: Co-operation and Competition in Information Technology in the Japanese System.* Cambridge, Mass.: Cambridge University Press.

Glasmeier, A. K. 1991. *The High-Tech Potential: Economic Development in Rural America.* New Brunswick, N.J.: Center for Urban Policy Research.

Harrington, J.W., and Warf, B. 1995. *Industrial Location.* New York: Routledge.

Henderson, J. 1989. *The Globalization of High Technology Production.* London: Routledge.

Hoare, A. G. 1993. *The Location of Industry in Britain.* New York: Cambridge University Press.

Hogan, W. T. 1991. *Global Steel in the 1990s: Growth or Decline.* New York: Lexington Books.

Knox, P.L., and Agnew, J.A. 1994. *The Geography of the Economy.* 2nd ed. London: Edward Arnold.

Krugman, P. 1991. *Geography and Trade.* Cambridge, Mass.: Cambridge University Press.

Law, C. M., ed. 1991. *Restructuring the Global Automobile Industry.* London: Routledge.

McDermott, M. C. 1989. *Multinationals: Foreign Divestment and Disclosure.* London: McGraw-Hill.

Pitelis, C., and Suugden, R., eds. 1991. *The Nature of the Transnational Firm.* London: Routledge.

Porter, M. E. 1990. *The Competitive Advantage of Nations.* New York: Free Press.

Sklair, L. 1991. *Sociology of the Global System: Social Change in Global Perspective.* Hemel Hempstead, England: Harvester Wheatsheaf.

Spero, J. E. 1990. *The Politics of International Economic Relations,* 3rd ed. London: Allen and Unwin.

Taylor, P. J. 1990. *Political Geography: World Economy, Nation State and Locality,* 2nd ed. London: Longman.

WORLD WIDE WEB SITES

WORLD FACTBOOK ON COUNTRIES
http://www.odci.gov/cia/publications/pubs.html

THE WORLD TRADE ORGANIZATION
http://www.wto.org/
The principal agency of the world's multilateral trading system. Its home page includes access to documents discussing international conferences and agreements, reviewing its publications, and summarizing the current state of world trade.

THE WORLD BANK
http://www.worldbank.org/
A leading source for country studies, research, and statistics covering all aspects of economic development and world trade. Its home page provides access to the contents of its publications, to its research areas, and to related websites.

U.S. DEPARTMENT OF COMMERCE
http://www.doc.gov/

Charged with promoting American business, manufacturing, and trade. Its home page connects with the web sites of its constituent agencies.

BUREAU OF LABOR STATISTICS WEBSITE
http://stats.bls.gov/
Contains economic data, including unemployment rates, worker productivity, employment surveys and statistical summaries.

ECONOLINK
http://www.progress.org/econolink
Econolink has "the best web sites that have anything to do with economics." Its listings are selective with descriptions of side content; many referenced sites themselves have web links.

WEBEC: WORLD WIDE WEB RESOURCES
http://netec.wustl.edu/WebEc/WebEc.html
A more extensive set of site listings—though more purely "economic" than "economic geographic."

INTERNATIONAL BUSINESS I: DYNAMICS

O B J E C T I V E S

- To consider the nature of international business
- To explain the bases of international trade and factor flows
- To examine the effects of natural and artificial barriers on international business
- To acquaint you with the major international institutions that deal with problems of trade, investment, and development
- To examine the effects of tariffs and quotas on international trade
- To understand the financing of international trade

Multinational corporate advertising in Piccadilly Circus, London, England. *(Photo: Michael Coyne, The Image Bank)*

Since World War II, world economies have become more integrated than ever. *International integration* refers to the concept of international specialization or division of labor. Contributing to this widespread integration have been technological breakthroughs in transportation and communications and massive transformations in business behavior. These innovations and developments have greatly enhanced the role of *international business*, which is any form of business activity that crosses a national border. International business includes the international transmission of merchandise, capital, and services. International trade is expanding, and its composition and patterns are changing. But in many respects, it is now less significant in the global business structure than the international movement of capital and services. More and more companies are investing in foreign countries to acquire raw materials, to penetrate markets, and to exploit cheap labor. The expansion of production overseas has been matched by a parallel, symbiotic expansion of service enterprises, which now account for an increasing share of foreign direct investments (FDIs).

The actors in the international business arena have become more numerous and relationships among them more complex in the postwar years. The immediate postwar decade was primarily a *two-actor era*, the firm and its foreign commercial constituents—customers, suppliers, joint-venture partners, licensees. During these years, the United States spent billions of dollars on the reconstruction of Japan and west European countries. The economic revival of these countries represented huge markets for U.S. capital, equipment, and consumer goods, and few political barriers impeded their flow.

By the mid-1950s, with postwar reconstruction virtually complete, Japan and several west European countries challenged the economic dominance of the United States. The high value of the dollar in relation to other currencies dampened demand for U.S. goods and services and encouraged other countries to aggressively seek export markets. Between 1955 and 1970, the years of the long postwar boom, an increasing volume of trade was associated with multinational corporations, domiciled in the United States, Western Europe, and Japan. These giant companies intensified the sensitivities of host governments, which became a *third actor* in international decision making. Corporate strategists had to contend not only with host-country policies in developed countries, but with policies in the new, self-conscious underdeveloped countries as well.

The early 1970s marked another watershed as the world economy entered a prolonged period of disorder. This time of crisis stemmed from four structural changes: First, the industrialized countries experienced a marked slowdown in their economic growth rates. Second, competitive rivalry among industrialized countries increased significantly. This competition was stimulated by the slower and more unstable growth rates and, in turn,

Foreign direct investment by U.S. firms in the Pacific. *(Photo: Greg Davis, The Stock Market)*

contributed to them. The rivalry gave rise to an increase in restrictive policies as each country sought to overcome its national crisis at the expense of other countries. Third, in 1973, the collapse of the Bretton Woods monetary arrangements, involving the replacement of fixed exchange rates and the convertibility of the dollar into gold with a system of more or less freely fluctuating exchange rates, permitted the United States to devalue its currency in an effort to retrieve lost competitiveness with its trading rivals. The new world monetary system, in which exchange rates shift almost daily, fueled inflation and destabilized commodity markets. The fourth structural change, increases in oil prices by 400% in 1973–1974 and 165% between 1979 and 1981, reduced real income in the advanced countries and dealt a particularly harsh blow to the oil-importing Third World countries. The oil crisis induced worldwide recession in the mid-1970s and again in the early 1980s. Recovery from the recessions has been weak in all countries except the western Pacific nations. In fact, Japan and other East Asian nations emerged from the recessions stronger than ever, their industrial growth fed by an expanding market in the Western Hemisphere.

While these events unfolded, the activities of multinational corporations grew substantially. In pursuit of larger profits than could be obtained domestically, multinationals built up industrial capacity in "offshore" production centers. Parent governments recognized that the ability of multinationals to transfer capital, production, and labor across national boundaries was a major cause of instability in the global economy. So began the *four-actor era* as parent governments sought to regulate the activities of multinationals.

In the late 1970s, international business became more and more politicized at home and abroad. An increase in government regulations at both ends occurred in response to mounting public concerns about pollution, natural-resource allocation, income and wealth distribution, consumer protection, energy, and the governance of corporations. These concerns were expressed by a variety of special-interest groups—ethnic, religious, occupational, and political.

Because these special-interest groups are now developing international linkages and loyalties as they strive to create international orders in their own image, we live in the *multiactor era*, in which a variety of actors are relevant to corporate decision making. In the new environment, multinational corporations find it difficult to react swiftly enough to changing opportunities and constraints. As a result, new forms of business entities such as trading companies, financial groups, and service companies are developing, and they have become preeminent features in the international business climate of the 1990s.

The increasingly complex international business environment warrants the attention of geographers. Knowledge of the international sphere of business helps us to understand what is occurring in the world as well as within our own countries. This chapter examines the concepts and patterns that underlie the expanding world of international business. It seeks answers to two questions: What theories shed light on the processes of international interaction? What are the dynamics of world trade and investment?

INTERNATIONAL TRADE

Why International Trade Occurs

Why has world trade jumped from $2 trillion annually in 1980 to $4 billion today? Why are so many countries, large and small, rich and poor, deeply involved in international trade? (Figure 10.1.) One answer lies in the unequal distribution of productive resources among countries. Trade offsets disparity with regard to the availability of productive resources. However, whether a country can export successfully depends not only on its resources, but also on the conditions of the economic environment; the opportunity, ability, and effort of producers to trade, and the capacity of local producers to compete abroad.

NATURAL AND HUMAN-MADE RESOURCES
Production factors—labor, capital, technology, entrepreneurship, and land containing raw materials—vary from country to country. One country is rich in iron ore, and another has tremendous oil deposits. Some countries have populations large enough to support industrial complexes; others do not. People are not only a natural resource, but also a precious human-made resource with differential skills. One country is home to an enormous pool of workers adept at running modern machinery; another abounds with scientists and engineers specializing in research-laden products. In some countries, entrepreneurs are more

FIGURE 10.1
World growth of export classes by value, 1965–1994. After the world recession in the early 1980s, oil prices declined, but machinery, transport equipment, and other manufactures, the products on which industrial growth is based, have continued to climb rapidly. Despite the concerns of some in the labor movement, the reality is that most U.S. manufacturing employees work in plants that export. Over 1.5 million export-related jobs have been created in the United States in the past four years alone. These jobs pay 10 percent to 15 percent more than non-export jobs. Export firms expand employment nearly 20 percent faster than other firms. Moreover, productivity in export firms is one-fifth higher than in non-export firms. Companies that export are 10 percent less likely to fail than others. *(Source:* United Nations Yearbook of International Trade Statistics, *1995, 1996)*

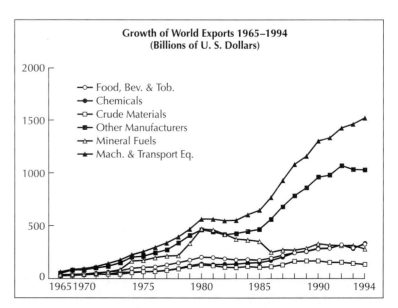

Growth of World Exports 1965–1994
(Billions of U. S. Dollars)

- Food, Bev. & Tob.
- Chemicals
- Crude Materials
- Other Manufacturers
- Mineral Fuels
- Mach. & Transport Eq.

able and knowledgeable than in others. The imbalance in natural and human-made resources accounts for much of the international interchange of production factors and the products and services that the factors can be used to produce.

CONDITIONS OF THE ECONOMIC ENVIRONMENT

A country that is well endowed in natural and human-made resources has an advantage over countries that lack these assets. But the assets in and of themselves are insufficient to guarantee success in the export market. American producers, for example, tend to be less successful exporters than their Japanese or European counterparts. Numerous economic environmental factors may reduce the ability of countries to best use their resources and productive advantages. These factors include inflation, exchange rates, labor conditions, governmental attitudes, and laws.

Inflation, which is a rise in the general level of the prices of goods and services, can be detrimental to a country's ability to compete domestically or internationally. Exchange rates, the prices of currencies in foreign-exchange markets, can also influence competitiveness. For example, if a currency is overvalued in relation to other currencies, local producers may find it difficult not only to compete with foreign imports, but also to export successfully. In addition, recurring labor disputes that interrupt production can create serious obstacles for exporters, and governments can encourage or discourage their export sectors. Finally, the competitiveness of exporters is affected by laws—labor laws, tax laws, and patent laws.

OPPORTUNITY, ABILITY, AND EFFORT OF THE PRODUCER TO TRADE

Success in the international trade arena hinges on demand for the good or service produced, an awareness of the demand by suppliers, the availability of appropriate foreign distribution channels, and a minimum of governmental controls. But even if all these conditions are met, many producers fail to respond. Some may lack the desire to try, whereas others may have the desire but never make the effort. The importance of desire and effort in determining success in international trade is exemplified by Japan. Although this island nation lacks raw materials, it has successfully imported materials and components for use in manufacturing goods for domestic and foreign markets.

COMPETITIVENESS OF LOCAL PRODUCERS ABROAD

The ability of a country to export is affected by the capacity of foreign producers to compete. A crucial element in the ability of domestic producers to compete is the relative cost of production. Countries with high labor, capital, and energy costs can expect strong competition from abroad. The existence of strong competitors may act as a disincentive to successful and profitable exporting.

Trade by Barter and Money

At one time, trade was simple and was conducted on a barter basis. *Barter*, or *countertrade*, is the direct exchange of goods or services for other goods or services. It still occurs in some traditional markets in underdeveloped countries and is of increasing importance in the modern world economy. Roughly 30% of world commerce is now countertrade. Russia and eastern European countries use barter to trade among themselves and with underdeveloped countries. Major oil-exporting countries such as Iran and Nigeria barter oil and gas for manufactured goods.

Despite its widespread use, particularly by governments that have turned toward national economic protectionism, barter is a cumbersome way of conducting international exchange. Even within a country, consumers would find it difficult to barter the goods or services that their families produce for goods and services to satisfy their daily needs. In fact, more time would be devoted to exchange than to production. Money provides a means for simplifying the domestic exchange procedure and trade between countries. Money does present some problems, such as those associated with exchange rates, but introducing it as an exchange medium does not alter the theoretical bases for international trade.

Classical Trade Theory

The flow of trade arises from an economic advantage that one country has over another in the output of a good or a service. It may occur because of the *absolute advantage* of one country in the production of a good. For example, Country A produces cloth twice as efficiently as Country B does, whereas Country B is 50% more productive than Country A is in the output of wheat; thus, Country A will be inclined to exchange its cloth for Country B's wheat, and vice versa. Each country has an absolute advantage in producing its good.

However, both countries need not have absolute advantages for trade to occur. Indeed, a country can successfully export a good in which it has an absolute cost *disadvantage*. This phenomenon is called the *theory of comparative advantage*, or the *theory of comparative cost*. Developed by 18th- and 19th-century English economists, notably David Ricardo (1912), the theory states that all countries have comparative advantages and that countries will export the goods that they can produce at the lowest *relative* cost. For example, Country A specializes in the export of raisins because it can produce them more cheaply, compared with other goods, than Country *B* can, and Country *B* exports paper because it can produce paper more cheaply than other goods. Raisins may be cheaper than paper in both countries, but the cost differential between raisins and paper is wider in Country *A* than in Country *B*. To explain the foreign-trade structure of a particular country, we must identify its comparative advantage, which involves a study of its productive resources.

Heckscher-Ohlin Trade Theory

David Ricardo's two-country, two-product, theory of corporation advantage provides the basis for other traditional trade theories. For example, it can be expanded by taking into account several countries and commodities and by allowing several production factors. The multifactor approach to trade theory derives from work by two Swedish economists, Eli Heckscher (1919) and Bertil Ohlin (1933). In brief, the *Heckscher-Ohlin theory* takes the view that a country should specialize in producing those goods that demand the least from its scarce production factors and that it should export its specialties in order to obtain the goods that it is ill-equipped to make (Olsen, 1971). Thus, if Countries *A* and *B* have different endowments of labor and machinery, both can gain from trade. Country *A* with many laborers and few machines can concentrate on, say, corn production and export its specialty in order to import cloth. Similarly, Country *B* with few laborers and many machines can specialize in cloth and export some cloth to import corn. Again, free trade is best from a global standpoint. When specialization is fostered, world output is maximized.

The Heckscher-Ohlin theory argues not only that trade results in gains, but also that wage rates will tend to equalize as the trade pattern develops. The reasoning behind this *factor-price equalization*, as it came to be called, is as follows: As Country *A* specializes in corn production and diverts production from cloth, its production pattern becomes more labor intensive. As a result, Country *A*'s abundance of labor diminishes, the marginal productivity of labor rises, and wages also increase. Conversely, in Country *B*, as cloth production replaces corn production, labor becomes less scarce, the marginal productivity of labor falls, and wages also fall.

Some economists find the notion that foreign trade evens out the production factors too simplistic. But others cling to the Heckscher-Ohlin theory with considerable tenacity. The theory has ideological importance, and in 1919, when Heckscher first put it forward, it was particularly opportune. After World War I, the United States introduced restrictions on immigration, and there was also a growing interest in protectionism. With the interests of free traders under threat, a strong case for free trade needed to be made. The Heckscher-Ohlin theory showed not only that free trade was desirable, but

also that it could compensate for restrictions on labor migration.

Inadequacies of Trade Theories

Trade theories are based on restrictive assumptions that limit their validity. They generally ignore considerations such as scale economies and transportation costs. Scale economies improve the ability of a country to compete even in the face of higher factor costs. The cost of moving a product greatly affects its "tradability." Brick, which has a high transportation cost relative to value and therefore is not extensively traded internationally, is a good example.

Trade theories assume perfect knowledge of international trading opportunities, an active interest in trading, and a rapid response by managers when opportunities arise. However, executives are often ignorant of their trading opportunities. Even if they are aware, they may fear the complexities of international trade.

Other inadequacies of trade theories include the assumptions of homogeneous products, perfect competition, the immobility of production factors, and freedom from governmental interference. Products are not homogeneous. Oligopolies exist in many industries. Production factors such as capital, technology, management, and labor are mobile. Governments interfere with trade; they can raise formidable barriers to the movement of goods and services.

The most important shortcoming of trade theories, however, is their failure to incorporate the role of firms, especially that of multinational corporations. Trading decisions are usually made on the microeconomic level by managers, not by governments in Country *A* or Country *B*. Multinational corporations also operate from a multinational perspective rather than from a national perspective. When international trade occurs between different affiliates of the same company, it is referred to as *intramultinational trade*. Special considerations, such as tax incentives or no competition from other affiliates of the same company, can often play a pivotal role in a company's international decisions.

Despite their limitations, traditional trade theories provide an essential basis for our understanding of international business. They still underlie the thinking of many scholars, managers, labor leaders, and government officials. They offer a background for understanding the barriers to international business. They also frequently explain commodity trade, such as the international movement of wheat.

A New Theory of International Trade and Transactions

Evidence that MNCs are responsible for a major share of world trade, and also for a variety of other kinds of international economic transactions that do not fit the conventional mold, are central issues fueling the recent dissatisfaction with the explanatory powers of traditional theory. This dissatisfaction generated alternative approaches from the 1970s onward, each offered as a replacement for existing theory.

The appearance of a theory of industrial structure arose, seemingly establishing a relationship between trade and industrial organization—more specifically, the multinational enterprise. With this new theory of foreign direct investment, writers set about simultaneously explaining both the overseas production of multinationals and the international trade conducted by these enterprises (Dunning, 1995).

A distinct set of assumptions has been the backbone for what these theorists have relied on, straying from those rooted in conventional trade theory. They have concluded that:

1. Individual firms possess unique competitive advantages.
2. Firm assets are mobile among branches of the enterprise.
3. The economic functions performed by various branches of a firm are decided in accordance with the spatial distribution of firm assets.

The new trade theory allows trade between countries even when they have identical factor endowments. Because of imperfect competition, there is a possibility of increasing returns to scale (scale economies) which offers gains from trade over and above those attainable from conventional theory alone. To summarize these gains from Helpman and Crudman (1985):

1. Countries will benefit from trade if they are able to generate *increasing returns to scale* (agglomeration or scale economies)—they are able to increase their output thereby reducing the unit cost and reducing prices, therefore, increasing trade.
2. Increasing returns (scale economy) industries will result and arise in productivity efficiencies in the host country, thereby enlarging global output and reducing prices worldwide—machinery, transport equipment, and other manufactures are good examples (Figure 10.1).
3. Increased competition will reduce profits, which will thin out competing companies from trade. With increasing returns, this effect will result in greater overall productivity and sufficiency and lower costs and prices.
4. With many different countries increasing their productivity efficiencies through increasing returns, and exchanging goods with each other, the world as a whole will gain from the resulting

increase and the variety of goods and the cost of goods over and above what any individual country is capable of supplying.

Fairness of Free Trade

Free trade is best from the standpoint of efficiency, but is it fair given the relationship of *unequal exchange* between developed and developing countries? This question is raised by theorists such as I. Wallerstein (1974), for whom imperialism is associated with relatively free trade. Their argument is that an artificial division of labor has made earning a good income from free trade difficult for most LDCs.

AN ARTIFICIAL DIVISION OF LABOR
The British were instrumental in creating an unfair division of labor. Implicit in the early 19th century argument for free trade was the notion that what was good for Great Britain was good for the world. But free trade was established within a framework of inequality among countries. Great Britain found free trade and competition agreeable only after becoming established as the world's most technically advanced industrial nation. Having gained an initial advantage over other manufacturers, Great Britain then threw open its markets to the rest of the world in 1849. Other countries

A tea-buying center in Kenya. About 16% of Kenya's export earnings come from tea, and 34% come from coffee. Kenya depends largely on these two commodities for export income. *(Photo: Jim Pickerell, World Bank)*

were instructed or lured to do the same. The pattern of specialization that resulted was obvious. Great Britain concentrated on producing manufactured goods, such as vehicles, engines, machine tools, paper, and textile yarns and fabrics, and exporting them in exchange for a variety of primary products. Imports included specialized cargoes such as Persian carpets, furs, wines, silks, and bulk imports such as timber, grains, fruit, and meat. Although many countries gained from the application of this artificial division of labor, none gained more than Great Britain did.

The only way other countries could break out of this artificial division of labor was by interfering with free trade. The United States and Germany did so in the 1870s by adopting protectionist policies. France and a few other European countries with embryonic industries did the same. Dependent countries, however, failed to escape, either because of colonialism, or because it was not in the interest of their ruling groups to do so.

The original division of labor changed little until after World War II, when a new structure began to evolve. Some underdeveloped countries were given a limited license to industrialize. As we saw in Chapter 9, the basic trend was export-led industrialization, concentrated in a few countries. For the best-off poor countries, industrial growth is geared to the needs of the old imperial powers. Thus, the growth of manufacture in the Third World, under multinational corporate auspices, is not a portent of its emancipation from an artificial division of labor.

THE WORSENING TERMS OF TRADE

A deterioration in the *terms of trade*—the prices received for exports relative to the prices paid for imports—exemplifies the problem of unequal exchange for LDCs. By and large, LDCs export raw and semiprocessed primary goods—agricultural commodities (cocoa, tea, coffee, palm oil, spices, bananas, seafood, sugar, jute, and cotton) and minerals (tin, iron ore, bauxite, aluminum, phosphate, diamonds, oil, copper, and uranium). Primary commodities account for about 70% and 47% of the total exports of low- and middle-income countries (excluding China and India), respectively. The proceeds from these exports are needed to pay for imports of manufactures, which are vital for continuing industrialization and technological progress. Shifts in the relative prices of commodities and manufactures can therefore change the purchasing power of the exports of LDCs dramatically (World Bank, 1988). The situation is exacerbated because many of these low- and middle-income countries are vulnerable, single-commodity-dependent countries (Table 10.1).

Between 1970 and 1990, LDCs experienced a worsening in their terms of trade. This was caused by a decline in the prices of primary commodities and an increase in the prices of manufactures. In some years,

the adverse shift was offset by an increased volume of LDCs' exports. For the period as a whole, however, import volume growth exceeded export volume growth and, given the overall deterioration in the terms of trade, the result was large current account deficits. Maintenance of these deficits was possible only because the LDCs had access to external finance sources.

Economist Keith Griffin (1969) provided an account of the theoretical causes for the decline in the terms of trade of LDCs for export commodities, especially food crops. According to Griffin, the cause of a decline in the terms of trade for any country depends on (1) the nature of the product exported, (2) the degree of structural rigidity in the economy, and (3) the bias of technical progress. Let us now look at how Engel's law applies to each of these factors.

We can use a statistical finding known as *Engel's law* to account for a deterioration in the terms of trade. Nineteenth-century German statistician Ernst Engel, who examined the income and spending patterns of wage-earning families in several European countries, arrived at the following conclusion: As incomes rise beyond a certain point, the proportion of disposable income spent on food declines (Figure 10.2). In the United States, for example, the percentage of disposable income spent filling supermarket carts is greater for families earning $20,000 a year than it is for families earning $60,000 a year. If we extended the concept of Engel's law to international trade, we could argue that as consumption of manufactured goods increased, agriculture would form a decreasing proportion of total trade, and income elasticity of demand would increase for manufactured goods. Thus, we would have a built-in structural disequilibrium, which would result in worsening terms of trade for exporters of primarily agricultural produce.

The economies of many LDCs are characterized by structural rigidity. They cannot alter the composition of exports rapidly in response to changing relative prices. Thus, if their commodity export prices decrease, they have no alternative but to accept declines in their terms of trade (Figure 10.3, Table 10.2).

A third factor that may lead to worsening terms of trade is technological advances in developed countries. Advanced technology (1) enables the industrial economies to reduce the primary content of final products, (2) enables the wealthy nations to produce high-quality finished products from less valuable or lower-quality primary products, and (3) enables the advanced economies to produce entirely new manufactured products, which are substitutes for existing primary products. These three technological developments are irreversible. The demand for many primary products may be inelastic for price decreases, but in the long run it may be very elastic for price increases. A rise in the price of a raw material provides an incentive for indus-

◈

T a b l e 10.1
Vulnerable, Single-Commodity Dependent Countries

Product's Percentage of Total Export Earnings		
40 to 59	*60 to 80*	*More than 80*

Agriculture and Fishing

40 to 59	60 to 80	More than 80
Benin (cotton)	Bhutan (spices)	Dominica (bananas)
Burkina Faso (cotton)	Burundi (coffee)	Rwanda (coffee)
Burma (lumber, opium[a])	Ethiopia (coffee)	Uganda (coffee)
Chad (cotton)	French Guiana (seafood)	
Cocos (Keeling) Islands (copra)	Guadeloupe (bananas)	
Comoros (spices)	Malawi (tobacco)	
El Salvador (coffee)	Maldive (seafood)	
Equatorial Guinea (cocoa, lumber)	Martinique (bananas)	
Finland (wood products)	St. Lucia (bananas)	
Ghana (cocoa)	Seychelles (seafood)	
Grenada (spices)		
Honduras (bananas)		
Iceland (seafood)		
Kiribati (copra, seafood)		
Mali (cotton)		
Mauritania (seafood)		
Nicaragua (seafood)		
Sudan (cotton)		

Crude Oil and Petroleum Products

40 to 59	60 to 80	More than 80
Congo	Bahrain	Algeria
Ecuador	Gabon	Angola
Syria	Libya	Brunei
Yemen	Venezuela	Iran
		Iraq Kuwait
		Nigeria
		Oman
		Qatar
		Saudi Arabia
		Trinidad & Tobago
		United Arab Emirates

Metals and Minerals

40 to 59	60 to 80	More than 80
Central African Republic (diamond)	Botswana (diamonds)	Nauru (phosphates)
Chile (copper)	Guinea (aluminum)	Zambia (copper)
Jamaica (aluminum)	Niger (uranium)	
Liberia (iron ore)	Papua New Guinea (copper)	
Mauritania (iron ore)	Suriname (aluminum)	
Togo (phosphates)		

[a]Although impossible to quantify, Myanmar's opium exports may exceed 40% of total exports.

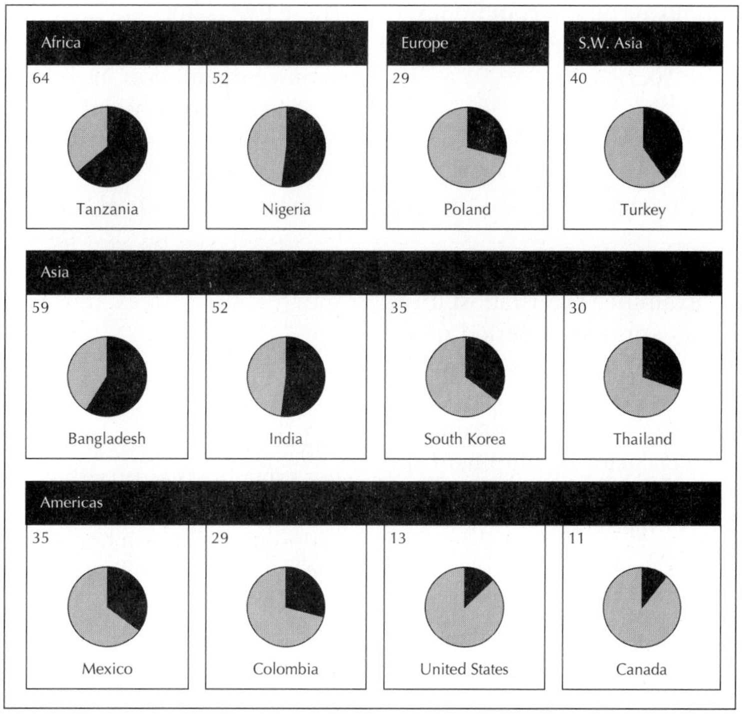

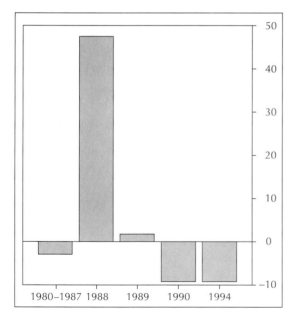

FIGURE 10.2

Food consumption as a percentage of total household expenditure for selected countries, 1981–1986. *(Source: United Nations, 1992, p. 140)*

FIGURE 10.3

The worsening terms of trade for ores, minerals, and nonferrous metals, 1980–1990. Many underdeveloped countries cannot alter the composition of exports rapidly in response to changing relative prices. *(Source:* Handbook of International Trade Statistics, *1993, CIA, p. 170)*

◈

T a b l e 10.2

Prices of Selected Ores, Minerals, and Nonferrous Metals, 1988–1991

	Annual Percentage Change			
	1988	*1989*	*1990*	*1991*
Copper	46	9	−7	−12
Aluminum	63	−23	−16	−14
Iron ore	4	14	12	7
Tin	5	19	−29	−11
Nickel	183	−3	−33	−1
Zinc	55	33	−8	−33
Lead	10	3	20	−32
Phosphate rock	16	13	−1	4
Ores, minerals, and nonferrous metals	48	2	−9	−11
Primary products	18	2	−3	−4

Source: Handbook of International Trade Statistics, *1993, C.I.A., p. 170.*

trial research geared to economizing on the commodity, or substituting something else for it, or producing it in the importing country (Figure 10.3).

COMPETITIVE ADVANTAGE OF A NATION

Michael Porter (1990), in his seminal study, *The Comparative Advantage of Nations*, examined the 10 most important trading nations in the world for a period of 5 years. Porter, from the Harvard University School of Business, concluded that four attributes of a nation combine to increase or decrease its global competitive advantage and world trade: (1) factor conditions; (2) demand conditions; (3) supporting industries; and (4) firm strategy, structure, and competition.

FACTOR CONDITIONS

To Porter, *factor conditions* include what we call in this text the *production factors*—land, labor, capital, technology, and entrepreneurial skill. Porter structured them slightly differently and elaborated on them as follows:

1. *Human resources:* the quantity of labor, the skill and cost of labor, the cultural factors that determine the motivation for the work ethic, and so forth.
2. *Physical resources:* (a) raw materials and their costs and (b) site and situation factors of the land and location itself, including physical geography, international time distance, and space conveyance.
3. *Capital resources:* all aspects of money supply and availability to finance the industry and trade from a particular country. Capital resources include the amount of capital available; the savings rate of the population; the health of capital, money markets, and banking in the host country; government policies that affect interest rates, savings rates, and the money supply; levels of indebtedness; trade deficits; public and international debt; and so forth.
4. *Knowledge-based resources:* research, development, the scientific and technical community within the country, its achievements and levels of understanding, and the likelihood for future technological support and innovation.
5. *Infrastructure:* all aspects of public services available to develop the conditions necessary for producing the goods and services that provide a country with a competitive advantage. Included are transportation systems, communications and information systems, housing, cultural and social institutions, education, welfare, retirement, pensions, and national policies on health care and child care.

These five factors are identified in current international and economic circles as the keys to the competitive advantage of a nation in the 1990s.

Porter also grouped the five factors into *basic factors*, which are fixed and have an availability limit, such as natural resources, location, distance from markets, and so forth, and *advanced factors*, which have no upper limits and can change dramatically. Examples of advanced factors are knowledge-based resources, capital resources, and human resources.

Demand Conditions

Porter defined *demand conditions* as the market conditions in a host country that aid the production processes in achieving better products, cheaper products, scale economies, and higher standards in terms of quality, service, and durability. Demand conditions cause firms to become innovative and therefore produce products that will sell not only in the local market, but also in the world market. For example, in the United States, consumers are well read and have a sophistication and demand that require producers to make products that are high value, durable, compact, and attractive. The result is a series of technologically advanced products that sell not only locally, but also internationally. Japan provides another example. Because of Japan's demand conditions—the compact nature of the country and its high population density—the Japanese have honed their techniques to produce high-value, attractive items that are smaller than many products found in America.

Different portions of a national market are in different stages of a life cycle. For example, mass-produced, identical items are no longer in high demand in the United States. Instead, individualistic applications of technology to produce goods for small market segments have led to flexible manufacturing and, consequently, because of demanding domestic buyers, high-quality, upscale items.

Supporting Industries

Related and supporting industries supply the parts and semifinished goods needed to complete and market a high-quality product. Supporting industries are vital to the overall global competitive advantage of industries in the domestic market.

Fiber-optic cables, for example, have aided the space-shrinking technologies of a global communications industry. Without them, today's complex global economic system could not function as it does, nor could it expand. Likewise, various means of propulsion, including the steam engine, revolutionized transportation technology. The use of the internal combustion engine and different parts of the automobile transformed the way that we live, move, and settle. In addition, the semiconductor and microprocessor industry allowed the extremely successful American computer manufacturing industry to be born and to grow. For example, in 1965, the memory capacity of a computer chip was 1,024 bytes of information (1K). By 1970, the capacity had increased to 10K; by 1977, 16K; and by 1980, 64K. Since 1970, a new high-capacity memory chip has appeared approximately every 3 years, each with quadruple the capacity of the last one. The chip that appeared in 1983 had a capacity of 256K, and through the years 1989, 1992, and 1996, chip capacity increased from 4 to 16 to 64 megabytes (MB).

Firm Strategy, Structure, and Competition

Firm strategy, structure, and competition relates to the conditions under which firms originate, grow, and mature. For example, because stockholders demand U.S. companies to show short-term profits, U.S. corporate performance may be less successful in the long run than it would be if it were judged for a much longer time, as Japanese and German corporate performance is.

State support of corporate strategy and performance is important. For example, a country can regulate taxes and incentives so that investment by a firm is high or low. In addition, competition within a country can impose demands on company performance; new business formations often pressure existing firms to improve products and lower prices and thus increase competitiveness. In Japan, for example, no fewer than 200 firms produce portable cassette player earphones. The market is extremely competitive, so quality is high and price is low. U.S. industries have adopted the so-called developmentalist stance of the Japanese government; doing so has fostered competitiveness in Japan but halted U.S. penetration into its markets.

Why Production-Factor Flows Occur

Besides trade, movement of international production factors can overcome differences in resource availability among countries. Profit and economic efficiency are the basic forces underlying the demand and supply incentives for international flows of production factors. From the supply side, the search for more profitable employment of production factors is the primary motive for most factor flows. For example, *capital* commonly goes to places where interest, profits, and capital gains promise to be favorable. Labor—unskilled, technical, or managerial—often flows to where opportunity and potential returns are better. The incentive in international *technology transfer* results from a desire to tap additional markets. From the demand side, international flows of production factors are frequently initiated in a similar way: The company that needs capital, labor, and technology starts the search in response to its need to lower costs, improve its productive efficiency, and introduce new products.

Let us look at capital flows in detail and then look briefly at labor and technology flows. The market for capital flows is far more efficient and integrated than the markets for labor and technology. The labor and technology markets generally suffer from poor information flows. In addition, the labor market is the most politically and socially sensitive of the international markets—and the most heavily regulated.

Harvesting coffee in Colombia. Coffee plantations have survived independence movements, having been originally established by European colonization. Coffee plantations are mainly run by multinational corporations. Plantations were established in the most agriculturally rich areas of the tropics and the workforce was originally slave labor. The world's largest coffee exporter, Colombia, also exports bananas and flowers, not to mention illegal cocaine. The capital, Bogota, looks very Americanized, but the rest of the country is struggling with important social, economic, and political problems. *(Photo: Diego Goldberg, Sygma)*

Forms of Capital Flow

Capital movement takes two major forms. The first type involves lending and borrowing money. Lenders and borrowers may be in either the private or the public sector. The public sector includes governments or international institutions such as the World Bank and agencies of the United Nations.

The second type of capital movement involves investments in the equity of a company. If a long-term investment does not involve managerial control of a foreign company, it is called *portfolio investment.* If the investment is sufficient to obtain managerial control, it is called *direct investment.* Multinational corporations are the epitome of direct investors.

Sources of Capital Flow

Monetary capital is the result of historical development. Unlike a natural resource like iron ore, it must be accumulated with time as a result of the willingness of a society to defer consumption. Low-income countries have low capacities to generate investment capital; all the capital that they do generate is usually employed domestically. Developed countries have much greater capacities for generating investment capital. They provide most of the world's private-sector capital, although a few fast-growing underdeveloped countries, such as South Korea, are also capital exporters.

Optimally, financial markets should produce an efficient distribution of money and capital throughout the world. However, there are many barriers to optimal distribution. Personal preferences of investors, practices of investment banking houses, and governmental intervention and controls confine money and capital movements to well-worn paths. They flow to some areas and not to others, even though the need in neglected areas may be greater.

International Money and Capital Markets

The global expansion of the financial system has three components: the internationalization of (1) domestic currency, (2) banking, and (3) capital markets. *International currency markets* developed with the establishment of floating exchange rates in 1973 and with the growth in private international liquidity, mostly in the form of Eurocurrencies. The growth of Eurocurrencies was only partly a reflection of the decline of the dollar, because the dollar remains the major trading currency. It was more the result of continued integration of the world economy—with the growing internationalization of productive and financial capital—and the result of increased competi-

tion among financial institutions, especially the commercial banks.

What are Eurocurrencies? *Eurocurrencies* are bank deposits that are not subject to domestic banking legislation. With relatively few exceptions, they are held in outside countries, "offshore" from the country in which they serve as legal tender. They have accommodated a large part of the growth of world trade since the late 1960s. The Eurocurrency market is attractive because it provides funds to borrowers with few conditions; it also offers investors higher interest rates than can be found in comparable domestic markets.

EURODOLLARS

At first, Euromarkets involved U.S. dollars deposited in Europe; hence, they were called *Eurodollar* markets. Although the dollar still represents about 80% of all Eurocurrencies, other currencies, such as the deutsche mark and yen, are also vehicles of international transactions. Therefore, *Eurocurrencies* is preferred to the less accurate term *Eurodollar.* However, even *Eurocurrencies* is a misnomer. Only 50% of the market is in Europe, the major center of which is London. Other Eurocenters have developed in the Bahamas, Panama, Singapore, and Bahrain.

Eurocurrencies first became significant with the growth in the Eurodollar deposits of the former Soviet Union. In the immediate postwar years, the Soviets doubted the safety of holding dollar reserves in the United States (where they could be confiscated) and transferred them to banks in Paris and London. However, three occurrences added impetus to the market in Eurocurrencies. First, during 1963 and 1964, President Kennedy, worried by the increased flow of dollars abroad, announced a program of capital control that lasted until 1973. As a consequence, international borrowers looked to Europe and the Eurocurrencies market. The main borrowers were U.S. multinational corporations raising loans to continue their expansion abroad. Second, in 1971, the United States began to finance its budget deficit by paying its own currency and flooded the world with dollars that helped to fuel worldwide inflation. Third, the oil crises of the 1970s were a major stimulus to the growth of Euromarkets. *Petrodollars* (OPEC oil surpluses) poured into the major Eurobanks.

THE DEBT CRISIS OF DEVELOPING COUNTRIES

The banks had to find outlets for all the money that they suddenly found in their coffers. One outlet was to send money to developing countries. Commercial lending to LDCs—along with official lending and aid—grew rapidly between 1974 and 1982. As a result, the total debt of underdeveloped countries rose fourfold, from about $140 billion at the end of 1974 to about $560 bil-

The city of London is a major center of Eurocurrencies, international banking, and capital markets. Because American banks were not happy with government controls in the post-war period, they began setting up branches overseas. This was called moving offshore. These offshore deposits, whether they came from the United States or Japan, were called Eurodollars. They were gladly received by most international banks because they were not regulated by national banking controls. London became the center of Eurocurrencies. Because banks were able to make loans on the basis of their Eurodollar accumulations, offshore banking business expanded rapidly during the 1970s and 1980s, leading to the internationalization of world finance. Eurocurrency deposits increased much more rapidly than official country reserves, allowing international money markets and international banks to take over international finance from their own governments. Major banks, such as Citibank or Chase Manhattan, expanded rapidly in the 1960s and 1970s, returning high rates of profit. *(Photo: Matthew Weinreb, The Image Bank)*

lion in 1982. Developing countries were happy to take advantage of this unaccustomed access to cheap loans with few strings attached. The borrowing enabled them to maintain domestic growth. However, these countries could not pay off the debts that they incurred, so now a widespread *debt crisis* exists among them, especially in Central and South American countries such as Mexico, Brazil, and Argentina (Figure 10.4, Table 10.3).

FIGURE 10.4
The external debt of developing countries grew substantially between 1970 and 1992. It consists of two parts: the public debt, which is owed to foreign governments, and the private debt, which is owed to private banks. *(Source: International Monetary Fund, 1991, p. 24)*

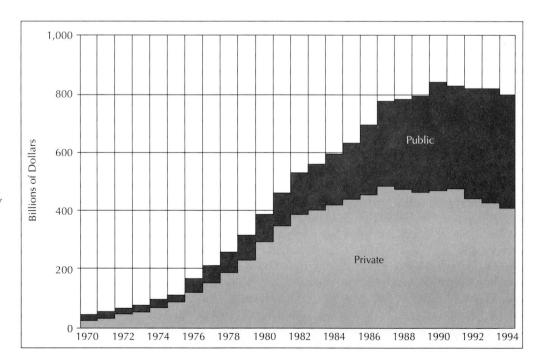

◈

T a b l e 10.3

Selected Heavily Indebted Less Developed
Countries, 1990

Country	Total External Debt (Billions of Dollars)	External Debt as a Percentage of GNP
Brazil	111	35
Mexico	96	48
Argentina	65	123
India	63	27
Indonesia	53	56
Venezuela	33	75
Nigeria	33	114
Philippines	29	66
Thailand	23	32
Chile	18	72
Colombia	17	44

Source: World Bank, 1991, pp. 244–245, 250–251.

In 1982, top-level bankers expressed concern about the stability of the international monetary system. The cause of this instability was the overexpansion of credit, particularly through the Eurocurrency market in the 1960s and 1970s, which led to a crisis that had its roots in the overaccumulation of capital and the declining rate of profit. The general crisis of a declining rate of profit was exacerbated by the imbalances caused by the oil-price hikes of the 1970s.

The financial crisis came to a head in 1982. A sharp rise in bankruptcies involving industrial capital put pressure on the banks and other financial institutions. In May 1982, when the American brokerage Drysdale went bankrupt, the heavy losses sustained by Chase Manhattan Bank forced the Federal Reserve Bank to pump approximately $3 billion into the banking system. Worse followed. In August 1982, Mexico ran into difficulties meeting interest and capital payments on its debts. Brazil and Argentina also appeared ready to default. A collapse of the financial system was forestalled by a series of emergency measures designed to prevent large debtor countries from defaulting on their loans. These measures involved banks, the International Monetary Fund (IMF), the Bank for International Settlements, and the governments of lending countries in massive bail-out exercises that accompanied debt reschedulings. The debt overhang persists in the 1990s because debt-service ratios—annual interest and amortization payments as a percentage of total exports—remain at dangerously high levels. In 1990, it was estimated that LDC external debt was equal to 45% of LDC gross national product (GNP) (World Bank, 1992). Consequently, the developing world is extremely vulnerable to changes in the world economy.

INTERNATIONAL BANKING

Paralleling the internationalization of domestic currency is the *internationalization of banking.* International banks have existed for centuries; for example, banking houses such as the Foggers, the Medici, and the Rothschilds helped to finance companies, governments, voyages of discovery, and colonial operations. The banks of the great colonial powers—Great Britain and France—have long been established overseas. American, Japanese, and other European banks went international much later. Major American banks—Bank of America, Citicorp, Chase Manhattan—moved into international banking in the 1960s, and the Japanese banks and their European counterparts in the 1970s. Today, Japanese-based banks, such as Sumitomo Bank, Fuji Bank, and Mitsubishi Bank, surpass the U.S.-based banks in global banking.

Banks were enticed into international banking because of the explosion of foreign investment by industrial corporations in the 1950s and 1960s. The banks of different countries "followed the flag" of their domestic customers abroad. Once established overseas, many of the banks found international banking highly profitable. From their original focus on serving their domestic customers' international activities, banks evolved to service foreign customers as well, including foreign governments.

As we said, a major problem for the banks is that their lending decisions to LDCs were often imprudent and resulted in excessive indebtedness. They lent too much money. For example, in 1981, the ratio of the banks' capital to total assets was around 4%, whereas in the early 19th century it was about 40%. Loans to Mexico from the Bank of America amounted to more than 70% of the bank's capital. Thus, the banks have a collective interest in the debts of other countries. In 1990, Mexico's foreign debt was more than $100 billion. Mexico owed $26 billion to U.S. banks and the rest to multilateral lending institutions, Western Europe, and Japan. If indebted countries such as Mexico were to default, a number of global banks would fail. The result would be dim prospects for further internationalization of banking and capital and for continued growth in the world economy in the 1990s.

CAPITAL MARKETS

Capital markets, or long-term financial markets, form the third component of the international financial system. Stock exchanges, futures exchanges, and tax havens have proliferated. American, European, and Asian multinational corporations take advantage of *tax-haven countries,* countries near their home coun-

Table 10.4
Tax and Banking Havens

No Taxation
Bermuda, Bahamas, Monaco, Cayman Islands, New Hebrides, Macao, Andorra, Anguilla, Maldive, Brunei

Low Taxation
Virgin Islands, United Arab Emirates, Hong Kong, Isle of Man

Strict Banking Secrecy
Switzerland, Costa Rica, Andorra

Tax Benefits for Companies
Jamaica, Antigua, Barbados, Liberia, Bahrain, Philippines, United Kingdom, Luxembourg, Greece, Liechtenstein, Cyprus

Tax Benefits for Shipping Companies
Panama, United States, Malta, Cyprus

Tax Benefits for Certain People
Israel, Liberia, Sri Lanka

tries where taxes on foreign-source income or capital gains are low or nonexistent (Table 10.4). The internationalization of capital markets has made international finance a round-the-clock business, with trade in currencies, stocks, and bonds transversing the world with the passage of the sun.

Financing International Trade

International trade is not primarily a barter transaction. Money for products is transacted as opposed to products for products. Thus, the buyer country must swap its currency, in proportion to the value of the product, for the currency of the exporting country. If a Silo or Circuit City retail chain in the United States wants to buy televisions, video camcorders, or AM/FM stereo cassettes from Japanese firms, the buyer must, unless otherwise agreed upon, convert U.S. dollars to Japanese yen in order to satisfy the terms of the purchase. As in the case with most international transactions—exports and imports—the seller receives payment in the currency of his or her own country, not in the currency of the purchasing country.

The value of the U.S. dollar, compared with foreign currency, is called the dollar's *exchange rate.* An exchange rate is the number of dollars required to purchase one unit of foreign money. For example, in early 1997, the price of Japanese yen, as stated in U.S. currency, was 108. That is, 1 American dollar would purchase 101 Japanese yen. A Canadian dollar would be purchased for $0.80 in U.S. currency, or 1 U.S. dollar could purchase 1.27 Canadian dollars. Table 10.5 shows that the value of the U.S. dollar strengthened in Europe

Table 10.5
1997 Foreign Exchange Rates

Country (Unit)	Per One U.S. $	1992	Country (Unit)	Per One U.S. $	1992
Africa			**Europe**		
Kenya (shilling)	41.32	20.83	Austria (schilling)	10.14	11.310
Morocco (dirham)	7.80	8.97	Great Britain (pound)	.55	.547
South Africa (rand)	4.25	3.17	Denmark (krone)	5.51	6.176
The Americas			Finland (mark)	5.33	4.340
Brazil (eruzeiro)	0.86	3745.00	France (franc)	4.99	5.400
Canada (dollar)	1.27	1.20	Germany (mark)	1.46	1.610
Mexico (nuevo peso)	6.30	3096.00	Greece (drachma)	225.30	111.900
Asia-Pacific			Ireland (pound)	.55	.590
Australia (dollar)	1.18	1.33	Italy (lira)	1429.00	1190.000
Hong Kong (dollar)	7.19	7.73	Spain (peseta)	121.30	100.900
India (rupee)	30.27	30.77	Switzerland (franc)	1.26	1.420
Japan (yen)	108.40	127.20			
South Korea (won)	733.40	805.80			
Tahiti (franc)	85.80	100.90			

From Foreign exchange rates, San Diego Union, January, 1997.

Located on Broad Street between Wall Street and Exchange Place, the New York Stock Exchange is a hub of financial activity—not only of the city and nation but of the world as well. The New York Stock Exchange is a domestic financial market open to participation by foreign corporations, governments, and international institutions, as both users and suppliers of funds. Foreign companies also have shares of stock listed on the New York Stock Exchange, just as they do on the London and Tokyo exchanges. *(Photo: New York Convention and Visitors Bureau)*

during 1992–1993, especially in Great Britain, Greece, Italy, Spain, and the Scandinavian countries, as a result of economic stagnation abroad and currency devaluation by European governments. However, the value of the dollar decreased in Asia and the Pacific during the same period. For American travelers, this meant taking a cruise to the Greek Islands or examining the ruins of Rome but postponing the hiking trip to Mount Fuji.

DETERMINING EXCHANGE RATES
If the value of a currency fluctuates according to changes in supply and demand for the currency on the international market, a *floating exchange rate* is in effect. Figure 10.5 shows the relationship between Mexican pesos and U.S. dollars. The demand curve, D, shows dollars sloping downward to the right. U.S. citizens will demand more Mexican pesos if they can be purchased with fewer dollars. Point p_0 suggests that fewer pesos can be purchased with a dollar, whereas Point p_1 suggests that many more pesos are available per dollar. For example, 1 dollar could purchase 3,096 pesos in June 1992.

The demand for Mexican pesos in the United States is based on the amount of goods and services that a U.S. citizen wants to purchase in Mexico. A lower exchange rate for the peso makes Mexican goods less expensive to Americans.

In Figure 10.5, the supply of pesos is upward sloping to the right, which suggests that, as the number of dollars increases per 10,000 pesos, more pesos are offered in the marketplace. Mexican residents desire more goods from the United States when the dollar exchange rate (price) for the peso is high. The more dollars per 10,000 pesos, the relatively cheaper American products are for Mexicans. Therefore, Mexican residents will demand more dollars with which to purchase American goods and will consequently supply more pesos to foreign-ex-

change markets when the exchange rate for the peso increases. The equilibrium position is reached when supply and demand conditions for foreign exchange is, therefore, based on supply and demand of international goods produced in Mexico demanded by Americans and American goods demanded by Mexicans.

Line S_2 represents devaluation of the peso because of economic restructuring in Mexico in 1986. This restructuring effectively reduced the number of dollars per 10,000 pesos on the international market. The result was that fewer American products could be purchased for the same amount of pesos because American goods and services became relatively more expensive. Cross-border purchases by Mexican border residents decreased dramatically, as did the international flow of goods and services from America to Mexico. At the

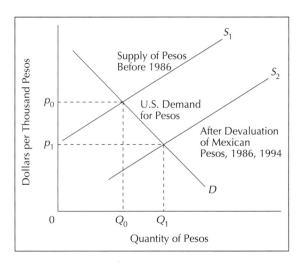

FIGURE 10.5
Determining the exchange rate of the U.S. dollar and the Mexican peso.

same time, the quantity of pesos available to Americans increased from q_0 to q_1. American purchasers poured into border communities and increased their travel to the main tourist destinations within Mexico. The international flow of goods from Mexico to the United States increased because Mexican goods and services were relatively less expensive.

When the dollar appreciates in foreign-exchange markets relative to other currencies, it can buy more foreign currency and, therefore, more goods and services from other countries (Figure 10.6). As a result, American retailers import more goods to the United States. At the same time, the appreciated U.S. dollar means more costly American goods and therefore less demand for them. Exports from America decline under these circumstances. Thus, a so-called strong dollar, which means that the dollar can buy more units of foreign currency, is not always desirable for U.S. trade. Conversely, when the dollar depreciates and is a "weak" dollar, it can buy fewer units of foreign currency and therefore fewer goods and services. Imports usually decline under these circumstances.

WHY EXCHANGE RATES FLUCTUATE

Exchange rates fluctuate for five reasons. First, as a country becomes wealthier and increases its real output and efficiency compared with that of other countries, it imports goods from abroad. The result is that the increased demand for foreign currency raises the exchange rate of the currency and decreases the value of the dollar internationally.

A second factor that influences the exchange rate is the inflation rate of a nation. If the inflation rate of one nation increases faster than that of its trading-partner nations, the currency of the nation with high inflation will depreciate compared with the currency of its trading-partner nations. Consequently, the products of the trading-partner nations will be more attractive to consumers in the country with high inflation. For example, when the U.S. inflation rate increases, the demand for Canadian dollars increases. This demand raises the U.S. dollar price of both the Canadian dollar and Canada's goods and services. At the same time, Canadians demand fewer of the comparatively higher-priced U.S. goods. Thus, Canada supplies fewer Canadian dollars to the U.S. foreign-exchange market. As a result, the U.S. dollar depreciates.

Third, domestic demand is a factor in determining exchange rates. Real income growth and the relative price levels between countries affect domestic demand. However, domestic demand also depends on consumer tastes and preferences. Americans will pay higher prices for specialty items and technologically advanced foreign goods than for comparable products at home. Examples of these foreign goods are VCRs and electronic consumer products from Japan, French wines and perfumes, German automobiles, and Italian shoes. A shift in the direction of foreign goods decreases domestic demand and causes the dollar to depreciate.

Fourth, the dollar may appreciate on foreign exchange markets if interest rates in America increase and, therefore, provide a higher yield to foreign investors who are interested in U.S. assets. Foreign investors increase their demand for dollars in order to purchase American companies and thus supply more of their foreign currency in exchange for U.S. dollars in the world foreign-exchange markets. In the early 1990s, U.S. interest rates were the lowest in 20 years, so rates on investments were lower than those in foreign nations. Foreign investors, although still drawn to companies, land, and fixed assets,

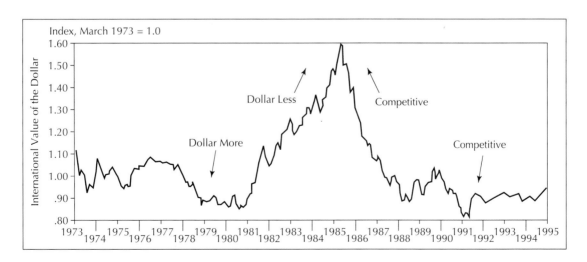

FIGURE 10.6

International value of the dollar, 1973–1995. From 1981 to 1986, the value of the dollar was relatively high compared to other major currencies. Since foreign goods were cheap, the U.S. imports increased, and their exports decreased. The dollar fell abruptly from 1986 to 1988 and has remained relatively weak on international currency exchanges, but the trade deficits continue. *(Source: Expanded from McConnell and Brue, 1993, p. 425)*

◈

Table 10.6
Trade in the U.S. Economy, 1960–1995*

	1960		1975		1995	
	Amount†	Percent of GDP	Amount†	Percent of GDP	Amount†	Percent of GDP
Exports of goods and services	$23.3	4.9	$136.3	8.6	$667	10.4
Imports of goods and services	22.3	4.4	122.7	7.7	817	11.0
Net exports	4.4	0.5	13.6	0.9	−150	0.6

*Data are on a national income accounts basis.
† In billions of dollars.
Source: Statistical Abstracts of the United States, *1996, p. 81.*

were not drawn to U.S. government securities, and the associated real rates of return were low. Therefore, the dollar continued to depreciate in the early 1990s.

Fifth, currency speculation helps determine exchange rates worldwide. Real events are important in determining exchange rates, just as they are in determining home prices in any U.S. city. However, the expectation of future events is almost as important as actual events. The expectation that economic events will cause the dollar to appreciate or depreciate promotes currency speculation that may, in the not-too-distant future, be a self-fulfilling prophecy. If a major event such as a sudden war or the assassination of a major political figure occurs in a foreign country, it can trigger fear, which encourages individuals to sell the currency of that

country to buy dollars. If all people reacted in the same manner to such events, the market would be driven down for that currency against the dollar, and the anticipated depreciation would actually occur.

U.S. Trade Deficit

The United States enjoyed a trade surplus throughout most of its history. Starting in 1976, however, the volume of imports began to exceed the volume of exports (Table 10.6). The merchandise trade deficit was $25 billion in 1980, but by 1987, it had jumped to $160 billion, then decreased to $74 billion in 1991. What was the cause of the American merchandise trade deficit and the

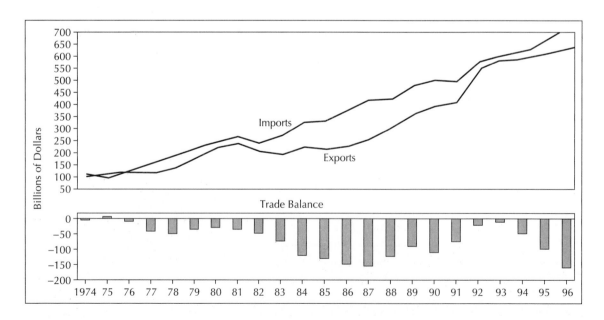

FIGURE 10.7
U.S. merchandise trade, 1974–1996. Until 1970, U.S. export value exceeded the value of imported goods. Since that time, the value of imports has exceeded exports, creating a trade imbalance. The greatest level of trade imbalance occurred between 1984 and 1988, mirroring the strong international value of the dollar. *(Source:* Statistical Abstracts of the United States, *1997, p. 81)*

negative trade balance? How has this deficit affected the United States?

CAUSES OF THE U.S. TRADE DEFICIT
Three causes of the U.S. trade deficit are generally cited: the increase in the value of the dollar, the rapid growth of the American economy, and the decrease in the volume of goods exported to less developed countries.

As shown in Figure 10.6, the international value of the dollar increased substantially between 1980 and 1985. This increase meant that American currency and products were expensive to foreign nations, whereas foreign goods were less expensive to Americans. Consequently, Americans imported a large number of products from other countries—more than they exported. The American government borrowed heavily to finance its deficit. This increased the demand for money locally and boosted interest rates. High interest rates lured foreign investors to American securities. These investors expected high returns on their investments.

During these Reagan years, the United States shifted to a more stringent policy to control inflation. The money supply was reduced relative to its demand, which kept interest rates high. The low inflation and high interest rates again increased foreign demand for U.S. investments because the real rate of return on investments was high.

By 1985, when the value of the dollar was 60% more than its 1980 average, the deficit was $212 billion. What caused the dollar to decline, and what effect did this have on the trade balance?

As the value of the dollar increased sharply and the volume of foreign imports to America increased, the demand for foreign currency to pay for the expanding volume of imports began to lower the value of the dollar to 1980 levels. In the United States, the demand for yen, marks, francs, pesos, and other currencies increased. At the same time, Great Britain, France, Japan, Germany, and the United States agreed to increase the supply of dollars in foreign-exchange markets to reduce the value of the dollar. The value of the dollar began to shift downward rapidly, but the annual federal deficit persisted through 1994, even though the export-import trade imbalance decreased by 50% from 1987 to 1994 (Figure 10.7).

The second reason for the large trade deficit of the 1980s and 1990s is that the economic growth in America outpaced that of the rest of the world by a large

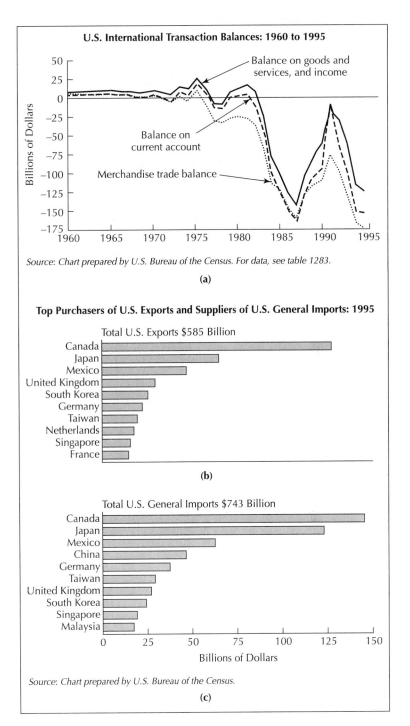

FIGURE 10.8
(a) U.S. international transaction balances, 1960–1995. In 1997, the United States had a $188 billion trade deficit, while the United Kingdom had a $22 billion trade deficit. **Canada** enjoyed a $33 billion trade surplus, along with Japan at $98 billion, Germany at $86 billion, Sweden at $22 billion, and Mexico at $5.6 billion. (b) Top purchasers of U.S. exports and suppliers of U.S. general imports, 1995. (c) **Canada** leads both as the top U.S. export recipient and total U.S. general import donor.

amount. For example, America's growth was three times European growth in 1984, and American growth continued to outpace European growth between 1985 and 1990 (Figure 10.8).

Rapid growth affected domestic consumption, so Americans purchased foreign goods in proportion to their relatively high domestic income. One result was that America, by 1990, had a substantially larger real per capita purchasing power than that of any European country, even though nominal rates in Switzerland, Sweden, and Germany were higher than those in America. This slower European growth meant slower importation of U.S. goods.

The third factor that caused the American trade deficit was the reduction in exports to less developed countries. As the international debt of less developed countries increased in the 1980s, many of those countries, including Mexico, Brazil, and Argentina, restricted the amount of foreign imports and restructured their debts, agreeing to lessen their trade deficits. These austerity programs reduced U.S. exports. With devalued currencies, these countries could not afford U.S. goods. The combined effect was that the United States could export fewer goods to developing countries, while importing a large amount of goods from those countries.

RESULTS OF THE U.S. TRADE DEFICIT

The U.S. deficit had three results: First, a reduction in aggregate demand occurred in the United States. Furthermore, an increase in the volume of imports pushed prices of domestic goods downward as domestic goods competed with imports for consumer demand.

Second, the trade deficit hurt the industries that are highly dependent on international trade. In the 1980s, for example, automobile manufacturers, steel manufacturers, and the American farmer struggled the most because of the trade deficit.

Third, America is now a debtor nation instead of a creditor nation, owing foreign governments more than they owe us. The U.S. foreign debt climbed to $721 billion in 1991, which made the United States the largest debtor nation in the world. In other words, American consumers have been subsidized because more goods and services flowed into the country than flowed out of the country. The reverse is true for Japan and its consumers. America has been living above its means, and consumers have received an economic boost during this period. This boost in domestic living standards is only temporary and not without cost to America. The federal budget deficit and the balance of trade deficit have led to the so-called selling of America to foreign investors. Foreign investors now own 20% of America's total domestic assets.

Classical Capital Theory

In the 19th and early 20th centuries, foreign portfolio investment overshadowed *foreign direct investment* (FDI).

Theorists concentrated therefore on foreign portfolio investment, which was directed toward raw-material extraction, agricultural plantations, and trade facilities. The theory of FDI received relatively little attention and remained underdeveloped. Given the massive scale of FDI today, however, an understanding of the rationale for such investment is important. The only help that classical capital theory gives us is that under free-market conditions capital will flow from where it is abundant to where it is in short supply or, in other words, from where the rates of return are low to where they are high.

As with classical trade theory, classical capital theory is macroeconomic. However, in reality international money and capital flows are dependent on managerial decisions. Therefore, it is necessary to examine international capital movements from a microeconomic perspective. Even a brief examination reveals that although foreign investment is a simple process conceptually, a complex of motivations is involved.

Motivation for Foreign Direct Investment

The primary reason for a firm to go international is profit. Three strategic profit motives drive a firm's decision to operate abroad. One motive for many direct investments is to obtain natural or human resources. *Resource seekers* look for raw materials or low-cost labor that is also sufficiently productive. A second motive is to penetrate markets, especially when *market seekers* have been prevented from exporting to a particular country. The third goal is to increase operating efficiency. *Efficiency seekers* look for the most economic sources of production to serve a worldwide, standardized market. These three motives are not mutually exclusive. Some segments of a corporation's operation may be aimed at obtaining raw materials, whereas other segments may be aimed at penetrating markets for the products made from the raw materials. These operations may also result in some productive and market efficiencies.

There may be a strong motivation for a firm to internationalize, but there may also be compelling constraints. Prominent among these are the uncertainties of investing or operating in a foreign environment. Consumers' incomes, tastes, and preferences vary from country to country. Japanese consumers, for example, are wary of foreign products, at least those that are not name brands. Cultural differences in business ethics and protocol, attitudes regarding time, and even body language in interpersonal relationships complicate the task of conducting business in two or more languages (McConnell and Erickson, 1986). Added to these barriers are problems relating to currencies, laws, taxation, and governmental restrictions (Stutz, 1992).

Bias in Foreign Direct Investment

Although managers may have the initiative and ability to implement rational investment scenarios, they often take a path of least resistance, which results in a less than ideal allocation of the world's investment capital and of the investors' capital. For example, the direct investment patterns of American companies overseas exhibit considerable bias. In the 1980s, 45% of FDI by American companies was in Europe, 20% in Canada, and 13% in Great Britain, but only 3% was in Japan and less than 2% in Africa (excluding South Africa). These patterns indicate a geographic bias (investing close to home), a cultural bias (investing in countries with similar cultures, especially the same language), and a historical bias (investing in countries to which they are tied historically). Historical bias is often encouraged by the investor's government. This bias maintains a strong national presence in these countries. The

trend in the 1990s is for the United States to increase investment in the Far East, countering past FDI biases.

Origin and Destination of Foreign Direct Investment

The period since World War II has been characterized by a massive flow of FDI associated with the growth of multinational corporations. Capital export was five times more in 1986 than it was in 1970 (United Nations, 1988). The postwar years have seen the development of the international car (Figure 10.9), television set, cassette recorder, and home computer, with different components produced in different countries under the same corporate control.

In addition to manufacturing multinationals, the growth of international banks and of service multinationals has been strong. Service multinationals sell services related to business and professional activities—medical, publishing, agrotechnology, warehousing, distribution, computer science, laboratory testing, hotel management, education, entertainment—and personal and social activities. The share of America's outward

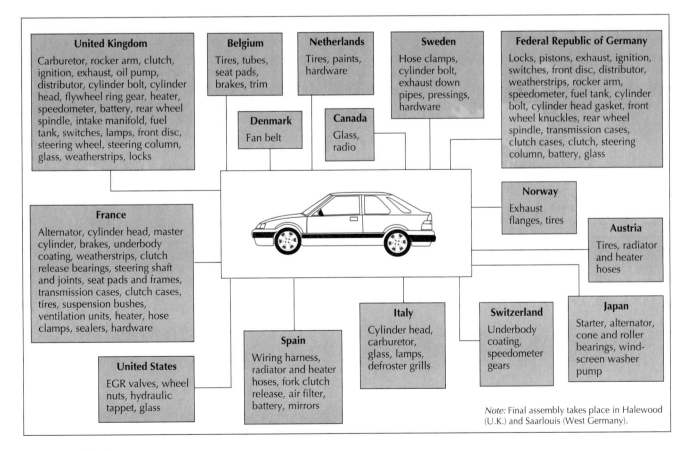

FIGURE 10.9
The international car: the component network for the Ford Escort (Europe). *(Source: Dicken, 1992, p. 301)*

investment in services increased from 24% in 1975 to 34% in 1990; Great Britain's share, from 29% to 35%; and Japan's share, from 36% to 52%. In the 1990s, a still larger share of outward investment has been devoted to exportable services, particularly those related to improving international markets, such as information. In fact, "the firm may simply be the supplier of information to those essentially national firms responsible for physically moving the goods and services and for investment in fixed industrial assets" (Robinson, 1981, p. 17).

World Investment by Multinationals

American firms lead the world in FDI, but their share of the total is slipping. Until the early 1970s, U.S.–based multinationals accounted for nearly two-thirds of the world's corporate investment abroad. This figure decreased to 45% in 1977, and it continued to fall throughout the 1980s as corporations headquartered in other countries stepped up their rates of FDI. The rate of increase has been most rapid for companies domiciled in Western Europe and Japan; however, some underdeveloped countries have also increased their outflow of FDI (Figure 10.10). Major bases are Hong Kong, Brazil, Singapore, South Korea, Taiwan, Argentina, Mexico, and Venezuela.

In 1990, the share of stocks held abroad by U.S. companies fell to approximately one-third of the world total. The United Kingdom, a country of only 60 million, ranks in second place, with approximately 15% of world FDI, followed by Japan, a country of 120 million, at 13%. Germany is not far behind with approximately 10%, followed by Switzerland and the Netherlands, each at approximately 6.5%. Other industrialized countries make up 15%, whereas less developed countries make up only 2%.

Significant changes in the destination of FDI are the increased flow to the United States and to the LDCs (Figure 10.10). In 1975, the United States accounted for only a small proportion of the stocks held by foreign companies; 10 years later, the United States emerged as a major host country. Investment in the developing world has focused mainly on eight countries—Brazil, Mexico, Singapore, Indonesia, Malaysia, Argentina, Venezuela, and Hong Kong—which accounted for more than one-half the stock for foreign investment in underdeveloped countries in the early 1980s. Availability of natural resources, recent strong growth, and political and economic sta-

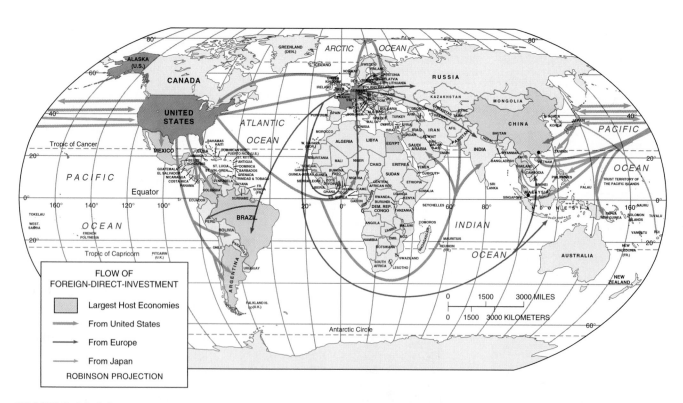

FIGURE 10.10

Flow of foreign direct investment. Multinationals' foreign direct investment originates, for the most part, in the United States, Japan, the United Kingdom, Germany, and France. These transnational corporations have invested most of their resources in other developed countries. In addition, U.S. multinational corporations are more likely than the Japanese or Europeans to invest in Latin America. European multinational corporations are more likely to invest in eastern Europe and the Middle East, while Japanese transnationals are more likely to invest resources in the Far East.

Foreign direct investment. Ford Motor Company headquarters in Britain. The Ford Motor Company is a true multinational corporation with 330,000 employees scattered throughout the world in 30 countries. Although accused of exploiting local labor in developing countries, Ford and other multinationals helped to stimulate the economy and, therefore, provide foreign sources of revenue for a struggling economy. *(Photo: Ford Motor Company United Kingdom)*

bility were among the factors that attracted foreign investment to these LDCs.

Investment by U.S. Multinationals in Foreign Countries

The pace of FDI by U.S. multinational corporations has increased since the 1960s. In 1960, almost two-thirds of foreign investment by U.S. companies was in nearby Canadian ventures. Other foreign investment was located in Europe and in Latin America. The Far East saw little investment by U.S. multinationals, partly because of distance, cultural, and political barriers. The pattern changed slightly, and by 1980, U.S. FDI was positioned chiefly in Europe in principally manufacturing operations.

Figure 10.11a shows that by 1995 Europe still maintained the preeminent area for U.S. FDI. During this period investment in Europe increased by 210%, whereas U.S. investment in Canada increased only 65%. The single greatest proportional increase, however, occurred in the Far East, with nearly a 370% increase in Japan and a 600–700% increase in Asia and the Pacific, primarily in the Four Tigers of South Korea, Taiwan, Singapore, and Hong Kong, but also in Malaysia and the Philippines. Part of the reason for the increase in investment in the Far East is that the labor-intensive semiconductor assembly plants are located in these cheap-labor countries.

Because of strong Japanese barriers to foreign investment, Japan entertains a relatively small proportion of total FDI by U.S. firms, but numerous overseas affiliates (Figure 10.11b). Another notable and disappointing world region is Africa, which during the 1980s received no increase in U.S. FDI, further hurting its opportunities for development.

Until the 1980s, FDI by U.S. firms centered on manufacturing and mining activity; more recently, investment has targeted real estate service activities (Figure 10.11c). By 1995, manufacturing and mining investment by U.S. firms worldwide had decreased to less than 50%, whereas investment in services amounted to 35% and was on the increase.

Investment by Foreign Multinationals in the United States

FDI in the United States grew at an incredible rate from 1970, when it was a skimpy $13 billion, to 1990, when it amounted to $550.7 billion. In fact, between 1980 and 1989, it increased more rapidly than did U.S. foreign investment (Figure 10.12). The popularity of investing in the United States was a result of the power and stability of the country economically and politically and the relatively inexpensive dollar in world markets. As discussed previously, during the 1970s and 1980s, the United States accumulated an increasing trade deficit. A chain of events created the intense FDI in the United States as a result of this deficit.

The dollar peaked in value internationally in 1985. This peak increased the amount of foreign imports to the United States and decreased the amount of domestic exports. The result was a trade deficit. Between 1985 and 1992, the value of the dollar began to fall. The trade deficit

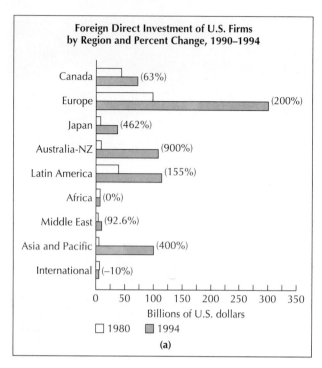

Foreign Direct Investment of U.S. Firms by Region and Percent Change, 1990–1994

Canada (63%)
Europe (200%)
Japan (462%)
Australia-NZ (900%)
Latin America (155%)
Africa (0%)
Middle East (92.6%)
Asia and Pacific (400%)
International (–10%)

Billions of U.S. dollars

□ 1980   ▪ 1994

(a)

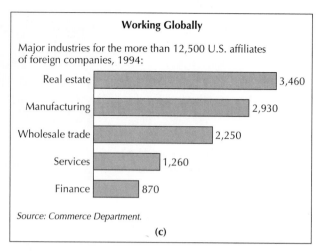

Working Globally

Major industries for the more than 12,500 U.S. affiliates of foreign companies, 1994:

Real estate 3,460
Manufacturing 2,930
Wholesale trade 2,250
Services 1,260
Finance 870

Source: Commerce Department.

(c)

FIGURE 10.11

(a) Foreign direct investment of U.S. firms by region and percent change, 1980–1994. Europe remained the favorite destination for U.S.–based multinational investments. Change from 1980 to 1994 was an increase of an investment by 210%. Latin America remains in second place, Australia is the only contender for third place, and Canada remains a steady fourth. The largest increase proportionally for foreign direct investment of U.S. companies occurred in Australia and the Pacific Rim countries, especially the Four Tigers, where the pace of U.S. investment accelerated sharply in the 1990s. *(Source:* Statistical Abstracts of the United States, *1996)* (b) Partners in industry. (c) Partners working globally.

Partners in Industry

Of the more than 12,500 U.S. affiliates of foreign companies, most were located in these nations in 1994:

Japan 3,281
Germany 1,281
U.K. 1,240
Latin America 1,078
France 661

Source: Commerce Department.

(b)

was reduced, but a large negative differential still exists. The deficit led to the outflow of American dollars into foreign hands. This money allowed foreign governments and corporations to buy American real estate, as Japan has done, and American factories, as Great Britain has done. Ironically, many foreign firms that sell to U.S. markets found it cheaper to produce products from plants that they own and operate *in* the United States. For instance, foreign multinational corporations such as Honda and Mazda opened plants in America to build automobiles that the U.S. formerly imported from Japan. Land, labor, and capital were cheaper in the United States; therefore, the product was less expensive to produce there.

As a result, foreign investment in the United States increased sharply. The average increase was 33% between 1980 and 1994 (Figure 10.12). Almost 1,000 new foreign enterprises were founded each year until 1994.

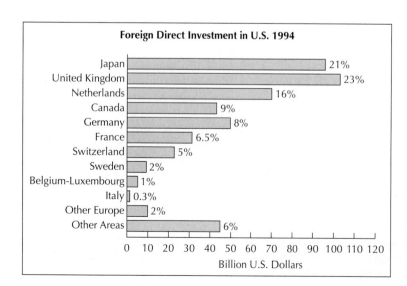

Foreign Direct Investment in U.S. 1994

Japan 21%
United Kingdom 23%
Netherlands 16%
Canada 9%
Germany 8%
France 6.5%
Switzerland 5%
Sweden 2%
Belgium-Luxembourg 1%
Italy 0.3%
Other Europe 2%
Other Areas 6%

Billion U.S. Dollars

FIGURE 10.12

Foreign direct investment (FDI) in the United States by source area, 1994. FDI in the United States is defined by the U.S. Bureau of the Census as all U.S. companies in which foreign interest or ownership is 10% or more.

U.S. Ford plant in Europe. General Motors and the Japanese manufacturers of automobiles produce two-thirds to nine-tenths of their products in their home countries. The Ford Motor Company produces two-thirds of its more than 5 million manufactured vehicles abroad. Ford Motor Company has been one of the most prolific of the American multinational corporations. The post-war years were the years of most rapid expansion with over 300 new subsidiaries being set up annually between 1960 and 1965 by American companies in foreign countries. American firms saw the advantage of multinational

expansion because post-war markets were developing rapidly for their goods in Europe, Japan, and throughout the world. By establishing manufacturing plants in those foreign countries, companies could reduce expensive transportation costs of finished products as well as import duties. A major factor is sometimes overlooked, and that is transfer pricing. Japan and European countries have higher rates of corporate tax than does America. Shrewd multinationals can invoice their subsidiaries in these foreign countries in such a way to show low profits in high tax countries and therefore shelter their income. *(Photo: Ford Motor Company United Kingdom)*

As of 1995, because of cultural affinities and the lack of language barriers, the United Kingdom maintained its lead as America's chief foreign investor, boasting 23% (Figure 10.12). Other European nations, especially the Netherlands, Germany, Switzerland, France, and Sweden, in that order, are strong investors in the United States. The Netherlands is especially noteworthy, if we consider its relatively small size. It commands a 16% share of FDI in the United States. Japan, however, had the largest proportional increase in the last 20 years. In 1970, Japan owned only 2% of foreign investment in the United States, but, by 1995, its share had risen more than tenfold to 21%.

FDI by Japanese firms is a relatively recent occurrence because of Japanese governmental policy, which regulates capital outflows in an effort to restrict foreign exchange claimed by other countries. The Japanese have been slow to invent new technology. They have found it more economical to invest in U.S. companies and thereby export knowledge-intensive activities and technology from the United States to Japan. Many of these knowledge-intensive industries are now well placed in Japan, but because of import restrictions and the high price of Japanese currency compared with the dollar, Japanese firms have found it even more profitable to sell to their chief buyer, the United States, from Japanese-owned manufacturing plants in Amer-

ica. The most notable example is the accumulation of Japanese-owned automobile-assembly plants and autoparts industries in the U.S. Midwest.

The Japanese have exported their labor-intensive industries to developing countries, where the labor rate is much lower. Notable locations include Australia, the Middle East, Brazil, and East Asia.

FDI in the United States is centered on manufacturing, chemicals, electrical machinery, electronics, pharmaceuticals, and services. Japanese investment also has controlling interests in 60 U.S. steel operations, 25 rubber and tire factories, 10 automobile-assembly plants, and 300 autoparts distributors. U.S. public attention has been heightened by recent Japanese purchases of retail and service industries that have been considered, until now, all-American industries—owned by Americans for Americans. Examples include the purchase of Columbia Records; the Music Company of America, which owns several motion-picture studios in Hollywood; Rockefeller Center; Federated Department Stores; the famous Spyglass Hill and Pebble Beach golf courses in Monterey, California; Yosemite Lodge and National Park Company. When Japanese interests anticipated buying the Seattle Mariners, an American professional baseball franchise, a public outcry was heard across the United States (Figures 10.13 and 10.14).

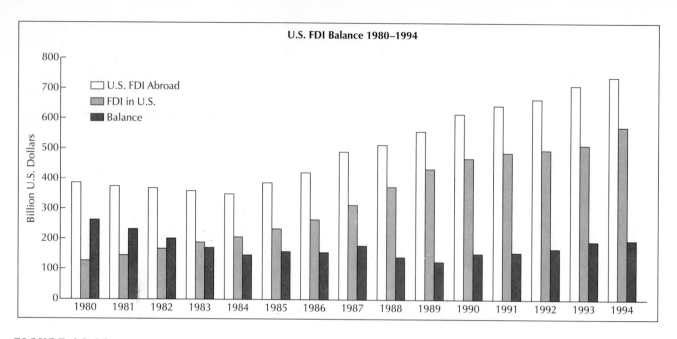

FIGURE 10.13
U.S. foreign direct investment position, 1980–1994. Investments abroad by U.S. multinationals is $130 billion greater per year than investment by foreign companies in the U.S. *(Source:* Statistical Abstracts of the U.S., *1997)*

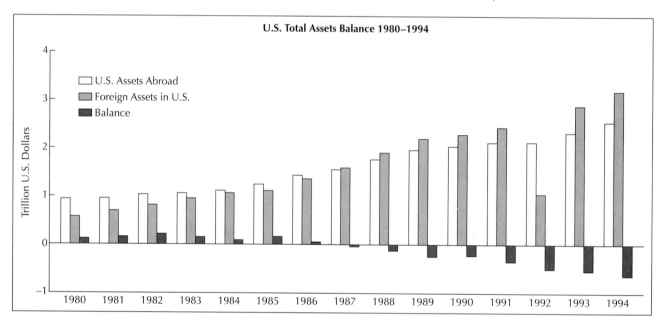

FIGURE 10.14
U.S. international investment position: U.S. assets abroad compared with foreign assets in the United States, 1980–1994. Foreign held assets in the U.S. have accumulated as a result of the mounting U.S. indebtedness to the rest of the world. This is shown by a comparison of total U.S. overseas assets, both governmental and private, with total assets held in the United States by foreign interests. After many years of positive asset balances with the world, the United States became a net debtor nation in the 1980s. *(Source:* Statistical Abstracts of the United States, *1997)*

Foreign Direct Investment in the United States by Region

Similar to the U.S. population and U.S. employment, FDI in the United States has moved west. The original pattern of investment in the Middle Atlantic and Great Lakes states of the north central United States has been historically strong by British and other Eu-

ropean nations, including Canada. Canadian investment has been the strongest, not unexpectedly, along the border states from the western north central region through New England, including the South Atlantic states. But the Pacific region of the United States leads overall in FDI, primarily because of recent heavy investment by the Japanese and other Far Eastern countries (Figure 10.15).

FIGURE 10.15
Foreign direct investment in the United States, 1981–1995. The value of foreign direct investment (FDI) during this period has quadrupled. The top four states accounted for more than a third of the total FDI, and California, Texas, and New York led all states. For the 25 states shown here, 85% of the total FDI is accounted for. A large influx of foreign direct investment has brought prosperity to the Rust Belt states of Michigan, Illinois, Indiana, and Ohio, with new extensions into Kentucky and Tennessee. *(Source:* Statistical Abstracts of the United States, *1997)*

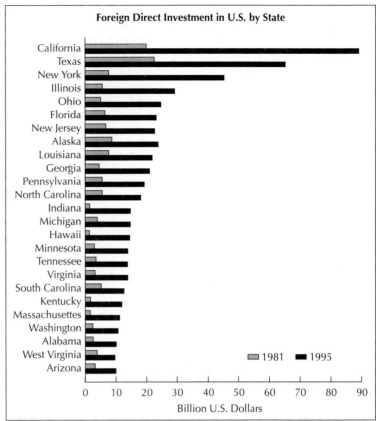

Overall, California was the FDI leader with approximately $90 billion in 1995. Texas was the second largest state for FDI in 1995, with $70 billion, followed by New York as a distant third, with $45 billion. In the Pacific region, Alaska was second with $25 billion.

Balance Between U.S. FDI and FDI in the United States

In the 1970s and early 1980s, U.S. FDI far outweighed foreigners' investment in the United States. However, in 1988 and 1989, U.S. FDI was slightly less than investment by foreign companies in America. Since 1989, however, U.S. foreign investment has surpassed FDI in the United States, so the balance is now a positive $20 billion rather than a deficit as in 1989.

Effects of Foreign Direct Investment

Are the effects of widespread FDI desirable? Should the operations of multinationals be controlled? There is no unanimity of opinion, particularly when LDC development is the issue.

There are two polar attitudes among scholars regarding the presence of multinationals in LDCs. Those on the right of the political spectrum argue that the multinational firm has a high potential to aid the economic development process. According to Herman Kahn (1973, p. 2):

> The transnational corporation (TNC) is probably the most efficient social, economic and political institution ever devised to accomplish the following tasks for the less developed nations:
>
> (1) Raising, investing, and reallocating capital.
>
> (2) Creating and managing organizations.
>
> (3) Innovating, adopting, perfecting and transferring technology.

(4) Distribution, maintenance, marketing, and sales (including trained personnel and providing financing).

(5) Furnishing local elites with suitable—perhaps ideal—career choices.

(6) Educating and upgrading both blue collar and white collar labor (and elites).

(7) In many areas, and in the not-so-distant future, serving as a major source of local savings and taxes and in furnishing skilled cadres (i.e., graduates) of all kinds to the local economy (including the future local competition of the TNC).

(8) Facilitating the creation of vertical organizations or vertical arrangements which allow for the smooth, reliable, and effective progression of goods from one stage of production to another. In many cases, while such organization is a partial negation of the classical free market, it is still often a very efficient and useful method of stable and growing production and distribution.

(9) Finally, and almost by themselves, providing both a market and a mechanism for satellite services and industries that can stimulate local development much more effectively than most (official) aid programmes.

This view is in marked contrast to that of scholars who argue that the multinational corporation is counterproductive to development. In the view of *dependency theorists,* modern capital-intensive industry does not result in rapidly increasing employment. Multinationals engender balance-of-payments problems because of heavy profit repatriations. Although the balance-of-payments problem could be avoided in part if multinational firms reinvested more of their profits in the host country, it is uncertain that the national interest would be served. Reinvestment causes growing foreign control of the economy and the denationalization of local industry. The multinational firm is an assault on political sovereignty. Moreover, the transnational system internationalizes the tendency to unequal development and to unequal income.

To be sure, multinationals are imperfect organs of development in the developing countries, and their potential for the exploitation of poor countries is tremendous. There is, therefore, an inherent tension between the multinational's desire to integrate its activities on a global basis and the host country's desire to integrate an affiliate with its national economy. Maximizing corporate profits does not necessarily maximize national economic objectives.

In general, the relative bargaining power of host countries has increased with time because the number of multinationals has grown; thus host countries have a wider range of choice. Larger and wealthier LDCs have more bargaining leverage. A consumer-products manufacturing corporation will accept more controls to gain access to a country with a large market. Finally, the degree of host-country control varies across industries. Manufacturing industries with advanced and dynamic technologies are more difficult to control than firms in the raw-materials area.

Developing countries have changed the rules of the game for FDI. The new rules, however, have made corporations hesitant to risk large amounts of capital in the raw-materials area and in smaller and poorer developing countries. Corporations prefer to invest in countries that follow an outward-oriented development strategy, impose few controls, offer incentives, and therefore appreciate the employment, skills, exports, and import substitutes that foreign investment can bring. The newly industrializing countries (NICs) exemplify this posture, and they do not align themselves with the LDCs that want to see multinational activity regulated.

BARRIERS TO INTERNATIONAL BUSINESS

Just as trade can, international flows of the production factors can help to reduce imbalances in the distribution of natural resources. Whereas trade *offsets* differences in factor endowments, factor movements *reduce* these differences. International trade and factor flow would occur more commonly if barriers did not exist. The main barriers relate to management, distance, and government.

Management Barriers

A number of managerial characteristics reduce trade and investment expansion. These characteristics include limited ambition, unawareness of opportunity, lack of skills, fear, and inertia.

LIMITED AMBITION
Firms may have the potential to expand but fail to do so because they are *satisficers*—settle for less than the optimal. Until the economic crisis of the 1970s, many

Manufacturing electronics in Shinjuku, near Tokyo, Japan. Shinjuku is one of the busiest manufacturing locales in Japan. Japan enjoys a $60 billion export surplus. Its largest trading partner is the United States. Electronics and automobiles have been its most important export products. Because Japan is in a severe recession, having lost 50% of its equity on the Japanese stock exchange, Japan refuses to lower trade barriers to U.S. products. It has had a long history of such seemingly unfair trade practices. Consequently, the stage is set for a new trade war between Japan and the United States. *(Photo: Kuninori Takahashi, Impact Visuals)*

U.S. firms paid little attention to foreign markets. They were satisfied with the large domestic market.

UNRECOGNIZED OPPORTUNITY

Firms may have the will to go international but may lack knowledge of potential markets. "Even the largest multinational firms cannot possibly access, much less interpret, all of the potential information about other countries that may be available. Managers' mental maps and images of world order, including such aspects as size, shape, proximity, intensity of activities, or geopolitical affiliations, are relative and reflect considerable amounts of distortion from reality" (McConnell and Erickson, 1986, p. 101). The burden of recognizing export opportunities rarely falls on the managers of individual companies, however. Most national and local governments are actively involved in increasing international awareness and in promoting exports.

LACK OF SKILLS

Firms may have the potential and will to go international *and* an awareness of the opportunities, but they may be thwarted by the complexity of international business and ignorance of foreign cultures. Governments and universities can aid companies by providing education. Knowledge of intermodal rate structures, freight forwarders, shipping conferences, and customs-house brokers (firms that contract to bring other companies' imported goods through local customs) is vital for the conduct of international business. Just as necessary is a knowledge of foreign cultures. According to McConnell and Erickson (1986), "More than a few international business 'blunders' can be attributed to a lack of information or misinterpretation of basic cultural patterns" (p. 101). In the past, for example, U.S. firms could not penetrate the Japanese market, partly because of a failure to appreciate the Japanese culture. This is changing. American entrepreneurs are no longer likely to lose deals by forgetting respectful bows at the beginning of meetings.

FEAR AND INERTIA

Firms may avoid international business activities because of a fear of the unknown or what is foreign—different currencies, laws, documentation requirements, taxation, political systems, languages, and customs. They may also fail to capitalize on opportunities because of inertia.

Distance as a Barrier

Companies that go international confront the geographic distance barrier. Obviously, distance affects international business in that all movement costs money. But the international movement of goods is not restricted to line-haul costs. Shipments of goods to foreign destinations incur bank collection charges, freight forwarders' and customs-house brokers' fees, consular charges, and cartage expenses. These costs entail extra clerical costs for preparing bills of lading (receipts given by carriers to exporters), customs declarations, and other shipping documents. Adding these terminal costs to line-haul costs yields a total outlay called *transfer costs*, which make goods more expensive to importers and less valuable to exporters. As we saw in Chapter 4, total transportation costs vary from commodity to commodity. They are higher on finished goods than on bulk shipments of raw materials requiring less care and less special handling. Small firms must pay what the market will bear. However, multinationals, through intrafirm trade, have the opportunity to practice *transfer pricing*—fixing prices for the movement of goods between affiliates.

Distance also influences trade in ways other than cost. Commercial relations are often smoother and less complicated between neighboring countries, if only because managers are more aware of export opportunities. Propinquity brings with it the possibility of frequent contact. A high level of interaction between neighboring countries enables each to better understand the other's economic and political system and culture.

Government Barriers to Trade

No country permits a free flow of trade across its borders. Governments have erected barriers to achieve objectives regarding trade relationships and indigenous economic development. Trade barriers include *tariffs*—schedules of taxes or duties levied on products as they cross national borders—and *nontariff barriers*—quotas, subsidies, licenses, and other restrictions on imports and exports. These kinds of obstacles (apart from *political bloc prohibitions*) are the most pervasive barriers to trade.

COSTS OF PROTECTION

Free marketeers advocate free trade because it promotes increased economic efficiency and productivity as a result of international specialization. They argue that trade, a substitute for factor movements, benefits each participating nation and that deviation from free trade will inhibit production. It follows, then, that *protectionism* will adversely affect the welfare of the majority. For example, it is estimated that in 1983 every dollar spent to preserve employment in the U.S. steel industry cost consumers $35 and amounted to a net loss of $25 for the economy (World Bank, 1988).

What are some of the major arguments in favor of protectionism? One of the most common is the cheap foreign-labor argument, which suggests that a country such as the United States with its high union wages must protect itself against a country such as Taiwan with its low-paid workers. This argument contradicts the principle of comparative advantage. Other more compelling arguments appeal to national gain. One argument asserts that a country with market power can improve its terms of trade with a tax that forces down the price at which other countries sell to it. Another argument is that tariffs can be used to divert demand from foreign to domestic goods so as to shift a country's employment problem onto foreign nations. Still another argument is that tariffs can be used to protect an infant industry that is less efficient than a well-established industry in another country. The *infant-industry argument* was invoked to justify protectionist policies in 19th- and 20th-century America and 19th-century Germany (Figure 10.16). It was also used to justify the protection that developed in the LDCs in the 1960s. Although these arguments have some merit, free marketeers recommend other approaches to attain desired goals. For example, they suggest that if grounds exist for protecting an infant industry until it has grown large enough to take advantage of scale economies, protection could be given through a subsidy rather than through a protective tariff.

The arguments for free trade and low tariffs have been accumulating over the years, yet one could argue that throughout the history of the United States, tariff rates have been unquestionably high (Figure 10.16). The Compromise Tariff of 1833 and the Smoot-Hawley Tariff of 1930 put rates at almost 70% of dutiable imports; that is, 70% of imports were subject to the tariff. The reasons for tariffs are clear. Special-interest groups who stand to gain economically from tariffs and quotas press the government for protection through the use of high-powered, politically savvy lobbyists and indoctrinators. The public, which must then absorb these tariffs and quotas as a surcharge on all imported products, are politically uninformed and not represented in Washington, D.C.

In 1947, however, the United States and 25 other nations signed the *General Agreement on Tariffs and Trade* (GATT). GATT established multinational reductions of tariffs and import quotas and now has more than 110

signatories (World Trade Organization [WTO]). It is discussed in more detail later in this chapter. The *Uruguay Round*, which began in 1986 and has now ended, once again reduced tariffs, so U.S. tariff rates are currently approximately 5% of dutiable imports.

TARIFF AND NONTARIFF BARRIERS
Tariffs are the most visible of all trade barriers, and they can be levied on a product when it is exported, imported, or in transit through a country. The tariff structure established by the developed countries in the post–World War II period works to the detriment of underdeveloped countries. The underdeveloped countries encounter low tariffs on traditional primary commodities, higher tariffs on semimanufactured products, and still higher tariffs on manufactures. These higher rates are, of course, intended to encourage firms in industrial countries to import raw materials and process them at home. They also discourage the development of processing industries in the developing world.

In recent years, the relative importance of tariff barriers has decreased, whereas nontariff barriers have gained significance. The simplest form of nontariff barrier is the *quota*—a quantitative limit in the volume of trade permitted. A prominent example of a product group subject to import quotas in developed countries is textiles and clothing. Since the early 1970s, textiles and clothing have been subject to quotas under successive Multifibre Arrangements (MFAs). These arrangements have created a worldwide system of managed trade in textiles and clothing in which the quotas severely curtail underdeveloped-country exports. Another common nontariff barrier is the *export-restraint agreement*. Governments increasingly coerce other governments to accept "voluntary" export-restraint agreements, through which the government of an exporting country is induced to limit the volume or value of exports to the importing country. The United States has employed this special

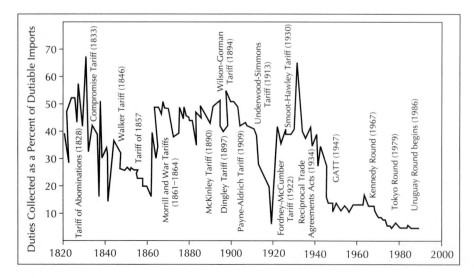

FIGURE 10.16
U.S. tariff fluctuations, 1820–1994. Compared to the inception of the General Agreement on Tariffs and Trades (GATT) in 1947, U.S. tariff rates have been historically high. They reached a peak during this century in the mid-1930s, with the Smoot-Hawley tariff, designed to protect U.S. markets for U.S. goods. The U.S. has the lowest barriers to trade in the world. Without new agreements lowering foreign barriers to our goods and services, we will continue to suffer. Remember, 95 percent of the world's consumers live somewhere other than America.

type of quota—*an export quota*—extensively. It has concluded "bilateral export agreements with . . . the Republic of China with respect to footwear and color television sets, Hong Kong with respect to textiles, Korea with respect to footwear, textiles and color television sets and with Japan as regards color television sets."

Other nontariff barriers include discretionary licensing standards; labeling and certificate-of-origin regulations; health and safety regulations, especially on foodstuffs; calendriers, which allow foodstuffs to be imported only during certain seasons to avoid competing with the peak production of the importers; and packaging requirements. Increasingly, loose, or break-bulk, cargo is unacceptable to developed-country mechanized transportation handlers. Dockers and longshoremen often demand bonuses for handling such items as unpacked skins and hides. Consumers, too, demand agricultural products in packaging that requires more investment on the part of the exporting country. These examples represent only a few of the hundreds of nontariff barriers devised by governments. The evidence indicates that these barriers in developed countries are higher for exports from developing countries than they are for exports from rich, developed countries.

EFFECTS OF TARIFFS AND QUOTAS

The economic effect of tariffs and quotas in the host country is the development and expansion of inefficient industries that do not have comparative advantages. At the world level, tariffs and quotas penalize industries that are relatively efficient and that do have comparative advantages. The result is less international trade and penalized consumers.

Figure 10.17 shows the economic effects of a protective tariff and an import quota. Let's first deal with the case of a protective tariff. Line D_d represents domestic demand in, for example, the United States for AM/FM auto cassette players, whereas line S_d is the domestic supply. (Disregard the S_d + Quota line for now.) The domestic equilibrium position is at price p_3 and quantity q_3. Now assume that the United States market for AM/FM auto cassette players is open to world trade. The world price is lower than the domestic price because, compared with Japan, Malaysia, or Taiwan, the United States has the comparative disadvantage of high labor costs.

The world equilibrium price is p_1. At this price, Americans will consume the quantity q_5 at Point N. With the low price, the domestic supply is only q_1 at Point P, with the quantity q_1–q_5 supplied by foreign imports (Figure 10.17).

Next, let's say that the United States imposes a tariff on the import of AM/FM auto cassette players. This tariff raises the price from p_1 to p_2. The equilibrium price and quantity are now p_2, q_4, respectively, at Point Q. The first reaction will be a decline in quantity demanded by American consumers, from q_5 to q_4, as they

lock up their demand curve toward the higher price (Figure 10.17).

American consumers are hurt by the tariff because they can buy fewer goods at a much higher price. While the consumers move back up the demand curve to Point Q, domestic producers, with a higher price opportunity, increase their production and move up their supply curve from Point P to Point R. Domestic production has increased from q_1 to q_2. Consequently, we can easily understand why domestic producers send lobbyists to Washington, D.C., to invoke tariffs that give the producers a relative advantage in the market.

The increased tariff reduces the number of Far East Asian imports from q_1–q_5 to q_2–q_4. The U.S. government, not the East Asian supplier, receives the tariff monies, p_1–p_2. At the same time, the market decreases because of reduced demand, and domestic supply increases. The shaded area represents the amount of tariff revenue paid to the U.S. government. This revenue is an economic transfer from the consumers of the country to the government.

The result of levying a tariff is reduced world trade and reduced efficiency of the international economic system, which hurts foreign suppliers, aids domestic producers, and costs the consumer. The indirect effect is that the supplying countries have a smaller market in America and thus earn fewer dollars with which to exchange or invest in American resources. As FDI decreases, trade deficits may increase.

Next, let us consider the effects of levying of an import quota. The difference between a tariff and an import quota is that a tariff yields extra revenue to the host government, whereas a quota produces revenue for the foreign suppliers. Imagine that the United States subjects a foreign nation to an import quota, rather

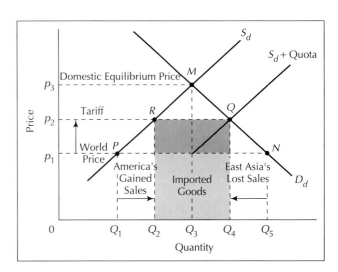

FIGURE 10.17
The economic effects of tariffs and quotas.

than imposing a tariff (Figure 10.17). The import quota in this case is q_2–q_4. The quantity q_2–q_4 is the number of auto cassette players that foreign producers are allowed to supply. Note that for easy comparison this example limits the quota to the exact amount of imported goods in our tariff example. The quota establishes a new supply curve, S_d + Quota, with an equilibrium position at Point Q. The new supply of auto cassette players is the result of domestic supply, plus a constant amount, q_2–q_4, which is supplied by importers.

The chief economic outcomes are the same as with the tariff example. The price is p_2 rather than p_1, and domestic consumption is reduced from q_5 to q_4. The American manufacturers enjoy a higher price for their goods, p_2 rather than p_1, and increased sales, q_2 rather than q_1. But the main difference is that the shaded box, paid by the domestic consumer on imports of q_2–q_4, is not paid to the U.S. government. Instead, the extra revenue in the shaded box is paid to the foreign supplier. That is, no tariff exists, and consequently, the foreign supplier keeps all the revenue in the box q_2–q_4–Q–R.

The result is that for local consumers, and their government, a tariff produces a better revenue situation than a quota does. Tariff money can be used to lower the overall tax rate and provide social services and infrastructure for the population as a whole. However, either case is detrimental to international trade and economic efficiency and takes away from the comparative advantages of supplying nations.

Stimulants to Trade

Not only do governments attempt to control trade, but they also attempt to stimulate trade. Free marketeers believe that government intervention to promote trade is yet another obstacle to free trade. In their view, gains from trade should result from economic efficiencies, not from government support. Examples of governmental assistance to promote exports include market research, provision of information about export opportunities to exporters, international trade shows, trade-promotion offices in foreign countries, and free-trade zones, areas where imported goods can be processed for reexport without payment of duties.

THE REDUCTION OF TRADE BARRIERS

Several efforts have been made to eliminate some of the trade barriers that were erected in the past. GATT and the Uruguay round were two of these efforts. In addition, the United States continues to try to penetrate Japan's closed markets.

GENERAL AGREEMENT ON TARIFFS AND TRADE

The most notable effort to reduce trade barriers has been a multilateral effort known as GATT. GATT was put into operation in 1947. When 23 developed and developing countries signed the agreement, they thought that they were putting in place one part of a future World Trade Organization (WTO). The organization would have wide powers to police its trading charter and regulate international competition in such areas as restrictive business practices, investments, commodities, and employment. It was to be the third in the triad of Bretton Woods institutions charged with overseeing the postwar economic order—along with the International Monetary Fund and the World Bank. But the draft charter of the ITO encountered trouble in the U.S. Congress and was never ratified. Only GATT remained—a treaty without an organization.

GATT is now administered on behalf of more than 100 member countries, which make decisions by a process of negotiation and consensus. This process has resulted in a substantial reduction of tariffs. However, GATT's rules have proved inadequate to cope with new forms of nontariff barriers such as export-restraint agreements. Such areas as services, which now account for about 30% of world trade, are not covered by GATT at all. GATT has also been of little help to developing countries with limited trading muscle. True, developing countries did gain acceptance of the General System of Preferences (GSP). Under GSP, which developed countries adopted in 1971, tariffs charged on imports of manufactured and semimanufactured products are granted preferential treatment. However, one of the more striking features of GSP schemes is the low proportion of manufactured goods that are eligible for preferential treatment. Because most GSP schemes are restrictive, few underdeveloped countries have benefited.

Since the mid-1970s, and especially since 1980, the liberal trading order that GATT helped to uphold has been steadily eroding. The resurgence of protectionism, especially in the guise of nontariff barriers, is a reflection of the world economic crisis. Between 1981 and 1990, the proportion of imports to North America and the European Union (EU) that were affected by nontariff barriers increased by more than 20%. Trade between developed and underdeveloped countries is increasingly affected by nontariff barriers; roughly 20% of LDC exports were covered by such measures in 1990. In the coming years, pressure on governments in developed countries to protect domestic jobs through trade barriers is likely to mount.

URUGUAY ROUND

In 1986, the eighth round of GATT opened in Montevideo, Uruguay. This Uruguay round of GATT negotiations centered on (1) removing barriers to international trade and services, which now account for 21% of all international trade; (2) ending limits on foreign economic investments; (3) establishing and policing patent, copyright, and trademark rights (intellectual property rights) on an international basis; and (4) reducing agricultural trade barriers and domestic subsidies.

The problem in 1990 with the Uruguay round was European opposition to reducing and phasing out agricultural subsidies. The EU wanted to maintain *export subsidies*, which are government payments that reduce the prices of goods to buyers abroad. Another type, the *domestic farm subsidy*, constitutes direct payments to farmers according to their production levels in order to subsidize their output. The result of these subsidies is increased domestic food output, which provides unfair competition for U.S. and LDC agricultural products on the world market. Both types of subsidies are artificial barriers that reduce prices on the world market and provide advantages to local farmers. French farmers roadblocked Paris in October 1993 to pressure the government into negotiating a better farm deal with America. The United States had pressed the EU in early 1993 to phase out all subsidies over a 10-year period, but settled for a deal that reduced EU farm subsidies by 21% over 6 years. France reluctantly agreed to that, but the farmers refused to go along.

In 1993, the Uruguay round of negotiations reconvened in an attempt to further resolve trade disagreements. The U.S. effort was an attempt to pry open foreign markets, especially those considered to be unfair traders. Because most of the industrialized world was in the midst of an economic slowdown or an outright recession in 1992, there was much interest in the international trade measures of GATT at Uruguay.

GATT CONCLUDES IN VICTORY

Finally, in Geneva on December 15, 1993, 117 nations agreed to reduce worldwide tariffs, lower subsidies, and eliminate other barriers to trade. In reducing worldwide protectionism, the GATT nations have chosen to improve resource allocation, which will increase trade and employment and raise wages and standards of living.

The United States obtained most of what they sought: the opening of agriculture markets, cuts in industrial tariffs, intellectual property rights, and the opening of markets in the world's service industries. These were both losers and gainers. America's heavy equipment manufacturers, toy makers, and beer brewers were joyful, but the pharmaceutical industry, Hollywood film makers, and textile manufacturers were not too pleased.

WORLD TRADE ORGANIZATION (WTO)

The last round of GATT established a World Trade Organization (WTO). Until the WTO was enacted, countries with conflict over trade had to resolve their own problems. WTO describes arbitration by a third party to settle conflicts between a pair of nations over trade disagreements. The judgments will be enforced through retaliatory trade sanctions by all members of the WTO, meaning that a part of a country's sovereignty is given up to this multilateral world organization. With the enactment of the WTO, a country's power to control flows across its borders is to some extent lost. For some countries not abiding by international trade agreements, this is a scary proposition.

Under the WTO, quantitative limits on imports become illegal. For example, with this provision, Japanese and Korean import bans on rice from the United States, peanuts, dairy products, and sugar are now banned. Intellectual property rights is perhaps the most important aspect of the WTO. All signatories of the WTO are required to protect patents, copyrights, trade secrets, and trademarks. This measure is designed to end the wholesale pirating of computer programs, video cassettes, musical recordings, books, and prescription drugs widely practiced in some developing countries (e.g., China). The WTO also prohibits members from requiring a certain proportion of content in products manufactured within their borders. This practice was widely employed as a device to limit the use of imported parts and components and, thus, to bolster local employment.

Criticism of the WTO has centered around the lack of environmental production standards. A good can be entered into international trade without meeting production standards of another country. Many signatories have argued that environmental standards should be a consideration for a good on the international market.

The WTO calls for free trade and financial services, shipping, and audio-visual product—movies, television, programs, and musical recordings. The Uruguay Round of GATT, which created the WTO, estimated that the increased world commerce resulting from the WTO agreement would add $500 billion annually to the global commerce. Of this total, the United States would be expected to gain $125 billion, the European Union $150 billion, and Japan $27 billion. The less developed and former communist nations would make up the difference. Consequently, the new WTO specification was predicted to raise the world economy to substantially higher levels of wealth and prosperity.

CLOSED JAPANESE MARKETS

The accusation of closed Japanese markets is a contentious issue that is difficult to prove. A number of econometric studies have failed to support unequivocally the conclusion that Japan's imports of American products are less than can be expected given its level of income and resources. In the recent past, Japan increased its imports and was the third largest importer of world merchandise as of 1990. It ranked first in importation of commercial services.

President Bush's visit to Japan in 1992 and President Clinton's trip in 1993 were efforts by the United States to pry open Japanese markets. The results were minimal at best, with an agreement by Japanese automakers to import automobile parts worth $10 million and 22,000 Chrysler, Ford, or General Motors vehicles by 1995.

Other agreements were made to boost the sales of computers and computer products by American companies in Japan. However, part of Japan's reluctance to open its markets may be because it is experiencing a lingering recession and continuing political turmoil and scandal, and it is not as economically robust as it was 5 years earlier.

Government Barriers to Production-Factor Flows

Although not as complex as trade barriers, obstacles to the free flow of capital, labor, and technology constrain international managerial freedom. Exchange controls and capital controls are the main types of controls that interfere with the movement of money and capital across national borders. *Exchange controls*, which restrict free dealings in foreign exchange, include multiple exchange rates and rationing. In multiple-exchange-rate systems, rates vary for different kinds of transactions. For example, a particular commodity may be granted an unfavorable rate. Foreign exchange may also be rationed on a priority basis or on a first-come, first-served basis. Thus, exchange rates are political tools bearing little relationship to economic reality. *Capital controls* are restrictions on the movement of money or capital across national borders. They are typically designed to discourage the outflow of funds.

All countries regulate migration, but the movement of workers from poorer to richer countries was the dominant pattern during the long postwar boom. When the boom ended, jobs moved to the workers. One reason for this change was tighter immigration laws in the advanced industrial countries. These tighter laws strengthened the position of labor and resulted in a growth in managed trade and a decline in managed migration.

Technology, which is highly mobile, can be transferred in many ways: export of equipment, provision of scientific and managerial training, provision of books and journals, personal visits, and the licensing of patents. Political and military controls regulate the export of technology. Although these controls are not yet terribly onerous, demands for more stringent controls are on the increase. One source of demand for control is labor unions in advanced industrial countries. These unions attribute domestic job loss to the export of high technology.

MULTINATIONAL ECONOMIC ORGANIZATIONS

In the world today, as nations turn inward to concentrate on problems of economic growth and stability, we are witnessing a resurgence of protectionism. But also in evidence is a strong, simultaneous countermovement toward international interdependence. This movement is exemplified by scores of multinational organizations, which for the most part are loosely connected leagues entailing little or no surrender of sovereignty on the part of member nations.

Some of these international organizations are global in scale. The most inclusive is the United Nations (UN), with 159 member nations that account for more than 98% of humankind. Much of the UN's work is accomplished through approximately two dozen specialized agencies such as the World Health Organization (WHO) and the International Labor Organization (ILO). Other international organizations have a regional character; for example, the Association of South-East Asian Nations (ASEAN) and the Asian Development Bank (ADB). Many international organizations are relatively narrow in focus—mostly military, such as the North Atlantic Treaty Organization (NATO), or economic, such as the Organization of Petroleum Exporting Countries (OPEC). Some international organizations are discussion forums with little authority to operate either independently or on behalf of member states; for example, GATT and the Organization of Economic Cooperation and Development (OECD).

Others, such as the IMF and the World Bank, have independent, multinational authority and power, performing functions that individual states cannot or will not perform on their own. Some international organizations integrate a portion of the economic or political activities of member countries—as, for example, the EU. International organizations to promote regional integration are the most ambitious of all. Some observers believe that regional federations are necessary to the process of weakening nationalism and developing wider communities of interest. However, if a rigid, inward-looking regionalism is substituted for nationalism, the ultimate form of international integration—world federation—will be difficult to achieve (Schwartzberg, 1987).

This section examines international economic organizations that affect the environment in which firms operate and thus influence the development of developing countries. We look at international financial institutions, groups that foster regional economic integration, and groups such as commodity cartels, which deliberately manipulate international commodity markets.

International Financial Institutions

International financial institutions are a phenomenon of the post–World War II period. The IMF and the International Bank for Reconstruction and Development (IBRD), or World Bank, were established in 1945. Regional development banks—the Inter-American Development Bank (IADB), Asian Development Bank (ADB), and African Development Bank (AFDB)—were established in the 1960s. Various other multilateral facilities, of which the United Nations Development Program (UNDP) is the

The atrium of the International Monetary Fund (IMF) headquarters in Washington, D.C. The IMF attempts to maintain foreign exchange balances and to promote economic modernization and growth in the Third World. Adjustment programs generally include measures to manage demand, improve the incentive system, increase market efficiency, and promote investment. *(Photo: International Monetary Fund)*

most important, were also established after 1960. These institutions are significant sources of multilateral capital, especially aid (i.e., capital provided on concessional terms) for developing countries. Multilateral capital is particularly important for the poorer developing countries that do not have access to private capital markets.

The IMF is an international central bank that provides short- to medium-term loans to member countries, and the IBRD is an international development bank that provides longer-term loans for particular projects. Both institutions are clusters of governments, each government paying a subscription or quota determined by the size of its economy. Because quotas determine a member's voting power, the banks are dominated by the most powerful economies—particularly, by the United States.

The IMF and the World Bank were established to prevent a recurrence of the crisis of the 1930s. They embody Keynesian principles, which offer a rationale for state intervention in the market. The right of the state to act in the economy is a principle of great importance to many developing countries, which prefer an authoritative to a market allocation of resources. Despite weighted voting, both institutions give LDC members a degree of formal power in excess of their share of actual financial contributions.

With the passage of time, developing countries have obtained more resources on better terms, especially as a result of the creation of two subsidiary World Bank organizations: the International Finance Corporation (IFC), founded in 1956, and the International Development Association (IDA), founded in 1960. These organizations provide loans with stipulations less stringent than those of the IBRD. For example, the IDA may pro-

vide loans with no interest charges, 10-year grace periods (no repayment of principal for the first 10 years), or 50-year repayment schedules for poorer developing countries. Because the IDA is much less credit-worthy than the IBRD, all its resources must come from member-government contributions.

Although the IMF and the World Bank have become more solicitous of LDC opinions and preferences, these institutions remain firmly wedded to liberal, as opposed to dependency, interpretations of development. The market-oriented position is consistent with the basic orientation of the industrialized world, especially that of the United States. Loans from the IMF and the World Bank, therefore, tend to uphold the basis of U.S. economic and foreign policy.

The developing world has never been satisfied with the IMF and the World Bank, particularly with regard to the level of available resources and the conditions imposed on their use. One counter has been the regional development banks—the IADB, ADB, and AFDB. These banks reflect the desire of developing countries to enhance their control. Of the three banks, the ADB is most under the control of developed countries, in particular Japan. The AFDB has been the most independent, but it is also the smallest.

Another counter was the creation of the UNDP in 1965. In terms of formal voting structure, the UNDP provides the developing world with dominant influence, even though approximately 90% of the contributions are from advanced industrial countries. The UNDP provides assistance to a wider variety of countries than the World Bank does.

International financial institutions are important for international business. The IMF, the World Bank, and regional development banks annually finance billions of dollars of the import portion of development projects. This can be valuable business for foreign companies involved in the projects, either occasionally as part owners or, more commonly, as contractors or suppliers.

Although international financial institutions facilitate international business, their project aid may not promote development. Conservative and radical critics agree that development must be primarily an indigenous process. Foreign aid has served many underdeveloped countries as an easy substitute for devising means to generate domestic development. It is also apparent that aid from institutions heavily influenced by developed countries is a palliative designed to ensure the continuity of unequal exchange in the world economy.

Regional Economic Integration

Regional integration is the international grouping of sovereign nations to form a single economic region. It is a form of selective discrimination in which both free-trade and protectionist policies are operative: free trade among members and restrictions on trade with nonmembers. According to economist Bela Balassa (1961), five degrees of economic integration are possible. At progressively higher levels, members must make more concessions and surrender more sovereignty. The lowest level of economic integration is the *free-trade area*, in which members agree to remove trade barriers among themselves but continue to retain their own trade practices with nonmembers. A *customs union* is the next higher degree of integration. Members agree not only to eliminate trade barriers among themselves, but also to impose a common set of trade barriers on nonmembers. The third type is the *common market*, which, like the customs union, eliminates internal trade barriers and imposes common external trade barriers; this regional grouping, however, permits free production factor mobility. At a still higher level, an *economic union* has the common-market characteristics plus a common currency and a common international economic policy. The highest form of regional grouping is full *economic integration*, which requires the surrender of most of the international sovereignty of its members.

Regional economic integration is evident within three major trade blocs in Figure 10.18. Western Europe has more than 70% of its imported items originating from within the region, and the percentage reached 75% by 1993. North American trade regionalization has fallen from 45% to 35% between 1970 and 1990, while East Asia's has increased from 35% to 45%.

Table 10.7 lists the different levels of integration of a variety of economic groups. In part B, the main

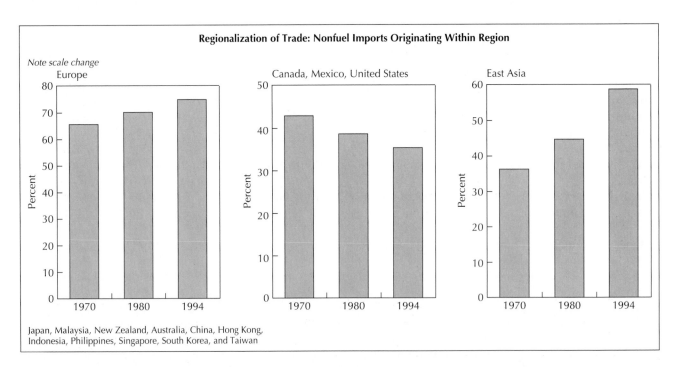

Regionalization of Trade: Nonfuel Imports Originating Within Region

Japan, Malaysia, New Zealand, Australia, China, Hong Kong, Indonesia, Philippines, Singapore, South Korea, and Taiwan

FIGURE 10.18

Regionalization of trade: nonfuel imports originating within the regions of Western Europe, North America, and East Asia. Of the three principal world trade blocs, Western Europe has the greatest regionalization with almost 75% of its trade originating with Western Europe itself. North America has seen its intraregional trade slip from 45% to 35%, while intraregional trade for East Asia has increased from 35% to 45% between 1970 and 1994. What effect will NAFTA have on the percentage of intraregional trade in North America? *(Source: United Nations, 1997, p. 138)*

🔲

T a b l e 10.7

Levels and Examples of Regional Integration

A. Levels of Regional Integration

Level	Characteristics	Examples
Free-trade area	Common internal tariffs but differing external tariffs	EFTA, LAFTA
Customs union	Common internal and external tariffs	UDEAC, CEAO
Common market	Common tariffs and few restrictions on production-factor mobility	EEC, COMECON

B. Examples of Regional Integration

Year	Grouping	Main Provisions
1949	Council for Mutual Economic Assistance (COMECON): USSR, Bulgaria, Czechoslovakia, German Democratic Republic, Hungary, Poland, Romania, Mongolia, and Cuba	Trade and development agreements and contracts
1957 (1973: plus U.K., Ireland, Denmark) (1981: plus Greece) (1986: plus Spain and Portugal)	European Economic Community (EEC, now EU): France, West Germany, Italy, Belgium, the Netherlands, and Luxembourg	Common market (common internal and external tariffs and common agricultural and industrial policies)
1960 (1961: plus Finland as an associate member) (1970: plus Iceland) (1973: minus U.K. and Denmark)	European Free Trade Association (EFTA): U.K., Switzerland, Austria, Denmark, Norway, and Sweden	Common internal tariffs but not common external tariffs
1960	Latin America Free Trade Association (LAFTA): Mexico and all of South America, except the Guyanas	Free-trade area (no common external tariff) and sectoral agreements
1966	Union Douanière et Economique de l'Afrique Centrale (UDEAC): Cameroon, Central African Republic, Congo, Gabon	Customs union with common central bank
1967	Association of South-East Asian Nations (ASEAN): Indonesia, Malaysia, Philippines, Singapore, and Thailand	Some regional trade preferences and sectoral agreements
1969 (1973: plus Venezuela) (1976: minus Chile)	Andean Common Market (ACM): Bolivia, Chile, Colombia, Ecuador, and Peru	Common market envisaged, common policy on foreign investment
1979	Southern Africa Development Co-ordination Conference (SADCC): Angola, Botswana, Lesotho, Malawi, Mozambique, Swaziland, Tanzania, Zambia, Zimbabwe	Sectoral integration
1989	United States-Canada Free Trade Agreement: Canada and the United States	Free trade area
1994	North American Free Trade Agreement (NAFTA): Canada, the United States, Mexico	Free trade area

provisions of the groups are described. These groups range from loosely integrated free-trade areas such as the Latin American Free Trade Association (LAFTA) to common markets such as the EEC. There are also links between members of regional blocs. Thus, East-West ties exist between EEC countries and the Eastern countries' Council for Mutual Economic Assistance (COMECON). By 1997, most of the COMECON countries had applied for membership in the EU. There are North-South ties between the EU and LAFTA, South-South ties between LAFTA and ASEAN, and some East-South ties between COMECON and various developing countries. Most of these links are bilateral; that is, agreements between nations within different regions. Fully fledged interregional integration has yet to be achieved. Indeed, regional groups are more concerned with closer economic integration *within* regions than *among* regions (Figure 10.19).

Barriers to successful regional integration are stronger in developing countries than in developed countries. The most significant barriers are political—an unwillingness to make concessions. Without concessions to the weaker partners of a regional group, the benefits from cooperation pile up in the economically more prosperous and powerful countries. Another difficulty is that developing countries have not historically traded extensively among themselves. Still another obstacle to integration is the issue of integrating transportation and power networks. Nonetheless, much potential for integration exists in developed countries, particularly because many are too small and too poor to grow rapidly as individual units.

Many developing countries turned to regional integration schemes in the 1960s and 1970s. Reasons for integration included a need to gain access to larger markets, to obtain more bargaining power with the developed countries than they could if they adopted a "go it alone" policy, to create an identity for themselves, to strengthen their base for controlling multinational corporations, and to promote cohesion solidarity. The effect of regional groupings differs from one company to another. Companies that enjoy a secure and highly profitable position behind national tariff walls are unlikely to favor removal of these barriers. Conversely, companies that see the removal of trade barriers as an opportunity to expand their markets see integration as a favorable development. Similarly, companies that traditionally exported to markets absorbed by a regional grouping have a strong interest in integration. They perceive these enlarged markets to be more attractive than they were in the past. But as outsiders, the shipments of these companies will be subject to trade controls, whereas barriers for internal competitors will decrease. Thus, foreign companies may face the prospect of losing their traditional markets because they are outside the integrated group of countries. As a result, there is an incentive to invest inside the regional grouping. This is why many U.S. firms invested directly in the EEC countries in the late 1950s and early 1960s, and again in the late 1980s.

THE EUROPEAN UNION (EU)

The most successful example of economic integration is the European Union (EU). It began with six nations: France, West Germany, Italy, Belgium, the Netherlands, and Luxembourg, in 1957, and since then has added most western European countries; for example, the United Kingdom, Ireland, Denmark, Greece, Spain, and Portugal. In addition, most eastern European nations have applied for membership in the EU. The population of potential EU countries numbers almost 350 million. The EU today is the largest single trade bloc in the world and, along with the EFTA, accounts for 46.5% of international trade, which is three times its fair world share based on its population. Western Europe is noted as a region of high world trade as a proportion of a country's GDP (Table 10.8).

The intent of the EU was to give its members freer trade advantages while limiting the importation of goods from outside Europe. It called for (1) the establishment of a common system of tariffs applicable to imports from outside nations; (2) the removal of tariffs and import quotas on all products traded among the participating nations; (3) the establishment of common policies with regard to major economic matters such as agriculture, transportation, and

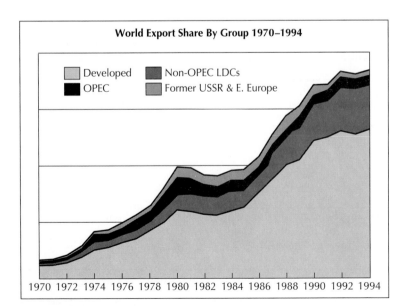

FIGURE 10.19
Changing shares of world exports, 1970–1994.

◈

Table 10.8

Exports of Goods and Services as a Percentage of Gross Domestic Product, Selected Countries, 1993

Country	Exports (% of GDP)
The Netherlands	57
Germany	36
New Zealand	27
Canada	**25**
United Kingdom	24
France	22
Italy	19
Japan	11
United States	**11**

Source: International Monetary Fund, 1992, p. 31.

so forth; (4) free movement and access of capital, labor, and currency within the market countries; (5) transportation of goods and commodities across borders with no inspection or passport examination; and (6) a common currency.

The EU has made tremendous progress toward its stated goals. The member nations have been afforded more efficient, large-scale production because of potentially larger markets within the EU, permitting them to achieve scale economies and lower costs per manufactured unit, something that pre-EU economic conditions had denied them. However, the current stumbling block seems to be the lack of a common economic unit of currency. In addition, certain countries, such as the socialized north, especially Denmark and Germany, have scoffed at opening borders that would allow southern Europeans to immigrate and thereby take advantage of very liberal social welfare programs of economic assistance from cradle to grave.

Americans are concerned about the economic power of the EU. Present tariffs have been reduced to zero among EU nations, whereas tariffs for American-made products have been maintained. Consequently, importing goods to EU nations is relatively more difficult. At the same time, increased prosperity among EU nations because of increased trade and comparative advantage allows these nations to become potential customers for more American export products.

The pressure placed on the United States by the EU has led the United States to promote freer trade through GATT and the U.S.–Canadian Free Trade Agreement. In the fall of 1992, the North American Free Trade Agreement (NAFTA) was signed by the U.S. and Mexican presidents and the Canadian Prime Minister, and in 1994 the U.S. Congress passed it. The

United States, Canada, and Mexico have become the largest free-trade zone in the world.

THE EU SINGLE CURRENCY

If members of the European Union decide to adopt a single currency, the sacrifice will be great. Each country will surrender the right to independently balance its own budget and manage its own debt. Each country will relinquish its individual monetary identity. The French franc, the British pound, and the Spanish peseta will all cease to exist.

There is no question that when the economic efforts of 15 countries and 400 million people are combined, European goods and services will be better represented in the world economy by a bank currency capable of maintaining price stability in all member countries at home and single, stable currency abroad. Presently, European national levels of inflation vary widely, and the respective currencies are at odds with one another. Blocking unforeseen difficulties that could arise, the Euro, as it is called, will become official as of New Year's Day 1999. From that time on, countries that meet the criteria will have the option of paying their debts in Euros. Companies will be able to make transfers into Euros. The exchange rate of participating countries' currencies will be locked in against one another and against the Euro from that day onward.

U.S.–CANADIAN FREE TRADE AGREEMENT

Even before signing the U.S.–Canadian Free Trade Agreement in 1989, Canada and the United States were each other's largest trading partners. According to the agreement, tariffs, quotas, and nontariff barriers will be totally eliminated by 1999. The agreement helps producers on both sides of the border. For example, for Canadian producers, their potential markets increase by a factor of 10. The population of Canada is about 25 million, but that of the United States is 265 million. At the same time, the Canadian market becomes a large group of consumers for U.S. producers. The reduced Canadian tariffs will help American producers gain access to the Canadian consumer. The estimate is that when the U.S.–Canadian Free Trade Agreement is fully enacted by 1999, $3 billion worth of annual gains will accrue to each country in 1990 dollars.

NORTH AMERICAN FREE TRADE AGREEMENT

In December 1993, President Clinton signed into law NAFTA, which could create one of the two most powerful trade blocs in the world. (The Mexican population is 90 million and estimated to be 100 million by the year 2000. Consequently, the total population in the trade bloc would be 380 million, slightly more than that of the EU.) Access to the North American market is coveted by EU countries and Japan. Proponents of NAFTA argued that EU nations would negotiate a free-trade agreement between the two blocs. Japan would then be left out of the world's wealthiest trading markets. The argument is that Japan would then be forced to reduce its tariffs and barriers to world international trade.

The problem with NAFTA is that it is not well received by all members of North American society. The critics' main argument was that it would rob lower-skilled assembly and manufacturing jobs from America and transplant them to Mexico, where the labor rate is one-fifth to one-eighth as much as it is in America. In addition, companies would flee America's more stringent climate regarding environmental pollution and workplace safety controls. No one is certain how many workers NAFTA will replace in already hard-hit North America. Critics of NAFTA also suggest that Japan, Korea, Taiwan, and other Far East Asian countries would build plants in Mexico and import goods duty free to the American and Canadian markets, which would hurt U.S. firms and workers even further.

The principal argument in favor of NAFTA was that free trade would enhance both American (U.S. and Canadian) and Mexican comparative advantages: raise per capita income in Mexico, and increase Mexican demand for goods from the United States and Canada. Another argument suggests that higher living standards in Mexico would help control the flow of undocumented aliens crossing the U.S. border, which is now estimated at 1 million per year. With free trade, wages would rise in Mexico; therefore, undocumented aliens could stay home and work in their native country.

NAFTA frees up auto trade with its elimination of performance standards on foreign producers in Mexico, who are often required to export a certain proportion of their Mexican output. NAFTA also frees up financial and investment trade by reducing restrictions on establishment of United States and Canadian financial service subsidiaries. Thus, NAFTA deregulates Canadian and United States banking, securities brokering, and insurance operations in Mexico. Truck transport service allows free access to the Mexican market opened to Canadian and American trucking companies, as well as free access to markets north for Mexican truckers. Before NAFTA, the United States' tariff rates on imports from Mexico averaged approximately five percent. Therefore, NAFTA did not substantially raise the incentive for Canadians and United States firms to invest in Mexico. Before NAFTA, a free trade zone, 100 miles south of the border, called "the maquilladora zone," had been in operation and NAFTA has in effect extended the maquilladora zone to all of Mexico, allowing imports from manufacturing plants to America with duties imposed only on the value added by manufacture in the maquilladora plants.

CANADA AND NAFTA

NAFTA partner Canada has done a better job than Mexico in auto exports. Yet for a time, Mexico was to create

a "giant sucking sound" of industries moving south to reap high profits and low wages, the term coined by Texas billionaire and presidential candidate, Ross Perot. Now, the number one exporter of cars and trucks to United States is no longer Japan, but Canada. Almost three years into the NAFTA Agreement, which many feared would send United States industry to Mexico, free trade is having the opposite effect of what was expected. Mexico, which most Americans think is gaining jobs through trade, according to a recent associated press poll, steadily slashed its auto manufacturing workforce until 1997, and replaced those jobs with dollar-an-hour employees on the maquilladora assembly lines. The United States actually gained jobs in the auto industry, though not the higher paying jobs everyone wants. Meanwhile, Canada has been gaining jobs. Canada has doubled the number of cars and trucks it exports to the United States since 1989. In 1986, it sold $11.1 billion more in vehicles to the United States than it bought. Canada employees almost 60,000 workers assembling mostly United States parts into motor vehicles—about 85 percent of which were exported to United States (Figure 10.20).

Mexico's labor is cheap. However, Canadian labor is low cost too—one-third less than the United States—and the quality and productivity are equal to or better than that of the United States. American automakers save $300 for each car they make in Canada because of Canada's national health care system, which effectively subsidizes the average auto worker, giving free medical care that United States companies must pay for separately for each employee. GM spends about $5,000 a year to provide health care to each current and retired employee, but less than $1,000 for each current Canadian worker. Canada's cheap dollar, worth about $.75 U.S., allows the dollars of United States manufacturers of automobiles to go farther. Government loans and other subsidies amounting to billions of dollars during the 1970s and 1980s also persuaded American companies to choose Canada—for plants in Branalea, Ontario, and Ste. Therese, Quebec.

By 1995, Mexico had lost 13% of its 137,000 auto manufacturing jobs when foreign competition wiped out Mexican parts suppliers. One year after NAFTA, the peso collapsed and auto sales plummeted 70%. But due to the peso devaluation and reduced labor costs, the recovery has begun, and Mexican production by Nissan, Volkswagen, GM, Ford, and Chrysler hit 1.2 million vehicles in 1997, approximately 85% of which was exported to the United States. Yet new equipment and more efficient production has meant fewer workers than were expected to produce cars in Mexico.

The lower valued peso, however, has meant a boom for maquilladora plants. The peso devaluation of 1994 has brought Mexico an additional 30,000 new jobs in this sector by 1997—low-paid, low-skilled labor producing items like circuit boards, wiring, and electrical systems for motor vehicles, toys, and athletic gear. The pay is about $1.25 U.S. per hour, or $2.00 with benefits.

NAFTA has been a winning situation in the long term for the auto industry's competitiveness because the trade agreement allowed United States automakers to win back a share of the United States market, a loss to cheaper Asian and European cars. In order to do this, they had to send some jobs to neighboring countries to keep costs down, but large numbers of United States workers have been laid off. The pressure to keep reducing the cost of labor to produce a car is continuing.

OPEC

A *cartel* is an agreement among producers that seeks to artificially increase prices by arbitrarily raising them, by reducing supplies, or by allocating markets. The most successful commodity cartel is the Organization of Petroleum Exporting Countries (*OPEC*). Founded in 1960, OPEC consists of 13 countries—Saudi Arabia, Iran, Venezuela, Kuwait, Libya, Nigeria, Iraq, Indonesia, Algeria, Gabon, Qatar, Ecuador, and the United Arab Emirates. In the 1970s, it forced acceptance of authoritative rather than market-oriented principles. The success of OPEC at raising oil prices encouraged other underdeveloped countries to create new regimes.

The first oil shock occurred in 1973, followed by another in 1978. The price for a barrel of crude petroleum oil peaked in 1981 at $36 (Figure 1.15). Major new exploration by American oil companies commenced, and billions of dollars of infrastructure were erected in territories of North America that were rich in oil, low-grade oil, shales, and tar sands, notably Colorado, Wyoming, Alberta, and Montana. With the decline in world oil prices, these new oil operations and oil explorations likewise declined, which sent shock waves through the oil industry and depreciated home and business values throughout Houston, Texas, and other oil-dependent cities.

By 1986, the price of a barrel of crude petroleum on the world market had declined to $15, but it was pushed up again in 1990 to $21 because of the Persian Gulf War and the closing of wells in Iraq and Kuwait. In 1991 and 1992, however, prices again gyrated downward because of the sluggish world economy and the overproduction of

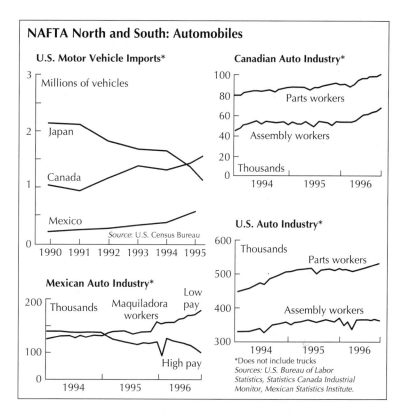

FIGURE 10.20
NAFTA north and south. U.S. motor vehicle imports, Canadian auto industry, Mexican auto industry, U.S. auto industry. Major restructuring has occurred as the United States embraced Mexico and Canada as partners in the North American Free Trade Agreement in 1994. The auto industry is shifting production across southern and northern borders to lower-cost locations.

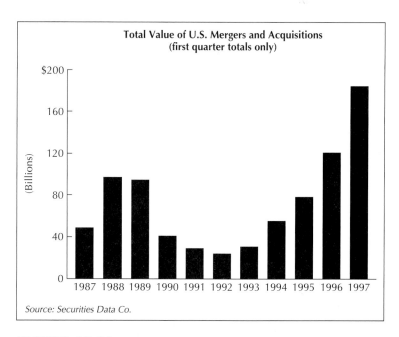

FIGURE 10.21
Total value of U.S. mergers. Mergers have increased in recent years due to technological changes, regulatory reform, and the 1996 telecommunication acquisitions that complement existing business.

OPEC nations. By 1994, the price was again $16 a barrel. Several years have past since the Persian Gulf War, but the oil industries' return to prewar production levels has not occurred because of the devastation in Iraq and Kuwait. With the lower output by Iraq and Kuwait as of 1994, higher levels of production or refining capacity are required by other OPEC nations.

World oil prices depend on the resumption of economic growth in not only the developed nations, but also the less developed nations, and the demand for crude petroleum. The resumption of production and refining in Russia and the former Soviet Union will also affect world petroleum prices.

The future demand for OPEC oil in the mid to late 1990s is likely to be substantial and increasing because of the expected recovery of world economic output and OPEC's large reserves offered at relatively low prices. At present, OPEC produces less than 50% of the world's output but controls 80% of the proven reserves. Despite the tremendous oil reserves of several trillion barrels, generating enough money to maintain its production and increased capacity will be a problem for OPEC nations because, surprisingly, most have serious financial debt problems. Much of their petrodollar accumulation is spent on military equipment to keep the wealthy sheiks in power.

Corporate Mergers

Corporations rarely pursued takeovers with the zeal they exhibit today, perhaps because they were too busy defending themselves. The realities of competition 1990s style began to change that as companies came to believe they had to be more efficient—and bigger—to survive in a smaller world. They began to merge, and their mergers begat more mergers, and so on, and so on. Now, companies in merger-frenzied industries not only view combining as an option, but as a necessity. If they do not become buyers or sellers, they stand to slowly shrivel as others grow (see Table 10.9 and Figure 10.21).

Numerous deals driven by fear of being left behind helped push 1996 into the record books as the Year of the Merger, with about $650 billion in announced U.S. deals, according to Securities Data. Some of the most deal-frenzied industries were telecommunications, utilities and broadcasting. Banking, which saw heated action in 1995, also inked some notable combos.

Surprisingly friendly deals like Boeing buying McDonnell Douglas and British Telcom acquiring MCI also highlight the new era. Who could have imagined McDonnell agreeing to be purchased by its archival or feisty MCI bought by a British competitor with a staid reputation?

◈

T a b l e 10.9

Top U.S. Corporate Mergers

Value in Dollars Unadjusted for Inflation

RJR Nabisco merger with Kohlberg Kravis Roberts completed in 1989	$25 billion	Kraft merger with Philip Morris, completed in 1988	13.44 billion
Bell Atlantic agrees to combine with Nynex in an exchange of stock, announced April 22, 1996	22.7 billion	Gulf merger with Standard Oil of California, completed in 1984	13.4 billion
Walt Disney buys Capital Cities/ABC with cash and stock, completed in 1996	19 billion	Boeing agrees to buy McDonnell Douglas, announced Dec. 15, 1996	13.3 billion
SBC Communications agrees to buy Pacific Telesis Group with stock, announced April 1, 1996	16.7 billion	Chase Manhattan merger with Chemical Banking, completed in 1996	13 billion
WorldCom agrees to acquire MFS Communications, announced Aug. 26, 1996	14.4 billion	Squibb merger with Bristol-Myers, completed in 1989	12.09 billion
Wells Fargo buys First Interstate Bancorp, completed in 1996	14.2 billion	AT&T buys McCaw Cellular Communications, completed in 1994	11.5 billion
Warner Communications merger with Time, completed in 1990	14.11 billion	Getty Oil merger with Texaco, completed in 1984	10.12 billion
		Martin Marietta merger with Lockheed, completed in 1995	10 billion

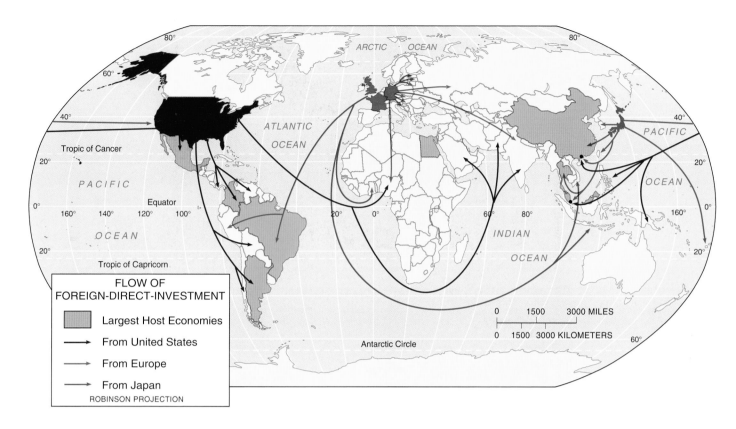

FLOW OF
FOREIGN-DIRECT-INVESTMENT

Largest Host Economies

From United States

From Europe

From Japan

ROBINSON PROJECTION

ABOVE: The great majority of the world's transitional corporations are based in the United States, Europe, or Japan. Most transitional corporation investment has been in companies and resources of other relatively developed countries. The U.S. transitionals make most of their investments in European and Japanese companies. Third World investment is channelized, with the United States investing in Latin America, Europe investing in Eastern Europe, and Japan investing in East Asia.

BELOW: Many less developed countries of the world have borrowed heavily from developed countries in order to finance economic development. Some countries, especially in Africa, have larger debts than their total GNP, resulting in little revenue left over to invest in their economies. Some countries are so heavily in debt that they cannot repay loans to banks or governments of developed countries, and must renegotiate or default, adding fuel to the North/South debate.

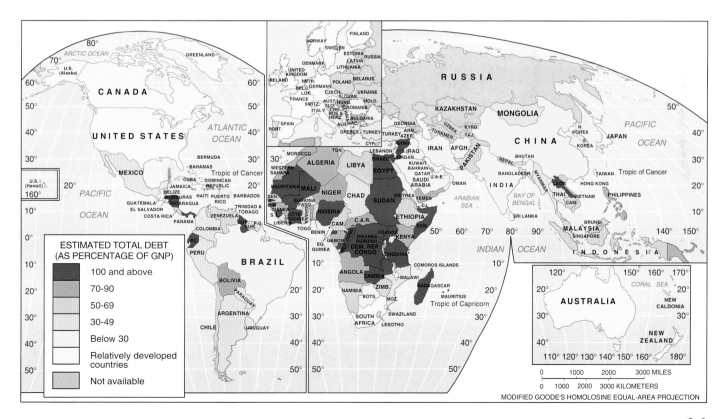

ESTIMATED TOTAL DEBT
(AS PERCENTAGE OF GNP)

100 and above

70-90

50-69

30-49

Below 30

Relatively developed countries

Not available

MODIFIED GOODE'S HOMOLOSINE EQUAL-AREA PROJECTION

3.1

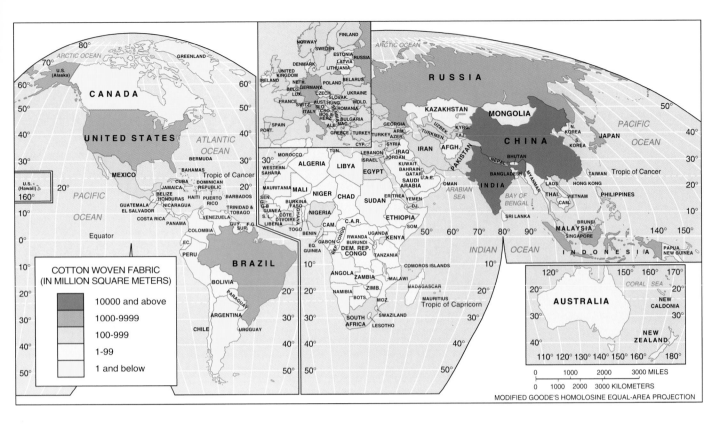

ABOVE: Cotton woven fabric is produced in both countries that have an abundant supply of inexpensive labor, such as China and India, and in countries that have an abundant supply of cotton or where demand is high, such as in the United States and Brazil. In spite of the return to woven cotton fabric as the preferred material worldwide, a depression in world prices throughout the 1980s and 1990s contributed to a crisis in the industry. Attempts are now being made to introduce new marketing policies in the hope that prices will recover as the developed world turns away from synthetic materials, back to the use of more popular natural fibers.

BELOW: Unlike the production of cotton, yarn, and woven cotton fabric, the production of men's and boy's shirts takes place almost exclusively in developed countries. Here, the principle locational factor is proximity to market. Inexpensive labor is still important and is supplied by locations such as the southeastern United States, Southern Europe, the Ukraine, Brazil, and Indonesia.

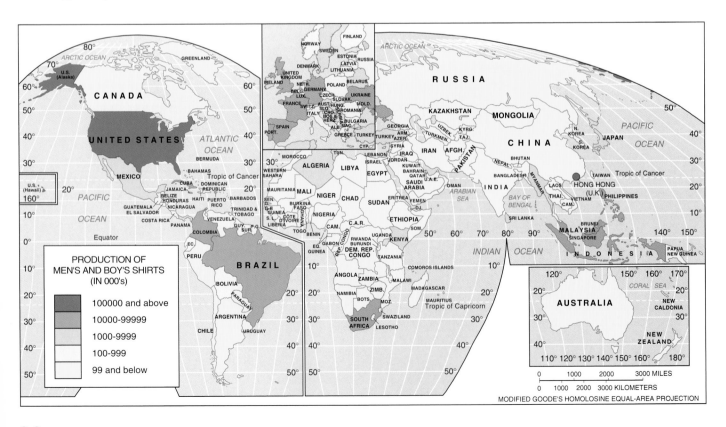

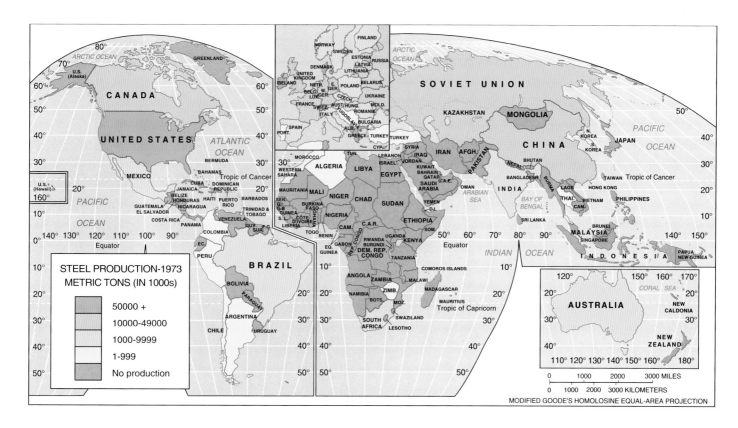

STEEL PRODUCTION-1973 METRIC TONS (IN 1000s)

- 50000 +
- 10000-49000
- 1000-9999
- 1-999
- No production

MODIFIED GOODE'S HOMOLOSINE EQUAL-AREA PROJECTION

ABOVE AND BELOW: The production of steel has been called the backbone of industrialization. In 1973, the United States was the clear world leader in the production of steel, but developing countries were interested in building their own steel plants to foster their economic development process. By 1991, Brazil, the former Soviet Union countries, and China had become major steel producers, and Mexico, Argentina, South Africa, India, Iran, and several Eastern European nations had become important steel producers. The United States, Western Europe, and Japan actually reduced their production of steel due to dwindling world markets. Furnaces using metal scrap led to plants being located close to markets. New ore refining and metal concentrating techniques that raise the iron content to 95% make the ore easily transportable, permit a vast increase in blast furnace capacity, and also favor market locations, such as the Great Lakes concentration of steel making in North America, which stretches from Chicago, Illinois, to Hamilton, Ontario. Electric arc furnaces have brought high quality steel making to areas deficient in coal, but with cheap hydroelectricity or gas supplies.

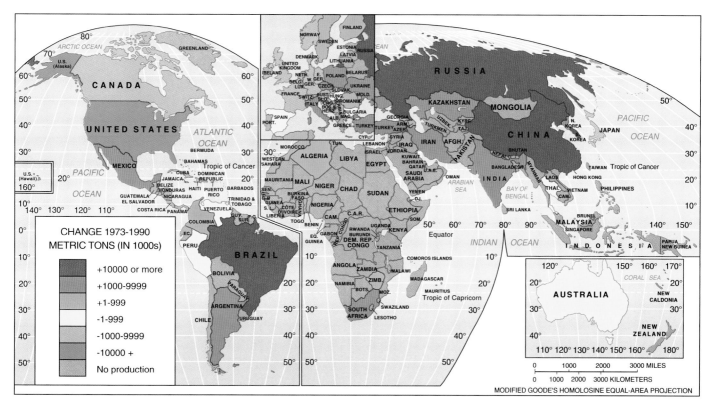

CHANGE 1973-1990 METRIC TONS (IN 1000s)

- +10000 or more
- +1000-9999
- +1-999
- -1-999
- -1000-9999
- -10000 +
- No production

MODIFIED GOODE'S HOMOLOSINE EQUAL-AREA PROJECTION

3.3

ABOVE: Per capita GNP varies widely among the republics of the former Soviet Union. Levels are much higher in the European portion of the region. Credit ratings, provided by the World Bank, are shown for Eastern Europe and the former Soviet Union, from 1888 to 1992, in Table 13.12. Presently there is tremendous instability and uncertainty surrounding every aspect of Russian life including the political, economic, social, legal, and fiscal systems. The result has been a reluctance on the part of U.S. businesses to make substantial investments. The best case scenario is that Russia and other former republics would evolve into peaceful free-enterprise democracies.

BELOW: The United States, Western Europe, Argentina, and Australia are leading wine-producing regions of the world. Income, social preference, religious custom, and climate are important factors. Why is wine production a good mirror of a country's economic development?

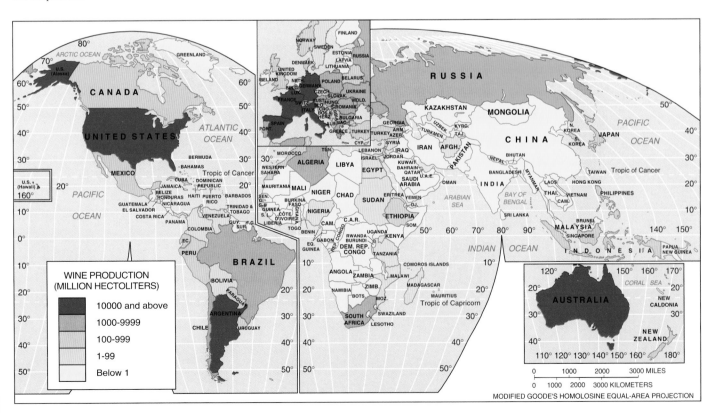

3.4

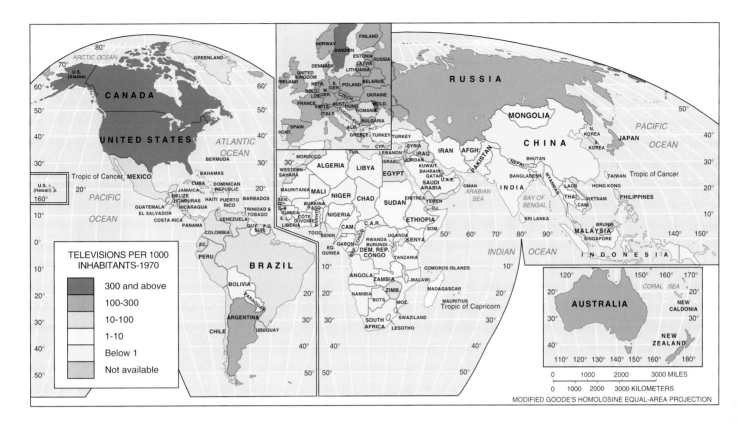

ABOVE AND BELOW: Televisions have diffused from North America to the rest of the world. The diffusion is not complete, however, as some countries in Africa and South Asia have less than 10 televisions per 1000 people. In 1954, the United States had almost 200 television sets per 1000 inhabitants, while Argentina, Germany, Italy, Japan, the Netherlands, Spain, Sweden, and Switzerland had less than one. But what is this television technology doing to us? Television changes the way those who use it view and understand themselves and the way in which they live. Television is mobilizing people throughout the world against harsh government policies, by allowing them to see how their lives compare with those who are better off. In India, Asian and U.S. satellite television programming is the chief inspiration for major changes in Indian culture, such as fashion and popular music, has increased the demand for brand name products and has engendered demands to liberate the Indian economy from state control.

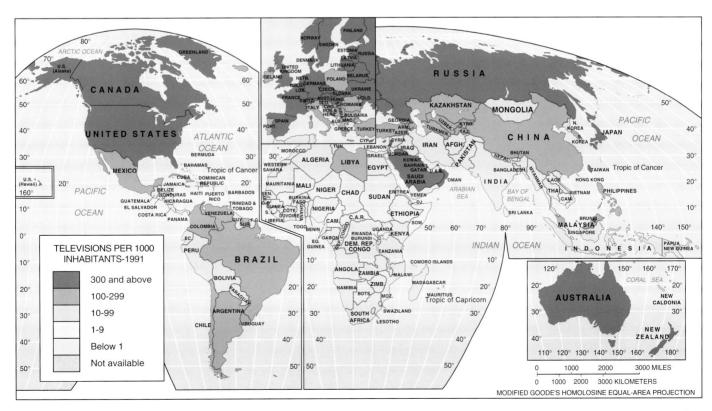

3.5

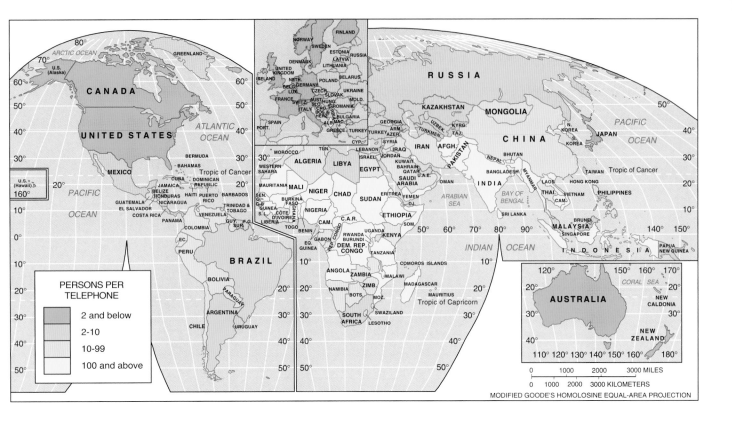

ABOVE: Developing countries have several hundred persons per telephone, while relatively developed countries have two persons per telephone. The United States is approaching one person per telephone and will soon surpass this ratio with the proliferation of cellular phones. By 1994, there were 1.1 million voice circuits between the United States and Europe. Private global satellite constellations are being created. Large scale, low earth orbit systems such as Motorola's 66-satellite Iridium system provide telephone communications to and from any place on earth.

BELOW: In developing countries, the percentage of women in secondary schools is far below that of men. This fosters lower educational development among women and lowers prospects for economic development and gender equality.

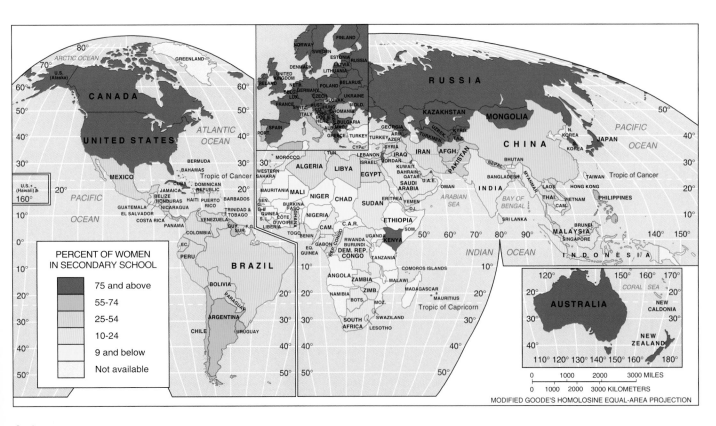

3.6

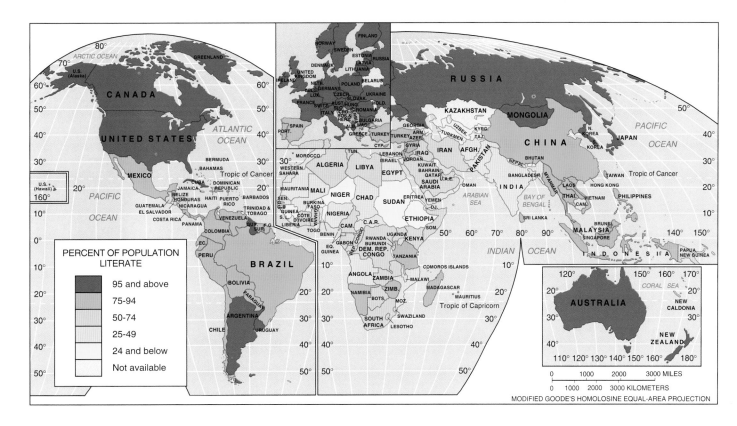

ABOVE: A great disparity in literacy rates exists between inhabitants of developed and less developed countries. In the United States and other highly developed countries, the literacy rate is greater than 98%. Notice the large number of countries in Africa and South Asia that have a literacy rate of less than 50%. As the global information age matures, education becomes even more paramount. In order to compete effectively in world trade, countries must learn how to invest in human capital. Educated workers in other countries are an incentive to transnational corporations. The transnational business and hyperfluidity of information assures that the business community will find the workers that they need at lower costs.

BELOW: The gender gap for literacy is most pronounced in the developing world, especially in Africa, South Asia, and the Middle East. In most developing countries, the female literacy rate is far below the male literacy rate.

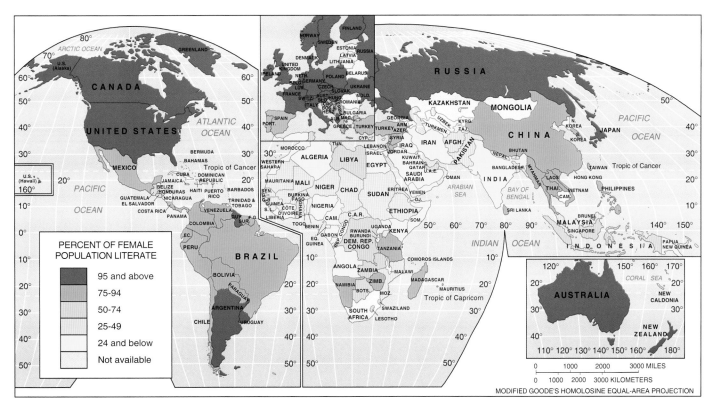

3.7

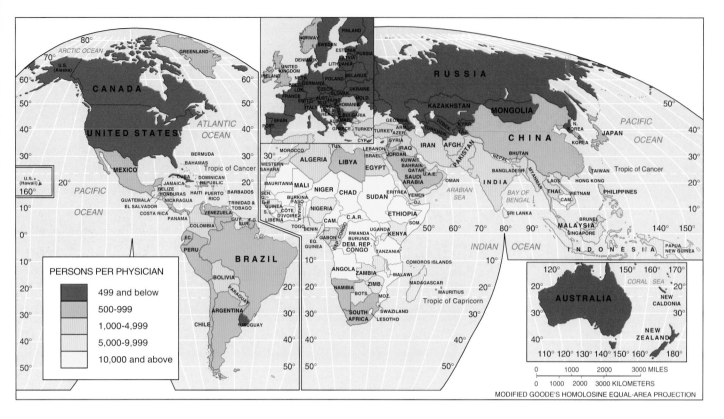

ABOVE: Persons per physician. An important measure of economic development is the number of persons per physician in a country. This measure is a surrogate for health care access, which includes hospital beds, medicine, nurses, and doctors. Most of Africa exhibits 10,000 people per physician or more, while Europe and the United States average approximately one doctor for every 200 people. Advances in medical and biotechnology will significantly lengthen the average life of humans in the near future. New ideas in medicine that revolve around holism are coming into the mainstream of thought, but the global AIDS epidemic will strike 100 million people in the next two decades. Entire countries in Africa and Asia are being ravaged by AIDS, which threatens to destroy the merchant class of some societies.

BELOW: The Human Development Index is perhaps the best overall measure of economic development. It combines three measures, including life expectancy, adjusted GDP per capita, and education (years of schooling literacy).

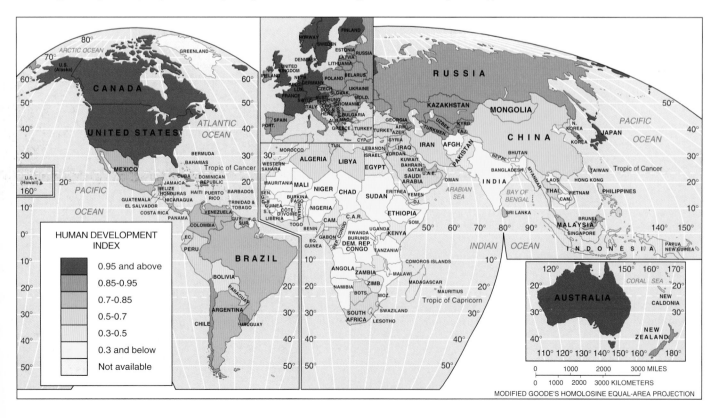

3.8

Globalization Smooths Business Cycles

Globalization in the restructured economy may be less vulnerable to the normal fluctuation of the business cycles. Companies today, large and medium size, are emerging as global players procuring parts and materials worldwide, as well as selling their products overseas. The result may be an economy that responds in new ways to swings in supply and demand.

Companies have restructured and are also more efficient because the labor market has changed so dramatically. Today, businesses are relying more heavily on outsourcing, cross-training, two-tier wage systems, temporary labor, and taking advantage of lower unionization levels. Ninety percent of TRW's North American employees, for example, receive variable pay, which means their take-home pay increases if the company does well, but declines if it does poorly.

The companies are managing their inventories much more efficiently than in the past because they have adopted just-in-time techniques and new computerized systems to better track the delivery of products. The ability of producers to expand abroad, at the same time that non–North American countries establish a presence in North America, obviates the decade old notions of how the regional economy operates. More than a quarter of the economy in North America now depends in some way on exports and imports.

Globalization has also changed the relationship between American producers and consumers. Before foreign competitors, such as the Japanese, saturated the North American market, North American companies had more power to make products of less quality and to charge a price that they wanted. The poor quality of many American cars prior to the Japanese onslaught is testimony to this fact. In the late 1990s, however, customers have the upper hand in demanding new products, better quality, and lower prices because of the globalized market.

Because of globalized competition, booms and busts are more difficult. If Dow or Dupont tries to raise prices, customers will now reflexively seek price quotes from Germany's companies or other suppliers. If a foreign competitor underbids a North American supplier, components can be air-freighted for delivery within 28 to 48 hours. In short, raising prices has never been tougher. This suggests that the economic cycles may not carry the inflationary risk they once did.

The decline in the proportion of American workers represented by unions—from 34% to 15% since 1980—goes a long way towards explaining why wages have not increased in real terms recently. Some economists blame unions for pushing up wages in good times and for not being flexible enough when bad times hit, thus causing wilder fluctuations in the business cycle.

Thanks to technology innovation, it is possible for companies to refrain from making a product until it is ordered. A company's production can be recalibrated on a daily or weekly basis. Inventory build-ups will play a small role in future recessions. Additionally, the economy is far more diverse today. Technological producing companies are growing, not just in California, but throughout the Midwest and the Northeast. As a result, the job market dependence on traditional big employers like steel and auto companies is fast declining. Additionally, the growth of the service sector builds resilience into the overall economy because health care, finance, business services, and government are not nearly as vulnerable to classic business cycles.

Globalization of the Space Economy by 2000

Early in this chapter we discussed the identification of three inward-seeking and competing trade blocs: (1) the European Union augmented as of 1997 to 15 members, but likely to include many more in the future; (2) the North American Free Trade Agreement of the United States, Canada, and Mexico; and (3) the Japanese-dominated Asian block of Four Tigers and five newly industrializing countries. We have just described the EU and NAFTA blocs in the previous section. The Asian block will continue to show the largest gains in growth (see Chapter 11). Much of this growth is intra-Asian trade and also foreign investment being produced by two groups: the overseas Chinese entrepreneurs and the Japanese multinationals. Large numbers of overseas Chinese are investing in the Chinese economic area. Capital flows are emanating from Singapore, Hong Kong, and Taiwan, as well as the United States and Canada. Large numbers of Japanese industries, especially small ones, have moved operations to neighboring agent countries as the rising Japanese yen has caused manufacturing inside Japan to be unprofitable. Asian countries have recently reduced tariffs, deregulated their financial sectors, and thrown open telecommunications and power/energy monopolies to private investment (see Chapter 11).

However, rather than seeing three great regional trading blocs competing for advantage in global trade and commerce and limiting imports from one another, a new pattern is emerging as we converge on the year 2000.

Rather than competing with one another, trade blocs are showing signs of closer economic integration because of the liberalization of trade, produced by the GATT negotiations in 1994, which created the World Trade Organization after 50 years of struggle. The result is a new globalization of world trade. Whereas before, the individual separate trade blocs were competing with one another, now the new players are powerful multinational corporations which operate in an almost borderless world of their own making. Today, the multinational corporation (MNC) follows the *triad corporate strategy* where each firm attempts to maintain a substantial corporate presence within each of the three major world regions, or trade blocs (EU, NAFTA, Asia), as a way of hedging its market share against loss to its competitors.

The year 2000 and beyond will see the overall levels of prosperity rise worldwide, thanks largely to the immense increase in economic integration among multinational corporations representative of the three triad trade blocs. There is now general agreement and understanding that the reversal of recent WTO progress in reducing tariffs and approving trades would cost the world and its citizens some or all of its post–World War II gains in living standards and political cooperation. The future growth of large and small economies, rich and poor ones included, depends on continuing movement towards more trade liberalization, not less.

SUMMARY

This chapter examined aspects of the international sphere of business. *International business* is any form of business activity that crosses a national border. It includes the international movement of almost any type of economic resource—merchandise, capital, or services.

Our discussion of traditional trade theory pointed to *comparative advantage* rather than *absolute advantage* as the underlying explanation for much of the trade that occurs. Beyond predicting that everyone gains something from trade, classical trade theory neglects to consider the distribution of benefits. Free trade was established in the 19th century within a framework of inequality among countries. An artificial division of labor was established between developed and developing countries. Developing world countries as primary producers became dependent on foreign demand and, therefore, vulnerable to the business cycle of expansion and contraction in developed countries.

The theoretical basis for international business was extended in a consideration of the basis of *production-factor flows*. Production factors that are most readily movable are capital, technology, and labor. We focused primarily on *capital movements*, enhancing understanding of FDI by using a managerial perspective. In many respects, the international movement of capital, technology, and managerial know-how is now more important than international trade is. In fact, the preeminent international business organizations

of the future will be those "devoted to improving international markets in terms of special inputs—services, skills, knowledge, capital. . . . The guts of such firms will be information network and data banks" (Robinson, 1981, pp. 18–19).

Theories of international trade and production-factor movements emphasize the benefits of a liberal, market-oriented business environment. However, a number of obstacles significantly impede the flow of merchandise, capital, technology, and labor. These obstacles include distance barriers, managerial barriers, and governmental barriers. Much progress was made in reducing governmental barriers—*tariffs* and *nontariff barriers*—during the long postwar boom. But the 1970s and 1980s saw a steady erosion in the liberal trading order as governments started to erect new barriers to international business.

International trade is not normally a barter arrangement. Money must be exchanged on international markets for goods and services. Determination of exchange rates allows world trade to function. Explanation of why exchange rates fluctuate is no easy matter but is related to levels of real output, inflation rates, demand factors, and currency speculation in trading-partner countries.

Despite the chillier economic climate since the mid 1970s, countries continue to participate in myriad multinational operations. Major organizations that can be important to international business are *international financial institutions* and groups that promote *regional integration*. Groups that obstruct market forces include *commodity cartels*. Leaders in developed countries view commodity cartels as an unfortunate departure from market-oriented principles. In contrast, most LDC leaders view commodity cartels as a means to reduce their vulnerability in a world of unequal exchange.

The General Agreement on Tariffs and Trade (GATT) has reduced trade barriers worldwide. The most recent Uruguay round of GATT has made progress in a number of difficult issues—farm policy, intellectual property rights, and trade barriers related to a growing international provision of services. The United States' current, most troublesome world trade problems, however, are the sluggish world economy and the closed Japanese markets. The Japanese own a $60 billion trade surplus.

The most widely acclaimed regional integration to date is the EU of Western Europe. By the year 2000, it is likely that most eastern and western European nations will be associated in a powerful trade bloc that manages almost 50% of worldwide trade. The United States and Canada followed suit in 1989, with the U.S.–Canadian Free Trade Agreement. Final approval for the momentous North American Free Trade Agreement (NAFTA), involving Canada, Mexico, and the United States, to challenge the EEC head-on occurred in 1994.

KEY TERMS

absolute advantage
brain drain
capital controls
cartel
common market
comparative advantage
countertrade
development bank
direct investment
domestic farm subsidies
economic integration
economic union
Engel's law
Eurocurrency
European Union (EU)

exchange controls
exchange rate
export quota
export subsidies
export-restraint agreement
factors of production
floating exchange rates
foreign direct investment
(FDI)
free-trade area
General Agreement on Tariffs and Trade (GATT)
import quota
infant industry
intellectual property rights

international currency
markets
International Monetary
Fund
intramultinational trade
market seekers
multinational corporation
(MNC)
NAFTA
nontariff barrier
Organization of Petroleum
Exporting Countries
(OPEC)
production factors
petrodollars

portfolio investment
protectionism
regional integration
tariff
tax-haven country
terms of trade
trade deficit
transfer costs
transfer pricing
unequal exchange
United Nations Conference
on Trade and Development (UNCTAD)
World Bank
WTO

SUGGESTED READINGS

Bagchi-Sen, S., and Pigozzi, B. M. 1993. Occupational and industrial diversification in the metropolitan space economy in the United States, 1985–1990. *Professional Geographer*, 45:44–54.

Britton, J.N.H., ed. 1996. *Canada and the Global Economy.* Montreal: McGill-Queens University Press.

Brotchie, J. F., Hall, P., and Newton, P. W., eds. 1987. *The Spatial Impact of Technological Change.* London: Croom Helm.

Dicken, P. 1991. *Global Shift: Industrial Change in a Turbulent World*, 2nd ed. London: Harper & Row.

Dunning, J. H., ed. 1985. *Multinational Enterprises, Economic Structure and International Competitiveness.* New York: John Wiley.

Glasmeier, A. K. 1991. *The High-Tech Potential: Economic Development in Rural America.* New Brunswick, N.J.: Center for Urban Policy Research.

Hoare, A. G. 1993. *The Location of Industry in Britain.* New York: Cambridge University Press.

Hogan, W. T. 1991. *Global Steel in the 1990s: Growth or Decline.* New York: Lexington Books.

Korth, C. M. 1985. *International Business: Environment and Management*, 2nd ed. Englewood Cliffs, N.J.: Prentice Hall.

Law, C. M., ed. 1991. *Restructuring the Global Automobile Industry.* London: Routledge.

Massey, D., and Jess, D. 1996. *A Place in the World: Places, Cultures, and Globalization.* New York: Oxford University Press.

Porter, M. E. 1990. *The Competitive Advantage of Nations.* New York: Free Press.

Root, F. R. 1990. *International Trade and Investment*, 6th ed. Cincinnati: South-Western Publishing.

Salvatore, D. 1990. *International Economics*, 3rd ed. New York: Macmillan.

Taoka, G., and Beeman, D. R. 1991. *International Business: Environments, Institutions, and Operations.* New York: HarperCollins.

WORLD WIDE WEB SITES

WEBEC—WWW RESOURCES IN ECONOMICS
Frames: http://netec.mcc/ac.uk/%7eadnetec/WebEc/framed.html
No frames: http://www.helsinki.fi/WebEc/NetEc
http://cs6400.mcc.ac.uk/NetEc.html
WebEc is ". . . an effort to categorize free information in economics on the WWW." Bill Goffe, author of Resources for Economists on the Internet notes WebEc is ". . . a particularly good place to look for a broader array of business and economic resources."

THE WORLD TRADE ORGANIZATION
http://www.wto.org/
The principal agency of the world's multilateral trading system. Its home page includes access to documents discussing international conferences and agreements, reviewing its publications, and summarizing the current state of world trade.

THE WORLD BANK
http://www.worldbank.org/
A leading source for country studies, research, and statistics covering all aspects of economic development and world trade. Its home page provides access to the contents of its publications, to its research areas, and to related websites.

U.S. DEPARTMENT OF COMMERCE
http://www.doc.gov/
Charged with promoting American business, manufacturing, and trade. Its home page connects with the web sites of its constituent agencies.

BUREAU OF LABOR STATISTICS WEBSITE
http://stats.bls.gov/
Contains economic data, including unemployment rates, worker productivity, employment surveys and statistical summaries.

INTERNATIONAL BUSINESS II: WORLD PATTERNS

OBJECTIVES

- To describe the evolving pattern of international commerce
- To understand the important position of North America within the pattern of international trade
- To document the giant emerging markets (GEMs) for North America
- To assess the effect of East Asia as an economic powerhouse on the industrial democracies of North America and Western Europe
- To examine global trade flows of six commodities and goods: microelectronics, automobiles, steel, nonoil commodities, grains and feed, textiles and clothing

The G-7 economic summit leaders.
(Source: Reuters/Peter Jones, Archive Photos)

COMPOSITION OF WORLD TRADE

The turbulent decades of the 1970s and 1980s saw major changes in the volume and composition of trade. World trade grew throughout the period, reaching about $2 trillion in 1986. Rates of trade growth, however, declined after 1973. Manufacturing exports, with the exception of a dip in the mid-1970s, continued their rapid growth (see Figure 11.1). They now account for about 60% of world exports by value. Fuel exports doubled their value share in the 1970s, from less than 10% to more than 20%, but they declined with an oversupply of oil and weakening demand in the 1980s. The export value share of other primary commodities—food, beverages, and crude materials—decreased from nearly 30% in the mid-1960s to less than 15% in the mid-1980s.

The changing structure of trade has affected different types of countries differently. OPEC countries recorded a meteoric rise in the value of their exports in the 1970s and a precipitous decline after 1980. The industrial countries of North America, Western Europe, and East Asia experienced a drop in their export earnings after the oil crisis. But as a group they recovered and now account for 80% of the value of world trade. With the exception of the major oil exporters, LDCs that depend heavily on the export of a few primary commodities fared badly. For them, the growth in volume of primary commodity exports has been negative since 1980.

Increased diversification of trade ties represents one of the most significant developments in the contemporary world economy. Advanced industrial countries still trade primarily among themselves, but the proportion has declined from more than 75% in 1970 to around 66% today. They have increased their share of exports to LDCs and their imports from LDCs have increased still more. Another major development has been the growth of manufacturing exports from LDCs to developed countries and, to a lesser extent, to other LDCs. Manufacturing exports now account for about 40% of total nonfuel exports of these countries, compared with 20% in 1963, and LDCs now supply 13% of the imports of manufactures by developed countries, compared with only 7% in 1973. Yet only a handful of Asian and Western Hemisphere countries are involved in this development.

The State of the World Economy

The global gross domestic product and the total output of goods and services across the world actually dipped below zero for the first time in 1991. It dipped only −0.4%, but a negative world output is a rare occurrence. The main reasons for this decline were a world recession and a major economic interruption in Eastern Europe and the former Soviet Union (Figure 11.2).

World output per capita fell more than 2% below zero in 1991, after an almost complete stagnation in 1989 and 1990. East Asia remained the fastest-growing area despite a slowdown in total world trade. Major military activity in the Persian Gulf disrupted financial markets and ruined the oil exports of Kuwait and Iraq. The Uruguay Round of GATT negotiating was postponed, and controversy among the EEC countries over the Maastricht Agreement also derailed European integration tem-

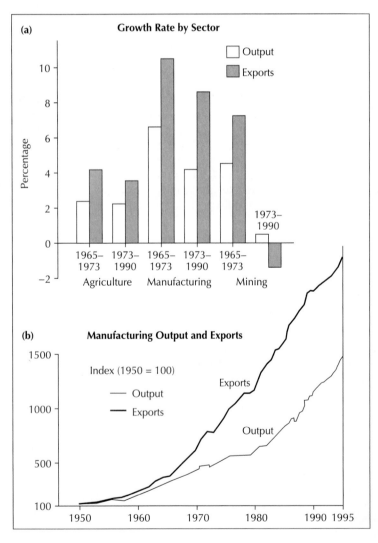

FIGURE 11.1
Postwar growth in world output and exports. *(Source: World Bank, 1997, p. 134)*

porarily. Democracy in South Africa was at a standstill, and Moscow announced the dissolution of the Soviet Union. Some of these world events created uncertainties in the international world economy and the world of trade (Tables 11.1 and 11.2).

Growth of world trade slowed substantially by 1991 to about 3%, the smallest gain since 1985. But even this small gain was welcomed because of the negative growth rate of world output. World merchandise exports actually increased 2% in value in 1991 to a record $3.47 trillion, as prices declined on the average (Figure 11.2).

By 1993 and 1994, there was high unemployment worldwide, disenchantment and withdrawal from the labor force that had been restructured, widespread frustration, homelessness, drug trafficking, and ethnic tensions and riots in the developed world. All this turbulence caused the electorate to express dissatisfaction, most notably in the United States with the election of the 1993 Clinton administration. While the entire world waited for recovery from recession and for new growth, heavy indebtedness by households and firms, by governments, notably that of the United States, as well as others, continues into 1994. High debts and high real long-term interest rates cooled the supply and demand for credit and international trade.

The World Economy is recovering from a recession that bottomed in 1991–1993. World GDP growth was 3.5% in 1995, up from 3% in 1994, and 2.5% in 1993. This is the fastest rate of economic growth since 1988

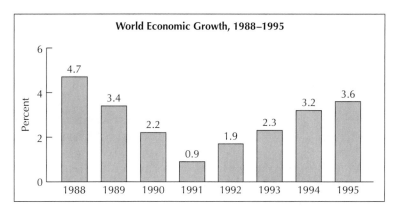

FIGURE 11.2
World economic growth, 1988–1995. Recovery is now progressing in industrialized countries. Developing countries were averaging 5.6% growth in 1995, while developed countries were averaging 3%.

(see Figure 11.2). World economic growth has been slow since 1990 because the industrialized countries have been in the midst of a recession. However, economic recovery in the industrialized countries is now progressing rapidly (Figure 11.3). Eighty percent of the world output comes from North America, Japan, and Europe (triad countries). North America's economic growth was 3% in 1995, compared to only 2% in 1994. The European Union (EU) grew by 3% in 1995, up from 2% in 1994. Japan, still in the early stages of recovery, grew 2% in 1995 and only 1% in 1994. By 1995, world exports have increased from 12% to more than 30% of world GNP, thus documenting world globalization of business. Exports jumped from $2 trillion in

Table 11.1
Growth of the World Economy, 1988–1993

	Annual Percentage Change					
	1988	*1989*	*1990*	*1991**	*1992†*	*1993†*
World output	4.4	3.2	1.8	−0.4	1.0	3.0
Developed market economies	4.4	3.3	2.6	0.9§	1.7	3.2
Economies in transition‡	4.5	2.3	−5.0	−15.9§	−12.0	−4.1
Developing countries	4.4	3.3	3.2	3.4	4.5	5.5
World trade (export volume)	8.5	7.2	4.7	3.4	4.5	6.5
Memo item						
Growth of world output per capita	2.7	1.5	—	−2.1	−0.7	1.3

* Preliminary.
† Forecast, based on Project LINK and DESD estimates.
‡ The former Soviet Union and Eastern Europe.
§ After 1990, the former German Democratic Republic is included in Germany.
Source: U.N. World Economic Survey, 1992/93, p. 1.

◈

T a b l e 11.2

Distribution of World Output ($ Billion)

Item	1980	1985	1990	1994	% Total	% Change 1980–1994
World Total	**11,982.7**	**17,951.2**	**25,442.3**	**28,202.7**	**100.00**	**235**
North America	3,234.1	4,791.6	6,495.4	7,473.7	26.5	234
Other Western Hemisphere	842.7	1,131.7	1,505.3	1,698.2	6.0	231
Middle East	471.4	657.7	876.2	964.7	3.4	201
Asia and Oceania	2,497.3	4,267.1	6,839.5	8,626.8	30.6	20
Western Europe	2,999.7	4,230.2	5,939.1	6,565.2	23.2	219
E.Europe and Ex-USSR	1,399.2	2,110.1	2,759.4	1,784.9	6.3	127
Africa	538.4	762.8	1,027.3	1,089.1	3.8	202

Note: Figures for each region's share of world output are purchasing power parity estimates based on the U.N. International Comparison Project.
Source: International Monetary Fund.

1980 to $4 trillion in 1996. Pacific Rim Asia NICs lead the way (Figure 11.4).

Giant Emerging Markets (GEMs)

Export policy of the triad countries has been focused on themselves—Europe, Japan, and North America. These markets will continue to grow throughout the decade, but another category of countries holds more promise for large increases in world exports by percentage. These countries can be called giant emerging markets (GEMs). In Latin America, they are Argentina, Brazil, and Mexico. In Asia, they are the Chinese economic area (China, Hong Kong, and Taiwan), Indonesia, Singapore, South Korea, and India. Others GEMs include Poland, South Africa, and Turkey. These 12 countries account for half of the world's population—three billion people (Tables 11.2 and 11.3).

The GEMs share important attributes in that they are all physically large with massive markets offering a vast array of products to serve their huge populations. All of the GEMs are of major political importance within their world regions, where they serve as regional economic drivers—their growth will cause further economic expansion in neighboring markets.

The U.S. Department of Commerce estimates GDP in the GEMs as averaging 7% annual growth from 1995–2000. This compares with an average of only 3% among the industrialized countries. More importantly, for the triad countries, GEM imports of goods and services will increase by an estimated 75% from 1995 to the year 2000, compared to an only 8% rise among industrialized countries' imports. In essence, the GEMs will become the fastest growing

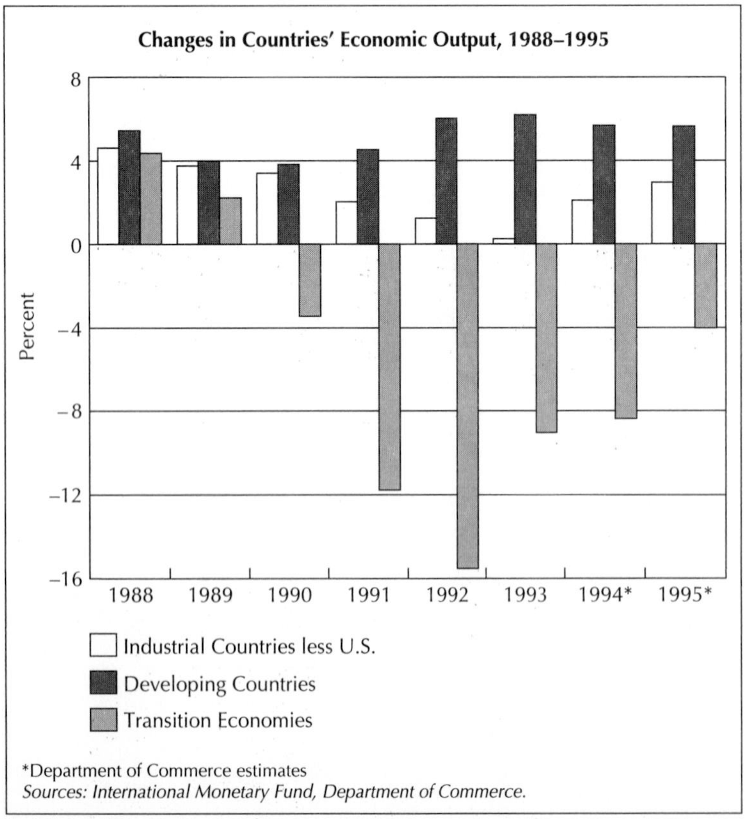

Changes in Countries' Economic Output, 1988–1995

☐ Industrial Countries less U.S.
■ Developing Countries
▨ Transition Economies

*Department of Commerce estimates
Sources: International Monetary Fund, Department of Commerce.*

FIGURE 11.3
Changes in countries' economic output, 1988–1995.

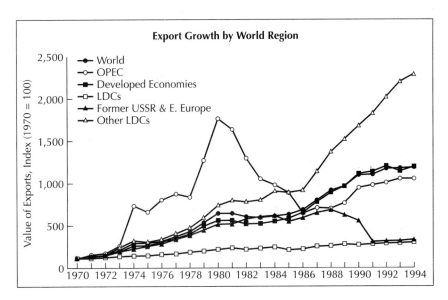

FIGURE 11.4

Exports by world region. OPEC's stranglehold on the world ended in 1980 and oil prices plummeted. Newly developing countries, especially the Four Tigers and other Asian countries, have grown the most rapidly since 1986, while the former Soviet Union and eastern European countries have also plummeted. *(Source: United Nations, 1996)*

🔶

T a b l e 11.3

Economic Growth in the Giant Emerging Markets, 1976–1995

Item	Average Growth 1976–85	Annual Percentage Change in Real GDP		
		1986	1990	1995
World	3.4	3.6	2.2	3.6
Industrial Countries	2.8	2.9	2.4	3.0
Developing Countries	4.5	4.8	3.8	5.6
Giant Emerging Markets[1]	6.0	6.8	3.3	6.3
Chinese Economic Area	7.9	8.8	3.9	8.7
China (12)	7.8	8.5	3.9	9.0
Hong Kong (11)	8.9	11.1	3.2	5.4
Taiwan (6)	8.6	11.6	4.9	6.4
India (29)	4.6	4.9	4.9	5.5
Indonesia (30)	5.7	5.9	7.2	7.0
South Korea (9)	8.0	11.6	9.5	7.5
Turkey (27)	4.0	7.5	9.3	3.0
South Africa (35)	2.1	0.0	−0.3	3.0
Argentina (22)	−0.5	7.3	0.1	5.5
Brazil (18)	4.0	7.6	−4.1	4.0
Mexico (2)	4.3	−3.6	4.5	−2.0
Poland (50)	1.8	4.2	−11.6	4.5
Other Developing Countries	0.3	−0.5	5.2	2.9
Eastern Europe and Former USSR[2]	3.9	3.5	−2.7	−4.9

[1]Numbers in parentheses indicate the country's ranking for U.S. exports of manufactured goods in 1993.

[2]Excludes Poland.

Sources: International Monetary Fund, Department of Commerce.

markets throughout the world up to the turn of the century, and perhaps beyond. Such developing countries make up 40% of the market for U.S. exports.

Of all the world's regions, Asia has been growing the fastest for the last several years, experiencing 8–10% annual growth (see Chapter 1). For example, in 1995, fiscal growth slowed a bit to 7% because of a slowdown in China. China had grown at double digit rates since 1992, but slowed to 9% (Figure 11.4). Never before in the history of humanity have so many people been raised from poverty in such a short period of time thanks to free-market mechanisms. Since 1970, free-market mechanisms and technological revolution have expanded the coverage of the market economies of the world from 1 billion to more than four billion. Three-quarters of them are in Asia. The free market, a Western idea, and what it can do, is a triumph of markets for both East and West.

The epitome of private enterprise in Asia are the Four Tigers (Singapore, South Korea, Taiwan, and Hong Kong). The tigers demonstrated that poor countries could industrialize quickly and enter the world manufacturing by employing market economic policies and thus rise quickly on the ladder of economic development. Not too far behind are the achievements of Thailand, Malaysia, and Indonesia, which have also seen accelerated growth into the late 1990s. The high economic growth of these Asian countries is contrasted markedly with the economic stagnation of India, Pakistan, Bangladesh, and other Asian low-income countries that pursued policies of import substitution and isolationism, producing excessive bureaucratic barriers and tariffs and fostering inefficiency and low growth. While India is changing its policies in this regard, elsewhere and especially in Africa, inefficient state-run manufacturing and agricultural enterprises continue to be the dominant mode of economic activity, thus penalizing its citizens.

The reduction in government-operated programs was not limited to the Third World. The 1980s and 1990s in North America and Europe saw the planned reduction of "big government" with the benefits of the laissez-faire market policies. Western conservatism was put into motion in the United States under Ronald Reagan, in England under Margaret Thatcher, and proliferated to Germany and Scandinavia, where the Christian Democratic parties strengthened their government position over the Social Democrats and attempted to reduce the welfare state policies and social expenditures and the high tax rates of former governments. In France, however, the country's telecommunications, defense, utilities, airlines, trains, and banking remained government operated, in contrast to most of the remainder of western European nations. In Latin America, economic programs took a tilt toward laissez-faire economic policies and allowed rapid progress of economic development for a few countries, notably Chile and Argentina.

Economic growth for several important countries in Latin America are in the 4–5% range. For example, Argentina has had 4 years of steady growth, averaging 7% annually. The outlook for 1997 growth is 6%. Brazil has had growth of 5% in 1994 and 1995, and should average between 3–5% in 1997. Mexico's economy contracted somewhat after the peso devaluation of 1994, resulting in a GDP decline of 2% in 1995. However, for 1996, the country began an economic expansion that appears to be continuing through to the end of the century. The low value of the peso has made Mexican goods more competitive on world markets, reversing the trend toward an increasing balance of payments and deficits for Mexico. Mexico, alone, was the only country to double its manufacturing exports between 1989–1994 (Table 11.1).

Central Europe and the newly independent states (NIS) of the former U.S.S.R. are continuing to experience negative growth and have been since 1990 (see Figure 11.3, Table 11.3). Eastern Europe and 15 former U.S.S.R. countries experienced 17% negative growth in 1992, which has eased a bit to approximately 5% negative growth in 1995. These "economies in transition" will continue to have negative growth rates, but they should ease towards the end of the century. Several of the countries, such as the Czech Republic, Hungary, and Poland, are now showing positive growth. Russia and the Ukraine, two leading former Soviet countries, have taken longer than expected to recover from the transition from Communism to a market economy. Among the newly independent states, only the Baltic Republics—Latvia, Lithuania, and Estonia—have been able to post positive growth by 1996.

The economic growth in Asia over the past three decades, first in Japan, followed by the Four Tigers (Asian newly industrialized economies of South Korea, Taiwan, Hong Kong, and Singapore), and now more recently by China, Indonesia, Malaysia, and Thailand, has brought with it a boost of international trade. It is impossible to separate increasing trade and economic growth. Economic growth leads to expanded levels of domestic consumption and investment, which result in higher levels of imports of consumer goods, raw materials, and capital goods.

In 1994 Asia, including Japan, generated one-third of total world output, up from less than a quarter 15 years ago (see Table 11.2). In contrast, the contribution of North America to international trade growth remained constant at approximately a quarter of the world output. At the same time, Asian international trade has accelerated (Figure 11.5). In 1980 Asia, including Japan, accounted for 16% of world exports. Over the next 14 years, the region increased its share to 26%. North America (the United States and Canada) expanded its combined share slightly from 15.1% to 15.5%.

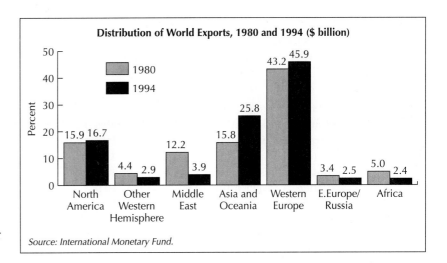

FIGURE 11.5
Distribution of world exports. Western Europe still leads the world in the distribution of world exports. Many small countries supply neighbors with manufactured goods, taking advantage of scale economies and satisfying the need for aerial specialization.

Western Europe still shows the greatest proportion of trade because of many small states in close proximity, specializing in the production of goods and services and exchanging with one another, much like the American states and Canadian provinces trade among themselves (Figure 11.5). The value of Asia's output in 1995 is estimated by the U.S. Department of Commerce at $8.6 trillion. Meanwhile the U.S. was producing $6.6 trillion and Western Europe about the same amount of output. Japan alone accounts for 9% of world output, valued at $2.5 trillion in 1994.

The *purchasing power parity* is used to measure China's output; the results indicate that its output is nearly equal to that of Japan. GDP estimates using purchasing power parity (PPP) exchange rates are higher than estimates calculated with market exchange rates for most developing countries because of the developing countries' low prices of nontraded goods and services on the international market. On the other hand, the PPP exchange rate estimates of GDP are lower for some countries, including Japan, where many nontraded goods, such as housing and services, were very high priced when valued at market exchange rates. The same applies for Switzerland. Because the cost of living is so high in Japan and Switzerland, their per capita incomes are grossly exaggerated and range in 1997 between $30,000 and $40,000. (See box: Purchasing Power Parity.)

The burgeoning markets of the GEMs represent the greatest North American export market of the future. The GEMs—Argentina, Brazil, Mexico, China, Hong Kong, Taiwan, India, South Africa, and Russia—are discussed in detail after treatment of North American (U.S. and Canada) and European countries' trade patterns. While Russia is a fairly small market for North American trade today, it is examined because it could constitute an important role in the future of the world free-market economy.

WORLD PATTERNS OF TRADE

The United States

The United States is the world's largest trading nation, accounting for more than $1.3 trillion worth of exports and imports in 1995. During the 1950s, the United States accounted for 25% of total world trade but now accounts for only approximately 10% (Figure 11.5). From 1960 to 1970, the United States enjoyed a net trade surplus as a result of OPEC price rises and the weak value of the dollar. However, this surplus turned into a deficit by 1985. Throughout much of the 1970s, the United States sold only 4% of its goods overseas but raised the overall level to approximately 7% during the late 1970s and the 1980s (see Figure 10.8).

Figure 11.6 shows the composition of U.S. merchandise trade with the world in 1995. Machinery and transportation equipment accounted for the largest single proportion of exports (45%). Chemicals and other manufactures added another 34%, whereas agricultural products amounted to 9%. More than 80% of U.S. exports are manufactured items. The United States, with its varying productivity and physical conditions, as well as its competitive edge in technology, is the chief world manufacturer today (Table 11.4).

Manufacturing goods also account for approximately 80% of the traded goods flowing to the United States from foreign companies. America has essentially farmed out much of its labor-intensive manufacturing to developing countries, shipping semifinished goods to Mexico for manufacturing and reimport (Table 11.5). At the same time, it has acquired an expensive taste for foreign-made luxury items such as automobiles from Germany and Japan, shoes from Italy and Brazil, electronic items from the Far East, and perfume and wine from France. Other imports include fuels for which the United States is not self-sufficient.

◻

PURCHASING POWER PARITY

The usual measure of a country's economy and strength is its GDP—that is, the value of all goods and services produced. When one divides the GDP by the population, the result is GDP per person. The following figure, however, shows GDP per person adjusted for the cost of living by the United Nations—in other words, GDP per person with *purchasing power parity.*

Some experts argue that European nations, notably Germany, Switzerland, and Sweden, have higher GDPs per capita than the U.S. GDP per capita. However, countries such as Norway, Sweden, Germany, Switzerland, and Great Britain have some of the highest costs of living in the world, with Japan topping the list. A Burger King Whopper, now $0.99 in the United States, costs from $6 to $9 in the capitals of these countries. The figure below shows that in 1990, at least, the United States had a clear superiority over other nations with regard to GDP per person, which averaged just about $20,000. Switzerland ranked below the United States at seven-eighths the U.S. level, whereas Kuwait, West Germany, and Singapore were rated at approximately three-quarters the U.S. level. Japan ranked at only seven-tenths the U.S. level, and the United Kingdom and France at two-thirds. Southern European countries fared even more poorly; Spain ranked at less than 50% of the U.S. level, and eastern European countries, the most prosperous of which is Hungary, ranked at less than one-third the U.S. level.

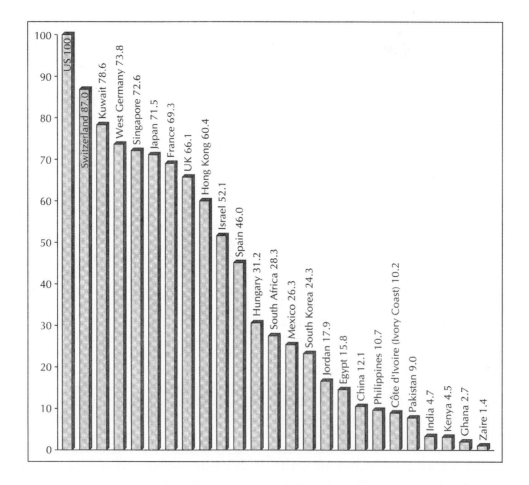

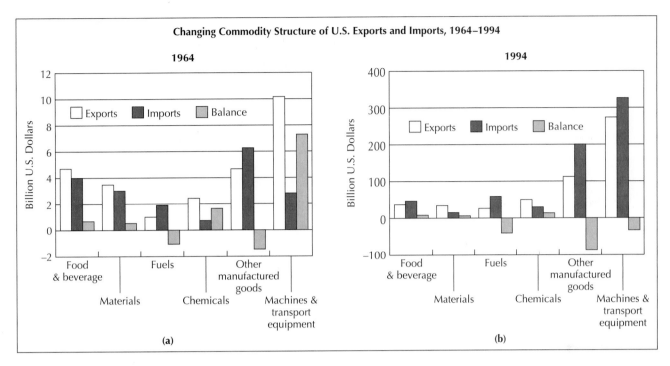

FIGURE 11.6
Changing structure of commodity exports and imports, 1964–1995. In the mid-1960s, the United States enjoyed a very large trade surplus in machinery and transport equipment. By the mid-1990s, however, the trade deficit, at least $160 billion, and imports and manufactured goods in both low and high technology categories exceeded exports. *(Source: United Nations, 1996)*

The direction of U.S. merchandise trade with the world is shown in Figures 11.7 and 11.8. Trade with GEMs between 1993 and 2000 is expected to double. Canada, with its 26 million people, located on the northern border of the United States, represents the single largest trading partner of the United States. Twenty percent of its exports and 19% of its imports go and come from that country. As an economic bloc, the EU or EC-15, accounts for 24% of U.S. exports and 18% of U.S. imports. The United States has a competitive disadvantage and trade deficit with Japan, and Asia, however.

During the last 30 years, the pattern of trade linkages between the United States and the world has shifted away from Western Europe toward the Asian and Pacific regions. Ties with Latin America, although they have contracted somewhat, remain strong, and Mexico ranks third behind Canada and Japan as the leading trade partners of the United States. Mexico is followed by Germany, the United Kingdom, Taiwan, South Korea, France, China, and Italy, which each had more than $20 billion worth of trade with the United States in 1991.

The continued strong performance by the U.S. economy and the continuation of controlled inflation with low unemployment rates, coupled with a favor-

able dollar exchange rate, will insure that U.S. goods and services remain highly competitive in world markets (see Figure 11.9). U.S. firms continued to restructure and downsize to reduce costs and are continuing to reengineer with information technologies at a rapid pace.

The U.S. trade deficit is increasing and will remain large. In 1994, the trade deficit was 1.8 billion compared with 75 billion in 1993. The 1995 deficit was 150 billion, and almost 200 billion by 1996 (see Figure 10.8). The growth of U.S. exports and imports remains high, and the deficit in merchandise trade continues to grow as Americans have an increasing taste for foreign goods, especially automobiles and electronics.

Trade in services will keep the trade deficit from ballooning. In 1995, the U.S. had a surplus of about 60 billion in services trade, up from a mere 7.5 billion in 1987. The growning services sector surplus has kept the overall trade deficit from exceeding over $200 billion annually (see Table 11.5). The largest net positive item in the services trade category is "other private services." These services include financial and business services worldwide. The second most important category is royalties/license fees, which include entertainment, videos, cable TV, compact disks, and recordings.

▣

T a b l e 11.4

Economic Growth in Top 20 Markets for Exports of U.S. Manufactured Goods, 1980–95
(percent except as noted)

Country	Rank 1980	Rank 1995	1993 Exports ($ Billion)	1993 Share of U.S. Exports	1993 Share of Top 20 Markets	Average Growth 1980–89	Growth in GDP 1995
Top 20 Markets			**319.0**	**82.1**	**100.0**	**3.6**	**2.8–4.1**
Canada	1	1	90.2	23.2	28.3	3.1	3.5–4.5
Mexico	2	2	36.0	9.3	11.3	2.1	–3.0–0.0
Japan	3	3	30.3	7.8	9.5	4.0	1.7–2.7
United Kingdom	4	4	24.7	6.3	7.7	2.4	3.0–3.8
Germany[1]	5	5	17.2	4.4	5.4	1.8	2.6–3.4
Taiwan	6	7	13.1	3.4	4.1	8.3	6.0–6.8
France	7	6	12.1	3.1	3.8	2.3	2.8–3.6
Singapore	8	10	11.0	2.8	3.4	7.3	7.0–7.8
South Korea	9	9	10.7	2.8	3.4	8.3	7.0–8.0
Netherlands	10	8	10.3	2.6	3.2	1.6	2.4–3.2
Hong Kong	11	12	8.5	2.2	2.7	7.6	5.0–5.8
China	12	16	7.7	2.0	2.4	9.5	7.0–11.0
Australia	13	11	7.6	2.0	2.4	3.3	3.6–4.4
Belgium/Luxembourg	14	13	7.2	1.9	2.3	2.0	2.3–3.1
Switzerland	15	19	6.6	1.7	2.1	2.3	2.2–2.9
Saudi Arabia	16	15	5.9	1.5	1.9	0.3	1.5–3.0
Malaysia	17	20	5.7	1.5	1.8	5.8	7.8–8.6
Brazil	18	17	5.3	1.4	1.7	3.0	3.0–5.0
Italy	19	14	5.0	1.3	1.6	2.4	2.7–3.3
Venezuela	20	18	3.9	1.0	1.2	–0.2	–2.0–2.0
United States						2.5	2.7–3.3
World			**388.7**			3.0	3.2–4.0

[1]Data for years through 1990 apply only to the former West Germany.
Sources: International Monetary Fund, World Bank, Bank for International Settlements, United Nations, Organization for Economic Cooperation and Development, Department of Commerce, and official country sources.

Growth of U.S. trade in service categories should increase rapidly after 1997, helping to reduce the trade deficit.

Continuing to the end of the century, economic growth and improved world market access due to the reduction of tariffs will offer great potential for U.S. exports. Between 1996 and 2000, world GDP is expected to increase about 5% annually, compared with approximately 2% annually from 1990–1995. Growth in the GEMS will average 6.5–7% during the last half of the decade, with the strongest growth coming from East Asia. By the year 2000, the U.S. exports to GEMS is expected to amount to the combined total exports to the EU and Japan (see Figure 11.5).

Shifting U.S. Trade Patterns

Since 1980, the United States has undergone an important global shift in its exports—away from the traditional European markets and toward Asia, Mexico, and Canada (Figure 11.5). The reasons behind this global shift include the faster growing economies of Asia as expanded markets for U.S. products. Another reason is the great progress made with GATT and the World Trade Organization in the process of liberalization of tariffs in many countries. The North American Free Trade Agreement (NAFTA) of 1994 was such a liberalization agreement, and it has increased trade between Mexico, the Unites States, and Canada.

⧫

Table 11.5

Distribution of U.S. Manufactured
Exports (Percent)

Item	1983	1988	1993
Canada	21.1	22.2	21.6
Japan	10.6	11.7	10.3
Mexico	4.4	6.4	8.9
Western Europe	27.3	27.3	24.4
Asia*	13.7	15.8	17.6
Western Hemisphere**	8.1	7.2	7.9
Other	14.8	9.4	9.3
Total	**100.0**	**100.0**	**100.0**

*Less Japan.
**Less Mexico and Canada.
Source: Department of Commerce.

Thirdly, there have been increasing amounts of U.S. and foreign investment in the Asian economies, and this has produced rapid increases of trade in capital and intermediate goods. Investment has also come from the overseas Chinese and Japanese multinational corporations.

Another shift in trade for the United States has been an expansion of the Mexican market for U.S. goods. U.S. exports have increased significantly over the past 10 years such that, in 1990, Mexico became the second largest market for U.S. manufactured goods after Canada and Japan dropped to number three in rank (Japan continues as the second largest for U.S. goods exports principally because of the large proportion of agricultural products and raw materials that are shipped to that country from the U.S.: meat, corn, wheat, and timber products). The value of U.S. manufactured goods exported to Mexico increased four-fold between 1983 and 1994 (see Table 11.5). This represented an increase of 9%, but during the same period,

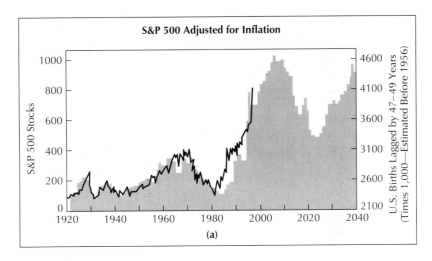

(a)

FIGURE 11.7

(a) Standard & Poor's 500 compared to the baby boom lag 49 years forward shows important growth for the U.S. economy to the year 2010. (b) U.S. exports to giant emerging markets, European Union, and Japan. Exports to giant emerging markets have increased more rapidly than exports to other developed regions, the EU, and Japan. Asia has been experiencing the strongest economic growth, over 8% annually for the last 7 years. The Asian emerging markets will be the fastest growing for U.S. exports. So far the big economic expansion in the U.S. economy has lasted 8 years, and inflation and unemployment have gone down. Gains from technological change, the rise of the service sector, and globalized markets for U.S. products are penetrating deeply into the economy. The Standard & Poor's 500 Stocks Index has risen from 300 in 1991 to over 800 in 1997, a meteoric rise. For the first time in 15 years, the top 500 companies in 1997 showed employment gains.

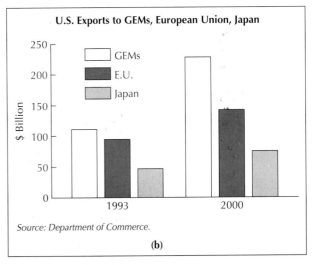

(b)

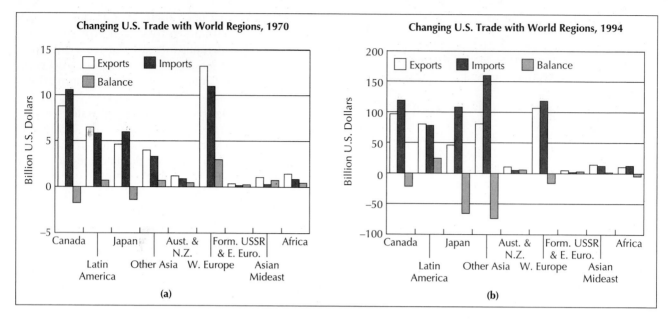

FIGURE 11.8
The changing U.S. trade with world regions, 1970 and 1994. Europe has been supplanted as the most important trade region for the United States between 1970 and 1994. More important now is Asia, while trade with Japan, especially imports, has grown rapidly, as has trade with Canada.

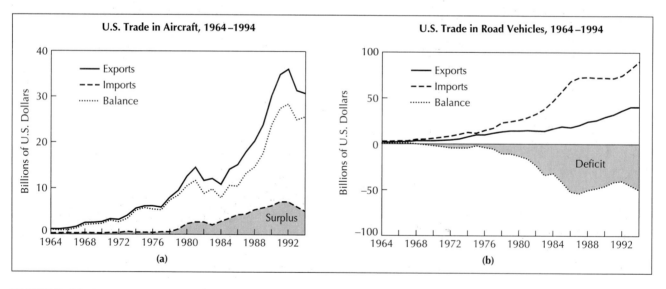

FIGURE 11.9
(a) U.S. balance of trade in aircraft, 1964–1994. The United States is the world leader in aircraft manufacturing and has maintained an important surplus since 1992. At the end of the Cold War, both exports and imports dipped slightly. *(Source: United Nations, 1996)* (b) U.S. trade in road vehicles, 1964–1994. Road vehicles have been a bright spot in U.S. trade up until 1970. From that point onward a trade deficit has developed. Foreign auto companies sell more cars in the United States than U.S. companies sell abroad. *(Source: United Nations, 1996)*

U.S. manufactured exports to developing economies of Asia increased 18% (Figure 11.6). During the same period, U.S. exports bound for Europe dropped slightly, and the share exported to Japan and Canada held constant. Despite this global shift, the most important trading partners for the United States remain almost unchanged from 1983 to 1994: Canada, Japan, Mexico, UK, and Germany (Figure 11.8).

Export Sectors

MEDICAL EQUIPMENT

The medical equipment sector is one of the most competitive sectors of the U.S. economy, with export growth averaging 15% per year over the last 5 years. In 1994, exports were 9 billion, far exceeding the 5 billion of medical equipment imports. A substantial trade surplus in medical equipment exists and is expected to reach 20 billion by the year 2000. Medical equipment markets are shifting, however, and industrialized countries have traditionally been the best markets, with Europe leading the list. Rising incomes in developing countries will shift the market for medical equipment in the future to the GEMS. GEMS are expected to grow 20% annually over the next 5 years in demand for U.S. medical equipment, compared to 12% growth for Western Europe, according to the U.S. Department of Commerce (see Figure 11.7).

MOTOR VEHICLES

The export of motor vehicles and aircraft accounts for approximately $30 billion each in export earnings. Excluding Canada, U.S. exports are concentrated in developing countries, which will provide most of the growth in the future. U.S. exports of motor vehicles and aircraft have been increasing at 11% annually, 16% if one excludes Canada, since 1989. The auto sector had a U.S. trade deficit of $51 billion in 1994, while the aircraft sector had a $6 billion surplus in 1994. The deficit in autos with Japan, 24 billion in 1994, accounted for more than half of the auto deficit (Figures 11.9a and 11.9b). U.S. trade in aircraft remains positive.

Half of the motor vehicle exports are shipped outside of North America. Japan, Taiwan, and Saudi Arabia are the three largest motor vehicle markets for U.S. road vehicles outside of North America. The European markets are saturated. World sales of motor vehicles has grown just 1.2% annually over the last 10 years in the developed markets because of the prolific European manufacturers. In real terms, export sales to these markets should be much higher. The growth of new markets in developing countries is leading to the globalization of U.S. production in these developing markets.

AUTOMOTIVE PARTS

The export of automotive parts has increased 16% annually since 1989, and the automotive parts industry has become more global in outlook. Exports are crucial to the auto parts industry. Exports as a share of output have jumped from 18% in 1989 to 27% in 1991 as the U.S. auto parts industry restructures itself. Most of the competition is from Japan. Over one-third of U.S. auto parts imports comes from Japan. The world market is anticipated to grow at 2.5% annually between 1995 and the year 2000 because of slow growth in the industrialized countries.

COMPUTER EQUIPMENT

The U.S. computer equipment industry commands more than 75% of the world's computer sales through global operations, but the United States has had a trade deficit since 1992 (Figures 11.10a and 11.10b). In 1995, the U.S. trade in computer equipment was $15 billion deficits, up from $3 billion in 1989. The import growth

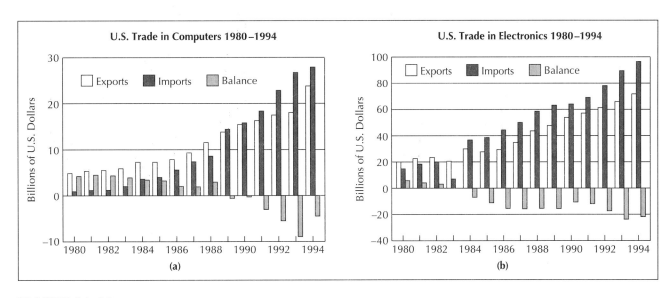

FIGURE 11.10

(a) U.S. trade balance in computers, 1980–1994. Most computer components and peripheral equipment sold in the United States are imported. The U.S. deficit has shrunk between 1993–1994 however. The United States leads the world in software manufacture and development. *(Source: United Nations, 1996)* (b) U.S. trade in electronics, 1980–1994. There has been a continuing deterioration of the trade balance in electronics since 1983. Foreign companies in the Far East can manufacture electronic items more inexpensively than can the United States.

has averaged 16% annually since 1989, and this is faster than the 7% export growth during the same period. Foreign sales now account for more than half of the total revenues of many leading U.S. computer equipment suppliers. With economic recoveries is Western Europe and Japan, and the robust growth of the GEMS, exports are anticipated to grow 8% in 1996, and 7% annually through the year 2000.

COMPUTER SOFTWARE

The United States is the world leader in computer software, supplying 75% of the $80 billion world market for packaged software per year (Table 11.6). The computer software industry is moving towards *multimedia*, which combine video, animation, voice, music, and text. Implementation of the Uruguay Round of GATT will help U.S. manufacturers because of new controls on *intellectual property rights* (IPR), reducing piracy, which has been a major problem for the industry. World market for software is expected to grow 12% annually between 1995 and 2000, benefiting world suppliers, reaching 160 billion per year in market demand by the year 2000. Asia and Latin America are expected to be the fastest growing markets for computer software.

INFORMATION SERVICES

The United States is the world leader in the production, use, and export of information services, commanding 50% of the world market by 1997 (Table 11.7). U.S. firms produce and use the most advanced software, and exports have grown rapidly. U.S. export informa-

Table 11.6
Packaged Software Markets, 1993–2000 ($ Million)

Item	1993	1994	1995	2000
World	**69,918**	**77,416**	**86,060**	**152,816**
United States	**31,400**	**35,600**	**40,000**	**74,263**
Western Europe[1]	24,703	26,563	28,970	45,243
Japan	7,008	7,499	8,365	14,562
Canada	**1,276**	**1,457**	**1,613**	**2,694**
Australia	1,044	1,183	1,345	2,551
Latin America[2]	1,640	1,770	1,840	4,250
Asia[3]	1,079	1,334	1,647	4,786
Other	1,768	2,010	2,279	4,466

[1]Austria, Belgium, Denmark, Finland, France, Germany, Italy, Greece, Ireland, Netherlands, Norway, Portugal, Spain, Sweden, Switzerland, and United Kingdom.
[2]Argentina, Brazil, Chile, Mexico, and Venezuela.
[3]China, Hong Kong, India, Malaysia, Singapore, South Korea, Taiwan, and Thailand.
Source: International Data Corporation.

tion services increased 22% annually during the 5 years from 1991–1996, averaging 3.2 billion per year. Imports are only a half billion per year. The information industry is especially affected by government policy regarding market access, intellectual property rights, privacy protection, data security, and telecommunications services. Export sales are expected to rise 13% an-

Table 11.7
Top 10 U.S. Information Services Companies

Rank	Company	Revenue ($ Million)		Percent Change 1992–1993	Share of U.S. Market 1993
		1992	1993		
1	IBM	7,352.0	9,711.0	32	21
2	EDS	8,155.2	8,507.3	4	18
3	Anderson Consulting	2,445.0	2,588.7	6	5
4	CSC	2,474.0	2,502.0	1	5
5	ADP	2,075.0	2,339.2	13	5
6	TRW	1,800.0	1,900.0	6	4
7	Digital	1,570.3	1,875.0	19	4
8	Unisys	1,336.0	1,593.1	19	3
9	First Data	1,205.0	1,500.0	24	3
10	AT&T	1,198.7	1,235.0	3	3

Source: Datamation.

nually during the next 5 years, according to the U.S. Department of Commerce. The largest markets for U.S. information services will be the industrialized countries, especially the United Kingdom, Japan, and Canada. The most important market among the GEMS will be the Chinese economic area and Korea, followed by Mexico, Brazil, and Argentina.

SERVICES: SOFT POWER

Services represent a hyperdynamic sector of the U.S. economy. Services are not just food and travel, but information software, telecommunications, advertising, and entertainment. Services in North America account for about a quarter of gross domestic product. They account for 30% of total exports. Canada and the United States have 150 of the world's top 500 service corporations. On the cutting edge of industry—software, information technology, and entertainment, which accounts for one-third of the world services sales—the United States and Canada's position is unchallenged. For example, U.S. service exports dwarf auto exports, $200 billion to $60 billion in 1995. There is a rising influence in the world economy of "soft power," defined as the ability to achieve desired outcomes in international affairs through attraction rather than coercion. *Soft power* is services power. It is the power of Microsoft to write programs, of Hollywood to make movies, and of U.S. cultural ideals, products, and practices to become known around the world through the information revolution. Soft power gives North America the edge over every other region of the world. Services will replace heavy industry as the motor of the world economy of the future.

U.S. Stocks Lead World Economy

The U.S. stock market's sixth straight annual gain led a 1996 rally in most equities worldwide, spurring companies to set records for mergers, acquisitions, and initial public offerings (Figure 11.7a). U.S. stocks rose more than 26% in 1996—on top of a 34% gain in 1995—helped by an expanding economy that drove the dollar 10% higher against the yen and up 8% versus the deutsche mark. Brighter prospects in the United States boosted trade worldwide, producing cross-border acquisitions and helping other stock markets.

The most expensive cross-border acquisition ever took place in November when British Telecommunications reached across the Atlantic and bid $25.5 billion for MCI Communications, America's second biggest long-distance telephone company.

The U.S. economy has reemerged as a growth magnet for the world's surplus savings because of its superior growth performance compared to other industrial na-

tions. Japan and Europe have been so concerned about their flagging economies that they want to knock their currencies down to promote growth. Dollar-denominated assets get a boost from competitive devaluation. One reason for the better U.S. economic performance is that labor is not demanding its piece of the pie as it is in Europe and Japan. In 1996, the hourly compensation cost in manufacturing was $17.20 in the U.S., $19.34 in France, $23.66 in Japan, and a staggering $31.88 in Germany.

Europe's higher minimum wage, greater degree of trade union power, and government support for a social market economy have all conspired to put a floor under the wages of the lowest paid. To be sure, this means that income polarity is far more pronounced in the United States than elsewhere. In the United States, male wage earners in the top 10% of incomes make 4.4 times what those in the bottom 10% make, compared with 2.5 in Europe and Japan. Between 1979 and 1995, the poorest one-fifth of American families saw their income drop 9%, while the wealthiest one-fifth enjoyed a 26% increase.

In the United States, capital clearly dominates labor. Overseas, that's not so. Profit ratios of publicly held U.S. companies have sprinted to near record levels in the 1990s, while family incomes have suffered. Similarly, while U.S. public sector entitlement problems are severe, they pale in contrast to those abroad. Taxes are around 32% of total economic output in the United States, versus 45% in France, Italy, and Germany.

Canada

The United States is Canada's most significant trading partner and accounts for 75% of exports and 65% of imports (Figures 11.11 and 11.15). Canada exports automobiles and transportation equipment, industrial supplies, and industrial plant and machinery parts. These products compose 60% of Canada's total exports. Canada also has vast supplies of natural resources, including forest products, iron ores, metals, oil, natural gas, and coal. The United States is in need of all these products. On the other hand, Canada imports industrial plant and machinery parts, transportation equipment, and industrial supplies from the United States. In addition, because of a longer growing season, balmy climates, and temperate agricultural territories, the United States can produce subtropical fruits and winter vegetables for colder Canada. High-tech manufactured goods are also a chief import from the United States.

Whereas the United States exports approximately 11% of its output, Canada exports approximately 20% of its total output. Because of its small population of approximately 30 million, Canada cannot attain the large-scale economy necessary for super efficient plants and,

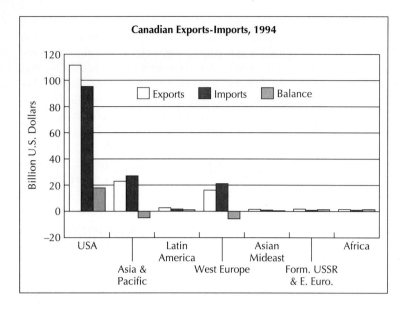

FIGURE 11.11
Canadian exports and imports by world region, 1994. Most of the foreign trade produced by the Canadian economy is with the United States. In 1997, Canada also enjoyed a $30 billion trade surplus with the world. In 1994, the United States took 82% of Canadian exports and supplied 68% of its imports. The second most important trade partner for Canada is Asia and the Pacific, followed by Western Europe. *(Source: United Nations, 1996)*

therefore, must import many of its goods. As with many small countries, especially in Europe, the smaller the country, the more dependent it is on foreign markets for imports and exports.

Canada can export energy resources because of its vast amounts of hydroelectric power and its ability to manufacture hydraulic turbines and electric generators. It also produces high-tech communications equipment, including fiber-optic cables. Transportation and telecommunications equipment are required because of the vast territories that must be overcome to interconnect with the second largest country in the world. For Canada, automobiles and automobile parts represent the largest category of exports to the United States. This is a result of the U.S.–Canadian Free Trade Agreement, which favored the export of automotive industrial goods from Canada to the United States.

Canada's second most important regional partner, according to Figure 11.11, is Asia, followed by Western Europe. Japan actually ranks second as an individual country, ahead of Great Britain, Germany, and other EU countries, which as a bloc constitute approximately 8% of Canadian exports and imports.

Canada's economy is among the leading major industrial countries, with 2.4% growth of GDP in 1993, and more than 4% in 1994 (see Table 11.4). The economic news is good for Canada, considering its lingering recession that began in 1990. The recession affected every sector of the Canadian economy, and manufacturing in particular. The U.S.–Canadian Free Trade Agreement (USCFTA), enacted in 1989, lowered trade barriers between the United States and Canada as part of the 10-year phasing out process. Today, Canadian exports and imports to the United States outpace every other world region four to one (see Figure 11.11). The overwhelming

share of Canada's foreign trade is with the U.S. Approximately 80% of Canadian exports go to the United States, and the United States supplies about 70% of its imports.

Canada's competitiveness has been improved by industrial restructuring that has laid off many manufacturing workers. In 1995 industrial production was above its long-term trend, although the recovery was fueled to a large degree by exports to the United States. Unemployment, while declining, still remains high, at approximately 10% in 1995, making job creation the government's chief objective. The public sector Canadian deficit increased 6.7% of GDP in 1994. Total public debt is now nearly 100% of GDP.

Canada and the United States have the largest bilateral (between two countries) trade relationship in the world. Canadian exports have grown rapidly in recent years and held about $170 billion in 1995. The overall trade surplus amounts to over $10 billion annually. Trade with the United States amounted to over $250 billion in goods and services in 1995. For a country roughly one-tenth the size of the United States, Canada accounts for 21% of U.S. exports and 20% of U.S. imports. The United States runs a large merchandise deficit with Canada; in 1995, it was $11 billion.

NAFTA has sparked U.S. and Canadian merchandise trade, which saw an increase of 14% over the first 2 years of its enactment in 1994–1995. Canada's merchandise trade to the United States includes motor vehicles, motor vehicle parts, engines, office machines, timber, and newsprint (Figure 11.12). U.S. exports to Canada include primarily automobiles, trucks, special vehicles, vehicle parts, paper and paper board, computers, and software. The best prospects for U.S. exports to Canada include computers and peripherals, automotive parts, telecommunications equipment and

FIGURE 11.12
U.S. trade with Canada by major trade category, 1994. The United States maintained a trade surplus with Canada as long as the country was a supplier of oil, ores, and bran based commodities. In recent years, however, the bilateral balance has moved to the negative side with new imports of manufactured items from Canada. Canada is now a major importer of automobiles to the United States, as General Motors and Ford have plants throughout southern Canada. By 1994, 91% of U.S. shipments to Canada and 75% of Canadian exports to the United States were industrial goods, with motor vehicles and motor vehicle parts dominating the movements in both directions. U.S. automakers find it more inexpensive to produce cars in Canada than in Detroit.

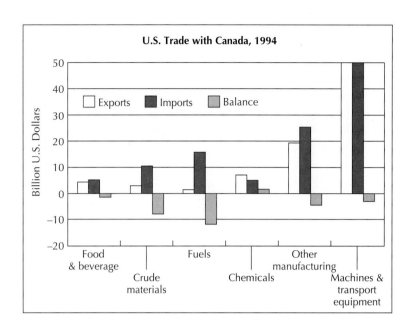

automobiles. In 1994, trade between company affiliates (intracompany trade) and current companies accounted for 45% of U.S. exports to Canada and roughly the same proportion of U.S. imports from Canada. Canada doubled the number of cars and trucks it exports to the U.S. (Figure 11.3) since 1989 as a result of operation for 3 years under NAFTA. In 1996, it sold $11.1 billion more in vehicles to the United States than it bought from the United States. Canada has overtaken Japan as the top auto exporter to the United States. Canadian labor is one-third less than the cost of U.S. labor, and the quality and productivity are equal to or better than that of the United States. Automakers save $300 for each car made in Canada because of its national health care system, which offers for free most of the medical care that U.S. companies buy for their employees. GM spends $5,000 per year to provide health care to each current and retired employee, and less than $1,000 per current Canadian worker. Canada's cheap dollar ($.75 U.S.) allows the dollar of the U.S. manufacturers to go farther (see Figure 10.20).

A fast-growing sector of U.S.—Canadian trade is U.S. nongovernment service exports, which jumped from $10 billion in 1988 to $17 billion in 1994. Canada's financial service market continues to expand as a result of the 1987 accord between the Toronto Securities Commission and the U.S. Department of Finance. This agreement allows the deregulation and integration of the financial securities industry, removing the distinctions among banks, trusts, insurance companies, and brokerages. U.S. direct foreign investment in Canada was more than 70 billion in 1994. About one-half of it was is the manufacturing sector. Canadian direct foreign investment in the U.S. totaled about 45 billion in 1994.

Therefore, Canada's investment income balance with the United States is the single largest deficit in Canadian nonmerchandise trade.

No other market in the world is as open to U.S. goods and services as is the Canadian market. Almost 98% of all bilateral trade passes freely without tariff, and U.S. and Mexican products will continue to have an advantage in the Canadian market as a direct result of the NAFTA legislation, which continues to remove trade barriers that were begun by USCFTA. Import needs of Canada for application and computer-based training software are expected to jump 16% annually between 1995 and the year 2000. Canadian government agencies will account for nearly half of this demand. Additionally, the market for CD-ROM drives and software also will experience rapid growth. Presently, U.S. vendors supply 95% of Canada's imported CD-ROM software and control 60% of the total Canadian market.

Trade between the United States and Canada is weighted heavily toward industrial goods, with 90% of U.S. shipment to Canada and 74% of Canadian exports to the U.S. in such goods. Machinery, transportation equipment (autos), and other manufactured products (auto parts) lead the way (see Figure 11.12). Since 1960, when Canada's exports were made up of 50% primary products—forest, mine, and field products—Canada has shifted its emphasis towards industrial merchandise (see Figure 11.13).

Like the United States, Canada has found more recent trade growth with Asia and the Pacific Rim than with its traditional trading partner, Western Europe. Since 1980, the Pacific has eclipsed the Atlantic as the leading avenue of North American commerce (see Figures 11.14 and 11.15).

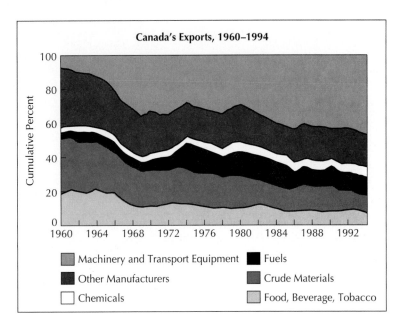

FIGURE 11.13
Canada's changing export composition, 1960–1994. In 1960, 50% of Canada's exports were made up of primary products: from the mine, forest, and field. Since that time, Canada has shifted its export predominance to industrial and manufactured products. In 1994, these products reached 73% of the total.

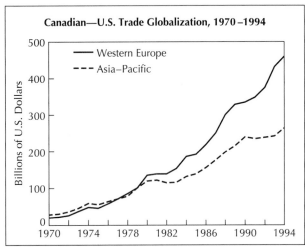

FIGURE 11.14
Canadian–U.S. trade globalization (exports and imports), 1970–1994. Asia and the Pacific became the leading export destination for both Canadian and U.S. trade by 1980, following a long period of dominance by Europe and the Atlantic countries. Countries most responsible for this shift are Japan, Hong Kong, Singapore, Taiwan, and South Korea. To this list we now must add Australia, Indonesia, Malaysia, Thailand, and the Philippines.

European Union (EU)

Western Europe's trade, as a proportion of total world trade, is disproportionately large compared with its population, a mere one-third of a billion. Although it possesses only one-fifteenth of the world population, it accounts for between 40% and 50% of world trade because of (1) the strength of the EU; (2) short distances; (3) good river, canal, and motorway transportation; and (4) relatively small countries that need complementary trade flows with one another in order to exist and flourish. Western Europe is barely half the size of the United States; some countries are comparable in population and size to individual U.S. states. Italy, France, the United Kingdom, and the former West Germany have a population of almost 60 million each (with the addition of East Germany, Germany is now

pushing toward 80 million). Other countries are much smaller. Some have food resources, such as Denmark and France; some have energy resources, such as Norway, the Netherlands, and the United Kingdom; some produce iron, steel, and heavy equipment, such as Italy, Germany, France, and Spain; and others produce high-value consumer goods.

The proximities, geographies, and areal complementarities of Europe have made intraregional trade ideal. This type of trade has increased from 55% of all western European exports to 73% in 1992. That is, 73% of all exports from western European nations go to other western European nations. If the EU grows from its present size to include the EFTA countries as well as associate members from eastern Europe and the Mediterranean, European intraregional trade will continue to increase.

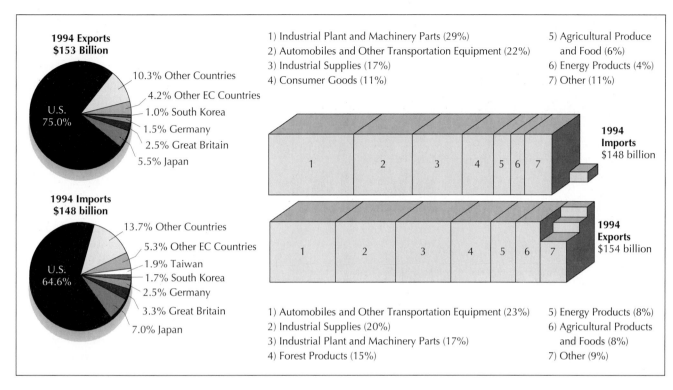

1) Industrial Plant and Machinery Parts (29%)
2) Automobiles and Other Transportation Equipment (22%)
3) Industrial Supplies (17%)
4) Consumer Goods (11%)
5) Agricultural Produce and Food (6%)
6) Energy Products (4%)
7) Other (11%)

1994 Exports $153 Billion

U.S. 75.0%

10.3% Other Countries
4.2% Other EC Countries
1.0% South Korea
1.5% Germany
2.5% Great Britain
5.5% Japan

1994 Imports $148 billion

1994 Imports $148 billion

U.S. 64.6%

13.7% Other Countries
5.3% Other EC Countries
1.9% Taiwan
1.7% South Korea
2.5% Germany
3.3% Great Britain
7.0% Japan

1994 Exports $154 billion

1) Automobiles and Other Transportation Equipment (23%)
2) Industrial Supplies (20%)
3) Industrial Plant and Machinery Parts (17%)
4) Forest Products (15%)
5) Energy Products (8%)
6) Agricultural Products and Foods (8%)
7) Other (9%)

FIGURE 11.15
Composition of **Canada's** world merchandise trade, 1994.

The economies of the EU countries grew at 2.5% in 1994, a notable increase over the prior year's contraction. Unemployment has been high recently in the EU, and as of 1995 was 11%, but has begun to decline through 1996. World recession, the reunification of East and West Germany, and the failure of communism all sparked recession in Europe. Western Europe accounts for 40% of all goods and services traded. The EU is the largest overseas market for U.S. multinationals.

The EU membership expanded to 15 countries in 1995 with the addition of Austria, Finland, and Sweden. Their economies totaled 400 billion. While this is only 5% of the GDP of the EU, they added 20 million new consumers, which results in a 15-member EU with a total GDP of $7.5 trillion as of 1995 and a population of close to 400 million.

While the EU is in the midst of its economic recovery, the structural problems, such as excessive labor costs, economic rigidity that stall the growth of smaller companies, industrial obsolescence, difficulty in making use of new technologies, and costly social welfare programs tend to restrain growth below the EU economic potential and is keeping unemployment above 10%. The EU is the largest trading block in the world, with exports totaling 1.5 trillion in 1994, about the same as imports. The EU ranks second as an export market to the U.S., after Asia (Figure 11.16). However, it is the fastest growing market for U.S. high-technology exports and remains the principal destination for U.S. foreign direct investment. The EU is as well the largest source area of foreign direct investment (FDI) in the United States.

The leading U.S. exports to the EU in 1994 were aircraft ($10 billion), data processing and office equipment ($7 billion), engines and motors ($4.2 billion), measuring, checking, and analyzing equipment ($3.2 billion), and thermoic and cathode valves ($3 billion). The U.S. Department of Commerce estimates the most promising sectors for U.S. exports to the EU to include telecommunications equipment, computer peripherals, software, electronic items, pollution control, machinery, medical equipment supplies, and aircraft.

The mounting trends toward (1) deregulation, (2) privatization, and (3) market decentralization in the EU make sales prospects for the above mentioned products good for 1995–2000. Sales of U.S. companies through European affiliates reached $2 trillion in 1994; total merchandise that year valued $500 billion, or one-quarter that amount. Most international trade takes the form of *intra-industry trade*—investment in foreign affiliates that produce abroad, rather than shipments of U.S. produced goods to target export markets. In 1994, U.S. affiliate sales in Europe exceeded $1 trillion. These sales accounted for more than half of

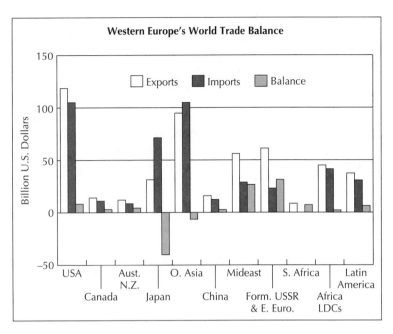

FIGURE 11.16
Western Europe's world trade balance.

U.S. affiliate sales worldwide, and were four times as large as U.S. affiliate sales in Canada or Asia. Consequently, Europe's overall importance for U.S. companies and the U.S. economy is much greater than trade statistics indicate. These U.S. companies' overseas affil-

iates are major importers of products manufactured in the United States. In 1995, sales by U.S. parent companies to the European affiliates made up 40% of U.S. exports to Europe and generated $25 billion in trade surplus for the United States. U.S. affiliates in Europe contributed 30 billion to overall U.S. trade surplus and services (fees, royalties, airfare, dividends, and other earnings), yielding a total U.S. surplus in affiliate trade of 55 billion in 1995.

THE UNITED KINGDOM
Great Britain's balance of merchandise trade was slightly negative in 1990. Similar to France, its export profile suggested a per capita level of $3,000 per person per year, approximately 20% of its GDP. Fifty percent of both imports and exports were manufactured products, and another 25% of both exports and imports were semimanufactured products (Figure 11.17). Great Britain is a net exporter of fuels because of its North Sea oil bonanza, but a net importer of agricultural products and foodstuffs. It has a trading-partner profile similar to that of France and Germany.

Birthplace of the Industrial Revolution, Great Britain became a net importer of manufactured goods

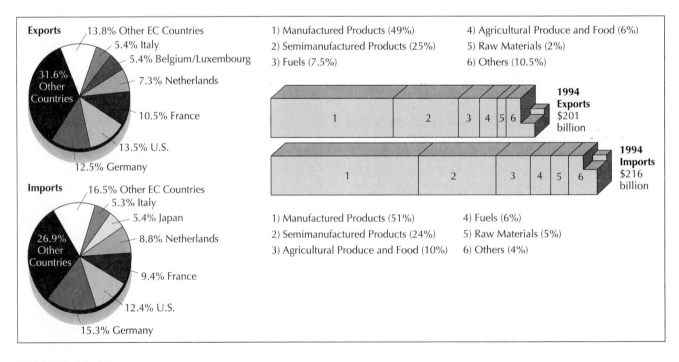

FIGURE 11.17
Composition of **Great Britain's** world merchandise trade in 1994. Considering its size, Britain is a leader in world trade, in fifth place nationally. Its most important trade partners include Germany, France, and the United States.

by 1990. The restructuring of the British economy, away from manufacturing, caused the country to depend on North Sea oil profits and overseas investments and services. The U.K. firms invested heavily outside the United Kingdom, especially in the United States. Relative to its population and economy, the United Kingdom owns more foreign assets than any other OECD nation does, including Japan and the United States. In addition, FDI by Japan in the United Kingdom has been greater than in any other western European nation.

As with the United States and France, a current problem in Great Britain is the overcapitalized defense research, development, and manufacturing industry. Attempts to find commercial applications of defense plants and weapons are currently being pursued.

Recently, Great Britain has suffered balance-of-payment problems and a declining standard of living. Referring to the box, we can see the United Kingdom's purchasing power parity, which shows its true standard of living as compared with those of other countries.

Like the rest of Europe, the economy of the United Kingdom (U.K.) slipped into recession in 1991, but it began to recover in 1993, sooner than the other EU countries. In 1994, GDP increased to over 3%. This was the best performance of any EU economy. In spite of major restructuring of manufacturing in the Midlands industrial belt, industrial competitiveness has increased in recent years. Output per person in the manufacturing sector has risen faster than wages, mainly because the 1995 hourly compensation of manufacturing workers was only half that of Germany's (see Table 8.4).

The economy is expected to continue with a strong rebound at approximately 3% per annum to the year 2000. A continuation of a ceasefire in Northern Ireland has promoted economic confidence, investment and development. The United Kingdom is the United States' largest export destination in Europe and the fourth largest export market worldwide (behind Canada, Japan, and Mexico). In 1994, the United States owned a 14% share of the U.K. import market. The highest in the EU. The United States had a 5.0 billion bilateral trade surplus in 1995. Between 1987 and 1994, U.S. exports to the United Kingdom increased a robust 11% per annum. Manufactured goods accounted for 95% of the $27 billion total.

The United Kingdom will remain the United States' main export market and principal destination for foreign direct investment by U.S. companies in Europe. It is an important market for computer software, aircraft computers and peripherals, electronic components, medical instruments, oil and gas, field equipment and machinery, apparel, nonelectric motors, and paracommunications equipment.

The United Kingdom is a base from which Canadian and U.S. companies can expand to the rest of Europe because of the familiarity of language and custom. Pollution control equipment markets are expected to grow at 10% annually, from $3 billion in 1993 to $5 billion in 1996. U.K. imports of U.S. equipment in this category are projected to increase 50% annually because the United Kingdom has raised its air pollution standards. The latest demand for pollution control equipment will be from companies in heavy manufacturing that must comply with new strict emissions limits to keep their operating licenses. Additionally, the telecommunications equipment and services market in the United Kingdom is expected to grow 5% annually to the year 2000. The chief beneficiary of sales will be U.S. exports. Various utilities have existing communications networks and own rights-of-way to lay cable, but many need infusions of external capital and technology to expand into telecommunications fields and are looking for joint ventures with U.S. partners.

GERMANY

Germany is the third leading country in the world in terms of international trade; it is also the third most important trading partner of the United States. It exports a large percentage of its GDP and is the world's leading exporter of merchandise. In addition, it has a trade surplus of more than $55 billion but has a heavy dependency on imported energy. Motor vehicles and mechanical engineering products are its chief exports, but it also exports a notable amount of chemicals and electrical engineering products (Figure 11.18). It imports many of the same items that it exports—motor vehicles, chemicals, and electrical engineering products—and is unable at the present time to provide the necessary amount of agricultural produce, which is its largest single import, 11.5%. Most of its trading partners are other EU countries.

Germany's exports are worth approximately $6,600 per capita per year, whereas those of the United States are worth approximately $1,500 per capita per year. Germany exports as many goods as the United States does but has only a quarter of the population.

Germany also enjoys a remarkably healthy balance of trade, exporting almost 16% more than it imports. Its industrial sector has been paramount in the manufacture of high-value products, including automobiles, chemicals, and machinery. Recently, however, its economy and trade have shifted from that of manufactured goods to the service sector. Frankfurt is now Europe's second largest banking center, after Switzerland and prosperous Liechtenstein. Both of the latter two are important investment and banking centers, not only because of Liechtenstein's tax-free status and Switzerland's extreme secrecy (recently under fire from Holocaust survivors) about foreign bank deposits, but also because of the keen political stability of the two countries.

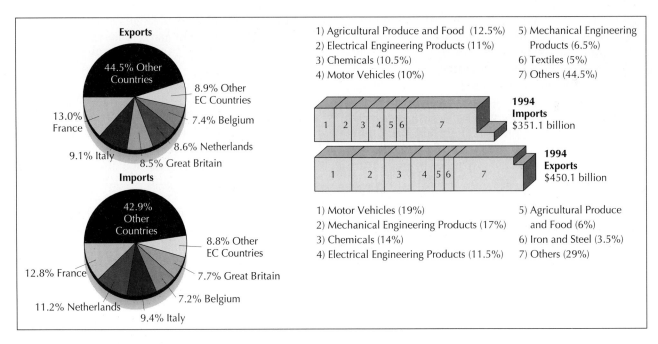

FIGURE 11.18
Composition of **Germany's** world merchandise trade in 1994. Germany has less than one-third of the United States' population; its exports in 1990 were $400 billion, almost equal to that of the United States. When exports and imports are added, Germany becomes the second largest trading nation in the world, second to the United States. Its most important trade partners are other EU nations. In addition, Germany has a sizable trade surplus.

The problem now facing former West Germany is the integration of former East Germany into its prosperous borders. During the first 2 years of unification, GDP in what was formerly East Germany decreased by more than 40%. Only now, in 1994, is recovery and growth beginning to occur. Unification has meant a drop in economic activity for Germany as a whole and has necessitated extreme financial support for the social programs of the east. But former East Germany has had the luxury of shock therapy with a safety cushion. Neither Russia, the former republics of the Soviet Union, nor any other eastern European nation has had such an easy transformation.

As of 1996, Germany's $2 trillion economy is by far the largest in Europe and accounts for 28% of all EU production of goods and services. Germany consumes $19 billion worth of products and is the second most important European market for U.S. companies. Germany is the world's second largest exporter after the U.S. Germany rebounded from a 1% decline in 1993 to post a 3% gain in 1994 and 1995. The economy bottomed in 1993 when the burden of reconstruction and reunification of East Germany made its impact on the German economy. The reunification of East Germany and its reconstruction is costing Germany $100 billion annually.

Unemployment stabilized in 1996 around 9%, but joblessness remains a major concern for industry and government. The structural weakness in the German economy is due to very high wages—50% higher in dollar terms than U.S. labor costs—and restrictive labor legislation (Table 8.4). Germany, like most EU countries, has been slow to move into robotics, high technology industries, information technologies, microelectronics and biotechnology. This is starting to hamper Germany's competitiveness in the world markets.

Germany is an economic powerhouse, and total exports exceeded 450 billion in 1995. Imports increased from 350 billion to 390 billion between 1994 and 1995 alone. Between 1987 and 1994, U.S. exports to Germany grew 8% per annum, while U.S. imports rose less than 1% per year. German products remain popular in America, especially automobiles (Figure 11.18). As a result, the long-standing U.S. deficit with Germany has increased to about $15 billion annually. The leading U.S. exports to Germany are computers, aircraft, environmental management technologies, and niche items (e.g., mountain bikes, sportswear, and works of art).

FRANCE

France exports approximately $3,000 per capita per year and suffers from a slight trade imbalance, with more imports than exports. Nonetheless, it is the fourth largest trader in the world. Exports include industrial plant and machinery parts and other manufactured goods such as chemicals, automobiles, iron, and steel (Figure 11.19). But unlike other European nations, it exports a much

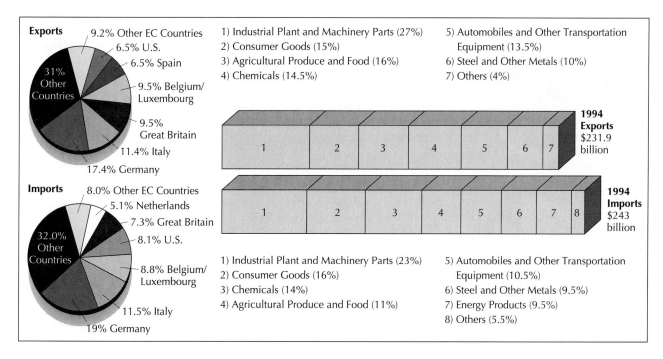

FIGURE 11.19

Composition of **France's** world merchandise trade in 1994. France is the world's fourth largest trading nation. Considering its size of approximately 60 million, less than a quarter of the population of the United States, it is surprising that its exports and imports are almost one-half as great as the latter. France is energy poor and therefore imports fuels, while exporting high-value industrial products, including automobiles and aircraft.

higher proportion of agricultural products and consumer goods. In fact, France produces more food than any other European nation does. As most other technologically advanced nations do, France also imports industrial plant and machinery parts, consumer goods, and chemicals. Agricultural products and transportation equipment account for another 21% of imports.

Whereas the United States exports 11% of its GDP, France exported almost 30% of its GDP in 1996. During the last 20 years, France's trade with Western Europe increased, while trade with French colonial markets in Africa and East Asia declined substantially. Trade with less developed countries amounts to only 15% of France's international flow, whereas trade with the EU amounts to 60%. Energy imports have been a particular stumbling block because France is not endowed with coal, as Germany is, or North Sea oil, as the Netherlands, Norway, and the United Kingdom are. More than 50% of France's electricity is generated from nuclear power.

Similar to Great Britain and the United States, France has been forced to restructure industrial jobs, which have decreased by 2 million during the last 15 years. France has also experienced economic difficulty in its balance of trade and suffered from a $7.5 million deficit in 1990. This deficit has impeded its overall trade and economic development.

France, as with other Organization of Economic Co-operation and Development (OECD) countries, has shifted more of its workforce toward the services. Employment in restaurants, retailing, publishing, banking, personal services, finance, and government-run social services is on the increase.

The French economy is starting to recover from the major downturn in 1993, which saw a negative growth in gross domestic product. Unemployment still remains high at over 12%. Economic output grew at 3% in 1995, and the government pursues a policy to maintain a strong currency. The United States and France produce many of the same types of goods and services, export them to one another, and mutually compete in world markets. France is the United States' third largest European trading partner. The U.S., for example, is the world's largest exporter of agricultural goods and services, and France is the second largest.

ITALY

The Italian government heavily subsidizes Italian manufacturing (15 billion in 1993, twice the EU average). Thus, manufactured goods, including machinery and metal-engineered goods, textiles and clothing, shoes, and automobiles are Italy's most important export products (Figure 11.20).

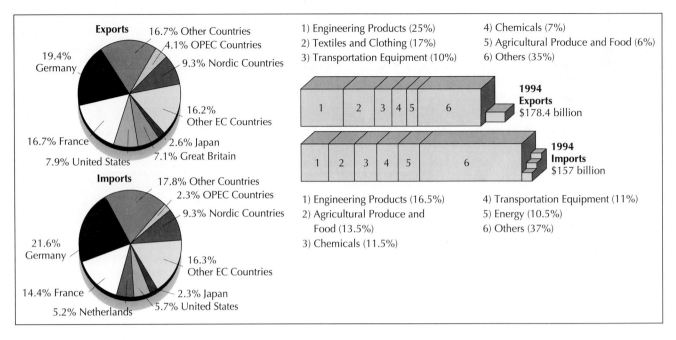

FIGURE 11.20

Composition of **Italy's** world merchandise trade in 1994. In 1990, Italy accounted for $193 billion worth of imports and $182 billion worth of exports, making it one of the world's trading leaders. Italy exports industrialized products, chemicals, machinery, textiles, and automobiles.

Italy's economic woes bottomed out in 1993, with a negative GDP growth of 0.7%. As of 1995, GDP grew at 2%. GDP is expected to grow to 3% in 1996. The Italian markets for U.S. and Canadian exports grew stronger in 1995 as the government continued with its process of privatization of companies. Italian privatized firms are opportunities for North American companies because they seek products and assistance to replace their anti-quated, inefficient systems. The best sales prospects for North American exporters are in the following sectors: medical and laboratory equipment, telecommunications equipment, process control instruments, environmental technology, computers and peripherals. Italian buyers value the design and engineering features provided by North American companies.

Latin America

Latin America comprises the South American countries of Colombia, Ecuador, Bolivia, Chile, Venezuela, Brazil, Argentina, Uruguay, and Paraguay, as well as the Middle American countries, which are dominated by Mexico. All these countries are considered developing countries, and their traditional role since the time of exploration and colonization has been to provide primary materials, namely agricultural exports and mineral resources, to the developed world of Europe and more recently to North America.

Latin America has economically advanced countries, including gigantic Brazil, with a population of 161 million, as well as some of the poorest countries in the world (e.g., Bolivia and Paraguay in South America and Haiti and Guatemala in Central America). Latin American countries are diverse not only with regard to population and size, but also with respect to development and natural resources. Some countries, such as those in the Caribbean and Central America (Dominican Republic, Costa Rica), have agricultural plantation surpluses. Others, as in Argentina, have midlatitude grain surpluses, whereas still others, such as Venezuela and Mexico, are rich in iron ore and oil. Brazil has a wealth of minerals and is a strong producer of manufactured goods.

The balance of merchandise trade and trading partners of Brazil and Mexico and the fastest growing sectors are shown in Figures 11.21 to 11.24. Once again, the chief trading partner of Mexico is the United States. Mexico's proximity and 2,000 miles of common border suggest that the United States is a much stronger trading partner with Mexico than with Brazil. Brazil continues to export primary products, such as soybeans, iron ore, and coffee, but also exports transportation equipment and metallurgical products. Imports include industrial plant and machinery parts, fuel and lubricants, chemicals, and iron and steel.

Mexico's balance of merchandise trade is centered on labor-intensive manufactured products (52%), many of which flow from plants along the border that are owned by U.S. MNCs (*maquilladoras*) back to the United States (South, 1992; Stutz, 1992a) (Figure 11.25). Petroleum and by-products, as well as agricultural products, account for 45% of Mexico's exports. Mexico is one of the world's largest exporters of energy, and oil provided more than 35% of its export revenue in 1994. Semifinished industrial supplies that act as input materials for final production compose 60% of Mexican imports, and manufacturing and plant equipment another 23%. These types of imports are necessary for Mexico to maintain its level as a rapidly industrializing Third World nation.

In 1994, for Latin America as a whole, 85% of the region's exports, mainly food, minerals, and fuels, went to the United States. This pattern was typical for Third World nations. For a long time, Latin America had an import-substitution policy for industrialized products. In 1990, Latin American imports from the United States consisted of 85% manufactured goods and only 15% consumer goods. However, today, Latin America's new hope to achieve wealth and a prominent place in the world economy is centered on export-led industrialization, and a growing variety of manufactures are exported, led by Brazil, Argentina, and Mexico.

The southern countries of Argentina, Uruguay, Chile, and Bolivia, farther from the United States, have had stronger ties to Western Europe. The East Asian NICs are currently strengthening their economic ties with Latin America. Unfortunately, Latin American trade within the region is not nearly so strong as that for North America, Western Europe, or the Pacific Basin. Each country seems to be tied more politically and economically to Europe, the United States, and the Far East than to one another.

One problem in Latin America has been negative growth rates. During the 1980s, only Colombia, Chile, and Paraguay increased their per capita incomes, whereas the remaining countries experienced a decline. Between 1981 and 1991, per capita GDP actually fell 30% in Peru, 29% in Guyana, 23% in Argentina, 23% in Bolivia, and 20% in Venezuela. Inflation was out of control in Latin America as well, running at 11,000% in Bolivia, 5,000% in Argentina, and 8,000% in Peru from 1981 to 1991. Under such circumstances, export producers were discovering that it was not profitable to concentrate on world markets because exchange rates and government taxes were putting them out of business.

At the present time, some Latin American countries can hardly keep up with debt service on their international loans. As previously discussed, Mexico, Brazil, and Argentina owe $100 billion to the developed world, and several other Latin American countries are following closely behind. Most of the loan money was put into urban infrastructures, but high world interest rates, oil prices, and international recessions have minimized exports. Brazil is a case in point. Exports are a little

more than one-tenth those of the United States, but the population is approximately two-thirds that of the United States. In general, the early 1990s was a disadvantageous time for Latin America because Latin American governments were asked by the International Monetary Fund (IMF) to devalue their currencies, and to invoke austerity programs by restructuring their economies, raising taxes, decreasing public expenditures, and selling unprofitable government-owned business enterprises, such as state banks, power companies, metal refineries, and transportation and airline companies. The latter 1990s have proven somewhat better for the GEMs.

ARGENTINA

Argentina's famous economic past was troubled by a growing pool of unemployed workers unable to find good jobs because of high wage costs. This brought class warfare in the streets in Argentina (and Brazil), marches by idealistic students, and strikes and demonstrations by activist union leaders and workers. In reaction and to find favor with worker voters, General Juan Peron outlawed outsourcing, developed legislation against employers dismissing employees, and strengthened unions' ability to strike. That was the pattern that operated in South America during the 1940s through the 1960s. The Peronist populist actions, put forward as humanistic measures to reduce the gaping inequalities in Argentine incomes, did not achieve their stated purposes: reduction of poverty and real income disparities. Instead, they resulted in hyperinflation lasting for decades (Argentina cried for itself, while Evita Peron, dying of cancer, sang, "Don't cry for me, Argentina").

Since 1990, president Menem's government moved to correct a floundering economy hit by hyperinflation, high budget deficits, and an over-capitalized public sector. Argentina has made good progress towards balanced economic growth with an annual percent increase of 6% from 1993 to 1996. Cumulative GDP growth between 1990–1996 has been more than 25%, and inflation has been slashed from 3,000% in 1990 to less than 7% in 1995. Recently, new tax reforms have placed public finances on a sustainable basis, resulting in budget surpluses. Argentina's government has instituted economic restructuring to reduce the role of the state and to reduce privatization. The labor force has been pushed into flexibility, and the result has been tens of thousands of new jobs. The state telephone company, the airlines, and the utilities have all been privatized. The share of the GDP of government control declined from 50% to 25% since 1989.

Argentina's economic resurgence has encouraged private investment. Confidence in the country's credit worthiness is resulting in record capital flows to finance trade deficits. The private sector is being relied on to increase productivity, while foreign investments play a major role in raising productivity standards. U.S., Canadian, European, and South American countries will be major sources of new technology and capital to the year 2000. Between 1990 and 1995, U.S. investment in Argentina doubled to $4.5 billion, for example. U.S. exports to Argentina are now third largest in Latin America after Mexico and Brazil. Twenty-three percent of Argentina's exports are headed for U.S. markets. Argentina's trade liberalization has eliminated almost all nontariff barriers and lowered duties. Brazilian exports to Argentina are more competitive as a result of the *Mercosur Trade Agreement*, which eliminated tariffs on most goods traded among the four member countries (Argentina, Brazil, Paraguay, Uruguay). The average tariff of Argentina is now 9%. The key sectors of import are computers, communications, and capital goods. The new export opportunities for the Argentina market include telecommunications equipment, electric power generation, transmission equipment, and medical equipment supplies and services.

BRAZIL

GDP averaged 2% between 1987 and 1992 in Brazil. More recently, economic growth has returned and averaged 5% between 1993 and 1996. Brazil is still not on sound economic footing because of erratic domestic policies and high inflation. Inflation rose more than 2,000% in 1994, and climbed more than 50% monthly by 1995. To counterbalance high inflation, Brazil introduced a new national currency, the *real*, in 1995.

With it came strict monetary controls, and consumer price increases have averaged 3% per month since then. Part of Brazil's policy continues to be pressured by the desire for high wages and the need to maintain high interest rates to finance domestic government debt and borrowing and to prevent capital outflows from the country. New measures in 1996 to implement sound economic policies included privatization of state industries, tax and social security reform, and a downsizing of government functions.

Through all of this turmoil, Brazil has maintained its trade competitiveness; imports rose 25% to $25 billion in 1994, and the EU was the most important regional market for Brazilian exports receiving 25% of the total (Figure 11.21). The largest single country market for Brazilian exports was the United States, accounting for $8 billion, or 20% of total exports in 1995. Brazil remains the United States' largest market in South America as well, and the third largest market in the Western Hemisphere after Canada and Mexico. Between 1987 and 1995, U.S. exports to Brazil grew 8% annually, and amounted to 8.5 billion in 1995.

The U.S. Department of Commerce projects that U.S. exports to Brazil will increase 18–25% through the year 2000. This is due to the fact that Brazil's trade regime is now more open and competitive. Imports are increasing in response to lower tariffs or reduced nontariff barriers and to the strength of the Brazilian *real* relative to the U.S. dollar. This has been accomplished through abolishing quotas and import licensing procedures. Trade lib-

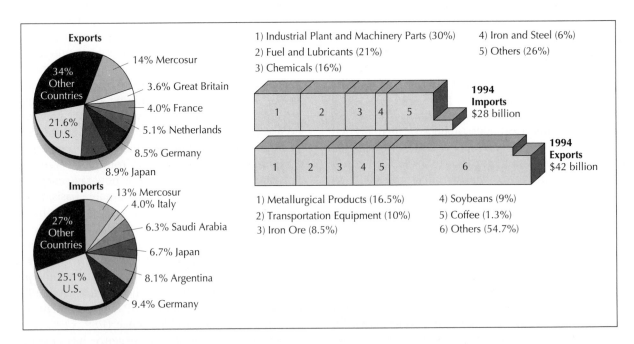

FIGURE 11.21

Composition of **Brazil's** world merchandise trade, 1994. Although Brazil is two-thirds the size of the United States in population, its imports are only 5% of the United States' and its exports are only 7%. Because it is a highly indebted nation, exports outnumber imports 3 to 2. The United States accounts for 21% of its exports and almost 26% of its imports.

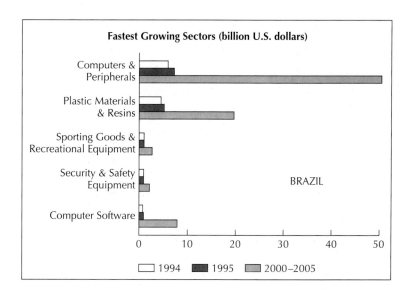

FIGURE 11.22
Brazil's fastest-growing sectors of the economy. For the period 2000–2005, it is forecast that computers and peripherals will be the fastest-growing sectors of the Brazilian economy, followed by plastic materials and resins. *(Source: U.S. Department of Commerce, 1997)*

eralization combined with Brazil's strong base of natural resources and industry, and its large market, will result in continued expansion of North American export sales. Unfortunately, Brazil has discriminatory government policies that prohibit or restrict much trade and services. Government procurement impedes U.S. exports of computers, software, and telecommunications equipment, the fastest growing sectors (Figure 11.22)

MEXICO

Mexico's economic woes were punctuated with the devaluation of the peso in early 1995. With insufficient capital inflows to finance its account deficit, the Mexican government let the peso devalue in the face of rapidly diminishing foreign reserves. Major government programs included reduced government spending, tax and price increases, and strict control of credit. The new government programs included accelerated privatization and liberalization of key industries.

Mexican markets continued to open to foreign competition, following the enactment of the North American Free Trade Agreement (NAFTA) on January 1, 1994. NAFTA has turned the Mexican economy around. Inflation has dropped to about 7%, down from its peak at more than 150% in 1987. Cooling of the 1995 uprising in the state of Chiapas has helped to calm the political situation, even though high level political assassinations and kidnappings continue to contribute to investor weariness. Mexico is experiencing a modernizing economy that is no longer protected from foreign competition, and an improved investment environment. Measures have facilitated the resurgence of growth. Growth bottomed at 0.4% in 1993, but has crept up to 6% in 1996. In 1997, rather than default on a $30 billion foreign debt, Mexico announced the return of its credit worthiness by the prepayment of the remaining $4 bil-

lion owed to the United States from the $13 billion emergency aid package negotiated in February of 1995. The debt will be repaid 4 years ahead of schedule, and this represents a victory for the Clinton administration and the pro-NAFTA analysts. The Mexican economy has rebounded, and its exports have surged since 1994, especially those northbound to the United States and Canada (see Figure 11.23). Mexico will save about $500 million in interest charges by replacing its higher-cost U.S. debt with cheaper currencies like the Japanese yen.

Mexico is among the fastest growing export markets for U.S. products. In 1986, Mexico became the third largest market for U.S. exports after Canada and Japan. Mexico is actually the second leading export market for U.S. manufactured goods. Foreign investment opportunities are particularly strong in Mexico in infrastructure development, where the government will invest $35 billion between 1995–2000. This will include airline privatization, highway construction, railroad services, and water and energy projects.

Under NAFTA, preferential duty treatment of U.S. origin goods gives them an edge over products from European or Japanese firms, which are often the principal competitors. U.S. and Canadian manufacturers must pay duty only on the value added by manufacturers in Mexico, not the products or parts shipped as semifinished or raw materials to plants in Mexico from North America. By 1997, 50% of the total trade that occurred between the United States and Mexico was created in 1994 with the passage of NAFTA. This proportion will increase towards the year 2000. Under NAFTA, half of U.S. exports to Mexico have been eligible for no Mexican tariffs—semiconductors, computers, machine tools, aerospace equipment, telecommunication and electronic equipment, and medical supplies. Another important feature of NAFTA was the gradual phase out of the Mexican

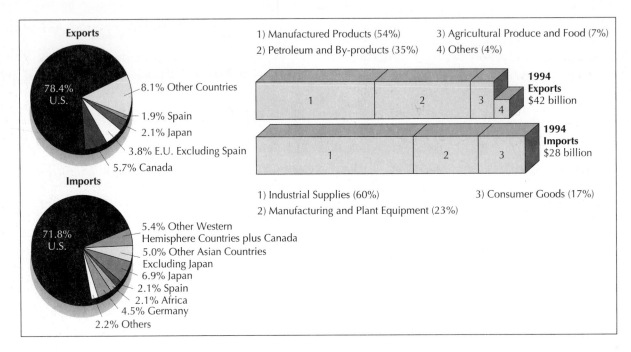

Exports

78.4% U.S.

8.1% Other Countries

1.9% Spain

2.1% Japan

3.8% E.U. Excluding Spain

5.7% Canada

Imports

71.8% U.S.

5.4% Other Western Hemisphere Countries plus Canada

5.0% Other Asian Countries Excluding Japan

6.9% Japan

2.1% Spain

2.1% Africa

4.5% Germany

2.2% Others

1) Manufactured Products (54%)   3) Agricultural Produce and Food (7%)
2) Petroleum and By-products (35%)   4) Others (4%)

1994 Exports $42 billion

1994 Imports $28 billion

1) Industrial Supplies (60%)   3) Consumer Goods (17%)
2) Manufacturing and Plant Equipment (23%)

FIGURE 11.23
Composition of **Mexico's** world merchandise trade, 1994. Mexico's proximity and the fact that it is a large supplier of energy are reasons that the United States dominates its exports. In addition, Mexico imports 68% of its products from the United States. NAFTA should increase both exports and imports with the United States.

Auto Decree, which helped the United States triple exports of passenger cars to Mexico since 1994. Since 1994, autos, auto parts, semiconductors, machine tools, and certain fruits and vegetables, including apples, realized export increases of between 100 and 10,000% when Mexican barriers were sharply reduced or eliminated under NAFTA (Figures 11.24, 11.25).

Japan and East Asia

The fastest-growing world trade region is East Asia and the Pacific. After Western Europe, this region has the largest amount of internal world trade. Exports and imports in 1990 amounted to $1.6 trillion. In 1970, this re-

gion accounted for approximately 10% of all world trade, but by 1990, its share had increased to 25%.

JAPAN, THE FOUR TIGERS, AND THE FIVE LITTLE DRAGONS

Japan has taken the lead role in the development of East Asia and the Pacific. The growth of its flattened economy, which occurred after World War II, is nothing short of an economic miracle and parallels that of Germany. It has been joined by the Four Tigers of Taiwan, South Korea, Singapore, and Hong Kong. However, five new emerging *Little Dragons* have followed suit with rapidly growing economies: the Philippines, Thailand, Malaysia, Indonesia, and the People's Republic of China.

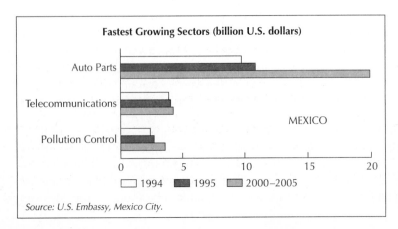

Fastest Growing Sectors (billion U.S. dollars)

Auto Parts

Telecommunications

Pollution Control

MEXICO

0   5   10   15   20

☐ 1994 ■ 1995 ▨ 2000–2005

Source: U.S. Embassy, Mexico City.

FIGURE 11.24
Mexico's fastest-growing sectors of the economy. For the period 2000–2005, it is predicted that auto parts will be the fastest-growing sector of the Mexican economy, followed by pollution control equipment and telecommunications. Because of NAFTA, U.S. exports of cars and trucks in 1995 rose 1,000% over 1993 levels. The U.S. import share of the Mexican market also increased tenfold. The United States currently dominates the Mexican market, but the Europeans and Japanese are looking at Mexico as a possible springboard to the NAFTA and South American markets. *(Source: U.S. Department of Commerce, 1997)*

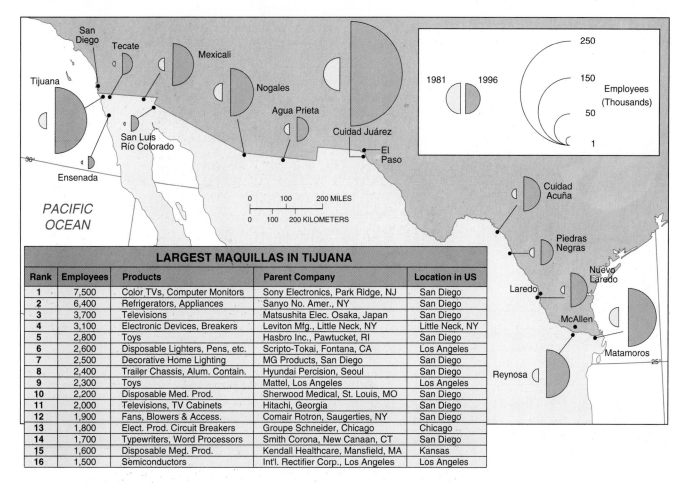

\multicolumn{6}{c}{**LARGEST MAQUILLAS IN TIJUANA**}
Rank	**Employees**	**Products**	**Parent Company**	**Location in US**
1	7,500	Color TVs, Computer Monitors	Sony Electronics, Park Ridge, NJ	San Diego
2	6,400	Refrigerators, Appliances	Sanyo No. Amer., NY	San Diego
3	3,700	Televisions	Matsushita Elec. Osaka, Japan	San Diego
4	3,100	Electronic Devices, Breakers	Leviton Mfg., Little Neck, NY	Little Neck, NY
5	2,800	Toys	Hasbro Inc., Pawtucket, RI	San Diego
6	2,600	Disposable Lighters, Pens, etc.	Scripto-Tokai, Fontana, CA	Los Angeles
7	2,500	Decorative Home Lighting	MG Products, San Diego	San Diego
8	2,400	Trailer Chassis, Alum. Contain.	Hyundai Percision, Seoul	San Diego
9	2,300	Toys	Mattel, Los Angeles	Los Angeles
10	2,200	Disposable Med. Prod.	Sherwood Medical, St. Louis, MO	San Diego
11	2,000	Televisions, TV Cabinets	Hitachi, Georgia	San Diego
12	1,900	Fans, Blowers & Access.	Comair Rotron, Saugerties, NY	San Diego
13	1,800	Elect. Prod. Circuit Breakers	Groupe Schneider, Chicago	Chicago
14	1,700	Typewriters, Word Processors	Smith Corona, New Canaan, CT	San Diego
15	1,600	Disposable Med. Prod.	Kendall Healthcare, Mansfield, MA	Kansas
16	1,500	Semiconductors	Int'l. Rectifier Corp., Los Angeles	Los Angeles

FIGURE 11.25

Maquilladora employment in the major border cities of Mexico, 1981 and 1988. Because of America's need for inexpensive labor, it signed the Border Industrialization Program with Mexico in the 1960s. Raw materials and semifinished goods may be transported from the United States to Mexico for fabrication and back to the United States with a much reduced tariff. Wage rates for comparable work in Mexico are from one-fifth to one-tenth of what would be paid for labor in America. The Mexican maquilladora program is a tremendous growth industry, with most plants hovering near the border to minimize transportation costs. By 1994, Tijuana, Mexico, had 550 plants, with 75,000 employees. Critics of NAFTA fear that the total removal of tariffs will escalate the maquilladora program even faster and displace hand-assembly workers from Los Angeles to Detroit. *(Source: Stutz, 1994; San Diego Economic Development Corporation, 1993)*

Forklifts awaiting export to America from Yokohama, Japan. Japan is the world's most prolific producer of motorized vehicles, automobiles, and commercial vehicles. More spectacular than the sheer volume of automobiles produced is the rate of industrial growth in this area. From 1960 to 1990, Japanese auto production increased by more than 5,000%. *(Photo: Pam Haragawa, Impact Visuals Photo & Graphics, Inc.)*

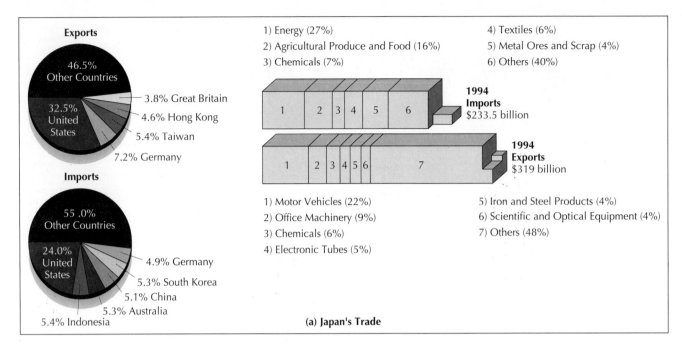

Exports

46.5% Other Countries

32.5% United States

— 3.8% Great Britain
— 4.6% Hong Kong
— 5.4% Taiwan
— 7.2% Germany

Imports

55 .0% Other Countries

24.0% United States

— 4.9% Germany
— 5.3% South Korea
— 5.1% China
— 5.3% Australia

5.4% Indonesia

1) Energy (27%)   4) Textiles (6%)
2) Agricultural Produce and Food (16%)   5) Metal Ores and Scrap (4%)
3) Chemicals (7%)   6) Others (40%)

1994 Imports $233.5 billion

1994 Exports $319 billion

1) Motor Vehicles (22%)   5) Iron and Steel Products (4%)
2) Office Machinery (9%)   6) Scientific and Optical Equipment (4%)
3) Chemicals (6%)   7) Others (48%)
4) Electronic Tubes (5%)

(a) Japan's Trade

FIGURE 11.26

(a) Composition of **Japan's** world merchandise trade, 1994. Japan has shown prodigious growth in manufacturing and trade and now stands as the world's second largest economy, after that of the United States. Most important, Japan has a huge trade surplus. Japan's leading trade partner for both exports and imports is the United States. (b) The global distribution of Japanese automobile exports, 1991. (c) Composition of **Australia's** world merchandise trade, 1994. Australia's small economy exports primary products: ores, minerals, coal, gold, wool, and cereals. Its territory and natural resources are vast compared to its small population of 17 million. Exports and imports are divided between Japan, the United States, and the EU.

While the rest of the world reeled from two major oil-price hikes and experienced a major recession, lower productivity, high unemployment, economic restructuring of manufacturing jobs, and a shift toward the service sector of the economy, East Asia and the Pacific forged ahead with unprecedented growth. It is difficult to identify the factors that account for this growth. However, diligence and hard work, a culture dedicated to succeeding competitively, and the work ethic that weds the laborer to the firm and to the government are part of the answer. In addition, Japan and other countries have protected home markets with extremely high import duties, while continuing to encourage rivalry among domestic companies to prepare them for international export wars (see East Asian Miracle, Chapter 9).

Two other factors contributed to the prosperity of these East Asian and Pacific countries. First, unlike America, where short-term profits were important to satisfy stockholders, banks, and financial institutions, Japan encouraged reinvestment and long-term growth cycles. These long cycles allowed firms time to develop products and to reinvest in the highest-quality production systems before the owners or employees could reap any of the profits. Second, many of the Asian/Pacific countries acted as resource supply centers for the United States from

1965 to 1975, during the Vietnam War, which allowed them to collect a heavy inflow of U.S. dollars.

All these factors combined allowed Japan to develop the world's second largest economy, after the U.S. economy, by 1980. Japan's economy supplanted that of the Soviet Union, which has almost three times the population of Japan. Japan's economy is also tremendously more prosperous than that of China, which has nearly nine times the population of Japan.

Between 1960 and 1990, the combined domestic product of East Asia and the Pacific increased 20 times over, which changed the economy from one of developing nations to NICs. The economy moved toward a new emphasis on electronics, automobiles, steel, textiles, and consumer goods. At the same time, internal policies required these countries to manage their population growth and natural resources, whereas other developing countries in Latin America, Africa, and South Asia did not have such policies. From 1970 to 1990, foreign investment in the region, especially in Japan and the Four Tigers, grew tenfold. This investment was led not only by U.S., British, German, Canadian, and Australian firms, but also by the Japanese.

The United States has taken the lead in trade with Pacific Rim nations, even though it is separated from them by 10,000 miles or more of ocean. Former U.S.

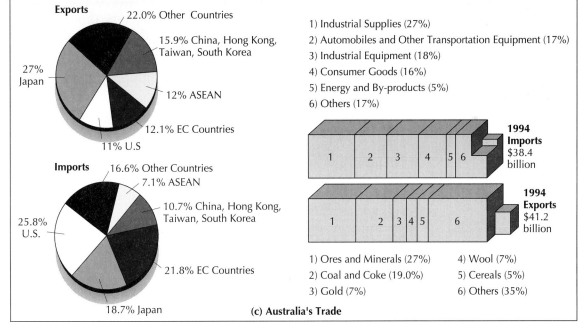

(b)

(c) Australia's Trade

trade ties were with the North Atlantic, but the focus has changed to the Pacific. Figure 11.14 shows North American (U.S. and Canadian) trade ties with East Asia and the Pacific versus ties to Western Europe. Until 1970, Western Europe was North America's chief trading partner. From 1970 to 1980, however, Asia and the Pacific had caught up with Western Europe in terms of total trade with North America, and since 1979, North American trade with Asia and the Pacific increased more rapidly than trade with Western Eu-

rope. The trade gap continued to increase until 1990, when North American trade with Asia and the Pacific outpaced trade with Western Europe by 50%, nearly hitting the $350 billion mark.

JAPAN

Japan is the second leading international trading nation. Figure 11.26 shows the Japanese balance of merchandise trade. Unlike most of Western Europe and North America, Japan shows a huge export trade surplus. Although the United States is its principal trading partner, both for exports and for imports, Japan has a tremendously diversified trading base, with almost 50% of exports and imports going to countries each composing less than 4% individually. Second-, third-, fourth-, and fifth-place countries relying on Japanese exports are Germany, Taiwan, Hong Kong, and Great Britain—an interesting mix of Asian/Pacific and EU nations. The other 50% of Japanese world trade is with Pacific Rim nations.

The United States provides fewer imports to Japan than the exports that it receives from Japan. However, U.S. products still account for the largest single proportion of goods imported by Japan. Indonesia, Australia, China, South Korea, and Germany follow. The United States and Indonesia fill a large need for energy that Japan cannot meet domestically. In addition, because of Japan's mountainous terrain, agricultural and food products compose 15% of total imports. Chemicals, textiles, and metals are also imported.

The world dominance of Japan in the manufacture of motor vehicles is truly phenomenal (Figure 11.26). Fully 22% of its exports are motor vehicles, followed by high-tech office machinery, chemicals, electronic tubes, iron and steel products, and scientific and optical equipment. Diversification is a key word used to describe the breadth of Japanese exports. Fifty-two percent are products that do not individually account for more than 4% in any one industrial category. Per capita exports amount to $2,500 in Japan, comparable to those of many of the EU nations, which are much smaller.

Similar to America's exportation of manufacturing jobs to Mexico and East Asia, Japan has done the same with automobile-assembly plants and autoparts firms in the United States, which now number near 400. In addition, with regard to grains, feed, and nonoil commodities, the United States ships these primary products to Japan as a chief way of accounting for its reception of high-tech manufactured items—microelectronics and automobiles, primarily. There is a reverse flow of high-tech goods from North America to Japan, most notably Boeing aircraft.

Japan, by 1980, had become the world's leading creditor nation and a dominant player in the world financial scene. Currently, Japanese banks account for all but 2 of the 10 largest in the world and dominate the 500 largest banks in the world. Tokyo became one of the main financial centers, alongside London and New York. Brokerage firms and banks from Japan occupy cities not only on the West Coast of America, but throughout the world as well. Japan has managed all this with almost a total lack of food, mineral, and energy resources. It has shown the way for other East Asian and Pacific countries to follow suit. Singapore and Hong Kong are also major players in the world banking scene and have important money markets.

From 1980 to 1994, Japan attained a trade surplus of $700 billion, by far the largest in the world. It enjoyed a $60 billion bilateral trade surplus with the United States, $35 billion with Western Europe, and $25 billion with Asia and the Pacific. Both the EU, which received 20% of Japanese exports, and the United States, which received 30%, contributed to this trade surplus. As a result, protectionist voices in Europe and in America can be heard periodically accusing Japan of establishing new markets for itself by temporarily undercutting prices of foreign competitors. This monopolistic approach to competition is illegal in North America. The Japanese have also been accused of selling goods by dumping them on local markets to weaken rivals and force them to sell their market shares to Japanese firms. At the same time, it is argued, the Japanese have restricted foreign firms from selling in Japan by a huge amalgam of import duties, tariffs, and regulations.

As mentioned earlier, in 1985, the United States, Japan, Great Britain, France, and Germany began selling dollars on the world market to try to drive down the currency's value. The result was a big success, and the dollar is now quite weak on world markets and is creating a trade surplus for the United States. The surplus has not eliminated the U.S. trade deficit yet, but movement is in that direction. At the same time, the yen soared, which caused the prices of Japanese goods in Europe and America to skyrocket. Japanese sales dropped, and the deficit narrowed even more. However, although the American and Japanese deficits have been narrowing from 1988 to 1994, they still represent a huge surplus for the Japanese.

The Japanese response to the decreased demand for their goods was to establish manufacturing plants in America and in Europe to reduce the prices of their marketed items. They also established manufacturing locations in Southeast Asia, where costs were lower. Both approaches—in North America and in Southeast Asia—allowed the Japanese to continue trading competitively with cheaper products. Thus, Japan has relegated Southeast Asia and the United States to Third World status by using its vast amounts of inexpensive labor to produce its manufactured products.

The selling of America to Japan certainly is no more true than on the North American West Coast, where, from San Diego to Vancouver, Japanese banks were more numerous than American banks and own 30% of the major buildings and real estate downtown by 1995.

JAPAN TODAY

The Japanese economy is in its fifth year of serious recession. GDP, which was negative in 1993, recorded only 1% gain in 1996. The collapse of Japan's "bubble economy" at the beginning of 1990 was the main reason for the recession. Equity values plunged to 50% of their paper value; heavy investment in North American properties tapered off to almost zero. The declining asset values left Japanese banks, which number seven of the largest 10 in the world, with nonperforming loans and precarious balance sheets. GDP is expected to increase to 2% in 1997. Until the year 2000, growth in Japan is expected to be well below the average 4% growth over the last 20 years per annum. Japan experiences low rates of labor force growth and excess capacity in manufacturing. The economy is experiencing economic restructuring such as that of North America (away from manufacturing and into services and communications). The sharp appreciation of the yen with respect to the U.S. dollar sent the price of Japanese automobiles higher in American markets, boosting North American production from 1994–1998.

The United States is by far Japan's largest trading partner, and in 1994, 30% of Japanese exports went to the United States and Canada, while 25% of imports came from Canada and the United States. After Canada, Japan has been the second largest market for U.S. exports for many years, but in 1995, the U.S.–Japan trade deficit widened to 60 billion. This constituted more than half of the U.S. merchandise trade deficit (the other largest U.S. trade deficit was with China). However, exports to Japan did expand at a 9% annual clip between 1987 and 1997. While U.S. total exports are 84% manufactured items, Japan's imports are only 64% manufactures because of their need for food and timber products from America. A relatively large share of U.S. exports to Japan consist of intracompany shipments from Japanese subsidiaries in the United States to their parent firms in Japan.

Japan's trade surplus with Southeast Asian countries exceeded its surplus with that of the United States in 1993–1997. As a group, Southeast Asian countries and the new industrializing countries (NICs) are now a more important export market for Japan than is the United States. To counterbalance the impact of the high priced yen, Japanese companies have been moving labor-intensive assembly manufacturing operations to offshore Southeast Asian cities—mostly to Thailand and Malaysia, but more recently to China and Hong Kong. The primary contention between Japan and the United States has been closed markets. The primary policy of U.S. trade with Japan has been to open Japanese markets for imports and foreign investment. Informal obstacles such as testing standards, certification requirements, intellectual property regulations, and impenetrable distribution channels have limited U.S. exports to Japan. Japanese exports manufactured goods and imports raw materials, food, and industrial components. Japanese operates an *exchange economy*—exchanging raw materials into high-value-added products, with high inputs of technology and labor.

Goods imported from Japan to the United States include automobiles ($20 billion), household appliances, televisions, computer chips ($13 billion), computers and office equipment ($12 billion), telephone and other telecommunications equipment ($6 billion), auto engines, jet engines, and electric motors ($5 billion), and industrial machinery ($4 billion).

Australia

Australia's balance of merchandise trade is shown in Figure 11.26. Australia's main trading partners are the United States, the EU, and Japan. Exports go primarily to Japan, which accounts for 27.5%. The next largest share, 15%, goes to China, Hong Kong, Taiwan, and South Korea. The Association of South-East Asian Nations (ASEAN)—Indonesia, Malaysia, the Philippines, Singapore, and Thailand—account for an additional 15%, whereas the United States and the EU account for approximately 12% each. The United States leads the list of importers, providing 23.5%, while EU countries supply 21.8%. Japan follows with 18%.

Australia is a nation that exports *primary products*—mainly ores and minerals, coal and coke, gold, wool, and cereals ("rocks and crops"). Almost all of its exports are from the vast wealth of land and resources that it enjoys. Because of its small population, 19 million, industrial supplies, automobiles, and industrial equipment account for more than 60% of imports. Japan has made its greatest market penetration into Australia and accounts for 50% of all vehicles purchased.

Australia is one of the leading raw-material suppliers in the world. It is the largest exporter of iron ore and aluminum and the second or third largest exporter of nickel, coal, zinc, lead, gold, tin, tungsten, and uranium. Consequently, Australia's current problem is to withstand the declining world prices of raw materials. To cushion against fluctuations in these prices, Australia needs to industrialize itself so that it can transform its raw materials into finished products and become an exporter of higher-value items. However, doing so is nearly impossible with a small industrial base that demands consumer products before industrial products.

China

Where does China stand in world trade? China enjoys a favorable trade balance, but its exports amount to only $60 per capita, one-hundredth of the German level of $6,000 per capita. China's primary trading partners are Hong Kong and Macao, which uses Hong Kong as a port of entry and export for its goods (Figure 11.27 to 11.29). Exports and imports are further connected to countries beyond the tiny borders of Hong

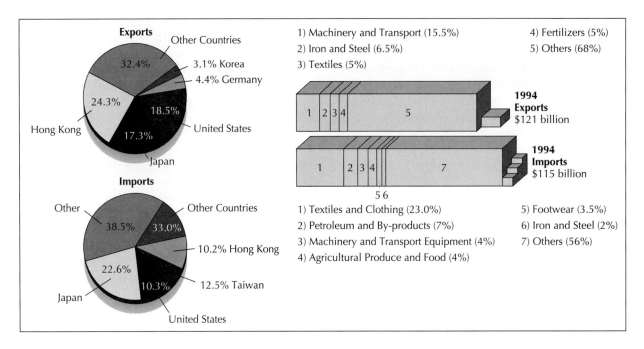

Exports

Other Countries 32.4%

3.1% Korea

4.4% Germany

18.5% United States

24.3% Hong Kong

17.3% Japan

1) Machinery and Transport (15.5%)
2) Iron and Steel (6.5%)
3) Textiles (5%)
4) Fertilizers (5%)
5) Others (68%)

1994 Exports $121 billion

Imports

Other 38.5%

Other Countries 33.0%

10.2% Hong Kong

12.5% Taiwan

10.3% United States

22.6% Japan

1994 Imports $115 billion

1) Textiles and Clothing (23.0%)
2) Petroleum and By-products (7%)
3) Machinery and Transport Equipment (4%)
4) Agricultural Produce and Food (4%)
5) Footwear (3.5%)
6) Iron and Steel (2%)
7) Others (56%)

FIGURE 11.27
Composition of **China's** world merchandise trade, 1994. World trade is tiny compared to its population size of more than 1.2 billion people, yet it enjoys a trade surplus, exporting more than it imports, including textiles, clothing, petroleum, and by-products.

Kong and Macao. Japan is China's second leading trade nation, followed by the United States.

China suffers from the communist ideology of *autarky*, which suggests that countries should be self-sufficient economically. This view was part of Mao Tse-tung's utopian vision of the nation. China opened its door of trade quite late, after the death of Mao in 1976. Shortly after President Nixon visited China, the first American president to do so, he granted them *Most Favored Nation* status as a trading partner. During the early 1980s, China finally allowed foreign companies to set up joint ventures there. *Special economic zones* (SEZs) were created near Hong Kong to produce goods for world markets. These economic zones received tax incentives but were subject to a host of legal red tape that was typical of the Communist government (see Figure 9.16).

China, the largest nation in the world, with 1.2 billion people, has the potential to become a major actor in world trade. It has a large worker base, low wages, and, because of the East Asian work ethic, relatively high levels of worker productivity. In addition, some Chinese who fled during the Maoist cultural revolution of 1966 to 1976 (the *overseas Chinese*), and who settled in the capitalistically successful countries of the Four Tigers or the Five Little Dragons, have now returned to China with new entrepreneurial skills and some investment capital. The result has been a dramatic increase in foreign trade. For example, exports increased from only 5 billion in 1976 to almost 62 billion in 1990.

By all measures, however, China is a poor country. It still struggles to provide its many people with sufficient food and housing. Almost no capital is available for start-up programs; therefore, China has open doors to foreign companies, especially in the industrial sector. Foreign investment has flooded in to establish factories, to mass produce items in the areas of oil exploration, to manufacture motor vehicles, and to construct commercial buildings and hotels in the major cities. State-owned manufacturing plants account for approximately 50% of manufacturing exports. The other 50%, and an increasing proportion, are small-scale industrial plants that are owned by rural townships but leased to private individuals for profit. Textiles, clothing, and industrial products accounted for 70% of exports in 1990.

In the early 1990s, the United States took issue with China's Most Favored Nation trade status. China has a very poor record of human rights violations, most notably the Tiananmen Square Massacre of hundreds of students in the summer of 1989. For this reason, U.S. protectionist sympathies in Congress have asked for China to be stripped of its Most Favored Nation status. Although China seems to have weathered this storm and is moving ahead under communist leadership, it has been reluctant to allow Chinese manufacturing to privatize as it has in part allowed the agricultural sector to do. Until privatization occurs, China's vast human resources will not be used efficiently to produce items for world trade, even with Hong Kong on board.

CHINA TODAY

China is currently watching Russia and former East Germany. High levels of unemployment that would initially be caused by privatization of state-owned industry in China may cause major social unrest, it is feared, and thus lead to a revolt that could topple the Communist Party. The safe approach is to allow state-owned industry to be heavily subsidized for the near term.

While the developed world was undergoing a major recession in the early 1990s, China recorded a GDP growth of 8% between 1987 and 1992, peaking at almost 14% in 1993. GDP increased to 12% in 1995, continuing the surge in investment-led growth that began in 1992. Industrial output soared 21% in 1994, and 26% in 1995, due in part to collectives and firms that are partially foreign owned. China's principal economic plan has been the continuation of high growth to generate jobs and to spread the benefits of economic reform from the cities to the countryside, in addition to controlling inflation. It is expected to continue to outpace other regions (Figure 11.27).

China and the developed world seem unsure if the government will continue its economic globalization program after the Deng Xiao Ping era. Deng was a chief architect of economic liberalization, which has brought wealth to many Chinese, but also brought controversy. The government would like to reduce inflation to 15% or less by 1997. Other problems include great disparities of income among Chinese citizens, greater regional autonomy, especially in the southern coastal special economic zones (SEZs), and a shortage of key commodities and services (see Figures 11.28 and 11.29).

Chinese manufactured goods lead export growth, whereas agricultural and primary products lead import

FIGURE 11.28
China's fastest-growing sectors of the economy. By the year 2000, China's fastest-growing sectors will be development of power systems to produce electricity, followed by telecommunications equipment and automobile parts and service equipment. What used to be a Chinese unified national market has now separated into a number of regional markets, each pursuing its own development plan, and each competing with the others to attract foreign investment and technology. Chinese per capita GDP is $550 U.S. dollars in 1996, with an inflation rate of 24% for the 1.2 billion people.
(Source: U.S. Department of Commerce, 1997)

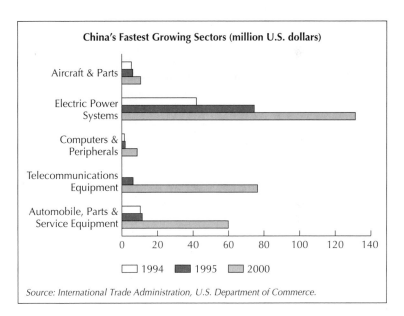

FIGURE 11.29
Chinese economic area growth relative to other regions, 1993–1999. Projected average annual increases in GDP. The Chinese economic area, which includes North and South Korea, is expected to lead the world in average annual GDP increases. In the decade prior to 1997, China experienced double digit growth. This growth is expected to continue at a robust rate in the vicinity of 7% to 8%. U.S. growth is expected to be between 2% and 2.5%, and Europe and the EU community near 2%.

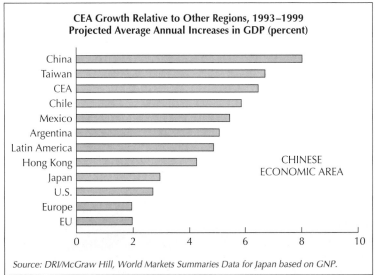

growth. SEZs in southern and eastern coastal areas continue to outpace the rest of the nation in economic growth and trade. Japan continues to be China's top trading partner and source of imports. Hong Kong is the main port for exports and imports of China. Imports from the United States have been growing rapidly, but the United States ranks fourth largest source of imports to China after Japan, EU, and Taiwan. China's major imports from the United States include cotton, fats and oils, manufactured fibers, fertilizer, aircraft, wood pulp, and leather. The U.S. trade deficit with China has increased steadily since 1987, and stood at 50 billion in 1996. Imports to the United States from China have grown 30% per year and exports 15%. Part of this imbalance is because China maintains an intricate system of import controls. Most products are subject to quotas, licensing requirements, or other restrictive measures.

Three out of four toys sold in America are foreign made, and 60% of those imports come from China. Sixty percent of all shoes sold in America come from China. In 1997, China is expected to dislodge Japan as the country giving America its biggest trade deficit. Other major goods imported into the United States from China include clothing, telephone and other telecommunication equipment, household appliances, televisions and computer chips, computers, and office equipment. China's higher trade barriers protect inefficient state-run companies that still employ two-thirds of the urban workers. China might decide that the pain of ending that would not be worth the smaller gain of joining the World Trade Organization (WTO). China

has been trying for more than two years to join the WTO, the Geneva group that sets the rules for global trade, but the United States has blocked the effort. It is China's large trade barriers that U.S. officials blame for America's poor showing in exports to China. Hong Kong is the entrepot of China (Figure 11.30).

HONG KONG

Economically, Hong Kong continues to prosper. Politically, China's resumption of sovereignty will result in a subtle erosion of civil liberties that neither the British nor the United States are in a position to arrest. Those focusing on Hong Kong's economic future are generally unperturbed by the change of political masters. Under the formula of "one country, two systems," Beijing has promised to maintain a capitalist economy in Hong Kong for 50 years following the handover. Hong Kong is too important to China for Beijing to alter the status quo.

Indeed, Hong Kong is already deeply entwined with the rapidly growing provinces of the South China coast. The city's dynamic entrepreneurs were quick to exploit Guangdong's low-wage special economic zones, and they are now among the largest investors in China. The flows have not been one way. Hong Kong has always been a major port for Chinese exports; now, investment from the mainland is pouring into Hong Kong as well. China's largest state-owned enterprises already operate in Hong Kong, using the city as their international headquarters.

At present, economic optimists are ascendent. Firms and banks are signing long-term contracts, and the

Samsung VCR assembly line in Seoul, South Korea. Foreign national corporations, like Samsung, have raised foreign exchange and have helped pay off Korea's massive debts, a total of $45 billion by 1990, making Korea one of the most indebted nations of the world for its size. Because of competitive exports, inexpensive labor, and a booming economy, its debt service ratio is now only 13% of its GNP. By contrast, some African nations are at 100% debt payment of their GNP per year, and the debt service ratio for South American nations is 37%. Recent downturns in the global economy have slowed South Korea's journey out of heavy debt. New problems now exist with North Korea's nuclear capability and sullen response to trade. *(Photo: T. Matsumoto, Sygma)*

stock and property markets stand at all-time highs. Fears of Hong Kong talent exiting in a debilitating "brain drain" have not materialized. Surveys show foreign investors more confident in Hong Kong's future now than they were 2 years ago.

But for those looking at Hong Kong through a political lens, the future looks somewhat less rosy. Hong Kong under the British was never democratic. Politics was dominated by a governor and a hand-picked group of local advisers. But the colonial administration generally respected civil liberties and provided an efficient and even-handed legal system. China has started to erode the civil liberties of Hong Kong already.

Taiwan

Taiwan, with a population of 21 million and a GDP of over $200 billion, has sustained a 6–7% economic growth rate since 1987. Exports account for an amazing 35% of GDP in Taiwan's export-oriented economy—

FIGURE 11.30
(a) **Hong Kong's** leading sources of imports and destination of exports, 1995. China controls well over a third of Hong Kong's exports and imports, and Hong Kong acts as an entrepot port and gateway for the export and import of goods from the China mainland. Hong Kong functions as a gateway to and from the mainland for regional and international traders, investors, and tourists. Singapore's distance and Taipei's regulations will likely give Hong Kong the edge in becoming East Asia's financial center after Tokyo. (b) Hong Kong's fastest-growing sectors of the economy. By the year 2000, electronic components will outdistance other sectors of the economy, followed by telecommunications equipment and plastic materials and resins. *(Source: U.S. Department of Commerce, 1997)*

one of the four newly developing NICs, or Four Tigers of the Orient (Figure 11.31). Similar to Hong Kong, there has been a restructuring in Taiwan's economy with long-sustained economic prosperity and increasing land and labor costs. Manufacturers of labor-intensive products such as toys, apparel and shoes, and circuit boards have moved offshore, mainly to Southeast Asia and to China. Manufacturing growth is now concentrated in

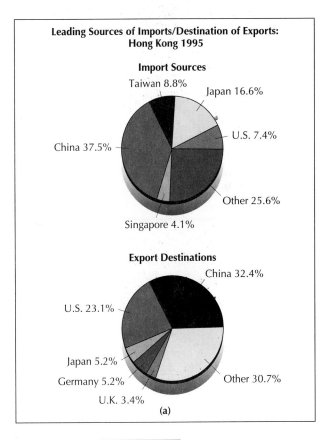

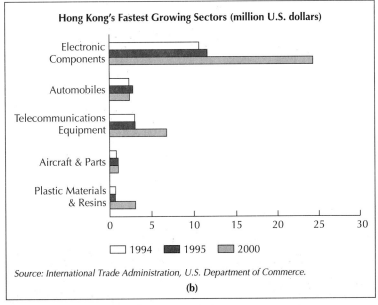

technology-intensive industries including petrochemicals, computers, and electronic components. Large state-run enterprises still account for one third of Taiwan's GDP, but major privatization efforts are underway to release power generation. During the 1996–2000 period, Taiwan's growth is expected to continue at a 6.5–7% clip annually.

Between 1987 and 1985, export growth to Taiwan proceeded at a 14% annual increase. Historical trade patterns in the east Pacific Rim are likely to keep Taiwan

dependent on capital goods from imports from Japan. Over the last 10 years, Taiwan has made great progress in lowering trade barriers and improving market access of foreign goods. The average tariff on industrial products coming into Taiwan is 9%, but for agricultural products, the tariff is 20%. Taiwan has committed to tariff practices specified under WTO provision; this should help resolve many of its trade problems. Korea and Singapore are heavily linked to Japan and the United States for exports and imports, but are also widely spread to many other countries (Figure 11.32).

South Korea

Another miracle of the Pacific Rim is South Korea, a country of 45 million, with a GDP of $370 billion in 1995. Its GDP has averaged over 8% over the last 10 years, and its GDP per capita is $9,000 (which compares favorably with the $300 GDP per capita in North Korea). South Korea's rapid advancement has turned this nation

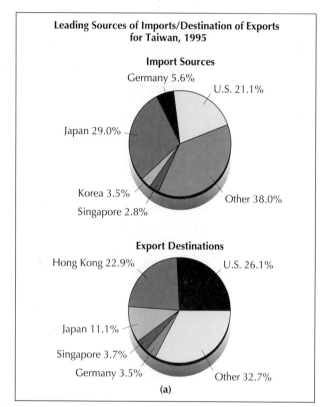

Leading Sources of Imports/Destination of Exports for Taiwan, 1995

Import Sources

Germany 5.6%
U.S. 21.1%
Japan 29.0%
Korea 3.5%
Singapore 2.8%
Other 38.0%

Export Destinations

Hong Kong 22.9%
U.S. 26.1%
Japan 11.1%
Singapore 3.7%
Germany 3.5%
Other 32.7%

(a)

FIGURE 11.31

(a) **Taiwan's** leading sources of imports and destination of exports, 1995. While Japan and the United States rank number one and two for Taiwan imports, the United States and Hong Kong are the chief destinations for Taiwanese exports. (b) Taiwan's fastest-growing sectors of the economy include computer peripherals, laboratory and scientific equipment, telecommunications, and pollution equipment. The consumer market, from cars and computers to insurance and world travel, is not only expanding in Taiwan in overall size, but new niches continually open up as Taiwan's consumers become more discerning and more sophisticated in the world economy. *(Source: U.S. Department of Commerce, 1997)*

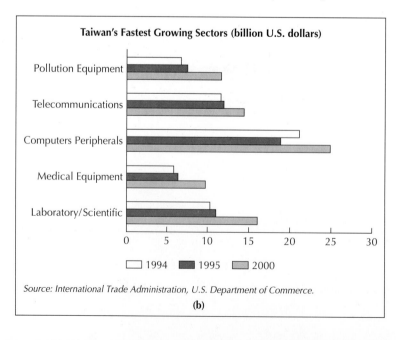

Taiwan's Fastest Growing Sectors (billion U.S. dollars)

Pollution Equipment
Telecommunications
Computers Peripherals
Medical Equipment
Laboratory/Scientific

☐ 1994 ◼ 1995 ▦ 2000

Source: International Trade Administration, U.S. Department of Commerce.

(b)

into one of the most economically powerful in the world. In 1995, South Korea opened power production to the private sector. These areas are limited to the production of electricity by coal, LNG (liquid natural gas), and water power, benefiting Canadian and U.S. companies. In the next 5 years, continuing with its privatization program, the South Korean government is expected to spend $5 billion on environmental improvements. South Korea is implementing an ambitious transportation infrastructure development program that includes major high-speed rail and transit programs, airport development, and highway construction. The fastest growing sectors in South Korea include transportation services and computers, and peripherals between the year 1995 and 2005 (see Figure 11.31).

While South Korean production and manufacturing technology have reached nearly the level of advanced countries in key heavy and high-tech industries, they still lag somewhat behind. South Korea remains a regional economic driver, especially as a potential gateway to China, one of the most difficult markets in Asia in which to invest and to which to export. This probably stems from the legacy of the Japanese occupation

FIGURE 11.32

(a) **South Korea's** leading sources of imports and destination of exports, 1995. Leading sources of imports included Japan and the United States, accounting for roughly 45% of imports. The leading export destinations included the United States and Japan, accounting for 36% of destinations, but a wide variety of other countries contributed to both import and export destinations for South Korea. (b) The fastest-growing sectors in South Korea include transportation services and computers and peripherals. South Korea is implementing an ambitious transportation infrastructure development program which includes major high-speed rail and transit programs, airport construction, and highway development. While South Korea's production technologies have reached nearly the level of advanced countries in key heavy and high-tech industries, they still lag somewhat behind. The legacy of Japanese occupation and the fear of being dominated have lead South Korea to implement product diversification. Because of this, South Korea is one of the most difficult markets in Asia in which to invest and to which to export.

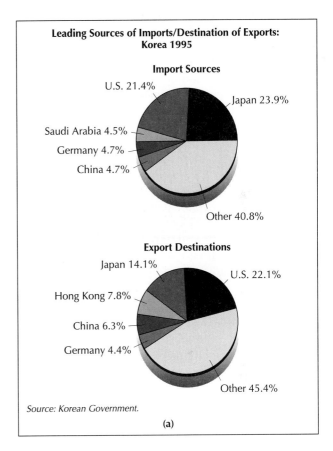

(a)

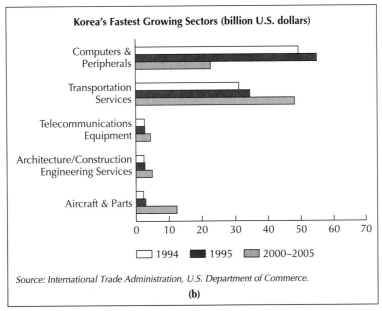

(b)

and the fear of being dominated by Japanese imports. The strongest competition that any foreign company faces in South Korea is domestic competition from South Korean companies.

India

In South Asia, India, with its 860 million people, is like a human flood of population in a land smaller than that of the United States. Its world trade is minuscule but growing, at about $15 per capita. India is, at long last, self-sufficient in food production and is an exporter of primary products, including gems and jewelry, textiles, clothing, and engineering products (Figure 11.33). In order for its factories to operate, it must import industrial equipment and machinery, and crude oil and by-products, as well as chemicals, iron, and steel. Twelve percent of its imports include uncut gems for its expanding jewelry trade.

In terms of trading partners, India is no longer dominated by its former colonial overseer, Great Britain. Its leading countries of export include the United States and Japan. Great Britain and other EU countries account for 19% of total exports and 25% of total imports. Because India represents such a large pool of demand, most manufactured goods and consumer goods are consumed locally, not exported. Since 1990, India has also become an exporter of cereals and grains, and textiles and clothing are now the chief exports.

India's GDP has continued at a 5–6% increase between 1987 and 1995. Record levels of foreign capital ($6 billion in portfolio and direct investment) stimulated a capital market over the last decade. But Indian public investment and infrastructure continues to be insufficient for a country with developmental goals. The economy is expected to grow 6% annually between 1996–2000.

Prime minister Rao and the Congress party began economic reforms in 1991 to liberalize the economy, privatize government-owned industry, and open India to international competition. The International Monetary Fund (IMF) persuaded India to turn its back on a policy of trade protection and import substitution that had been in place since the country became independent in 1949. Since 1991, India has moved forward with traditional macroeconomic tools to handle deficits, inflation, and a balance of payments. Import tariffs were slashed, and the government loosened its hold on business by dismantling the licensing system that governed all economic activity, moving strongly towards deregulation of the private sector. One of the results is the accelerating development of information technology centers and industrial parks around the city of Banglore, where companies produce software for international markets. Banglore is India's Silicon Valley, and

foreign multinationals such as IBM, Texas Instruments, Digital Equipment, Hewlett Packard, Motorola, 3M, and QUALCOMM have set up operations in a $150 million science park.

The United States is India's single largest trading partner and total bilateral trade is $8 billion in 1985. Tariffs on those capital goods and equipment have been reduced from 35% to 25%, and imports of consumer goods continue to be banned. Patent protection is lacking new corporate laws, but pending trademark legislation should significantly enhance intellectual property rights protection in the near future. The biggest export opportunities to India from developed regions of the world include large electronic components. In 1996, the market will total $2 billion, a 20% increase over 1995. Because of increasingly stringent environmental regulations and growing industry awareness, markets for pollution control equipment are increasing at an annual rate of 40%. Food processing and packaging equipment is in demand as India's agricultural sector employs 70% of the country's workforce.

South Africa

The Republic of South Africa, with a population of 41 million and a GDP of $115 billion, is the most productive economy in all of Africa and accounts for 75% of the GDP of the southern African region. It accounts for almost 50% of the entire continent's output (Figure 11.34). Manufacturing now accounts for 26% of the GDP, indicating a diversification from the traditional dependence on gold and diamond exporting. Finance and business services account for 16% of the GDP as the nation moves towards the tertiary and quaternary sectors. The remarkable succession and peaceful transfer of power to the new government headed by Nelson Mandela produced an upswing in business confidence and a relaxation of foreign embargoes against South Africa. Job growth has been sluggish, however, and unemployment is 40% among blacks. GDP in the near future is expected to average 4% per year, until the end of the century.

Apartheid, the policy of racial separation, has kept a shortage of skilled labor and a concentration of economic power among few large economic enterprises and a relatively high tax rate. It has also caused widespread illiteracy, unemployment, and social problems which will be expensive to remedy and take years. The triad countries are finding South Africa to have an attractive market because of the pent-up demand for goods and services after nearly a decade of sanctions. Leading U.S. exports to South Africa include aircraft and parts, industrial chemicals, computer software, pharmaceuticals, medical equipment, telecommunications equipment, and building and housing products. The United States is South Africa's largest trading partner. South Africa recently has paid attention to reorienting its own economy away from import substitution towards an international competition.

The Former Soviet Sphere

The most momentous occurrence in our lifetimes is certainly the breakup of the former Soviet Union and the return of eastern European nations to market

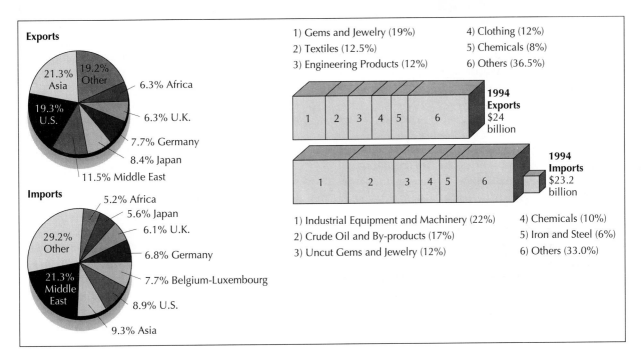

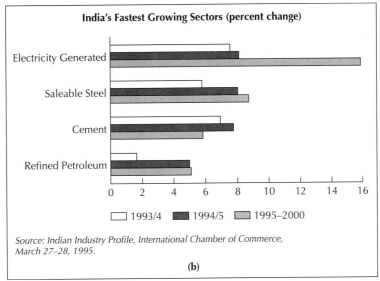

(b)

Source: Indian Industry Profile, International Chamber of Commerce, March 27–28, 1995.

FIGURE 11.33

(a) Composition of **India's** world merchandise trade, 1994. India exports gems, jewelry, and textiles. The country imports badly needed industrial equipment and machinery, as well as crude oil for fuel. Its total trade is tiny compared to its 900 million population. (b) India's leading sources of imports and destination of exports. In the words of India's minister of commerce, P. Chidambaram: "India's road to reform will not be the German autobahn, or the North American interstate. We'll take the Indian road—with potholes, twists and turns, and slowing down at crossing . . . we will reach our goal." India's fastest-growing sectors in 1995–2000 include electricity-generated steel and cement production. India's $9 billion infotech market has had a 25% annual growth rate during the early 1990s, making it one of the fastest-growing in the world. Canadian and U.S. firms are looking at secondary urban areas and at infrastructure projects in cities and regions that are likely to grow, rather than major metropolitan areas in India. For most North American firms, India's huge and growing population is the primary attraction.

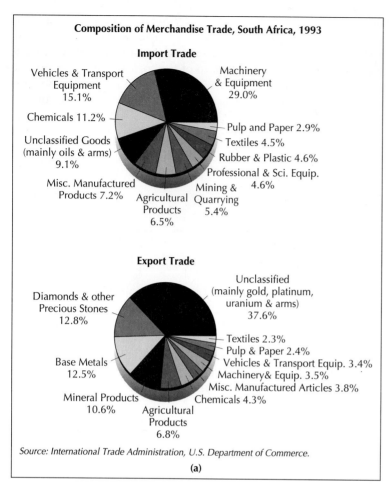

Composition of Merchandise Trade, South Africa, 1993

Import Trade

Vehicles & Transport Equipment 15.1%
Chemicals 11.2%
Unclassified Goods (mainly oils & arms) 9.1%
Misc. Manufactured Products 7.2%
Agricultural Products 6.5%
Mining & Quarrying 5.4%
Professional & Sci. Equip. 4.6%
Rubber & Plastic 4.6%
Textiles 4.5%
Pulp and Paper 2.9%
Machinery & Equipment 29.0%

Export Trade

Diamonds & other Precious Stones 12.8%
Base Metals 12.5%
Mineral Products 10.6%
Agricultural Products 6.8%
Chemicals 4.3%
Misc. Manufactured Articles 3.8%
Machinery & Equip. 3.5%
Vehicles & Transport Equip. 3.4%
Pulp & Paper 2.4%
Textiles 2.3%
Unclassified (mainly gold, platinum, uranium & arms) 37.6%

Source: International Trade Administration, U.S. Department of Commerce.

(a)

economies. As of 1997, the former Soviet Union's 15 republics—Russia and the 14 republics—and the eastern European nations were still in transition. The old central-planning systems have clearly broken up, but new systems that will replace them are not yet fully in place.

Before the breakup of the Soviet Union in 1991 and the return to market economies by the eastern European nations, beginning in 1989, this communistic, centrally planned region accounted for only 8% of world trade. The communist ideology was based on autarky, or self-sufficiency, which meant that these nations kept their distance from the rest of the world in terms of trade. This policy was disadvantageous because Western Europe produced many items that eastern European nations and the Soviet Union needed. At the same time, Western Europe had a great demand for energy resources and raw materials from the Soviet Union. Only near the end of the empire did a major pipeline provide natural gas from the Ural Mountains to Germany and other western European nations.

It is impossible to gauge what the future will hold for world trade from this region.

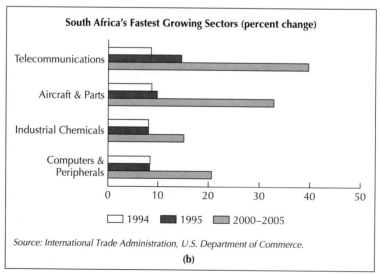

South Africa's Fastest Growing Sectors (percent change)

Telecommunications
Aircraft & Parts
Industrial Chemicals
Computers & Peripherals

0   10   20   30   40   50

☐ 1994   ◼ 1995   ☐ 2000–2005

Source: International Trade Administration, U.S. Department of Commerce.

(b)

FIGURE 11.34

(a) **South Africa's** composition of merchandise trade, imports and exports. In Nelson Mandela's September 1993 address to the United Nations, when he called for the lifting of U.S. sanctions, he said: "We have together walked a very long road. We have traveled together to reach a common destination." Former U.S. secretary of commerce Ronald H. Brown has stated: "South Africa is the key to the economic success of southern Africa, and probably the key to the renewal of the rest of the continent." (b) South Africa's fastest-growing sectors of the economy include telecommunications, aircraft and parts, computers and peripherals, and industrial chemicals. U.S. and Canadian firms involved in South Africa's redevelopment will contribute to the country's revitalization process while gaining a strong commercial foothold in the new South Africa and the Africa beyond.

In 1992, the national income was 20% less in the former Soviet Union than it was in the previous year. In 1993, output decreased 16% in the region as a whole because the economies were in major transition. However, not all countries were suffering equally, as shown by the credit ratings of Eastern Europe and the republics of the former Soviet Union. These credit ratings were based on a United Nations/Department of Economic and Social Development (UN/DESD) survey of 100 banks, and credit was rated according to analysis of economic conditions, and the chance of default within the countries. With 100 being good, repressive Albania, closed for most of its history, rated 41, whereas Bulgaria, Poland, and Romania hovered around the 50s to 60s. The former Soviet Union, which had a credit rating of 168 in 1988, could barely register 79 in 1992, whereas Hungary and Czechoslovakia were rated 111 and 126, respectively.

Comparable ratings for the republics of the former Soviet Union were much, much worse than those of Eastern Europe. Armenia rated 39; Estonia, 69; Georgia, 36; Kazakhstan, 47; the Kirghiz Republic, 34; Latvia, 64; Lithuania, 63; Moldavia, 38; the Tajik Republic, 34; the Turkmen Republic, 34; and the Uzbek Republic, 37.

The output of this region, country by country, was less than it was 10 years earlier, and investment fell to pre-1970 levels. It was clear that the former Soviet Union and Eastern Europe were in the throes of a savage economic restructuring and contraction. With the collapse of trade under the Council for Mutual Economic Assistance (COMECON), orders from nearby COMECON countries decreased by more than 50%, seriously hurting internal production and trade. There is a slowdown in the production of necessary raw materials, equipment, and replacement parts, and a massive shortage of food, pharmaceuticals, textiles, garments, footwear, and machinery. Bulgaria, formerly a food exporter, was forced to import large quantities of food to feed its population during the winter of 1992 and 1993. In the former Soviet Union itself, agricultural production decreased 25% since 1991, and the area under cultivation decreased by 5 million hectares. The former Soviet Union is wrecked by inflation, which increased 400% between 1990 and 1994 (see Russia's Transition, Chapter 12).

The 14 republics of the former Soviet Union lag most in the transition to a market economy, and Russia is not far ahead. Eastern European nations vary, but the central ones, the Czech Republic and Hungary, are farther ahead than Poland in the north and the southern republics of Romania, Albania, and Bulgaria.

What does the future hold for this region with regard to international trade? There is a certain complementarity in Euro-Asia. Western Europe needs the minerals, oil, natural gas, and other raw materials that are in vast supply in Russia and the republics of the former Soviet Union. At the same time, the eastern bloc nations need foodstuffs and industrial equipment and machinery to resume their powerhouse of economic production.

Multinationals from every OECD country are investigating the potential for investment in the former Soviet Union. For example, Pizza Hut, a Chicago-based company, has unveiled plans to open 1,000 new restaurants within Russia and the 14 republics of the former Soviet Union. Automobile manufacturers from Western Europe, Japan, and the United States are also investigating their opportunities, as are consumer electronics producers. However, uncertainties still exist, and these multinationals must be cautious because of the gigantic economic uncertainties in the political and economic transformation of the former Soviet Union and its former member states.

The process of reorienting Russia's economy to market-driven enterprise through the privatization process is not yet complete (see Chapter 1, and Chapter 12 on privatization in Russia). Although much business and property have been privatized, agriculture remains in state hands as well as a third of the manufacturing plants. Since the dissolve of the Soviet Union in 1991, economic production has plummeted. As of 1994, Russia's population was 150 million and it generated a GDP of around $850 billion. Russia is still experiencing declining output, but not to the 14% level it did in 1993. Unemployment has leveled off, but difficult inflationary battles continue, and infrastructures continue to decay. Russian privatization has attracted investment targets including Aerofloat, the United Utility of Russia (National Electric Utility), and Gazpron (Russia's gas company). Russian's real disposable income grew between 1993–1995 by 12%. Some companies producing consumer goods are experiencing rapidly growing sales.

Russia is one of the GEMs and constitutes one of the 10 fastest-growing world markets for manufactured goods. Normally manufacturing exports to Russia continue to be infrastructure-related: engineering equipment, and automatic data processing and telecommunications equipment. In 1995, the largest single export to Russia was aircraft and related equipment. Total U.S. exports to Russia have been increasing steadily, and were $3.4 billion in 1995. Hindrances to increased world exports to Russia are substantial value added taxes, high import duties, and high excise levees. The two most promising prospects for triad country exports to Russia from 1995–2000 are telecommunications equipment, computers and computer peripherals, pollution control equipment, oil and gas field machinery, construction equipment, medical equipment, electrical power systems equipment, automotive parts and services, building products, and food processing and packaging equipment.

The Suez Canal remains a vital link for distribution of oil from the Persian Gulf to Europe. Despite the fact that the Middle East was one of the early hearths of civilization and city development, it did not share in the capitalist expansion and prosperity of the last 300 years that was centered on Europe and North America. It was only with the discovery of vast oil deposits by U.S. companies in the Persian Gulf area in the second half of the 20th century that the focus of world attention returned to the Middle East. Today, the Middle East contains the greatest extremes of wealth, versus poverty, to be found anywhere in the world, all based on have and have-not oil supplies. *(Photo: Frederic Neema/Reuters, Archive Photos)*

The Middle East

The Middle East contains approximately 60% of the world's oil reserves, with Saudi Arabia containing 200 billion barrels, more than one-third of the world's total. Other oil producers and exporters are Bahrain, Kuwait, the United Arab Emirates, Oman, and Qatar. Four other countries also produce oil but have either interrupted supplies or very small amounts of oil—Iraq, Iran, Egypt, and Syria. Countries in the region without oil supplies include Afghanistan, Israel, Jordan, Lebanon, Turkey, Yemen, Morocco, and Tunisia.

Inexpensive oil from the Middle East has fueled the world for a long time. In fact, the United States and the western European nations have enjoyed a large supply. At $4 to $5 a barrel, Middle Eastern oil helped rebuild Europe after World War II. However, the 1970s and 1980s were tumultuous with regard to international oil prices. In 1973 and 1979, oil supplies were interrupted by Arab boycotts of Western political maneuvering, by Israel's Western bloc support in the Yom Kippur War, and by the overthrow of the Shah of Iran. Prices increased dramatically to $25 a barrel. In 1980, the Iraq–Iran War reduced world flow further, and

Iranian oil field and peasant herder. Iran remains an important source of world oil supplies, but squabbles among OPEC members have kept world prices low. *(Photo: Paolo Koch, Photo Researchers, Inc.)*

prices continued to climb toward $30 a barrel. OPEC's revenues reached $300 billion, and a worldwide recession was triggered. However, because of squabbling among OPEC members and because U.S.–backed Saudi Arabia decided to undercut the market to provide Western stability, oil prices decreased to less than $10 a barrel in 1986. Revenues plummeted. In January 1994, prices were again at $16 a barrel, and OPEC nations had lost much of their stranglehold on the world market, dropping to approximately equal to OPEC's pre-1973 levels.

While all the fluctuations were occurring, other sources of oil, synthetic fuels, and solar, geothermal, and nuclear power sources had been explored and embraced. OPEC's largest stranglehold was on Western Europe, which traditionally had poor supplies of fossil fuels. Even before the oil crisis, the Middle Eastern nations fulfilled 45% of Western Europe's energy needs. By 1994, however, that proportion had dropped to 20% as a result of not only the exploitation of the North Sea oil fields, but also increased coal production in central Europe.

The Persian Gulf War between Kuwait and Iraq further suggested to the world that Middle Eastern oil supplies would be erratic. Nonetheless, because the former Soviet Union ceased its oil production while changing over to a market economy, the demand for international trade in oil is still significant.

The region's second most important activity after oil is agriculture, but water is scarce and the few sources that do exist are heavily tapped. The Tigris and Euphrates rivers, for example, as well as the Jordan River, are argued over by countries such as Turkey, Lebanon, Syria, Israel, and Iraq. In any case, wheat, barley, vegetables, cotton, and citrus fruits can be grown and supplied to Europe, which is a short distance away. Agriculture is also the basis of Turkey's export economy. Turkey is a large exporter of wheat and mineral resources, including iron, copper, and zinc.

Egypt is now one of the world's leading exporters of cotton, yarns, textiles, and denim. Egypt has been criticized because much of its agricultural base is devoted to cotton at the expense of food crops needed to feed its people. Israel exports cut and polished diamonds, machinery, chemicals produced from Dead Sea salts, and phosphates from the Sinai Desert. In addition, countries of the Middle East, ranging from Morocco to Jordan, especially Egypt and Israel, have an invisible trade in tourist dollars. Tourism has been one of the mainstays of the economies of these two countries, but recent uprisings by Islamic fundamentalists in Iran have shocked the tourist industry. Recently, several tourist buses have been attacked with machine-gun fire near the town of Luxor. These attacks created a scare throughout Western countries and was detrimental to foreign exchange.

MAJOR GLOBAL TRADE FLOWS IN THE 1990S

Although we do not have space in this text to identify every major international trade-flow pattern of most manufactured items and services, six have been selected for examination: microelectronics, automobiles, steel, textiles and clothing, grains and feed, and nonoil commodities. Each of these goods shows major patterns of international trade and has been selected for examination on the basis of its representativeness and importance.

Global Trade Flow of Microelectronics

Microelectronics includes semiconductors, integrated circuits and parts for integrated circuits, and electronic components and parts. Japan and the Far East countries, especially the Four Tigers of South Korea, Taiwan, Hong Kong, and Singapore, but including the 14 Pacific Basin nations, together account for the predominant flow of microelectronics in the world (Figure 11.35).

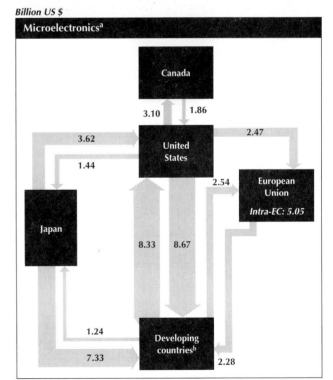

[a]Including integrated circuits and IC parts, semiconductors, and electronic component parts.
[b]Including Western hemisphere other than the United States and Canada; Africa and the Middle East; South Asia; and East Asia other than Japan, Mongolia, North Korea, Laos, and Cambodia.
Note: Width of arrows scaled to dollar volume. Trade flows less than $1 billion not shown.

FIGURE 11.35
Global trade flows of microelectronics, 1994, in billions of U.S. dollars. *(Source: Data from United Nations, 1994)*

The single largest flow from this group is from the developing countries of the Western Hemisphere, other than the United States and Canada, most notably Mexico, and the countries in South Asia and East Asia, which send more than $8.33 billion worth of microelectronics to the United States. Japan sends another $7.33 billion worth of microelectronics to developing countries and more than $3.62 billion worth to the United States. The single largest flow of microelectronics in the world is from the United States to developing countries and is worth $8.67 billion.

Although the United States no longer leads the world in the manufacture of semiconductors, it is still a major player in the global trade flow of microelectronics. Canada and the western European nations of the EU and the European Free Trade Area (EFTA) account for a much smaller proportion of overseas trade in this category. However, intra-European trade in microelectronics accounts for $5 billion.

Global Trade Flow of Automobiles

Global trade within the EU accounts for the largest single flow of automobiles. More than $72 billion worth of automobiles were shipped among EU countries in 1990, and more than another $8.5 billion worth were sent to EFTA countries (Figure 11.36). The United States imported $3.8 billion worth as well, including Mercedes, Audis, Porsches, Volkswagens, Peugeots, Fiats, and Renaults. If we consider the value and production of automobiles in Europe, it is surprising that halfway around the world Japan made a major inroad into the European market, shipping $8.5 billion worth to Western Europe in 1992. The single largest volume of flow is also accounted for by Japan and its shipment to the United States, $21 billion worth in 1997. Eastern Europe and the U.S.S.R. accounted for a minuscule amount of global trade flow in automobiles.

Japanese automobile manufacturers made major penetrations in the world automobile market between 1960 and 1994 (Figure 11.37). During the same time, the big three automakers in America scaled down operations substantially. However, General Motors and

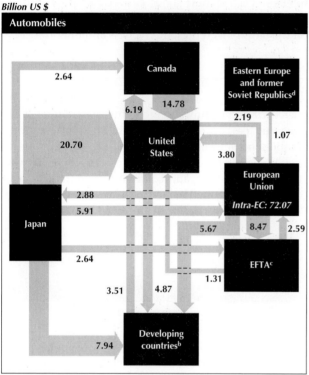

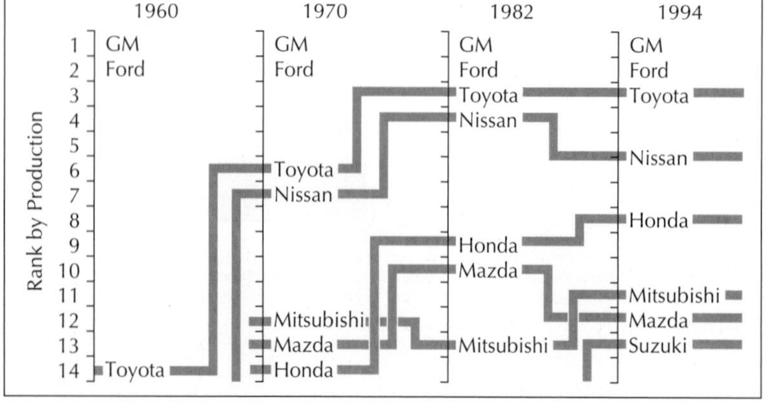

Billion US $

Including Austria, Switzerland, Sweden, Norway, Finland, Iceland, and Liechtenstein.
Including Poland, Hungary, Czech Republic, Slovakia, Albania, Romania, Bulgaria, and former Soviet Republics.
Note: Width of arrows scaled to dollar volume. Trade flows less than $1 billion not shown.

FIGURE 11.36
Global trade flow of automobiles, 1994, in billions of U.S. dollars. *(Source: Data from United Nations, 1994)*

FIGURE 11.37
The rise of Japanese automobile manufacturers, 1960–1994. *(Source: World Motor Vehicle Data, 1996)*

🔲

T a b l e 11.8
Top Automobile Manufacturers Worldwide, 1990

Rank	Company	Country of Origin	Passenger Car Production*	Share of World Total (%)	Percentage Produced Outside Home Country
1	General Motors	USA	5,523,134	15.6	41.8
2	Ford	USA	4,060,586	11.5	58.7
3	Toyota	Japan	3,330,380	9.4	8.3
4	Volkswagen	Germany	2,713,671	7.7	30.5
5	Peugeot-Citroën	France	2,320,266	6.5	15.4
6	Nissan	Japan	2,289,123	6.5	13.8
7	Fiat	Italy	2,108,622	6.0	7.1
8	Renault	France	1,755,510	5.0	17.6
9	Honda	Japan	1,604,430	4.5	28.0
10	Mazda	Japan	1,184,166	3.3	18.3
11	Chrysler	USA	1,052,537	3.0	13.0
12	VAZ	USSR	724,740†	2.0	—
13	Mitsubishi	Japan	708,418	2.0	—
14	Daimler-Benz	Germany	536,993	1.5	—
15	BMW	Germany	489,742	1.4	—
16	Rover	UK	466,619	1.3	—
17	Volvo	Sweden	423,385	1.2	33.8

*Excludes commercial vehicles.
†1987.
Source: Calculated from SMMT, 1990, pp. 28–33.

Ford remained the world's largest automobile manufacturers by 1990 (Table 11.8), even though the Japanese had captured almost 26% of the American automobile and light truck market. In 1990, the American competitors appeared to be losing out to the high-quality, beautifully engineered subcompacts and luxury models from Japan. But then, slowly, between 1990 and 1994, the market share situation in the United States began to reverse itself. By 1993, Japan's share was 22%, and by 1994, 20%.

The problem for Japan was the increased value of the yen compared with the dollar. As we learned in Chapter 10, the dollar is now weak on world international markets. In contrast, the yen has appreciated more than 16% in each of the last 2 years and now stands at approximately 100 to the dollar. The high value of the yen adversely affects Japanese sales in America because comparable vehicles are now priced $2,000 more than American vehicles.

Another factor is that American car builders, after years of struggle, are finally turning out higher-quality competitive products with a flair for style not seen since the 1960s. The Honda Accord has fallen from grace in

America because, analysts believe, the styling is too bland and conservative. Public taste has changed. Porter (1990) described demand conditions and taste as an important factor in global comparative advantage. American consumers are bypassing ordinary passenger cars for light trucks, minivans, four-wheel drives, and sporty utility vehicles. This preference gives American companies a strong edge because Japan is just entering these other product lines.

The final reason for the reversal is that some consumers look at the hostile relationship between America and Japan and choose to buy American. Only two of the top 10 cars and trucks sold in America were imported from Japan in 1993. Honda's compact-sized Accord was the nation's best-selling import car for several years. However, the Ford Taurus supplanted it in the 1993 list of 10 best-selling cars and light trucks in America (Table 11.9).

The situation for Japanese MNCs, Toyota, Nissan, Mitsubishi, Mazda, and Honda, would be much worse if it had not been for political pressure to build U.S. factories, starting with the Honda plant in 1982 in Marysville, Ohio (Table 11.10). Japanese-made vehicles

T a b l e 11.9
Top United States Vehicle Sellers, Second Half of 1996

	Type of Vehicle	1996 Vehicle Sales	Rank 1996
1	Ford F-series pickup	263,574	1
2	Chevrolet C/K pickup	258,234	2
3	Ford Taurus	173,819	4
4	Ford Ranger	165,136	9
5	Toyota Camry	149,499	7
6	Chevrolet Cavalier	146,612	11
7	Dodge Caravan	145,442	8
8	Ford Explorer	145,270	3
9	Ford Escort	141,816	10
10	Honda Accord	133,005	6
11	Dodge Ram pickup	122,059	5

Source: Automotive News Data Center, Detroit, Michigan, 1996.

that are assembled and built in America account for nearly half the Japanese–U.S. car sales. Locally made Japanese cars are cheaper than those imported from Japan because they escape the high-priced parts and labor associated with the yen. Nissan's newest plants are built in low-wage Smyrna, Tennessee, producing Sentras and Ultimas. These American-made Japanese cars are called *Japanese transplants*. Transplant sales are down.

Exports of automobiles from major European producing countries have been extremely heavy. Although France produces innocuously designed cars—the Peugeot, Renault, and Citroën—it is second in Western Europe in terms of passenger car exports, with Germany number one at 2,451,251 vehicles in 1989 and Belgium and Spain third and fourth. U.S. companies have diversified, so both Ford and General Motors now have substantial international automobile production (Figure 11.38).

Global Trade Flow of Steel

Whereas America has lost as much as two-thirds of its steelworker employment in the last 20 years and now

T a b l e 11.10
Final Assembly Foreign Automobile Plants in North America

Company	Plant Location	Date Opened	Current Planned Capacity	Projected Plant Employment
United States				
BMW	Greer, So. Carolina	1996	150,000	1000
Honda	Marysville, Ohio	1982	360,000	2600
	East Liberty, Ohio	1989	150,000	1800
Nissan	Smyrna, Tennessee	1983	480,000	3300
	Decherd, Tennessee	1996	300,000 engines	1000
Toyota/GM (NUMMI)	Fremont, California	1984	340,000	2500
Toyota	Georgetown, Kentucky	1988	400,000	3000
Mazda	Flat Rock, Michigan	1987	240,000	3500
Mercedes Benz	Vance, Alabama	1997	50,000	600
Mitsubishi/Chrysler (Diamond Star)	Normal, Illinois	1988	240,000	2900
Subaru/Isuzu	Lafayette, Indiana	1989	120,000	1700
Canada				
Honda	Alliston, Ontario	1987	100,000	700
Hyundai	Bromont, Quebec	1989	20,000	400
Toyota	Cambridge, Ontario	1988	50,000	1000
Suzuki/GM (CAMI)	Ingersoll, Ontario	1989	200,000	2000
Volvo	Halifax, Nova Scotia	1990	40,000	300

Source: Press and company reports, and Dicken, 1992, p. 294.

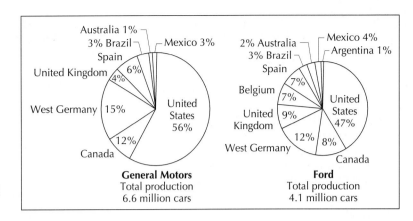

FIGURE 11.38
International automobile production by General Motors and Ford, 1990.

is a net importer of steel, Western Europe continues to lead the world in steel production and in the steel trade. The EU accounted for more than $31 billion worth of steel traded internally in 1992, and trade with the EFTA accounted for another $10 billion worth (Figure 11.39).

The single largest flow of steel was from Japan to developing countries. In addition, the EEC sent $6.68 billion worth of steel to developing countries in 1992. Steel requires large and highly efficient plants, which are possible only with tremendous capital investments and large-scale economies. Africa, for example, has only a tiny portion of world steelmaking, less than 1% of the world's total output.

In the post–World War II period, steel made by traditional producers in Europe and North America became overpriced. New production centers began to emerge in Brazil, South Korea, Taiwan, and Japan. The migration of steel production to the Third World reflected the growing importance of labor costs, government subsidies, and taxes to the delivered cost of steel. In Chapters 10 and 11, we discussed the British and U.S. steel industries: too many employees, reluctant unions, demoralized management, inefficiency, and lack of government support of an ailing industry. Competition has made the international steel industry highly mobile. In the past, competition was national; now it is global.

Global Trade Flow of Textiles and Clothing

As discussed in Chapter 9, labor-intensive textile and clothing manufacture has dramatically shifted to developing countries (Figure 11.40). These countries include South American and Central American countries; Africa and the Middle East; South Asia; and East Asia other than Japan, Mongolia, North Korea, Laos, and Cambodia. Major global shifts have occurred in textile production in the last 40 years with a decline in the

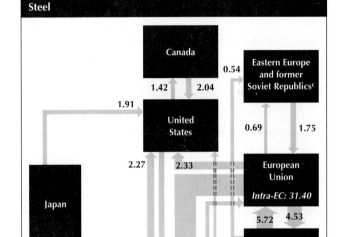

[a]Including Western Hemisphere other than the United States and Canada; Africa and the Middle East; South Asia; and East Asia other than Japan, Mongolia, North Korea, Laos, and Cambodia.
[b]Including Austria, Switzerland, Sweden, Norway, Finland, Iceland, and Liechtenstein.
[c]Including Poland, Hungary, Albania, Romania, Bulgaria, Czech Republic, Slovakia, and former Soviet Republics.
Note: Width of arrows scaled to dollar volume. Trade flows less than $500 million not shown.

FIGURE 11.39
Global trade flow of steel, 1994, in billions of U.S. dollars.
(Source: United Nations, 1994)

United States and Western Europe, which were the dominant producers as late as 1950. New production centers include the Far East, India, and China. International trade in textiles reflects these shifts in production. Developing countries accounted for more than $78 billion worth of exports in 1992.

Surprisingly, Germany and Italy lead the world in textile export, even though the main manufacturing centers of older industrialized nations have given ground to the Third World in textile and clothing manufacture. Tiny Hong Kong leads the world in clothing export. Some of the top six leading world exporters are from East Asia and the Pacific (Table 11.11). Again, Western Europe accounts for more than $70 billion worth of textile and clothing trade among nations within Western Europe (EU and EFTA).

Major gainers during the last 30 years include the East Asian countries of China, Hong Kong, South Korea, and Taiwan. By 1990, European nations accounted for 47% of the world trade flow, whereas East Asian countries accounted for another 43%. However, a much larger proportion of textiles and clothing flowed from developing countries to the United States than they did from Western Europe to the United States. Eastern Europe, the former Soviet Union, Japan, and Canada are relatively small players in the world textile and clothing trade.

Global Trade Flow of Grains and Feed

The primary products of wheat, corn, rice, other cereals, feed grains, and soybeans are included in the category of grains and feed. The United States exports more than $20 billion worth of feed and grains and thus is a world leader in this category (Figure 11.41). Considering its huge area, Canada is also a major exporter.

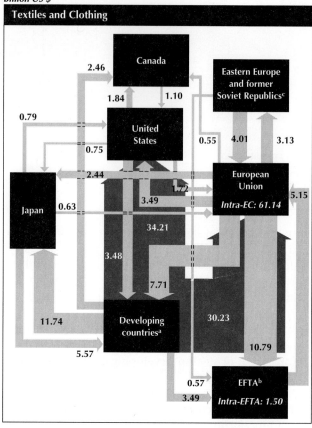

Billion US $

aIncluding Western Hemisphere other than the United States and Canada; Africa and the Middle East; South Asia; and East Asia other than Japan, Mongolia, North Korea, Laos, and Cambodia.
bIncluding Austria, Switzerland, Sweden, Norway, Finland, Iceland, and Liechtenstein.
cIncluding Poland, Hungary, Albania, Romania, Bulgaria, Czech Republic, Slovakia, and former Soviet Republics.
Note: Width of arrows scaled to dollar volume. Trade flows less than $500 million not shown.

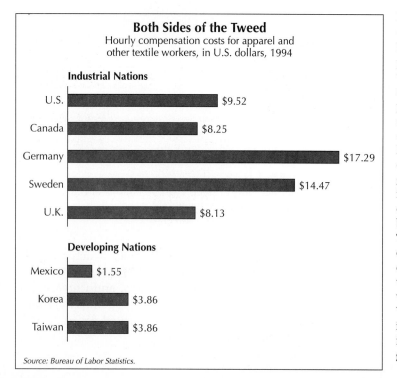

Both Sides of the Tweed
Hourly compensation costs for apparel and other textile workers, in U.S. dollars, 1994

Industrial Nations

U.S.	$9.52
Canada	$8.25
Germany	$17.29
Sweden	$14.47
U.K.	$8.13

Developing Nations

Mexico	$1.55
Korea	$3.86
Taiwan	$3.86

Source: Bureau of Labor Statistics.

FIGURE 11.40

Global trade flow of textiles and clothing, 1996, in billions of U.S. dollars. Canadian and U.S. productivity is growing every year. That means that we can produce more goods and services with the same amount of labor hours. Which means that people should be able to buy more goods and services—or more leisure—each year. But they can't, or at least the majority of them can't. Median wages have actually declined over the past 24 years. That one statistic speaks volumes about the net benefits of globalization. The globalizers claim that cheaper imported goods will more than compensate for the disruption and downward pressure on wages caused by increasing international trade and investment. This trade includes, of course, what is produced overseas by U.S. and Canadian multinational corporations like Nike, who pay $1.60 a day to their workers in Vietnam. But it hasn't worked out that way. A declining real wage means exactly that: Whatever benefit we have gotten from cheaper imported goods has been marginalized by other forces, including runaway factories and increased global competition. *(Source: United Nations, 1996)*

◻

Table 11.11

The World's Leading Clothing-Exporting Countries, 1991

Rank	Exporter	Share of World Clothing Exports (%)		Average Annual Growth (%)		
		1980	1991	1980–1988	1988	1989
1	Hong Kong*	12.0	14.5	11.5	10.0	18.5
2	Italy	11.0	9.5	9.0	1.0	4.0
3	South Korea	7.0	9.5	14.5	17.5	4.5
4	China	4.0	6.5	14.0	30.0	26.0
5	West Germany	7.0	5.5	8.0	7.5	5.0
6	Taiwan	6.0	5.0	8.5	−5.5	0.5
7	France	5.5	3.5	4.5	8.0	10.0
8	Turkey	0.5	3.0	43.5	7.0	18.0
9	Portugal	1.5	2.5	17.5	11.5	12.6
10	United Kingdom	4.5	2.5	3.5	7.5	−6.0
11	Thailand	0.5	2.5	27.0	21.5	28.0
12	United States	3.0	2.5	3.5	36.0	34.5
13	India	1.5	2.0	13.0	7.5	24.5
14	Netherlands	2.0	1.5	7.0	9.5	4.0
15	Greece	1.0	1.5	15.0	−16.5	22.5
Above 15 countries		67.0	72.0			

*Includes substantial reexports.

Source: GATT, 1992, p. 21.

Japan, with its small base of agriculture and arable land, is a net importer, as is Eastern Europe and the former Soviet Union. Trade within the EU is large and was $15.15 billion in 1992. Developing countries such as India, Egypt, and Argentina are some of the largest net exporting developing countries in this category.

World trade in grains, feeds, and food products had been as high as 30% in 1965, but these commodities slipped to only 15% by 1992. Some of this reduction is because Western seeds, grains, and fertilizers are now commonplace in Third World nations, and technology from the grain revolution has taught developing countries to provide for themselves. Another portion of the reduction from 1965 to 1992 reflects the worsening terms of trade (Chapter 12) for primary goods, as prices of manufactures and energy rose rapidly and gave producers of feed, grains, and agricultural products less leverage in world international commerce. The 1973 and 1978 Middle East oil crises and the sharp increase in oil prices in 1979 signaled the beginning of the falling prices of such energy-intensive products.

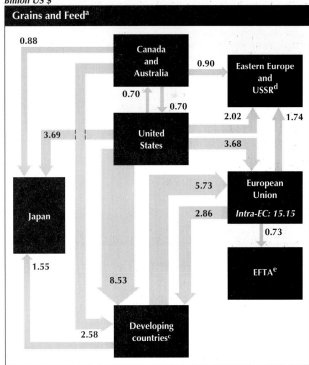

[a]Including wheat, corn, rice, other cereals, feedgrains, and soybeans.
[b]Including unwrought copper, aluminum, nickel, zinc, and tin, and ores thereof; iron ore and pig iron; uranium ore and alloys; other nonferrous ores, unwrought metals, and crude minerals; crude rubber, wood and pulp; hides; cotton fiber and other textile fibers; crude animal and vegetable materials.
[c]Including Western Hemisphere other than the United States and Canada; Africa and the Middle East; South Asia; and East Asia other than Japan, Mongolia, North Korea, Laos, and Cambodia.
[d]Including Poland, Hungary, Czech Republic, Slovakia, Albania, Romania, Bulgaria, and former Soviet Republics.
[e]Including Austria, Switzerland, Sweden, Norway, Finalnd, Iceland, and Liechtenstein.
Note: Width of arrows scaled to dollar volume. Trade flows less than $500 million not shown for grains/feed; trade flows less than $2 billion not shown for nonoil commodities.

FIGURE 11.41

Global trade flow of grains and feed, 1994, in billions of U.S. dollars. *(Source: United Nations, 1994)*

Billion US $

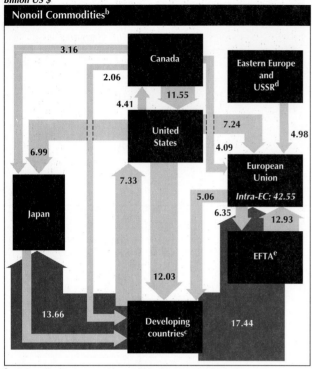

Nonoil Commodities[b]

3.16 · 2.06 · 4.41 · 11.55 · 6.99 · 7.33 · 7.24 · 4.98 · 4.09 · 5.06 · 6.35 · 12.93 · 12.03 · 13.66 · 17.44 · Intra-EC: 42.55

Canada · Eastern Europe and USSR[d] · United States · European Union · Japan · EFTA[e] · Developing countries[c]

[a]Including wheat, corn, rice, other cereals, feedgrains, and soybeans.

[b]Including unwrought copper, aluminum, nickel, zinc, and tin, and ores thereof; iron ore and pig iron; uranium ore and alloys; other nonferrous ores, unwrought metals, and crude minerals; crude rubber, wood and pulp; hides; cotton fiber and other textile fibers; crude animal and vegetable materials.

[c]Including Western Hemisphere other than the United States and Canada; Africa and the Middle East; South Asia; and East Asia other than Japan, Mongolia, North Korea, Laos, and Cambodia.

[d]Including Poland, Hungary, Czech Republic, Slovakia, Albania, Romania, Bulgaria, and former Soviet Republics.

[e]Including Austria, Switzerland, Sweden, Norway, Finalnd, Iceland, and Liechtenstein.

Note: Width of arrows scaled to dollar volume. Trade flows less than $500 million not shown for grains/feed; trade flows less than $2 billion not shown for nonoil commodities.

FIGURE 11.42
Global trade flow of nonoil commodities, 1994, in billions of U.S. dollars. *(Source: United Nations, 1994)*

Global Trade Flow of Nonoil Commodities

Nonoil commodities include copper, aluminum, nickel, zinc, tin, iron ore, pig iron, uranium ore, and alloys. Crude rubber, wood and pulp, hides, cotton fiber, and animal and vegetable minerals and oils are included in this category. Although the United States, for a highly developed country, shows an amazing export potential, sending $7 billion worth to Japan, $12 billion worth to developing countries, and another $7.24 billion worth to EU countries, the developing countries of the world lead in the export of nonoil commodities (Figure 11.42). The largest single flow of nonoil commodities outside the more than $42 billion worth exchanged within the EU is from developing countries to Japan ($13.66 billion) and from developing countries to Western Europe ($17.44 billion).

Since the early days of Europe, the international division of labor was based on international trade. Under this unfair program, less developed countries traded their nonoil commodities—grains, feed, food stocks, and energy sources—for industrialized goods from primarily Europe and other developed nations. This pattern still existed in 1992 but had weakened somewhat with the internationalization of manufactures.

Combined grain is off-loaded on a farm near Wataga, Illinois. The United States is a world supplier of primary products, including grains, timber, and other agricultural products. Most every country in the world grows grain, and approximately a quarter of the 600 million tons of wheat produced each year enters world trade. By 1960, overpopulation in the Third World had begun to outstrip its own food production, and Europe had long been a grain importer by then. The United States and Canada became the only large grain exporters, providing 80% of the world's exports. No doubt the Richard Nixon grain deals with the Soviet Union countries in the mid-1970s helped establish detente and ease tensions during the Cold War. *(Photo: Martha Tabor, Impact Visuals Photo & Graphics, Inc.)*

KEY TERMS

capital resources
demand conditions
factor conditions
factor-driven stage
Five Little Dragons
Four Tigers
General Agreement on Tariffs and
    Trade (GATT)

human resources
infrastructure
innovation-driven stage
investment-driven stage
Japanese transplants
knowledge-based resources
Little Dragons
microelectronics

physical resources
purchasing power parity
Organization of Economic Cooperation
    and Development (OECD)
supporting industries
wealth-driven stage

SUGGESTED READINGS

Berry, B. J. L., Conkling, E. C., and Ray, D. M. 1997. *Global Economy.* Upper Saddle River, N.J.: Prentice Hall.

Brett, E. A. 1985. *The World Economy Since the War.* London: Macmillan.

Corbridge, S. 1993. *Debt and Development.* Oxford: Blackwell.

Daniels, P. W. 1996. *The Global Economy in Transition.* White Plains, N.Y.: Longman.

Daniels, P. W. 1993. *Service Industries in the World Economy.* Series: Institute of British Geographers Studies in Geography. Cambridge, Mass.: Blackwell Publishers.

Dicken, P. 1992. *Global Shift: The Internationalization of Economic Activity.* 2d ed. New York: Guilford Press.

Dicken, P., and Lloyd, P. E. 1990. *Location in Space: Theoretical Perspectives in Economic Geography.* 3d ed. New York: Harper & Row.

Ettlinger, N. 1991. The roots of competitive advantage in California and Japan. *Annals of the Association of American Geographers* 81, no. 3:391–407.

Freeman, M. 1991. *Atlas of the World Economy.* New York: Simon & Schuster.

Janelle, D. G., ed. 1992. *Geographical Snapshots of North America.* New York: Guilford.

Korth, C. M. 1985. *International Business: Environment and Management,* 2d ed. Englewood Cliffs, N.J.: Prentice Hall.

Krugman, P. 1992. *The Age of Diminished Expectations.* Cambridge, Mass.: MIT Press.

Momsen, J. H.. 1991. *Women and Development in the Third World.* London: Routledge.

Porter, D., Allen, B., and Thompson, G. 1991. *Development in Practice: Paved with Good Intentions.* London and New York: Routledge.

Porter, M. E. 1990. *The Competitive Advantage of Nations.* New York: Free Press.

Root, F. R. 1990. *International Trade and Investment,* 6th ed. Cincinnati: South-Western Publishing.

Thurow, L. 1992. *Head to Head: The Coming Economic Battle Among Japan, Europe, and America.* New York: William Morrow.

Wallace, I. 1990. *The Global Economic System.* London: Unwin Hyman.

Wallerstein, I. 1991. *Geopolitics and Geoculture: Essays on the Changing World-System.* Cambridge: Cambridge University Press.

WORLD WIDE WEB SITES

TRADE, STATISTICS, MAPS
http://lcweb.loc.gov/homepage/lchp.html
Library of Congress, links to other libraries & online catalogs.
http://econwpa.wustl.edu/EconFAQ/EconFAQ.html
Resources for Economists on the Internet.

SPACESTATE—SOFTWARE FOR SPATIAL DATA ANALYSIS
http://lambik2.rri.wvu.edu/spacestat
http://www.geom.umn.edu/docs.snell/chance/ welcome.html
http://www.worldbank.org/
World Bank

http://www.lib.virginia.edu/gic
GIS and other map generation; links to NCGIA, University geography departments, government agencies, environmental resources, and digital electronic maps.

COMPLETE COUNTRY INFORMATION FROM THE UNITED NATIONS
http://www.un.org/pubs/cyberSchoolBus/infonation/

TRADE STATISTICS
http://www.census.gov/indicator/www/ustrade.html
http://www.odci.gov/cia/publications/nsolo/factbook/so.html

C H A P T E R

1 2

DEVELOPMENT

OBJECTIVES

- To outline the goals of development
- To acquaint you with the attributes of less developed nations
- To examine major theories and perspectives on development
- To examine the causes of Soviet collapse and the formidable transition process

Collecting water from a well in Ethiopia.
(Photo: World Bank)

The modern world has its origin in the European societies of the late 15th and early 16th centuries. One of its most striking characteristics is the division between rich and poor countries. Early on, this division was achieved through an international system in which the wealthy minority industrialized, using primary products produced by the impoverished majority. More recently, as we have seen in Chapters 9 through 11, this original division of labor gave way to a new division of labor. Now, the wealthy minority are increasingly engaged in office work and the masses in hands-on manufacturing jobs on the global assembly line as well as in agriculture and raw-material production. The creation of today's world with a rich core and a poor periphery was not the result of conspiracy among developed countries, but of an "invisible hand." It was the outcome of a systemic process—that process by which the world's political economy functions.

This chapter deals with how this world of unequal development came about, how present structures are the result of the past. We begin with a discussion of the characteristics of LDCs and of some frequently propounded views on the nature of development. Goals for development are introduced, and development objectives that most people would subscribe to are listed.

WHAT'S IN A WORD? "DEVELOPING"

From Primitive to Underdeveloped

If the United States, with its high level of material consumption, is described as a developed country, then what adjective should we use to describe the poor countries of the world? Certainly, there are many from which to choose. In the past half-century, each of the following terms has flourished in succession: primitive, backward, undeveloped, underdeveloped, less developed, emerging, developing, and rapidly developing. Today, Western social scientists use the word *developing* and, increasingly, the phrase *less developed countries* (LDCs), but social scientists in the Marxist tradition favor the term *underdeveloped*.

Underdeveloped was formerly used by Western social scientists to describe situations in which resources were not yet developed. People and resources were seen as existing, respectively, in a traditional and natural state. Scholars in the Marxist tradition now use *underdeveloped* to describe not an initial state, but rather a condition arrived at through the agency of imperialism, which set up the inequality of political and economic dependence of poor countries on rich countries. Thus, instead of viewing underdevelopment as an initial or *passive state*, Marxists view it as an *active process* (Rodney, 1972).

THE GOALS OF DEVELOPMENT

In his plea for a definition of development based on human well-being, Dudley Seers (1972) asked, "Why do we confuse development with economic growth? . . . Development means the condition for the realization of the human personality. Its evaluation must therefore take into account three linked economic criteria: Whether there has been a reduction in (i) poverty; (ii) unemployment; (iii) inequality" (p. 21). He pointed out that some countries have experienced not only rapid growth of per capita income, but also increases in poverty, unemployment, and inequality. He urged measures of development at the family level based on nutrition, health, infant mortality, access to education, and political participation.

Seers (1972) cited works indicating that during the United Nations Development Decade of the 1960s "the growth of economic inequality and unemployment may actually have accelerated" (p. 34). For example, in India, the much heralded Green Revolution, which depends on fertilizer and water inputs, has benefited mainly the farmers in the Punjab who were already wealthy and who owned large tracts of land (Wharton, 1969). Seers (1972) also urged the use of measures that indicate a degree of national independence. Among them, "the proportion of capital inflows in exchange receipts, the proportion of the supply of capital goods (or intermediates) which is imported, the proportion of assets, especially subsoil assets owned by foreigners, and the extent to which one trading partner dominates the pattern of aid and trade" (p. 30).

A related view of development goals was expressed by Denis Goulet (1971), who echoed Seers's concern: "There may be considerable merit . . . in asking whether higher living standards, self-sustained growth, and modern institutions are good in themselves or necessarily constitute the highest priorities" (p. 85). Goulet argued for three general development goals: life sustenance, esteem, and freedom.

LIFE SUSTENANCE
There can be no dispute that food, health, adequate shelter, and protection are essential to human well-being. When they are sufficient to meet human needs, a state of development exists; when they are insufficient, a degree of underdevelopment prevails.

ESTEEM
All people value respect. The feeling that "one is being treated as an individual who has worth, rather than as a tool for the satisfaction of other individuals' purposes, is . . . the basic source of human contentment" (Hagen, 1968, pp. 411–412). Esteem or recognition is closely associated with material prosperity. Consequently, it is often difficult for those who are materially deprived or

African children watching television. Although literacy levels increased between 1970 and 1990 in all but one country, literacy rates for women are still below 50% in 45 of the countries for which data are available. Literacy rates for men are below 50% in only 17 countries. *(Photo: Raccah, United Nations)*

Freedom of expression and achievement of a humane lifestyle are ultimate, essentially unresolvable issues. More down-to-earth development goals include the following:

1. a balanced, healthful diet
2. adequate medical care
3. environmental sanitation and disease control
4. labor opportunities commensurate with individual talents
5. sufficient educational opportunities
6. individual freedom of conscience and freedom from fear
7. decent housing
8. economic activities in harmony with the natural environment
9. social and political milieus promoting equality

In conventional usage, *development* is a synonym for economic growth. But growth is not development, except insofar as it enables a country to achieve the nine goals. If these goals are not the objectives of development, if modernization is merely a process of technological diffusion, and if spatial integration of world power and world economy is devoid of human referents, then *development* should be redefined. The realities of the contemporary world, however, do not offer much hope for achieving these humane objectives anytime soon.

CHARACTERISTICS OF LESS DEVELOPED COUNTRIES

Rapid Population Growth

Can we ascribe the problems of LDCs to rapid population growth? Yes, to a degree. After all, the present rapid increase of population is most apparent in developing countries, many of which have average annual population growth rates of at least 2.5%. It seems there are just too many millions of people in developing countries who must be fed, housed, clothed, educated, employed, cared for in illness, and, finally, looked after in old age. Most people would argue that these populations should be controlled if development is to take place.

Table 12.1 shows population statistics for various countries. The most striking dimension is the *annual population increase*. Keep in mind that a steady 2% annual population increase will double the population in 35 years. A 3% annual population increase means that the population will double in about 17 years. Population *densities* are also high in LDCs. These statistics show why such a large income gap exists between LDCs and industrially advanced countries (IACs) (see Table 12.2).

In many LDCs, the rapid population growth rates tax the food supplies. Because more mouths must be fed,

"underdeveloped" to experience a sense of pride or self-worth. Mass poverty prevents people and societies from receiving due recognition or esteem. These people may even reject development. For example, if people are humiliated or disillusioned through their contacts with the "progress" introduced by foreigners, they may return to their traditional ways in order to regain a sense of self-respect.

FREEDOM

Freedom can be defined as "the capacity, the opportunity to develop and express one's potentialities" (Warwick, 1968, p. 498). As with life sustenance and esteem, the degree to which freedom exists in a society can be used to assess development.

Donald Warwick (1968) asked students of development to "devote explicit attention to the values used in gauging progress" of individuals or societies (p. 498). Moreover, he remarked that it is not useful to define development in terms of urbanization, commercialization, industrialization, or modernization. Instead, he advised that development be viewed as a coordinated series of changes from a phase of life perceived by a population as being less human to a phase perceived as being more human.

Table 12.1

Population Density and Increase, 1980–1990

	Population per Square Mile, 1990	Annual Role of Population Increase, 1980–1990
United States	69	0.8
Bangladesh	2130	3.0
Venezuela	56	2.7
Haiti	573	1.2
Kenya	110	3.9
Pakistan	369	3.0
Philippines	570	2.6
India	669	2.1
World	101	1.7

less food is available per person, and some LDCs approach levels of subsistence and starvation. Perhaps the standard of living could be raised by producing more consumer goods and food in particular. But any increase in consumer goods and food production is likely to increase the population first, before it increases the standard of living. One reason for this is that the death rate, or mortality rate, declines with increased production. Higher per capita food consumption leads to better chances of survival and longer life spans. In addition, medical and sanitation improvements usually accompany the early stages of increased productivity and reduce the infant mortality rate. Kenya is a prime example of this phenomenon. Consequently, the relationship between a population increase and an increase in goods and food is a complex one. In the long run, a population increase will accompany an increase in the standard of living and will probably stop only when the standard of living again plunges to the bare subsistence level.

Rapid population growth also reduces the ability of households to save; therefore, the economy cannot accumulate investment capital. In addition, with rapid population growth, more investment is required by the government to maintain a level of real capital per person. If government investment fails to keep pace with the population growth, each worker will be less productive, having fewer tools and equipment with which to produce goods. This declining productivity results in reduced per capita incomes and economic stagnation. (See Chapter 1: Vicious Cycle of Poverty.)

Rapid population growth in agriculturally dependent countries, such as China and India, means that the land must be further subdivided and used more heavily than ever. Smaller plots from subdivision inevitably lead to overgrazing, overplanting, and the pressing need to increase food production for a growing population from a limited amount of space.

Many LDCs are rapidly urbanizing. Rapid population growth means large flows of rural farmers to urban areas and more urban problems. Housing, congestion, pollution, crime, and lack of medical attention are all seriously worsened by the rapid urban population growth.

Assuredly, a rapid increase in population—especially the number and proportion of young dependents—creates serious problems in terms of food supply, public education, and health and social services; it also intensifies the employment problem. However, a high rate of population growth was once a characteristic of present-day developed countries, and it did not prevent their development. This observation makes it difficult to argue that population growth necessarily leads to underdevelopment or that population control necessarily aids development.

Table 12.2

GNP Per Capita, Population, and Growth Rates

	GNP Per Capita		Population	
	Dollars 1996	Annual Growth Rate (%) 1965–1996	Millions, 1996	Annual Growth Rate (%) 1980–1996
Industrially advanced countries IACs (19 nations)	21,330	2.4	1071	0.1
Less developed countries LDCs (97 nations)				
Middle income LDCs (56 nations)	2040	5.3	1505	1.4
Low income LDCs (41 nations)	330	2.9	3448	2.0

Source: World Bank, World Development Report, 1997.

Unemployment and Underemployment

Are the problems of LDCs a product of unemployment and underemployment? Unemployment and underemployment are major problems in LDCs. *Unemployment* is a condition in which people who want to work cannot find jobs. *Underemployment* means that those people who are working are not able to work as many hours as they would like, usually much less than 8 hours per day. Reliable statistics on unemployment and underemployment in LDCs are difficult to obtain, but many scientists suggest that unemployment in these countries is approximately 20%.

Many of the cities in LDCs have recently experienced rapid flows of migrants from rural areas as a result of poor agricultural output and lack of land reform. This large number of migrants is created by the expectation of jobs and higher salaries in the cities. However, the cities usually have much higher unemployment rates than their rural counterparts. Once in the cities, many of the migrants cannot find work and contribute to the unemployment situation. Other migrants find limited amounts of work as shop clerks, handicraft peddlers, or street vendors. If you visit an LDC, such as Mexico, you can easily see that idleness in the shop or on the street occupies a larger amount of time than work does. Although these people are not without jobs, they are without jobs that require substantial work, and therefore they are underemployed.

Certainly, unemployment and underemployment do not lend themselves to development. However, they are not the sole reasons for the problems of LDCs.

Low Labor Productivity

Are the problems of LDCs a result of low labor productivity? It is true that a day's toil in a developing country produces very little compared with a day's work in a developed country. This is particularly evident in agriculture. American farmers spread their labor over 30 to 60 hectares; African or Asian farmers pour their energies into a hectare or so. As a consequence, human productivity in a developing country may be as little as one-fiftieth that in a developed country. Why is this?

The populations of LDCs are not equipped for high productivity for several reasons. One reason is the small scale of operations. Another is that physical capital investment is extremely low. Rapid population growth has reduced the amount of investment available to maintain productivity. Most developing countries lack the machines, engines, power lines, and factories that enable people and resources to produce more than is possible with bare hands and simple tools. In addition, LDCs are less able to invest in *human capital* (Table 12.3). Investments in human capital—such as education, health, and other social services, including provision of food—prepare a population to be productive workers. When an LDC lacks human capital, productivity is low.

Low labor productivity in LDCs is exacerbated by a lack of organizational skills and the absence of a management class, both of which are necessary for increased productivity. Workers have low skill levels, and few are able to handle supervisory jobs. Many of the most intelligent workers immigrate to IACs, where they are

Table 12.3
Selected Socioeconomic Indicators of Development, 1996

Country	Per Capita GNP ($)	Life Expectancy at Birth in Years	Infant Mortality per 1000 Live Births	Adult Literacy Rate (%)	Daily per Capita Calorie Supply	Per Capita Energy Consumption*
Japan	34,810	80	4	99	2848	3484
United States	25,910	76	7.5	99	3666	7794
Canada	19,570	78	6.2	49	3500	7152
Brazil	3,540	66	58	78	2709	897
Mauritania	500	46	123	17	2528	114
Haiti	220	57	74	15	1911	51
India	340	59	95	43	2104	226
Bangladesh	230	51	88	33	1925	51
Ethiopia	130	50	123	5	1658	20
Mozambique	80	46	147	28	1632	84

*Kilograms of oil equivalent.
Source: World Bank, World Development Report, 1997, and U.S. Bureau of the Census, 1997.

more highly paid for their labor. The United States and Europe, for example, are replete with some of the best-trained labor from LDCs, including doctors, engineers, mathematicians, and scientists who have come looking for more challenging work and better support services from companies or the government. This immigration has contributed to a so-called *brain drain*, whereby LDCs lose talented people to IACs. The U.S. immigration policy has encouraged the brain drain as well.

Although it must be acknowledged that low labor productivity is a universal attribute of LDCs, it is not a *causative* factor. The important question to consider is this: What prevents labor productivity from improving in developing countries?

Adverse Climate and Lack of Natural Resources

Can the problem of LDCs be traced to an adverse climate, insufficient rainfall, poor soils, and a lack of mineral resources? Yes, to a degree. Obviously, the uneven allocation of the gifts of nature makes development more difficult in some areas than in others. East Africa is a case in point. Large areas of East Africa have poor soils and little, unpredictable rainfall. The dry, wooded steppe in the rain shadow of Mount Meru, Tanzania, is on the arid margins of agriculture, and supports Masai pastoralists at low population densities. In contrast, large numbers of East African farmers live in better-watered areas—along the coast, near the lakes, and in the highlands. For example, in the Kigezi district of southwest Uganda, fertile volcanic soils and ample rainfall make agriculture highly productive on carefully terraced hillsides. In many densely populated areas of East Africa, the main problems are not environmental but economic and political.

Some LDCs have sizable natural-resource endowments of minerals such as bauxite, copper, tin, tungsten, nitrates, and petroleum. The Organization of Petroleum Exporting Countries (OPEC) is an example of LDCs that have used their resource endowments for rapid economic growth. In other cases, natural resources in LDCs are controlled by multinational corporations (MNCs), who divert their profits abroad to their home countries. Still other LDCs just lack mineral and petroleum deposits and have little arable land.

Limited natural resources can pose a serious developmental problem for LDCs. An LDC may be able to receive grants from more prosperous nations and to train workers to increase output, but increasing the supply of natural resources is impossible in most cases. With few natural resources, certain LDCs may be unwise if they expect to achieve development levels such as those of Germany and Japan. On the other hand, we must be careful not to entirely rule out the potential for high levels of development just because an LDC lacks natural resources. For example, Japan, Israel, Switzerland, Ireland, Norway, and Singapore achieved high levels of development despite a dearth of natural resources. Likewise, the Netherlands has, for centuries, obtained its necessary resources through various long-distance trading connections.

It is untenable to attribute development problems solely to a lack of resources or a poor climate. Vagaries of weather, exposure to natural disasters such as flood or drought, and danger of soil erosion are not exclusive to the developing world. Despite technological advances in North America, Europe, and the former Soviet Union, climate still poses recurring problems for farmers in these countries. Furthermore, as we said, although some developed countries have unfavorable natural-resource endowments, they were able to develop.

Lack of Capital and Investment

Is the situation in LDCs attributable to a lack of capital and investment? Most LDCs suffer from a lack of capital accumulation in the form of machinery, equipment, factories, public utilities, and infrastructure in general. The more capital, the more tools available for each worker, thus a close relationship exists between output per worker and per capita income. If a nation expects to increase its output, it must find ways to increase per capita income. Furthermore, an increase in the investment of an individual country will increase its gross domestic product (GDP).

In most cases, increasing the amount of arable land for an LDC is no longer a possibility. Most cultivable land is already in use, as we learned in Chapter 3. Therefore, capital accumulation for an LDC must come from savings and investment. If an LDC can save, rather than spend all its income, and invest some of its earnings, resources will be released from the production of consumer goods and be available for the production of capital goods, as we saw with the production possibilities curve analysis in Chapter 1. But barriers to saving and investing are high in LDCs. The United States has had notable problems in the recent past with savings and investing. An LDC has even less margin for savings and investing, particularly when domestic output is so low that all of it must be used to support the many needs of the country. Ethiopia, Bangladesh, Uganda, Haiti, and Madagascar save between 2% and 3% of their domestic outputs. In 1989, India and China managed to save an average of 27% of their domestic outputs, compared with 33% for Japan, 25% for West Germany, and 10% for the United States.

Many LDCs have suffered from *capital flight*, which means that individuals in these countries have invested and deposited their monies in overseas ventures and in banks in IACs for safekeeping. They have done so for fear of expropriation by politically unstable governments, future unfavorable exchange rates brought on by hyperinflation in the LDCs, high levels of taxation, and the possibility of business and bank failure. World Bank statistics for 1992 suggest that inflows of foreign aid and bank loans to LDCs were almost completely offset by capital flight. By 1990, for example, Mexicans were estimated to have held about $100 billion in assets abroad. This amount is roughly equal to their international debt. Venezuela, Argentina, and Brazil also have foreign holdings between $30 and $60 billion each.

Finally, investment obstacles in LDCs have impeded capital accumulation. The two main problems with investment in LDCs are (1) lack of investment opportunities comparable to those available in IACs and (2) lack of incentives to invest locally. We just discussed investment abroad, but what deters local investment? Usually LDCs have lower levels of domestic spending per person, so their markets are weak compared with those of advanced nations. Factors that keep the markets weak are a lack of demand, a lack of trained personnel to manage and sell products at the local level, and a lack of government support to ensure stability. There is also a lack of infrastructure to provide transportation, management, energy production, and community services—housing, education, public health—which are needed to improve the environment for investment activity.

Again, a shortage of capital and low investment are not causative factors of underdevelopment. But the important point is to understand what prevents capital from accumulating in LDCs.

Lack of Technology

Can the situation in LDCs be attributed to the absence of technology? Yes, to a degree. LDCs lack the basic technological advances necessary for creating capital and for applying methods to increase productivity and accumulate wealth. The IACs already have a large body of research and high-level technology from which the LDCs could draw. In the past, technological borrowing from advanced nations occurred in the Pacific Rim countries of Hong Kong, Singapore, South Korea, and Taiwan. In addition, OPEC nations benefited from the IACs' advanced technology in oil exploration, drilling, and refining. Today, the former Soviet Union countries desire technology from North America to aid their fledgling economies. Unfortunately, for LDCs to put this available technology to use, they must have a certain level of capital goods (machinery, factories, etc.), which they by and large do not possess. The need is to channel the flow of technologically superior capital goods that have high levels of reliability to the developing nations so that they can improve their output.

Information is the capital commodity of the 21st Century. Individual access will increase, and information will be used as a means to gain wealth and power. Most of the peoples of the world will not participate in this revolution. The implications will be profound.

In IACs, technology has been developed primarily to save labor and to increase output. This technology has been capital intensive. However, in LDCs the reverse is needed. The large pool of unskilled labor must not be supplanted by new technology. LDCs need capital-saving technology that is labor intensive. Therefore, much of the mid-latitude technology of the IACs is unsuited for low-latitude agricultural systems. The same is true for advanced technologies in the manufacturing and service sectors.

In LDCs, there is a strong inertia to maintain traditional production methods in agriculture and in industry, so that a basic level of food can be maintained and starvation can be avoided. If a new technological application fails, even only temporarily, malnutrition and starvation will not hesitate to overcome these countries.

Consequently, LDCs must be very traditional and conservative when they borrow technology.

Cultural Factors

Can cultural factors account for development problems of LDCs? Perhaps in part. Economic factors and forces of nature are not sufficient in themselves to explain the poverty cycle and attributes of LDCs. Economic development frequently involves a new world view and a willingness to accept changes in resource utilization, and changes in customs and traditions that frequently hamper the developmental process. Whether we like it or not, we must acknowledge that economic development is a change of the person—the way he or she thinks, does business, operates, and maintains relationships with other members of society. If the will to change is strong, economic development is more likely to take place. If there is little will to develop and a particular group within society is unwilling to change its ways of doing things, economic development is less likely to occur.

Tribal allegiances in Africa, for example, are stronger than national allegiances are. Warring tribes dilute economic efficiency and hamper economic development because of the political and economic disorganization associated with defending territories and ways of livelihood. Yugoslavia has likewise disintegrated because of ethnic clashes.

Religious beliefs in some nations hamper economic development because of a built-in resistance to change. For example, in India, a highly religious nation, it has been estimated that 8% of the per capita income is spent on religious ceremonial activities, to the economic detriment of most families who are desperately struggling to provide for themselves. Another impediment to economic development is *fatalism*, which dominates the populations of some LDCs. Fatalism is the religious or philosophical belief that future life conditions will not improve, regardless of the amount of work or effort involved. Fatalism is widespread in India. In Africa, a similar condition could be called the "capricious cosmos." The individual sees no correlation between personal actions and the experiences or outcome in the future. Both attitudes impinge on all aspects of life, including family planning, work, savings, investment, effort, and relationships with other individuals.

Finally, the existence of a strong social stratification system or a caste system works against people, especially women, in developing countries. If better jobs, education, and positions in society are allocated according to one's class, a certain feeling of futility develops. People are less likely to work hard or to attempt to improve their status with so little incentive (Figure 12.1).

Political Factors

Do institutional and political factors play a certain role in limiting the development of LDCs? Yes. Petty politics and inept social service administration create certain problems. In Nigeria, for example, 20% of the oil profits are stolen every year. The institutional situation in other LDCs such as Somalia is totally out of control. Political corruption and bribery are commonplace in Latin America, for example, and are actually expected as a way of life. Political decisions and even laws are interpreted according to individual whim and personal gain.

Throughout much of the developing world, land reform, which is vitally important to increased agricultural output, is lacking because the government is too inept to redistribute the land owned by a few wealthy families. For some nations, such as the Philippines, land reform is the single most pressing problem deterring them from economic development. In contrast, strong action taken by the South Korean government after the Korean War allowed for increased productivity and the development of an industrial and commercial middle class that made South Korea one of the true success stories of the Pacific Rim.

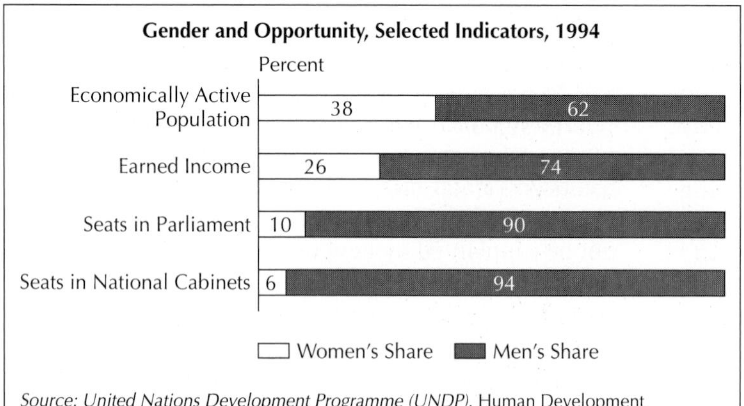

Vicious Cycle of Poverty

Is development of LDCs hindered by a catalog of causally related internal factors? The idea of causal links between attributes of un-

Gender and Opportunity, Selected Indicators, 1994

Percent

Indicator	Women's Share	Men's Share
Economically Active Population	38	62
Earned Income	26	74
Seats in Parliament	10	90
Seats in National Cabinets	6	94

☐ Women's Share   ■ Men's Share

Source: United Nations Development Programme (UNDP). Human Development Report 1995 (UNDP, New York, 1995). Figure 2.4. p. 31.

FIGURE 12.1
Gender and opportunity. Gender plays a significant role in business opportunity for women. The percentage of businesses owned by women by country is also illuminating. In 1996, for the United States this was 36%, Australia 33%, Canada 31%, Germany 28%, Japan 23%, Mexico 16%.

Consumers in relatively developed countries of the world obtain their food in supermarkets, such as the one shown. *(Photo: Richard Hayman, Photo Researchers, Inc.)*

derdevelopment is summarized in the often-used expression *vicious cycle of poverty.* Vicious-cycle explanations emphasize the multicausality of underdevelopment. These explanations suggest that it is not "just" a lack of ambition, or "just" an absence of specialization, or "just" a low output per capita, or "just" a population problem, or even "just" a political problem that holds back underdeveloped countries. Rather, a combination of interwoven limiting factors thwarts development. An example of a vicious cycle is shown in Figure 1.7a. According to Hungarian economist Ta'mas Szentes (1971), "the main weakness of the vicious cycle theories is that they reveal neither the historical circumstances out of which the assumed 'magic' circle originated, nor the underlying socioeconomic relations and the fundamental, determinant causes" (p. 54).

THE LESS DEVELOPED COUNTRIES' DEBT CRISIS

The foreign debt of LDCs has grown significantly during the last two decades. The debt is owed to foreign governments and to private banks. The total debt of nations in trouble as of 1993 exceeded $900 billion, and most LDCs that are heavily in debt are experiencing difficulties in repaying their debts. All LDCs, including those who are not experiencing debt-payback difficulties, owe IAC governments and banks more than $1.5 trillion. According to the World Bank, this represents approximately 30% of the combined GNPs of the LDCs. Table 12.3 lists some of the more heavily indebted LDCs as of 1990. This table lists the absolute amount of each country's debt as well as the proportion of debt compared with annual GNP. Some nations, notably Argentina and Nigeria, have debts in excess of 100% of their annual GNPs.

Causes of the Debt Crisis

In the 1970s and 1980s, LDCs took out large loans from IACs. These loans were granted with the expectation of future growth of the LDC economies. But a series of major economic changes in the world scene meant that many of these loans could not be repaid.

The Arab oil embargoes of 1973 and 1978 sent the price of crude oil from $8 a barrel to $35 a barrel. It has since dropped significantly and currently hovers at $20 a barrel on the world markets. However, the phenomenal oil prices created a cash-flow crisis for oil-importing LDCs. Oil-importing LDCs, as well as some of the oil-producing LDCs, based their expectations of future development on inexpensive oil. From 1973 onward, these nations faced increased foreign deficits because they had to continue to borrow money so that oil could flow into their nations. These borrowed funds increased their debt crisis. Furthermore, monies spent for oil could have been used for economic development, such as for increased infrastructure, improved education, and needed agricultural reforms. The debt of oil-importing LDCs grew from $150 billion in 1973 to $800 billion by 1985. Even oil-exporting nations such as Mexico, Libya, and Nigeria overborrowed to build economic strength based on their expectations of rapidly inflated oil prices, and, therefore, profits. When oil prices fell from $35 a barrel to $15 a barrel in the mid-1980s, these nations found themselves with debts that they could not afford.

Another blow to LDCs occurred between 1980 and 1985, when the U.S. dollar appreciated in value on world markets. Because many of the goods imported by LDCs were based on dollar value, the prices of these goods increased and foreign exchange decreased. This meant that the LDCs had to produce more domestic goods in order to import the same level of foreign goods. They also needed to export more goods to compensate for each dollar's worth of debt—and interest on the debt—because most LDC loans were denominated in U.S. dollars.

Finally, some of the LDCs, notably Mexico, Brazil, and Argentina, have been on the verge of bankruptcy. These countries required the IACs to rewrite their loans and cancel or *write-down* a portion of the principal and interest. Creditor banks in the United States were required to rewrite loans and even to increase loans. The result was a loss of confidence in the future ability of many LDCs to repay. The stocks of the creditor banks fell, and some CEOs were replaced. At the

same time, in the early to mid-1980s, the United States experienced a large budget deficit that continued into 1994. To finance the payback of this deficit, the U.S. government issued bonds. The issuance of bonds on the world market absorbed a large portion of investor revenue that would have been absorbed by LDCs as loans and private capital flows; thus, they were further deprived of the ability to grow and pay off their debts.

Higher oil prices, a decreased ability to produce goods and sell them on the world market, higher world interest rates, a valuation of the U.S. dollar, and a decline in public and private lending to LDCs because of loss of confidence all contributed to the enormous LDC debt crisis that prevails today. As mentioned, these difficulties triggered banking problems in the IACs, particularly in America. For example, in the mid-1980s, Citicorp and Chase Manhattan Bank determined that a large portion of the LDC debt would never be repaid. This threatened Citicorp's banking and financial system because a large portion of the LDC debt had to be written off as uncollectible. The International Monetary Fund (IMF) chose to attack this problem and to deal with each nation on an individual basis in an attempt to solve the debt problems. Many nations were able to rewrite their debts and schedule them for longer periods as long as their governments agreed to implement austerity programs. In order to invoke domestic austerity programs, these nations had to reduce imports and expand exports, which further reduced their living standards. Even so, from 1980 to 1990, the GDP for LDCs as a whole increased 3.8% per year, partly as a result of the agreements with the IMF. In comparison, the rate for IACs was only 2%. This enormous growth amid economic difficulty is proof that, in certain cases, LDCs can improve their plight and, in time, join the ranks of the advanced nations.

HOW ECONOMIC DEVELOPMENT IS MEASURED

Geographers measure economic development through a number of social, economic, and demographic indexes.

Abundance, availability, and great variety are the norm at grocery stores throughout the United States. However, for three-fourths of the people of the world, food is much less available. If developing countries produce a food surplus, it may be sold at a market such as this one near Bangkok, Thailand. *(Photo: Susan McCartney, Photo Researchers, Inc.)*

Per Capita Income

Per capita income is a statistic that is seldom readily available to economic geographers. However, the GNP and the population of a country are more easily acquired. Consequently, by dividing GNP by the number of people in a country, the economic geographer can determine GNP per capita. As shown in Figure 12.2, GNP per capita is more than $15,000 in most highly developed nations. At the same time, the United Nations estimates that 1 billion people live on less than $1 a day, or $365 per capita per year. Japan, North America, Western Europe, Australia, and New Zealand have the highest per capita incomes in the world. The Middle East, Latin America, South Asia, East Asia, Southeast Asia, and sub-Saharan Africa have the lowest. Income and GNP figures for former Soviet Union countries, portions of Eastern Europe, Iraq, Iran, Afghanistan, and portions of Southeast Asia were unavailable in 1990.

Per capita purchasing power is a more meaningful measure of actual income per person (Figure 12.2). The relative purchasing power in developed nations is more than $10,000 per capita per year, whereas in Africa it is much less than $1000 per capita per year. Per capita purchasing power includes not only income, but the price of goods in a country. The United States is surpassed by Japan, Switzerland, and Germany in per capita income. However, it surpasses all countries in per capita purchasing power because goods and services are relatively inexpensive in America, compared with those in other IACs. Although a Big Mac at McDonald's may cost as little as 99¢ in America, it costs as much as $6 in Europe and $10 in Japan. Prices in IACs, especially for food, are generally higher outside the United States. (See Purchasing Power Parity box in Chapter 11.)

Economic Structure of the Labor Force

The economic structure of a country also bespeaks its economic development. Economic geographers divide employment into five categories:

1. The *primary sector* mainly involves the extraction of materials from the earth—mining, lumbering, agriculture, and fishing.
2. The *secondary sector* includes assembling raw materials and manufacturing.
3. The *tertiary sector* is devoted to the provision of services—most notably wholesaling, retailing, and professional and personal services, including medical, legal, and entertainment.
4. The *quaternary sector* of the economy includes information processing, such as finance, insurance, real estate, and computer related fields.
5. The *quinary sector* includes medical care, research, education, arts, and recreation.

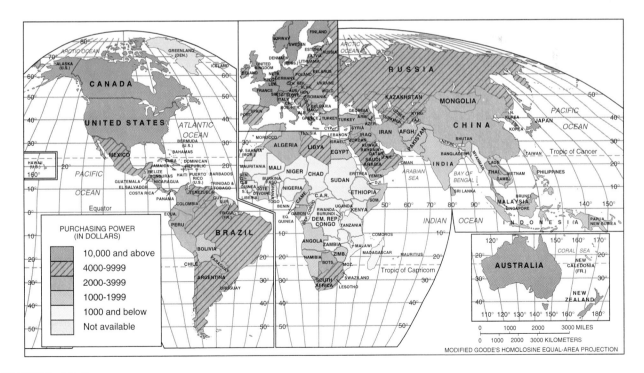

FIGURE 12.2

Per capita purchasing power. Per capita purchasing power is a better measure of a country's relative wealth than is GNP per capita because it includes the relative prices of products. For example, Switzerland, Sweden, and Japan have higher per capita GNPs than the United States. However, the United States has the world's highest per capita purchasing power because of relatively low prices for food, housing, fuel, general merchandise, and services (see Purchasing Power Parity box in Chapter 11).

We learned in Chapter 7 that fewer than 2% of the workers in America and Western Europe are engaged in agriculture, whereas in certain African nations, India, and China, more than 70% of the laborers are in the primary, or agricultural, sector. More than 75% of U.S. laborers are in the tertiary and quaternary sectors (Figure 12.3).

Consumer Goods Produced

The quantity and quality of consumer goods purchased and distributed in a society is also a good measure of the level of economic development in that society. A large amount of consumer goods means that a country's economic resources have fulfilled the basic human needs of shelter, clothing, and food, and more resources are left over to provide nonessential household goods and services. Televisions, automobiles, home electronics, jewelry, watches, refrigerators, and washing machines are some of the major consumer goods produced worldwide on varying scales. Figure 12.4 shows telephones per 100 persons worldwide for 1994. In IACs, more than one television, telephone, or automobile exists for every two people. In developing nations, only a few of these products exist for a thousand people. For instance, the ratio of persons to television sets in developing countries is 150 to 1, and population to automobiles is 400 to 1. In California, the ratio for these consumer

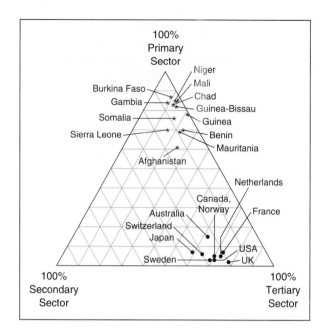

FIGURE 12.3
Portion of labor force in 20 countries that work in the three largest sectors of the economy. The countries near the top have labor forces primarily engaged in agriculture in the primary sector. The cluster of countries located in the bottom right show a concentration in the tertiary and quaternary sectors of the economy. The 20 countries shown here are the top 10 and bottom 10 ranked by the Human Development Index.

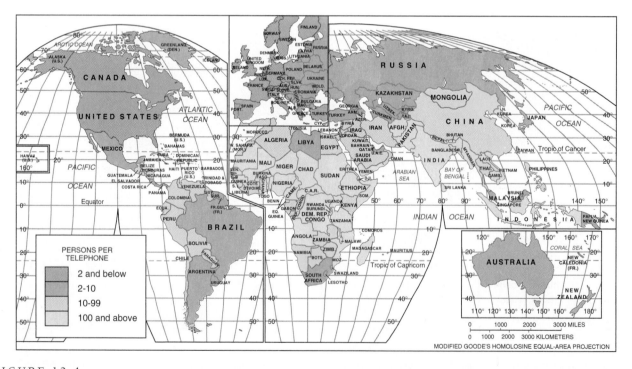

FIGURE 12.4
Persons per telephone. Developing countries have several hundred persons per telephone, while relatively developed countries have two persons per telephone. The United States is approaching one person per telephone and will soon surpass this ratio with the proliferation of cellular phones.

items is almost 1 to 1. The number of consumer goods such as telephones and televisions per capita is a good indicator of a country's level of economic development.

Education and Literacy of a Population

The more men and women who attend school, usually the higher the level of economic development in a country. Unfortunately, in the world today, only 75 women attend school for every 100 men who attend. Some areas of the world are particularly disadvantageous to women's education (Figure 12.5). The regions that have low percentages of women attending secondary school also generally have poor social and economic conditions for women.

The *literacy rate* of a country is the proportion of people in the society who can read and write. Figure 12.6 shows the literacy rates of men and women worldwide, and Figure 12.7 shows the literacy rates of women only. Comparing these two figures, one can clearly see that in some countries the percentage of women who can read and write is much lower than that of men. In many nations, the literacy rate of women is less than 25%, whereas the literacy rate of men is between 25% and 75%. The Middle East and South Asia, where the role of women is clearly subservient to men, show the greatest disparities. How-

ever, in the highly developed world of the IACs, the literacy rates of men and women are almost identical. In addition, because more people can read and write, a proliferation of newspapers, magazines, and scholarly journals improve and foster communication and exchange, which leads to further development.

Health and Welfare of a Population

Measures of health and welfare, in general, are much higher in developed nations than in LDCs. One measure of health and welfare is diet. Most people in Africa do not receive the U.N. daily recommended allowance. However, in developed nations, the population consumes approximately one-third more than the minimum daily requirement and are therefore able to maintain a higher level of health. In all cases, the figures represent averages. Naturally, in some areas of each country, calories and food supplies are insufficient, even in the United States.

People in developed nations also have better access to doctors, hospitals, and medical specialists. Figure 12.8 shows worldwide access to physicians and health care. For relatively developed nations, there is one doctor for 1,000 people, but in developing countries, each person shares a doctor with many thousands of others. Africa by far has the worst access to health care, followed by Southeast Asia and the East Indies. Portions

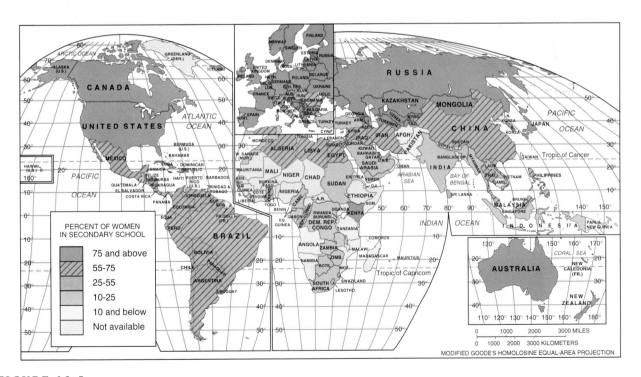

FIGURE 12.5

Percentage of women in secondary schools. In developing countries, the percentage of women in secondary schools is far below that of men and women in developed countries. This fosters lower educational development among women and lowers the prospects for economic development and gender equality.

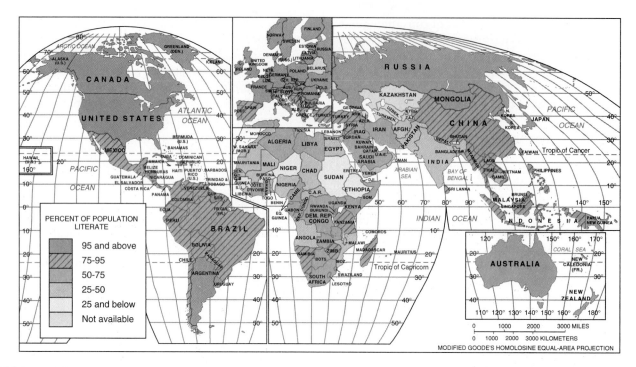

FIGURE 12.6

Literacy rate. A great disparity in literacy rates exists between inhabitants of developed countries and less developed countries. In the United States and other highly developed countries, the literacy rate is more than 98%. Notice the large number of countries in Africa and South Asia where the literacy rate is less than 50%.

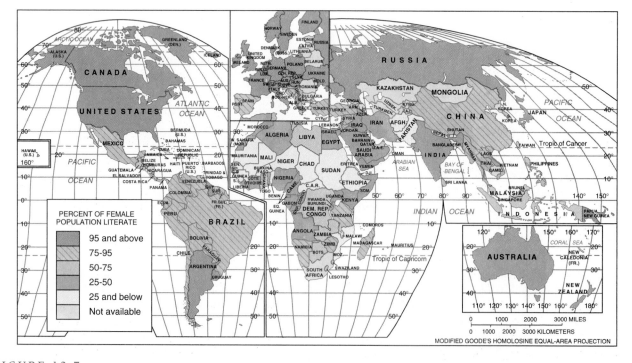

FIGURE 12.7

Literacy rate of women. The gender gap for literacy is most pronounced in the developing world, especially Africa, South Asia, and the Middle East. In most developing countries, the female literacy rate is much below the rate for men.

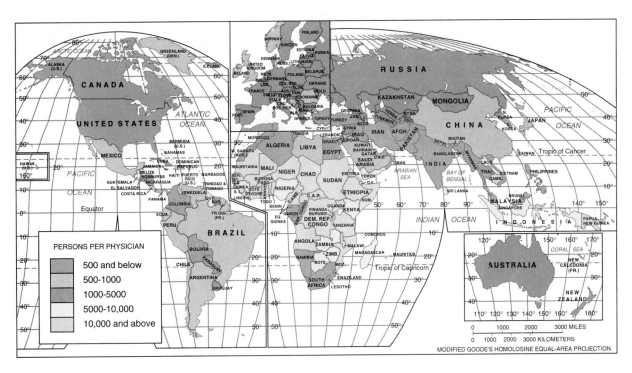

FIGURE 12.8
Persons per physician. An important measure of economic development is the number of persons per physician in a country. This measure is a surrogate for health care access, which includes hospital beds, medicine, and nurses and doctors. Most of Africa exhibits 10,000 people per physician, or more, while Europe and the United States average one doctor for every 200 people.

of the Middle East are also lacking in medical care. AIDS is out of control (Table 12.4).

AIDS is tightening its grip outside the United States and Western Europe. In India, researchers estimate that by the year 2000, anywhere from 15 million to 50 million people could be HIV positive. Half the prostitutes in Bombay are already infected, and doctors report that the disease is spreading along major truck routes and into rural areas, as migrant workers bring the virus home. In central and eastern Europe, countries that had largely escaped the epidemic are seeing an explosion in the number of cases, mainly among IV drug users and their heterosexual contacts.

Then there is Africa, across much of which the disease continues to rage unchecked. Already the sub-Saharan region accounts for more than 60% of the people living with HIV worldwide, or some 14 million men, women, and children. As many people will die there this year from the disease as were massacred 2 years ago in the Rwanda holocaust. The social consequences of this die-off are catastrophic. By the year 2000, nearly 2 million children in Kenya, Rwanda, Uganda, and Zambia will have lost their parents to the disease.

Virtually every AIDS expert agrees that only an effective vaccine can halt the epidemic. Yet, after more than a decade of trying to develop one, scientists have made little headway. The failure can be traced in part to the notorious craftiness of the virus itself. Researchers still aren't sure what part of the immune system confers protection against HIV, and until they are, they won't be able to induce immunity with a vaccine.

The Epidemiological Transition

As the populations of industrialized countries moved from Stage 2 to Stage 3 of the demographic transition, the main causes of illness and death changed from infectious and parasitic diseases to circulatory and degenerative diseases (see Figure 12.9). In the industrialized countries today, more people die of heart disease, stroke, and cancer, and few die from infectious and parasitic diseases, such as malaria, AIDS, typhoid dysentery, sleeping sickness, and cholera. This study of the occurrence, distribution, and control of disease is called *epidemiology*. The transition from infectious and parasitic diseases (Stages 1 and 2 of the demographic transition) to circulatory and degenerative diseases (Stages 3 and 4) in a society is called the *epidemiological transition*. Diseases are grouped into three broad categories: (1) infectious diseases caused by the entry, growth, and dissemination of foreign organisms within the body; (2) chronic and degenerative diseases that result from long-term deterioration of the body; (3) genetic or inherited diseases

✦

T a b l e 12.4

AIDS: The Global Epidemic

Hot Spots	North America	Latin America	Caribbean	Sub-Saharan Africa	North Africa and Middle East	Western Europe	Central and Eastern Europe and Central Asia	South and Southeast Asia	East Asia and Pacific	Australia and New Zealand
People living with HIV/AIDS	750,000	1,300,000	270,000	14,000,000	200,000	510,000	50,000	5,200,000	100,000	13,000
Percent change 1992 to 1996	–13%	43%	47%	37%	46%	2%	238%	261%	658%	–14%
Percent who are women	20%	20%	over 40%	over 50%	20%	20%	20%	over 30%	20%	20%
Percent of population age 15–45 with AIDS	0.50%	0.60%	1.70%	5.60%	0.10%	0.20%	0.02%	0.60%	0.00%	0.10%
Deaths in 1996	61,300	70,900	14,500	783,700	10,800	21,000	1,000	143,700	1,200	1,000
Main modes of transmission	1. Male homo-sexual 2. Intra-venous drug use and hetero-sexual	1. Male homo-sexual 2. Intra-venous drug use and hetero-sexual	Hetero-sexual	Hetero-sexual	1. Intra-venous drug use 2. Hetero-sexual	1. Male homo-sexual 2. Intra-venous drug use and hetero-sexual	1. Intra-venous drug use 2. Male homo-sexual	Hetero-sexual	1. Intra-venous drug use 2. Male homo-sexual	1. Male homo-sexual 2. Intra-venous drug use and hetero-sexual

All numbers are estimates as of December 1996.
Source: UNAIDS.

that are caused by the characteristics of the genes and chromosomes inherited from parents. Modern medical science can lower the death rate by 60% in developing countries by attacking infectious and parasitic diseases, and maternal and perinatal causes of diseases.

Demographic Characteristics

IACs have much lower infant mortality rates than those of LDCs. In developed nations, on the average, fewer than 10 babies in 1,000 die within the first 100 days. In many less developed nations, more than 100 babies die per 1,000 live births. Paradoxically, as noted in Chapter 3, even though the infant mortality rate is higher in developing nations, the natural rate of increase and the crude birth rate are higher as well. Some developing nations show a 3% per annum growth, which means that the population doubles in 17 years, although the average for LDCs is closer to 2%, with a doubling time of 35 years. In contrast, most developed nations have a less than 0.8% relative increase per year, and a few nations are at zero population growth (ZPG). LDCs average 40 live births per 1,000 population, while IACs average 10.

Age structures also differ substantially. In developing nations, as many as 50% of the people are younger than age 15. In IACs, however, median ages range from 30 to 40, and life expectancy is longer (Figure 12.10).

Total Human Development

The six developing regions of the world are Latin America, the Middle East, East Asia, Southeast Asia, South Asia, and sub-Saharan Africa. The six developed regions of the world are Japan, Western Europe, the South Pacific, Anglo-America, Eastern Europe/former Soviet Union, and South Africa. The boundaries of each of these so-called *world cultural realms* follow continental outlines and major political boundaries or climatic zones. Each region has more highly and less highly developed regions within it. For example, Latin America varies dramatically, from the more advanced nations of Argentina and Venezuela to the less advanced nations of Bolivia, Paraguay, and Guyana. Within each cultural realm, similarities of language, religion, race, population characteristics, and economic development characteristics exist. But with all the differences among

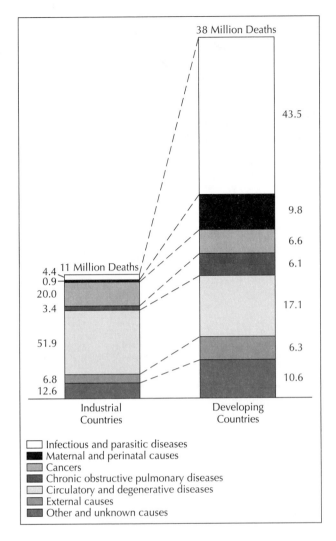

FIGURE 12.9
The epidemiological transition: percentages of deaths by cause in industrial and developing countries.

nations and realms, how can their development levels be compared?

The United Nations devised a way to measure total human development, a *Human Development Index* (HDI), which combines three development factors: life expectancy (a demographic factor), literacy rate (a social factor), and per capita purchasing power (an economic factor). Because it includes all three factors—social, demographic, and economic—it is a useful index, even though the variables selected are somewhat arbitrary (Figure 12.11). The index is a standardized scale, where the highest possible value is 1.0 and the lowest is zero. The minimum for each index was set as the lowest level actually observed. For example, the minimum index for literacy was 12%, which was Somalia's rate. Overall, Anglo-America and Japan had the highest human development score, 0.98, followed by Australia and New Zealand with 0.97, and Western Europe with

0.95. The lowest scores were in sub-Saharan Africa, which averaged 0.23. Nigeria, for example, had a literacy rate of 14%, a per capita purchasing power of $450, and a life expectancy of 45 years (Table 12.5).

Major Perspectives on Development

"Theories" of development have existed for many years. The earliest of them can be traced to the classical economists. But discussion of the term *development* in the social sciences is fairly recent. It was not required before the collapse of the colonial system or before the onset of the Cold War between the United States and the Soviet Union for power and influence in the world. Since the late 1940s, however, the problem of how to accelerate the pace of development in roughly 100 ex-colonial countries has generated intense interest among planners and in the academic world.

Three groups of perspectives on development can be identified in terms of the scale upon which they concentrate (national, global) and whether or not development is defined exclusively in terms of economic growth. The first and most widely accepted theories of development are *modernization theories*. These concentrate on the national scale and define development in terms of economic growth and Westernization. Their underlying assumption is that modernization influences are projected to peripheral regions from Western Europe and North America; hence, the path to progress from traditional to modern is unidirectional. In this view, rich industrial countries, without rival in social, economic, and political development, are modern, whereas poor countries must undergo the modernization process to acquire these traits.

The second group of theories are *world political economy theories*. They focus on the structure of political and economic relationships between dominant and dominated countries. They pay special attention to the global history of economic growth that brought poor countries to their present position (Agnew, 1982). Their underlying assumption is that the poverty of the Third World is the outcome of a worldwide network of intrusion by the rich countries into the poor.

Theories that make up a third group come from scholars who wonder whether certain kinds of development are desirable, and if, indeed, development will ever materialize for people in underdeveloped countries. These *ecopolitical economy theories* concentrate on the ecological and cultural consequences of economic growth. Their proponents are disenchanted with research that (1) neglects the diverse value systems and world views of societies engaged in the process of development, (2) equates development with economic growth, (3) advocates the trickle-down theory of benefits to the poor instead of the channeling of resources directly into basic

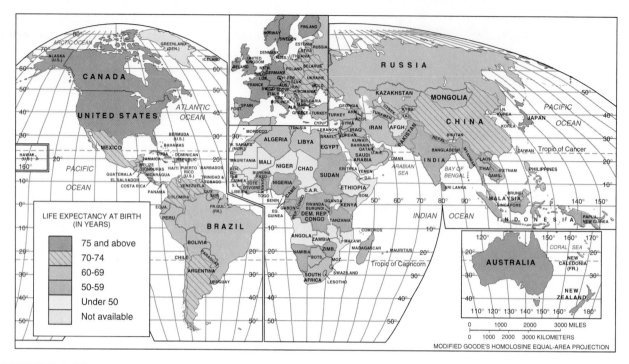

FIGURE 12.10

Life expectancy at birth reflects access to health care and food. For the world as a whole, babies born this year will live to their mid-60s. Life expectancy at birth actually ranges from the high 70s in Canada, the United States, Europe, and Japan, to the mid-60s in most Latin American and Middle Eastern countries, Russia, and China, to the mid-40s in most African nations.

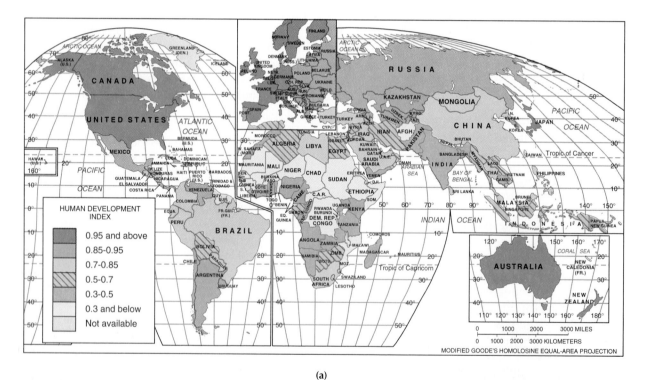

(a)

FIGURE 12.11

(a) The Human Development Index is perhaps the best overall measure of economic development. The United Nations constructs a single index measuring life expectancy at birth, per capita purchasing power, years of schooling, and literacy rates (see Table 12.4 for actual values by cultural realm). Northwestern Europe is clearly distinguished from the remainder of Europe. Central America and Bolivia have the lowest scores for Latin America. Sub-Saharan Africa, South Asia, and Southeast Asia have the lowest scores overall. (b) Canada and the United States lead the world.

human needs, and (4) ascribes no merit whatsoever to the contributions of indigenous systems to the development process (Yapa, 1980).

World Political Economy Theories

World-system theory provides a framework for understanding the development of the capitalist system and its three components—the core, the semiperiphery, and the periphery—from the start of the 16th century (Wallerstein, 1974). The *development-of-underdevelopment theory* owes much to the work of André Gunder Frank (1969). He drew on the Latin American experience to argue that the development of the West depended on the impoverishment of the periphery. The develop-

ment-of-underdevelopment theory pays special attention to a theory of dependence that relates the effects of imperialism on developing countries.

An attractive theory for geographers is Johan Galtung's (1971) *structural theory of imperialism* based on *center-periphery relations*. The Center is represented by wealthy industrial nations, the Periphery by underdeveloped nations. Both the Center and the Periphery have centers of their own (mainly urban elites and some rural elites) and peripheries (essentially the rural poor and also the urban powerless). The dominant economic and political power lies with the center in the Center (i.e., the elites in the industrialized countries); the dominant strength in numbers lies with the periphery in the Periphery (i.e., the majority of Third World people). For Galtung, the center-periphery theory of imperialism provides a fruitful basis for empirical research within liberal and radical schools of thought.

Another way to examine imperialistic relations is to focus on the *role of the MNC* as a source of political and economic control (Barnet and Müller, 1975). The buildup of crises in the capitalist core, frequently expressed as conflicts between capital and labor, has forced firms to emphasize multinational production and to channel capital in the direction of peripheral regions, where hourly wage rates and fringe benefits are low. Despite the economic benefits derived from the transfer of capital, the foreign enclaves of multinationals, their "export platforms," create few jobs relative to the size of the massive industrial reserve army in developing countries.

Ecopolitical Economy Theories

Ecopolitical economy theories stem from concern over the meaning of development and the costs and benefits of economic growth and cultural change. Geographer Lakshman Yapa (1980), who coined the term *ecopolitical economy*, pointed to insights derived from world political economy theory in criticizing the conventional wisdom of striving for higher and higher rates of economic growth in developing countries in the hope that benefits will diffuse to the poor. The benefits of economic growth fail to reach the poor because the economic surplus is diverted to national and foreign elites, in effect preempting the basic-goods fund. The only way that economic development can improve the living conditions of the poor is by using resources directly in the production and distribution of basic goods.

Human Development Index
Country ranking*, 1995

Legend:
- Human development index ranking
- GDP per person† ranking

Countries (top to bottom): Canada, United States, France, Japan, Britain, Germany, Greece, Hong Kong, Israel, Singapore, Chile, Portugal, South Korea, Argentina, Czech Republic, Venezuela, Hungary, Mexico, Colombia, Poland, Thailand, Malaysia, Russia, Brazil, Turkey, South Africa, Philippines, Indonesia, China, India

Axis: 0, 50, 100, 150

*Selected countries
†At PPP exchange rates
Source: United Nations Development Program.

(b)

◈

T a b l e   12.5
Human Development Index, 1995

Less Developed Regions		Industrially Advanced Regions	
Latin America	.76	Japan	.98
East Asia	.61	Anglo-America	.98
Southeast Asia	.52	Australia/New	.97
Middle East	.51	Zealand	
South Asia	.29	Western Europe	.95
Sub-Saharan	.23	Eastern Europe	.87
Africa		South Africa	.67

COLONIALISM AND GLOBAL CORE-PERIPHERY RELATIONS

Thus far, we have dealt with attributes of LDCs and ideas about development that find expression in the social sciences. The balance of the chapter examines the themes of inequality and unequal development in more detail. We begin with the development of the capitalist world economy, viewed in terms of waves of colonialism and core-periphery relations.

Cycles of Colonialism

In modern history, there have been two waves of colonialism (Figure 12.12). The first wave began in 1415, when the Portuguese seized control of the commercial naval base of Ceuta on the Strait of Gibraltar, and ended soon after 1800. The second wave began in 1825 and ended in 1969. During the first wave, European power centered on the Americas; during the second wave, the focus switched to Africa, Asia, and the Pacific. Colonies of the first wave were mainly settlement colonies where quasi-European societies were created by immigrants. The second wave involved colonies of occupation, in which a small number of Europeans exercised political control. Exceptions to the latter included 19th-century settler colonies in Australasia and in southern and eastern Africa.

In each wave, a few colonial powers overshadowed the rest. During the first wave, Spain and Portugal stood apart from the Netherlands, Great Britain, and France. In the second wave, when the number of major colonial powers increased from 5 to 10, Great Britain and France were far ahead of their contemporaries. At its peak in 1933, the British Empire covered more than 24% of the world's land surface and included nearly one-quarter of the world's population (502 million people).

The first wave of colonialism involved conquest, plunder, slavery, and the annihilation of indigenous people. The Spaniards, for example, virtually exterminated the Carib population of Hispaniola; in 1492, the Caribs numbered 300,000, but by 1548 the figure was down to 500 (Griffin, 1969). The arrival of the Spaniards in Mexico led to the destruction of the Aztec civilization and a population decline from 13 to 2 million. In Africa, the slave trade greatly reduced the population in large parts of the Congo Basin and in the West African forest.

During the second wave, there was less destruction and disruption of local societies. Conditions varied from colony to colony, however. For example, the effects of colonialism were generally stronger in southern Africa and East Africa than in West Africa. In southern Africa and East Africa, the British alienated land to build an export-oriented economy.

During the second wave of colonialism, the imperialist countries saw the developing regions as immense supply depots for the cheap production of raw materials from which their economies could profit. The economies of developing countries were often deformed into subsidiaries of the colonial powers: Jamaica became a sugar plantation, Sri Lanka a tea plantation, Zambia a copper mine, and Arabia an oil field.

Core Structure and Global Center-Periphery Relations

Why did colonialism expand at a particular time and contract at another? World-system theory suggests that the key to understanding waves of colonialism is the changing structure of the core: periods of instability and stability in the core coincide with periods of colonial expansion and contraction in the periphery. During periods of instability, when there is competition among rival core countries, colonialism expands. In the presence of a single, hegemonic core country, colonialism contracts. A hegemonic power can control the world without the expensive encumbrances that colonies represent.

Hegemony exists when one core power enjoys supremacy in production, commerce, and finance and occupies a position of political leadership. The hegemonic power owns and controls the largest share of the world's production apparatus. It is the leading trading and investment country, its currency is the universal medium of exchange, and its city of primacy is the financial center of the world. Because of political and military superiority, the dominant core country maintains order in the world system and imposes solutions to international conflicts that serve its self-interests. Consequently, hegemonic situations are characterized by periods of peace as well as by universal ideologies such as freedom to trade and freedom to invest.

During a core power's rise to hegemony, core periphery relations become more informal. Economic

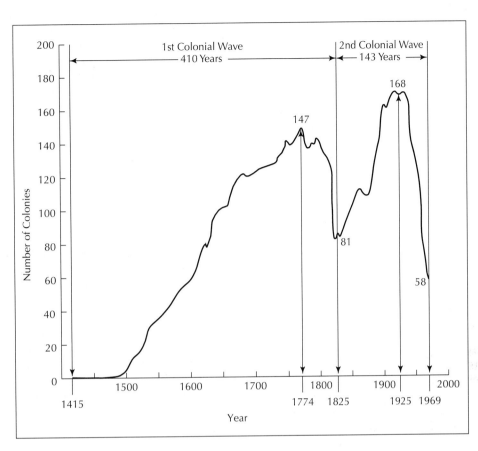

FIGURE 12.12
Waves of colonialism. The first wave of colonialism, such as that by the Spanish and Portuguese in Latin America, peaked in the mid-18th century and involved conquest, plunder, slavery, and annihilation of indigenous peoples. The second colonialism wave, which peaked in the early 20th century, involved less destruction and disruption of local societies. *(Source: deSouza, 1986)*

Nairobi, capital of Kenya, is a city created by Europeans out of the African "wilderness." Home to more than 2 million people, Nairobi originated in 1899 as a railway camp during the building of the Mombasa–Lake Victoria railway line. To the north and west of the city, the fertile "White Highlands" were alienated to European farmers who specialized in wheat, corn, pyrethrum, sheep, and cattle. Land was also alienated along the coast for European use—often as a vacation spot on the Indian Ocean. Since independence, colonial land ownership and use has been modified. Land in the "White Highlands," for example, has been acquired by Africans and divided into small holdings. *(Photo: Oldrich Karasek, Peter Arnold, Inc.)*

linkages between center and periphery increasingly focus on the hegemonic power. This reorientation results in decolonization. The hegemonic power relies on economic mechanisms to extract the surplus value of the periphery. Lopsided development between the core and the periphery flourishes, and terms of trade deteriorate for the colonized periphery. During a power's fall from hegemony, rival core states, which can focus on capital accumulation without the burden of maintaining the political and military apparatus of supremacy, catch up and challenge the hegemonic power.

Competition exists when power in the center of the world system is dispersed among several countries. With pluralization, competing centers control and own a larger part of the world's production apparatus. They increase their shares of world trade, enclose national economic areas behind tariff and nontariff barriers, and use their national currencies in a growing volume of transactions. Without a hegemonic power dominating the world system, political tensions increase and may develop into armed conflicts.

Competition forces increasingly more formal core-periphery relationships. Competing core countries rely on the mechanism of colonialism to extract the surplus of the periphery. Economic linkages between the colonized and the colonizers become multilateral to a greater extent, and economic transactions with core countries become more frequent.

World-system theory maintains that periodic fluctuations from a single, hegemonic power to a group of competing countries are essential to the survival of the world system. The system would break into separate empires if competition at the center were to persist, or mutate into a world empire if hegemony were to last. Moreover, this cyclic realignment of core-periphery relations is not limited to colonialism; it is manifested in all the ways that the world system binds itself together. Trade, for example, is more formally structured during periods of core instability.

REGIONAL DISPARITIES WITHIN DEVELOPING COUNTRIES

Inherited colonial structures inhibit the development efforts of LDCs in the post-independence era. Major cities of developing countries are still "export platforms." They link the rich industrial countries and their sources of raw materials. Under the new industrial division of labor, they supply a small core of developed countries with manufactures: engineering and metal products, clothing, and miscellaneous light manufacturing. As a result, modern large-scale enterprise remains concentrated in capital and port cities. Injections of capital into urban economics attract new migrants from rural areas and provincial towns to principal cities. Urban primacy increases. Migrants, absorbed by the system, are maintained at minimal levels. There is little incentive to decentralize urban economic activities. The markedly hierarchical, authoritarian nature of political and social organization retards the diffusion of ideas throughout the urban hierarchy.

The relationships between the different parts of the capitalist system accentuate inequality. Polarizing effects within former colonies concentrate services and innovations at the center, promoting the imbalance between center and periphery. Capital movements, trade flows, internal migration, and institutional controls all tend to have an absolute negative effect on the growth rate and development of the periphery. In this section, we survey the persistence of disequilibrium within developing countries.

The Center-Periphery Concept

The center-periphery concept is one of the most geographic ideas presented by regional analysts. It echoes the Marxist argument that the center appropriates to itself the surplus of the periphery for its own development. The center-periphery phenomenon may be regarded as a multiple system of nested centers and peripheries, like a Chinese puzzle box. At the world level, the global center (rich industrial countries) drains the global periphery (most of the underdeveloped countries). But within any part of the international system, within any national unit, other centers and peripheries exist. Centers at this level, although considerably less powerful, still have sufficient strength to appropriate to themselves a smaller, yet sizable, fraction of remaining surplus value. A center may be a single urban area or a region encompassing several towns that stand in an advantageous relation to the hinterland. Even in remote peripheral areas local, regional imbalances are likely to exist, with some areas growing and others stagnating or declining. (See Figure 12.15 on p. 558.)

There are reverse flows from the various centers to the peripheries—to peripheral nations, to peripheral rural areas. Yet these flows, themselves, may further accentuate center-periphery differences. For example, World Bank, United States Aid for International Development (USAID), and International Development Association (IDA) loans generally support major infrastructural projects such as roads and power stations, which are proven money earners, and which reinforce the centrality and drawing power of the cities and the modern export sector of agriculture. AID strongly supports projects dealing with agriculture, health and family planning, school construction, and road building; industrialization projects are seldom financed.

Yet many Western social scientists see core regions in underdeveloped countries as " 'beachheads' . . . the centres from which the benefits of modernization flow outwards to revitalize the stagnating agricultural sector"

(McGee, 1971, p. 13). Such a vision could cast social scientists in the role of augmenting national, regional, and individual inequalities in developing countries. Let us see why by describing and analyzing some of the major center-dominant models of regional development.

Center-Dominant Models of Regional Development

One view of regional development stems from the studies of neoclassical theorists. They hypothesize that differences between center and periphery are only temporary within a free-market system. Regional development inequalities that occur at first will be corrected by the mobility of factors under pure competition. For example, if wages are higher in Region *A* than in Region *B*, labor will move from the lower-paid area to the higher, thereby leading to an adjustment in relative wage rates. Because the mechanism in a free-market economy is self-regulating, no government intervention is necessary: Regional differences in wage rates and income will occur automatically. Unfortunately, little evidence exists to support such a view—especially in developing countries, where perfectly mobile production factors are hardly characteristic features.

Two models offering an alternative explanation of regional differentials emerged in the late 1950s. Both models suggest that, with time, interaction *increases* rather than *decreases* inequalities between rich and poor regions. The models of Swedish economist Gunnar Myrdal (1957) and American economist Albert Hirschman (1958) indicated that in a developing country operating under a capitalist system, *deviation-amplifying forces*, rather than *deviation-counteracting forces*, increase and rigidify the differences between center and periphery. *Deviation-amplification* (as opposed to *deviation reduction*) refers to any process that amplifies an initial "kick" and increases divergence from an initial condition.

MYRDAL'S CIRCULAR AND CUMULATIVE CAUSATION MODEL

Myrdal (1957) argued that during early stages of development, economic inequalities are increased through the operation of *circular and cumulative causation*. He reasoned that "change does not call forth contradicting forces [as equalization models suggest] but, instead, supporting changes, which move the system in the same direction as the first change but much further" (p. 13). According to Myrdal, once growth has been initiated in favored locations in a free economy, inflows of labor, skills, capital, and commodities develop spontaneously to support these locations. The flows, however, induce *backwash effects* that amplify inequalities between expanding and other regions. Myrdal argued that if events follow an uncontrolled course, backwash effects perpetuate growth in expanding regions and retard

The periphery in the Center is illustrated by these shacks occupied by North African migrant workers in Paris. *(Photo: International Labor Office)*

growth elsewhere. For development to occur throughout a country, *spread effects* must, on the average, be stronger than backwash effects.

In developing countries, spread effects often refer to the benefits that trickle down from a major city to surrounding areas. These *trickle-down effects* may include increased demand for primary commodities, increased investment, and the diffusion of modern technology. Conversely, backwash effects are the demands by the city on surrounding areas. These effects can include an inward flow of commodities, capital, and skilled workers. Even a new transportation route can initiate backwash. It can permit industrial plants in the growing city to supply stagnating areas with goods formerly supplied by the poor region's own craft industries.

HIRSCHMAN'S CIRCULAR MODEL

Hirschman (1958) advanced a similar model of polarized development. His model shows that after an initial decision is made to locate a particular industry at a specific point, it has an initial multiplier effect, as shown in Figure 12.13. New local demands are generated by the factory and by the purchasing power of its labor force. The labor force creates a demand for housing and for a

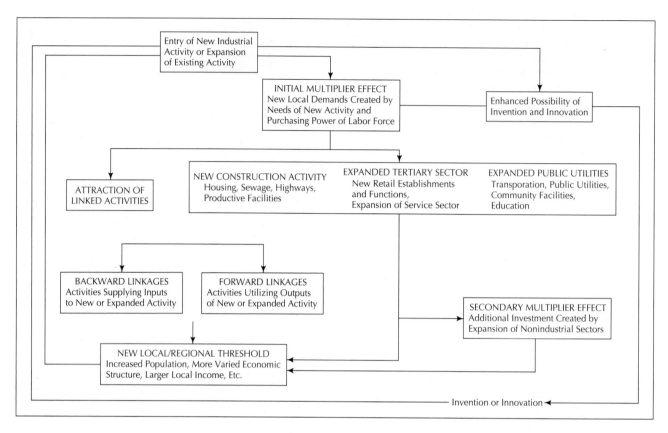

FIGURE 12.13
Initial multiplier effect and the process of cumulative upward causation. *(Source: Pred, 1966, p. 25)*

set of services. The new factory attracts industries producing complementary goods. Linked industries either provide needed inputs or purchase semifinished products from the initial factory. The entire process has a cumulative self-generating momentum; after the first cycle of growth is completed, a new spiral of growth is initiated at a higher threshold. The diagram of circular and accelerative growth illustrates that cycles of growth can be carried through *any* number of times—at least until the process is arrested by diseconomies or interrupted by the competitive advantages of other growth points. It also shows that the initial multiplier stimulates expansion of nonindustrial activities; they, in turn, trigger a second multiplier that induces further growth and still higher thresholds.

Thus, through the operation of unrestrained market forces, the center grows, feeding on more and more resources from the hinterland. The result is unbalanced or polarized growth. According to Hirschman, *polarization effects* are offset eventually by *trickle-down effects*—the equivalent of Myrdal's spread effects. However, Hirschman thought that only governments can provide enough incentives for positive trickle-down effects to outweigh negative polarization effects and ensure sustained peripheral growth.

THE COLLAPSE OF COMMUNISM IN THE SOVIET UNION

Soviet communism was based on the *labor theory of capital*. It viewed capitalism as a way for the few elite in the capitalist system to gather extra value from the laborers. Communism, on the other hand, was to extract the extra value of labor, which was the value of production minus wages, and redirect it to the society as a whole in the form of subsidized food, subsidized housing and education, and subsidized medical attention. Unfortunately, much of this capital went into an unnecessarily large military machine. The primary features of the Soviet system were state ownership of all the means of production—land, factories, and raw materials—and state-directed economic planning, which established prices, set quotas, and directed the distribution of all wealth.

In the 1950s and 1960s, economic growth was an impressive 6% in the Soviet Union, compared with a mere 3% in the United States. This led premier Nikita Khrushchev to state in the United Nations, "We will bury you with the superiority of our economic system through world competition and trade." However, 6% growth rates could not continue; by 1990, the rate was –4%, and by 1991, –14%. By 1996, growth rates returned to zero.

An abortive attempt in 1991 to push President Mikhail Gorbachev into the background failed, but the break in the Soviet Union had already occurred. Boris Yeltsin emerged as the strong leader of the Russian Republic. The former Soviet Union today is loosely called the *Commonwealth of Independent States*, or CIS, but this commonwealth is extremely weak. The position and quality of the economy continues to deteriorate. Production is declining, prices have risen, and shortages are massive. The political situation also continues to be completely unstable, and civil war could break out at any time, reverting the country to the old authoritarian ways. Many members of the Russian congress would like to see the authoritarian system reinstated because they have everything to gain if it is, but Yeltsin dissolved congress, stormed the parliament, and arrested the policymakers.

Economic development potential exists in the former Soviet countries (Figure 12.14); the eastern European nations of Czechoslovakia, Hungary, Poland, Romania, Bulgaria, and Albania; and the fractured republics of former Yugoslavia are also readying themselves for development. These countries provide major lessons for LDCs that are pursuing or contemplating pursuing the economic development model given by the former Soviet Union's central-planning programs.

What caused the decline of the former Soviet Union in the 1990s? What created the long queues of consumers on the streets of St. Petersburg (Leningrad), Kiev, and Moscow? In this section, we first attempt to answer these questions. Then we discuss how the Soviet Union can embrace a market economy and what the IACs can do to help.

Why Soviet Communism Failed

Soviet-style communism failed in the early 1990s after 70 years of struggle. The factors that led to its downfall were an inability to meet consumer demand, lack of proper motivation and labor in-

centives in the economic system, central-planning problems and mismanagement, inadequate agriculture, and too much military investment.

INABILITY TO MEET CONSUMER DEMAND
Before the collapse of communism, the poor quality and shortages of Soviet manufactured goods suggested that failure was imminent. What caused the shortages? Why were the goods substandard?

After World War II, rapid Soviet growth was possible with more labor, more capital, and more land. However, by the 1980s, these resources were being maximized, so no growth occurred, and in the 1990s growth was negative.

The military machine and capital goods were all-important, so attempts were made to increase the production of these goods. However, this emphasis had several costly consequences. First, the percentage of output devoted to consumer products decreased. Store shelves became empty, and the standard of living dwindled. Second, the World Bank estimated that the manufacturing sectors of the economy were so heavily subsidized and overemphasized by the central economic planning groups that the Soviet Union lagged 12 to 15 years behind the United States and Germany in technological expertise and machinery. Refrigerators, televisions, radios,

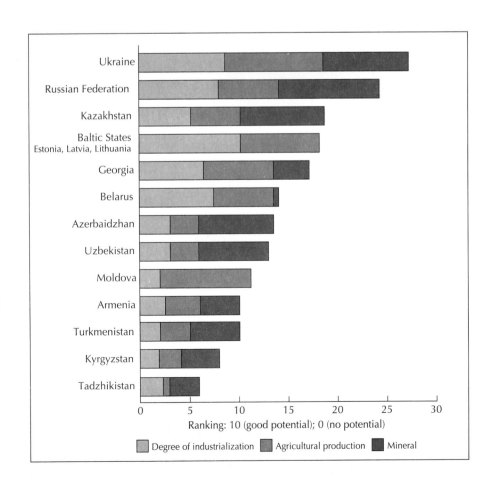

FIGURE 12.14
Economic development potential of Russia and the former Soviet Republics, which show large variation in their potential for economic development. *(Source: Fisher, 1992, p. 320)*

and automobiles were all extremely poor quality—primitive by German, Japanese, and U.S. standards.

Both these factors had an adverse effect on the citizens. Incentive problems, discussed in more detail next, already existed. According to estimates, the labor productivity of the average Soviet worker was only 35% that of his or her American counterpart. The poor attitude of citizens was aptly expressed by one worker's response to an inquiring journalist: "We pretend to work, and the government pretends to pay us."

Added to the incentive problems was the inability of the system to supply the goods and services that the consumers wanted. Consumer expectations always outstripped supply. Long lines existed for almost every product, and many times stores ran out of goods before the lines dwindled. Empty shelves became the hallmark of the Soviet Union. Morale was lowered even more when the Soviet citizens quickly discovered—through satellite communications and television broadcasts from Western Europe and Radio-Free Europe—the gross difference between their standard of living and that of the IACs.

Understandably, the result of all these problems was extreme frustration and deteriorating confidence in the system. This unrest finally led to rebellion and the collapse of Soviet communism in 1991.

The poor standard of living in the Soviet Union continues today. Whereas in 1995 the United States produced 600 cars, 800 television sets, and 1,000 telephones for every 1,000 people, the former Soviet Union had only 40 cars, 50 television sets, and 100 telephones for each of its 1,000 citizens. Overall, the total consumption per capita by Soviet citizens was less than 25% that of Americans.

LACK OF INCENTIVES AND MOTIVATION

The Soviet system lacked incentives—the key ingredient of the capitalist system—for workers, for managers, and for competition. In a capitalist system, more work effort, greater creativity, and an improved product mean higher wages and promotions for workers. However, in the Soviet Union, work effort was not necessarily rewarded with higher wages or promotions. For the few workers who did receive promotions, even though their wages were higher, increased purchasing power never materialized because of what is known in economics as the *ruble overhang*: the accumulation of rubles by a society with no real wealth or purchasing power attached to them. If extra rubles are paid but no consumer goods are available for purchase, the ruble is worthless.

Another problem was that the output and mix of products and prices were determined by Soviet central planners. In a capitalist system, if a product is in short supply, its price increases. The incentive is to produce more items to make more profit. When production increases, prices are forced back down because of the larger supply. These self-correcting mechanisms were absent in the Soviet system. Without such price changes, Soviet planners had no way to gauge whether their decisions were correct or incorrect. Thus, many products that were in high demand were in short supply and remained so. Other items remained stored in warehouses because of an overabundance and lack of demand. Because the planners determined the output, there were no incentives for managers to increase their supplies.

MISMANAGEMENT AND CENTRAL-PLANNING PROBLEMS

Richard Ericson (1992) pointed out that the Soviet planning system that worked in the 1930s and 1940s, wherein production goals, quotas, prices, and supplies were set individually and separately from one another, could not work in the 1980s and 1990s because of the increased complexity of the international market. In the later years, there were far more industries for which to plan, and many more components and supplies for each item produced. In 1990, 48,000 industrial enterprises in the Soviet Union produced goods that were demanded at various stages in the economy. The planners had to ensure that all resource demands were met and that final production targets were achieved. Literally billions of supplies and demands had to be met. The outputs of any one industry were inputs to other industries, and these were yet inputs to the final products demanded in stores across the Soviet Union. If a supplier experienced a slowdown in production because of a closed mine or a breakdown in the transportation system, supplies along the way would not be met, and the entire system would break down. Then, no one would be able to fulfill production goals and quotas. The bottleneck or chain reaction would be passed on to the consumer and to the state, which consumed a large portion of its own products. The result would be resource immobilization.

The capitalist economy is a powerfully organizing force that sets millions of laborers, companies, nations, and prices in motion and allocates resources in an efficient manner. Coordination is accomplished according to consumer wants, resource availability, and business efficiency and practices. However, duplicating this coordination with state-operated economic planning units run by individuals making arbitrary decisions on allocations, prices, and resources was impossible.

AGRICULTURAL FAILURE

As discussed in Chapter 5, Soviet agricultural programs were essentially disappointing. One-fourth the Soviet Union's annual GNP and 30% of its labor force were engaged in agriculture, yet it became the world's largest importer of food. According to one estimate, the average Soviet agricultural worker was only 10% as efficient as an agricultural worker in the United States.

There is no question that the Soviet Union possessed a relatively small amount of good agricultural land, but it had enough land to be more productive.

The main problems discussed so far also apply to agriculture: inadequate administration and planning, insufficient incentives for workers, and an increasing import drain on the foreign currency reserve, which should have been used for investment, technology, and infrastructure.

MILITARY WASTE

Throughout history, Russia and the Soviet Union have been repeatedly invaded from the outside. Most recently were invasions by Napoleon in 1805 and Hitler's army in 1942. Prior to this, a series of other dictators and warring tribes had disrupted the region. Consequently, the Soviet Union became paranoid of its neighbors, especially its most powerful neighbors. In the later stages of the Cold War, fear was focused on the United States. This fear led to the arms escalation of the 1970s and 1980s, with thousands of nuclear warheads stockpiled on each side. The Soviet military expenditure accounted for 15% to 20% of the GNP, while in America it accounted for 3% to 6%. Consequently, much money that could have been used for investment, development, and growth was funneled into the military. At the same time, the most talented people were also channeled into the military as administrators and planners, including research scientists and engineers who created plans and devices for defense and attack military systems. This talent could have been best used in other sectors of the economy.

The Transition to a Market Economy

Most of the 15 former Soviet socialist republics, and notably Russia, have committed themselves to forming capitalist-type market economies. The political and economic undertones of maintaining the status quo, however, are still quite powerful. What do these countries require to establish strong market economies? Many factors are required, most of which Americans take for granted.

PRIVATIZATION

If the former Soviet Union countries are to move toward a market system, private ownership of the means of production is necessary. This means private ownership of land, factories, houses, and machinery. Department stores must also be privatized. Private property must be established and rules for its governance maintained. How this will occur is not altogether clear, but there has been a mad scramble to resurrect titles, deeds, and ownership papers from the period of 1900 to 1917, when most property was expropriated.

INTRODUCTION OF COMPETITION

If a market economy is to develop and flourish, competition must be introduced. Once again, this is a tall order. The 48,000 state-owned enterprises that were the backbone of the Soviet economy must be free to operate as independent agents, competing with one another for improved products and lower prices. Because most of these stable enterprises are operating as monopolies, competition was all but absent in the former Soviet Union. But competition is necessary for products to improve, for efficiency to take hold, and for profits to accrue to the laborers, who now have a vested interest in these enterprises. Competition is already budding as a result of joint ventures between the former Soviet republics and foreign governments or MNCs. For example, Pizza Hut is opening 1,000 restaurants from the Ukraine to the Baltic republics, and many American firms are attempting to establish a foothold in what will be one of the world's largest markets.

A REDUCED ROLE FOR GOVERNMENT

A reduced role for government in the control and operation of enterprise needs to occur. At the same time, the government must establish and maintain law and order in a new economy where privateers and black markets flourish. The government's first order of business is to establish private ownership and some level of equality with regard to opportunities for advancement in order to set the tone and to guarantee competition. At the same time, it must realize that all government-controlled enterprises may fail in the privatization process and that unemployment will flourish. In the old Soviet system, unemployment was quite low, so in order to prevent civil unrest with the higher unemployment levels that capitalism will bring, some measure of unemployment insurance must be provided immediately.

DEREGULATION OF PRICES

In a market economy, prices are not regulated the way they are in a communist system. Consequently, prices must be deregulated in the former Soviet Union. Because most prices were set artificially low, without regard for market demand and supply, the decontrol of prices will have a serious inflationary effect. With increased prices, workers will demand and receive higher wages. In 1992, Boris Yeltsin decontrolled prices in the former Soviet republic of Russia. Immediately, prices tripled.

Participation in the World Economy

For most of its 70 years of existence, the Soviet economy was isolated from the rest of the world. So that it can convert to a market system, it will have to join the world economy and take part in international trade, making its currency convertible to world standards, such as the deutsche mark, the yen, and the dollar. Firms that want

Russian shoppers are at a disadvantage compared with their American counterparts. With ideology diminishing, the Iron Curtain fallen, and the empire collapsed, Russia must now operate in an open, competitive world, as other developed nations do. Russia's decision to join NATO's Partners for Peace is a sharp contrast to Stalin's decision to keep the Soviet Union from participating in the Marshall Plan in 1948. Nonetheless, Russia's transition from totalitarianism to democracy is a huge undertaking, and there are no guarantees. Russian nationalism is on the rise, and free market economic reforms are on the wane.

to do business in the former Soviet Union currently cannot buy and sell because the ruble does not exchange in world monetary markets. Buying and selling can occur only with in-kind types of real-money products. In-kind, or nonfinancial, investment is a means of capital exchange without converting to world monetary standards. However, this method is cumbersome and not attractive to many foreign investors.

DECREASED RESISTANCE BY HARDLINERS

Hardline Soviet economists currently outnumber reform moderates in the Russian congress. Many former Communist Party members who held office in the dissolved congress would like to see the return of Soviet communism to protect their status and position. These older Communist Party members are most reluctant to admit that Soviet communism has failed. Currently, the economy is lagging, and promises made by reform leaders such as Yeltsin have not materialized, so these bureaucrats' position remains strong. Decreased resistance by the hardliners would allow a market economy to take root more easily.

WORKER INCENTIVES

In order for the former Soviet Union to convert to a market system, the workers, the former *proletariat*, will have to be strongly in favor of the move. Currently, workers who want change outnumber those who want to return to the old communist ways. But under the new capitalist system, new challenges await these workers. Harder work, longer hours, and increased productivity are needed. Workers in the former Soviet Union are accustomed to little or no responsibility for the quality and quantity of their work. All this will change. In addition, harder work should result in higher wages, which should yield more consumer goods: housing, food, appliances,

and even automobiles. If these increases do not result from harder work in the short run, reform may not be given enough time to succeed. Soviet Union laborers for 70 years have been trained to abhor private property. Now that they suddenly have the opportunity, can 70 years of hardened attitudes be changed overnight? Nobody seems to know the answer to this question.

How to Help the Former Soviet Union Embrace a Capitalist Economy

Assistance for Russia and the former Soviet federated socialist republics in their move toward a market economy can take public and private avenues: *direct foreign aid* and *private investment*.

DIRECT FOREIGN AID

In 1992, the IMF approved $24 billion in emergency aid to Russia. By 1993, about $12 billion had been spent, and another $6 billion went toward writing down the Russian debt. The remaining $6 billion for currency stabilization has not yet been dispersed because of policy differences between the IMF and Russia. The U.S. share of IMF and World Bank funding is large. In addition, many other IACs claim that they will contribute more toward direct foreign aid than Gorbachev asked for in 1991, but only after the substantial costs of the Soviet military are redirected toward economic development.

In 1994, the Russian economy deteriorated still more. Hyperinflation, a collapsing ruble, rising unemployment, decreased production, and a huge budget deficit pointed toward political disaster. Moscow's total debt, including the debt taken over from the former Soviet Union, is about $100 billion. All that Western creditors can hope for in the short term is for renegotiated debt-restructuring deals, similar to the one negotiated with Mexico and Argentina 3 years ago. The long-term solution requires converting some loans and credit to Russia into direct aid, as America has done with LDCs. The money that America gives to Egypt, Israel, Pakistan, and others each year is not repaid. Why is America's investment in Russia any less important?

The situation in Russia is likened to that of post–World War II years, when Western Europe lay prostrate. "The patient is sinking while the doctor deliberates" was the famous statement by Secretary of State George Marshall in 1947. At first, Americans thought that Europe could rebuild itself after the war but were proven wrong. By 1947, America knew that democracy would not spring automatically from the tired soil of continental Europe, particularly with communist propagators in the region. For democracy to grow, it first had to be planted. Unfortunately, the eastern European nations were already out of reach.

The Marshall plan was not cheap; it cost America about $60 billion in today's market, in the form of direct foreign-aid grants, not repayable loans. Public money seeded the ground for private investment money, which began flowing heavily into Europe when the Marshall plan was terminated in 1952. What was the payoff?

Today, every nation in Western Europe has a stable democratic government in alliance with the United States. NATO is a successful alliance. The population flourishes by every social and economic measure. Because of NATO's solidarity, the Cold War ended successfully several years ago. The EU is America's most significant trading partner and among the few with which the United States has a trade surplus.

Without the Marshall plan, Western Europe, defeated and destitute, could easily have followed Eastern Europe down the Communist path that it is currently trying to retrace. Thanks to America's help, the nations of Western Europe are fully recovered and were able to help in the current endeavor, which is to rebuild Eastern Europe and Russia.

PRIVATE INVESTMENT

The former Soviet Union's 15 republics compose a market comparable to that of the North American Free Trade Association—Canada, the United States, and Mexico. The total EU market is just 25% larger. The former Soviet Union is such a large market that one would expect it to attract a vast amount of foreign investment and to use this investment to strengthen its economy. The hope is that not only private investment and capital will flow into the former Soviet Union, but also the managerial skills and investment behavior necessary to put the nations back on their feet economically.

Deep problems exist, however. No one is sure exactly who is setting the standards for government and business reform. No one is exactly sure who holds the

power. No one is sure how quickly the political and economic climate will change. No one is sure of the eminence of suppliers for products that are needed to produce consumer goods for such a large market in the present Soviet system. Another major problem is lack of convertibility of the ruble. How will American firms be paid? How will IACs receive a profit? Answers to these questions and others remain ominous and can be sorted out only with time.

Economy and Privatization, 1997

The demise of the Soviet Union and the move towards a market economy has created massive economic problems (hyperinflation, shortages, crime). To remedy the economic situation, the Russian government reduced agricultural subsidies and raised the price of bread and other basic foodstuffs. Defense expenditures and the production of weapons for sale on the world market were reduced initially. The result was a steady drop in the total sale of weapons abroad, and thus, a reduction in the needed inflow of hard currency. Because Russia had few manufactured goods, food products, or other exports, arms exports were again increased in 1994 and 1995. The defense industry has tried to convert its production to nonmilitary goods, similar to such defense conversion programs in the United States. However, thus far, too little progress has been made in this structural economic changeover.

The Russian government has sold off small enterprises, such as cafes, restaurants, markets, shops, and small service establishments in a move towards privatization of state-owned companies. The privatization of medium and small-scale factories, warehouses, and service industries has proven to be much more complicated, however. Since it is almost impossible to calculate the future value of these large enterprises, and since many such enterprises have such antiquated technology and infrastructure and are presently in a state of great disrepair, domestic and foreign investors have been difficult to find.

Two approaches toward privatization have been practiced so far. The first is "immediate privatization," whereby a group of investors buys an establishment or enterprise and begins operating it. The second approach entails selling or giving certificates to the employees of various enterprises, giving them first chance to acquire partial ownership. A number of small stock exchanges and commodity exchanges are now operating within Russia, and some joint stock companies exist. Russia is moving toward the marketing of all shares of former state-owned businesses and enterprises, and each Russian citizen has been given the equivalent of $100 worth of vouchers to buy stock.

Nowhere has so much been privatized so quickly. Between 60% and 70% of Russia's businesses are now in private party hands. But there have been major problems associated with the privatization scheme. What had been a Soviet integrated economy fell apart into 15 separate parts almost overnight. Lines of authority, interindustry linkages, and supplies of customers quickly disappeared. Rigid control gave way to anarchy and left an economic vacuum. The results have led to chaos. Massive corruption associated with the transfer of asset ownership is occurring, especially the factories for durable goods and the military-industrial sectors. Most of these privatization schemes are viewed locally as shams in that former communist party bosses and industry managers have received the major shares of stock and ownership for themselves and their families.

Private ownership does not mean that a plant manager will be able to effectively operate a firm according to market principles. The new market system needs vigorous competition to make it operate efficiently, and this is still lacking in Russia. Most enterprises are still not competitive on the world market. The managers who are holdovers from the old communist regime (*nomenklatura*) still seek and find government subsidies that prop up inefficient operations. One indication that these old practices are still being used is that there are almost no bankruptcies among the largest enterprises. The Russian government holds the largest share of stock in these largest 10,000 enterprises, especially those that control the natural resources of the country. Russia remains a country of monopolies. For example, some 31 ferrous metallurgical enterprises produce two-thirds of this economic sector's output.

Most large enterprises operate much as they did before 1992, relying on state subsidies, offering no research and development, and almost nothing in the way of new products or manufacturing efficiency. Rapid privatization has also created increased crime, inflation, and unemployment. The Russian mafia controls 60–80% of all private banking activities, and 25% of Russia's entrepreneurs make regular extortion payments, including American company chains (Subway Sandwiches, Estee Lauder, and Radisson Hotels). Inflation remains at 5% per month and unemployment is high (20%). As former Soviet factories and enterprise are shut down and people lose their jobs, other factories and enterprises are starting to run more efficiently under the new system of privatization.

In 1993, the free purchase and sale of agricultural land began for the first time since the Bolshevik Revolution, and outright private land ownership is now being practiced. Serious food shortages still exist. As the new Russian government began to phase out price controls of the communist distribution system, staple food items, such as bread and milk, have shot up in price, more rapidly than wages. Russia also has a high level of debt because it assumed the bulk of the Soviet Union's international shortfall.

Boris Yeltsin solicited aid from the International Monetary Fund and the World Bank. The Russian ruble on world currency markets has tumbled. In 1995, Russia was unable to continue paying principal and interest on its foreign debt and negotiated a moratorium. Without large bailouts from these international organizations, or foreign aid from Western nations such as Germany, France, and the United States, and without money to invest and modernize its oil and natural gas industries, inflation and economic decline will continue and the debt will not be paid back.

The Russian government established new ties with the West in June 1994 by joining the NATO (North Atlantic Treaty Organization) Partnership for Peace military cooperation program, and also by signing an agreement promoting food trade with the EU. But differences do exist between Russia and its new Western allies. Now, Russians face an expanding NATO alliance into neighboring states, which were formerly under Soviet control. In 1996, 9 countries—Poland, Hungary, the Czech Republic, Slovakia, Albania, Romania, Lithuania, Latvia, and Ukraine—applied for membership in the NATO alliance. Russians now wonder how much of a threat NATO expansion might be. NATO expansion will cause more friction between Russia and the West, and Russia may seek a military alliance with a former nemesis, China or Japan. The latter alliance is altogether possible because of the complementarity of needs—Russia needs capital investment and technology, and Japan wants the return of the Kuril Islands, taken by the U.S.S.R. in World War II.

In the first multiparty presidential election in Russia, Boris Yeltsin in 1996 won the majority of votes, strengthening Russian democracy. Most Russians, while unhappy with the Yeltsin administration, apparently prefer the continuation of market reforms and individual freedoms, even under Yeltsin's faltering leadership, rather than a return to the communist past. The Russians did not heed the communist call for a "new empire" because they are worried about local economic issues, not about foreign policy.

Granted, the problems between Russia and the West, including NATO expansion into Eastern Europe, the supply of nuclear reactors to Iran, and Russia's assertiveness in places like Chechnya, are unlikely to vanish soon. At least the Western nations, and the United States in particular, have a chance to manage them without a strong likelihood of a new communist-evoked cold war anytime soon. The people of Russia appear to be making their choice for freedom, in spite of the probability that painful economic and political reforms may take a generation or longer to mature.

In Russia today, new personal freedoms may outweigh the drawbacks of reform. However, Russia is not only trying to establish a reformed system, it is also trying to dismantle the 1,000-year-old Russian culture of "unfreedom." Many substitute freedom of action for freedom of thought. Some Russians see freedom as license. They don't realize freedom requires self-discipline. They fear that freedom leads to anarchy. They view it as the ability, if one can, to lord it over those weaker than they are. This may explain why, in a survey of almost 2,500 Russians conducted in 1996 by Richard Rose, a professor at the University of Strathclyde in Scotland, 77% of respondents answered "order" when asked whether order or democracy was "more important for Russia."

Nor will it be easy to undo 70 years of a command economy. The Soviet government dismantled free-market commercial codes, contract etiquette, legal and accounting practices that are essential to conducting business, codes of honor, business customs, and ethics. In spite of new and promising reforms, the changeover will require decades. The question remains, do the Russians have enough patience? Previous reforms in Russia have not worked well, and there is a growing discontent with Russia's present form of corrupted capitalism. It may be a market, but not a market that most people of the world could tolerate.

Help for Less Developed Countries from Advanced Nations

The IACs must come to the aid of LDCs today. How can this occur peacefully? Three methods are generally cited for IACs to help LDCs: (1) expand trade with LDCs, (2) invest private capital in LDCs, and (3) provide foreign aid to LDCs. As a last resort, the World Bank may also step in to help.

Expansion of Trade With Less Developed Countries

Some economists have suggested that expanding trade with LDCs is one way to help them. It is true that reducing tariffs and trade barriers with LDCs will improve the situation somewhat. With the North American Free Trade Agreement (NAFTA), the United States is attempting to remove all trade barriers with Mexico, for example. However, increasing trade with LDCs means that they must produce something for export, something of value that an IAC wants. Not all developing nations have abundant raw materials and agricultural surpluses. Furthermore, when a major trade relation exists between an LDC and an IAC, the LDC is often tied to the developed nation, and as a result of the economic vitality of the IAC, the LDC experiences downturns that it would not normally experience. This relationship could have dire consequences for the price of exported goods from the LDC. For example, the world price of copper dropped from $1.50 an ounce in 1974 to $0.50

an ounce in 1975. Thus the export markets were destroyed for some LDCs who depended on copper for survival.

Private Capital Flows to Less Developed Countries

LDCs are also a destination for investments from MNCs, private banks, and large corporations. For example, major U.S. automakers have now built numerous plants in Brazil and Mexico. In Tijuana alone, 500 U.S. labor-intensive manufacturing plants now take advantage of the average $0.90 per hour worker wage. As mentioned earlier, Citicorp and Chase Manhattan Bank have made loans to the Philippine and Argentine governments. Private capital flows have been increasing to LDCs from advanced nations since 1950. They reached a high point in the early 1980s near the time of the LDC debt crisis of $30 billion per year. Since the debt crisis, however, investments and private capital flows have decreased substantially because of concerns about returns on investment.

All too often, however, the LDC must not only provide the investing corporation or country with some pledge of a financial return, but also guarantee that a politically stable environment will prevail. An international trade climate must also be supported by financial and marketing systems, a favorable tax rate, maintenance of infrastructure, and some measurement of a reliable labor flow. These latter guarantees are frequently lacking in poor LDCs and preempt the major capital flows that would otherwise exist. African nations in particular have not been able to tap private capital flows from large corporations and commercial banks because of problems with these conditions.

Foreign Aid From Advanced Nations

In order to reverse the vicious cycle of poverty shown in Figure 1.17a, foreign aid is needed in the form of direct grants, gifts, and public loans to LDCs. Capital accumulation is necessary to retool the workplace, to increase productivity, and to retrain the labor pool. Capital is needed for improvement of the sadly lacking infrastructures in developing nations that are necessary to attract private capital flows. These infrastructures include transportation facilities, communications systems, educational programs, irrigation for agriculture, and public health not only for populations, but also for visiting MNCs and their entrepreneurial classes.

The United States has been a major world player in foreign-aid programs. U.S. foreign aid averaged $15 billion per year, for example, from 1990 to 1994. The majority of this aid was administered through the U.S. State Department's Aid for International

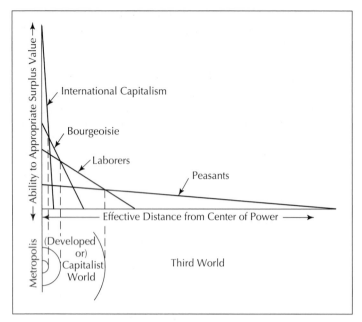

FIGURE 12.15
The world's stratified society. In a world viewed as a set of von Thünen rings, the peasants exist at its outer fringe—poor and discriminated against with respect to the use and control of world resources. Their access to credit, insurance, education, manufactured goods, technical knowledge, and infrastructure is virtually nonexistent. *(Source: deSouza and Porter, 1974, p.81)*

World Bank Aid

The IACs, including the United States, have established a *World Bank*, whose major goal is to support developing nations in their quest for improved economic status. The World Bank is generally considered a "basket case" funding agency. When all hope has faded for private capital flows and for expanding trade and direct foreign aid from IACs, the World Bank sells bonds to finance and fund bailout programs for LDCs. World Bank activities center primarily on infrastructure. Examples of World Bank projects are large dams, irrigation projects, transportation systems, roads, highways, airports, and basic communications systems. Sanitation programs, health programs, and housing have also been funded. The World Bank's efforts are to improve the economic setting of a country so that private capital flows will be forthcoming and will be self-perpetuating. The World Bank has also provided consultancy expertise to developing nations to determine not only their internal strengths and weaknesses, but also how to reinforce their positive aspects of comparative advantage and cumulative causation.

SUMMARY

In this chapter, we consider problems of Third World development. We begin by discussing goals for development and by listing objectives that are by and large universally endorsed. We then explore typical characteristics of LDCs—overpopulation, lack of resources, capital shortage—considering whether these attributes can be properly construed as causative factors.

In discussing major perspectives on development, we see how modernization theories, which stress economic growth and Westernization, have obscured important aspects of underdevelopment. World political economy theories explain why the Third World does not develop. Ecopolitical economy theories question the desirability of the income-oriented approach, instead emphasizing the feasibility of the basic-needs approach to development.

As the world system expanded, it became differentiated into a core of rich countries and a periphery of poor countries. One distinctive linkage between the

Development (AID). Additional direct aid included food programs to needy countries under the U.S. government's Food for Peace program. Other nations have also rallied, and in 1990, IACs as a group contributed a total of $50 billion. In addition, OPEC nations contributed $2.5 billion.

Unfortunately, to the discredit of the United States, most of its aid has stipulations such as purchase requirements that make the LDCs patronize American products and services. Furthermore, in the past, foreign aid has been distributed on the basis of political ties, as opposed to economic need. For example, Israel, Egypt, and Turkey each receive nearly $3 billion in aid per year. These nations are neither the most populous nor the most needy, but they do occupy a strategic area of the Middle East where vast oil deposits exist and where America is fighting a cold war against Islamic fundamentalism. In addition, the United States has guaranteed its support of Israel in its hostile environment.

Also, unfortunately, IAC contributions amount to only 0.5% of their collective GDPs. This amount is too small. To make matters worse, the former Soviet Union and eastern European nations are now making strong pleas for increased aid from the West. Many LDCs fear that aid that would normally be channeled to them will now go toward supporting privatization in former communist areas of the world. The IACs' fear is that the cost of failure (because of IAC neglect) of democratization in Russia will be far greater than the cost (as a result of IAC aid) of success. If Western Europe and America agree, the larger portion of their foreign aid (through grants, loans, and direct aid) will be syphoned from the LDCs.

core and the periphery was colonialism. Our discussion of waves of colonialism demonstrates that underdevelopment is not a state but a process.

When colonial holdings disappeared after World War II, experts drew up plans to accelerate the pace of development in developing countries. Regional scientists emphasized the center-dominant model. Although this approach tends to concentrate income and wealth, especially in the early stages, some scholars contend that eventual convergence between rich and poor regions is the norm. Other scholars believe that only progressive taxation, social services, and other government actions can spread the benefits of growth. In contrast, social theorists argue that because developing countries are dependent on core regions, and the spread from center to periphery is limited, institutional reform must precede planning.

KEY TERMS

backwash effects
basic needs
bazaar economy
capital flight
center-periphery
circular and cumulative causation
Commonwealth of Independent States (CIS)
colonial division of labor
colonial organization of space
dependency
development
deviation amplification
deviation counteraction
direct foreign aid

dual economy
ecopolitical economy
fatalism
firm-centered economy
growth-stage theories
involution
land alienation
less developed countries (LDCs)
limiting factors
lower circuit
modernization theories
multiplier
polarization effects
primary sector
privatization

quaternary sector
secondary sector
spread effects
squatter settlements
tertiary sector
trickle-down effects
underdevelopment
underemployment
upper circuit
urban peasants
vicious cycle of poverty
world cultural realms
world political economy theories
world system theory
write-down

SUGGESTED READINGS

Anderson, J., Brooke, C., and Cochrane, A. 1996. *A Global World*. New York: Oxford University Press.

Armstrong, W., and McGee, T. G. 1985. *Theatres of Accumulation: Studies in Asian and Latin American Urbanization.* New York: Methuen.

Bingham, R. D., and Hill, E. W. 1997. *Global Perspective on Economic Development.*

Corbridge, S. 1986. *Capitalist World Development: A Critique of Radical Development Geography.* Totowa, N.J.: Rowan and Littlefield.

Johnston, R. J., Taylor, G. P., and Watts, M. J., eds. 1996. *Geographies of Global Change.* New York: Blackwell.

WORLD WIDE WEB SITES

TERRA: BRAZIL'S LANDLESS MOVEMENT—NEW YORK TIMES [REALPLAYER]
http://www.nytimes.com/specials/salgado/home/
The New York Times has recently opened this web special by documenting the plight of Brazil's landless in both words and pictures.

1997: COUNTDOWN TO HISTORY—SOUTH CHINA MORNING POST
http://www.scmp.com/1997/

HONG KONG 1997: LIVES IN TRANSITION—PBS
http://www.pbs.org/hongkong

THE YELLOW PAGES: GEOGRAPHY
http://theyellowpages.com/geography.html
Lists links of geography sites and search tools.

GEOPEDIA
http://www.geopedia.com
Geopedia Online contains key information on every country of the world. Each country profile provides facts and data on geography, climate, people, religion, language, history, and economy, making it ideal for students of all ages.

UN HUMAN DEVELOPMENT
http://www.undp.org/undp/rbec/nhdr/1996/
Report includes excellent information for developing countries.

WASHINGTON POST NEWSPAPER SITE
http://www.washingtonpost.com
Within this site, under "International," is a choice called "World by Region." This connects you to an index of all recent Washington Post articles on the country of your choice. This site also connects students to AP wire service stories and other sources.

CIA FACT BOOK
http://www.odci.gov/cia/publications/95fact/
This site includes data from every country in the world with a map.

LIBRARY OF CONGRESS
http://www.loc/.gov
Astounding variety of resources and exhibits.

GLOSSARY

absolute advantage   The ability of one country to produce a product at a lower cost than another country.

absolute location   Fixed position in relation to a standard grid system.

abstract space   A geographic space, homogeneous in all respects. Movement over this space is equally easy in all directions.

accessibility   A measure of aggregate nearness. It refers to the nearness of a given point to other points.

accessibility index   A measure of the shortest path between one vertex and all others.

achieved characteristics   Sociodemographic characteristics, such as education, occupation, income, marital status, and labor force participation, over which we have some degree of control.

acid rain   Acid rain, snow, or fog derives from the combustion of coal, releasing sulfur and nitrogen oxides that react with water in the earth's atmosphere.

adaptive economic behavior   Behavior based on systematic analysis of alternative locations that leads to the development of rational industrial location.

administrative principle   The spatial organization of central places in which a higher-order administrative place is surrounded by six lower-order administrative places.

adoptive economic behavior   Economic location decision making based on copying past, successful patterns.

African Development Bank   An international financial institution that extends loans to African countries for purposes of development.

age distribution   The age structure of a population divided into five-year age cohorts.

agglomeration   A measure of aggregate nearness. It refers to total aggregate nearness among a number of points.

agglomeration economies   The savings in cost that result from the clustering of firms.

agribusiness   Food production by commercial farms, input industries, and marketing and processing firms that contribute to the total food sector.

agricultural involution   The ability of the agricultural system in densely populated areas of Asia to absorb increasing numbers of people and still maintain minimal subsistence levels in rural communities.

agriculture   The livelihood of farming and cattle ranching by cultivation of the soil and the rearing of livestock.

arable land   Land that can be used to grow food crops.

ARC/INFO   A GIS software package produced by Environmental Systems Research Institute of Redlands, California.

areal differentiation   The study of geographic areas for purposes of comparing their similarities and differences.

areal integration   An approach to the world economy that searches for theory and broad-sweeping understanding, rather than unique characteristics of every place or country.

ascribed characteristics   Sociodemographic characteristics such as gender, race, and ethnicity, with which we are born and over which we have basically no control.

Asian Development Bank   An international financial institution that extends loans to Asian countries for development projects.

assembly costs   The costs of bringing raw materials together at a factory.

Association of Southeast Asian Nations (ASEAN)   Economic association of seven Southeast Asian countries.

Atlantic Fruit and Vegetable Belt   An area of high agricultural production per acre, stretching from southern Virginia through the eastern half of North and South Carolina to coastal Georgia and Florida.

attributes   Non-graphic descriptors of point, line and area entities in a GIS.

automatic vehicle identification (AVI)   Automatic surveillance of highway vehicles aimed at adjusting tolls, based on levels of congestion.

automobile component plants   Plants manufacturing parts required in the assembly of an automobile.

average product   The total output divided by the number of units of input used to produce it.

average total costs   Total costs divided by the quantity of output.

back office accounts   Business functions not requiring direct customer contact.

baby boom   The dramatic rise in the U.S. birth rate following World War II, between the years 1946 and 1964.

baby bust   The years immediately following the baby boom in which U.S. fertility rates fell dramatically, or any period of low population growth.

backhaul   A carrier's return trip.

backward integration   A firm takes over operations that were previously the responsibility of its suppliers.

backward linkages   Integration of production firms with their supply firms or sources.

balance-oriented life-style   A mind-set that insists that because resources are finite, they must be recycled, and input rates slowed down to prevent ecological overload.

balance of payments   For a country, the difference between payments for exports and payments for imports during a year. A negative balance occurs when imports exceed exports.

balance of trade   For a country, the difference in the total value of exports from the total value of imports for a given year.

basic cost   The cost of an input for a firm at its least-cost location.

basic firms   Firms that produce goods or services for export outside the local region.

bazaar economy   The traditional and labor-intensive sector of a Third World city.

behavioral matrix   A device used for analyzing nonoptimal decision making. It shows the location of decision makers with respect to information and their ability to use information.

beta index   A measure of linkage intensity or connectivity.

bid rent   The amount of money offered to purchase a piece of land.

bilateral trade agreement   Agreements between two countries calling for favorable concessions of trade.

birth rate   The number of live births per 1,000 population per year.

boomdocks   Boom towns in remote, boomdock areas that provide inexpensive housing.

brain drain   Less developed countries lose talented people to industrially advanced nations through immigration politics.

break-of-bulk   The stage at which a shipment is divided into parts. This typically occurs at a port where the shipment is transferred from water transport to land transport.

Bretton Woods   The New Hampshire location of a 1944 international meeting of treasury and bank officials of the Allied countries. The meeting designed the International Monetary Fund (IMF) and the International Bank for Reconstruction and Development. It also led indirectly to the creation of the General Agreement on Tariffs and Trade (GATT).

browsers   Remote controllers that assembly text, facts, figures and photos on the web pages to appear on computers screens around the country and around the world.

business cycles   Periods of economic expansion and contraction.

business process reengineering (BPR)   It includes management realignments, mergers, consolidations, operation integration, downsizing and reorienting distribution practices to become more efficient and more profitable in the world economy.

California Environmental Quality Act (CEQA)   Legislation enacted in 1971 requiring environmental impact reports and review to be prepared for all major developmental and land-use change.

Canada/U.S. Free Trade Agreement (CFTA)   The agreement in 1989 that lifted tariffs and trade barriers between these two countries.

capital   A factor of production, including tools, buildings, and machines used by labor to fashion goods from raw material.

capital accumulation   The engine that drives economic growth under the capitalist system.

capital controls   Restrictions on the movement of money or capital across national borders.

capital flight   Local individuals and whole countries have invested their monies in overseas ventures and in foreign banks for safekeeping.

capital goods   Manufactured items that can be used to create wealth or other goods. For example, a home washing machine is a consumer good, but the machines used to make that washing machine are capital goods.

capital resources   Money supply and availability to finance an industry.

capital-intensive   A term that applies to an industry in which a high proportion of capital is used relative to the amount of labor employed.

capitalism   The political-economic system based on private property and profit.

carrying capacity   The maximum population an ecosystem can support.

cartel   An organization of buyers and sellers, capable of manipulating price and/or supply.

cartograms   Maps that show relationships often modifying actual geographic space in deference to phenomena being displayed.

ceiling rent   Maximum rental that a particular user pays for a site.

central business district   The downtown area of a city.

Central Florida Manufacturing Region   Includes the cities of Jacksonville, Tampa, St. Petersburg, Orlando, and Miami.

central function   A good or service offered by a central place.

central place   Center for local exchange of goods and services, including stores and population settlement.

central-place theory   A theory that attempts to explain the size and spacing of settlements and the arrangement of their market areas.

centrifical drift   The outward movement from the city of industrial properties due to the attention of less expensive land.

child dependency ratio   The proportion of a population for a country under the age of 15.

Chi-Pitts   Megalopolis centered on Chicago, Detroit, Cleveland, and Pittsburgh.

circular and cumulative causation   Myrdal's theory that continuing changes move the system in the same direction as initial changes, but mush farther, resulting in industrial and urban agglomeration and innovation.

circular flowing market system   The flow of labor and resources from households to firms and the flow of products and wages from firms to households.

cloning spatial structure   A headquarters/branch plant structure allowing for the complete production process to take place at each site.

colonial division of labor   An artificial form of labor specialization imposed on underdeveloped countries by colonial powers.

colonial organization of space   The European organization and zoning of land at all scales—urban, regional, national—to serve Europe's own interests during the colonial period.

command economy   A society in which a central authority or government establishes the rules of economic behavior and decision making and usually owns the means of production.

commercial agriculture   Agricultural goods produced for sale in the city or on the international market.

commercial geography   The study of products and exports of the main regions of the world.

common market   A form of regional economic integration among member countries that disallows internal trade barriers, provides for common external trade barriers, and permits free factor mobility.

Commonwealth of Independent States (CIS)   The former Soviet Union's 15 socialist republics, which today are independent countries, the largest of which is Russia.

comparative advantage   The theory that stresses relative advantage, rather than absolute advantage, as the true basis for trade. Comparative advantage is gained when countries focus on exporting the goods they can produce at the lowest relative cost.

compensation   Wages, salaries, or bartered items paid by employers to workers, plus fringe benefits, such as unemployment compensation and medical payments.

compensation demand   A table that illustrates the total amount of a good or service that all consumers in a country, taken together, are willing to buy at each possible price.

competitive-bidding process   The aspect of classical location theory in which those people willing and able to pay the highest price for a particular site win the competition and put the site to the highest and best use.

complementarity   A concept in which two places interact based on a demand in one place and a supply in the other, the demand and the supply being specifically complementary.

computer network   Interlinking of computers, regionally or internationally, via hardwire, fiber optics, or telecommunications satellite for information transfer and decision making.

concentric zone model   The city is viewed as a series of concentric zones, with the focus of economic activity on the central business district (CBD). City land use expands symmetrically in all directions.

concrete space   The actual surface of the earth in all its geographic complexity.

conglomerate   A widely diversified corporation that controls the production and marketing of dissimilar products.

conglomerate merger   The control, production, and marketing of diverse products. (See Diversification, Horizontal Integration, Vertical Integration.)

connectivity   A measure of the interrelation between places in a network.

constant returns to scale   As outputs double, total costs double, but the cost per unit of output remains unchanged.

consumer   Individuals that purchase goods or services from a firm or market.

conurbation   A continuously urban area formed by the expansion and consequent coalescence of previously separate urban areas.

convenience goods   Central functions that are low in price, uniform in quality and style, and purchased frequently. They include goods and services needed on a day-to-day basis, such as groceries, gasoline, and drugstore items.

core-periphery model   Periphery countries and regions send raw materials to core or industrialized regions at the center of the world economy. From here they are processed and redistributed to all regions of the world. Peripheral regions are dependent on core regions, but not the reverse.

Corn Belt   The midwestern region of the United States, centered on the state of Iowa, that predominates in the growing of corn and cattle raising.

cost-insurance-freight (CIF) pricing   A policy whereby each consumer is charged production costs plus a flat markup to cover transportation charges.

cost-space convergence   The reduction of travel costs between places as a result of transport improvements.

countertrade   The direct exchange of goods or services for other goods and services.

crude birth rate   The number of raw births per 1,000 population per year in a given country.

crude death rate   The number of deaths per 1,000 population in a country in a given year.

culture   The total way of life of a population, including beliefs, traditions and lifestyles.

cumulative causation   The process by which economic activity leading to increasing economic development tends to concentrate in an area with an initial advantage, draining investment capital and skilled labor from the peripheral area.

customs union   A form of regional economic integration among countries that disallows internal trade barriers and provides for common external trade barriers.

cyberspace   Information infrastructure provided by linked networks, databases, servers, browsers and the Internet.

daily urban system   Trade areas around cities or employment centers that have a large area of dominant influence.

death rate   Annual number of deaths per thousand population.

decentralization   The transfer of control from national to local and regional levels of economic activity or from the core to the periphery.

deindustrialization   The economic transformation of manufacturing industry to service occupations.

demand and supply   Demand is the quantity of a good that buyers would like to purchase during a given period at a given price in a competitive market economy. Supply is the quantity of a good that sellers would like to sell during a given period at a given price.

demand conditions   The market conditions in a host country that aid the production process.

dematerialization   The reduction in raw materials used to make finished products.

demographic equation   Population at time 2 equals population at time 1, plus births, minus deaths, plus in-immigration, minus out-migration.

demographic transition   The historical shift of birth and death rates from high to low levels in a population.

dependence   A conditioning situation in which the economies of one group of countries are underdeveloped by the development and expansion of other groups.

dependency ratios   The portion of elderly or young people in a society.

dependency theorists   The theory of economic development that argues that the Third World is dependent on, and has been exploited by, their former colonial overseers.

depletion curves   Graphs used to project lifetimes of resources.

deregulation   The reduction of government controls on economic activity within a country.

desertification   Overuse, overpopulation and drought causes the expansion of desert lands and non-cultivatable regions.

developing countries   Countries experiencing economic growth and modernization but are, heretofore, less developed than advanced industrialized countries.

development   A historical process that encompasses the entire economic and social life of a nation, resulting in change for the better. Development is related to, but not synonymous with, economic growth.

development bank   An investment and/or loan fund that aids the development of underdeveloped countries.

deviation reduction   Policies or processes that decrease in the quality between rich and poor regions.

deviation-amplification   Any process that amplifies an initial inequality and increases divergence from an initial condition.

diffusion   The spread of innovation outward from a core region or hearth.

direct investment   The purchase of enough of the equity shares of a company to gain some degree of managerial control.

direction   The orientation of places toward each other.

diseconomies of scale   Diminishing marginal returns to scale that occur when a firm becomes too large to manage and operate efficiently.

disguised unemployment   A term used by some economists to describe the surplus of labor that is thought to exist in the traditional sector of underdeveloped countries.

disorganic development   A form of "development" at odds with local cultural and political institutions.

distance   A measure of the cost to overcome the space between two places.

distance-decay effect   With increase in distance, the decline in the level of interaction between two places.

distance learning   Virtual instruction, class attendance or participation via telecommunications allowing instruction source and student to be separated from one another.

diversification   A company produces an increasing number of unrelated products, each with elements of horizontal and vertical integration.

division of labor   The specialization of workers in particular operations of a production process. Labor specialization is a source of scale economies, a necessary ingredient in the evolution of a market-exchange economy.

domestic farm subsidy   Direct payments to farmers according to their production levels in order to subsidize their output.

double cropping   More than one crop is produced from the same plot throughout the year.

doubling time   The time in years required for the population of a country to double.

dual economy   In the study of industrial location, a term used to refer to two types of business enterprises, fundamentally different from one another: large, organically complex center firms and small, simply structured periphery firms. In the study of development, the term is used to refer to two types of social and economic systems existing simultaneously within the same territory: one system modern, the other traditional.

dual-sourcing   A strategy of cloning used by multilocational companies to guarantee continuity of production, usually by undermining the potential monopoly control of a work force in one place over a particular production process.

dumping   The practice of firms selling off goods in markets at lower than domestic market prices in order to clear reserves.

East Asia Miracle   The rise of certain East Asian countries into an era of double digit growth rates and progressing from the status of developing countries to newly industrialized countries since the 1950s, including Twain, South Korea, Hong Kong, Singapore, China, Malaysia, Indonesia and Thailand.

economic integration   The ultimate form of regional integration. It involves removing all barriers to interbloc movement of merchandise and factors of production, and unifying the social and economic policies of member nations. All members are subject to the binding decisions of a supranational authority consisting of executive, judicial, and legislative branches.

economic liberalism   Sometimes used as a synonym for capitalism.

economic person   The behavior of any individual who consistently does those things that will enable him or her to achieve the desired economic objective and, therefore, yields the maximum amount of utility.

economic rent   The monetary return from the use of land after the costs of production and marketing have been deducted.

economic restructuring   The transformation of an economy by discarding inefficient and obsolete systems and bringing new, efficient and competitive systems on line through technological innovation, human resource development and privatization.

economic union   A form of regional economic integration having all the features of a common market, as well as a common central bank, unified monetary and tax systems, and a common foreign economic policy.

ecumenopolis   The global urban network of the twenty first century, consisting of most of the world's population.

edge   A link; a route in a topological diagram of a network.

egalitarian society   A society that has as many positions of prestige in any given age-sex sector as there are persons capable of filling them. It is forged through cooperative behavior.

elasticity   The responsiveness of prices to changes in supply and/or demand for a good.

electronic data interchange   The electronic movement of standard business documents between and within firms.

E-mail   A computerized message network.

Engel's Law   The principle according to which, with given tastes or preferences, the proportion of income spent on food decreases as income increases.

entrepreneurial skill   The human know-how and skill that combines other resources to create a product or service using innovations and decisions.

environmental determinism   The notion that human behavior is environmentally prescribed.

environmental perception   The ways in which people form images of other places, and how these images influence decision making.

environmental sustainability   The level of the environment for which renewable resources can resupply and, therefore, sustain themselves in a steady state.

equilibrium price   The price at which the quantity supplied equals the quantity demanded and, therefore, no surplus or shortage is produced.

EU   In 1992, the EEC, or Common Market, changed its name to the European Union and expanded membership to include 15 countries that are associated economically, mainly through tariff reductions on exports and imports.

Eurocurrency   A currency deposited in a commercial bank outside the country of origin.

Eurodollar   The major form of Eurocurrency.

Euromarkets   The international financial markets that usually exist outside of the country whose currency is utilized.

European Economic Community (EEC) A group of European countries established in 1958 on the basis of a treaty signed in Rome in 1957. The community consists of 12 members—France, West Germany, Italy, Belgium, the Netherlands, Luxembourg, Britain, Ireland, Denmark, Greece, Spain, and Portugal—whose aim is to establish a United States of Europe; sometimes called simply, E.C.

European free-trade association A group of European countries established in 1960 for the purposes of trade, aiming to abolish tariffs on imports of goods originating in the group. The original members were Britain, Norway, Sweden, Denmark, Portugal, Austria, and Switzerland. Finland joined as an associate member in 1961. Iceland joined in 1970. Britain and Denmark left the free-trade area in 1973 upon joining the European Economic Community (EEC).

exchange controls Restriction of free dealings in foreign exchange, including multiple exchange rates and rationing.

exchange rate The value of the U.S. dollar compared with foreign currency.

exchange value The value at which a commodity can be exchanged for another commodity.

expert systems Rule-based instruction in the form of computer software enabling computers and machines to help make decisions.

expert systems Software systems that can control operations or manufacturing using programmed knowledge from experts.

export quota The government of an exporting country limits the volume or value of exports to an importing country voluntarily.

export restraint agreement A nontariff barrier whereby governments coerce other governments to accept voluntary trade export restraint agreements.

export-led industrialization A development strategy emphasizing the production and export of manufactured items. Its success depends on a rising world economy.

externality An external effect in the environment or region surrounding a firm or business, impacting it in a positive or negative way.

exurbs Small, semi-rural towns, at least 100 miles away from major metropolitan regions that house suburbanites seeking cheaper housing, a slower pace of life and desire to be near nature. Smart cities: A city that is linked via computer network and information technology to itself and other sources of information across the country and world. Complete with infrastructure—the medium over which information travels; access points—the terminals which users can interact with the network; and applications—the use to which the network's information and resources can be put.

factor of production One of the economic inputs—land, labor, capital, entrepreneurship, technology—essential to a production effort.

factor conditions Land, labor, capital, enterprise and other factors of production.

factor-driven stage Processing and exporting natural resources and primary products.

factor endowments The combination of land, labor, capital and enterprise that each country naturally possesses.

fatalism The philosophical belief that future life conditions will not improve, regardless of the amount of work or effort involved.

feed lots Penned corrals near cities used for fattening cattle; forage is trucked into these fattening pens.

Fertile Crescent Present-day countries of Israel, Lebanon, Syria, Iraq, and parts of Turkey, Iran, and Jordan, where early plant and animal domestication occurred 10,000 years ago.

fertility The actual reproductive performance of an individual, a couple, a group, or a population.

fertility rate The number of live births per 1,000 people in a country per year.

fiber-optic cable A new high-carrying capacity technology in telecommunications.

fifth wave Technology period that is characterized by growth in advanced services, cyberspace and the creation of information technologies.

filtering process The relocation of people within a city under free-market conditions. Filtering may be upward or downward. Upward filtering refers to the movement of people into higher quality housing. Downward filtering refers to the movement of people into lower quality housing.

firm-centered economy The modern and capital-intensive sector of a Third World city.

first world Western market economies.

fixed cost The cost of the investment in land, plant, and equipment that must be paid, even if nothing is subsequently produced.

flexible economy Microtechnologies increasingly allow the economy to custom design and produce products and services at lower cost for niche markets. The flexible economy will become more and more customized.

flexible factories A manufacturing process that allows speedy redesign and retooling to produce small runs of niche products using compact efficient teams that allow little or no inventory supervision or coordination.

flexible manufacturing systems Use of computer controlled machinery and robotics to minimize costs and deliver products to fast changing market demands.

floating exchange rate The value of a currency fluctuates according to changes in supply and demand for the currency on the international market.

Food for Peace programs U.S. program that allowed agricultural surpluses to be distributed to starving nations instead of being liquidated.

Food Stamp program U.S. government program that allowed agricultural surpluses to be distributed to America's poor, instead of being liquidated.

forces of production In materialist science, forces including living labor power, appropriated natural resources, and capital equipment provided by past generations of workers.

Fordism A mode of capital accumulation based on integrated production and assembly.

foreign direct investment (FDI) Investing in companies in a foreign county, with the purpose of managerial and production control.

foreign sourcing An arrangement whereby firms based in advanced industrial countries provide design specifications to producers in underdeveloped countries, purchase the finished products, and then sell them at home or abroad.

forward integration　A firm begins to control the outlets for its products.

four-field rotation system　Rotating three crops amongst four fields over a period of years, while allowing a fourth rotated field to remain fallow, thus rusting the soil for that year.

four questions of economic geography　What will be produced? How will it be produced? Where will it be produced? For whom will it be produced?

Four Tigers　South Korea, Taiwan, Hong Kong, and Singapore.

four worlds　First World—most developed countries; Second World—former command economies of Eastern Europe and the USSR; Third World—the remaining areas of the world that are less developed than the first two; Fourth World—the poorest of the Third World nations.

franchising　A license venture allowing a licensee to manufacture and sell under the original company's name.

free-trade area　A form of regional economic integration in which member countries agree to eliminate trade barriers among themselves, but continue to pursue their independent trade policies with respect to nonmember countries.

free-trade zones　Areas where imported goods can be processed for reexport without paying duties, since the goods will not be used locally.

freight rates　Payment to a carrier for the loading, transporting, and unloading of goods.

freight-on-board (FOB) pricing　A policy whereby a consumer pays the plant price plus the cost of transportation.

friction of distance　There are time and cost factors associated with movement across space, which exerts a friction to movement and flow.

functional disintegration　Locational dispersal of functions among firms.

functional region　An area differentiated by the activity within it; that is, by the interdependence and organization of its features.

game theory　Developed primarily by John von Neumann, a mathematical approach to decision making in the face of uncertainty.

General Agreement on Tariffs and Trade (GATT)　An international agency, headquartered in Geneva, Switzerland, supportive of efforts to reduce barriers to international trade.

general systems theory　A theory that applies the principles of organization, interaction, hierarchy, and growth to any system.

gentrification　Intercity neighborhood redevelopment and property upgrading by high-income settlers, thus displacing lower income residents.

geocoding　The process of entering spatial data into a GIS database by assigning Cartesian coordinates to each point, line and area entity.

geographical database　A digital information set developed from field-based or satellite spatial observation.

geographic inertia　The tendency of a place with established infrastructure to maintain its importance as a focus of activity after the original conditions influencing its development have altered, ceased to be relevant, or no longer exist.

geographic information system (GIS)　A technique for storing and transforming geographic data on a wide variety of variables to produce digital maps with hardware and associated computer software.

geothermal energy　Energy produced by heat from deep inside the earth as water interfaces with heated rocks from the earth's core, producing steam.

giant emerging markets (GEMs)　Countries of the world showing the greatest potential for high economic growth rates and which constitutes new regions for export and import from the developed world—East Asia, South America, Eastern Europe, Russia and India.

global office　Workplaces interlinked by satellite communications offering telecommunications and computer-based methods of information transfer.

global positioning systems (GPS)　A hand-held device that communicates with satellites to determine extremely accurate terrestrial location.

global shift of manufacturing　The pattern of employment and production shift from developed nations to developing nations of Eastern Europe and the Pacific Rim.

grain elevators　Large storage devices dotting the Great Plains of America that are used for wheat.

graph theory　The branch of mathematics concerned with the properties of graphs; that is, with the vertices and edges. Graph theory is used to describe and evaluate networks.

graphs　Idealized transportation networks that are comprised of *vertices*, which represent points or intersections, and *edges*, which represent interconnecting routes.

gravity model　The product of the masses or the populations of two cities, divided by the distance squared, is proportional to the traffic flow between them.

Green Revolution　A popular term for the greatly increased yield per hectare that followed the introduction of new, scientifically bred and selected varieties of such food crops as wheat, maize, and rice.

greenhouse effect　The warming of the atmosphere due to increased amounts of carbon dioxide, nitrous oxides, methane, and chloroflourocarbons.

gross domestic product (GDP)　The monetary value of final output produced by U.S. businesses, individuals, and government inside the United States once a year.

gross national product (GNP)　The market value of all goods and services produced by U.S. businesses, individuals, and government inside and outside of the U.S. in a given year.

growth stage theories　A developmental sequence of the national economic improvement of underdeveloped countries based on sufficient surplus to generate and sustain economic growth.

growth-oriented life-style　A mind-set that insists on maximum production and consumption. It assumes an environment of unlimited waste and pollution reservoirs and indestructible ecosystems.

guilds　Craft, professional, or early trade associations.

Gulf Coast Manufacturing Region　Region that stretches from southeastern Texas through southern Louisiana, Mississippi, and Alabama to the tip of Florida panhandle.

hamlet　The smallest of central places, usually less than 200 in population.

Hecksher-Ohlin Theory　Theory that a country should specialize in producing those goods that demand the least from its scarce production factors and that it should export specialties in order to obtain goods that it is too ill-equipped to make.

hierarchical marginal good   The highest order good offered by a center at a given level of the central-place hierarchy.

hierarchy   In central-place theory, the arrangement of settlements in a series of discrete classes, the rank of each determined by the level of specialization of functions.

hierarchy of central places   Small central places and their trade areas, nested within medium-sized central places and their trade areas. Medium-sized central places are nested with a few large central places within large trade areas.

high value added   Commodities, which due to the manufacturing process, are much more valuable than the raw materials used to produce them.

higher order of goods   Goods or services with higher thresholds, such as furniture or hospital care.

highest and best use   The notion that land is allocated to the use that earns the highest location rent.

horizontal integration   A business strategy to increase a firm's scale by buying, building, or merging with another firm at the same stage of production of a product.

HOV   High occupancy vehicle (HOV) lanes used for express traffic of vehicles with more than one rider.

hub-and-spoke networks   Hubs are major cities that collect passengers from small cities, in the local vicinity, via spoke lines. Hubs redistribute passengers between sets of original major cities.

human capital   Educational attainment, labor force participation, occupation, and income.

human development index (HDI)   Life expectancy, literacy rate, and per capita purchasing power.

human resources   The skill and cost of labor, the cultural factors of the term, and the motivation for the work ethic.

human suffering index   A useful descriptive measure of the differences in living conditions among countries, including 10 measures of human well-being.

hypermedia/hypertext   Incorporates innovative mechanisms interacting with a computer. It is based on hypertext navigation. Hypertext navigation can move through a database where information is linked together and is accessible in any form desired, regardless of its location. Hypertext links lead to other links which further link to other topics. Hypermedia improves on hypertext by linking together not only text, but also graphic, sound, animation and video.

import-substitution industrialization   A development strategy to replace imports of final manufactures through domestic production. Subsidies, loans, and protective tariff regulations are often the means of assuring local production.

industrial inertia   Location decisions for industrial activity, once made, lead to a continuation of that pattern.

industrial restructuring   A term used to refer to the alternating phases of growth and decline in industrial activity. It emphasizes changes in employment between regions, and links these with change in the world economy.

industry life cycle   The typical sequence of developmental stages in the evolution of an industry.

industry life cycle model   Industries experience periods of rapid growth, diminishing growth, and stability or decline.

inelastic supply   A supply schedule that does not show large increases due to a price increase.

infant industry   A young industry which, it is argued, requires tariff protection until it matures to the point where it is efficient enough to compete successfully with imports.

infant mortality rate   Number of deaths during the first year of life per 1,000 live births.

information technology   Microelectronics technologies, including microprocessors, computers, robotics, satellites, and fiberoptic cables.

information warehouse   A decision support tool collecting information from multiple sources and making that information available to users in a consolidated, consistent manner.

infrastructure   The services and supporting activities necessary for an economy to function; for example, transportation, banking, education, health care, and government.

innovation   A new idea applicable to something useful for humankind.

innovation drive stage   Innovation in new product design derived from high levels of technology and skill.

innovation wave   Innovations are adopted into the mainstream by the generation that innovated the technologies—as they accumulate the purchasing power.

integrated circuit   Transistors connected on a single, small silicon chip acting as a semiconductor of electrical current.

intellectual property rights   Establishing and policing patent, copyright, and trademark rights on an international basis.

intelligent vehicle highway systems (IVHS)   In-vehicle computers and navigation systems to help drivers avoid accidents and congestion while offering information on trip destinations.

intensive subsidence agriculture   A high-intensity type of primitive agriculture practiced by densely populated areas of the developing world.

Inter-American Development Bank   An international financial institution that extends loans for development to countries in Latin America and the Caribbean.

interest   Payments made to owners of capital as a compensation for its contribution to production.

intermediate technology   Low-cost, small-scale technologies "intermediate" between "primitive" stick-farming methods and complex Western agri-industrial technical packages.

International Bank for Reconstruction and Development   An international financial institution that extends loans to underdeveloped countries at commercial rates of interest; also called the World Bank.

international business   Any form of business activity that crosses a national border.

international currency markets   The internationalization of domestic currency, banking, and capital markets.

International Development Association   An adjunct of the International Bank for Reconstruction and Development. It extends loans with generous interest and repayment terms to the poorer underdeveloped countries.

International Finance Corporation   An adjunct of the International Bank for Reconstruction and Development. It provides either loans or equity investments to private-sector companies in underdeveloped countries.

International Monetary Fund (IMF)   An international financial agency that attempts to promote international monetary cooperation, facilitate international trade, promote exchange stability, assist in the establishment of a multilateral system of payments without restrictions on foreign currency exchange, make loans to help countries adjust to temporary international payment problems, and lessen the severity of international payments disequilibrium.

international subcontracting　The arrangement by multinational corporations (MNCs) to use Third World firms to produce entire products, components, or services in order to cover markets in an advanced industrial country.

Internet　A computing and data network that allows any source of stored information throughout the World Wide Web to be connected by modem that converts data, like text, statistics, and photos into electronic signals and is shared amongst computers.

intervening opportunity　An alteration in the complementarity of places.

investment-driven stage　Using foreign technology and scale economies to produce standardized products with mass labor inputs provided by the local population.

involution　The ability of the peasantry or the protoproletariat in the Third World to absorb an unusual number of people. The process of involution is characterized by a tenacity of basic pattern, internal elaboration and ornateness, technical hairsplitting, and unending virtuosity.

isodapane　The locus of points of equal transport cost from a factory.

isotims　Isocost lines of transportation, assembly, and distribution costs for each potential plant location. (See Isodapanes.)

isotropic environment　An imaginary plain or surface with uniform environmental conditions, including transport costs.

isotropic surface　A plain that is homogeneous in all respects, with equal ease of movement in all directions from every point.

Japan Incorporated　Sometimes used by Japan's competitors, an appellation acknowledging the successful marriage between Japanese businesses and government.

jobs/housing balance　The notion of expanding the supply of housing in job-rich areas, and the quantity of jobs in housing-rich areas.

joint venture　An enterprise undertaken by two or more parties. It may be a jointly owned subsidiary, a consortium, or a syndicate.

journey to work　Travel by individuals to work, yielding the largest proportion of travel by American households.

just-in-time manufacturing　Quick delivery and response of parts and inventory delivery from component plants to final assembly operations.

knowledge-based resources　Research, development, and scientific and technical skill within the country.

kolkhozes　Collective farms in the former Soviet Union.

Kondratiev cycles　Successive cycles of growth and decline in industrial economies, occurring with a periodicity of some 50 to 60 years duration.

Kuzbass　Short for Kuznetsk Basin, centered on the town of Novosibirsk, located on the Transiberian Railroad, east of the Euro Mountains in Russia/Asia.

labor　A factor of production that includes human physical exertion performed in the creation of a good or service.

labor economies　The reduction in costs that are due to a finer division of labor and its increased efficiencies.

labor exchange　The market for workers seeking gainful employment.

labor force　Those in society that produce work for pay, both employed and unemployed.

labor intensive industries　Those industries that use a large share of labor as inputs to the final product as opposed to some other form of capital.

labor migration theory　Theories to explain the process of changing residences from one geographic locale to another, due to economic factors.

labor process　The nature and degree of the division of labor.

labor-intensive　A term that applies to an industry in which a high proportion of labor is used relative to the amount of capital employed.

laissez-faire　The relative absence of government control and intervention in an economy.

land　A factor of production that includes not only a geographic portion of the earth's surface, but also the raw materials from this region.

land alienation　A term referring to the land taken away from indigenous people by Europeans for their own use.

law of demand　An empirical regularity of consumer behavior that presumes the quantity of a good demanded and the price of the good are inversely related.

law of diminishing returns　The law according to which, when factors of production (land, labor, and capital) are doubled, output doubles; but if one factor of production or only some factors are doubled, output increases, but fails to double. The law assumes given levels of technological knowledge.

law of retail gravitation　This law uses the gravity model concept to identify the breaking point between two cities' spheres of influence.

law of supply　An empirical regularity of producer behavior that presumes the quantity of a good produced and the price of that good are directly related.

lean producers　Producers using just-in-time delivery of supplies to keep costs of production low.

least-cost-to-build network　A transport system designed to keep the cost to the builder as low as possible.

least-cost-to-use network　A transport system designed to keep the cost to the user as low as possible.

less developed countries (LDCs)　The Third and Fourth Worlds, characterized by high rates of population growth and low per capita income.

level three alert　At this level, urban pollution is near its highest, and residents are advised to stay indoors, cease from strenuous exercise and, automobiles are ordered off highways.

licensing venture　The rental of patents, trademarks, or technology by a company in exchange for royalty payments.

life expectancy　The average number of years an individual can be expected to live given current socioeconomic and medical conditions.

limits to growth　The optimum population size for the world, provided by the club of Rome, shows that growth must be limited. A gloomy forecast by Paul Ehrlich suggests worldwide famine and war as the inevitable results of continued increases in world population.

line-haul costs　Costs involved in moving commodities along a route.

linear market   The spatial organization of central places into a K = 2 hierarchy.

literacy rate   The percentage of a population 15 years and over that can read and write.

Little Dragons   Philippines, Thailand, Malaysia, Indonesia, and China.

localization economies   The declining average costs for firms as the output of the industries, of which they are part, increases.

localized raw material   A material that is not available everywhere; thus, it exerts a specific influence on industrial location.

location   (See absolute location; relative location.)

location allocation model   A spatial model used in the world economy to identify the best location for a public facility, industrial plant, or transportation route.

location rent   The advantage of one parcel of land over another because of its location; the concept of declining rents with an increase in distance from the market.

location theory   A compilation of ideas and methods dealing with questions of accessibility.

locational costs   Costs over and above the basic cost of an input or the least-cost price.

locational inertia   The stabilizing effect of invested capital in a region.

locational interdependence   A concept that implies that competition from rival producers can lower potential revenues at a given point and space.

logical positivism   The formation of hypotheses, data collection, and a search for theory through the scientific method.

Lome Convention   A 1974 trade agreement signed by the European Economic Community (EEC) and 46 countries in Africa, the Caribbean, and the Pacific.

long run   A period long enough to permit a firm to vary its production function and the quantity of all of the inputs.

long-run average cost (LRAC)   The graphical relationship between the average costs and the level of output over the long run.

long wave   The expansion of an economy from innovation, expansion, decline and bust over a period of 50 to 60 years that reoccur in the techno-economic system.

lower circuit   The traditional and labor-intensive sector of a Third World city.

lower order of goods   Goods or services with low thresholds, such as a loaf of bread or a gallon of gas.

machinofacture   The phase of the developing division of labor where mechanization and division of labor within production occurs.

Maglev   A magnetically levitated train that operates with a linear induction engine and cruises on a cushion of air at high speeds on a detached right-of-way and heralded as the state of the future in ground transportation systems.

Maine Street   The Canadian megalopolis extending from Windsor, Hamilton, Toronto, Quebec, to Montreal, along the northern shores of the eastern Great Lakes.

malnutrition   A state of poor health in which an individual does not obtain enough essential vitamins and nutrients, especially proteins.

manufacture   Workers and machines producing goods in a factory setting.

manufacturing belt   The occurrence of numerous manufacturing firms in a particular geographical region of a country.

maquiladora program   The border industrialization program whereby U.S. companies establish factories in Mexico and enjoy tariff breaks on finished products returned to America.

margin of cultivation   The location at a certain distance from the market at which marginal products are produced and, therefore, the edge of profitability is reached.

marginal cost   The extra cost caused by the production of an additional unit of output.

marginal product   The addition to output resulting from an increase of one unit of increased input.

Marine fisheries   Species of fish populations found in oceans, seas, and other saltwater bodies.

market area   The territory surrounding any central point of exchange. It includes all potential customers for whom market price plus transport cost will still be sufficiently low to justify their purchases at that price in the central place.

market basket   A batch of housing, goods, and services that individuals consume on a given income with a given set of preferences.

market economy   A free economy in which prices are based on supply and demand.

market equilibrium   The quantity price relationship where supply and demand curves intersect.

market exchange   An economic system that establishes market prices. The prices are the mechanism for connecting economic activity among a large number of individuals and for controlling a large number of decentralized decisions.

market linkage   The connection resulting from the sale of a firm's output to nearby firms.

market orientation   Plant location is oriented to the market, rather than the source of raw materials, because of a savings in transportation costs.

market share   The percent of total consumer demand for a particular product that is held by a single firm.

marketing principle   The spatial organization of central places when a central place of any order is at the midpoint of each set of three neighboring places of the next higher order.

Marxist   The view that the world economy is the product of exploitation by capital of labor following the principles and teachings of Karl Marx.

massing of reserves   The principle that states that large firms can maintain smaller inventories of spare machines or machine parts than can small firms.

material index   In Weberian analysis, a measure of the weight a raw material loses in processing; the weight of raw materials divided by the weight of the finished product.

maximum sustainable yield   Maximum production consistent with maintaining future productivity of a renewable resource.

Mediterranean cropping   Agriculture producing specialty crops because of mild climates, including citrus, grapes, nuts, avocados, tomatoes, and flowers.

megalopolis   A giant, sprawling urban region encompassing many cities, towns, and villages. The term was coined by geographer Jean Gottman to describe the Atlantic Urban Region that extends from Boston to Washington, DC.

megatrend   Future directions of the economy, according to John Naisbitt, which have the capacity to transform our lives.

mercantile model   A model that attempts to explain the wholesale trade relationships that link regions.

mercantilism   A theory popular among European nations in the early modern period stating that the economic and political strength of a country lay in its acquiring gold and silver, to be achieved by restricting imports, developing production for exports, and prohibiting the export of gold and silver.

merger   The consolidation of two or more companies.

microelectronics   Semiconductors, integrated circuits, and electronic components and parts.

microprocessor   A computer the size of one's fingernail used for applications, including calculators, electric typewriters, computers, industrial robots, and aircraft guiding systems.

MicroVision   A GIS software package used for target marketing.

migration   Movement of a population, resulting in a change of permanent residence.

milk shed   The area around the city from which milk can be shipped daily, without spoilage.

milpa farming   Temporary use of rain-forest land for agriculture by cutting and burning the overgrowth.

mine tailings   Leftover ore wastes from which minerals have been extracted.

minicity   A multifunctional urban node that is the focal point of the outer city, especially in North America. Suburban minicities include a variety of land uses—retailing, wholesaling, manufacturing, entertainment, and medical functions.

mixed crop and livestock farming   The raising of beef cattle and hogs as the primary revenue source, with the production of crops fed to the cattle.

mixed economic systems   Economic systems that are a hybrid form of capitalists, command economy, or a traditional system, usually where both government and private decision determine how resources are allocated.

modernization   A replacement of traditional approaches to production with new technologies and techniques.

money capital   Items of exchange used to purchase capital goods, such as money.

monopoly   A single firm dominates the market sales.

Monte Carlo simulation   A probabilistic model that accounts for sheer chance and reproduces a particular process by discovering the major rules of the game.

Moscow Industrial Region   This industrial region, also known as the Central Industrial Region, is located near the population center of Russia, west of the Euro Mountains.

most-favored nation   All nations in GATT are to be given the same favorable treatment with regard to exports and imports—no one is to be given preferential treatment.

multimedia   Computers with TVs, VCRs, CD players and other entertainment devices in a human-machine communications media with combined applications.

multinational   A company with established operations in several host countries, usually headquartered in one parent country.

multinational corporations (MNCs)   Companies based on one country that do business in one or more other countries.

multiplant enterprises   Companies with factories and offices in widely scattered locations, sometimes in other countries.

multipurpose travel   Bypassing the closest central place in favor of the next largest central place, which offers all needs of a particular shopper (e.g., groceries, gasoline, basic services, etc.).

multiple-nuclei model   Besides the traditional central business district (CBD), the modern city has a number of outlying high-intensity nucleations of commercial, industrial, and residential land uses.

multiplier   An "injection" into the spending stream in the belief that total output will increase as a result. The opening of a new factory in a region is an example of an injection. New funds flow into the region from the outside, thereby raising the level of regional income.

National Environmental Policy Act (NEPA)   Legislation enacted in 1969 by the federal government requiring environmental impact reports and review to be prepared and conducted on every major public land development.

nation-state   A sovereign country.

Nationwide Personal Transportation Survey (NPTS)   A major national travel survey conducted every 7 years, giving information regarding mobility of the American population.

NATO   North Atlantic Treaty Organization.

need creation   A process in which luxuries are marketed as necessities.

negative population growth   A falling level of population where out-migration and death exceed in-migration and births.

neo-classical economies   The school of economic practice that emphasizes the efficiency of markets and advocates non-intervention of governments.

NeoFordism   Fragmentation of the labor force and the distilling of the traditional blue-collar class through the introduction of electronic information systems.

neo-Malthusian   Someone who accepts the Malthusian principle, but hopes to avoid famine with the intervention of government controls.

net energy   The amount of energy available minus the quantity used to find, concentrate, and deliver energy to the consumer.

net migration   The net effect of immigration and emigration on an area's population in a given period, expressed as an increase or decrease.

network   Any set of interlinking routes that cross or meet one another at nodes, junctions, or terminals.

networked computers   The linking of computers, servers and databases from around the country and around the world allowing each computer work station more power and information than it contains in a stand-alone mode.

New International Economic Order (NIEO)   A 1974 U.N. resolution originating with the underdeveloped countries and calling for a more equal distribution of the world's income.

newly industrializing countries (NICs)   Countries that have recently progressed from traditional economies to manufacturing economies, such as the East Asian Miracle countries.

nomadic capital   The switching of production from place to place because of innovation in transportation and communication savings.

nonbasic firms Firms producing a good or service for a consumption inside the region.

non-ferrous metals Metals that do not contain iron.

non-tariff barriers Restrictions that limit entry into an industry by competitive firms or countries.

nonrenewable resources Resources that are fixed in amount—that cannot regenerate—such as fossil fuels and metals.

normal lapse rate The rate at which the atmosphere cools with increasing elevation (3.6 degrees per 1000 feet).

normative model A model that attempts to describe how people should behave and make decisions if they wish to achieve certain well-defined objectives.

North American Free Trade Agreement (NAFTA) Agreement between Canada, the United States, and Mexico passed by the U.S. Congress and signed by President Clinton in December 1993.

offshore assembly An arrangement whereby firms based in advanced industrial countries provide design specifications to producers in underdeveloped countries, purchase the finished products and then sell them at home or abroad.

oligopoly The control of a market by a small number of firms or producers.

opportunity cost The cost of foregone alternatives given up in order to produce other activities or goods.

opportunity costs The cost of time that is lost due to participating in a particular activity and foregoing another activity.

optimizers Economic persons who organize themselves and their activities in space so as to optimize utility.

optimum population size The theoretical number of people that would provide the best balance of population and resources for a desirable standard of living.

Organization of Petroleum Exporting Countries (OPEC) The international cartel of oil-producing countries.

organizational ecology Businesses are born, they grow and mature, they die, and between birth and death, some of them migrate.

organizational structure of capital A term that is often applied to the size and associated characteristics of firms.

output The quantity of a good or service that is produced by an economy or a firm.

outsourcing Subcontracting and the shifting of work to other locations and firms outside the principal corporation.

overpopulation A level of population in excess of the "optimum" level relative to the food supply or rate of consumption of energy and resources.

Pacific Rim The 14 countries bordering the Pacific Ocean, in North America, South America, East Asia and Australia/New Zealand.

paddy A flooded field planted with rice seedlings.

parity price Equality between the prices at which farmers could sell their products, and the prices they could spend on goods and services to run the farm.

part-process spatial structure A headquarters/branch plant structure in which the production process is geographically fragmented or differentiated.

pastoral nomadism Animal herds used for subsidence, moved from one forage area to another, in a cyclical pattern of migration.

Paul Revere network The shortest network interconnecting a set of points.

peasant agriculture Subsidence agriculture, using little mechanical equipment and producing meager, labor-intensive crops.

per capita income GDP or GNP divided by the population of a country in a given year.

perestroika During the Gorbachev years in the Soviet Union, the restructuring and opening of the Soviet economy.

Peter's projection A map projection attempting to give area to the Third World in proportion to its true size.

petrol dollars OPEC oil surpluses poured into the major Euro banks.

phenomenological approach Life takes on meaning only through individual experiences and needs; little need for models or theory.

physical resources These are raw materials, site and situation factors, international time distance, and space conveyance.

physiologic density The number of people per square mile of arable or farmable land.

plantation A large landholding or estate devoted to the production of export crops, such as coffee, tea, sugar cane, sisal, and hemp. Plantations are usually located in underdeveloped countries and depend on foreign capital for their operation.

Po River Valley The principal industrial district of southern Europe, including Torino, Milan, Genoa, Cremona, Verona, and Venice.

polarization effects The negative influences prosperous regions exert on less prosperous regions.

political economies The economic form of countries and nation-states.

population density The number of people per unit of land, normally a square kilometer or square mile.

population distribution The spatial pattern of population across space.

population growth rate The difference between the birth rate and death rate; generally expressed as so many persons per hundred.

population hurdle The rate of investment in a society must be greater than population growth or it will be stuck in a vicious, Malthusian cycle of poverty.

population potential The degree of nearness or accessibility of a place to the population of a country or a region based on distance.

population pyramid A special type of bar chart indicating the distribution of a population by age and sex.

portfolio investment Capital investment in the equity of a company, not involving managerial control of a foreign country.

post-industrial A country's economy based on service and industry and information-based businesses.

postindustrial society The stage of an evolving society in which traditional manufacturing activity has given way to the growth of high-tech industry and an employment emphasis on services, government, and management-information activities.

pre-industrial economies Economies that have not experienced an industrial revolution and, thus, are traditional in scope.

price ceiling   A legally mandated price level below the typical market price.

price elasticity of demand   Changes in the quantity demanded of a good or service in response to the changes in the price of that good or service.

price elasticity of supply   Changes in the quantity supplied of a good or service in response to the changes in the price of that good or service.

price floor   A guaranteed price above the market price.

primary economic activity   An economic pursuit mainly involving natural or culturally improved resources, such as agriculture, livestock raising, forestry, fishing, and mining.

primary sector of the economy   Mining, lumbering, agriculture, and fishing.

primate city   A country's leading city economically, culturally, historically, and politically, and much larger than competing cities in population, wealth, and power.

privatization   Government-owned companies are transferred to private ownership and management.

product life cycle   The typical sequence through which a product passes, from its introduction into the market to when it is replaced by a new product.

product market   The market where households buy and firms sell the products and services they have produced.

production factors   Labor, capital, technology, entrepreneurship, and land containing raw materials.

production function   The technological and organizational characteristics of a firm that transform inputs into outputs. In the short run, at least one input is fixed in amount. In the long run, all the inputs are variable.

production linkages   Economies that accrue to firms that locate near other producers manufacturing their basic raw materials.

production possibilities analysis   A table or curve that shows various combinations of goods or services that can be produced given employment, resources, and technology are held constant.

profits   The reward, in monetary terms, paid to management or entrepreneurial skill for its supply to the economy over and above its costs.

protectionism   An effort to protect domestic producers by means of controls on imports.

protoproletariat   An urban class engaged within a peasant system of production. Most of its income is gained from informal income opportunities.

psychic income   Nonmonetary rewards gained from operating at a particular point.

purchasing power parity   The amount of income per capita based not on dollars, but on goods and services that the dollars will purchase.

pure capitalism   An economic system in which the means of production are privately owned, and markets and prices are used to direct and coordinate all economic activity.

pure competition model   A market structure of industry made up of many small firms that produce homogeneous products and that have no real influence on the market price of their products.

pure material   Materials that bear their full weight into the finished product.

pure raw material   A material that does not lose weight in processing.

quaternary activity   Those sectors of the economy associated with research; the gathering and disseminating of information; the administration and transmission of information, including radio, television, newspaper and magazine publishing; educational systems; and the computer information technologies.

quaternary economic activity   An information-oriented economic activity, as pursued, for example, in research units, think tanks, and management-information services.

quaternary sector of the economy   Information processing, finance, insurance, real estate, education, and computer and telecommunication fields.

quinery industries   According to Barry, Conkling and Ray, medical care research, education, arts and recreation. According to de Blij and Muller, corporate decision-making.

radical humanist   The theory that labor, the environment and the means of reduction have for centuries been dominated by white male power establishment.

Randstad   The ring city in the western Netherlands, including Rotterdam, the Hague, Amsterdam, and Utrecht megalopolis.

range   The average minimum distance consumers are willing and able to travel to purchase a good (or service) at a particular price in a central place.

range of the good   The distance people are willing to travel to purchase a good or a service of a particular threshold.

rank society   A society in which positions of valued status are circumscribed, so that not all with sufficient talent to hold such positions actually achieve them.

rank-size rule   An empirical rule describing the distribution of city sizes in an area. It states that the population of any given city tends to be equal to the population of the largest city in the set divided by the rank of the given city. For example, if the population of the largest city numbers 10,000, the population of the fifth largest city will be 2000—that is, 10,000 divided by 5.

rate of natural increase   The excess of births over deaths, or the difference between the crude birth rate and the crude death rate.

raw material orientation   The processing plant is oriented to the site of the raw materials, as opposed to the market, because of savings and transportation costs.

raw materials   A substance in the physical environment considered to have value of usefulness in the production process.

real capital   Human-made resources used to produce goods and services that do not directly satisfy human wants and needs.

real time information systems   Systems that allow data retrieval and data information access in an instantaneous format, allowing processing of information as the information is gathered and recorded: instantaneous communication.

reciprocity   A mutually beneficial form of economic exchange common in egalitarian societies.

redistribution   A form of economic exchange in which equity is maintained by a central authority that redistributes production.

regional growth forest   Long-range urban forecasting of population housing and the economic activity for the region and small geographic areas within it.

relative location Position with respect to other locations.

renewable resources Resources capable of yielding output indefinitely if used wisely, such as water and biomass.

rent Payments made to land owners as a productive factor for their contribution to the production process and the operation of the economy.

rent gradient A sloping net-profit line. The intersection of the line and the point of zero profit indicates the limit of commercial crop production.

reserve A known and identified deposit of earth materials that can be tapped profitably with existing technology under prevailing economic and legal conditions.

reserve deficiency minerals Those minerals for which U.S. reserves are not sufficient to meet anticipated near-term industrial needs.

resource A naturally occurring substance of potential profit that can be extracted under prevailing conditions.

resource dependent Countries in early stages of economic development depended on natural resources endowments and their own cheap labor.

resource market A place where households sell and firms buy resources of the services of resources.

returns to scale The degree to which output increases as inputs are increased.

Rhine-Alsace-Lorraine Region The second most industrial region of Europe, located in southwestern Germany and eastern France.

Ruhr Located on the Rhine in northeastern Germany, Europe's most important industrial district.

rule of 70 Dividing the average annual rate of growth by 70, yields the doubling time of population for a country.

S-curve Products and technologies go through three stages of growth—the innovation phase, the growth phase and the maturity phase.

saddle point The minimum-maximum point in game theory.

satellite firms Small organizations that manage to survive in the market by minimizing their labor cost and maximizing labor exploitation (also called peripheral firms).

satisficers Decision makers who make choices that are satisfactory rather than optimal.

sawah A flooded field planted with rice seedlings.

scale The size of operation that will determine the volume of industrial output.

scale economies The cost-reducing changes that lower the average costs of firms as they grow in size. These changes may be internal or external to firms.

scarcity The fact that the world's resources are limited in their supply and, therefore, have a value.

second law of thermodynamics The law according to which any voluntary process has as a consequence a net increase in disorder or entropy. It can also be expressed as the degradation of energy into a less useful form, such as low-grade heat.

second wave The second wave of the industrial revolution was the age of steam, Circa 1815-1875.

Second World The communist and socialist countries.

secondary economic activity The processing of materials to render them more directly useful to people; manufacturing.

secondary sector of the economy Manufacturing and assembling of raw materials.

servers Special computers that distribute data to computer users.

service industries Wholesale and retail trade, transportation services, public administration, hotels, restaurants, personal and professional services.

service linkages Economies that occur when a cluster of firms becomes large enough to support specialized services.

settlement-building function Sales of goods and services beyond the local retail and service hinterland of a central place.

settlement-forming function Sales of goods and services that occur totally within the hinterland of a central place.

settlement-serving function Sales of goods and services to residents of a central place.

shifting cultivation Temporary use of rain-forest land for agriculture by cutting and burning the overgrowth.

Shimble index A graph-theoretic measure of the compactness of a network; sometimes called the dispersion index.

shopping goods Central functions that are normally higher in price than convenience goods. They vary in quality and style, and are purchased infrequently.

short run A period short enough for at least one input to be fixed in an amount and invariable.

Silicon Valley The world's largest manufacturing area for semiconductors, microprocessors, and computer equipment, located south of San Francisco.

slash-and-burn agriculture Temporary use of rain-forest land for agriculture by cutting and burning the overgrowth.

smart cars Cars equipped with microcomputers and video screens to take the frustration out of driving.

smart highways Vehicle sensing systems to monitor traffic, volume, and speed, resulting in optimum traffic flow control and radar collision warning systems.

social relations of production Class relations between owners of the means of production and the workers employed to operate these means.

social surplus The portion of annual production of any society that is neither consumed by direct producer nor used for the reproduction of the stock of means of production available at the start of the year. In a class-divided society, the social surplus is always appropriated by the ruling class.

soft technology Innovative management techniques of the multi-national corporation.

soil bank program Government program that paid farmers to keep acreage out of production because of market overproduction.

solar energy Energy in the form of radiation received from the sun and changed into heat as it strikes the earth's surface.

Southeastern Manufacturing Region Also known as the Peidmont Region, it stretches south from Central Virginia, through North Carolina, South Carolina, and Northern Georgia, and into Alabama and Tennessee.

sovkhozes State-owned large collective farms in the former Soviet Union.

space preference structure Measures the trade off between the distance travelled to shop for goods and services and the town size.

spatial association Connections between households, cities or industries in the same region or country.

spatial decision support systems (SDSS) Prescription decision-making principles laden with data and rules for data use, designed to help decision makers reach decisions about spatial problems that confront them.

spatial diffusion The spread of information, goods, or people across space.

spatial equilibrium The equilibrium price where supply and demand curves intersect, allowing suppliers the good or service price minus transportation cost and the buyer the good or service price plus transportation costs.

spatial fetishism Attributing the cause of an event to locational factors.

spatial interaction The movement, contact, and linkage between points in space; for example, the movement of people, goods, traffic, information, and capital between one place and another.

spatial margins to profitability The intersection of a space-cost curve, and the market price of a finished product.

spatial mismatch theory Reverse commuting, traveling to work against the main flow of traffic from the intercity to the suburbs.

spatial monopoly A situation in which a single firm controls a given area of the market by virtue of its location.

spatial oligopoly A situation in which a few firms compete for a given segment of the total market space.

spatial organization A theme in geography emphasizing how space is organized by individuals and societies to suit their own designs. It provides a framework for analyzing and interpreting location decisions and spatial structures in a mobile, interconnected world.

spatial process A movement or location strategy.

spatial regularity The arrangement of economic activity on the earth's surface in a discernable pattern.

spatial structure The internal organization of a distribution that limits, channels, or controls a spatial process.

spending wave The owning and spending cycles of people as they age and raise families.

spot markets Commodity markets in which goods are traded for immediate purchase and delivery.

spread city A term that usually refers to the contemporary suburban or multifunctional American metropolis. The spread city encompasses more territory and has less "centrality" than the compact nineteenth-century industrial city.

spread effects The beneficial influences prosperous regions exert on less prosperous regions.

spring wheat Wheat planted in the spring and harvested in the fall. Fields are fallow in the winter.

squatter settlements Residential areas that are home to the urban poor in underdeveloped countries. The various terms used to identify squatter settlements include the following: calampas, tugurios, favelas, mocambos, ranchos, and barriadas in Latin America; bidonvilles and gourbivilles in North Africa; bustees, jhoupris, jhuggis, kampongs, and barung barong in south and southeast Asia.

stages of production According to the law of diminishing returns, the three stages that total product passes through as successive units of variable input are applied to a fixed input. In Stage 1, the average product curve rises to its peak; in Stage 2, it declines; and in Stage 3, the total product curve declines.

stagflation Economic periods of low economic growth accompanied by high unemployment and high inflation.

standardized economy Economy that consists of countless varieties of products resulting from a common economic formula of mass production, as goods and services are made cheaper by assemblyline manufacture.

strategic minerals Those minerals deemed critical to the economic and military well-being of the nation.

stratified society A society in which members of the same sex and equivalent age status do not have equal access to the basic resources that sustain life.

stationary state The dynamic state of a system in which input and output are balanced at a point below the maximum limits of the system and its surroundings.

stochastic model A model that assumes bounded rationality; it recognizes the major role of chance in the decision-making process.

subsidence agriculture Peasant agriculture, using little mechanical equipment, producing meager, and labor-intensive crops.

suburb An outlying residential district of a city.

suitability modeling A GIS map overlay technique to assess the ability of each increment of land under study to support a given use; a type of spatial decision support system (SDSS).

surplus value The difference between the value produced by a worker (value of units of labor produced) and the worker's wage (value of labor power).

sustainable communities A community that competes for businesses, jobs and residents in the years ahead in an effort to coexist with other cities and towns. They will be able to reduce the cost and burden of government while simultaneously increasing the quality and level of government services and thus, avoid being consigned to geopolitical obsolescence.

sustainable development The economic system and degree of output of a society that can be maintained in the long-run, without depletion of human or natural resources.

swidden Temporary use of rain-forest land for agriculture by cutting and burning the overgrowth.

synfuels Synthetic fuels produced by technological innovation for use in internal combustion engines.

target pricing The government pays a farmer directly the difference between the market selling price and the price that the government has set artificially.

tariff A schedule of duties placed on products. A tariff may be levied on an ad valorum basis (i.e., as a percentage of value) or on a specific basis (i.e., as an amount per unit). Tariffs are used to serve many functions—to make imports expensive relative to domestic substitutes; to retaliate against restrictive trade policies of other countries; to protect infant industries; and to protect strategic industries, such as agriculture, in times of war.

tariff protection A scheme of taxing imports that produces growth of domestic industry.

Taylorism The application of scientific management principles to production.

technique The method of procedure by which inputs are combined to produce a finished product.

technostructure Corporate technical personnel, including scientists and technicians.

telecommuting   The substitution of commuting to a worker's employment center by producing work from a telework center or from home via telecommunications innovations such as computers, modems, faxes and phone.

teleprescence   Also called teleworking or telecommuting which allows labors to be scattered widely from the production center via telecommunications innovation.

terms of trade   The relative price levels of exports to imports for a country.

terminal costs   Costs incurred in loading, packing, and unloading shipments, and preparing shipping documents.

tertiary activities   Those sectors of the economy that provide markets and exchange for commodities, including wholesale and retail trade and associated transportation government information services, as well as personal and professional services.

tertiary economic activity   An economic pursuit in which a service is performed, such as retailing, wholesaling, servicing, teaching, government, medicine, and recreation.

tertiary sector of the economy   The provision of goods and services, most notably wholesaling, retailing, and professional services.

thematic maps   Maps whose purpose is to display the locations of a single theme or attribute variation over space or the relationships amongst several selected attributes interrelated.

theory of contestable markets   The free entry of new transportation operators into the market to ensure efficiency.

Third World   Developing countries.

Third World debt crisis   The dangerous economic position of certain Third World nations that carry an enormous debt to overseas banks, private banks, and governments, the interest payments of which rob the host country of needed investment.

Thomas Malthus   During 1766-1834, Malthus theorized the geometric population growth, mixed with arithmetic growth in food supplies would lead to famine.

threshold   The minimum level of demand needed to support an economic activity.

Thunenization   According to Berry, Conkling and Ray, the outward expansion of agricultural production regions based on von Thunen's model of spatial rent.

TIGER files   Geocoded digital line graphs that are computer readable, representing roads, streets, and highways across America.

time-space convergence   The reduction in travel time between places that results from transport improvements.

trade area   The area dominated by a central place; sometimes called a hinterland or tributary area.

trade deficit   The excess of imports over exports for a country for any specific year.

traditional economies   Older, self-sustaining economies producing a subsistence level for family or village and not a major port of the world economy and its export and import.

traditional market   An economic system in which culture, tradition, and folkways determine how scarce resources will be used by the economy.

traffic principle   The spatial organization of central places when as many central places as possible lie on a traffic route between two important cities.

tragedy of the commons   Public resources are frequently ruined by the cumulative isolated actions of individuals in that overuse is practiced, rather than conservation, thus ruining resources held in common.

TRAILMAN   A microcomputer-based software package for logistics planning in Africa.

transfer of capital   The movement of the means of production to offshore locations by a company undertaking a foreign production center.

transfer costs   Terminal costs and other fixed costs, plus linehaul costs or over-the-road costs equal transfer costs.

transfer pricing   The transfer of taxable profits to lower tax assessing countries through price setting, thus minimizing corporate tax penalties and maximizing profit.

transferability   The condition that costs be acceptable in order for exchange of goods to occur between a supply area and a demand area.

transit oriented development (TOD)   Development guidelines to help communities reduce dependency on automobiles.

transmaterialization   Technology and innovation has reduced the demand for minerals as inputs to mature industries which undergo replacement by higher quality, smaller, and technologically more advanced materials producing high tech, less expensive and more durable products, such as fiberoptics, smart metals, transistors and computers, as coined by resource economist Lorna Waddell and Walter Labys.

transnational corporation   Companies that operate factories or service centers in countries other than the country of origin.

transport costs   The alternative output given up when inputs are committed to the movement of people, goods, information, and ideas over geographical space.

transport gradient   Increase in transportation costs with increasing distance from central markets.

transportation center   A port or city that performs break-of-bulk and associated services along transportation routes, such as railroad, highway, coastline or river.

traveling salesman network   The shortest network interconnecting a set of points where the origin and destination points are identical.

triage   The partitioning into three groups of nations, those so seriously deficient in food reserves that they cannot survive, those that can survive without food aid, and finally, those that can be saved by immediate food relief measures.

tributary areas   The area dominated by a central place; the area to which a central place services and from which it draws raw materials and labor supply.

trickling-down effects   The beneficial impact of prosperous regions on less prosperous regions.

truck farming   Sometimes called "market gardening," a specialization of intensively cultivated fruits, vegetables, and vines in developed economies.

turnkey project   A technique of competitive duplication of Western industrial facilities employed by multinationals. The contractor not only plans and builds the project, but also trains the buyer's personnel and initiates operation of the project.

ubiquitous raw material   A material that is available every-where; thus, it does not exert a specific influence on industrial location.

ubiquity   In Webber's model, raw material available everywhere and available at the same price.

Ukraine Industrial Region   Centered on the Donets Basin and the iron and steel industry of Krivoy Rog, formerly a most important industrial district of the Soviet Union.

undercapitalization   Lacking sufficient capital to operate a business or enterprise optimally.

underemployment   Shortage of job opportunities forcing people to accept less than full time employment or being employed well beneath one's training and ability.

undernutrition   A state of poor health in which an individual does not obtain enough calories.

United Nations Conference on Trade and Development   A UN organization that includes most underdeveloped countries. Although it has little statutory authority, it serves as a forum for discussion of common problems of its members.

upper circuit   The modern and capital-intensive sector of a Third World city.

urban field   The area dominated by a central place, to which a central place services, and from which it draws raw materials and labor supply.

urban hierarchy   System of cities according to types of businesses providing by each and the size correlating to the hinterland around each city.

urbanization   The process of a society changing from rural to urban and economic activities concentrating in cities.

urbanization economies   The declining average costs for firms as cities increase their scales of activity.

Uruguay Rounds   The last round of the GATT talks that began in 1986 and ended in December, 1993.

use value   The usefulness of a commodity to the person who possesses it.

value added   The difference between the revenue of a firm obtained from a given volume of output and the cost of the input (the materials, components, services) used in producing that output.

value free   Opinions and statements that are not laden with belief or innuendo regarding what ought to be or what ought not to be regarding the world economy.

variable costs   Expenditures firms incur as output changes. As output rises, variable costs rise; as output falls, variable costs fall.

varignon frame   Weights on strings running through holes in a sheet of plywood to represent the proportionality of attraction of each location, attempting to answer the question, "Where should economic activity be located?"

venture capital   Financial capital for the initiation for new industries or high risk businesses.

vertex   A point; a node in a topological diagram of a network.

vertical integration   A business strategy to increase a firm's scale by buying, building, or merging with another firm in a different stage of production of the same product; may be forward or backward.

vicious circle   A concept emphasizing the multicausality of underdevelopment; that is, a combination of interwoven limiting factors, rather than "just" a single factor, thwarts development.

vicious cycle of poverty   Explanations for the multiple causality of underdevelopment.

virgin and idle lands program   Khrushchev's hasty and grandiose dryland agricultural programs of Central Asia, which pulverized the soil, resulting in enormous dust storms.

Volga Industrial Region   Region along the Volga and Cama Rivers, the location of substantial oil and gas production and refining, east of Moscow.

voluntary export restraint   A restriction placed on a country's exports by their own companies to avoid retaliation by importing countries.

Von Thünen's Isolated State   A book that revealed laws that regulated the interaction of agricultural prices, distance and land use, as land users seek to maximize their income.

wage differentials   Difference in wages between and among groups on the basis of occupation, type of labor, region, race or gender.

wealth-driven stage   Stage of economic development described by a population that has a high overall standard of living and high levels of mass consumption and technological virtuosity.

Webber's Theory of Location   A minimization theory accounting for transportation costs, location of raw materials and a fixed location of labor.

weight-losing raw material   A raw material that undergoes a loss of weight in the process of manufacture.

welfare payments   The provision of government payments to low income members of society sustaining a minimum level of well-being.

welfare state   A political system in which the government assumes the primary role for the provision of social welfare

World Trade Organization (WTO)   The world trade union which came into existence following the Uruguay Round of the GATT Treaty. WTO enforces trade rules and assesses penalties against violators.

West Coast Manufacturing Region   Manufacturing region stretching from San Diego, north to Los Angeles and up to San Francisco.

wheat belts   Areas, such as those in North America, in which wheat predominates as an agricultural product.

wind farm   Capturing wind energy with wind turbines and converting it to electricity.

winnowing   Removing chafe from rice seeds by the blowing winds.

winter wheat   Wheat planted in the fall that sprouts in spring and is harvested in early summer.

World Bank   A group of international financial agencies including the International Bank for Reconstruction and Development, the International Finance Corporation, and the International Development Association.

world cultural realms   Giant world regions that possess similarities of culture, economy, and historical development.

world economy   A multistate economic system created in the late fifteenth and early sixteenth centuries by European capitalism and, later, its overseas progeny.

world political economy theories   The structure of political and economic relationships between dominant and dominated countries.

World Wide Web   The interconnected data sources via computer that one can access via the Internet.

world's breadbasket   The Midwest of the United States, with its rich agricultural potential.

write-down   A cancellation or reduction of a portion of the principal and/or interest by a creditor bank for less developed country's loans because of the possibility of default.

zaibatsu   A large Japanese financial enterprise, similar to a conglomerate in the West.

zero population growth (ZPG)   As a result of the combination of births, deaths, and migration, the population of a country is level, not rising or falling from year to year.

zero-sum game   A game in which the "payoff" to one player is exactly the value "lost" by the opponent.

Zuider Zee Reclamation   The Dutch program to reclaim portions of their country from the North Sea by building dikes across inlets and pumping water off the land.

REFERENCES

Abler, R. 1975. Effects of space-adjusting technologies on the human geography of the future. In *Human Geography in a Shrinking World,* edited by R. Abler, D. Janelle, A. Philbrick, & J. Sommer, pp. 35–56. North Scituate, MA: Duxbury Press.

Abler, R., J. S. Adams, and P. Gould. 1971. *Spatial Organization.* Englewood Cliffs, NJ: Prentice Hall.

Abu-Lughod, J. 1987–1988. The shape of the world system in the thirteenth century. *Studies in Comparative International Development* 22(4):3–25.

Aglietta, M. 1979. *A Theory of Capitalist Regulation: The U.S. Experience.* London: New Left Books.

Agnew, J. A. 1982. Sociologizing the geographical imagination: Spatial concepts in the world-system perspective. *Political Geography Quarterly* 1:159–166.

———. 1987. Bringing culture back in: Overcoming the economic-cultural split in development studies. *Journal of Geography* 86:276–281.

Aitken, S., F. Stutz, et al. 1993. Neighborhood integrity and residents' familiarity: Using a geographic information system to investigate place identity. *Tijdschrift Voor Economische en Sociale Geografie* 34:1–12.

Allen, B. J. 1985. Dynamics of fallow successions and introduction of Robusta coffee in shifting cultivation areas of the lowlands of Papua New Guinea. *Agroforestry Systems* 3:227–238.

Alonso, W. 1964. *Location and Land Use.* Cambridge, MA: Harvard University Press.

Amin, A., and I. Smith. 1986. The internationalization of production and its implications for the UK. In *Technological Change, Industrial Restructuring, and Regional Development,* edited by A. Amin and J. B. Goddard, pp. 41–76. Boston: Allen and Unwin.

Amin, S. 1976. *Unequal Development.* New York: Monthly Review Press.

Arrighi, G., and J. S. Saul. 1973. *Essays on the Political Economy of Africa.* New York: Monthly Review Press.

Asimov, I. 1978. *The Naked Sun.* London: Granada.

Augelli, J. P. 1985. Food, population, and dislocation in Latin America. *Journal of Geography* 84:274–281.

Averitt, R. T. 1968. *The Dual Economy: The Dynamics of American Industry Structure.* New York: W. W. Norton.

Ayeni, B., G. Rushton, and M. L. McNulty. 1987. Improving the geographical accessibility of health care in rural areas: A Nigerian case study. *Social Science and Medicine* 25:1083 1094.

Balassa, B. 1961. *The Theory of Economic Integration.* Homewood, IL: Richard D. Irwin.

Ballmer-Cao, T., and J. Scheidegger. 1979. *Compendium of Data for World System Analysis.* Zurick: Sociologisches Institut der Universitat.

Bannock, G. 1971. *The Juggernauts: The Age of the Big Corporation.* Harmondsworth, Eng: Penguin Books.

Baran, P. 1957. *The Political Economy of Growth.* New York: Monthly Review Press.

Barnet, R., and R. Muller. 1975. *Global Reach.* New York: Simon and Schuster.

Batty, M., and E. Saether. 1972. A note on the design of shopping models. *Journal of the Royal Town Planning Institute* 58:303–306.

Bauer, P. T. 1972. *Dissent on Development.* Cambridge, MA: Harvard University Press.

Baum, J., and C. Oliver. 1992. Institutional embeddedness and the dynamics of organizational populations. *American Sociological Review* 57:540–559.

Beer, S. 1968. *Management Science: The Business Use of Operations Research.* New York: Doubleday & Company, Inc.

Behrman, J. N. 1984. *Industrial Policies: International Restructuring and Transnationals.* Lexington, MA: D.C. Heath.

Bell, D. 1973. *The Coming of Postindustrial Society.* New York: Basic Books.

Bell, D. E., H. Raiffa, and A. Tversky, Eds. 1988. *Decision Making.* New York: Cambridge University Press.

Bell, T. L. 1973. *Central Place Theory as a Mixture of the Function Pattern Principles of Christaller and Lösch: Some Empirical Tests and Applications.* University of Iowa, Department of Geography, unpublished doctoral dissertation.

Berry, B. J. L. 1961. City size distributions and economic development. *Economic Development and Cultural Change* 9:573 588.

———. 1967. *Geography of Market Centers and Retail Distribution.* Englewood Cliffs, NJ: Prentice Hall.

———. 1968. Interdependency of spatial structure and spatial behavior: General field theory formulation. *Papers and Proceedings of the Regional Science Association* 21:205–227.

———. 1969a. Relationships between regional economic development and the urban system—The case of Chile. *Tijdschrift voor Economische en Sociale Geografie* 60:283–307.

———. 1969b. Policy implications of an urban location model for the Kanpur region. In *Regional Perspective of Industrial and Urban Growth—The Case of Kanpur,* edited by P. B. Desai, et al., pp. 203–219. Bombay: Macmillan.

Berry B. J. L., E. C. Conkling, and D. M. Ray. 1997. *Global Economy.* Englewood Cliffs, NJ: Prentice Hall.

Berry B. J. L., and H. M. Meyer. 1962. *Comparative Studies of Central Place Systems. Final Report NONR 2121-18 and NR 389-126.* Washington, DC: Geography Branch, U.S. Office of Naval Research.

Blaikie, P. M. 1971. Spatial organization of agriculture in some north Indian villages: Part 1. *Transactions, Institute of British Geographers* 52:1–40.

Blaikie, P. M., and H. Brookfield. 1987. *Land Degradation and Society.* New York: Methuen.

Blanchet, D. 1991. On interpreting observed relationships between population growth and economic growth: A graphical exposition. *Population and Development Review* 17(1):105 114.

Bloom, D., and R. Freeman. 1986. The effects of rapid population growth on labor supply and employment in developing countries. *Population and Development Review* 12(3):381 414.

Bluestone, B., and B. Harrison. 1987. The impact of private disinvestment on workers and their communities. In *International Capitalism and Industrial Restructuring*, edited by R. Peet, pp. 72–104. Boston: Allen and Unwin.

Bohannan, P., and P. Curtin. 1971. *Africa and Africans*. Garden City, NY: Natural History Press.

Borchert, J. R. 1961. The Twin Cities urbanized area: Past, present, and future. *Geographical Review* 51:47–70.

———. 1963. *The Urbanization of the Upper Midwest: 1930–1960*. Minneapolis: University of Minnesota, Upper Midwest Economic Study, Urban Report No. 2.

———. 1967. American metropolitan evolution. *Geographical Review* 57:301–331.

———. 1972. Banking linkages of Third Order countries. *Annals of the Association of American Geographers* 62:253.

———. 1987. *America's Northern Heartland: An Economic and Historical Geography of the Upper Midwest*. Minneapolis: University of Minnesota Press.

Borchert, J. R., and D. D. Carroll. 1971. *Minnesota Settlement and Land Use 1985*. Minneapolis: University of Minnesota, Center for Urban and Regional Affairs.

Boserup, E. 1965. *The Conditions of Agricultural Growth: The Economics of Agrarian Change Under Population Pressure*. Chicago: Aldine.

———. 1970. Present and potential food production in developing countries. In *Geography and a Crowding World*, edited by W. Zelinsky, L. A. Kosinski, and R. M. Prothero, pp. 100–110. New York: Oxford University Press.

———. 1970. *Woman's Role in Economic Development*. New York: St. Martin's Press.

———. 1981. *Population and Technology*. New York: Blackwell.

———. 1981. *Population and Technological Change: A Study of Long-Term Trends*. Chicago: University of Chicago Press.

Bradshaw, B., and W. P. Frisbie. 1983. Potential labor force supply and replacement in Mexico and the states of the Mexican Cession and Texas: 1980–2000. *International Migration Review* 17(3):394–409.

Brown, L. R. 1981. Eroding the base of civilization. *Journal of Soil and Water Conservation* 36:255–260.

Bunge, W. 1966. *Theoretical Geography*. Lund Studies in Geography, Series C1. Lund, Sweden: Gleerup.

———. 1971. *Fitzgerald: The Geography of a Revolution*. Cambridge, MA: Schenkman Publishing Company.

Bureau of Land Management (BLM). 1988. *Public Land Statistics 1987*. Washington, DC: GPO.

Burgess, E. W. 1925. Growth of the city. In *The City*, edited by R. E. Park, E. W. Burgess, and R. D. McKenzie, pp. 47–62. Chicago: University of Chicago Press.

Burns, A. F. 1934. *Production Trends in the United States*. New York: National Bureau of Economic Research.

Caldwell, J., and P. Caldwell. 1990. High fertility in sub-Saharan Africa. *Scientific American* 40:118–125.

Caldwell, J., I. O. Orubuloye, and P. Caldwell. 1992. Fertility decline in Africa: A new type of transition? *Population and Development Review* 18(2):211–242.

Carson, R. 1962. *Silent Spring*. Boston: Houghton Mifflin.

Cassen, R. W. 1976. Population and development: A survey. *World Development* 4:785–830.

Chapman, K., and D. Walker. 1987. *Industrial Location*. New York: Basil Blackwell.

Chisholm, G. G. 1899. *Handbook of Commercial Geography*. London: Longmans, Green.

Chisholm, M. 1979. *Rural Settlement and Land Use: An Essay in Location*, 3d ed. London: Hutchinson.

———. 1982. *Modern World Development*. Totowa, NJ: Barnes & Noble.

Christaller, W. 1966. *The Central Places of Southern Germany*. Translated by C. W. Baskin. Englewood Cliffs, NJ: Prentice Hall.

Clark, C. 1967. *Population Growth and Land Use*. New York: St. Martin's Press.

Clarke, B., and L. Bolwell. 1968. Attractiveness as part of retail potential models. *Journal of the Royal Town Planning Institute* 54:477–478.

Clarke, W. C. 1977. The structure of permanence: The relevance of self-subsistence communities for world ecosystem management. In *Subsistence and Survival: Rural Ecology in the Pacific*, edited by T. P. Bayliss-Smith and R. Feachem, pp. 363–384. London: Academic Press.

Coale, A. J., and E. M. Hoover. 1958. *Population Growth and Economic Development in Low-Income Countries*. Princeton, NJ: Princeton University Press.

Cochran, T. C. 1966. The entrepreneur in social change. *Explorations in Entrepreneurial History* (2d series) 4:25–38.

Cohen, R. B. 1981. The new international division of labor, multinational corporations and urban hierarchy. In *Urbanization and Urban Planning in Capitalist Society*, edited by M. Dear and A. J. Scott, pp. 287–315. New York: Methuen.

Cole, J. 1987. *Development and Underdevelopment*. London: Methuen.

Commoner, B. 1975. How poverty breeds overpopulation (and not the other way around). *Ramparts* 10:21–25, 58–59.

Conkling, E., and J. McConnell. 1985. The world's new economic powerhouse. *Focus* 35:2–7.

Corbridge, S., editor. 1993. *World Economy*. New York: Oxford University Press.

Council on Environmental Quality. 1980. *The Global 2000 Report to the President*, Vol 1. Washington, DC: GPO.

Curran, J., and J. Stanworth. 1986. Trends in small firm industrial relations and their implications for the role of the small firm in economic restructuring. In *Technological Change, Industrial Restructuring, and Regional Development*, edited by A. Amin and J. B. Goddard, pp. 233–257. Winchester, MA: Allen and Unwin.

Cyert, R. M., and J. G. March. 1963. *A Behavioral Theory of the Firm*. Englewood Cliffs, NJ: Prentice Hall.

Daly, H. 1986. Review of population growth and economic development: Policy questions. *Population and Development Review* 12(3):582–585.

Daniels, P. W. 1985. *Service Industries: A Geographical Appraisal*. New York: Methuen.

Darst, G. 1987. Energy worries fading: Conservation drive wanes in Washington. *Minneapolis Star and Tribune* 15 March, pp. 1, 3.

Datoo, B. A. 1976. *Toward a Reformulation of Boserup's Theory of Agricultural Change.* Dar es Salaam, Tanzania: Department of Geography, University of Dar es Salaam. Mimeographed.

de Blij, H. J., and P. O. Muller. 1985. *Geography: Regions and Concepts,* 4th ed. New York: John Wiley.

———. 1992. *Geography: Regions and Concepts,* revised 6th edition. New York: John Wiley & Sons.

de Souza, A. R. 1985. Dependency and economic growth. *Journal of Geography* 85:94.

———. 1986. To have and have not: Colonialism and coreperiphery relations. *Focus* 36(3):14–19.

de Souza, A. R., and J. B. Foust. 1979. *World Space Economy.* Columbus, OH: Merrill.

de Souza, A. R., and P. W. Porter. 1974. *The Underdevelopment and Modernization of the Third World.* Resource Paper No. 28. Washington, DC: Association of American Geographers.

Demeny, P. 1971. The economics of population control. In *National Academy of Science, Rapid Population Growth.* Baltimore: Johns Hopkins University Press.

———. 1981. The North–South income gap: A demographic perspective. *Population and Development Review* 7(2):297 310.

Demko, G. J., and W. B. Wood. 1987. International refugees: A geographical perspective. *Journal of Geography* 86:225 228.

Dent, H.S. and J.V. Smith, Jr. 1993. *The Great Boom Ahead.* New York: Hyperion.

Diaz-Briquets, S., and L. Perez. 1981. Cuba: The demography of revolution. *Population Bulletin* 36(1):101–121.

Dicken, P. 1986. *Global Shift: Industrial Change in a Turbulent World.* London: Harper and Row.

———. 1992. *Global Shift: Industrial Change in a Turbulent World,* 2d ed. London: Harper & Row.

Dickinson, R. E. 1964. *City and Region.* London: Routledge and Kegan Paul.

Directors of the World Resources Institute. 1992. *World Resources, 1992–93.* New York: Oxford University Press.

Dodson, R. F. 1991. *VT/GIS: The von Thünen GIS Package.* Santa Barbara, CA: National Center for Geographic Information & Analysis. (Technical paper 91-27.)

Doxiadis, C. A. 1970. Man's movements and his settlements. *Ekistics* 29:318.

Dumont, R., and M-F. Mottin. 1983. *Stranglehold on Africa.* London: André; Deutsch.

Dunn, E. S., Jr. 1954. *The Location of Agricultural Production.* Gainesville: University of Florida Press.

Edwards, C. 1985. *The Fragmented World.* New York: Methuen.

Ehrlich, P. 1968. *The Population Bomb.* New York: Ballantine Books.

———. 1971. *The Population Bomb,* 2d ed. New York: Sierra Club/Ballantine Books.

Electronics Industries Association of Japan. 1989. *Facts and Figures on the Japanese Electronics Industry.* Tokyo: Author.

Emmanuel, A. 1972. *Unequal Exchange: A Study of the Imperialism of Trade.* London: New Left Books.

Encyclopaedia Britannica. 1987. *1987 Britannica Book of the Year.* Chicago: Encyclopaedia Britannica.

Engels, F. 1958. *The Condition of the Working Class in England.* Stanford, CA: Stanford University Press.

Enke, S. 1960. The economics of government payments to limit population. *Economic Development and Cultural Change* 8(4): 339–348.

Environmental Systems Research Institute, Inc. 1989. Why GIS? (revised, 1992). *ARC News* 11(5):1–4.

Ericson, R. E. 1991. The classical Soviet-type economy: Nature of the system and applications for reform. *Journal of Economic Perspectives* 20:23.

Ewart, W. D. and W. Fullard. 1973. *World Atlas of Shipping.* London: Philip and Son.

Fanon, F. 1963. *The Wretched of the Earth.* New York: Grove Press.

Fellman, J., A. Getis, and J. Getis. 1992. *Human Geography: Landscapes of Human Activities.* Dubuque, IA: W. C. Brown.

Ferguson, R., and E. Carlson. 1990. The boondocks distant communities promise good homes but produce malaise: Census shows people moving so far from jobs they lack time to enjoy life. *The Wall Street Journal* 25 October, pp. Al, A6.

Fieldhouse, D. K. 1967. *The Colonial Empires.* New York: Delacorte Press.

Financial Times. 1989. Britain's regions: A test for Thatcherism. *Financial Times* 27 January:33–40.

Fisher, J. S. 1989. Anglo-America: Economic growth and transformation. In *Geography and Development,* edited by J. S. Fisher, pp. 146–166. Columbus, OH: Merrill.

Fisher, J. S. 1992. *Geography and Development: A World Regional Approach.* New York: Macmillan.

Fishlow, A. 1965. *American Railroads and the Transformation of the Antebellum Economy.* Cambridge, MA: Harvard University Press.

Food and Agriculture Organization. 1983. *Per Capita Dietary Energy Supplies in Relation to Nutritional Requirements.* World Food Report, Series 211. Rome: Author.

———. 1985. *The State of Food and Agriculture 1984.* Rome: Author.

Foreign exchange rates. *San Diego Union,* July 4, 1993.

Frank, A. G. 1969. *Capitalism and Underdevelopment in Latin America.* New York: Monthly Review Press.

———. 1981. *Crisis in the Third World.* London: Heineman.

Franklin, S. H. 1965. Systems of production: Systems of appropriation. *Pacific Viewpoint* 6:145–166.

Freeman, M. 1986. Transport. In *Atlas of Industrializing Britain, 1780–1914,* edited by J. Langton and R. J. Morris, pp. 80–93. New York: Methuen.

———. 1987. *The Challenge of New Technologies NOECD, Interdependence and Corporation in Tomorrow's World,* Paris, OECD.

———. 1988. Introduction. In *Technical Change and Economic Theory,* edited by G. Dosi et al., pp. 34–61. London: Pinter.

Fried, M. 1967. *The Evolution of Political Society.* New York: Random House.

Friedmann, J. 1966. *Regional Development Policy: A Case Study of Venezuela.* Cambridge, MA: The MIT Press.

Froebel, F., J. Heinrich, and O. Kreye. 1977. The tendency towards a new international division of labor. *Review* 1(1):77–88.

Fuentes, A., and B. Ehrenreich. 1987. Women in the global factory. In *International Capitalism and Industrial Restructuring,* edited by R. Peet, pp. 201–215. Boston: Allen and Unwin.

Fusfeld, D. R. 1986. *The Age of the Economist,* 5th ed. Glenview, IL: Scott, Foresman.

Gaffikin, F., and A. Nickson. 1984. *Job Crisis and the Multinationals: Deindustrialization in the West Midlands.* Chicago: Third World Books.

Galbraith, J. K. 1967. *The New Industrial State.* Boston: Houghton Mifflin.

———. 1969. *The Affluent Society.* Boston: Houghton Mifflin.

———. 1973. *Economics and the Public Purpose.* Boston: Houghton Mifflin.

Galtung, J. 1971. A structural theory of imperialism. *Journal of Peace Research* 21(2):81–107, 110–116.

Gamble, A. 1981. *Britain in Decline: Economic Policy, Political Strategy, and the British State.* London: Macmillan.

General Agreement on Tariffs and Trade (GATT). 1990. *International Trade, 1989–1990,* Vol II. Geneva: Author.

Geertz, C. 1963. *Agricultural Involution: The Processes of Ecological Changes in Indonesia.* Berkeley: University of California Press.

George, S. 1977. *How the Other Half Dies—The Real Reasons for World Hunger.* Montclair, NJ: Allanheld, Osmun.

Ginsburg, N. S. 1961. *Atlas of Economic Development.* Chicago: University of Chicago Press.

GIS World. 1992. Fort Collins: GIS World Inc. October, pp. 46–47.

Globe and Mail Newspaper. (Toronto), 1986, 6 May.

Godlund, S. 1961. *Population, Regional Hospitals, Transport Facilities and Regions: Planning the Location of Regional Hospitals in Sweden.* Lund, Sweden: Gleerup.

Goliber, T. J. 1985. *Sub-Saharan Africa: Population Pressures on Development.* Population Bulletin, Vol. 40, No. 1. Washington, DC: Population Reference Bureau.

Goodchild, M. F., and B. Massam. 1969. Some least-cost models of spatial administrative systems in southern Ontario. *Geografiska Annaler* 52, B-2:86–94.

Gordon, D. M. 1977. Class struggle and the stages of American urban development. In *The Rise of the Sunbelt Cities,* edited by D. C. Perry and A. J. Watkins, pp. 55–82. Beverly Hills: Sage Publications.

Gottman, J. 1964. *Megalopolis.* Cambridge, MA: The MIT Press.

Gould, P. R. 1960. *The Development of the Transportation Pattern in Ghana.* Evanston, IL: Northwestern University Press, Department of Geography, Studies in Geography, No. 5.

———. 1964. A note on research into the diffusion of development. *Journal of Modern African Studies* 2:123–125.

———. 1970. Tanzania 1920–63: The spatial impress of the modernization process. *World Politics* 22:149–170.

———. 1975. *Spatial Diffusion: The Spread of Ideas and Innovations in Geographic Space.* Learning Package Series No. 11. Columbus: The Ohio State University.

———. 1983. Getting involved in information and ignorance. *Journal of Geography* 82:158–162.

———. 1985. *The Geographer at Work.* London: Routledge and Kegan Paul.

Goulet, D. 1971. *The Cruel Choice.* New York: Atheneum.

Grandstaff, T. 1978. The development of swidden agriculture (shifting cultivation). *Development and Change* 9:547 579.

Greenhut, M. L. 1956. *Plant Location in Theory and Practice.* Chapel Hill: University of North Carolina Press.

Greenow, L., and V. Muniz. 1988. Market trade in decentralized development: The case of Cajamarca, Peru. *The Professional Geographer* 40:416–427.

Gribben, R. 1989. Economic divide will stay but shift northward. *The Daily Telegraph,* 3 January:4.

Griffin, E., and L. Ford. 1980. A model of Latin American city structure. *Geographical Review* 70:397–422.

Griffin, K. 1969. *Underdevelopment in Spanish America.* London: Allen and Unwin.

Hagen, E. E. 1962. *On the Theory of Social Change.* Homewood, IL: The Dorsey Press.

———. 1968. Are some things valued by all men? *Cross Currents* 18:406–414.

Hägerstrand, T. 1965. A Monte Carlo approach to diffusion. *European Journal of Sociology* 6:43–67.

Haggett, P. 1965. *Locational Analysis in Human Geography.* London: Edward Arnold.

Haggett, P., and R. J. Chorley. 1969. *Network Analysis in Human Geography.* London: Edward Arnold.

Håkauson, L. 1979. Towards a theory of location and corporate growth. In *Spatial Analysis, Industry and the Industrial Environment. Volume 1: Industrial Systems,* edited by F. E. I. Hamilton and G. J. R. Linge, pp. 115–138. New York: John Wiley.

Hall, P., Ed. 1966. *Von Thünen's Isolated State.* Translated by C. M. Wartenberg. Oxford, Eng: Pergamon.

———. 1971. *The World Cities.* New York: McGraw-Hill.

———. 1982. *Urban and Regional Planning.* London: Allen and Unwin.

Hammer, M. and R. Champy. 1993. *Reengineering the Corporation.* New York: Harper Collins.

Hannan, M., and G. Carroll. 1992. *Dynamics of Organizational Populations.* New York: Oxford University Press.

Hannan, M., and J. Freeman. 1989. *Organizational Ecology.* Cambridge, MA: Harvard University Press.

Hansen, J. 1970. *The Population Explosion: How Sociologists View It.* New York: Pathfinder Press.

Hanson, S., editor. 1986. *The Geography of Urban Transportation.* New York: Guilford Press.

Hardin, G. 1974. Living on a lifeboat. *Bioscience* 24:561–568.

———. 1968. The tragedy of the commons. *Science* 162: 1243–1248.

Harrington, M. 1977. *The Vast Majority: A Journey to the World's Poor.* New York: Touchstone.

Harris, C. 1954. The market as a factor in the localization of industry in the U.S. *Annals of the Association of American Geographers* 44:315–348.

Harris, C., and E. Ullman. 1945. The nature of cities. *Annals of the American Academy of Political and Social Science* 242:7–17.

Harris, M. 1966. The cultural ecology of India's sacred cattle. *Current Anthropology* 7:51–59.

Hartshorn, T. A. 1992. *Interpreting the City: An Urban Geography.* New York: John Wiley & Sons.

Harvey, D. 1972. *Society, the City, and the Space-Economy of Urbanism.* Resource Paper No. 18. Washington, DC: Association of American Geographers.

———. 1973. *Social Justice and the City.* London: Edward Arnold.

———. 1974. Population, resources, and the ideology of science. *Economic Geography* 50:256–277.

———. 1985. *The Urbanization of Capital.* Baltimore: Johns Hopkins University Press.

Havens, A. E., and W. F. Flinn. 1975. Green revolution technology and community development: The limits of action programs. *Economic Development and Cultural Change* 23:468–481.

Heckscher, E. 1919. The effect of foreign trade on the distribution of income. *Economisk Tidskrift* 21. Reprinted in *Readings in the Theory of International Trade,* edited by H. Ellis and L. Metzler, pp. 272–300. Homewood, IL: Richard D. Irwin, 1950.

Heilbronner, R. 1989. Reflections: The triumph of capitalism. *The New Yorker* 25 January:98–109.

Henderson, J. V. 1988. *Urban Development: Theory, Fact and Illusion.* New York: Oxford University Press.

Henderson, J., and Castells, M., editors. 1987. *Global Restructuring and Territorial Development.* London: Sage.

Henige, D. 1970. *Colonial Governors.* Madison: University of Wisconsin Press.

Hepworth, M. 1987. *The Geography of the Information Economy.* New York: Guilford Press.

———. 1990. *Geography of the Information Economy.* New York: Guilford Press.

Herbert, D. T. 1982. The changing face of the city. In *The Changing Geography of the United Kingdom,* edited by R. J. Johnston and J. C. Doornkamp, pp. 227–255. London: Methuen.

Higgins, B. 1956. The dualistic theory of underdeveloped areas. *Economic Development and Cultural Change* 4:99–115.

Hirschman, A. O. 1958. *The Strategy of Economic Development.* New Haven, CT: Yale University Press.

Hofstede, G. 1980. *Culture's Consequences: International Differences in Work-Related Values.* Beverly Hills, CA: Sage.

Hollerbach, P. 1980. Recent trends in fertility, abortion, and contraception in Cuba. *International Family Planning Perspectives* 6(3):97–106.

Hoover, E. M. 1984. *An Introduction to Regional Economies,* 3d ed. New York: Alfred A. Knopf.

Horvath, R. J. 1969. Von Thünen's isolated state and the area around Addis Ababa, Ethiopia. *Annals of the Association of American Geographers* 59:308–323.

Hotelling, H. 1929. Stability in competition. *Economic Journal* 39:41–57.

Howard, E. 1946. *Garden Cities of Tomorrow.* London: Faber.

Hoyle, B. S. and R. D. Knowles. 1993. *Modern Transport Geography.* London: Belhaven Press.

Hoyle, B. S., and D. A. Pinder. 1981. Seaports, cities and transport systems. In *Cityport Industrialization and Regional Development,* edited by B. S. Hoyle and D. A. Pinder, pp. 1–10. Oxford: Pergamon.

Hoyt, H. 1939. *The Structure and Growth of Residential Neighborhoods in American Cities.* Washington, DC: Federal Housing Administration.

Hubbert, M. K. 1962. *Energy Resources: A Report to the Committee on Natural Resources.* Washington, DC: National Academy of Sciences.

Huff, D. L. 1963. A probability analysis of shopping center trade areas. *Land Eeonomics* 53:81–89.

Hughes, A., and M. S. Kumar. 1984. Recent trends in aggregate concentration in the United Kingdom economy. *Cambridge Journal of Economics* 8:235–250.

Huke, R. E. 1985. The green revolution. *Journal of Geography* 84:248–254.

Humphrys, G. 1972. *South Wales.* Newton Abbott, Eng: David and Charles.

Hunker, H., and A. J. Wright. 1963. *Factors of Industrial Location in Ohio.* Columbus: The Ohio State University Press.

Huntington, E. 1924. *Civilization and Climate.* New Haven, CT: Yale University Press.

Hurni, H. 1983. Soil erosion and soil formation in agricultural systems, Ethiopia and northern Thailand. *Mountain Research and Development* 3:131–142.

IDRISI. 1992. *IDRISI: A Grid Based Geographic Analysis System.* Worcester, MA: Graduate School of Geography, Clark University.

Ihlanfeldt, K. R., D. L. Sjoquist. 1990. Job accessibility and racial differences in youth employment rates. *American Economic Review* 8:267–276.

Ihlanfeldt, K. R., and D. L. Sjoquist. 1990. Job accessibility and racial differences in youth employment rates. *American Economic Review* 8:267–276.

International Institute for Environment, Development and World Resources. 1987. *World Resources 1987.* New York: Basic Books.

———. 1993. *World Resources, 1993–1994.* New York: Oxford.

International Monetary Fund. 1978. *29th Annual Report on Exchange Restrictions.* Washington, DC: Author.

———. 1985. *IMF Survey, 21 January.* Washington, DC: Author.

———. 1991. *World Economic Outlook: October 1991.* Washington, DC: Author.

———. 1992. *International Financial Statistics.* Washington, DC: Author.

Isard, W. 1956. *Location and Space Economy.* Cambridge, MA: The MIT Press.

———. 1960. *Methods of Regional Analysis: An Introduction to Regional Science.* Englewood Cliffs, NJ: Prentice Hall.

Jackson, W. A. D. 1962. The Virgin and Idle Lands Program reappraised. *Annals of the Association of American Geographers* 52:69–79.

Janelle, D. G. 1968. Central place development in a time-space framework. *The Professional Geographer* 20:5–10.

Jefferson, M. 1921. *Recent Colonization in Chile.* Research Series No. 6. New York: American Geographical Society.

Johnston, W. 1991. Global work force 2000: The new world labor market. *Harvard Business Review* 41:115–129.

Junkerman, J. 1987. Blue-sky management: The Kawasaki story. In *International Capitalism and Industrial Restructuring,* edited by R. Peet, pp. 131–144. Boston: Allen and Unwin.

Kahn, H. 1973. If the rich stop aiding the poor... . *Development Forum* 1(2):1–3.

Kansky, K. J. 1963. *Structure of Transportation.* Chicago: University of Chicago, Department of Geography, Research Paper No. 84.

Keeble, D. E. 1967. Models of economic development. In *Models in Geography,* edited by R. J. Chorley and P. Haggett, pp. 243–302. London: Methuen.

Kelley, A. 1973. Population growth, the dependency rate, and the pace of economic development. *Population Studies* 27:405–414.

Kennedy, P. 1993. *Preparing for the 21st Century's Winners and Losers.* New York Review of Books, New York: Random House, pp. 32-44.

Kennelly, R. A. 1954. The location of the Mexican steel industry. *Revista Geografica* 15:109–129.

———. 1955. The location of the Mexican steel industry. *Revista Geografica* 16:60–77.

Keyfitz, N., and W. Flieger. 1990. *World Population Growth and Aging: Demographic Trends in the Late Twentieth Century*. Chicago: University of Chicago Press.

Keynes, J. M. 1936. *The General Theory of Employment Interest and Money*. New York: Harcourt, Brace.

Kidron, M. and R. Segal. 1984. *The New State of the World Atlas*. New York: Simon and Schuster.

Knight, C. G., and J. L. Newman, Eds. 1976. *Contemporary Africa: Geography and Change*. Englewood Cliffs, NJ: Prentice Hall.

Knight, C. G., and R. P. Wilcox. 1976. *Triumph or Triage: The Third World Food Problem in Geographical Perspective*. Resource Paper No. 75-3. Washington, DC: Association of American Geographers.

Knox, P. L. 1994. *Urbanization: An Introduction to Urban Geography*. Englewood Cliffs, New Jersey: Prentice Hall.

Kolars, J. F., and J. D. Nystuen. 1974. *Human Geography: Spatial Design in World Society*. New York: McGraw-Hill.

Komorov B. 1981. *The Destruction of Nature in the Soviet Union*. London: Pluto Press.

Kondratiev N. D. 1935. The long waves in economic life. *Review of Economic Statistics* 17:105–115.

Krasner, S. D. 1985. *Structural Conflict*. Berkeley and Los Angeles: University of California Press.

Kumar, K., and K. Y. Kim. 1984. The Korean manufacturing multinationals. *Journal of International Business Studies* 1:45–61.

Kuznets, S. 1930. *Secular Movements in Production and Prices*. Boston: Houghton Mifflin.

———. 1954. *Economic Change*. New York: W. W. Norton.

———. 1965. *Economic Growth of Nations*. Cambridge, MA: Harvard University Press.

———. 1972. Problems in comparing recent growth rates for developed and less developed countries. *Economic Development and Cultural Change* 20(2):195–209.

Landsberg, M. 1987. Export-led industrialization in the Third World: Manufacturing imperialism. In *International Capitalism and Industrial Restructuring*, edited by R. Peet, pp. 216–239. Boston: Allen and Unwin.

Lappé, F. M., and J. Collins. 1976. More food means more hunger. *Development Forum* 4(8):1–2.

———. 1977. *Food First*. Boston: Houghton Mifflin.

Lecomber, R. 1975. *Economic Growth Versus Environment*. New York: John Wiley.

Leibenstein, H. 1957. *Economic Backwardness and Economic Growth*. New York: John Wiley.

Leontief, W., et al. 1977. *The Future of the World Economy: A United Nations Study*. New York: Oxford University Press.

Lewis, P. 1983. The galactic metropolis. In *Beyond the Urban Fringe: Land-Use Issues of Nonmetropolitan America*, edited by R. Platt and G. Macinko, pp. 23–49. Minneapolis: University of Minnesota Press.

Lewis, W. A. 1954. Economic development with unlimited supplies of labour. *Manchester School of Economic and Social Studies* 22:139–191.

Lin Piao, L. 1965. Long live the victory of the People's War. *Peking Review* 3:9–30.

Linneman, H. 1966. *An Econometric Study of International Trade Flows*. Amsterdam: North-Holland Publishing.

Livingstone, I. 1971. Agriculture versus industry in economic development. In *Economic Policy for Development*, edited by I. Livingstone, pp. 235–249. Harmondsworth, Eng: Penguin Books.

Lloyd, P. E., and P. Dicken. 1972. *Location in Space: A Theoretical Approach to Economic Geography*. New York: Harper and Row.

Lord Ritchie-Calder. 1973/1974. UNICEF's grandchildren. *UNICEF News* 78.

Lösch, A. 1954. *The Economics of Location*. Translated by W. H. Woglom and W. F. Stolper. New Haven, CT: Yale University Press.

Lyle, J., and F. P. Stutz. 1983. Land use suitability modelling and mapping. *Cartographic Journal* 18:39–50.

Mabogunje, A. L. 1980. *The Development Process: A Spatial Perspective*. London: Hutchinson.

Mackay, J. R. 1958. The interactance hypothesis and boundaries in Canada. *Canadian Geographer* 11:1–8.

Magirier, G. 1983. The eighties: a second phase of crisis? *Capital and Class* 21:61–86.

Malecki, E. J. 1979. Locational trends in R&D by large U.S. corporations. *Economic Geography* 55:309–323.

———. 1980. Dimensions of R&D locations in the U.S. *Research Policy* 9:2–22.

Malthus, T. R. 1970. *An Essay on the Principle of Population and a Summary View of Principle of Population*. Harmondsworth, Eng: Penguin Books.

Manners, G. 1971. *The Changing World Market for Iron Ore 1950–1980*. Baltimore: Johns Hopkins University Press.

Mantoux, P. 1961. *The Industrial Revolution in the Eighteenth Century*. New York: Macmillan.

Marshall, J. N. 1979. Organization theory and industrial location. *Environment and Planning A* 14:1667–1683.

Marx, K. 1967. *Capital*, 1st ed. Volume 1. New York: International Publishers.

Mason, A. 1988. Savings, economic growth, and demographic change. *Population and Development Review* 14:113–144.

Mason, R. J., and M. T. Matson. 1990. *Atlas of United States Environmental Issues*. New York: Macmillan.

Mason, R. H., R. R. Miller, and D. R. Weigel. 1975. *The Economics of International Business*. New York: John Wiley.

Massey, D. 1973. Towards a critique of industrial location theory. *Antipode* 5(3):33–39.

———. 1977. *Industrial Location Theory Reconsidered*. Unit 26, Course D204. Milton Keynes, Eng: The Open University.

———. 1984. *Spatial Divisions of Labor: Social Structures and the Geography of Production*. New York: Methuen.

———. 1987. The shape of things to come. In *International Capitalism and Industrial Restructuring*, edited by R. Peet, pp. 105–122. Boston: Allen and Unwin.

Massey, D., and R. A. Meegan. 1978. Industrial restructuring versus the cities. *Urban Studies* 15:273–288.

———. 1979. The geography of industrial reorganization: The spatial effects of restructuring the electronical engineering sector under the Industrial Reorganization Corporation. *Progress in Planning* 10:155–237.

———. 1982. *The Anatomy of Job Loss*. London: Methuen.

Mather, C. 1986. The Midwest: Image and Reality. *Journal of Geography* 85:190–194.

McCall, M. K. 1977. Political economy and rural transport: An appraisal of Western misconceptions. *Antipode* 53:503–529.

McCarty, H. H., and J. B. Lindberg. 1966. *A Preface to Economic Geography.* Englewood Cliffs, NJ: Prentice Hall.

McClelland, D. 1961. *The Achieving Society.* New York: Van Nostrand.

McConnell, C. R., and S. L. Brue. 1993. *Macro-Economics: Principals, Problems, and Policies.* New York: McGraw-Hill.

McConnell, J. E. 1980. Foreign direct investment in the United States. *Annals of the Association of American Geographers* 70:259–270.

———. 1983. The international location of manufacturing investments: Recent behavior of foreign-owned corporations in the United States. In *Spatial Analysis, Industry and the Industrial Environment, Volume 3: Regional Economics and Industrial Systems,* edited by F. E. I. Hamilton and G. J. R. Linge, pp. 337–358. New York: John Wiley.

McConnell, J. E., and R. A. Erickson. 1986. Geobusiness: An international perspective for geographers. *Journal of Geography* 85:98–105.

McGee, T. G. 1967. *The Southeast Asian City.* London: Bell.

———. 1971. *The Urbanization Process in the Third World.* London: Bell.

———. 1974. *The Persistence of the Proto-Proletariat: Occupational Structures and Planning for the Future of Third World Cities.* Monash, Victoria: Department of Geography, Monash University.

McNee, R. B. 1960. Towards a more humanistic geography: The geography of enterprise. *Tijdscrift voor Economische en Sociale Geografie* 51:201–206.

———. 1974. A systems approach to understanding the geographic behavior of organizations, especially large corporations. In *Spatial Perspectives on Industrial Organization and Decision Making,* edited by F. E. I. Hamilton, pp. 47–76. New York: John Wiley.

Meadows, D. 1974. *Dynamics of Growth in a Finite World.* Cambridge, MA: Wright-Allen Press.

Meadows, D., et al. 1972. *The Limits to Growth.* New York: Universe Books.

Meijer, H. 1986. *Randstad Holland.* Utrecht/The Hague: Information and Documentation Centre for the Geography of the Netherlands.

Meinig, D. 1962. *On the Margins of the Good Earth: The South Australian Wheat Frontier 1869–1884.* Association of American Geographers, Monograph Series No. 2. Chicago: Rand McNally.

Mellor, J. W. 1962. Increasing agricultural production in early stages of economic development. *Indian Journal of Agricultural Economics* 17:29–46.

Mensch, G. 1979. *Stalemate in Technology: Innovations Overcome the Depression.* Cambridge, MA: Ballinger.

Merrick, T. W. 1986. *World Population in Transition.* Population Bulletin, Vol. 41, No. 1. Washington, DC: Population Reference Bureau.

Mexico slips into reverse. 1986. *Newsweek* 17 March:34–35.

Mikesell, M. W. 1969. Patterns and imprints of mankind. In *The International Atlas.* Chicago: Rand McNally.

Miller, G. T., Jr. 1975. *Living in the Environment: Concepts, Problems, and Alternatives.* Belmont, CA: Wadsworth.

Mills, E. S. 1972. The value of clean land. In *Quality of the Urban Environment,* edited by H. Perloff. Washington, DC: Resources for the Future.

Mishan, E. J. 1977. *The Economic Growth Debate: An Assessment.* London: Allen and Unwin.

Missen, G. I., and M. I. Logan. 1977. National and local distribution systems and regional development: The case of Kelantan in West Malaysia. *Antipode* 9:60–74.

Morello, T. 1983. Sweatshops in the sun? *Far Eastern Economic Review* 20:88–89.

Morgan, W. T. 1963. The ``White Highlands'' of Kenya. *Geographical Journal* 129:140–155.

Morrill, R. L. 1963. The development and spatial distribution of towns in Sweden. *Annals of the Association of American Geographers* 53:1–14.

———. 1970. *The Spatial Organization of Society.* Belmont, CA: Wadsworth.

Morris, M. D. 1979. *Measuring the Condition of the World's Poor: The Physical Quality of Life Index.* New York: Pergamon.

Muller, E. K., and P. A. Groves. 1979. The emergence of industrial districts in nineteenth century Baltimore. *Geographical Review* 69:159–178.

Muller, P. O. 1976. *The Outer City: Geographical Consequences of the Urbanization of the Suburbs.* Resource Paper 75-2. Washington, DC: Association of American Geographers.

———. 1981. *Contemporary Suburban America.* Englewood Cliffs, NJ: Prentice Hall.

Murdie, R. A. 1965. Cultural differences in consumer travel. *Economic Geography* 41:211–233.

Murphy, R. 1978. *Patterns on the Earth.* Chicago: Rand McNally.

Muth, R. 1969. *Cities and Housing.* Chicago: The University of Chicago Press.

Myrdal, G. 1957. *Economic Theory and Underdeveloped Regions.* London: Duckworth.

Naisbitt, J. 1982. *Megatrends: Ten New Directions Transforming Our Lives.* New York: Werner Books.

Naisbitt, J., and P. Aburdene. 1990. *Megatrends 2000.* New York: William Morrow & Co.

Nakamura, R. 1985. Agglomeration economies in urban manufacturing industries: A case of Japanese cities. *Journal of Urban Economies* 17:108–124.

National Geographic. 1993. Kazakhstan: A broken empire. *National Geographic* March:23–35.

National Research Council. 1986. *Population Growth and Economic Development: Policy Questions.* Washington, DC: National Academy Press.

———. 1993. *Toward a Coordinated Spatial Data Infrastructure for the Nation.* Washington, DC: National Academy Press.

Nations, J. D., and D. I. Komer. 1983. Central America's tropical rainforests: Positive steps for survival. *Ambio* 12:232–238.

Niemann, B. J., Jr., and S. S. Nieman. 1993, April. *Geo Info Systems.* Metuchen, New Jersey.

Norcliffe, G. B. 1975. A theory of manufacturing places. In *Locational Dynamics of Manufacturing Industry,* edited by L. Collins and D. F. Walker, pp. 19–59. New York: John Wiley.

Notestein, F. W. 1970. Zero population growth: What is it? *Family Planning Perspectives* 2:20–24.

O'Kelly, M. 1992. *Research problems in HUB networks.* Paper presented at the 88th annual meeting of the Association of American Geographers, San Diego, CA.

Organization for Economic Cooperation and Development (OECD). 1983. *Long-Term Outlook for the World Automobile Industry.* Paris: Author.

Ohlin, B. 1933. *Interregional and International Trade.* Cambridge, MA: Harvard University Press.

———. 1976. Economic theory confronts population growth. In *Economic Factors in Population Growth,* edited by A. Coale, pp. 121–143. New York: Wiley.

Ogden, P. 1984. *Migration and Geographical Change.* Cambridge, Eng: Cambridge University Press.

Olsen, E. 1971. *International Trade Theory and Regional Income Differences.* Amsterdam: North-Holland.

Osleeb, J., and R. G. Cromley. 1978. The location of plants of the uniform delivered price manufacturer: A case study of Coca Cola, Ltd. *Economic Geography* 54:40–52.

Paddock, W., and P. Paddock. 1976. *Time of Famines—America and the World Food Crisis.* Boston: Little, Brown.

Park, R. E., and C. Newcomb. 1933. Newspaper circulation and metropolitan regions. In *The Metropolitan Community,* edited by R. D. McKenzie, pp. 98–110. New York: McGraw-Hill.

Parrott, R., and Stutz, F. P. 1992. Urban GIS Applications. In Maguire, Goodchild and Rhind (eds.), *GIS: Principles and Applications.* London, England: Longman.

Partant, F. 1982. *La Fin du Développement: Naissance d'une Alternative?* Paris: La De;a;couverte.

Patel, N. R. 1979. Locating rural social service centers in India. *Management Science* 25:22–30.

Peet, R. 1985. The social origins of environmental determinism. *Annals of the Association of American Geographers* 75:309 333.

———. 1987a. Industrial restructuring and the crisis of international capitalism. In *International Capitalism and Industrial Restructuring,* edited by R. Peet, pp. 9–32. Boston: Allen and Unwin.

———. 1987b. The geography of class struggle and the relocation of United States manufacturing industry. In *International Capitalism and Industrial Restructuring,* edited by R. Peet, pp. 40–71. Boston: Allen and Unwin.

———. Ed. 1987c. *International Capitalism and Industrial Restructuring.* Boston: Allen and Unwin.

Pelzer, K. J. 1945. *Pioneer Settlement in the Asiatic Tropics.* Special Publication No. 219. New York: American Geographical Society.

Penrose, E. 1959. *The Theory of the Growth of the Firm.* Oxford: Basil Blackwell.

Perrons, D. 1981. The role of Ireland in the new industrial division of labor: A proposed framework for analysis. *Regional Studies* 15:81–100.

Pickard, J. P. 1972. U. S. metropolitan growth and expansion, 1970–2000, with population projections. In *Population Growth and the American Future.* Washington, DC: GPO.

Pickett, J., Ed. 1977. The choice of technology in developing countries. *World Development* 5(9/10):773–879.

Pletsch, C. E. 1982. The three worlds or the division of social scientific labor, circa 1950–1975. *Comparative Studies in Society and History* 23:565–590.

Polanyi, K. 1971. *Primitive, Archaic, and Modern Economics: Essays of Karl Polanyi.* Boston: Beacon Press.

Population Crisis Committee. 1987. *The International Human Suffering Index.* Washington, DC: Author.

Population Reference Bureau. 1986a. *The United States Population Data Sheet.* Washington, DC: Author.

———. 1986b. A potpourri of population puzzles. *Population Education Interchange* 15(2).

———. 1985. *World Population Data Sheet.* Washington, DC: Author.

———. 1996. *World Population Data Sheet.* Washington, DC: Author.

———. 1997. *World Population Data Sheet.* Washington, DC: Author.

Porter, D. R. 1985. The office/housing linkage issue. *Urban Land* 15:16–21.

Porter, M. E. 1990. *The Competitive Advantage of Nations.* New York: The Free Press.

Porter, P. W. 1965. Environmental potentials and economic opportunities: A background for cultural adaption. *American Anthropologist* 67:409–420.

———. 1979. *Food and Development in the Semi-Arid Zone of East Africa.* Syracuse, NY: Maxwell School of Citizenship and Public Affairs, Syracuse University.

Porter, D. R., and T. J. Lasser. 1988. The latest on linkage. *Urban Land* December:7–11.

Potter, J. 1986. Review of population growth and economic development: Policy questions. *Population and Development Review* 12(3):578–581.

Pred, A. R. 1966. *The Spatial Dynamics of U.S. Urban-Industrial Growth, 1800–1914.* Cambridge, MA: The MIT Press.

———. 1967. *Behavior and Location: Foundations for a Geographic and Dynamic Location Theory, Part 1.* Lund Studies in Geography, Series B. 27.

———. 1977. *City-Systems in Advanced Economies.* London: Hutchinson.

Preston, S. 1986. Are the economic consequences of population growth a sound basis for population policy? In *World Population and U.S. Policy: The Choices Ahead,* edited by J. Menken. New York: W. W. Norton & Co.

Price, A. G. 1939. *White Settlers in the Tropics.* Special Publication No. 23. New York: American Geographical Society.

Rae, J. B. 1965. *The American Automobile: A Brief History.* Chicago: University of Chicago Press.

Rand McNally Commercial Atlas and Marketing Guide. 1988. Chicago: Rand McNally.

Raporport, C. 1982. The FT European 500: Financial Times survey. *Financial Times* 21 October.

Ravenstein, E. G. 1885. The laws of migration. *Journal of the Royal Statistical Society* 48:167–227.

———. 1889. The laws of migration. *Journal of the Royal Statistical Society* 52:241–301.

Ray, D. M. 1971. The location of United States' manufacturing subsidiaries in Canada. *Economic Geography* 47:389–400.

Rees, J. 1972. The industrial corporation and location decision analysis. *Area* 4:199–205.

———. 1974. Decision-making, the growth of the firm and the business environment. In *Spatial Perspectives on Industrial Organization and Decision-making,* edited by F. E. I. Hamilton, pp. 189–212. New York: John Wiley.

Rees, J., and H. A. Stafford. 1986. Theories of regional growth and industrial location: Their relevance for understanding high-technology complexes. In *Technology, Regions, and*

Policy, edited by J. Rees, pp. 23–50. Totowa, NJ: Rowman and Littlefield.

Reilly, W. J. 1931. *The Law of Retail Gravitation.* New York: The Knickerbocker Press.

Reissman, L. 1964. *The Urban Process: Cities in Industrial Societies.* New York: Free Press.

Relph, E. 1976. *Place and Placelessness.* London: Ron.

Replogle, M. A. 1988. *Bicycles and Public Transportation: New Links to Suburban Transit Markets,* 2d ed. Washington, DC: The Bicycle Federation.

Ricardo, D. 1912. *The Principles of Political Economy and Taxation.* New York: E. P. Dutton.

Riddell, J. B. 1970. *The Spatial Dynamics of Modernization in Sierra Leone: Structure, Diffusion, and Response.* Evanston, IL: Northwestern University Press.

Rimmer, P. J. 1977. A conceptual framework for examining urban and regional transport needs in Southeast Asia. *Pacific Viewpoint* 18:133–147.

Robinson, R. D. 1981. Background concepts and philosophy of international business from World War II to the present. *Journal of International Business Studies* Spring/Summer: 13–21.

Rodney, W. 1972. *How Europe Underdeveloped Africa.* Dar es Salaam, Tanzania: Tanzania Publishing House and Bogle-L'Overture Publications.

Roepke, H. G. 1959. Changes in corn production on the northern margin of the Corn Belt. *Agricultural History* 33:126–132.

Ross, R., and K. Trachte. 1983. Global cities, global classes: The peripheralization of labor in New York City. *Review* 6:393–431.

Rostow, W. W. 1960. *The Stages of Economic Growth: A Non-Communist Manifesto.* Cambridge, MA: Cambridge University Press.

Rubenstein, J. M. 1992. *The Cultural Landscape: An Introduction to Human Geography,* 3d ed. New York: Macmillan.

———. 1994. *The Cultural Landscape: An Introduction to Human Geography,* 4th ed. New York: Macmillan.

Rushton, G. 1969. Central place theory. *Annals of the Association of American Geographers* 62:253.

———. 1969. A computer model for the study of agricultural land use patterns. In *Computer Assisted Instruction in Geography,* edited by the Association of American Geographers, pp. 141–150. Washington, DC: Commission on College Geography.

———. 1971. Preference and choice in different environments. *Annals of the Association of American Geographers* 3:146–149.

———. 1984. Use of location-allocation models for improving the geographical accessibility of rural services in developing countries. *International Regional Science Review* 9:217–240.

———. 1988. The Roepke lecture in economic geography: Location theory, location-allocation models, and service development planning in the Third World. *Economic Geography* 64:97–119.

Ryan, W. 1972. *Blaming the Victim.* New York: Vintage Books.

Saarinen, T. F. 1969. *Perception of Environment.* Resource Paper No. 5. Washington, DC: Association of American Geographers.

Sack, R. D. 1974. The spatial separatist theme in geography. *Economic Geography* 50:1–19.

Said, E. W. 1981. *Covering Islam: How the Media and the Experts Determine How We Shall See the Rest of the World.* New York: Pantheon.

Samuelson, R. J. 1989. Superpower sweepstakes. *Newsweek* 20 February:43.

Sanderson, S. W., and B. J. L. Berry. 1986. Robotics and regional development. In *Technology, Regions, and Policy,* edited by J. Rees, pp. 171–186. Totowa, NJ: Rowman and Littlefield.

Santos, M. 1971. *Les Villes du Tiers Monde.* Paris: Editions M-Th. Génin.

———. 1977. Spatial dialectics: The two circuits of urban economy in underdeveloped countries. *Antipode* 9:49–60.

Sayer, A., and R. Walker. 1993. *The New Social Economy: Reworking the Division of Labor.* Cambridge, MA: Blackwell.

Saxenian, A. 1985. The genesis of Silicon Valley. In *Silicon Landscapes,* edited by P. Hall and A. Markusen, pp. 20–34. Boston: Allen and Unwin.

Schaefer, F. 1953. Exceptionalism in geography: A methodological examination. *Annals of the Association of American Geographers* 43:226–249.

Schlebecker, J. T. 1960. The world metropolis and the history of American agriculture. *Journal of Economic History* 20:147–208.

Schumacher, E. F. 1973. *Small Is Beautiful.* London: Blond and Briggs.

Schumpeter, J. A. 1939. *Business Cycles: A Theoretical, Historical, and Statistical Account of the Capitalist Process.* 2 vols. New York: McGraw-Hill.

Schwartzberg, J. E. 1987. The U.S. Constitution, a model for global government. *Journal of Geography* 86:246–252.

Scobie, J. R. 1964. *Argentina: A City and a Nation.* New York: Oxford University Press.

Scott, A. J., and M. Storper, Eds. 1986. *Production, Work, Territory: The Geographical Anatomy of Industrial Capitalism.* Boston: Allen and Unwin.

Scott, E. 1972. The spatial structure of rural northern Nigeria: Farmers, periodic markets, and villages. *Economic Geography* 48:316–332.

Scott, A. J. and D. P. Angel. 1988. The global assembly of U.S. semiconductor firms. *Environment and Planning A* 20:1047 1067.

Seers, D. 1972. What are we trying to measure? *The Journal of Development Studies* 8:21–36.

Shabecoff, P. 1987. Peering into the energy future and sighting gas shortages. *The New York Times* 25 September, p. 26.

Shaw, S. 1993. HUB structures of major U.S. passenger airlines. *Journal of Transport Geography* 1(1):47–57.

Shipler, D. 1987. Reagan is preparing to sail into uncharted policy waters. *The New York Times* 31 May, Section 4, p. 1.

Simon, H. A. 1957. *Models of Man.* New York: John Wiley.

———. 1960. *The New Science of Management Decision.* New York: Harper and Row.

Simon, J. L. 1980. Resources, population, environment: An oversupply of false bad news. *Science* 208:1431–1437.

———. 1981. *The Ultimate Resource.* Princeton, NJ: Princeton University Press.

———. 1986. *Theory of Population and Economic Growth.* Oxford, Eng: Basil Blackwell.

———. 1989. On aggregate empirical studies relating population variables to economic development. *Population and Development Review* 15:323–332.

Simoons, F. J. 1961. *Eat Not This Flesh.* Madison: University of Wisconsin Press.

Sivanandan, A. 1987. Imperialism and disorganic development in the silicon age. In *International Capitalism and Industrial Restructuring,* edited by R. Peet, pp. 185–200. Boston: Allen and Unwin.

Skinner, G. W. 1964. Marketing and social structure in rural China. *Journal of Asian Studies* 24:3–43.

Smith, D. M. 1966. A theoretical framework for geographical studies of industrial location. *Economic Geography* 42:95 113.

———. 1977. *Human Geography: A Welfare Approach.* London: Edward Arnold.

———. 1981. *Industrial Location: An Economic Geographical Analysis,* 2d ed. New York: John Wiley.

Soja, E. W. 1968. *The Geography of Modernization in Kenya.* Syracuse, NY: Syracuse University Press.

Soja, E., R. Morales, and G. Wolff. 1987. Industrial restructuring: An analysis of social and spatial change in Los Angeles. In *International Capitalism and Industrial Restructuring,* edited by R. Peet, pp. 145–176. Boston: Allen and Unwin.

South, R. B. 1992. Transnational Maquiladora location. *Annals of the Association of American Geographers* 80(4):549 570.

Spengler, J. 1974. *Population Change, Modernization, and Welfare.* Englewood Cliffs, NJ: Prentice Hall.

Starbuck, W. H. 1965. Organizational growth and development. In *Handbook of Organizations,* edited by J. G. March, pp. 451–522. Skokie, IL: Rand McNally.

Stebelsky, I. 1983. Wheat yields and weather hazards in the Soviet Union. In *Interpretations of Calamity from the Viewpoint of Human Ecology,* edited by K. Hewitt, pp. 202–218. Boston: Allen and Unwin.

Stopford, I. M., and J. H. Dunning. 1983. *Multinationals: Company Performances and Global Trends.* London: Macmillan.

Storey, D. J. 1986. The economics of smaller businesses: Some implications for regional economic development. In *Technological Change, Industrial Restructuring, and Regional Development,* edited by A. Amin and J. B. Goddard, pp. 215–232. Winchester, MA: Allen and Unwin.

Storper, M., and R. Walker. 1989. *The Capitalist Imperative: Territory, Technology, and Industrial Growth.* New York: Basil Blackwell.

Streeten, P. 1981. *First Things First: Meeting Basic Human Needs in Developing Countries.* New York: Oxford University Press.

Stutz, F. P. 1992a. Urban and Regional Planning. In T. Hartshorn (ed.), *Interpreting the City: An Urban Geography.* New York: John Wiley & Sons.

Stutz, F. P. 1992b. Maquiladoras Branch Plants: Transportation—Labor Cost Substitution Along the U.S./Mexican Border. In *Snapshots of North America,* Washington, D.C.: 27th Congress of the International Geographical Union Official Book.

Stutz, F. P. 1992c. San Diego: The Next High Amenity Pacific Rim World City. In Blakeley and Stimson (eds.), *The New City of the Pacific Rim.* University of California, Berkeley: Institute of Urban and Regional Development.

Stutz, F. P. 1992d. The Labor Shed of Tijuana in Relation to the U.S. Mexican Border. In T. Hartshorn (ed.), *Interpreting the City: An Urban Geography.* New York: John Wiley & Sons.

Stutz, F. P., and S. Aitken. 1990. *Neighborhood Character/Community Disruption for Route 125/54 Middle Section.* Final report of the California Department of Transportation (Caltrans) funded project, Project No. 89-90-003.

Stutz, F. P., and P. Hinshaw. 1976. Churchgoer spatial travel behavior. *The Southeastern Geographer* 17:35–47.

Stutz, F. P., and A. Kartman. 1982. Spatial variations in housing prices in the United States. *Economic Geography* July.

Stutz, F. P., and Parrott, R. 1992. Urban GIS applications. In *GIS: Principles and Procedures,* edited by D. Maguire, M. Goodchild, and D. Rhind, pp. 247–260. London: Longman.

Stutz, F. P., Parrott, R., and Kavanaugh, P. 1992. Charting Urban Space–Time Population Shifts Using Trip Generation Models. *Urban Geography,* 13(5):468–475.

Stutz, F. P., and J. Supernak. 1992. *Understanding Land Use and Transportation Linkages.* Final report of the California Department of Transportation (Caltrans) funded project, Project No. 90/91-004.

Stutz, F.P. 1994 in T. de Souza and F.P. Stutz. *The World Economy: Resources, Location, Trade and Development.* Upper Saddle River: Prentice Hall.

Stutz, F.P. 1998 in T. de Souza and F.P. Stutz. *The World Economy: Resources, Location, Trade and Development.* Upper Saddle River: Prentice Hall.

Stycos, J. 1971. *Ideology, Faith, and Family Planning in Latin America.* New York: McGraw-Hill.

Sunday Times. 1987. London: The *Times* Newspaper, 6 December, p. 79.

Sunkel, O. 1982. Big business and dependency. *Foreign Affairs* 24:517–531.

Susman, P., and E. Schutz. 1983. Monopoly and competitive firm relations and regional development in global capitalism. *Economic Geography* 59:161–177.

Sweezy, P. 1942. *The Theory of Capitalist Development.* London: Dobson.

Symanski, R. 1971. *Market Cycles in Andean Columbia.* Ph.D. dissertation. Syracuse, NY: Syracuse University.

Szentes, T. 1971. *The Political Economy of Underdevelopment.* Budapest: Adademiai Kiado.

Taaffee, E.J., H.L. Gauthier and M.E. O'Kelly. 1996. *Geography of Transportation.* Upper Saddle River: Prentice Hall.

Taaffe, E. J., R. L. Morrill, and P. R. Gould. 1963. Transport expansion in underdeveloped countries. *Geographical Review* 53:503–529.

Takes, A. P., and A. J. Venstra. 1961. Zuyder Zee reclamation scheme. *Tijdschrift voor Economische en Sociale Geografie* 51:163.

Tata, R. J., and R. R. Schultz. 1988. World variation in human welfare: A new index of development status. *Annals of the Association of American Geographers* 78:580–593.

Taylor, M. J. 1975. Organizational growth, spatial interaction and location decision making. *Regional Studies* 9:313–323.

Taylor, M. J., and N. Thrift, Eds. 1982. *The Geography of Multinationals.* New York: St. Martin's Press.

Teitelbaum, M. S. 1975. Relevance of demographic transition theory for developing countries. *Science* 188:420–425.

The Economist. 1982. Money and finance, 16 October.

———. 1987a. Britain: the best of times, the worst of times, 21 February:1–26.

———. 1987b. Japanese property: A glittering sprawl, 3 October:25–28.

———. 1987c. Telecommunications supplement, 17 October: 23.

———. 1988a. Come back multinationals, 28 November:73.

———. 1988b. The regions revive, 2 April:45–46.

———. 1988c. The pleasures of three-part harmony, 24 December:41.

———. 1988d. Why it's still a triangle, 24 December:41–44.

Thrift, N. 1986. Geography of international economic disorder. In *A World in Crisis,* edited by R. J. Johnston and P. J. Taylor, pp. 12–67. New York: Blackwell.

Tipps, D. 1973. Modernization theory and the comparative study of societies: A critical perspective. *Comparative Studies in Society and History* 155:199–226.

Toffler, A. 1981. *The Third Wave.* New York: Bantam.

TRAILMAN. 1993. *TRAILMAN Version 1.1: Transportation and Inland Logistics Manager.* Knoxville: Department of Geography, The University of Tennessee.

Troughton, M. J. 1985. Industrialization of U.S. and Canadian agriculture. *Journal of Geography* 84:255–263.

U.K. Department of Employment. 1985. *Employment Gazette.* London: HMSO.

Ullman, E. L. 1940–1941. A theory of location for cities. *American Journal of Sociology* 46:853–864.

———. 1943. *Mobile: Industrial Seaport and Trade Center.* Chicago: University of Chicago Press.

United Nations. 1975. *Women in Africa.* New York: Author.

———. 1981. *1980 Report by the Executive Director of the United Nations Population Fund.* New York: Author.

———. 1982. *Estimates and Projections of Urban, Rural and City Populations, 1950–2025: The 1980 Assessment.* New York: Author.

———. 1983. *Demographic Yearbook, 1981.* New York: Author.

———. 1983. *Yearbook of National Accounts Statistics, 1981.* New York: Author.

———. 1985. *Statistical Yearbook.* New York: Author.

———. 1987. *Yearbook of World Energy Statistics.* New York: Author.

———. 1988. *National Accounts Statistics: Analysis of Main Aggregates, 1985.* New York: Author.

———. 1997. *Transnational Corporations in World Development: Trends and Prospects.* New York: Author.

———. 1997. *Industrial Statistics Yearbook, 1997.* New York: Author.

———. 1990. *Demographic Yearbook, 1988.* New York: Author.

———. 1991a. *World Population Prospects 1990.* New York: Author.

———. 1997b. Yearbook of International Trade Statistics. New York: United Nations.

———. 1997. *Yearbook of International Trade Statistics.* New York: Author.

———. 1997a. *World Economic Survey, 1992/93.* New York: Author.

———. 1997b. *Industrial Statistics Yearbook, 1990.* New York: Author.

United Nations Fund for Population Activities (UNFPA). 1997. *Report by the Executive Director.* New York: United Nations.

U.S. Bureau of Economic Analysis. 1997. *Survey of Current Business.* Washington, D.C.: U.S. Department of Commerce.

U.S. Bureau of the Census. 1997. *Statistical Abstract of the United States.* Washington, DC: GPO.

———. 1989. Geographical mobility: March 1986 to March 1987. *Current Population Reports,* Series P-20, No. 430, Table 5.

———. 1997. *Statistical Abstract of the United States.* Washington, DC: GPO.

———. 1992a. Educational attainment in the United States: March 1997. *Current Population Reports,* Series P-20, No. 462, Table 8.

———. 1997b. *Statistical Abstract of the United States.* Washington, DC: GPO.

———. 1997. *Statistical Abstract of the United States.* Washington, DC: GPO.

U.S. Bureau of Mines. 1997. *Minerals Yearbook.* Washington, DC: GPO.

———. 1997. *Mineral Commodity Summaries 1986:* An Up-to-Date Summary of 87 Nonfuel Mineral Commodities. Washington, DC: GPO.

———. 1997. *Mineral Commodity Summaries 1988: An Up-to-Date Summary of 87 Nonfuel Mineral Commodities.* Washington, DC: GPO.

U.S. Department of Agriculture. 1970. *The Marketing Challenge.* Foreign Economic Development Report No. 7. Washington, DC: USDA.

———. 1997. *Yearbook of Agriculture.* Washington, DC: USDA.

U.S. Department of Commerce. 1997. *U.S. Foreign Trade Highlights.* Washington, DC: International Trade Administration.

———. 1993. *World Trade Statistics.* Washington, DC: International Trade Administration.

Vale, T. R. 1985. What kind of conservationist? *Journal of Geography* 84:239–241.

Van de Walle, E. 1975. Foundations of the model of doom. *Science* 189:1077–1078.

van der Iaan, F. 1992. Raster GIS allows agricultural suitability modeling at a continental scale. *GIS World* 5(8):42–50.

Van Valkenburg, S., and C. C. Held. 1952. *Europe.* New York: John Wiley.

Vance, J. E., Jr. 1964. *Geography and Urban Evolution in the San Francisco Bay Area,* Institute of Government Studies, University of California, Berkeley.

———. 1970. *The Merchant's World: The Geography of Wholesaling.* Englewood Cliffs, NJ: Prentice Hall.

———. 1977. *This Scene of Man: The Role and Structure of the City in the Geography of Western Civilization.* New York: Harper & Row.

———. 1986. *Capturing the Horizon: The Historical Geography of Transportation.* New York: Harper & Row.

Vernon, R. 1966. International investment and international trade in the product cycle. *Quarterly Journal of Economics* 80:190–207.

———. 1979. The product cycle hypothesis is a new international environment. *Oxford Bulletin of Economics and Statistics* 41: 255–268.

Viner, J. 1950. *The Customs Union Issue.* New York: Carnegie Endowment for International Peace.

Vining, D. 1985. The growth of core regions in the Third World. *Scientific American* April:48.

Vogeler, I. 1981. *The Myth of the Family Farm: Agribusiness Dominance of U.S. Agriculture.* Boulder, CO: Westview.

Vogeler, I., and A. R. de Souza, Eds. 1980. *Dialectics of Third World Development.* Totowa, NJ: Rowman and Allanheld.

Von Thünen, J. H. 1826. *The Isolated State.* Hamburg: Perthes.

Waddell, L.M. and W.C. Labys. 1988. *Transmaterialization Technology and Materials Demand Cycles.* Morgantown: W. Virginia University.

Wallerstein, I. 1974. *The Modern World-System: Capitalist Agriculture and the Origins of the European World Economy in the Sixteenth Century.* New York: Academic Press.

Walsh, J. 1974. U.N. Conference: Topping any agenda is the question of development. *Science* 185:1144.

Warf, B. 1989. Telecommunications and the globalization of financial services. *Professional Geographer* 41:257–271.

Watson, M., P. Keller, and D. Mathieson. 1984. *International Capital Markets: Development and Prospects.* Washington, DC: IMF.

Watts, D. 1988. Thatcher's Britain—A manufacturing economy in decline? *Focus* Vol. 12.

Webb, W. P. 1931. *The Great Plains.* New York: Grosset and Dunlap.

———. 1929. *Alfred Weber's Theory of the Location of Industries.* Translated by C. J. Friedrich. Chicago: University of Chicago Press.

Weber, M. 1930. *The Protestant Ethic and the Spirit of Capitalism.* New York: Scribners.

Weeks, J. 1992. *Population: An Introduction to Concepts and Issues,* 4th ed. Belmont, CA: Wadsworth.

———. 1994. *Population: An Introduction to Concepts and Issues,* updated fifth edition. Belmont, CA: Wadsworth.

Wharton, C. R., Jr. 1963. Research on agricultural development in Southeast Asia. *Journal of Farm Economics* 45:1161–1174.

———. 1969. The green revolution. Cornucopia or Pandora's box? *Foreign Affairs* 47:464–476.

Wheaton, W. 1987. Income and urban residence: An analysis of consumer demand for location. *American Economic Review* 67:620–631.

Wheeler, J. O., and R. L. Mitchelson. 1989. Information flows among major metropolitan areas in the United States. *Annals of the Association of American Geographers* 79:523–543.

Wheeler, J. O., and F. P. Stutz. 1971. Spatial dimensions of urban social travel. *Annals of the Association of American Geographers* 61(2):371–386.

Williams, J. F. 1989a. Japan: Physical and human resources. In *Geography and Development,* edited by J. S. Fisher, pp. 330–343. Columbus, OH: Merrill.

———. 1989b. Japan: The economic giant. In *Geography and Development,* edited by J. S. Fisher, pp. 346–365. Columbus, OH: Merrill.

Williams, M. 1979. The perception of the hazard of soil degradation in South Australia: A review. In *Natural Hazards in Australia,* edited by R. L. Heathcote and B. L. Thom, pp. 275–289. Canberra: Australian Academy of Science.

Wise, M. J. 1949. On the evolution of the jewellery and gun quarters in Birmingham. *Transactions, Institute of British Geographers* 15:59–72.

Wolpert, J. 1964. The decision process in a spatial context. *Annals of the Association of American Geographers* 54:537–558.

Womack, J. R., D. T. Jones, and D. Roos. 1990. *The Machine that Changed the World.* New York: Rawson Associates.

Wong, K., and D. K. Y. Chu. 1984. Export processing zones and special economic zones as generators of economic development: The Asian experience. *Geografiska Annaler* 66:1–16.

World Bank. 1997. *World Tables 1971.* Washington, DC: Author.

———. 1990. *World Development Report.* Washington, DC: Author.

———. 1991. *World Development Report.* New York: Oxford University Press.

———. 1992. *World Development Report.* New York: Oxford University Press.

———. 1993. *World Development Report.* New York: Oxford University Press.

———. 1994. *World Development Report.* New York: Oxford University Press.

———. 1995. *World Development Report.* New York: Oxford University Press.

———. 1996. *World Development Report.* New York: Oxford University Press.

———. 1997. *World Development Report.* New York: Oxford University Press.

———. 1998. *World Development Report.* New York: Oxford University Press.

World Resources Institute. 1997. *World Resources 1997–98: A Guide to the Global Environment.* New York: Oxford University Press.

Yapa, L. S. 1980. Diffusion, development and ecopolitical economy. In *Innovation Research and Public Policy,* edited by J. A. Agnew, Geographical Series No. 5, pp. 101–41. Syracuse, NY: Syracuse University Press.

———. 1985. The population problem as economic disarticulation. *Journal of Geography* 84:242–247.

Yearbook of World Electronics Data. 1990. Oxford: Elsevier Advanced Technology.

Yeates, M. 1975. *Main Street: Windsor to Quebec City.* Toronto: Macmillan.

———. 1984. The Windsor-Quebec city axis: Basic characteristics. *Journal of Geography* 83:240–249.

Zipf, G. K. 1949. *Human Behavior and the Principle of Least Effort.* Reading, MA: Addison-Wesley.